INTRODUCTION TO UNIVERSITY MATHEMATICS

INTRODUCTION TO
UNIVERSITY MATHEMATICS

J. L. SMYRL
M.A., Ph.D., F.I.M.A.
Senior Lecturer, Department of Mathematics,
University of Strathclyde

HODDER AND STOUGHTON
LONDON SYDNEY AUCKLAND TORONTO

ISBN 0 340 20111 8 (Boards)
0 340 20112 6 (Unibook)

First published 1978

Text set in 11/12 pt Photon Imprint printed by photolithography and bound in Great Britain at The Pitman Press, Bath for Hodder & Stoughton Educational, a division of Hodder & Stoughton Limited, Mill Road, Dunton Green, Sevenoaks, Kent

PREFACE

Much attention has recently been focused on the problems of transition from school to university. In mathematics, following a period of very rapid change at school level, the step from a school to a university textbook presents special difficulties. 'Traditional' first-year university texts are by no means ideal for students accustomed to 'modern' school texts, while 'modern' university texts, in addition to being difficult for students with a 'traditional' background, tend to overlook the point of view of the engineer and the scientist.

This book covers the critical area of sixth-year school–early university mathematics. The approach used is one that should come easily to students with a modern background, but considerable care has been taken over the explanation of fundamental ideas so that the book should also be suitable for those with a traditional background. In the selection of material I have had the basic first-year course at Strathclyde very much in mind, and I have gone far enough to cover our second-year course for engineers. The knowledge assumed is that of the H-grade mathematics course in the Scottish Certificate of Education—most of the work for Papers I and II of the Scottish Sixth Year Studies examination is covered, and at this stage pupils could benefit considerably from using a book that enables them to look ahead. I am hopeful that the book, despite its Scottish origin, will likewise be suitable for GCE A-level work and for first-year courses in English universities and for corresponding courses in Polytechnics and Colleges of Education.

Following the tradition of the first-year course at Strathclyde, the book is not aimed too closely at any one particular specialisation but endeavours to cater for all students who like to relate mathematics to the physical world, providing a sound basis for later specialisation. Comprehensive sets of exercises are provided and include some exercises that require extra persistence and manipulative skill so that good students, particularly those who intend to specialise in mathematics, will be suitably challenged—however, it is important that others should not be unduly worried if they find some exercises too difficult.

Most readers will undoubtedly wish to study calculus, which is developed in Chapters 12, 14, 15, etc., simultaneously with the algebra topics of the other chapters. Some of the longer chapters, notably

Chapters 8, 9 and 18, are not intended to be taken all at once but rather split over a two-year course. However, the arrangement of material, being easy for reference and for revision, is that which I believe best suits students, and it allows maximum flexibility in the use of the book for various courses.

J. L. Smyrl

ACKNOWLEDGEMENTS

This book began as an attempted revision of Margaret M. Gow's book *A Course in Pure Mathematics*, to be undertaken jointly by Dr Gow and myself. Unfortunately Dr Gow was forced by pressure of other commitments to withdraw from this venture, and as I proceeded with the task on my own it became more and more obvious that I was writing a new book. In the event it was decided to publish the book as a new title *Introduction to University Mathematics*, but some debt to Dr Gow's book should be acknowledged. I have selected freely from the examples and exercises in Dr Gow's book and occasional traces of Dr Gow's text will be found in mine. For the examples and exercises my thanks are also due to the Senates of the Universities of Durham, Leeds, Liverpool, London, Sheffield and Strathclyde who gave permission for the use of questions from their examination papers.

Finally, I am most grateful to Professor D. C. Pack, O.B.E., M.A., D.Sc., who read the entire manuscript and made numerous helpful suggestions, and to Mrs G. Wybar (formerly Miss Lorna Ferrie), who performed with unfailing patience the mammoth task of typing the manuscript.

J. L. Smyrl

CONTENTS

CHAPTER 1

INTRODUCTION

1.1 Representation

Every schoolboy understands how a difficult investigation may sometimes be simplified by the use of a model. Shipbuilders, architects, civil engineers and town planners make frequent use of a scale model in which the true set-up is replaced by a miniature version. Scale models are particularly appropriate for investigations in which the shape of the objects involved is highly relevant.

However, the scale model is not the only kind of representative set-up which can be used in place of the true one. Consider, for example, the systems shown in Figs. 1.1 and 1.2. Figure 1.1 shows a mechanical system comprising a ball attached to one end of a spring, the other end of which is fixed. If the ball is held at rest, away from its natural position of equilibrium, and then released, it oscillates.

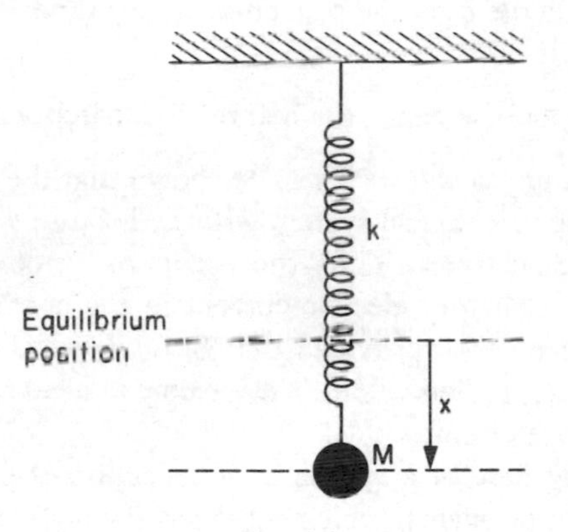

FIG. 1.1

The motion of the ball is affected by

(a) its mass M,
(b) the spring constant k, which measures the stiffness of the spring,

and the actual displacement x of the ball from its equilibrium position at any time after release also depends on

(c) the displacement x_0 at the moment of release.

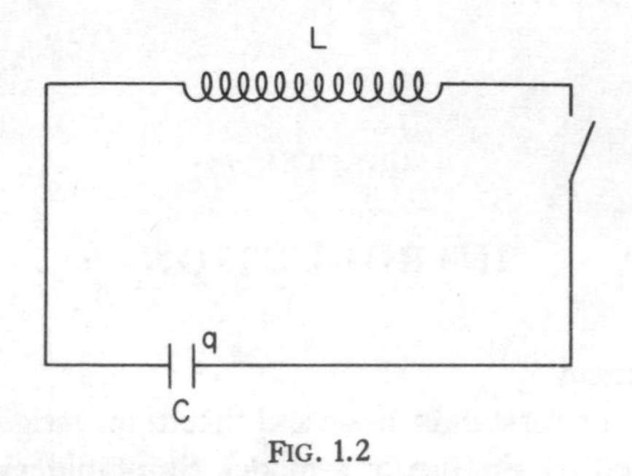

FIG. 1.2

Figure 1.2 shows an electrical system comprising a circuit with a switch, a coil and a capacitor. For readers unfamiliar with elementary circuit theory it is sufficient to realise that electric current, i.e. flow of electric charge, occurs in the circuit only when the switch is closed, and this current is affected by

(a′) a property of the coil, called its inductance, which we shall assume has the value L,
(b′) a property of the capacitor, called its capacitance, which we shall assume has the value C,

while the actual charge q on the capacitor at any time after the switch is closed also depends on

(c′) the charge q_0 on the capacitor before the switch was closed.

From the relevant physical laws it can be shown that the electrical system is analogous to the mechanical system with L, $1/C$, q_0, q playing the roles of M, k, x_0, x respectively. Thus the oscillatory motion of the ball is modelled by the oscillatory electric current in the corresponding circuit (and, of course, vice versa) provided that L, C, q_0 are chosen so that $L = M$, $C = 1/k$, $q_0 = x_0$; to find values of x we may instead measure values of q in the appropriate circuit.

We are thinking here of a *system* as a collection of parts, or *elements* (mass, spring, displacement), which together display some characteristic behaviour (oscillatory motion) in which we are interested. The representative system satisfies two important requirements:

(1) Each element in the original system has a counterpart in the representative system.
(Mass → inductance, spring → capacitor, displacement → charge)
(2) The behaviour of the original system is in some sense imitated by the behaviour of the representative system.
(Oscillatory motion → oscillatory current flow)

Clearly some features of the original system are irrelevant to the behaviour under discussion and so need have no counterpart in the representative system, e.g. shape and colour of the mass. Likewise there may be features of the representative system which do not correspond to anything in the original system.

1.1.1 Mathematical Systems The ideas discussed in Section 1.1 are of considerable importance in mathematics. As a very elementary example, suppose we wish to build a single street of new houses to rehouse all the families from several streets of old houses. Each street of houses can be represented by a group of (say) beads; by combining into a single group all the groups of beads representing streets of old houses we get the group of beads that represents the required street of new houses. Better still, we can dispense with material objects such as beads and instead use *numbers*; each group of objects can be represented by a number, and the process of combining several groups is then equivalent to the process of 'adding' representative numbers. This system of numbers, with abstract rather than material elements, will be called a *mathematical system*; the 'behaviour' of the system is the way the elements add together in accordance with well-known tables of addition. The system becomes more useful when further elements (e.g. negative numbers, fractional numbers) and other kinds of behaviour (e.g. multiplication, division) are introduced.

1.1.2 Operations The behaviour (i.e. the motion) of the mechanical system in Section 1.1 is activated by natural forces obeying certain natural laws; likewise for the behaviour of the electrical system. The behaviour (numbers adding, multiplying, etc.) of the above mathematical system is regulated by a human operator who imposes the constraint of certain arithmetic laws (addition tables, multiplication tables, etc.). We refer to addition, multiplication, etc. as *operations*.

Addition, multiplication, subtraction, division are called *binary* operations because two numbers are used to produce the answer: taking a square root, for example, is not a binary operation since the operation is performed on a single number.

1.2 The Real-Number System

In Chapter 3 we return to the subject of the cardinal or 'counting' numbers 1, 2, 3, ... For some purposes they provide an adequate system, but when dealing with, for example, the measurement of length, then an adequate set of numbers has to include fractions and irrational numbers such as $\sqrt{2}$. Further, when it is necessary to distinguish between, say, distance above and below a fixed level, then negative numbers are required.

Conversely, the idea of length may be used to provide a visual representation of numbers, and this is what is done when we represent numbers by points on a 'real axis' (Fig. 1.3). The set of numbers corresponding to all points on the real axis is called the set of *real numbers*.

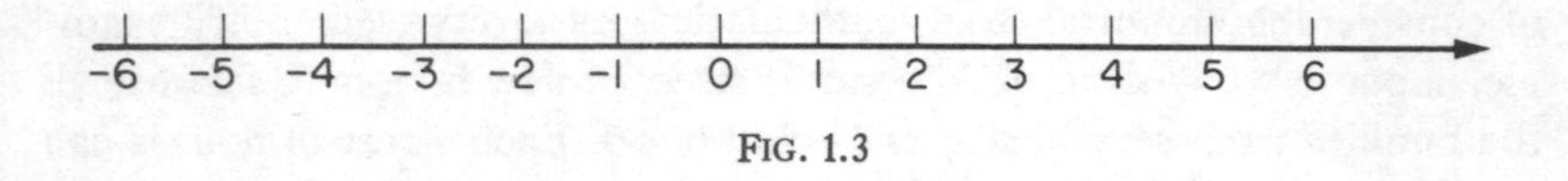

FIG. 1.3

This set of numbers, together with the usual arithmetic operations, constitutes the *real-number system.*

The *integers* are the whole numbers 1, 2, 3, . . ., together with zero and the negative numbers $-1, -2, -3, \ldots$.

The *rational numbers* are numbers of the form p/q, where p and q are both integers. The rational numbers include the integers.

All other numbers, e.g. $\sqrt{2}, \sqrt{3}, \pi$ are called *irrational numbers.*

1.2.1 Decimal Notation A number can be described directly (e.g. 5), or indirectly in terms of both numbers and operations (e.g. $3 + 2$). Since the same number can be described in many ways we require an 'equality' sign, e.g. $9 - 3 = 4 + 2 = 3 + 3 = 6$.

Economic use of symbols is achieved by decimal notation; for example, the number $1 \times 10^4 + 2 \times 10^3 + 3 \times 10^2 + 4 \times 10 + 5$ is denoted (directly) by 12345 and the number $2 \times 10 + 4 + 6 \times 10^{-1} + 3 \times 10^{-2} + 7 \times 10^{-3}$ by $24{\cdot}637$. Certain rational numbers (e.g. $\frac{1}{2}, \frac{3}{5}, \frac{173}{20}$) in which the denominator has no prime factors other than 2 or 5, can be readily expressed in decimal form. Certain others, however, like $\frac{1}{3}$ or $\frac{2}{11}$, require an unending decimal representation, and later we shall discuss an operation represented by an unending sequence of additions so that, for example,

$$\frac{3}{10} + \frac{3}{10^2} + \frac{3}{10^3} + \frac{3}{10^4} + \cdots = \frac{1}{3},$$

$$\frac{1}{10} + \frac{8}{10^2} + \frac{1}{10^3} + \frac{8}{10^4} + \cdots = \frac{2}{11};$$

hence we write $\frac{1}{3} = 0{\cdot}3333\ldots$ or $\frac{1}{3} = 0{\cdot}\dot{3}$, $\frac{2}{11} = 0{\cdot}1818\ldots$ or $\frac{2}{11} = 0{\cdot}\dot{1}\dot{8}$. Rational numbers in which the denominator has prime factors other than 2 or 5 can be represented as *non-terminating decimals* of this type, where the repeating pattern makes the full description clear.

Irrational numbers can be expressed as non-terminating decimals, in theory at least, but no repetitive pattern is found so that the complete description is impossible; in practice this does not affect us much, since we usually work to the accuracy of a specified number of decimal places.

1.2.2 Numbers to any Base The decimal notation is so called because of the special role played by the number ten. The notation for the entire set of real numbers could be similarly based on any positive integer b, i.e. we could write 12345 to mean $1.b^4 + 2.b^3 + 3.b^2 + 4.b + 5$ and $24{\cdot}637$ to mean $2.b + 4 + 6.b^{-1} + 3.b^{-2} + 7.b^{-3}$. We assume here that b is greater than any of the decimal digits used, e.g. if b is 8 then the digits 8, 9 are no longer necessary as these numbers are denoted by 10, 11 respectively in the notation based on b.

We call b the *base* for the notation, and say that numbers are expressed *to the base b*, or *in the scale of b*.

Example 1 Express the number 32·14 (scale of 6) in decimal scale.

$$32{\cdot}14 \text{ (scale of 6)} = 3.6 + 2 + \frac{1}{6} + \frac{4}{6^2} \text{ (scale of 10)}$$
$$= 18 + 2 + \frac{5}{18} = 20{\cdot}2777\ldots.$$

Note that the decimal form of the number is non-terminating.

The *duodecimal scale* (scale of 12) necessitates the use of two extra digits, say t and e, for the numbers ten and eleven respectively. Thus

$$3t5e \text{ (scale of 12)} = 3.12^3 + 10.12^2 + 5.12 + 11 \text{ (scale of 10)}$$
$$= 6695.$$

The *binary scale* (scale of 2) uses only the digits 0, 1. Electronic computers work with numbers in this form.

Example 2 Express 10101 (binary scale) in decimal scale.

$$10101 \text{ (binary scale)} = 1.2^4 + 1.2^2 + 1 \text{ (decimal scale)}$$
$$= 21.$$

Converting from decimal scale to another scale is easily done by successive divisions as illustrated in the following examples.

Example 3 Convert 1265 to the scale of 7.

$$\begin{array}{r|l} 7 & 1265 \\ \hline 7 & 180 + 5 \\ \hline 7 & 25 + 5 \\ \hline & 3 + 4 \end{array}$$

Hence 1265 (decimal scale) = 3455 (scale of 7). The answer is obtained from the final quotient (when we cannot divide by 7 any more times) together with all the remainders.

Example 4 Convert 1265 to binary scale.

$$\begin{array}{r|r}
2 & 1265 \\ \hline
2 & 632 + 1 \\ \hline
2 & 316 + 0 \\ \hline
2 & 158 + 0 \\ \hline
2 & 79 + 0 \\ \hline
2 & 39 + 1 \\ \hline
2 & 19 + 1 \\ \hline
2 & 9 + 1 \\ \hline
2 & 4 + 1 \\ \hline
2 & 2 + 0 \\ \hline
 & 1 + 0
\end{array}$$

Hence 1265 (decimal scale) = 10011110001 (binary scale).

Naturally addition tables, multiplication tables, etc. are different when numbers are expressed in a different scale. In practice these operations can easily be carried out by first converting the numbers to decimal scale.

Example 5 Multiply 15 by 14, where the numbers are expressed in the scale of 6.

$$\begin{aligned}
15\,(\text{scale of } 6) &= 11\,(\text{decimal scale}) \\
14\,(\text{scale of } 6) &= 10\,(\text{decimal scale}) \\
\text{Product} &= 110 \\
&= 302\,(\text{scale of } 6).
\end{aligned}$$

Binary scale arithmetic is so simple that conversion to decimal scale is unnecessary. The only strange result we need to remember is that 1 + 1 = 10.

Example 6 Multiply 110·1 by 10·1, where the numbers are in binary scale.

$$\begin{array}{r}
110{\cdot}1 \\
10{\cdot}1 \\ \hline
11{\cdot}01 \\
1\,101\phantom{{\cdot}01} \\ \hline
\mathbf{10\,000{\cdot}01}
\end{array}$$

1.2.3 Sets of Numbers, Variables, Identities We sometimes describe a set of numbers by listing the set inside curly brackets, e.g. $\{1, \frac{3}{2}, \frac{5}{7}\}$, $\{1, \frac{1}{2}, \frac{1}{3}, \frac{1}{4}, \ldots\}$ dots being used in the case of an infinite set when subsequent numbers follow the pattern established by those written down. For ease of reference an entire set may be denoted by a single letter, usually a capital, e.g. we often refer to R, *the set of all real numbers.*

If S is a set of numbers, then a *variable* on S is a symbol, usually a small letter, which may denote any number in S. Thus if x is a variable on the set $\{1, \frac{3}{2}, \frac{5}{7}\}$ we think of x as being capable of interpretation as either 1, or $\frac{3}{2}$, or $\frac{5}{7}$. Consider the following examples:

(a) A stone is released from a height of 100 m above the ground. Thereafter its height above the ground is h m, where h is a variable on the set of all real numbers from 0 to 100.

(b) A cone has height 12 cm, base radius 2 cm. The radius of cross-section perpendicular to its axis is r cm, where r is a variable on the set of all real numbers from 0 to 2.

(c) A rectangle is formed by bending a 4 m length of wire. Neglecting the thickness of the wire, the length of one side of the rectangle is x cm, where x is a variable on the set of all real numbers between 0 and 200.

(d) The result of throwing a die is x, where x is a variable on the set $\{1, 2, 3, 4, 5, 6\}$.

If the set S contains only one number, the variable is called a *constant.*

Corresponding to the alternative ways of writing the same number there are alternative ways of writing an expression involving variables. For example $\frac{1}{2} + \frac{1}{8} = \frac{10}{16}$, $\frac{1}{3} + \frac{1}{9} = \frac{12}{27}$ are particular cases of the equation

$$\frac{1}{x-3} + \frac{1}{x+3} = \frac{2x}{x^2-9}, \tag{1}$$

which is true for all x in R (the set of all real numbers) except $x = 3$ or $x = -3$; likewise

$$x^2 - y^2 = (x - y)(x + y) \tag{2}$$

is true for all x in R and all y in R.

Equations, like (1) and (2), that are true for all values of the variables—all values, that is, for which both sides of the equation make sense—are called *identities.* In such cases the equality sign is used to indicate that the two sides are merely alternative ways of writing the expression. This should be clearly distinguished from the use of the equality sign in equations like $x^2 - 4 = 0$ that are true only if x is a variable on some very limited set—in this case the set $\{-2, 2\}$.

Occasionally we shall replace the usual equality sign by the more emphatic symbol $\equiv$ to indicate an identity.

An expression involving a variable x, or involving several variables x, y, etc., is called a function of x, or a function of x, y, etc. In the usual function notation, such expressions are denoted by symbols like $f(x)$, $g(x, y, \ldots)$ where the variables appearing in the expression are enclosed in brackets and the letters f, g may be regarded as standing for the form of the expression, i.e. the precise sequence of operations to be performed on the variables. Thus if $f(x) = x^2 + 2x + 1$, then we would take $f(y)$ to mean $y^2 + 2y + 1$, and further

$$\begin{aligned} f(2x) &= (2x)^2 + 2(2x) + 1 = 4x^2 + 4x + 1, \\ f(x^2) &= (x^2)^2 + 2x^2 + 1 = x^4 + 2x^2 + 1, \\ f(x+1) &= (x+1)^2 + 2(x+1) + 1 = x^2 + 4x + 4, \\ f(2) &= 2^2 + 2\cdot 2 + 1 = 9. \end{aligned}$$

Likewise, if $g(x, y) = x^2 + 2xy^2 + y^3$, then we would take $g(u, v)$ to mean $u^2 + 2uv^2 + v^3$ and further

$$\begin{aligned} g(x+1, y^2) &= (x+1)^2 + 2(x+1)\,y^4 + y^6, \\ g(1, 2) &= 1^2 + 2.1.2^2 + 2^3 = 1 + 8 + 8 = 17. \end{aligned}$$

For a full discussion see Chapter 12.

1.3 Algebra

In discussing the behaviour of the mechanical system of Section 1.1, it is possible to make general statements like

(a) the time in seconds taken for one complete oscillation is $2\pi\sqrt{(M/k)}$,

or particular statements like

(b) if $M = 1$ and $k = 4$ the time taken for one complete oscillation is π s.

Obviously general statements such as (a) that are true for all values of M and k, not just for particular values, are especially useful.

With the real-number system we are mainly concerned with the kind of behaviour that can be expressed in the form of general statements about numbers (i.e. *identities*) rather than statements about particular numbers. This behaviour we call the *algebra* of the real-number system.

In the familiar situations where real numbers are used, the algebra of the real-number system models such aspects of the behaviour as are general rather than particular; for example, Fig. 1.4 illustrates two squares and two rectangles combining, regardless of the dimensions x

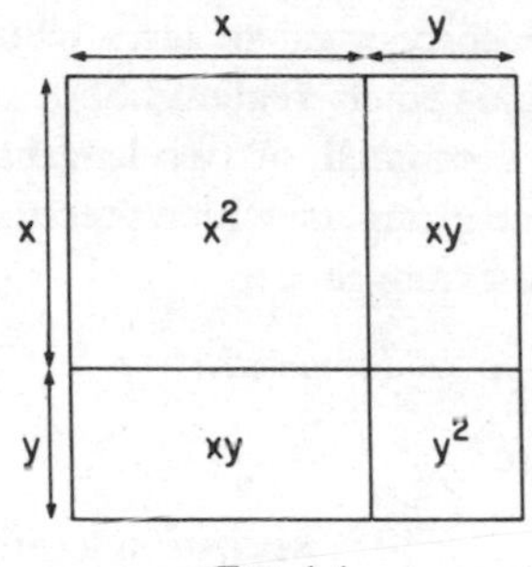

FIG. 1.4

and y, to form a square—this 'behaviour' is modelled in the real-number system by the identity

$$(x + y)^2 \equiv x^2 + y^2 + 2xy. \qquad (1.1)$$

1.3.1 Axioms Before we can make any general statements about the behaviour of a system we must first make some basic assumptions. In the case of the mechanical system in Section 1.1 we assume certain physical laws (Newton's laws of motion, Hooke's law for the spring); the statement of these laws need not concern us here, but it is important to realise that the behaviour of the system is to be regarded as the *consequence* of these laws. Similarly we regard the behaviour of the electrical system in Section 1.1 as the consequence of certain electrical laws which we accept as true.

Now if the assumptions we make about one system are directly analogous to those we make about another system, then we can conclude that the analogy is valid for all the consequent behaviour of the two systems. For example, once we have observed that the analogy described in Section 1.1 exists between the laws governing the mechanical and electrical systems respectively, then since the motion of the ball is determined by the mechanical laws, and the current flow is determined by analogous electrical laws, we conclude that the entire motion of the ball is modelled by the current flow. Thus an alternative statement of (2) in Section 1.1 is as follows:

(2′) The laws governing the behaviour of the representative system are analogous to the laws governing the behaviour of the original system.

In the algebra of the real-number system certain identities, called *axioms*, are assumed to be true. Other identities follow from these just as the motion of the ball follows inevitably from the relevant physical laws.

Moreover, the axioms correspond to laws which we would accept as true in the simple situations where real numbers are used. For example, if two groups of beads are combined, or two lengths of string are joined, it makes no difference which group, or which piece of string, is added to the other; the corresponding axiom is

$$x + y \equiv y + x \qquad \textit{(commutative law of addition)} \qquad (1.2)$$

Other familiar axioms are

$$
\begin{array}{lll}
x + 0 \equiv x, & \text{(0 is } \textit{neutral} \text{ in addition)} & (1.3) \\
(x + y) + z \equiv x + (y + z) & \textit{(associative law of addition)} & (1.4) \\
xy \equiv yx, & \textit{(commutative law of multiplication)} & (1.5) \\
x1 \equiv x, & \text{(1 is } \textit{neutral} \text{ in multiplication)} & (1.6) \\
(xy)z \equiv x(yz), & \textit{(associative law of multiplication)} & (1.7) \\
x(y + z) \equiv xy + xz & \textit{(distributive law)} & (1.8)
\end{array}
$$

Example 7 Show that the identity (1.1) follows from the identities (1.2)–(1.8).

$$
\begin{aligned}
(x + y)^2 &\equiv (x + y)(x + y) \\
&\equiv (x + y)\,x + (x + y)\,y && \text{(by (1.8))} \\
&\equiv x(x + y) + y(x + y) && \text{(by (1.5))} \\
&\equiv x^2 + xy + yx + y^2 && \text{(by (1.8))} \\
&\equiv x^2 + xy + xy + y^2 && \text{(by (1.5))} \\
&\equiv x^2 + 2xy + y^2.
\end{aligned}
$$

Note: steps for which no reason has been given follow as a direct consequence of the meaning of the operation involved.

Naturally there is some choice about which identities are taken as axioms, e.g. (1.1) could be taken as an axiom, but Example 7 shows that some of (1.2)–(1.8) would then be redundant. It is desirable to have a set of axioms in which none are redundant and which is complete in the sense that the axioms determine the entire algebra of the real-number system; such a set is given in Chapter 6.

The real-number system has grown up in connection with simple well-known physical situations to which we seldom question our right to apply it. Strictly speaking, the behaviour of a physical system is modelled by the algebra of the real-number system if

(a) the elements of the physical system can be represented by real numbers,
(b) the behaviour of the physical system is governed by laws that are analogous to the axioms of the real-number system.

1.4 Inequalities

Frequently we compare quantities, as when we say one is bigger, smaller, longer, shorter, etc. than another. This aspect of the physical world can be incorporated in our mathematical representation as follows: we say the number b is greater than the number a if on the real axis b lies to the right of a (written $b > a$)—alternatively we say a is less than b (written $a < b$).

The statement 'either $a < b$ or $a = b$' is abbreviated to $a \leqslant b$ or, alternatively, $b \geqslant a$.

The statement '$x > a$ and also $x < b$' is abbreviated to $a < x < b$; it implies that x lies between a and b as shown.

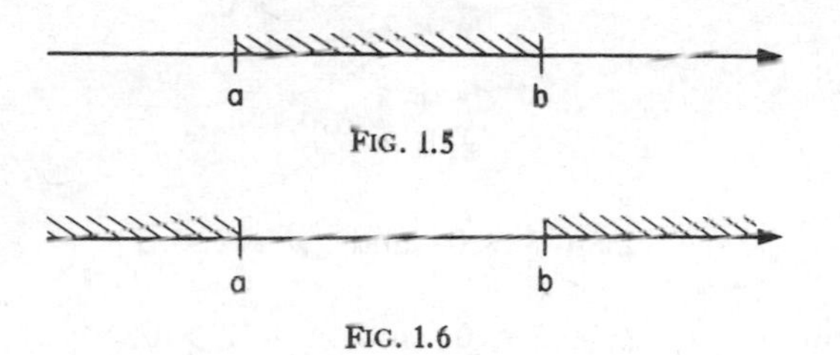

FIG. 1.5

FIG. 1.6

The statement 'either $x < a$ or $x > b$' must be written in full; it implies that x lies outside the portion of the real axis joining a and b. For example,

(a) $-1 < x < 3$ means that x lies between -1 and 3,
(b) $2 < x < 1$ is impossible,
(c) $x \leqslant -1$ or $x \geqslant 3$ means that x does not lie between -1 and 3.

The procedure when performing algebraic operations in the presence of inequality signs can be set out in rules very similar to those we apply to equations:

(a) to each side of an inequality we may add (or subtract) the same number—this means we may transpose terms from one side to the other provided we change their signs.
(b) both sides of an inequality may be multiplied (or divided) by the same positive number; if both sides are multiplied (or divided) by the same negative number, the inequality sign must be reversed, e.g. $5 > 3$, but $-5 < -3$.
(c) if $ab > 0$, then either $a > 0$ and $b > 0$ or else $a < 0$ and $b < 0$; likewise for a/b.
(d) if $ab < 0$, then either $a > 0$ and $b < 0$ or else $a < 0$ and $b > 0$; likewise for a/b.
(e) $a^2 \geqslant 0$ for any number a.

Example 8 Solve the inequality $x^2 - 7x + 10 > 0$.

Factorising, we have $(x - 2)(x - 5) > 0$. Hence $x - 2 > 0$ and $x - 5 > 0$ or else $x - 2 < 0$ and $x - 5 < 0$. That is $x > 2$ and $x > 5$ or else $x < 2$ and $x < 5$. Equivalently, $x > 5$ or else $x < 2$.

Example 9 Solve the inequality $(x + 1)/(x + 2) > 3$.

We have

$$\frac{x + 1}{x + 2} - 3 > 0;$$

that is

$$\frac{x + 1 - 3x - 6}{x + 2} > 0,$$

giving

$$\frac{2x + 5}{x + 2} < 0.$$

Hence

$$2x + 5 > 0 \quad \text{and} \quad x + 2 < 0$$

or else

$$2x + 5 < 0 \quad \text{and} \quad x + 2 > 0;$$

that is

$$x > -\tfrac{5}{2} \quad \text{and} \quad x < -2$$

or else

$$x < -\tfrac{5}{2} \quad \text{and} \quad x > -2 \quad \text{(impossible)}.$$

Hence finally

$$-\tfrac{5}{2} < x < -2.$$

Example 10 Solve the inequality

$$\frac{x^2 - 7x + 10}{x^2 + 5} < 0.$$

Since $x^2 + 5 > 0$ for all x, we require that $x^2 - 7x + 10 < 0$ and so (see Example 8) $2 < x < 5$.

Example 11 Determine the sign of

$$\frac{x^2 - x - 2}{x^2 - 5x + 4}.$$

Since $x^2 - x - 2 = (x + 1)(x - 2)$, its sign is positive if $x + 1 > 0$ and $x - 2 > 0$ or $x + 1 < 0$ and $x - 2 < 0$, i.e. if $x > 2$ or $x < -1$; likewise its sign is negative if $-1 < x < 2$.

Since $x^2 - 5x + 4 = (x - 4)(x - 1)$, its sign is positive if $x > 4$ or $x < 1$, negative if $1 < x < 4$.

It follows that the sign of $(x^2 - x - 2)/(x^2 - 5x + 4)$ is positive when $x < -1$ or when $1 < x < 2$ or when $x > 4$, negative when $-1 < x < 1$ or when $2 < x < 4$.

Example 12 Show that the arithmetic mean of two positive numbers is greater than or equal to their geometric mean, i.e. if $a > 0$, $b > 0$ then $\frac{1}{2}(a + b) \geqslant \sqrt{(ab)}$.

Since $(x - y)^2 \geqslant 0$ for all x and all y, we have

$$x^2 - 2xy + y^2 \geqslant 0$$
$$\text{i.e. } x^2 + y^2 \geqslant 2xy.$$

Substituting $x = \sqrt{a}$, $y = \sqrt{b}$, we get the required result.

The *modulus* of the real number x (sometimes called its *numerical* value or *absolute* value) is denoted by $|x|$ and is defined thus:

$$|x| = x \text{ if } x \geqslant 0, \quad |x| = -x \text{ if } x < 0.$$

In other words the modulus sign has no effect on a positive number and has the effect of removing the minus sign from a negative number. For example,

$$|5| = 5, \quad |-3| = 3.$$

We easily see that

(a) $|x| \geqslant 0$,
(b) $|xy| = |x||y|$,
(c) $\dfrac{|x|}{|y|} = \left|\dfrac{x}{y}\right|$,
(d) $|x| - |y| \leqslant |x + y| \leqslant |x| + |y|$.

Note that the distance between a and b on the real axis is $(b - a)$ or $(a - b)$ according as $b > a$ or $a > b$. This distance is given correctly by $|a - b|$ in either case.

Example 13 Solve the inequality $|x - 3| \leqslant 2$.

The distance between the points 3 and x must be at most 2. Hence $1 \leqslant x \leqslant 5$.

Example 14 Show that $x^2 + y^2 \geqslant 2|xy|$.

From Example 12 we have the result that $x^2 + y^2 \geqslant 2xy$ for all x and all y. Substituting $|x|$ for x and $|y|$ for y, we get that

$$|x|^2 + |y|^2 \geqslant 2|x||y|,$$

that is,

$$x^2 + y^2 \geqslant 2|xy|.$$

Example 15 Prove the following inequalities, in which a, b stand for unequal positive numbers:

(a) $a^3 - a^2 > a^{-2} - a^{-3}$, $\quad (a \neq 1)$;
(b) $a^{m+n} + b^{m+n} > a^m b^n + a^n b^m$ $\quad$ (m, n positive integers).

(a)

$$a^3 + a^{-3} - a^2 - a^{-2} = (a^3 - a^2)(1 - a^{-5}).$$

If $a > 1$, $a^3 - a^2$ and $1 - a^{-5}$ are both positive; if $a < 1$, $a^3 - a^2$ and $1 - a^{-5}$ are both negative. It follows that when $a \neq 1$,

$$a^3 + a^{-3} - a^2 - a^{-2} > 0,$$

that is,

$$a^3 - a^2 > a^{-2} - a^{-3}.$$

(b)

$$a^{m+n} + b^{m+n} - a^m b^n - a^n b^m = (a^m - b^m)(a^n - b^n).$$

$a^m - b^m$ and $a^n - b^n$ are both positive when $a > b$ and both negative when $a < b$. Hence

$$a^{m+n} + b^{m+n} > a^m b^n + a^n b^m.$$

Example 16 Given that a, b, c, d are real numbers, prove that

$$a^4 + b^4 \geqslant 2a^2b^2, \; a^4 + b^4 + c^4 + d^4 \geqslant 4abcd.$$

Prove also that

$$(a^2 + b^2)^2 + (c^2 + d^2)^2 \geqslant 2(ab + cd)^2.$$

Show that, if $a^4 + b^4 + c^4 + d^4 \leqslant 1$, then

$$1/a^4 + 1/b^4 + 1/c^4 + 1/d^4 \geqslant 16.$$

From Example 12,

$$a^4 + b^4 \geqslant 2a^2b^2 \text{ and } c^4 + d^4 \geqslant 2c^2d^2.$$

Hence

$$a^4 + b^4 + c^4 + d^4 \geqslant 2(a^2b^2 + c^2d^2) \tag{1}$$

and so, from Example 14,

$$a^4 + b^4 + c^4 + d^4 \geqslant 4|abcd| \geqslant 4abcd. \tag{2}$$

Also,

$$\begin{aligned}(a^2 + b^2)^2 + (c^2 + d^2)^2 &= a^4 + b^4 + c^4 + d^4 + 2(a^2b^2 + c^2d^2) \\ &\geqslant 4abcd + 2(a^2b^2 + c^2d^2) \qquad \text{(by (2))}.\end{aligned}$$

Hence

$$(a^2 + b^2)^2 + (c^2 + d^2)^2 \geqslant 2(ab + cd)^2.$$

Finally, from Example 12,

$$1/a^4 + 1/b^4 \geqslant 2/a^2b^2 \quad \text{and} \quad 1/c^4 + 1/d^4 \geqslant 2/c^2d^2.$$

Hence

$$\begin{aligned} 1/a^4 + 1/b^4 + 1/c^4 + 1/d^4 &\geqslant 2(1/a^2b^2 + 1/c^2d^2) \\ &\geqslant 4/|abcd| \\ &\geqslant 16/(a^4 + b^4 + c^4 + d^4) \qquad \text{(by (2))} \\ &\geqslant 16 \qquad \text{if } a^4 + b^4 + c^4 + d^4 \leqslant 1. \end{aligned}$$

Exercises 1

Solve the following inequalities:

1. $2x + 5 < x + 7$
2. $\dfrac{x + 3}{1 - 2x} > 0$
3. $\dfrac{x + 3}{1 - 2x} > 1$
4. $-2 < x^2 + 3x < 10$
5. $|x^3 + x| < 2$
6. $\dfrac{3x^2 - 6x + 6}{x^2 + x + 1} > 1$
7. $\dfrac{2x^2 - 3x + 1}{x^2 - 1} > 1.$
8. Given that a, b, c are positive, prove that

   (a) $(b + c)(c + a)(a + b) \geqslant 8abc$,
   (b) $(a^7 + b^7)(a^2 + b^2) \geqslant (a^5 + b^5)(a^4 + b^4)$.

   Given $y + z > x$, $z + x > y$, $x + y > z$, use (a) to prove that

   $$(y + z - x)(z + x - y)(x + y - z) \leqslant xyz.$$

9. Given that $x/y + y/x = a$, $y/z + z/y = b$, $x/z + z/x = c$ and x, y, z are all real, prove that

   (a) a^2, b^2, c^2 are each not less than 4,
   (b) if two of a, b, c are equal to -2, the other must be equal to $+2$,
   (c) $[a \pm \sqrt{(a^2 - 4)}]\,[b \pm \sqrt{(b^2 - 4)}]\,[c \pm \sqrt{(c^2 - 4)}] = 8$, where one of the ambiguous signs is opposite to the other two.

1.5 Further Developments

An important operation with numbers, the operation of taking a limit, is introduced in Chapter 14. Having introduced a new operation we have in effect introduced a new kind of behaviour about which the algebra of the real-number system was not concerned. The study of this behaviour and the problems it enables us to deal with is the main theme of the book from Chapter 14 to the end.

The set of real numbers under all the operations now at our disposal is a powerful tool for solving problems arising from science and engineering. We have three basic requirements:

(a) the physical entities involved should be capable of measurement so that they can be represented by real numbers,
(b) the physical behaviour under investigation must be modelled by operations within the real-number system, and
(c) we must know methods whereby the mathematical version of the problem can be solved.

In connection with (a) it should be pointed out that some quite common physical entities (e.g. force, displacement, velocity, acceleration) cannot be adequately represented by numbers. For this reason a new mathematical system (vectors) is introduced (see Chapter 7).

Other new mathematical systems (complex numbers, matrices) will be studied mainly in connection with (b) and (c).

In the light of (c) it may be necessary to introduce approximations, or to neglect factors believed to have only minor effects, in order to arrive at a mathematical problem which we can solve; in these circumstances it is important to appreciate the limitations of the mathematical solution, but such solutions can nevertheless be of immense value.

CHAPTER 2

TWO-DIMENSIONAL CO-ORDINATE GEOMETRY

In the early sections of this chapter a number of results are stated without proof. Most of these results are either revision or simple extension of school work. In the introduction to conic sections (Section 2.4) the reader need not be concerned with proof as this material is presented solely for information purposes.

2.1 The Straight Line (revision)

(a) *Distance, gradient and section formulae.* The distance between $A(x_1, y_1)$ and $B(x_2, y_2)$ is given by

$$AB = \sqrt{[(x_2 - x_1)^2 + (y_2 - y_1)^2]}$$

and the gradient of AB (denoted by m_{AB}) is given by

$$m_{AB} = \frac{y_2 - y_1}{x_2 - x_1}.$$

The point P which divides AB in the ratio $\lambda : \mu$ is

$$\left(\frac{\lambda x_2 + \mu x_1}{\lambda + \mu}, \frac{\lambda y_2 + \mu y_1}{\lambda + \mu}\right).$$

P divides AB internally or externally according as the ratio $\lambda : \mu$ is positive or negative.

The co-ordinates of the mid-point of AB are

$$\tfrac{1}{2}(x_1 + x_2), \quad \tfrac{1}{2}(y_1 + y_2).$$

(b) *The angle between two given lines.* If θ is the angle between two straight lines of gradients m_1, m_2, then

$$\tan\theta = \pm\frac{m_1 - m_2}{1 + m_1 m_2}.$$

This formula gives the tangent of the acute or obtuse angle between the lines according as the sign chosen makes the right-hand side positive or negative.

The lines are parallel if $m_1 = m_2$; the lines are perpendicular if $m_1 m_2 = -1$.

(*Hint*: If the lines make angles θ_1, θ_2 respectively with the positive x-axis, then $\tan\theta_1 = m_1$, $\tan\theta_2 = m_2$, and

$$\tan\theta = \pm\tan(\theta_1 - \theta_2) = \pm\frac{\tan\theta_1 - \tan\theta_2}{1 + \tan\theta_1\tan\theta_2}\Big).$$

(c) *Centroid of a triangle.* The centroid of the triangle whose vertices are (x_1, y_1), (x_2, y_2), (x_3, y_3) has co-ordinates

$$\tfrac{1}{3}(x_1 + x_2 + x_3), \quad \tfrac{1}{3}(y_1 + y_2 + y_3).$$

This result may be obtained by regarding the centroid as a point of trisection of a median of the triangle and using the section formula.

(d) *The equation of the straight line.* In cartesian co-ordinates any equation of the first degree in x and y, i.e. any equation of the form

$$ax + by + c = 0,$$

where a, b, c are constants, represents a straight line, but the reader should be familiar with the forms of this equation shown in Table 2.1.

TABLE 2.1

Equation	Description of line
$x =$ constant	line parallel to the y-axis
$y =$ constant	line parallel to the x-axis
$y = mx$	line of gradient m through the origin
$y = mx + c$	line of gradient m making an intercept c on Oy
$y - y_1 = m(x - x_1)$	line of gradient m passing through the point (x_1, y_1)
$\frac{y - y_1}{x - x_1} = \frac{y_2 - y_1}{x_2 - x_1}$	line joining the points (x_1, y_1), (x_2, y_2)
$\frac{x}{a} + \frac{y}{b} = 1$	line making intercepts a and b on Ox, Oy respectively
$x\cos\alpha + y\sin\alpha = p$	line such that the perpendicular to it from the origin is of length p and makes an angle α with Ox
$ax + by + c + \lambda(Ax + By + C) = 0$	line through the point of intersection of the lines $ax + by + c = 0$, $Ax + By + C = 0$

In the last case note that the equation is of the first degree and so represents a straight line; also, the equation is satisfied by any pair of co-ordinates (x, y) that satisfies *both* the equations $ax + by + c = 0$, $Ax + By + C = 0$.

(e) *Distance of a point from a straight line and the equations of the bisectors of the angles between two straight lines.* The perpendicular distance of the point $P(x_1, y_1)$ from the line $ax + by + c = 0$ is

$$\pm\frac{ax_1 + by_1 + c}{\sqrt{(a^2 + b^2)}}.$$

If the positive sign is chosen when c is positive and the negative sign when c is negative, this formula gives a positive result when P lies on the same side of the line as the origin, a negative result when P lies on the opposite side from the origin.

(*Hint*: Let the perpendicular from the point $P(x_1, y_1)$ meet the given line at the point $Q(x_1 + h, y_1 + k)$. The slope of PQ is k/h and so $k/h = b/a$; also the point Q lies on the given line so that $a(x_1 + h) + b(y_1 + k) + c = 0$. We may thus solve for h and k and hence find $\sqrt{(h^2 + k^2)}$, which gives the required distance PQ.

Note also that the given line divides the xy-plane into two regions, in one of which $ax + by + c > 0$, while in the other $ax + by + c < 0$. The value of $ax + by + c$ at the origin is c, and so the signs of $(ax_1 + by_1 + c)$ and c are the same or opposite according as the points (x_1, y_1) and $(0, 0)$ lie on the same or opposite sides of the given line.)

From the above formula we deduce that the equations of the bisectors of the angles between the lines $ax + by + c = 0, Ax + By + C = 0$ are

$$\frac{ax + by + c}{\sqrt{(a^2 + b^2)}} = \frac{Ax + By + C}{\pm\sqrt{(A^2 + B^2)}}$$

2.1.1 Parametric Equations of a Straight Line Let a straight line AB (Fig. 2.1) drawn through the fixed point $A(x_1, y_1)$ make an angle θ with Ox; let $P(x, y)$ be any point on AB, or on AB produced, and let $AP = r$. Then

$$x = x_1 + r\cos\theta, \quad y = y_1 + r\sin\theta.$$

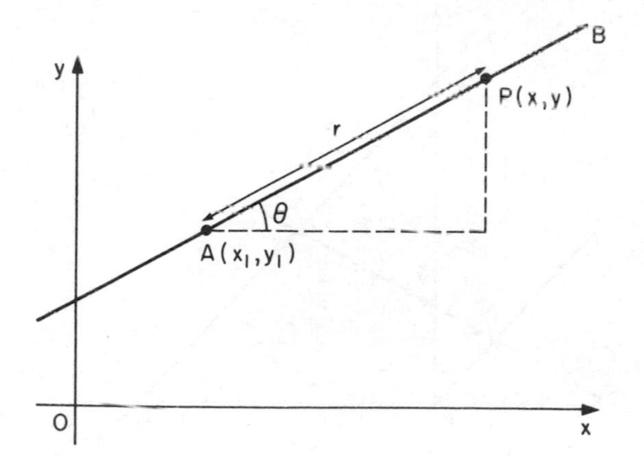

FIG. 2.1

These equations, which give the co-ordinates of any point on the line in terms of the single variable (parameter) r, are the parametric equations of the line. They may also be written in the form

$$\frac{x - x_1}{\cos\theta} = \frac{y - y_1}{\sin\theta} = r.$$

Note that points on BA produced correspond to negative values of r.

Example 1 Find the equation of the straight line drawn through the point $P(h, k)$ such that, if it meets the axes of co-ordinates in the points A and B, then P is the middle point of AB.

Show that if any straight line drawn through P meets the axis of x at the point X, the axis of y at the point Y and the parallel through the origin to the straight line AB at the point Q, then $2/PQ = 1/PX + 1/PY$, the lengths being measured algebraically.

If, in Fig. 2.2, $P(h, k)$ is the mid-point of AB, $A \equiv (2h, 0)$, $B \equiv (0, 2k)$, the equation of AB is

$$\frac{x}{2h} + \frac{y}{2k} = 1,$$

and the equation of the line drawn parallel to AB through the origin is

$$\frac{x}{h} + \frac{y}{k} = 0. \tag{1}$$

The equation of any line through P may be taken as

$$\frac{x - h}{\cos\theta} = \frac{y - k}{\sin\theta} = r. \tag{2}$$

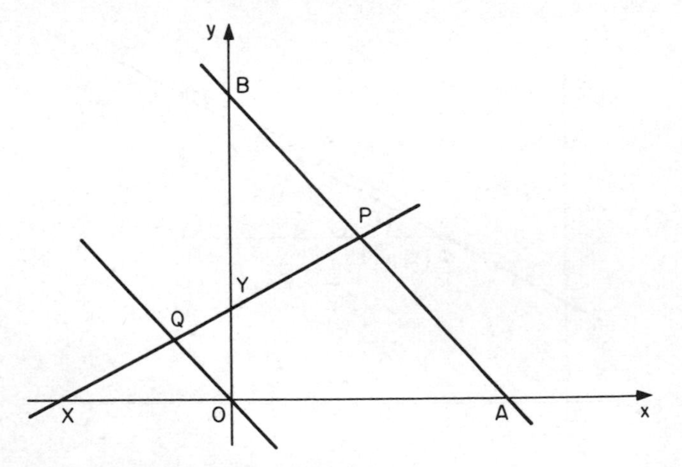

FIG. 2.2

This line meets Ox at X where $y = 0$ and $r = PX$.
Hence

$$PX = -k/\sin\theta. \tag{3}$$

Similarly,

$$PY = -h/\cos\theta. \tag{4}$$

Lines (1) and (2) meet at Q where

$$\frac{h + r\cos\theta}{h} + \frac{k + r\sin\theta}{k} = 0 \quad \text{and} \quad r = PQ.$$

Hence

$$\frac{\cos\theta}{h} + \frac{\sin\theta}{k} = -\frac{2}{PQ},$$

and so by (3) and (4)

$$\frac{1}{PX} + \frac{1}{PY} = \frac{2}{PQ}.$$

2.2 The Circle (revision)

(a) The equation of the circle with centre (α, β) and radius a is

$$(x-\alpha)^2 + (y-\beta)^2 = a^2.$$

(b) The equation $x^2 + y^2 + 2gx + 2fy + c = 0$ represents a circle with centre $(-g, -f)$ and radius $\sqrt{(g^2 + f^2 - c)}$. This circle is real if $g^2 + f^2 \geqslant c$.

(c) The equation of the circle whose diameter is the line joining the points (x_1, y_1) and (x_2, y_2) is $(x - x_1)(x - x_2) + (y - y_1)(y - y_2) = 0$.

(*Hint*: If P_1, P_2 are the points (x_1, y_1), (x_2, y_2) respectively, and P is any other point on the circle, then the lines PP_1 and PP_2 are perpendicular.)

(d) The circles

$$x^2 + y^2 + 2gx + 2fy + c = 0, \quad x^2 + y^2 + 2g'x + 2f'y + c' = 0$$

cut orthogonally if $2gg' + 2ff' = c + c'$.

(*Hint*: If O_1, O_2 are the centres and the circles intersect orthogonally at P, then the angle O_1PO_2 is a right-angle so that $O_1O_2^2$ equals the sum of the squares of the radii.)

(e) For any value of λ other than $\lambda = -1$ the equation

$$x^2 + y^2 + 2gx + 2fy + c + \lambda(x^2 + y^2 + 2g'x + 2f'y + c') = 0$$

represents a circle; for $\lambda = -1$ the equation represents a straight line. If the circles

$$x^2 + y^2 + 2gx + 2fy + c = 0, \quad x^2 + y^2 + 2g'x + 2f'y + c' = 0$$

intersect, then the circle or straight line represented by the above equation passes through the points of intersection.

2.3 Pairs of Straight Lines

The equation

$$(a_1x + b_1y + c_1)(a_2x + b_2y + c_2) = 0 \tag{1}$$

is satisfied by all points which satisfy *either* of the two equations

$$a_1x + b_1y + c_1 = 0, \quad a_2x + b_2y + c_2 = 0 \tag{2}$$

and by no other points. Hence eq. (1) represents the pair of lines given by (2). Thus any second-degree equation in x and y which factorises into two linear factors represents a pair of lines, e.g.

(a) $x^2 - 3y^2 - 2xy + 2x - 6y = 0$ gives $(x - 3y)(x + y + 2) = 0$ and so represents the pair of lines $x - 3y = 0$, $x + y + 2 = 0$,
(b) $x^2 - 2xy + y^2 + 4x - 4y + 4 = 0$ can be written as $(x - y + 2)^2 = 0$ and so represents the line $x - y + 2 = 0$, i.e. this second-degree equation is effectively the same as the linear equation $x - y + 2 = 0$.

Exercises 2(a)

1. Prove that if the points

$$(x_1, y_1), \quad (x_2, y_2), \quad (x_3, y_3), \quad (x_4, y_4)$$

are the vertices of a parallelogram taken in order, then

$$x_1 + x_3 = x_2 + x_4 \quad \text{and} \quad y_1 + y_3 = y_2 + y_4.$$

2. The fixed line $x/a + y/b = 1$ meets the axis of x at X and the axis of y at Y. Any straight line perpendicular to this straight line meets the axis of x at X' and the axis of y at Y'. Prove that the locus of the intersection of the straight lines XY' and $X'Y$ is the circle $x^2 + y^2 = ax + by$.
3. Show that the condition for the equation

$$ax^2 + 2hxy + by^2 + 2gx + 2fy + c = 0$$

to represent a pair of straight lines is

$$abc + 2fgh - af^2 - bg^2 - ch^2 = 0, \qquad h^2 - ab > 0.$$

(Treat the equation as a quadratic in y and solve for y in terms of x; the term inside the square root must be a perfect square to give solutions of the form $y = Ax + B$.)

Verify that the equation

$$y^2 - 4xy + x^2 - 10y + 8x + 13 = 0$$

represents a pair of straight lines.

4. Prove that, if $\mu < 169/56$, there are two finite real values of λ for which the equation

$$9x^2 + \lambda xy + \mu y^2 - 45x + 13y + 14 = 0$$

represents a pair of straight lines. If $\mu = -1$, find the separate equations of the lines constituting *one* of these pairs.

5. Show that the equation $ax^2 + 2hxy + by^2 = 0$ represents a pair of straight lines through the origin if $h^2 > ab$. Show also that if θ is an angle between these lines then

$$\tan\theta = \pm\frac{2\sqrt{(h^2 - ab)}}{a + b}.$$

6. Show that the equation

$$(x^2 - y^2)\cos\alpha + 2xy\sin\alpha = \lambda(x^2 + y^2), \qquad 0 < \lambda < 1,$$

represents a pair of straight lines, and find the acute angle between them.

7. Find the equations of the two circles of radius $\sqrt{2}$ with their centres on the x-axis which touch the line $x + y + 1 = 0$.

8. A circle passes through the points of intersection of the circles

$$3x^2 + 3y^2 - 6x - 1 = 0 \text{ and } x^2 + y^2 + 2x - 4y + 1 = 0$$

and also passes through the centre of the first circle. Find its equation and verify that it cuts the second circle orthogonally.

9. Find the equation of the circle on the join of (1, 3), (2, 4) as diameter, and obtain the equation of another circle with centre (−1, 2) which meets the first circle orthogonally.

2.4 Conic Sections

Suppose the plane π (Fig. 2.3) makes an angle α with the axis of the cone, vertex O, angle at vertex 2β. Then π intersects the cone along a curve which

(a) is circular if $\alpha = 90°$
(b) is called an *ellipse* if $\alpha > \beta$ (as in Fig. 2.3)
(c) is called a *hyperbola* if $\alpha < \beta$
(d) is called a *parabola* if $\alpha = \beta$.

The cone is regarded as continuing beyond the vertex O, i.e. it consists of two similar parts, each extending indefinitely. In case (c) the plane intersects both parts and so the hyperbola consists of two parts. By passing the plane π through O we see that, as a special case, the hyperbola may reduce to a pair of straight lines. It should also be noted that a circle may be regarded as a special case of an ellipse.

In Fig. 2.3 two spheres have been drawn so as to touch π and the cone. The upper sphere touches π at S and touches the cone along a circle; the

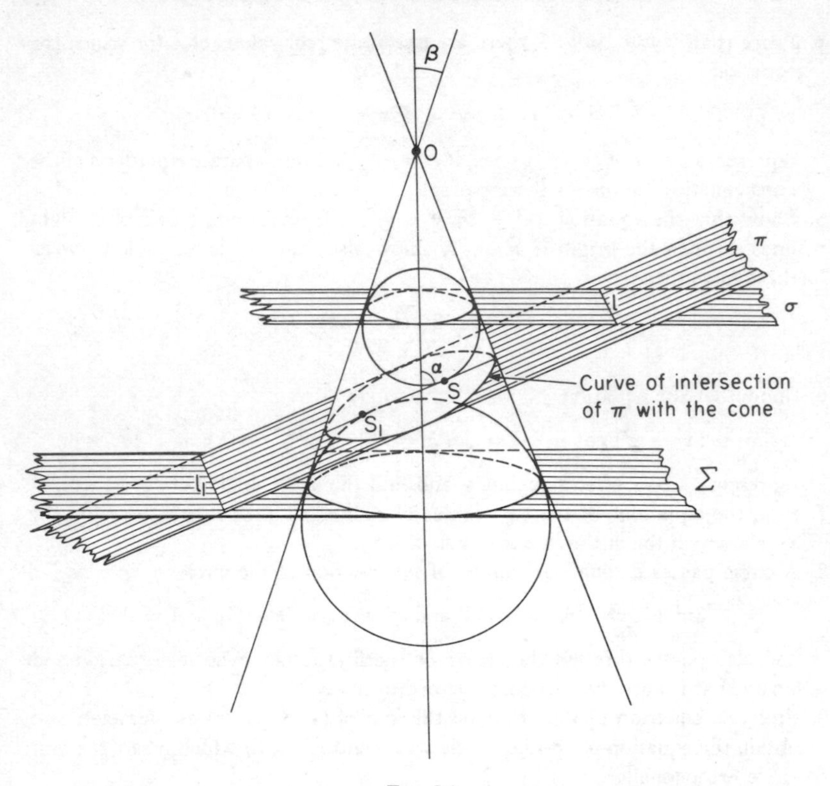

FIG. 2.3

plane of this circle of contact is denoted by σ. The lower sphere touches π at S_1 and touches the cone along a circle in the plane Σ. The planes σ, Σ intersect the plane π in lines l, l_1 respectively. If P is any point on the curve of intersection of π with the cone, then it can be shown that the distance of P from S is e times the perpendicular distance of P from l, where $e = \cos \alpha/\cos \beta$; the same result holds if we take S_1 and l_1 instead of S and l. Figure 2.3 has been drawn for case (b), but corresponding figures can be drawn for the other cases. Thus a conic section may be given the following alternative definition:

Definition. If S is a fixed point and l is a fixed line which does not pass through S, the locus of a point which moves in the plane of S and l so that its distance from S is in a constant ratio to its distance from l is called a *conic section* or simply a *conic*. The fixed point is called a *focus* of the conic, the fixed line a *directrix*, and the constant ratio (denoted by e) the *eccentricity* of the conic. The conic is called a ***parabola***, ***ellipse*** or ***hyperbola*** according as $e = 1$, $e < 1$ or $e > 1$.

From this definition the standard equations given in Sections 2.4.1–2.4.3 are easily derived.

2.4.1 The Parabola The standard equation of the parabola is $y^2 = 4ax$. When the equation of the curve is in this form,

(a) the focus S is the point $(a, 0)$,
(b) the directrix ZM is the line $x = -a$,
(c) the vertex O is the origin of co-ordinates,
(d) the x-axis is the axis of symmetry of the curve and is called the axis of the parabola,
(e) the y-axis is the tangent at the vertex,
(f) the latus rectum LL_1 (the double ordinate through the focus) is of length $4a$.

The parabola $y^2 = 4ax$ is shown in Fig. 2.4.

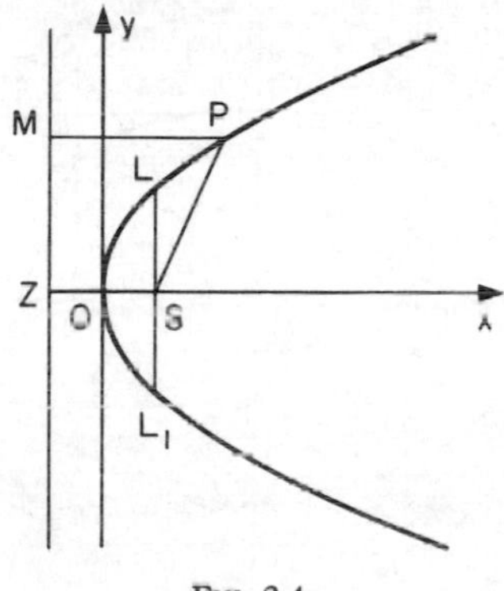

FIG. 2.4

The point whose co-ordinates are given in terms of the single variable t by the equations

$$x = at^2, \quad y = 2at \tag{1}$$

lies on the parabola $y^2 = 4ax$ for all values of t, and eqs. (1) may be taken as the parametric equations of the parabola $y^2 = 4ax$. As t varies from $-\infty$ to $+\infty$, the point given by (1) describes the parabola completely. The point $(at^2, 2at)$ is referred to as the point of parameter t or simply the point $[t]$.

2.4.2 The Ellipse The standard equation of the ellipse is

$$\frac{x^2}{a^2} + \frac{y^2}{b^2} = 1, \tag{2.1}$$

where $b^2 = a^2(1 - e^2)$.

When the equation of the curve is in this form,

(a) the foci S, S_1 are the points $(\pm ae, 0)$,
(b) the directrices ZM, Z_1M_1 are the lines $x = \pm a/e$,
(c) the eccentricity e, less than unity, is given by $e^2 = 1 - b^2/a^2$,
(d) the centre O is the origin of co-ordinates,
(e) the major axis AA_1 is of length $2a$ and lies along the x-axis,
(f) the minor axis BB_1 is of length $2b$ and lies along the y-axis,
(g) each latus rectum is of length $2b^2/a$.

The ellipse $x^2/a^2 + y^2/b^2 = 1$ is shown in Fig. 2.5. Since the major and minor axes (i.e. the principal axes) of the curve lie along the axes of co-ordinates, (2.1) is sometimes called the equation of the ellipse referred to its principal axes.

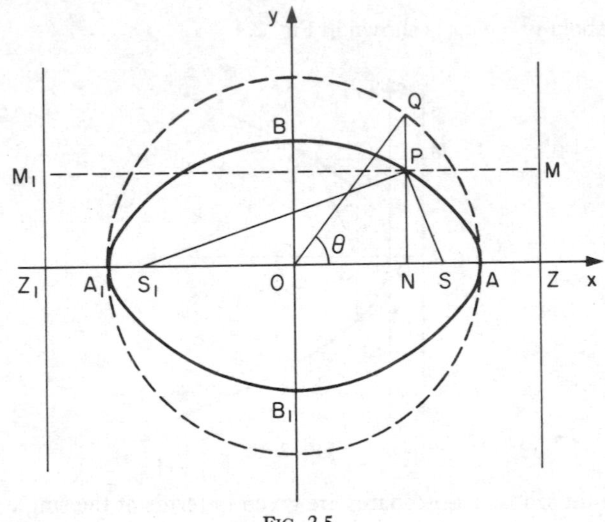

FIG. 2.5

The sum of the focal distances SP, S_1P of any point P on the ellipse is constant and equal to the length of the major axis. Hence the ellipse may also be defined as the locus of a point which moves in a plane so that the sum of its distances from two fixed points is constant.

The circle described on the major axis of an ellipse as diameter is called the *auxiliary circle* of the ellipse. If NP (Fig. 2.5), any ordinate of the ellipse, is produced to meet the auxiliary circle at Q and if $\angle\, x\,OQ = \theta$, then Q is the point $(a \cos \theta, a \sin \theta)$, and from (2.1) P is $(a \cos \theta, b \sin \theta)$. The angle θ is called the *eccentric angle* of P. Since $NP/NQ = b/a$, NP can be obtained as the orthogonal projection of NQ on a plane inclined at an angle ϕ to NQ, where $\cos \phi = b/a$. Thus an ellipse is the orthogonal projection of its auxiliary circle on a plane drawn through A_1A (or through any line parallel to A_1A) which makes an angle ϕ with the plane of the auxiliary circle. The equations

$$x = a \cos \theta, \quad y = b \sin \theta$$

may be taken as parametric equations of the ellipse.

2.4.3 The Hyperbola The standard equation of the hyperbola is

$$\frac{x^2}{a^2} - \frac{y^2}{b^2} = 1 \tag{2.2}$$

where $b^2 = a^2(e^2 - 1)$.

When the equation of the curve is in this form,

(a) the foci, S, S_1 are the points $(\pm ae, 0)$,
(b) the directrices ZM, Z_1M_1 are the lines $x = \pm a/e$,
(c) the eccentricity e, greater than unity, is given by $e^2 = 1 + b^2/a^2$,
(d) the centre O is the origin of co-ordinates,
(e) the transverse axis AA_1 is of length $2a$ and lies along the x-axis,
(f) the conjugate axis BB_1 is of length $2b$ and lies along the y-axis, with its centre at the origin,
(g) each latus rectum is of length $2b^2/a$.

The hyperbola $x^2/a^2 - y^2/b^2 = 1$ is shown in Fig. 2.6.

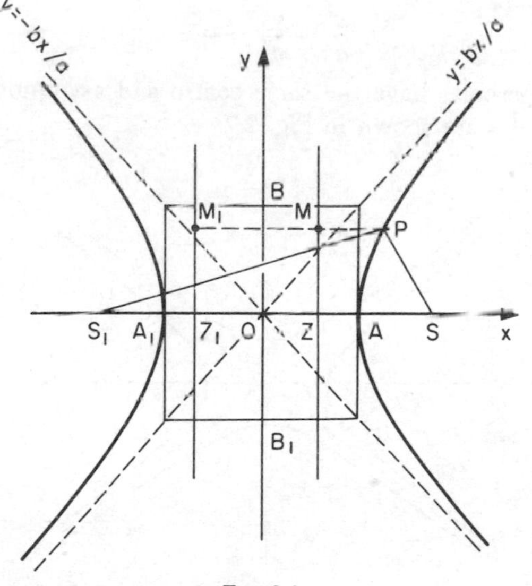

FIG. 2.6

Writing eq. (2.2) in the form

$$y/b = \pm(x/a)\sqrt{(1 - a^2/x^2)},$$

we see that $y/b \simeq \pm x/a$ if x^2 is large. The hyperbola gets closer and closer to the lines $y = \pm(b/a)\,x$ as x becomes larger and larger. These lines are called *asymptotes* of the hyperbola and are shown in Fig. 2.6.

The difference of the focal distances of a point on a hyperbola is constant and equal to the length of the transverse axis. Hence the hyperbola may also be defined as the locus of a point which moves in a plane so that the difference of its distances from two fixed points is constant.

The point whose co-ordinates are

$$x = a \sec\theta, \quad y = b \tan\theta \tag{1}$$

lies on the hyperbola $x^2/a^2 - y^2/b^2 = 1$ for all values of θ, and so eqs. (1) may be taken as the parametric equations of the hyperbola.

Other parametric equations of the hyperbola are

$$x = \tfrac{1}{2}a(t + 1/t), \quad y = \tfrac{1}{2}b(t - 1/t).$$

Two hyperbolas are said to be *conjugate* if the traverse and conjugate axes of the one coincide respectively with the conjugate and transverse axes of the other.

The hyperbola conjugate to

$$\frac{x^2}{a^2} - \frac{y^2}{b^2} = 1$$

is

$$\frac{x^2}{a^2} - \frac{y^2}{b^2} = -1.$$

Conjugate hyperbolas have the same centre and asymptotes. Two conjugate hyperbolas are shown in Fig. 2.7.

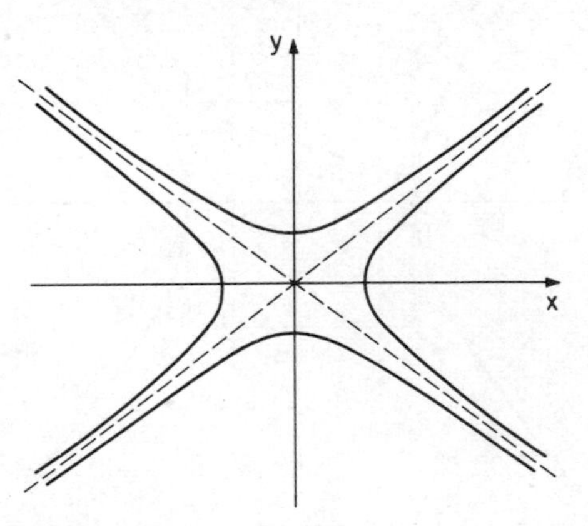

FIG. 2.7

A hyperbola is said to be *rectangular* if its asymptotes are perpendicular.

The asymptotes $y = \pm(b/a)x$ of the hyperbola $x^2/a^2 - y^2/b^2 = 1$ are perpendicular if $b^2 = a^2$. In this case, the eccentricity of the hyperbola is $\sqrt{2}$, its equation is $x^2 - y^2 = a^2$ and its asymptotes $y = \pm x$ bisect the angles between the axes. (See Example 4, Section 2.6, for a simpler form for the equation of a rectangular hyperbola.)

Exercises 2(b)

1. A point P moves on the parabola $y^2 = 4ax$, and B is the point $(2a, 0)$. Show that the locus of the middle point of BP is a parabola whose focus is the point $(3a/2, 0)$.
2. Show that if two variable points P, Q, on the parabola $y^2 = 4ax$ subtend a right angle at the vertex, then PQ meets the x-axis at a fixed point.

   Show also that the locus of the mid-point of PQ is a parabola whose vertex is at the point $(4a, 0)$.
3. The points $P\,(ap^2, 2ap)$ and $R\,(ar^2, 2ar)$ lie on the parabola γ with the equation $y^2 = 4ax$. Find the equation of the line PR in its simplest form, and obtain the condition which must be satisfied by p and r if PR is to pass through the focus of γ.

   PR and QS are focal chords of γ. Show that both pairs of opposite sides of the quadrilateral $PQRS$ intersect on the directrix of γ.
4. The circle $x^2 + y^2 = a^2$ intersects the x-axis at the points $A\,(a, 0)$ and $B\,(-a, 0)$ and P is the point $(a\cos\theta, a\sin\theta)$. Prove that the equations of the lines AP and BP are

$$(x - a)\cos(\theta/2) + y\sin(\theta/2) = 0$$

   and

$$(x + a)\sin(\theta/2) - y\cos(\theta/2) = 0.$$

   The lines AB and CD are two perpendicular diameters of a circle, and P is a variable point on its circumference. The straight line AP meets CD in Q and the straight line through Q parallel to AB meets BP in R. Prove that the locus of the point R is a parabola passing through the points B, C and D.
5. P and Q are points on the parabola $y^2 = 4ax$. The perpendicular bisector of PQ meets the axis of the parabola at R, and the perpendicular to the axis drawn through the mid-point V of PQ meets the axis at M. Prove that MR is of constant length for all positions of P and Q.
6. The foci of an ellipse of eccentricity $12/13$ are the points $(0, \pm 12)$. Show that the equation of the ellipse is $x^2/5^2 + y^2/13^2 = 1$.
7. Given that the point P on the ellipse $x^2/a^2 + y^2/b^2 = 1$ has the eccentric angle θ, find the equations of the lines PA, PA', joining P to the ends A, A' of the major axis.

   The lines through P perpendicular to PA, PA' meet the major axis in K, K'; show that length KK' is constant.

8. A straight rod PQ has length $a + b$, and R is the point on it such that $QR = a$, $RP = b$. The rod moves so that its ends P, Q slide on two perpendicular straight lines OX, OY respectively. Find with respect to axes OX, OY, the equation of the curve traced by R.
9. Given that two variable points $P[\theta]$, $Q[\phi]$ move on the ellipse

$$x^2/a^2 + y^2/b^2 = 1$$

in such a way that $(\theta - \phi)$ is constant and equal to 2α, show that the locus of the mid-point of PQ is the ellipse

$$x^2/a^2 + y^2/b^2 = \cos^2 \alpha.$$

10. The distance between the foci of an ellipse is 8 units and that between the directrices is 18 units. Find the equation of the ellipse referred to its principal axes.
11. PFQ and $PF'Q'$ are two focal chords of an ellipse, the foci being F, F', and the eccentric angles of Q and Q' are ϕ and ϕ'. Show that the ratio $\tan \frac{1}{2}\phi : \tan \frac{1}{2}\phi'$ is constant for all positions of P.
12. The line $lx + my = 1$ meets the ellipse $x^2/a^2 + y^2/b^2 = 1$ in points A, B. Show that the equation

$$x^2(l^2 - 1/a^2) + 2lmxy + y^2(m^2 - 1/b^2) = 0$$

represents the pair of straight lines joining A, B to the origin.
13. Show that the equation of the chord of the ellipse

$$b^2x^2 + a^2y^2 = a^2b^2$$

whose mid-point P is (α, β) is

$$x\alpha/a^2 + y\beta/b^2 = \alpha^2/a^2 + \beta^2/b^2.$$

14. Prove that the equation of the line joining the points $(ct_1, c/t_1)$, $(ct_2, c/t_2)$ on the rectangular hyperbola $xy = c^2$ is

$$x + t_1t_2y - c(t_1 + t_2) = 0.$$

A, B, C, D are four points on a rectangular hyperbola. Prove that if AB is perpendicular to CD, then AC is perpendicular to BD and AD is perpendicular to BC. Prove further that three of the points are on one branch of the hyperbola and one on the other branch.
15. Prove that the point P whose co-ordinates are

$$x = a \sec \theta, \quad y = a \tan \theta,$$

lies on the rectangular hyperbola $x^2 - y^2 = a^2$.

If Q is the point whose parameter is $(\theta + \pi/2)$ and $R(x_1, y_1)$ is the mid-point of PQ, prove that

$$y_1/x_1 = \sin \theta + \cos \theta$$

and find the locus of R.
16. A chord PQ of the rectangular hyperbola $xy = c^2$ meets the asymptotes at R and S. Prove that $PR = SQ$.

2.5 Change of Origin (translation of axes)

Suppose that new axes O_1X, O_1Y (Fig. 2.8) are drawn through O_1, parallel to the original axes and in the same sense. Then if P is the point (x, y) and O_1 is the point (α, β) referred to the original axes Ox, Oy, and if P is the point (X, Y) referred to the new axes, we have

$$x = X + \alpha, \quad y = Y + \beta.$$

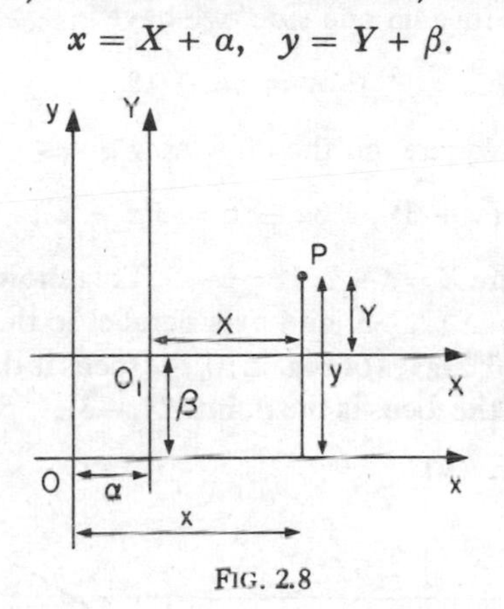

FIG. 2.8

Thus if the equation of a curve referred to Ox, Oy is $f(x, y) = 0$, the equation referred to O_1X, O_1Y is $f(X + \alpha, Y + \beta) = 0$.

For example, the parabola $y^2 = 4ax$ has equation $(Y + \beta)^2 = 4a(X + \alpha)$ when referred to O_1X, O_1Y.

On the other hand it may be that the equation of a curve is known with reference to O_1X, O_1Y; if this equation is $F(X, Y) = 0$, then the equation in the original co-ordinates is $F(x - \alpha, y - \beta) = 0$.

For example, suppose that a parabola has its vertex at (α, β) and its axis is parallel to the x-axis. Then if axes O_1X, O_1Y are drawn as shown (Fig. 2.9) the equation of the parabola referred to these axes is $Y^2 = 4aX$;

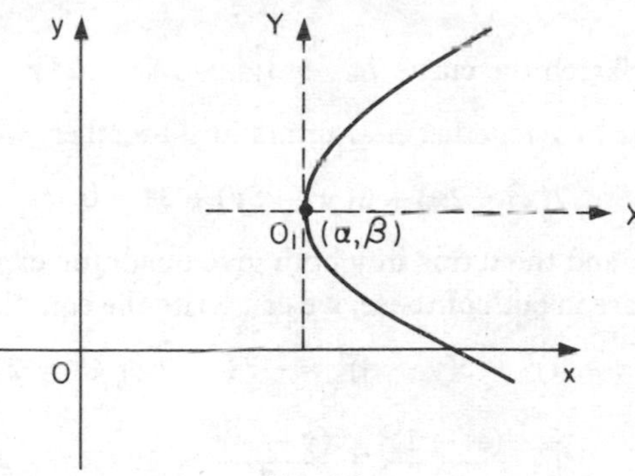

FIG. 2.9

it follows that the equation referred to the original axes is

$$(y-\beta)^2 = 4a(x-\alpha).$$

Example 2 Sketch the curve $y^2 - 3x + 6y + 15 = 0$.

Collecting the terms in y on one side, we have

$$y^2 + 6y = 3x - 15,$$

and 'completing the square' on the l.h.s. now gives

$$(y+3)^2 = 3x - 6 = 3(x-2);$$

that is, $Y^2 = 3X$, where $X = x - 2$, $Y = y + 3$. This shows that the curve is a parabola with vertex at $(2, -3)$ and axis parallel to the x-axis (Fig. 2.10). Referred to the X- and Y-axes of Fig. 2.10, the focus is the point $(a, 0)$ where $a = \frac{3}{4}$. It follows that the focus is the point $(2\frac{3}{4}, -3)$.

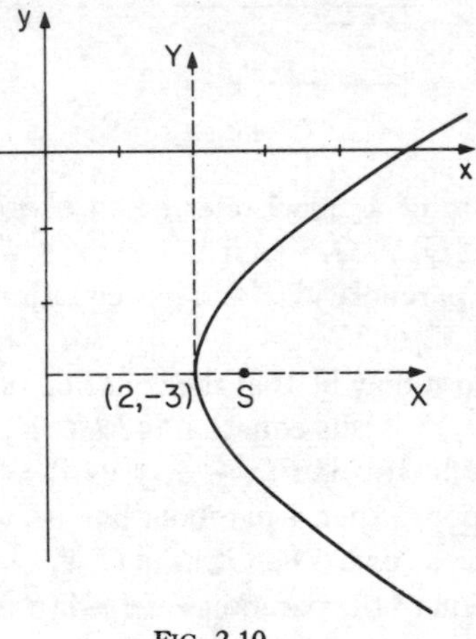

FIG. 2.10

Example 3 Sketch the curve $7x^2 + 3y^2 + 14x - 24y + 34 = 0$.

Grouping terms in x together and terms in y together, we have

$$7(x^2 + 2x) + 3(y^2 - 8y) + 34 = 0.$$

The terms in x and the terms in y both give quadratic expressions; completing the square in both of these, we can write the equation in the form

$$7(x+1)^2 + 3(y-4)^2 = -34 + 7 + 48 = 21;$$

that is,

$$\frac{(x+1)^2}{3} + \frac{(y-4)^2}{7} = 1.$$

This shows that the curve is an ellipse with centre at the point $(-1, 4)$, the axes of the ellipse having lengths $2\sqrt{3}$, $2\sqrt{7}$ (see Fig. 2.11).

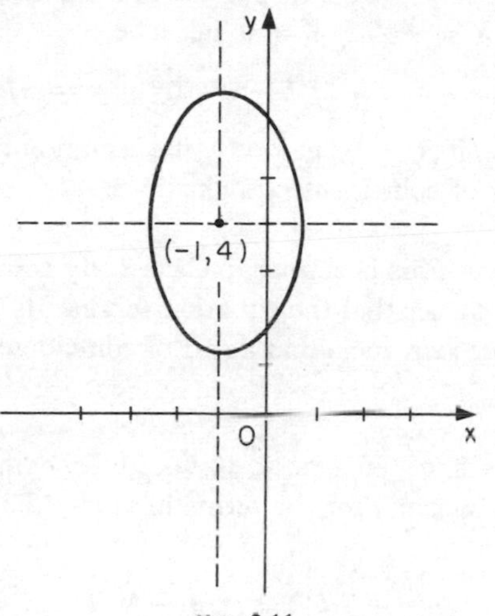

FIG. 2.11

The method illustrated in Examples 2 and 3 enables us to reduce any equation of the form

$$ax^2 + by^2 + 2gx + 2fy + c = 0 \qquad (2.3)$$

to a well-known standard form.

Case 1: $a = b$. Equation (2.3) can now be written in the form

$$(x^2 + 2g'x) + (y^2 + 2f'y) + c' = 0,$$

where $g' = g/a$, $f' = f/a$, $c' = c/a$. Completing the square in both brackets we get

$$(x + g')^2 + (y + f')^2 = g'^2 + f'^2 - c'$$

showing that the equation represents a circle, centre $(-g', -f')$ and radius $\sqrt{(g'^2 + f'^2 - c')}$.

Case 2: $a = 0$, $b \neq 0$. Equation (2.3) can now be written in the form

$$y^2 + 2f'y = -2g'x - c'$$

where $g' = g/b$, $f' = f/b$, $c' = c/b$. Completing the square on the l.h.s. we get

$$\begin{aligned}(y + f')^2 &= -2g'x - c' + f'^2 \\ &= -2g'[x - (f'^2 - c')/2g'].\end{aligned}$$

This is of the form $Y^2 = 4AX$ with $X = x - (f'^2 - c')/2g'$, $Y = y + f'$, $A = -g'/2$. Hence the equation represents a parabola with vertex at the point $((f'^2 - c')/2g', -f')$ and axis parallel to the x-axis. The focus of the parabola is where $X = -g'/2$, $Y = 0$, i.e. where

$$x = -g'/2 + (f'^2 - c')/2g', \quad y = -f'.$$

As a special case, if $A = 0$, i.e. $g = 0$, the equation reduces to $Y^2 = 0$, representing a pair of coincident straight lines.

Case 3: $a \neq 0$, $b = 0$. This is similar to Case 2. By completing the square for the terms in x, we see that the equation represents a parabola with its axis parallel to the y-axis, including a pair of coincident straight lines as a special case.

Case 4: Either $a > 0$, $b > 0$, *or* $a < 0$, $b < 0$. By completing the square both for the terms in x and for the terms in y, eq. (2.3) can be written in the form

$$a(x + g/a)^2 + b(y + f/b)^2 = g^2/a + f^2/b - c = c' \quad \text{(say)}.$$

If $c' = 0$ this equation is satisfied by the single point $(-g/a, -f/b)$. If $c' \neq 0$ we may write

$$\frac{(x + g/a)^2}{c'/a} + \frac{(y + f/b)^2}{c'/b} = 1.$$

Since a and b have the same sign it follows that c'/a, c'/b have the same sign. If $c'/a > 0$, $c'/b > 0$ write $A^2 = c'/a$, $B^2 = c'/b$ and the equation has the form

$$\frac{X^2}{A^2} + \frac{Y^2}{B^2} = 1$$

with $X = x + g/a$, $Y = y + f/b$; i.e. the equation represents an ellipse, centre at $(-g/a, -f/b)$, the axes having lengths $2A$, $2B$.

If $c'/a < 0$, $c'/b < 0$, the equation has the form

$$-\frac{X^2}{A^2} - \frac{Y^2}{B^2} = 1,$$

which is not satisfied by any real points.

Case 5: Either $a > 0$, $b < 0$, *or* $a < 0$, $b > 0$. Proceeding as in Case 4, we find that if $c' = 0$, i.e. $g^2/a + f^2/b - c = 0$, then the equation is of the form

$$\frac{X^2}{A^2} - \frac{Y^2}{B^2} = 0,$$

which represents a pair of straight lines. If $c' \neq 0$, the equation is of one or other of the forms

$$X^2/A^2 - Y^2/B^2 = 1 \quad \text{or} \quad X^2/A^2 - Y^2/B^2 = -1,$$

and represents a hyperbola.

Summary. The nature of the curve represented by eq. (2.3) is determined by the sign of ab.

(a) If $ab > 0$ the curve is an *ellipse* (including a circle as a special case) if it is a real curve.
(b) If $ab < 0$ the curve is a *hyperbola* (including a pair of straight lines as a special case).
(c) If $ab = 0$ the curve is a *parabola* (including a pair of coincident straight lines as a special case).

2.6 Rotation of Axes

Suppose that new axes OX, OY are drawn so that $\angle xOX = \theta$, the origin being unchanged. If P is the point (x, y) referred to the original axes, (X, Y) referred to the new axes, then writing $r = OP$ and $\phi = \angle XOP$ we have

$$\begin{aligned} x &= r \cos(\theta + \phi) \\ &= (r \cos\phi)\cos\theta - (r \sin\phi)\sin\theta. \end{aligned}$$

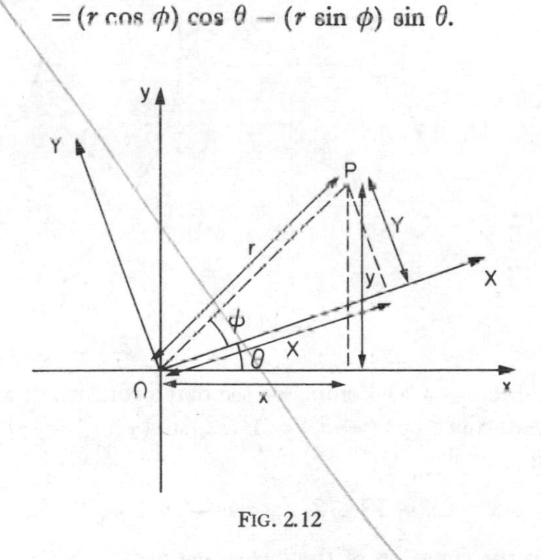

FIG. 2.12

Since $X = r\cos\phi$, $Y = r \sin\phi$, we get

$$x = X\cos\theta - Y\sin\theta;$$

similarly

$$y = r\sin(\theta + \phi) = X\sin\theta + Y\cos\theta. \tag{2.4}$$

Hence the curve with equation $f(x, y) = 0$ referred to the original axes has equation $f(X \cos \theta - Y \sin \theta, X \sin \theta + Y \cos \theta)$ when referred to the new axes.

Equations (2.4) may be solved for X, Y in terms of x, y; this gives

$$\begin{aligned} X &= x \cos \theta + y \sin \theta \\ Y &= -x \sin \theta + y \cos \theta \end{aligned} \tag{2.5}$$

Alternatively, eqs. (2.5) can be obtained from eqs. (2.4) by regarding the axes Ox, Oy as obtained by rotation through $-\theta$ of OX, OY.

Example 4 Find the equation of a rectangular hyperbola referred to its asymptotes as axes.

The equation of the rectangular hyperbola referred to the normal axes (see Fig. 2.13) is $x^2 - y^2 = a^2$. Choosing the asymptote $y = -x$ as X-axis

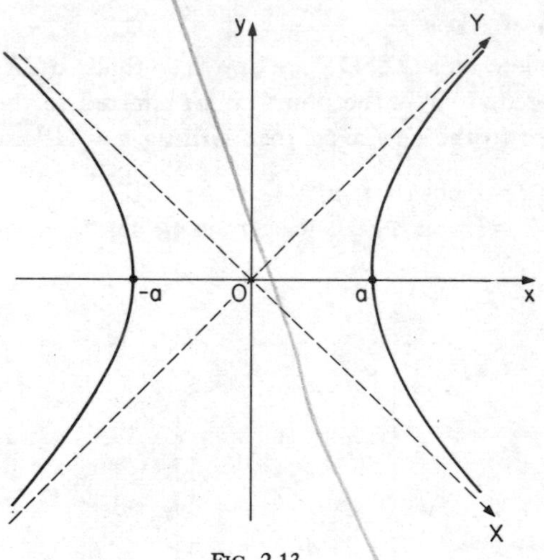

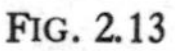
FIG. 2.13

and the asymptote $y = x$ as Y-axis, we see that a rotation of axes through $-45°$ is involved. Since $\cos(-45°) = 1/\sqrt{2}$, $\sin(-45°) = -1/\sqrt{2}$, we see from (2.4) that

$$x = (X + Y)/\sqrt{2}, \quad y = (-X + Y)/\sqrt{2}.$$

Substituting in the equation of the curve, we get

$$(X + Y)^2/2 - (-X + Y)^2/2 = a^2;$$

that is

$$2XY = a^2$$

or

$$XY = c^2, \quad \text{where } c^2 = a^2/2.$$

Example 5 Show that by a suitable rotation of axes, the equation of the curve $24x^2 - 24xy + 17y^2 = 1$ can be expressed in the form $aX^2 + bY^2 = 1$, and find a, b.

If the X, Y axes are obtained by rotation of the original axes through an angle θ, then writing $c = \cos\theta$, $s = \sin\theta$, it follows from (2.4) that the given equation becomes

$$24(Xc - Ys)^2 - 24(Xc - Ys)(Xs + Yc) + 17(Xs + Yc)^2 = 1;$$

that is

$$24(X^2c^2 + Y^2s^2 - 2XYsc) - 24[(X^2 - Y^2)sc + XY(c^2 - s^2)]$$
$$+ 17(X^2s^2 + Y^2c^2 + 2XYsc) = 1.$$

To obtain an equation of the required form the product term (XY-term) must vanish. The coefficient of XY is

$$-48sc - 24(c^2 - s^2) + 34sc,$$

that is

$$-14sc - 24(c^2 - s^2).$$

Thus we require

$$-7sc - 12(c^2 - s^2) = 0;$$

that is

$$12s^2 - 7sc - 12c^2 = 0,$$
$$(3s - 4c)(4s + 3c) = 0,$$
$$3s - 4c = 0 \quad \text{or} \quad 4s + 3c = 0.$$

The first of these equations gives $\tan\theta = \frac{4}{3}$; in other words $\sin\theta = \frac{4}{5}$, $\cos\theta = \frac{3}{5}$. (If the second equation is used we get a value of θ differing by $\pi/2$ from the value given by the first equation, but this means that the same pair of lines becomes the new axes, whichever equation is used.) For this rotation we have $x = (3X - 4Y)/5$, $y = (4X + 3Y)/5$, and substitution in the given equation gives

$$\tfrac{24}{25}(3X - 4Y)^2 - \tfrac{24}{25}(3X - 4Y)(4X + 3Y) + \tfrac{17}{25}(4X + 3Y)^2 = 1,$$

that is

$$8X^2 + 33Y^2 = 1.$$

This is of the required form with $a = 8$, $b = 33$.

Example 6 Show that by a suitable rotation of axes, the equation of the curve $5x^2 - 2xy + 7y^2 = 1$ can be expressed in the form $aX^2 + bY^2 = 1$, and find a, b.

Following the procedure used in Example 5, the given equation becomes

$$5(X\cos\theta - Y\sin\theta)^2 - 2(X\cos\theta - Y\sin\theta)(X\sin\theta + Y\cos\theta) + 7(X\sin\theta + Y\cos\theta)^2 = 1$$

and the XY-term vanishes provided that

$$-10\sin\theta\cos\theta - 2(\cos^2\theta - \sin^2\theta) + 14\sin\theta\cos\theta = 0,$$

that is

$$\sin^2\theta + 2\sin\theta\cos\theta - \cos^2\theta = 0.$$

Unfortunately the l.h.s. of this equation does not factorise simply, and instead of the procedure used in Example 5 we write the equation in the form

$$\tan^2\theta + 2\tan\theta - 1 = 0.$$

By the usual formula for the roots of a quadratic equation we get

$$\tan\theta = -1 \pm \sqrt{2}.$$

Taking the positive sign, we get $\tan\theta = \sqrt{2} - 1$, and so

$$\sin\theta = (\sqrt{2}-1)/\sqrt{(4-2\sqrt{2})}, \quad \cos\theta = 1/\sqrt{(4-2\sqrt{2})}$$

(see Fig. 2.14). Hence, from (2.4),

$$x = [1/\sqrt{(4-2\sqrt{2})}][X - (\sqrt{2}-1)Y],$$
$$y = [1/\sqrt{(4-2\sqrt{2})}][(\sqrt{2}-1)X + Y),$$

and the given equation becomes

$$5[X-(\sqrt{2}-1)Y]^2 - 2[X-(\sqrt{2}-1)Y][(\sqrt{2}-1)X+Y] + 7[(\sqrt{2}-1)X+Y]^2 = 4-2\sqrt{2},$$

that is

$$(6-\sqrt{2})X^2 + (6+\sqrt{2})Y^2 = 1.$$

This is of the required form with $a = 6-\sqrt{2}$, $b = 6+\sqrt{2}$.

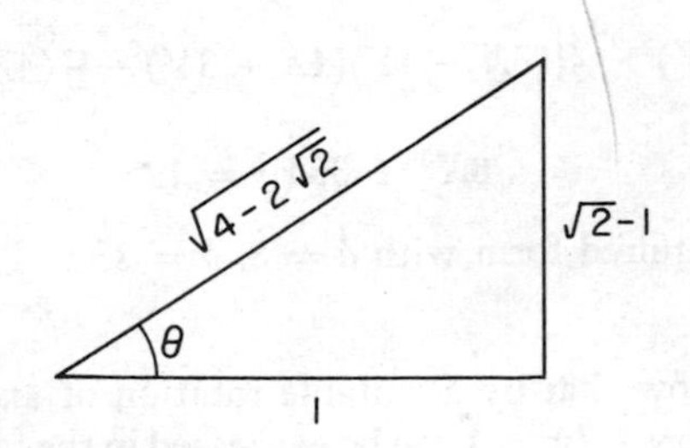

FIG. 2.14

2.7 The General Equation of the Second Degree

The general equation of the second degree is

$$ax^2 + 2hxy + by^2 + 2gx + 2fy + c = 0. \qquad (2.6)$$

The purpose of this section is threefold:

(a) It will be shown that any such equation represents a conic section if it represents a real curve (we have already met equations like $-3x^2 - 4y^2 = 1$ which is not satisfied by any values of x, y and so does not represent a real curve).
(b) It will be shown how to choose axes so that eq. (2.6) reduces to the standard, or *canonical,* form (see Section 2.4) of the equation of the appropriate conic section.
(c) It will be shown how to determine which type of conic section is represented by (2.6) without using the procedure in (b).

We shall do (a) and (b) simultaneously, for (a) can be done by showing that the procedure in (b) is always possible. We begin by showing that the method illustrated in Examples 5 and 6 can always be used to reduce (2.6) to a form in which the xy-term is missing; the effect of rotation of axes is found by substituting from (2.4) in (2.6), which gives

$$\begin{aligned} a(X\cos\theta - Y\sin\theta)^2 + 2h(X\cos\theta - Y\sin\theta)(X\sin\theta + Y\cos\theta) \\ + b(X\sin\theta + Y\cos\theta)^2 \\ + 2g(X\cos\theta - Y\sin\theta) + 2f(X\sin\theta + Y\cos\theta) + c = 0; \end{aligned}$$

that is

$$a'X^2 + 2h'XY + b'Y^2 + 2g'X + 2f'Y + c = 0, \qquad (2.7)$$

where
$$\begin{aligned} a' &= a\cos^2\theta + 2h\cos\theta\sin\theta + b\sin^2\theta, \\ b' &= a\sin^2\theta - 2h\sin\theta\cos\theta + b\cos^2\theta, \\ h' &= h(\cos^2\theta - \sin^2\theta) - (a-b)\sin\theta\cos\theta, \qquad (1) \\ g' &= g\cos\theta + f\sin\theta, \\ f' &= -g\sin\theta + f\cos\theta. \end{aligned}$$

We have $h' = 0$ provided that

$$h\sin^2\theta + (a-b)\sin\theta\cos\theta - h\cos^2\theta = 0. \qquad (2)$$

Equation (2) can be solved by factorising the l.h.s. (see Example 5) or by use of the general formula for the solution of the quadratic equation

$$\tan^2\theta + [(a-b)/h]\tan\theta - 1 = 0$$

(see Example 6); it may also be solved by writing

$$h\cos 2\theta = \tfrac{1}{2}(a-b)\sin 2\theta,$$

that is

$$\tan 2\theta = \frac{2h}{a - b},$$

and it is always possible to find a value of θ satisfying this equation. (When $b = a$ we take $2\theta = \frac{1}{2}\pi$.)

We have thus shown that it is always possible, by rotation of axes, to reduce (2.6) to a form with no xy-term, i.e. to the form (2.3). In Section 2.6 it has already been shown that (2.3) represents a conic section if it represents a real curve, and a method (change of origin) was given for reducing the equation to its canonical form. This completes (a) and (b).

From eqs. (1) it can be seen that $a' + b' = a + b$. This result is independent of θ and so is true for any rotation of axes. We say that $a + b$ is *invariant* under rotation of axes. It can also be shown that $a'b' - h'^2 = ab - h^2$ so that $ab - h^2$ is also invariant under rotation of axes. These invariants provide useful checks, e.g.

(i) in Example 5, $a = 24$, $b = 17$, $h = -12$ so that

$$a + b = 41, \quad ab - h^2 = 264,$$

while $a' = 8$, $b' = 33$, $h' = 0$, giving

$$a' + b' = 41, \quad a'b' - h'^2 = 264.$$

(ii) in Example 6, $a = 5$, $b = 7$, $h = -1$, so that

$$a + b = 12, \quad ab - h^2 = 34,$$

while $a' = 6 - \sqrt{2}$, $b = 6 + \sqrt{2}$, $h' = 0$, giving

$$a' + b' = 12, \quad a'b' - h'^2 = 34.$$

Example 7 Sketch the curve $x^2 + xy + y^2 + \sqrt{2}x + 2\sqrt{2}y = 1$ by first reducing its equation to canonical form.

If we substitute from (2.4) in the given equation, the equation corresponding to (2) is

$$\tfrac{1}{2}(\sin^2\theta - \cos^2\theta) = 0.$$

which is satisfied by taking

$$\sin\theta = \cos\theta = \frac{1}{\sqrt{2}};$$

that is, $\theta = \pi/4$. Thus the product term is eliminated by rotation of axes through $\pi/4$, which amounts to writing

$$x = \frac{1}{\sqrt{2}}(X - Y), \quad y = \frac{1}{\sqrt{2}}(X + Y),$$

giving

$$\tfrac{1}{2}(X - Y)^2 + \tfrac{1}{2}(X - Y)(X + Y) + \tfrac{1}{2}(X + Y)^2 + (X - Y) + 2(X + Y) = 1,$$

that is

$$\tfrac{3}{2}X^2 + \tfrac{1}{2}Y^2 + 3X + Y = 1$$
$$\tfrac{3}{2}(X^2 + 2X) + \tfrac{1}{2}(Y^2 + 2Y) = 1$$
$$\tfrac{3}{2}(X + 1)^2 + \tfrac{1}{2}(Y + 1)^2 = 1 + \tfrac{3}{2} + \tfrac{1}{2} = 3;$$

that is

$$\frac{\xi^2}{2} + \frac{\eta^2}{6} = 1, \text{ where } \xi = X + 1, \eta = Y + 1,$$

this being the canonical form of the equation of an ellipse, semi-axes $\sqrt{2}$, $\sqrt{6}$.

Thus if the axes OX, OY are inclined at 45° to the axes Ox, Oy (see Fig. 2.15), and if the axes $O'\xi$, $O'\eta$ are parallel to the axes OX, OY with origin O' at the point $(-1, -1)$ relative to OX, OY, i.e. O' is the point $(0, -\sqrt{2})$ relative to Ox, Oy, then relative to the axes $O'\xi$, $O'\eta$ the

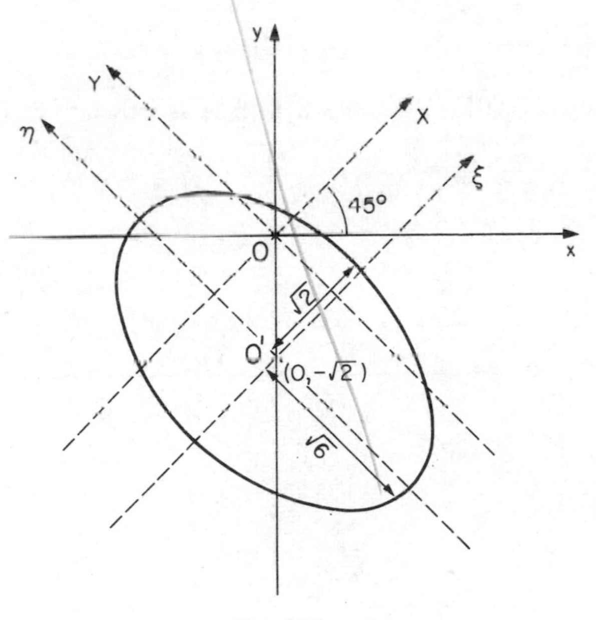

FIG. 2.15

equation becomes

$$\xi^2/2 + \eta^2/6 = 1$$

and the ellipse must therefore be placed as shown in Fig. 2.15.

Example 8 Show that $3x^2 + 4\sqrt{3}xy - y^2 = 7$ is a hyperbola, and find the equations of its asymptotes.

If we substitute from (2.4) in the given equation, the equation corresponding to (2) is

$$2\sqrt{3}\sin^2\theta + 4\sin\theta\cos\theta - 2\sqrt{3}\cos^2\theta = 0;$$

that is

$$\tan^2\theta + (2/\sqrt{3})\tan\theta - 1 = 0,$$

which is satisfied by $\tan\theta = 1/\sqrt{3}$. Thus by taking $\theta = \pi/6$, in which case $\cos\theta = \sqrt{3}/2$, $\sin\theta = \frac{1}{2}$, we eliminate the xy-term. We put

$$x = \tfrac{1}{2}(\sqrt{3}X - Y), \quad y = \tfrac{1}{2}(X + \sqrt{3}Y),$$

and the equation becomes

$$\tfrac{3}{4}(\sqrt{3}X - Y)^2 + \sqrt{3}(\sqrt{3}X - Y)(X + \sqrt{3}Y) - \tfrac{1}{4}(X + \sqrt{3}Y)^2 = 7,$$

that is

$$5X^2 - 3Y^2 = 7$$

or

$$\frac{X^2}{7/5} - \frac{Y^2}{7/3} = 1.$$

It follows that the curve is a hyperbola as shown in Fig. 2.16. (Note

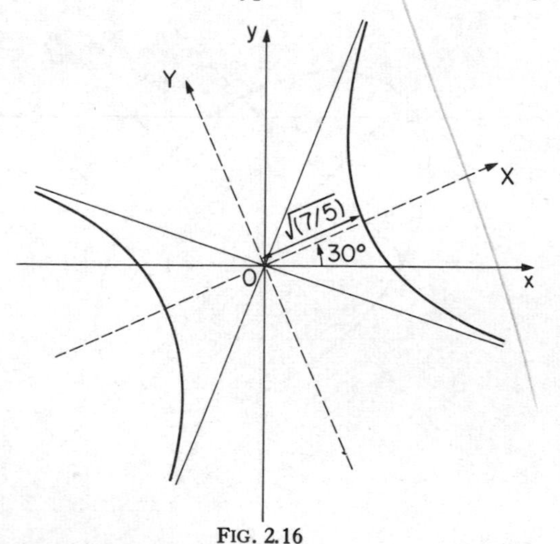

FIG. 2.16

that when there are no first-degree terms, a change of origin is not required.)

The asymptotes are the lines

$$Y = \pm \frac{\sqrt{(7/3)}}{\sqrt{(7/5)}} X = \pm \sqrt{(5/3)} X.$$

To find the equations of the asymptotes relative to the original axes, we first note (see eq. (2.5)) that

$$X = \tfrac{1}{2}(\sqrt{3}x + y), \quad Y = \tfrac{1}{2}(-x + \sqrt{3}y).$$

Hence the asymptotes are the lines

$$\tfrac{1}{2}(-x + \sqrt{3}y) = \pm\sqrt{(5/3)} \times \tfrac{1}{2}(\sqrt{3}x + y),$$

that is

$$-(\sqrt{5}+1)x + (\sqrt{3} - \sqrt{(5/3)})y = 0$$
$$\text{and} \quad (\sqrt{5}-1)x + (\sqrt{3} + \sqrt{(5/3)})y = 0.$$

If we wish to determine only the *type* of conic section represented by (2.6), this can be done simply by evaluating $ab - h^2$. To see this we note that under rotation of axes eq. (2.6) changes into eq. (2.7) and we have seen that $a'b' - h'^2 = ab - h^2$. If $h' = 0$, then $ab - h^2 = a'b'$ and moreover (2.7) is of the form (2.3); following the summary at the end of Section 2.6 we see that

(a) if $ab - h^2 > 0$, i.e. $a'b' > 0$, the curve is an *ellipse* (including a circle as a special case) if it is a real curve,
(b) if $ab - h^2 < 0$, i.e. $a'b' < 0$, the curve is a *hyperbola* (including a pair of straight lines as a special case),
(c) if $ab - h^2 = 0$, i.e. $a'b' = 0$, the curve is a *parabola* (including a pair of coincident straight lines as a special case).

For example in Example 7 we have $a = 1, b = 1, h = \frac{1}{2}$ so $ab - h^2 = \frac{3}{4} > 0$, showing that the equation represents an ellipse, and in Example 8 we have $a = 3, b = -1, h = 2\sqrt{3}$, so that $ab - h^2 = -15 < 0$, showing that the equation represents a hyperbola. The expression $ab - h^2$ is called the *discriminant* of eq. (2.6).

Exercises 2(c)

1. Find the vertices and axes of the parabolas

   (a) $y = x^2 + 2x + 2$, (b) $x + y^2 = 4.$,

   Draw a sketch to show the position of the curves.
2. Find the co-ordinates of the vertex of the parabola $y^2 = 2y + x$ and make a rough sketch of the curve.

3. Find the co-ordinates of the centre of the hyperbola $x^2 - 4y^2 + 4x - 8y = 4$. Find also the equation of each asymptote.
4. Find the equation of the parabola $y = a + bx + cx^2$ which passes through the points (0, 3), (1, 6), (2, 13). Find its axis and vertex. The parabola is reflected in the tangent at its vertex. What is the equation of the resulting parabola?
5. By using the discriminant, verify that the conic

$$x^2 + 2\sqrt{2}xy + 2y^2 + 6\sqrt{3}x + 3 = 0$$

is a parabola.

Show that the rotation

$$x = \frac{1}{\sqrt{3}}(x' - y'\sqrt{2}), \quad y = \frac{1}{\sqrt{3}}(x'\sqrt{2} + y')$$

reduces the equation to one containing no $x'y'$ term, and that a translation $x' = X + u$, $y' = Y + v$ reduces it further to the form $X^2 = 2\sqrt{2}Y$. Find u and v.

6. Reduce to canonical form the equations

(a) $3x^2 + 2xy + 3y^2 = 2$, (b) $5x^2 - 6xy + 5y^2 = 8$.

7. Find the canonical form of the equation of the hyperbola

$$7x^2 + 48xy - 7y^2 = 25.$$

What is the angle between the asymptotes?

8. By rotating the co-ordinate axes through 45°, find the canonical form of the equation

$$x^2 + 2hxy + y^2 = c^2.$$

Hence show that the curve it represents is an ellipse if and only if $|h| < 1$.

9. What is the value of c if the ellipse $x^2 + xy + cy^2 = 1$ has an axis along the line $x = y$? Substitute this value and reduce the equation to canonical form.

2.8 Polar Co-ordinates

Let O be a fixed point and Ox a fixed straight line in a plane. Then the position of any other point P in the plane is uniquely determined by the co-ordinates $r = OP$ and $\theta = \angle xOP$ which are called the *polar co-ordinates* of P. Ox is termed the *initial line* and O the *pole*; OP is called the *radius vector* and θ the *vectorial angle* of P.

The angle θ is positive or negative according as it is measured counter-clockwise or clockwise from Ox. If OM makes an angle θ with Ox, the point $P(r, \theta)$ lies on OM at a distance r from O when $r > 0$; P lies on MO produced at a distance $|r|$ from O when $r < 0$. The convention is sometimes made (notably in connection with the study of complex numbers in the Argand diagram) that only positive values of r shall be used, but this is not convenient in geometry.

In Fig. 2.17, the points $P(1, 60°)$, $Q(-2, 60°)$ and $R(-2, -60°)$ are shown. Note that each of these points may be expressed in terms of other polar co-ordinates. For example, R is the point $(2, 120°)$ and also $(-2, 300°)$.

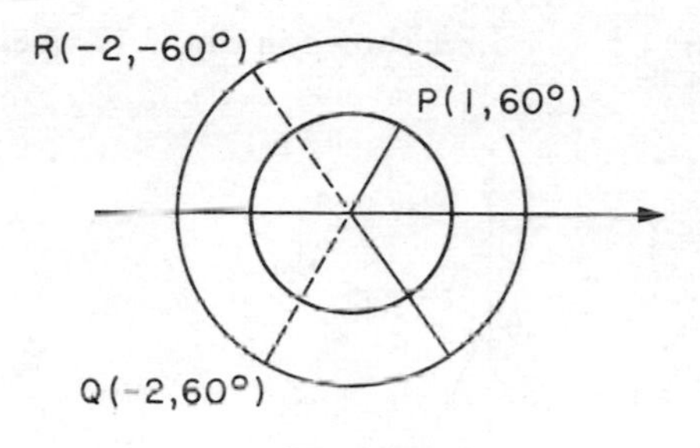

FIG. 2.17

2.8.1 Relations between Cartesian and Polar Co-ordinates If the cartesian origin and x-axis (Fig. 2.18) are chosen as the pole and initial

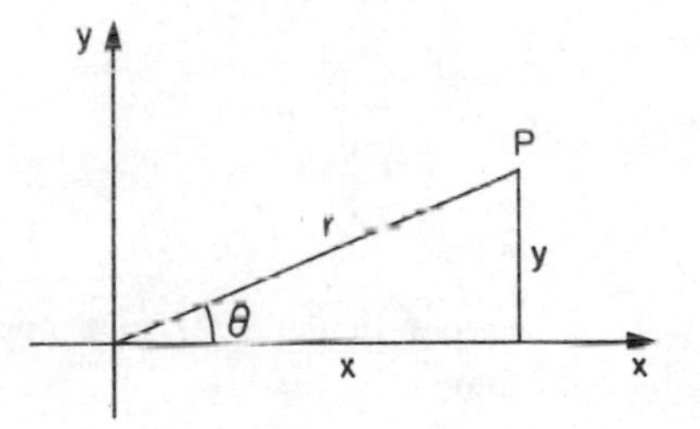

FIG. 2.18

line respectively of polar co-ordinates, the cartesian and polar co-ordinates of a point are connected by the relations

$$x = r\cos\theta, \quad y = r\sin\theta, \tag{2.8}$$

$$r^2 = x^2 + y^2, \quad \tan\theta = y/x. \tag{2.9}$$

Equations (2.9) do not determine r, θ uniquely. To obtain unique values of r and θ for a given x and y we may take $r = +\sqrt{(x^2 + y^2)}$ and θ as the angle which satisfies the relations $\cos\theta = x/r$, $\sin\theta = y/r$ such that $-\pi < \theta \leqslant \pi$.

2.8.2 The Straight Line in Polar Co-ordinates

Equation	*Description*
(1) $\theta = \alpha$	Line through the pole inclined at an angle α to the initial line
(2) $r = a \operatorname{cosec} \theta$	Line drawn parallel to the initial line at a distance a from the pole, i.e. $y = a$
(3) $r = b \sec \theta$	Line drawn perpendicular to the initial line at a distance b from the pole, i.e. $x = b$
(4) $r \cos(\theta - \alpha) = p$	Line shown in Fig. 2.19 where $P \equiv (r, \theta)$, the perpendicular $OM = p$ and $\angle xOM = \alpha$ so that $OP \cos(\theta - \alpha) = OM$. The corresponding cartesian equation is $x \cos \alpha + y \sin \alpha = p$.

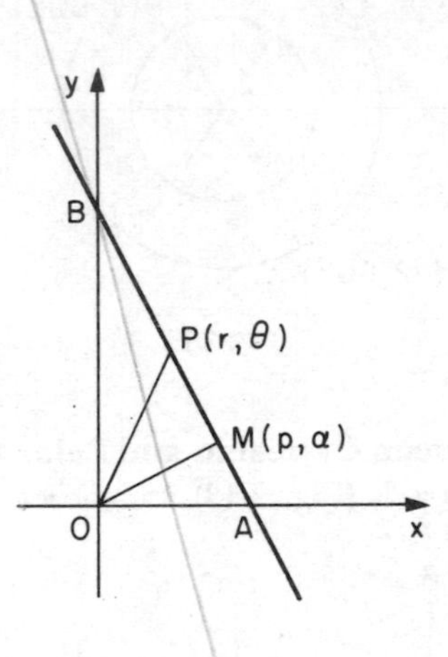

FIG. 2.19

2.8.3 The Circle in Polar Co-ordinates

Equation	*Description*
(1) $r = a =$ constant	Circle with centre at the pole and radius a, i.e. circle $x^2 + y^2 = a^2$

(2) $r = 2a \cos \theta$ — Circle of radius a with centre on the initial line and passing through the pole, i.e. circle $x^2 + y^2 = 2ax$

(3) $r = 2a \sin \theta$ — Circle of radius a touching the initial line at the pole, i.e. circle $x^2 + y^2 = 2ay$

(4) $r^2 - 2pr \cos (\theta - \alpha) + p^2 - a^2$ — Circle shown in Fig. 2.20 with centre $C(p, \alpha)$ and radius a. If $P(r, \theta)$ is any point on the circle, then

$$PC^2 = OP^2 + OC^2 - 2OP.OC \cos (\theta - \alpha).$$

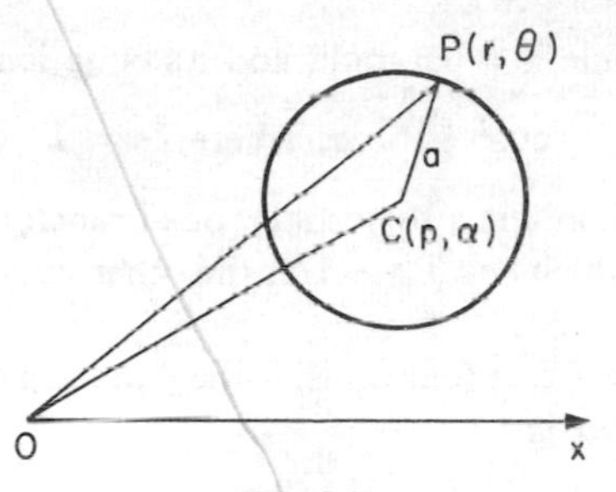

FIG. 2.20

2.8.4 The Conic in Polar Co-ordinates The simplest equation of a conic in polar co-ordinates is obtained by taking the pole at a focus S and taking as initial line the perpendicular SZ from S to the corresponding directrix DD' of the conic. In Fig. 2.21, $P(r, \theta)$ is any point on a conic which has eccentricity e and semi-latus rectum $LS = l$. The lines LK, PM are drawn perpendicular to DD', and PN is perpendicular to SZ.

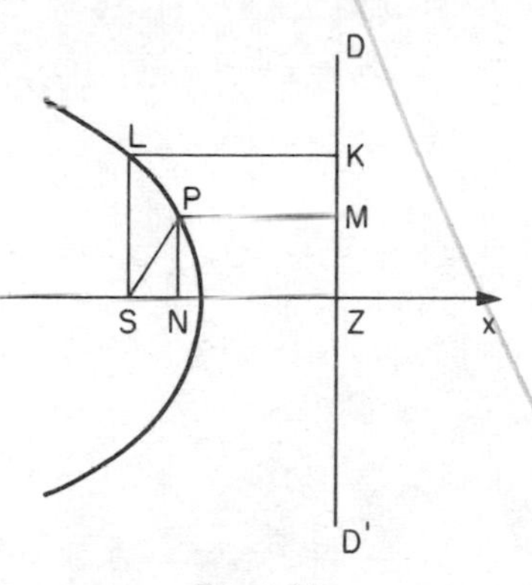

FIG. 2.21

Then, by the focus–directrix property of the conic

$$LS = l = eLK \tag{1}$$

and

$$PS = ePM, \tag{2}$$

that is

$$r = e(LK - SN),$$

and so

$$r = l - er\cos\theta, \text{ by (1).}$$

Hence

$$\frac{l}{r} = 1 + e\cos\theta.$$

When $e = 1$, the conic is a parabola and its equation is

$$r\cos^2\tfrac{1}{2}\theta = a, \text{ where } 2a = l.$$

When $e > 1$, the conic is a hyperbola, one branch of which is given by the values of θ for which $\cos\theta > -1/e$, the other by values of θ for which $\cos\theta < -1/e$.

Since $SZ = LK = l/e$, it follows that the equation of the directrix DD' is $r = (l/e)\sec\theta$, that is

$$\frac{l}{r} = e\cos\theta.$$

If, in Fig. 2.21, ZS produced is taken as initial line,

$$SN = r\cos(\pi - \theta) = -r\cos\theta,$$

and by (2) the equation of the conic is

$$\frac{l}{r} = 1 - e\cos\theta.$$

CHAPTER 3

IDEAS ABOUT SETS

3.1 Enumeration of Sets

In mathematics the word *set* is used in preference to similar words like group, collection, etc. It is interpreted broadly so that we might talk, for example, about the set of all planets or the set of all people (dead or alive) who have at some time played football, although we cannot easily visualise a 'group' in either case; in other words a set is taken as defined even though we cannot produce a membership list, provided there is a qualification which enables us to decide if any given object is a member or not.

If it is possible to set up a correspondence between the members of a set A and the members of a set B such that

(a) with each member of A is associated precisely *one* member of B,
(b) each member of B acts as the associate of precisely *one* member of A,

then we say that the sets A, B are in *(1, 1) correspondence*. The *cardinal number* of a set is n if the members of the set are in (1, 1) correspondence with the set of numbers $\{1, 2, 3, \ldots, n-1, n\}$. If the members of a set are not in (1, 1) correspondence with any terminating set of whole numbers the set is said to be *infinite*; it is *countably infinite* if the members are in (1, 1) correspondence with the non-terminating set $\{1, 2, 3, \ldots\}$ and the symbol ∞ is frequently used to indicate the cardinal number in this case.

It must be emphasised that ∞ is not to be included as a number within the real-number system because its behaviour is so different. For example, if $A = \{2, 4, 6, 8, \ldots\}$, then A is countably infinite as shown by the (1, 1) correspondence

$$\begin{array}{cccccccc} 2 & 4 & 6 & 8 & \ldots & 2n & \ldots \\ \downarrow & \downarrow & \downarrow & \downarrow & & \downarrow & \\ 1 & 2 & 3 & 4 & \ldots & n & \ldots \end{array}$$

Similarly, if $B = \{1, 3, 5, 7, \ldots\}$, then B is countably infinite, and moreover combining the sets A and B gives the countably infinite set $\{1, 2, 3, 4, \ldots\}$. This suggests that $\infty + \infty = \infty$ if we permit ∞ to be treated as an ordinary number!

Clearly, a set is countably infinite if its members can be arranged in some ordered sequence, there being a first, second, etc. but no last member.

Example 1 Show that the set of all rational numbers is countably infinite.

Consider the following arrangement of the positive rationals:

$$\begin{array}{cccccccc}
1 \rightarrow 2 & & 3 \rightarrow 4 & & 5 \rightarrow 6 & \ldots \\
\swarrow \quad \nearrow & & \swarrow \quad \nearrow & & \swarrow & \\
\frac{1}{2} \quad \frac{2}{2} & & \frac{3}{2} \quad \frac{4}{2} & & \frac{5}{2} \quad \frac{6}{2} & \\
\downarrow \; \nearrow \; \swarrow & & \nearrow \; \swarrow & & & \\
\frac{1}{3} \quad \frac{2}{3} & & \frac{3}{3} \quad \frac{4}{3} & & \frac{5}{3} \quad \frac{6}{3} & \\
\swarrow \quad \nearrow & & \swarrow & & & \\
\frac{1}{4} \quad \frac{2}{4} & & \frac{3}{4} \quad \frac{4}{4} & & \frac{5}{4} \quad \frac{6}{4} & \\
\downarrow \; \nearrow \; \swarrow & & & & & \\
\frac{1}{5} \quad \frac{2}{5} & & \frac{3}{5} \quad \frac{4}{5} & & \frac{5}{5} \quad \frac{6}{5} &
\end{array}$$

The nth row includes all positive rationals with denominator n and so all positive rationals have a place in this arrangement. Moving through the array as indicated by the arrows we meet each one in a definitive order. Since numbers are repeated many times (e.g. 2, $\frac{4}{2}$; $\frac{6}{3}$ etc.) we can ignore all but the first appearance of any number. Further, we could start with zero and proceed to $+1, -1, +2, -2, +\frac{1}{2}, -\frac{1}{2}$, etc. thereby defining an order for *all* rationals.

Example 2 Show that the set of all real numbers is not countably infinite.

It is sufficient to consider the set of real numbers between 0 and 1. If this set is countably infinite then it can be arranged in an ordered sequence. Suppose $x_1, x_2, x_3, \ldots$ is such an arrangement.

All these numbers can be expressed in decimal form, so we may write

$$\begin{aligned}
x_1 &= 0{\cdot}\alpha_1\alpha_2\alpha_3 \ldots \\
x_2 &= 0{\cdot}\beta_1\beta_2\beta_3 \ldots \\
x_3 &= 0{\cdot}\gamma_1\gamma_2\gamma_3 \ldots
\end{aligned}$$

where the α's, β's, γ's, ... are digits from the set $\{0, 1, 2, \ldots, 9\}$.

Now put $\alpha_1' = \alpha_1 - 1$, $\beta_2' = \beta_2 - 1$, $\gamma_3' = \gamma_3 - 1$, ... with the proviso that any of these values for $\alpha_1', \beta_2', \gamma_3', \ldots$ which equal -1 are replaced by $+1$, so that $\alpha_1', \beta_2', \gamma_3', \ldots$ are all digits from the set $\{0, 1, 2, \ldots, 8\}$.

The number $0{\cdot}\alpha_1'\beta_2'\gamma_3' \ldots$ lies between 0 and 1 and differs in one decimal place at least from each member of the set $\{x_1, x_2, x_3, \ldots\}$ which was supposed to include all real numbers between 0 and 1. We conclude that the real numbers between 0 and 1 cannot be arranged in an ordered sequence.

[The reader must accept for the moment that the decimal representation of a number is unique with exceptions like, for example,

$$0{\cdot}3456 = 0{\cdot}3455\dot{9}$$

where a repeating 9 is involved. This cannot be fully understood without the idea of an infinite series (see Chapter 22).]

3.1.1 Sets of Possibilities Special difficulties arise when enumerating the set of all possible ways in which something can happen or in which something can be done, e.g. finding the number of different arrangements that can result from shuffling a pack of 52 playing cards. In such situations we refer to the set of possible *results*, or *outcomes*, of an *experiment*; this economy of terms is convenient, but it means that we are interpreting the word 'experiment' rather widely, e.g. we might regard the act of selecting a football team as an 'experiment' the possible results of which are the various possible team selections that can be made from the available players.

Tossing a coin is a simple experiment with two possible results which may be denoted by H, T. Tossing several coins is a more complex experiment, but the results may be regarded as consisting of several component results, viz. the results for the individual coins. Starting from the 2 possible results for the first coin we can build up the set of possible results for any number of coins by drawing a 'tree' as shown below for the case of 3 coins.

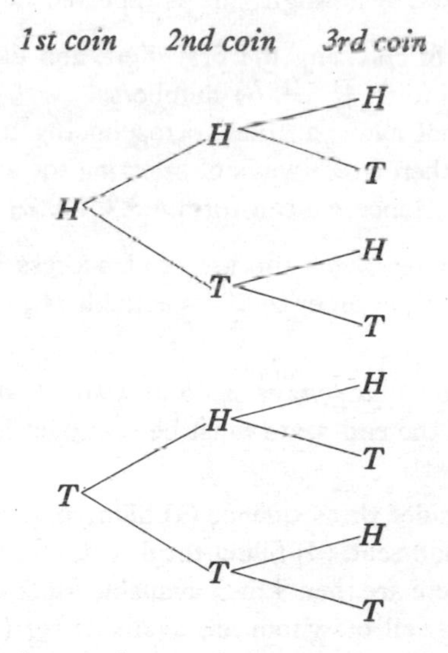

Thus for two coins the results can be represented by the set $\{HH, HT,$

TH, *TT*} (each result having two components) and for three coins by the set {*HHH*, *HHT*, *HTH*, *HTT*, *THH*, *THT*, *TTH*, *TTT*} (each result having three components).

This situation is typical of many where the experiment can be broken down into a sequence of simpler experiments and the following principle applied:

Basic principle. If an experiment Z_1 has n possible results and each of these results can be accompanied by m possible results from another experiment Z_2, then the combined experiment 'both Z_1 and Z_2' has mn possible results.

The examples show the application of this principle to more than two experiments.

Example 3 How many results are possible when (a) two dice, (b) three dice are thrown?

There are 6 possible results from the first throw, each of which can be accompanied by 6 possible results from the second throw, etc. Hence for (a) we get $6 \times 6 = 36$ and for (b) we get $6^3 = 216$.

Example 4 How many 3-digit positive numbers can be formed using the digits 3, 5, 7, 9?

What is the answer if no digit can be repeated in the number?

There are 4 ways of choosing the first digit, and likewise for the other two. Hence we can form 4^3, i.e. 64 numbers.

If repetition is not allowed, then corresponding to each way of selecting the first digit there are 3 ways of selecting the second and 2 ways of selecting the third. Hence we can form 4.3.2, i.e. 24 numbers.

With more difficult problems the secret of success is to try several approaches, if necessary, in order to find a suitable sequence of experiments.

Example 5 In how many ways can 5 men and 4 women be arranged in a row of 9 seats if the end seats must be occupied by men?

First method. Consider the sequence (a) filling first seat, (b) filling ninth seat, (c) filling second seat, (d) filling third seat, etc. Any of 5 men can be used for (a), and there are then 4 men available for (b). After this there are 7 people remaining, all of whom are available for (c), leaving 6 people available for (d), etc.

Applying the basic principle this leads to the answer

$$5 \times 4 \times 7 \times 6 \times 5 \times 4 \times 3 \times 2 \times 1 \quad \text{i.e. } 100\,800.$$

Note the difficulty that arises if we do not consider the ninth seat before going on to the second, third, etc.

Second method. Consider the sequence (a) placing first woman, (b) placing second woman, etc. until all the women, and then all the men, are placed.

There are 7 seats available for women, so (a) can be done in 7 ways, after which (b) can be done in 6 ways, etc. Once the women have all been seated, the remaining 5 seats are all available for men and so the first man can be placed in 5 ways, the second man in 4 ways, etc.

Applying the basic principle this approach leads to the answer

$$7 \times 6 \times 5 \times 4 \times 5 \times 4 \times 3 \times 2 \times 1 \quad \text{i.e. } 100\,800.$$

Note the difficulty that arises if we do not place the women before placing the men.

Example 6 A milkman delivers 14 bottles of milk in a street containing 6 houses. In how many ways can he do this?

The natural tendency is to take as component results the delivery at individual houses, but this leads to great difficulties. Consider instead the fate of individual bottles; the first bottle can be delivered in 6 ways, and for each of these ways there are 6 ways of delivering the second bottle, and so on, giving the answer 6^{14}.

3.1.2 Permutations and Combinations A rearrangement of a set which leaves the membership of the set unchanged but changes only the order in which the members appear is called a *permutation* of the set, e.g. $\{a, b, c\}$, $\{a, c, b\}$, $\{c, b, a\}$ are all permutations of the same set.

To find how many permutations of a set with n members are possible, consider the sequence (a) choosing a member for first position, (b) choosing a member for second position, and so on until the n positions have been filled; applying the basic principle we get the answer $n(n-1)(n-2) \cdots 2.1$, a product which is usually denoted by $n!$ and called *factorial n*.

For example, the number of ways of arranging 6 people in a row is $6!$, i.e. $6 \times 5 \times 4 \times 3 \times 2 \times 1$, i.e. 720.

If r is any positive integer not greater than n, then using the same argument as before we see that the number of ways of filling the first r

positions in a set of n members is $n(n-1)\cdots(n-r+1)$; this number is denoted by ${}^nP_r$, an abbreviation for 'the number of permutations of n objects r at a time'.

In other words, ${}^nP_r$ is the number of ways in which it is possible to select r members from a set of n members and then arrange the selected members in the order 1st, 2nd, ..., rth.

For example, the number of ways of filling a row of 6 seats from a group of 10 people is ${}^{10}P_6$, i.e. $10 \times 9 \times 8 \times 7 \times 6 \times 5$, i.e. 151 200. In practice we can evaluate ${}^nP_r$ simply as a product of r numbers, starting with n and decreasing by 1 each time; the general formula may be written

$$
\begin{aligned}
{}^nP_r &= n(n-1)\cdots(n-r+1) \\
&= \frac{n(n-1)\cdots(n-r+1)(n-r)(n-r-1)\cdots 2.1}{(n-r)(n-r-1)\cdots 2.1} = \frac{n!}{(n-r)!}
\end{aligned}
$$

We further define $0!$ to be 1. With this convention the last formula for ${}^nP_r$ gives ${}^nP_n = n!$ in agreement with the earlier formula. If a set S' has r members all of which also belong to a set S, we sometimes say that S' is a *combination* of r members from S, e.g. the England soccer team is a combination of 11 players from the set of all English players.

The symbol ${}^nC_r$ (i.e. number of combinations of r objects from n) is used to denote the number of ways in which a set of r members can be selected from a set of n members, no account being taken of the arrangement.

For example, the number of ways in which we can pick 6 men from 10 volunteers for an expedition is ${}^{10}C_6$ since, in the absence of further information, we would assume the arrangement of the 6 men to be unimportant; since each group of 6 can be permuted in $6!$ ways we see that there are ${}^{10}C_6 \times 6!$ permutations of 10 men taken six at a time, so that

$$
{}^{10}C_6 \times 6! = {}^{10}P_6 = 151\,200,
$$

that is

$$
{}^{10}C_6 = \frac{151\,200}{6!} = \frac{151\,200}{720} = 210.
$$

More generally the argument used in this example shows that

$$
{}^nC_r \times r! = {}^nP_r
$$

and so

$$
{}^nC_r = \frac{n(n-1)\cdots(n-r+1)}{1.2\cdots r} = \frac{n!}{r!(n-r)!}
$$

In practice ${}^nC_r$ can be remembered as a product of r numbers, starting from n and going down by 1 each time, divided by another product of r

numbers, starting from 1 and going up by 1 each time, e.g.

$$^{10}C_6 = \frac{10.9.8.7.6.5}{1.2.3.4.5.6} = 210.$$

Example 7 In how many ways can a committee of 4 men and 4 women be chosen from 9 men and 8 women? What is the answer if there are 4 different offices (e.g. secretary, treasurer, etc.) for men and 4 offices for women?

In the first case the answer is

$$^{9}C_4 \times {}^{8}C_4 \quad \text{i.e.} \frac{9.8.7.6}{1.2.3.4} \times \frac{8.7.6.5}{1.2.3.4} \quad \text{i.e. } 8820.$$

In the second case the answer is

$$^{9}P_4 \times {}^{8}P_4 \quad \text{i.e. } 9.8.7.6.8.7.6.5 \quad \text{i.e. } 5\,080\,320.$$

Example 8 How many anagrams of the word 'imitation' are possible?

The 9 letters can be arranged in 9! ways. Since the letter *i* occurs 3 times and the letter *t* occurs twice the 9! arrangements are not all different. For each different arrangement there are 3! × 2! ways of interchanging letters without changing the anagram and so the number of anagrams is 9!/(3!2!), i.e. 30 240.

Example 8 illustrates a general principle, viz. if in a set of n members there are n_1 alike of one kind, n_2 alike of another kind, . . ., n_s alike of another kind, then the number of *different* permutations of the set is

$$\frac{n!}{n_1!\, n_2! \ldots n_s!}$$

3.1.3 Properties of ^nC_r

(a) $^nC_r = {}^nC_{n-r}$. This result can be proved using the formula for ^nC_r or, more simply, by observing that any selection of r objects from n can equally well be made by selecting $(n - r)$ objects for exclusion. It is of practical value when evaluating ^nC_r for values of r close to n, e.g. to evaluate $^{100}C_{98}$ write

$$^{100}C_{98} = {}^{100}C_2 = \frac{100.99}{1.2} = 4950.$$

(b) $^nC_r + {}^nC_{r-1} = {}^{n+1}C_r$. We can argue as follows: if a set of n members has a new member added to it, then the number of selections of r

members from the set increases by the number of such selections that include the new member; now the number of selections including the new member is just the number of ways of selecting $(r - 1)$ of the original members, and hence the result. A proof using the formula for ${}^nC_r$ is more straightforward but gives less insight into the result.

Exercises 3(a)

1. Regarding a word as being any ordered succession of letters, how many three letter words are there of the type vowel–consonant–vowel?
2. How many 3-digit numbers can be formed from the digits 1, 2, 3, 4, 5, 6, 7 (a) if the numbers may be even or odd, (b) if the numbers are to be odd?
3. Express 90 000 as a product of primes and hence find how many factors it has (excluding 1 and 90 000). How many even factors does it have?
4. Show that n lines in a plane have at most $\frac{1}{2}n(n - 1)$ intersections.
5. If a regular polygon has n sides and lines are drawn joining all the vertices then $\frac{1}{2}n(n - 3)$ of these joining lines are not sides of the polygon.
6. Verify that ${}^nP_{n-1} = {}^nP_n$.
7. In how many ways can 7 persons be seated in a row if a certain two are (a) to sit together, (b) not to sit together?
8. In how many ways can two numbers be selected from 1, 2, 3, . . ., 12 so that their sum is even? Give a general formula for your answer if the numbers are selected from (a) 1, 2, . . ., $2n$, (b) 1, 2, . . ., $2n$, $2n + 1$.
9. How many different sequences is it possible to ring on 8 bells, each bell being rung once in each sequence? In how many of these does a particular bell ring last?
10. Three cards are taken at random from a full pack. How many different sets of three are possible (a) with no restrictions, (b) containing only spades, (c) having the same value (e.g. three sixes or three kings), (d) having values which add up to 4 (ace counting as 1, face cards > 10).
11. Four consecutive houses in a terrace row are to be painted using some or all of four given colours. Each house can be painted with any of the four colours but no two adjacent houses can be painted with the same colour. In how many ways can this be done?
12. An American city consists of p streets running east and west and q avenues running north and south. In how many ways can a man go from the north-west corner to the south-east corner, facing either south or east all the time?

3.2 Sets of Points

A set can frequently be described by means of a property common to all the members; for example, we denote by $\{x: x^2 - 5x + 6 = 0\}$ the set of numbers x that satisfy the equation $x^2 - 5x + 6 = 0$, i.e. the set $\{2, 3\}$; by $\{x : x > 2\}$ the set of numbers greater than 2; by $\{x : x^2 - 5x + 6 = 0, x > 2\}$ the set of numbers satisfying both these conditions, i.e. $\{3\}$. Any set of real numbers represents a set of points on the real axis, and the words *number* and *point* are frequently treated as synonymous.

Suppose $b > a$. Then the set of points on the real axis corresponding to $\{x: a < x < b\}$ is called an *open interval*; similarly $\{x: a \leqslant x \leqslant b\}$ denotes a *closed interval* and $\{x: a \leqslant x < b\}$, $\{x: a < x \leqslant b\}$ denote *half-open* (or *half-closed*) intervals, open *at the right* and open *at the left* respectively. The positive real axis is sometimes denoted by $\{x: 0 < x < \infty\}$ and the entire real axis by $\{x: -\infty < x < \infty\}$ or by R (see Section 1.2.3).

3.2.1 Points in a Plane A plane in which a pair of co-ordinate axes has been chosen may be regarded as the set of all points (x, y) where x and y are any two real numbers, i.e. the set of points $\{(x, y): x \text{ in } R, y \text{ in } R\}$. This set is commonly denoted by R^2.

The set of points $\{(x, y): F(x, y) = 0\}$ lies on some curve in the xy-plane. (The notation $F(x, y)$ for functions of two variables x, y was briefly introduced in Section 1.2.3. In this section we shall restrict our considerations to functions F that, in the sense of definitions given later, are *continuous*—however, it is not intended to discuss such subtleties at this stage.) The way in which the values of $F(x, y)$ change as we move from point to point in the plane can be illustrated by drawing a number of curves $F(x, y) = c$ with different values of c; this is similar to showing the rise and fall of ground level by the lines on a contour map and these curves are sometimes called 'level curves' for $F(x, y)$.

Example 9. What are the greatest and least values of the expression $2x + y$ for all points (x, y) within the circle $x^2 + y^2 = 5$?

The straight lines in Fig. 3.1 represent $2x + y = c$ with $c = 0, \pm 1, \pm 2,$

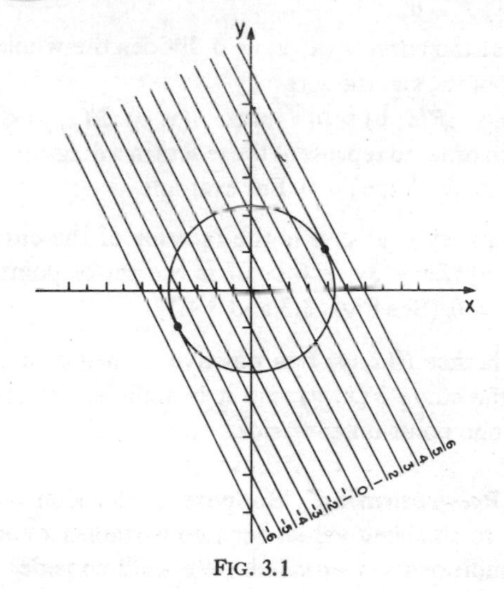

FIG. 3.1

± 3, ± 4, ± 5, ± 6 (as indicated). The lines that are tangential to the circle have the largest and smallest values of all lines having any points in the circle. It can be verified that $c = \pm 5$ give tangents to the circle and so these are the extreme values of $2x + y$.

Example 10 Draw level curves for $x^2 - 2y^2$.

The set of points where $x^2 - 2y^2 = 0$ forms a pair of straight lines $\sqrt{2}y = \pm x$. For $x^2 - 2y^2 = c$ where $c \neq 0$ we get hyperbolas.

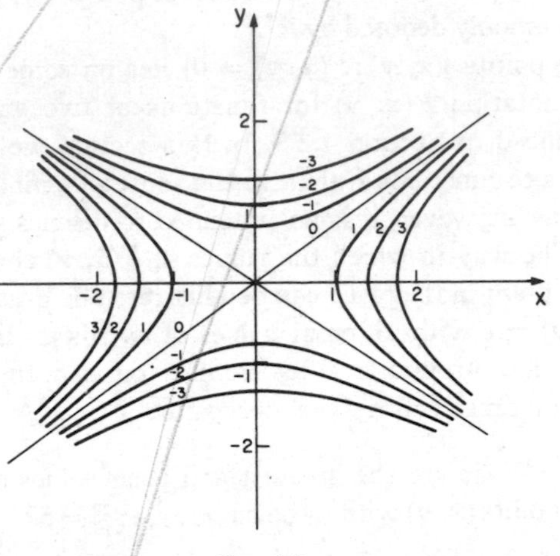

FIG. 3.2

We notice that the curve $F(x, y) = 0$ divides the whole plane into two distinct sets of points, viz. the sets

$$\{(x, y) : F(x, y) > 0\} \quad \text{and} \quad \{(x, y) : F(x, y) < 0\}.$$

It follows that, in order to represent these sets in a diagram, we should begin by drawing the curve $F(x, y) = 0$. For example,

(a) the set $\{(x, y) : x^2 + y^2 < 4$ is the *interior* of the circle $x^2 + y^2 = 4$,
(b) the set $\{(x, y) : 2x - 3y + 6 > 0\}$ is the set of points *below* the line $2x - 3y + 6 = 0$. (See Figs. 3.3 and 3.4.)

To decide whether $F(x, y)$ has positive or negative values on a particular side of the curve $F(x, y) = 0$ it is sufficient to check the value of $F(x, y)$ at any one point on that side.

3.2.2 Linear Programming Suppose a decision has to be made which amounts to choosing values for two variables x and y so that some condition or conditions are satisfied. (We shall consider only the case of

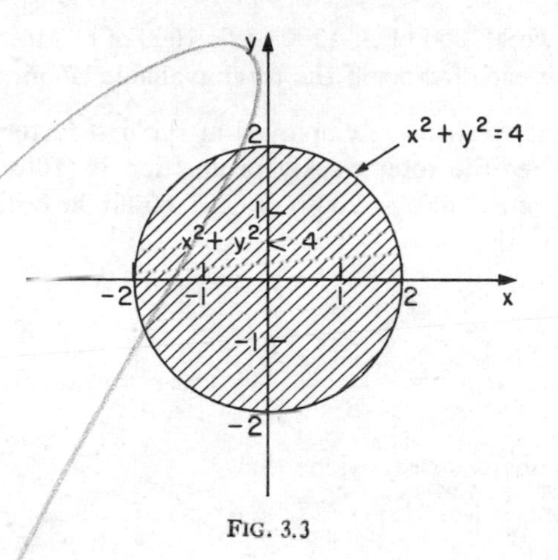

FIG. 3.3

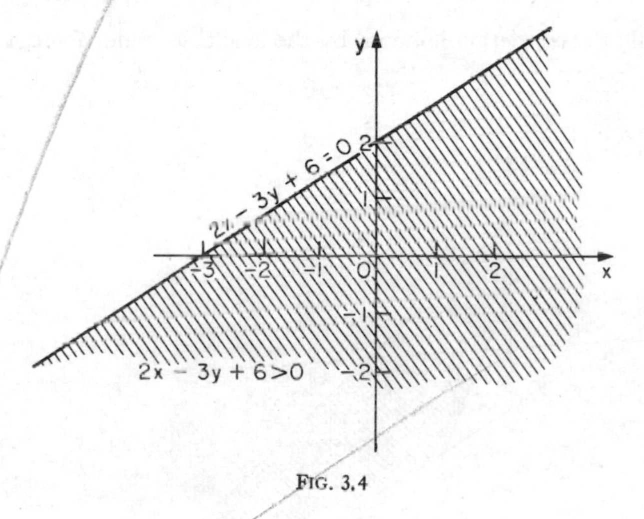

FIG. 3.4

two variables although the ideas can be extended quite generally.) Any particular decision can be represented by the corresponding point (x, y). Taking into account all the restrictions imposed on x and y we can then see the whole range of valid decisions as a set of points.

Example 11 A firm produces three articles A, B, C in two factories. For each 100 employees at the first factory the weekly production is 100 of A, 200 of B, 500 of C, while the corresponding production figures for the second factory are 500, 300, 200. To fulfil its orders the firm must produce each week 1000 of A, 1300 of B, 1600 of C. How many men can be employed at each factory if the total available labour force is 1000?

Suppose x hundred men are employed at the first factory, y hundred at the second. Then the total weekly production is $(100x + 500y)$ of A, $(200x + 300y)$ of B, $(500x + 200y)$ of C. To fulfil the orders we see (after division by 100) that

$$x + 5y \geqslant 10 \quad (1)$$

$$2x + 3y \geqslant 13 \quad (2)$$

$$5x + 2y \geqslant 16. \quad (3)$$

There are the obvious restrictions that

$$x \geqslant 0 \quad (4)$$

$$y \geqslant 0 \quad (5)$$

and finally the restriction imposed by the available labour force, viz.

$$x + y \leqslant 10. \quad (6)$$

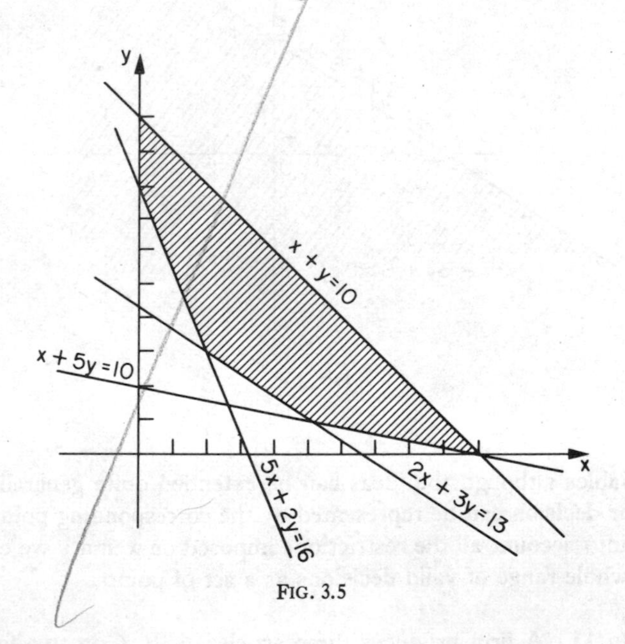

FIG. 3.5

Each of the conditions (1)–(6) implies that the point (x, y) is in the region to one side of a straight line, and the points that satisfy all six conditions must be common to all these regions, i.e. in the shaded region of the figure. Any point (x, y) in the shaded region (subject to the condition that $100x$ *and* $100y$ are whole numbers!) is a solution of the problem.

When we have a set of valid decisions at our disposal we can take some other factor into account and look for the *best* decision.

Example 12 What is the answer in Example 11 if total costs are to be minimised and these amount to £80 per week for each employee at the first factory and £40 per week for each employee at the second factory?

We must choose the point (x, y) in the required region so as to minimise $80x + 40y$, i.e. to minimise $2x + y$. The broken lines in Fig. 3.6 are the 'level' lines for $2x + y$ and scrutiny of these lines shows that $2x + y$ takes its least value in the region at P, i.e. at $(2, 3)$. Hence we would employ 200 men at the first factory, 300 at the second.

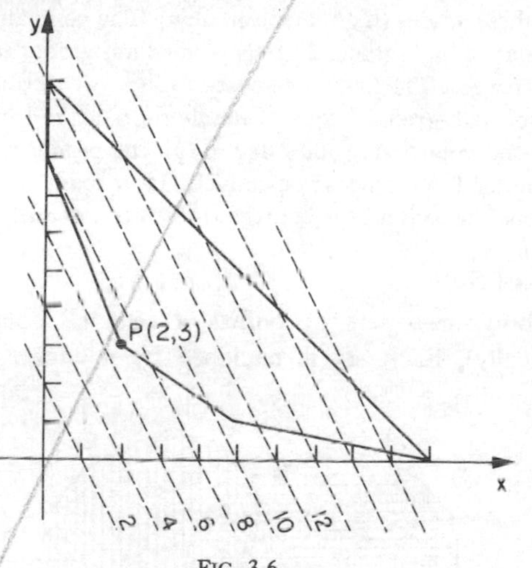

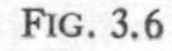

FIG. 3.6

Exercises 3(b)

1. Sketch the region in which all points (x, y) satisfy simultaneously the three inequalities

$$x - 2y + 4 \geqslant 0, \quad x - 3y + 4 \leqslant 0, \quad x + y - 8 \leqslant 0.$$

Find the point (x, y) in the region for which $5x - 12y$ has the greatest value. Find also the least value of $12x - 5y$ for points (x, y) in the region.

2. Sketch the region defined by the inequalities

$$x \geqslant 0, \quad y \geqslant 0, \quad x + 2y \leqslant 8, \quad x + y \leqslant 5, \quad 2x + y \leqslant 8.$$

Find the maximum value of $x + 3y$ if x, y satisfy the given inequalities.

3. In one figure, sketch the curves $y^2 = x$, $x^2 = y$, and shade the region R occupied by the set of points

$$\{(x, y); y^2 \leqslant x, x^2 \leqslant y\}.$$

Find the maximum of $x + y$ for points in R. Mark the point in R for which $y - x$ is greatest and find this greatest value.

4. Draw a sketch to illustrate the set of points R where

$$R = \{(x, y): y^2 \leqslant x, x + y \leqslant 2\},$$

and find the maximum values of $x + 2y$, $2x + y$ in R.

5. A farmer has 100 acres of land, £1000 capital and 160 man-days of labour available to him. Each acre of crop A requires 1 man-day of labour and £10 capital, and gives a return of £40. Each acre of crop B requires 4 man-days of labour and £20 capital, and gives a return of £120. How many acres of each crop should he have for maximum money return?

6. A company has two mines producing a certain ore which after crushing is sorted into three grades (high, medium, low). The company has a contract to supply 10 tons of high-grade, 22 tons of medium-grade, and 42 tons of low-grade ore per week. The first mine costs £400 a day to run and produces per day 4 tons of high-grade, 4 tons of medium-grade, and 6 tons of low-grade ore. The second mine costs £300 a day to run and produces 2, 6, 18 tons of the high, medium and low grades respectively. How many days per week should each mine operate to fulfil the contract most economically?

3.3 Algebra of Sets

Figure 3.7 shows two sets of points A (shaded horizontally) and B (shaded vertically). Each set is enclosed by a curve. We can use the

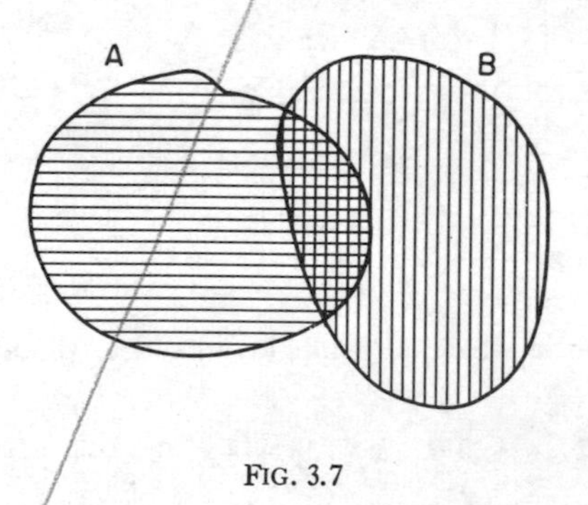

FIG. 3.7

diagram to give immediate answers to the questions

(a) do A and B have common members?
(b) does A have members which do not belong to B?
(c) does B have members which do not belong to A?

In this section our discussion of sets depends very much on the answers to (a), (b) and (c) for each pair of sets irrespective of the nature of the members or the numerical details (number of members in each set, number of common members). Consequently this diagram may be used to represent any pair of sets A and B, not necessarily sets of points, provided the answers to (a), (b) and (c) are the same for the actual sets as for the representative sets of points. Such a representative diagram is called a *Venn diagram*. There is no suggestion of a (1, 1) correspondence between the members of the actual set and the points in the representative set; in fact *any* closed curve is used for A and *any* closed curve for B subject to the answers to (a), (b) and (c) being unchanged. Similarly a Venn diagram for any number of sets is considered a correct representation if it is so for each pair of sets.

B is called a *subset* of A (denoted $B \subset A$, or $A \supset B$) if all members of B are members of A.

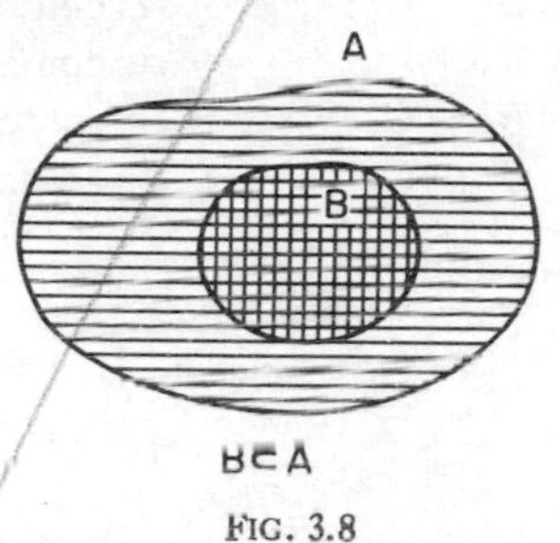

FIG. 3.8

The *union* of A and B (denoted $A \cup B$) is the set of all members which are in either A or B, i.e. in Fig. 3.7 $A \cup B$ is the set of all shaded points.

The *intersection* of A and B (denoted $A \cap B$) is the set of members common to both A and B, i.e. the set of points shaded both ways.

The *universal set* (chosen in relation to the sets we are discussing at a particular time) is a set such that all sets under discussion are subsets of it, e.g. for sets of real numbers we can take the universal set to be R (the set of all real numbers) or when only positive real numbers are involved then $\{x : x \text{ in } R, x > 0\}$ would be a suitable universal set.

The *complement* of A (denoted A') is the set of all members in the universal set that are not in A.

In a Venn diagram the universal set is usually represented by a rectangle large enough to enclose all other sets appearing. If the universal set is denoted by S then

$$S \cup A = S, \quad S \cap A = A, \quad A \cup A' = S \qquad \text{for any set } A, \quad (3.1)$$

where the equality sign now means 'is the same set as'.

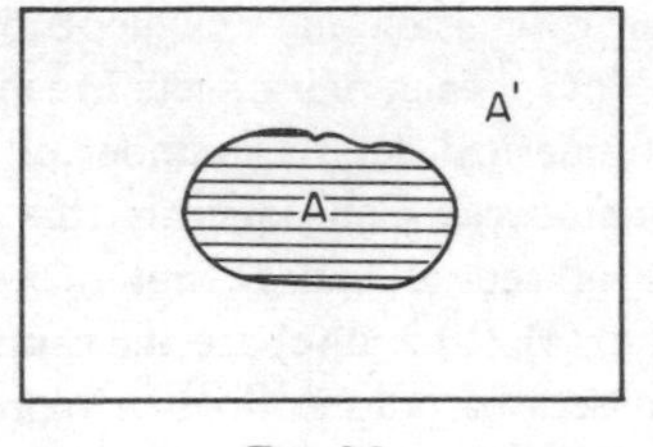

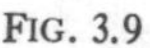
FIG. 3.9

A set with no members is called *empty* and denoted by ϕ. Thus

$$A \cup \phi = A, \quad A \cap \phi = \phi, \quad A \cap A' = \phi \qquad \text{for any } A. \qquad (3.2)$$

It is assumed that ϕ is a subset of any other set, i.e. $\phi \subset A$ for any A. From our definitions we see that

$A \cup B$	$= B \cup A$	($\cup$ is commutative)	(3.3)
$A \cap B$	$= B \cap A$	($\cap$ is commutative)	(3.4)
$A \cup (B \cup C)$	$= (A \cup B) \cup C$	($\cup$ is associative)	(3.5)
$A \cap (B \cap C)$	$= (A \cap B) \cap C$	($\cap$ is associative)	(3.6)

and the not-so-obvious results

$A \cap (B \cup C)$	$= (A \cap B) \cup (A \cap C)$	($\cap$ is distributive w.r.t. $\cup$)	(3.7)
$A \cup (B \cap C)$	$= (A \cup B) \cap (A \cup C)$	($\cup$ is distributive w.r.t. $\cap$)	(3.8)
$(A \cup B)'$	$= A' \cap B'$,		
$(A \cap B)'$	$= A' \cup B'$	(de Morgan's laws).	(3.9)

Example 13 Use Venn diagrams to demonstrate (3.7).

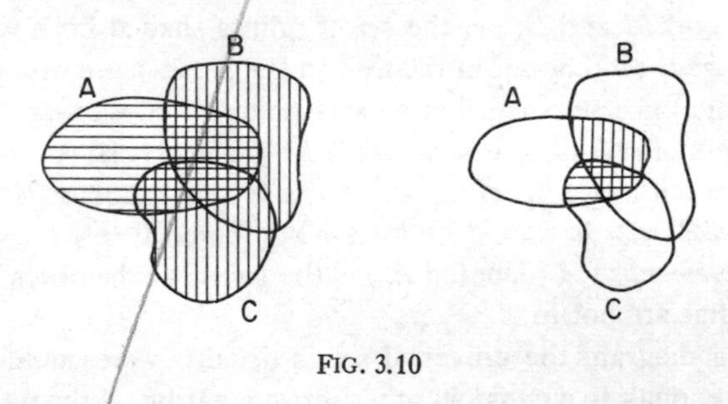

FIG. 3.10

In the first diagram $B \cup C$ is shaded vertically, A is shaded horizontally, and so $A \cap (B \cup C)$ is double-shaded. In the second diagram, $A \cap B$ is shaded vertically, $A \cap C$ is shaded horizontally, and so $(A \cap B) \cup (A \cap C)$ is the total shaded area.

Comparing the two diagrams we see that

$$A \cap (B \cup C) = (A \cap B) \cup (A \cap C).$$

The diagrams have been drawn for the case where any pair of sets have common members but the result holds in all cases.

Thinking of $\cup$, $\cap$, $'$ as *operations* for sets we have now established the basic laws for an algebra of sets. This algebra, usually called *Boolean algebra*, will not be pursued any further here, but it should be pointed out that such an algebra is a model for other entities which operate together in a similar way—we now look briefly at an example of this taken from electrical engineering.

3.3.1 Switching Circuits The basic entities to be discussed here may be described as 'switch arrangements'. Although we refer to an individual switch as (say) the switch A, the letter A denotes, strictly speaking, the *arrangement* consisting of this single switch.

Let $A \cup B$ denote the arrangement in which the switches A and B are connected in parallel, $A \cap B$ the arrangement in which they are connected in series. Denote by the same letter A any other switch linked to A so as to open when A opens and close when A closes, and by A' any other switch linked to A so as to close when A opens and open when A closes.

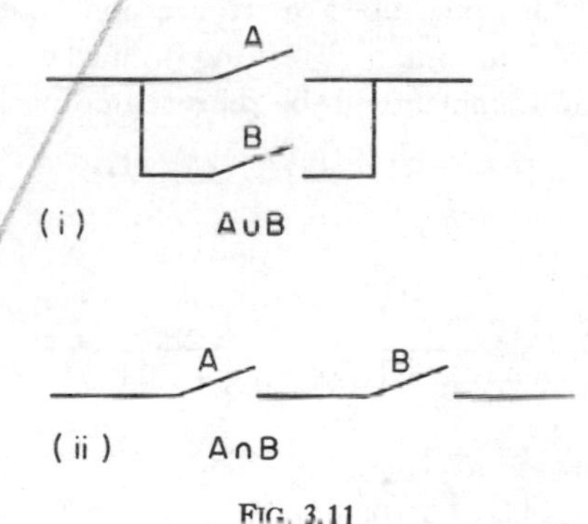

FIG. 3.11

Finally let S denote an arrangement which always provides a closed path, ϕ an arrangement which never provides a closed path, e.g. A and A' connected in parallel gives S, A and A' connected in series gives ϕ.

A specification indicating which switches in an arrangement are closed and which are open will be called a *setting* of the arrangement. Thus a single switch A has two possible settings (open and closed), an arrangement involving two switches A, B has four possible settings, an arrangement involving three switches A, B, C has eight possible settings, and so on.

Any setting of an arrangement has one of two possible *results*; either it provides a closed path and so allows current to pass, or else it does not.

Two arrangements involving switches A, B, . . . are equal if, whenever they have the same setting, they give the same result.

We can now verify that all the identities (3.1) to (3.9) are satisfied. The algebra of sets can thus be used in the design and analysis of complicated switching circuits.

Example 14 Find a simpler arrangement equivalent to that shown in Fig. 3.12.

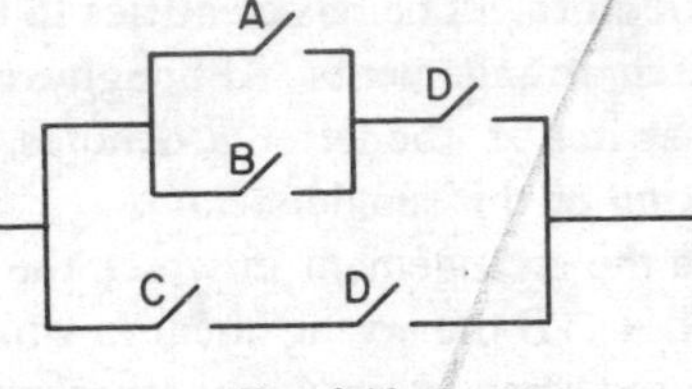

FIG. 3.12

The given arrangement can be broken down into two parts (a) and (b) as indicated in Fig. 3.13. Since these parts are connected in parallel, the given arrangement is their 'union'. Breaking (a) into two further parts, we see that the given arrangement can be represented by

$$[(A \cup B) \cap D] \cup (C \cap D).$$

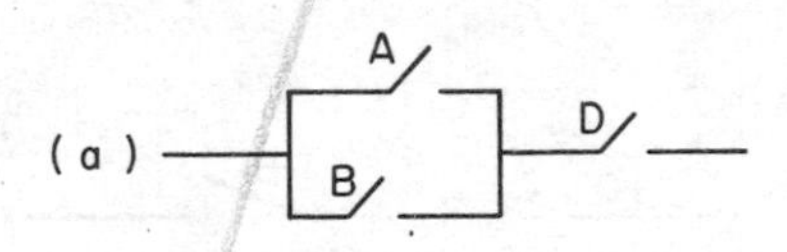

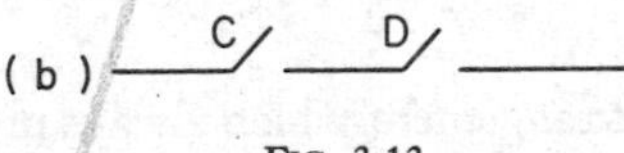

FIG. 3.13

From (3.4) this is equivalent to $[D \cap (A \cup B)] \cup (D \cap C)$, that is

$$D \cap [(A \cup B) \cup C]$$

on using (3.7).

From (3.5) we see that the parentheses are unimportant, and again using (3.4) we can write the expression in the form $(A \cup B \cup C) \cap D$. This gives the simple arrangement shown in Fig. 3.14.

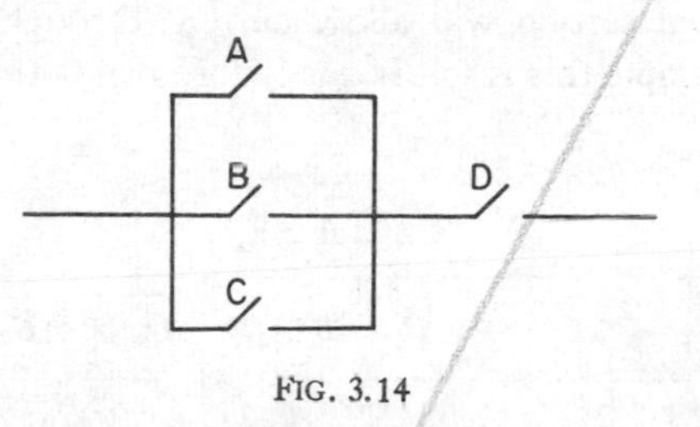

FIG. 3.14

We have seen that any setting of an arrangement has one of two possible results. Consequently all settings can be given a 'value' 1 or 0, according as they allow or do not allow the passage of current.

TABLE 3.1

A	B	$A \cup B$	$A \cap B$
1	1	1	1
1	0	1	0
0	1	1	0
0	0	0	0

The four possible settings of the arrangements $A \cup B$ and $A \cap B$ have their values listed in Table 3.1. For example, reading across the third row in the table we infer that if A is open and B is closed, then $A \cup B$ allows current through while $A \cap B$ does not; this can be abbreviated to $0 \cup 1 = 1$, $0 \cap 1 = 0$. In this way we can study the effect of any particular setting using a very simple mathematical model consisting of two numbers 0, 1 with two operations $\cup$, $\cap$ subject to 'arithmetic' as set out in the table above.

Example 15 Determine whether the arrangement in Example 14 allows current through when A is open, B closed, C open, D closed.

We give A and C the value 0, and B and D the value 1. The simplified form of the arrangement is $(A \cup B \cup C) \cap D$, so its value can be written as $(0 \cup 1 \cup 0) \cap 1$.

Using the table for $A \cup B$, we can put $0 \cup 1 = 1$, and so

$$0 \cup 1 \cup 0 = 1 \cup 0 = 1,$$

and finally

$$(0 \cup 1 \cup 0) \cap 1 = 1 \cap 1 = 1,$$

the last result coming from the table for $A \cap B$.

Hence the given setting will allow current through. Naturally, with such a simple example this result is easily foreseen without the use of the above method.

TABLE 3.2

A	B	C	$A \cup B$	$A \cup C$	$B \cap C$	$A \cup (B \cap C)$	$(A \cup B) \cap (A \cup C)$
0	0	0	0	0	0	0	0
0	0	1	0	1	0	0	0
0	1	0	1	0	0	0	0
0	1	1	1	1	1	1	1
1	0	0	1	1	0	1	1
1	0	1	1	1	0	1	1
1	1	0	1	1	0	1	1
1	1	1	1	1	1	1	1

By definition two arrangements are equal if the values corresponding to the possible settings are the same. For example, consider Table 3.2. This shows that the eight possible values of $A \cup (B \cap C)$ agree with the eight possible values of $(A \cup B) \cap (A \cup C)$ and so verifies that

$$A \cup (B \cap C) = (A \cup B) \cap (A \cup C).$$

In the algebra of real numbers an identity must be deduced from the axioms since it cannot be verified for all values of the variables. In the algebra of switch arrangements, however, each symbol A, B, . . . has only two possible values, and so an identity may readily be proved by direct verification as above.

Exercises 3(c)

1. In a sketch, show the following sets of points by shading:

$$A = \{(x, y) : y \geqslant x^2 - x + 1\}, \quad B = \{(x, y) : y \leqslant 1 + x\}, \quad C = A \cap B.$$

Find the maximum value of $x + y$ in C.

2. In a certain examination candidates are tested in Mathematics, Science, French, German, and Latin. The letters M, S, F, G, L denote the sets of successful candidates in these subjects respectively. In order to obtain a final certificate, a candidate must pass in Mathematics, Science and Latin or else in

Mathematics and two languages. Express in terms of unions and intersections of sets the set of candidates who

(a) obtain a final certificate,
(b) pass three subjects without obtaining a final certificate.

3. In a certain set of 100 students, a subset A are blondes and the rest are brunettes, B are left-handed, and C are left-handed blondes. Express in terms of A, B, C, the subset consisting of right-handed brunettes. If A contains 40 students, B contains 12 and C contains 5, how many right-handed brunettes are there?
4. Draw figures illustrating sets P, Q, X, Y for which

$$P \subset Q, \quad X \cap Y = \phi.$$

Draw a figure to illustrate

$$A \cup (B \cap C) = (A \cup B) \cap (A \cup C),$$

and state the other distributive law. If

$$(A \cup B) \cap (A \cup C) \cap (A \cup D) = A,$$

what can be said about the sets B, C, D?

5. The operation 'set subtraction' may be defined by the equation

$$A - B = A \cap B'$$

where A, B are subsets of a universal set S and B' denotes the complement of B with respect to S. Illustrate this operation by means of a Venn diagram.
Is set subtraction commutative? Prove your answer.
Is the result $A - (B - C) = (A - B) - C$ true or false? Prove your answer.
Prove that $(A - B)' = A' \cup B$.
(Proof by the use of Venn diagrams is sufficient in all cases.)

6. Complete Table 3.3 for the switch arrangement $(A \cup B)'$. Use tables of this kind to verify that

$$(A \cup B)' = A' \cap B', \quad (A \cap B)' = A' \cup B'.$$

TABLE 3.3

A	B	$(A \cup B)'$
0	0	1
0	1	
1	0	
1	1	

7. Verify the following identities for switch arrangements (a) by Boolean algebra, (b) by constructing a table of values:

$$A \cap (A \cup B) = A, \quad (A \cap B') \cap (A \cup B) = A \cap B',$$
$$(A \cup B') \cap (A' \cup B) = (A \cap B) \cup (A' \cap B').$$

8. Use tables of values to verify for switch arrangements that

$$(A \cup B) \cup C = A \cup (B \cup C), \quad A \cap (B \cup C) = (A \cap B) \cup (A \cap C).$$

3.4 Probability

It is very natural to make the assumption that certain events are 'equally likely'. For example, when normal unbiased coins are tossed there is no reason to suppose that heads will turn up more often than tails; we now express this numerically by saying that each result has a probability of $\frac{1}{2}$. Likewise when a die is cast there are six equally likely results to each of which we attribute a probability of $\frac{1}{6}$; when two dice are cast there are 36 equally likely results each with probability $\frac{1}{36}$.

In general if an experiment has N possible outcomes, all of which may be regarded as equally likely, we say each outcome has a ***probability*** of $1/N$. The set S of all possible outcomes has then N members, and is sometimes called the ***sample space*** for the experiment; we may write $S = \{r_1, r_2, \ldots, r_N\}$, where each of $r_1, r_2, \ldots$ refers to a possible result.

An *event* E will be taken to mean a subset of S, and we will say an event E takes place if any of the results in E occur. For example, when casting a die the sample space S is the set of possible results viz. $\{1, 2, 3, 4, 5, 6\}$; getting a 3 is the event $\{3\}$ which occurs only when the result 3 occurs, and getting an even number is the event $\{2, 4, 6\}$ which occurs when any one of the results 2, 4, 6 occurs. For the toss of two coins we can write $S = \{HH, HT, TH, TT\}$ and for both coins to give the same result we have the event $\{HH, TT\}$.

The definition of probability can be slightly extended as follows:

Definition. If the event E has n members out of a total sample space of N members, we define the ***probability of E***, denoted by $p(E)$, by the equation

$$p(E) = n/N.$$

In particular, the probability of an even number from the throw of a die is $\frac{3}{6}$, i.e. $\frac{1}{2}$, and the probability of similar results from two coins is $\frac{2}{4}$, i.e. $\frac{1}{2}$.

Example 16 Two men are selected at random from a group consisting of four married men and three single men. What is the probability that two married men are chosen?

The sample space is the set of possible ways of selecting two men from a group of seven men and so has ${}^7C_2$ members.

The event that two married men be chosen is a subset containing ${}^4C_2$ members. Hence the probability of this event is ${}^4C_2/{}^7C_2$, i.e. $\frac{2}{7}$.

From the definition we see that $p(S) = N/N = 1$; since one of the results of S *must* occur, this means that a probability of 1 implies *certainty*. Also, $p(\phi) = 0/N = 0$; since the event ϕ is the empty subset which contains none of the results in S and therefore cannot occur, we equate zero probability with *impossibility*.

Hence $0 \leqslant p(E) \leqslant 1$ for all events E; $p(E) = 0$ only if E is impossible and $p(E) = 1$ only if E is certain.

The event E' is the subset containing all results in S that are not in E. Thus to say that the event E' occurs is the same as saying that the event E does not occur. In addition if E has n members then E' has $N - n$ members so that

$$p(E') = \frac{N-n}{N} = 1 - \frac{n}{N} = 1 - p(E).$$

For example, the probability of not getting a six with a die is $1 - \frac{1}{6}$, i.e. $\frac{5}{6}$.

The event $E_1 \cup E_2$ is the subset of S that contains those results of S that are in E_1 together with those that are in E_2; in other words $E_1 \cup E_2$ occurs when either E_1 or E_2 occurs. For example, if E_1 is the event of getting a 3 with a die and E_2 is the event of getting a 5, then $E_1 \cup E_2$ is the event of getting either a 3 or a 5.

Likewise $E_1 \cap E_2$ is the event 'both E_1 and E_2'. For example if E_1 is the event of getting two even numbers with two dice and E_2 is the event of getting two numbers with sum 8, then

$$E_1 = \{(2,2), (2,4), (2,6), (4,2), (4,4), (4,6), (6,2), (6,4), (6,6)\},$$
$$E_2 = \{(2,6), (3,5), (4,4), (5,3), (6,2)\},$$

so that

$$E_1 \cap E_2 = \{(2,6), (4,4), (6,2)\};$$

in words, $E_1 \cap E_2$ is the event of getting two even numbers with sum 8.

Two events are called ***mutually exclusive*** if the occurrence of either precludes the occurrence of the other. Obviously E_1 and E_2 are mutually exclusive if, and only if, $E_1 \cap E_2 = \phi$.

Theorem 1 If E_1 and E_2 are mutually exclusive events, then

$$p(E_1 \cup E_2) = p(E_1) + p(E_2).$$

Proof. If E_1 has n_1 members, E_2 has n_2 members, and the sample space has N members, then since $E_1 \cap E_2 = \phi$ it follows that $E_1 \cup E_2$ has $n_1 + n_2$ members, so that

$$p(E_1 \cup E_2) = \frac{n_1 + n_2}{N} = \frac{n_1}{N} + \frac{n_2}{N} = p(E_1) + p(E_2).$$

Example 17 From a throw of 2 dice, what is the probability that we get either a total of 8 or else a total of 9?

Probability that we get a total of 8 is $\frac{5}{36}$.
Probability that we get a total of 9 is $\frac{4}{36}$.

Since these events are mutually exclusive the probability that we get one or other is $\frac{5}{36}+\frac{4}{36}$, i.e. $\frac{1}{4}$.

Example 18 From a throw of 2 dice what is the probability that we get a total of at least 8?

This event occurs if the total is either 8, 9, 10, 11, or 12, there being no possibility of a total higher than 12.

Hence the probability is the sum of the probability for totals of 8, 9, 10, 11, 12, that is

$$\frac{5}{36}+\frac{4}{36}+\frac{3}{36}+\frac{2}{36}+\frac{1}{36}=\frac{5}{12}.$$

Theorem 2 If E_1 and E_2 are events associated with two independent experiments then the probability that E_1 occurs in the first experiment and also E_2 occurs in the second experiment is $p(E_1) \times p(E_2)$.

Proof. Here E_1 and E_2 are subsets of separate sample spaces. Let S_1, the sample space for the first experiment, have N_1 members and let E_1 be a subset of S_1 with n_1 members. Then $p(E_1) = n_1/N_1$.

Similarly let S_2, the sample space for the second experiment, have N_2 members and let E_2 be a subset with n_2 members. Then $p(E_2) = n_2/N_2$.

If the experiments are independent the combined experiment can result in the combination of any result in S_1 with any result in S_2, giving a sample space of N_1N_2 members. Any result in E_1 combined with any result in E_2 gives the required event which has therefore n_1n_2 members. Hence the probability of this event is n_1n_2/N_1N_2, i.e. $p(E_1) \times p(E_2)$.

Example 19 Find the probability of getting two even numbers from two dice.

Probability of an even number with each die is $\frac{3}{6}$, i.e. $\frac{1}{2}$. Since the result from one die in no way affects the result from the other, we can apply Theorem 2 to get the answer $\frac{1}{2} \times \frac{1}{2}$, i.e. $\frac{1}{4}$.

Example 20 A bag contains 5 white and 3 black marbles, a second bag contains 6 black and 4 green marbles, and a third bag contains 3 red and 2 yellow marbles. A marble is drawn from each bag. What is the probability of obtaining a white, black and yellow marble in this way?

Applying Theorem 2 we get the answer $\frac{5}{8} \times \frac{6}{10} \times \frac{2}{5}$, i.e. $\frac{3}{20}$.

If the result of the first experiment affects the chances of the second event (E_2) occurring in the second experiment, we may be able to apply Theorem 2 in the revised form:

Theorem 2a If $p(E_1)$ is the probability that E_1 occurs in the first experiment and $p'(E_2)$ is the probability that when E_1 occurs, E_2 should occur in the second experiment, then the probability that E_1 and E_2 both occur is $p(E_1) \times p'(E_2)$.

Example 21 Solve the problem in Example 16 by use of Theorem 2a.

The probability of choosing a married man with first selection is $\frac{4}{7}$. If a married man has been chosen, the group now consists of 3 married and 3 single men and in that situation the probability of again choosing a married man is $\frac{3}{6}$, i.e. $\frac{1}{2}$.

By Theorem 2a, probability of choosing 2 married men is $\frac{4}{7} \times \frac{1}{2}$, i.e. $\frac{2}{7}$.

3.4.1 Application of Probability We now state the basis on which we are prepared to make practical use of the foregoing ideas, which at first sight may have seemed of somewhat limited application.

Basis for applications. If $p(E)$ is the probability of the event E in a certain experiment, then in a very large number of trials of that experiment the event E should occur in an approximate proportion $p(E)$ of the trials.

The emphasis here is on the very large number of trials; if we toss a coin 12 times or even 100 times we must not expect equal numbers of heads and tails, but if we toss the coin several thousand times then we expect a roughly equal distribution. If we are to play a game of guessing the sum of the 2 numbers obtained by throwing 2 dice, provided we are going to play the game a very large number of times it is correct strategy to guess 7 consistently since that is the sum with the highest probability and occurs most frequently in the long run.

The manner in which we apply the idea of probability to actual forecasting suggests an alternative definition of the probability $p(E)$ of the event E.

Empirical definition of $p(E)$. Suppose the event E occurs in n out of N trials of an experiment. If for all very large values of N the ratio n/N is approximately constant, say

$$\frac{n}{N} \simeq k \qquad \text{for large } N,$$

then $p(E)$, the probability that the event E should occur, is defined by

$$p(E) = k.$$

We are thinking here of the value of k as having been chosen on the basis of results from large numbers of trials of the experiment; hence there is a certain lack of precision about the value attached to $p(E)$ by this definition, but this is unimportant in the applications. The advantage of this definition is that it provides a working basis in cases where the previous definition, based on the idea of a sample space S consisting of a finite number of equally likely outcomes, does not. Examples of this are to be found in insurance (e.g. what is the probability that a man aged 25 survives until the age of 65?) and in industry (e.g. what is the probability that a machine mass-producing articles and in normal working order produces a defective article?). In these examples the idea of a finite sample space of equally likely outcomes simply does not apply; what can be done, however, is to use mortality tables or results from previous trials on the machine to estimate the probability in accordance with the empirical definition—in other words we find the proportion of times the event occurred in very large samples of results from the past.

The theorems in Section 3.4 are assumed to hold equally in cases where the empirical definition of $p(E)$ is used.

Example 22 A machine mass-producing cigarettes is known to give a slight deficiency in tobacco weight in 1% of the cigarettes. What is the probability that a packet of 10 cigarettes from this machine contains precisely 2 defectives?

We assume that each time a cigarette is produced the probability of its being defective is $0{\cdot}01$ and the probability of its being correct is $0{\cdot}99$. Hence in a packet of 10 cigarettes the probability of a particular 2 being defective (say the first and third) and the rest correct is $(0{\cdot}01)^2 \times (0{\cdot}99)^8$. Hence the probability of some 2 being defective and the rest correct is ${}^{10}C_2\,(0{\cdot}01)^2 \times (0{\cdot}99)^8$, i.e. $0{\cdot}00415$.

Exercises 3(d)

1. A drawer contains 5 pens and 4 pencils while another drawer contains 6 pens and 3 pencils. In how many ways can a group of 4 pens and 2 pencils be selected (a) if all come from the first drawer, (b) if the pens all come from one drawer and the pencils from the other, (c) if there is no restriction?

   What is the probability that a group of 6 selected at random from the first drawer consists of 4 pens and 2 pencils?
2. 3 cards are taken at random from a full pack. Find the probabilities that the 3 cards are (a) all spades, (b) of the same suit, (c) of the same value (e.g. all

sixes or all kings), (d) have values adding up to 4 (ace counting as 1).

3. Two cars each carrying three married couples collide. Of the twelve people involved, 9 have driving licenses and after the accident, 9 are incapacitated. What is the probability that exactly two drivers are not incapacitated?

   If the set L comprises the ladies; the set D the drivers; and the set I, the incapacitated; define in words the sets

$$L \cap (D \cup I), \quad I \cup (D \cap L).$$

4. In how many ways can 9 different books be divided among 3 people so that 1 person gets 4 books, 1 person gets 3 books, and 1 person gets 2 books? If 1 of the 3 people is a historian, and exactly 4 of the books are about history, what is the probability that, in a perfectly random distribution in the above manner, the historian gets all 4 history books?
5. There are 3 groups of students. Each group is to choose one out of n courses. If choice is random, find the probability that each group chooses a different course. Find how large n should be if this probability is to exceed $0{\cdot}9$.
6. What is the probability of finding in a family of 7 children (a) exactly 3 boys, (b) at least 5 boys, assuming that the probability of a boy is $\frac{1}{2}$?
7. A city council of 100 members has 50 members who own houses, 50 members who own business property, and 20 members who own neither. A committee of 5 is chosen at random. What is the probability that it contains

   (a) exactly 3 house-owners?
   (b) exactly 3 house-owners who also own business property?
   (c) at least 3 house-owners who also own business property?

8. Certain articles are packed in boxes of 20. A buyer receives a box in which 4 articles are defective. What is the probability that if he takes 2 articles at random (a) none is defective, (b) just one is defective, (c) at least one is defective?

   The buyer takes articles at random, one at a time, out of the box. He tests each article on taking it out and stops when he has found one which is not defective. What is the probability that he takes out precisely 3 articles?

9. A batch of 20 articles includes 3 which are defective, the rest being satisfactory. A sample of 5 articles is taken from the batch. Find the probability

   (a) that the sample includes all 3 defective articles,
   (b) that when the sample is taken from the batch the first 3 articles are the defective ones.

10. A man tests a batch of TV sets one at a time until he finds one in working order. If there are 10 sets in the batch, 3 of which are defective, what is the probability that he finds a good one (a) at the first attempt? (b) at the second attempt? (c) at the third attempt? (d) at the fourth attempt?

CHAPTER 4

SOME USEFUL RESULTS IN ALGEBRA

4.1 The Binomial Theorem

In the algebra of real numbers there are two distributive laws, viz.

$$x(y + z) = xy + xz, \tag{1}$$

$$(x + y)z = xz + yz. \tag{2}$$

We have already stated (1) as an axiom and (2) follows immediately from (1) together with the fact that multiplication is commutative.

It now follows that

$$(x + y)(z + t) = (x + y)z + (x + y)t \qquad \text{(using (1))}$$

$$= xz + yz + xt + yt \qquad \text{(using (2))}$$

Note that in the final form of the r.h.s.

(a) each term is a product involving one variable from each bracket on the l.h.s.

(b) every possible product involving one variable from each bracket occurs.

From (b) we expect 4 terms since there are 4 ways of selecting one variable from two in the first bracket, together with one variable from two in the second bracket.

When any number of brackets are multiplied together, each bracket being a sum of any number of variables, we find that (a) and (b) apply, e.g. from

$$(A + B)(a + b + c)(\alpha + \beta + \gamma + \delta)$$

we would get 2.3.4 terms, i.e. 24 terms, each term involving a capital letter, a lower-case letter, and a Greek letter.

In the case of the product

$$(x_1 + y_1)(x_2 + y_2) \cdots (x_n + y_n), \tag{3}$$

the statements (a) and (b) can be expressed in the form

(a′) each suffix 1, 2, . . ., n appears once and only once in each term

(b′) there is a term corresponding to every possible distribution of the suffices 1, 2, . . ., n between the two letters x and y.

Thus, from (b′) we have a term $x_1 x_2 \cdots x_n$ in which *all* the suffices have been given to x, a term $y_1 y_2 \ldots y_n$ in which *all* the suffices have been given to y, and terms corresponding to every possible distribution between these two extremes.

Consider those terms in which r suffices have been given to y and the remaining $(n - r)$ suffices given to x; since there are ${}^nC_r$ ways of selecting the r suffices given to y, there are ${}^nC_r$ such terms, i.e. ${}^nC_r$ terms in which the letter y appears r times and the letter x appears $(n - r)$ times. Further, if we put

$$x_1 = x_2 = \cdots = x_n = x, \quad y_1 = y_2 = \cdots = y_n = y,$$

each of these terms reduces to $x^{n-r} y^r$ while (3) becomes $(x + y)^n$. We have thus shown that

$$(x + y)^n = {}^nC_0 x^n + {}^nC_1 x^{n-1} y + \cdots + {}^nC_r x^{n-r} y^r + \cdots + {}^nC_n y^n.$$

This result is the *Binomial theorem for integral index.* Since

$$(x + y)^n = x^n(1 + y/x)^n$$

it is sufficient to remember the binomial theorem in the simpler form in which it is usually quoted, viz.

$$(1 + x)^n = {}^nC_0 + {}^nC_1 x + {}^nC_2 x^2 + \cdots + {}^nC_r x^r + \cdots + {}^nC_n x^n;$$

that is,

$$(1 + x)^n = 1 + nx + \frac{n(n-1)}{1.2} x^2 + \cdots$$
$$+ \frac{n(n-1)\cdots(n-r+1)}{1.2.\cdots.r} x^r + \cdots + x^n.$$

For example

$$(1 + x)^9 = 1 + 9x + \frac{9.8}{1.2} x^2 + \frac{9.8.7}{1.2.3} x^3 + \frac{9.8.7.6}{1.2.3.4} x^4 + \cdots + x^9$$

$$= 1 + 9x + 36x^2 + 84x^3 + 126x^4 + 126x^5 + 84x^6 + 36x^7$$
$$+ 9x^8 + x^9.$$

Note that in this example it is sufficient to calculate the coefficients up to the term x^4 since ${}^9C_5 = {}^9C_4$, ${}^9C_6 = {}^9C_3$, ${}^9C_7 = {}^9C_2$ etc. In general ${}^nC_r = {}^nC_{n-r}$ so a similar symmetry is shown by the coefficients in all binomial expansions.

4.1.1 Pascal's Triangle From the binomial expansions

$(1 + x)^1 = 1 + x,$
$(1 + x)^2 = 1 + 2x + x^2,$
$(1 + x)^3 = 1 + 3x + 3x^2 + x^3,$
$(1 + x)^4 = 1 + 4x + 6x^2 + 4x^3 + x^4$, etc.

we get a 'triangle' of coefficients

$$\begin{array}{ccccc} 1 & 1 & & & \\ 1 & 2 & 1 & & \\ 1 & 3 & 3 & 1 & \\ 1 & 4 & 6 & 4 & 1 \end{array}$$

which continues to build up according to the rules

(a) the first and last number in every row is 1
(b) any other number can be found by adding two numbers—the number directly above and its neighbour on the left—from the previous row (e.g. the arrow indicates that 6 = 3 + 3)

Rule (b) follows from the equation

$${}^{n+1}C_r = {}^{n}C_r + {}^{n}C_{r-1} \qquad \text{(see Section 3.1.3).}$$

Using these rules we can add a fifth row to the above triangle as follows:

$$\begin{array}{lcccccc} & 1 & (1+4) & (4+6) & (6+4) & (4+1) & 1 \\ \text{i.e.} & 1 & 5 & 10 & 10 & 5 & 1. \end{array}$$

4.1.2 Applications The following examples illustrate the use of the binomial theorem in performing algebraic expansions.

Example 1 Expand $(1 - 2x^3)^7$.

$$(1 - 2x^3)^7 = 1 + 7(-2x^3) + \frac{7.6}{1.2}(-2x^3)^2 + \frac{7.6.5}{1.2.3}(-2x^3)^3 + \cdots + (-2x^3)^7$$

$$= 1 + 7(-2x^3) + 21(2^2x^6) + 35(-2^3x^9) + 35(2^4x^{12}) + 21(-2^5x^{15}) + 7(2^6x^{18}) + (-2^7x^{21})$$

$$= 1 - 14x^3 + 84x^6 - 280x^9 + 560x^{12} - 672x^{15} + 448x^{18} - 128x^{21}.$$

Example 2 Expand $(x - 1/x)^5$.

$$\left(x - \frac{1}{x}\right)^5 = x^5\left(1 - \frac{1}{x^2}\right)^5$$

$$= x^5\left\{1 + 5\left(-\frac{1}{x^2}\right) + \frac{5.4}{1.2}\left(-\frac{1}{x^2}\right)^2 + \cdots + \left(-\frac{1}{x^2}\right)^5\right\}$$

$$= x^5\left\{1 - \frac{5}{x^2} + \frac{10}{x^4} - \frac{10}{x^6} + \frac{5}{x^8} - \frac{1}{x^{10}}\right\}$$

$$= x^5 - 5x^3 + 10x - \frac{10}{x} + \frac{5}{x^3} - \frac{1}{x^5}.$$

Example 3 Expand $(1 + x + x^2)^4$.

$$\begin{aligned}(1 + x + x^2)^4 &= 1 + 4(x + x^2) + 6(x + x^2)^2 + 4(x + x^2)^3 + (x + x^2)^4\\ &= 1 + 4(x + x^2) + 6x^2(1 + x)^2 + 4x^3(1 + x)^3 + x^4(1 + x)^4\\ &= 1 + 4(x + x^2) + 6x^2(1 + 2x + x^2)\\ &\quad + 4x^3(1 + 3x + 3x^2 + x^3) + x^4(1 + 4x + 6x^2 + 4x^3 + x^4)\\ &= 1 + 4x + 10x^2 + 16x^3 + 19x^4 + 16x^5 + 10x^6 + 4x^7 + x^8.\end{aligned}$$

Example 4 Find the constant term in the expansion of $(2x^2 + 3/4x^6)^{12}$.

$$\left(2x^2 + \frac{3}{4x^6}\right)^{12} = 2^{12}x^{24}\left(1 + \frac{3}{8x^8}\right)^{12}.$$

Each term in the expansion of $(1 + 3/8x^8)^{12}$ has the form ${}^{12}C_r(3/8x^8)^r$ and so when $r = 3$ this expansion has a term in $1/x^{24}$.

Hence the constant term in the expansion of $(2x^2 + 3/4x^6)^{12}$ is given by

$$2^{12}\ {}^{12}C_3\frac{3^3}{8^3} = 2^{12}\frac{12.11.10}{1.2.3}\frac{3^3}{2^9} = 47\,520.$$

The binomial theorem is also useful for numerical approximation.

Example 5 Evaluate $(1{\cdot}03)^{10}$ to two decimal places.

$$(1 + x)^{10} = 1 + 10x + 45x^2 + 120x^3 + 210x^4 + \cdots + x^{10}$$

and so

$$\begin{aligned}(1{\cdot}03)^{10} &= 1 + 10(0{\cdot}03) + 45(0{\cdot}03)^2 + 120(0{\cdot}03)^3 + 210(0{\cdot}03)^4\\ &\qquad + \cdots + (0{\cdot}03)^{10}\\ &= 1 + 0{\cdot}3 + 45(0{\cdot}0009) + 120(0{\cdot}000\,027) + 210(0{\cdot}000\,000\,81)\\ &\qquad + \cdots\\ &= 1 + 0{\cdot}3 + 0{\cdot}0405 + 0{\cdot}003\,24 + 0{\cdot}000\,170\,1 + \cdots\\ &= 1{\cdot}34 \text{ to two decimal places.}\end{aligned}$$

Exercises 4(a)

1. Find the coefficient of x^4 in the expansion of $(2 - 3x)^9$.
2. Find the independent term in the expansion of $(x + 1/x)^{20}$.
3. Write out the complete expansion in ascending powers of x of (a) $(2 + x)^5$, (b) $(1 - 2x)^5$.
4. Find the term independent of x in the expansion of (a) $(x^2 + 1/x)^{12}$, (b) $(x - 1/x^2)^9$.
5. Expand the binomial expression $(x + 1/2x)^6$ in a series of descending powers of x. By putting $x = 10$, or otherwise, evaluate $(10{\cdot}05)^6$ correct to the nearest integer.
6. Evaluate the coefficient of x^{-9} in the expansion of $(3x + 1/x^2)^9$.
7. Show that, in the binomial expansion of $(1 + x)^{12}$ when $x = 0{\cdot}03$, the ratio of the rth to the $(r - 1)$th term is less than $0{\cdot}1$ if $r > 4$. If, in the expansion of $(x - a/x^2)^6$ the term independent of x is 240, show that $a = \pm 4$.
8. Show that in the expansion of $\{x^\alpha + 1/x^\beta\}^{\alpha+\beta}$, where α, β are positive integers, there is a term independent of x and its value is $(\alpha + \beta)!/\alpha!\beta!$.
9. (a) Use a binomial expansion to evaluate $(1{\cdot}04)^5$ correct to four decimal places.
   (b) Show that the term independent of x in the expansion of $(x + 1/x)^{2n}$ is $(2n)!/(n!)^2$.
10. Expand $(1 + y)^{10}$ in ascending powers of y up to and including the term in y^5.

    Given that the first five terms in the expansion of $(1 + x + x^2)^{10}$ are

    $$1,\ 10x,\ ax^2,\ bx^3,\ \text{and } cx^4,$$

    find the values of a, b, c. What are the values of the coefficients of x^{20}, x^{19}, x^{18}?
11. Show that $(2 - \frac{1}{2}x)^8 = 256 - 512x + 448x^2 - 224x^3 + \cdots$, and that $(2 - \frac{1}{2}x - \frac{1}{2}x^2)^8 = 256 - 512x - 64x^2 + \cdots$.

    Hence, or otherwise, show that if x is so small that x^4 and higher powers of x may be neglected,

    $$(2 - \tfrac{1}{2}x - \tfrac{1}{2}x^2)^8 + (2 + \tfrac{1}{2}x - \tfrac{1}{2}x^2)^8 = 128(4 - x^2),$$

    approximately.
12. From the expansion of $(1 + x)^6$ in ascending powers of x deduce
    (a) the expansion of $(2 - x^2)^6$,
    (b) the constant term in the expansion of $(y^2 + 1/y)^6$,
    (c) the value of $(1{\cdot}03)^6$ to three decimal places,
    (d) the expansion of $(1 + x + x^2)^6$, up to and including the term in x^3. Would it be safe to take only the first three terms in (d) if $|x| < 0{\cdot}1$ and two-decimal accuracy is required?

4.2 Polynomials

If x is a variable, n is a non-negative integer and $p_0, p_1, p_2, \ldots, p_n$ are given constants of which p_0 is not zero, then

$$p_0x^n + p_1x^{n-1} + \cdots + p_{n-1}x + p_n$$

is a polynomial of degree n in x. We shall denote this polynomial by $P(x)$.

4.2.1 The Remainder and Factor Theorems Let $P(x)$, a polynomial of degree n in x, be divided by $x - a$, where a is a constant. Then the quotient $Q(x)$ is a polynomial of degree $n - 1$, the remainder R is a constant and

$$P(x) \equiv (x - a)Q(x) + R \tag{1}$$

Since (1) is an identity we can substitute any real number for x in (1). Putting $x = a$ in (1), we get $P(a) = R$.

Hence when $P(x)$ is divided by $x - a$, the remainder is $P(a)$. This result is known as the *remainder theorem*. The *factor theorem* follows immediately: if $P(x)$ is a polynomial, $x - a$ is a factor of $P(x)$ if, and only if, $P(a) = 0$.

4.2.2 Further Properties of Polynomials

Property 1 If the polynomial $P(x) \equiv p_0x^n + p_1x^{n-1} + \cdots + p_{n-1}x + p_n$ $(p_0 \neq 0)$ is equal to zero when x has any one of the n distinct values $a_1, a_2, \ldots, a_n$, then

$$p_0x^n + p_1x^{n-1} + \cdots + p_{n-1}x + p_n \equiv p_0(x - a_1)(x - a_2) \cdots (x - a_n).$$

By the factor theorem, since $P(a_1) = 0$, $x - a_1$ is a factor of $P(x)$ and the quotient when we divide $P(x)$ by $x - a_1$ is a polynomial of degree $n - 1$ whose first term is p_0x^{n-1}. Hence we write

$$P(x) \equiv (x - a_1)Q_{n-1}(x), \tag{1}$$

where $Q_{n-1}(x) \equiv p_0x^{n-1} + \cdots$. Since $P(a_2) = 0$, we have from (1)

$$0 = P(a_2) = (a_2 - a_1)Q_{n-1}(a_2),$$

that is, $Q_{n-1}(a_2) = 0$, since $a_2 \neq a_1$.

Hence $x - a_2$ is a factor of $Q_{n-1}(x)$, and we write

$$Q_{n-1}(x) \equiv (x - a_2)Q_{n-2}(x)$$

and, by (1),

$$P(x) \equiv (x - a_1)(x - a_2)Q_{n-2}(x),$$

where $Q_{n-2}(x) \equiv p_0x^{n-2} + \cdots$.

Proceeding in this way, we see that

$$P(x) \equiv (x - a_1)(x - a_2) \cdots (x - a_n)Q_0(x),$$

where $Q_0(x) \equiv p_0$. Hence

$$P(x) \equiv p_0(x - a_1)(x - a_2) \cdots (x - a_n).$$

Property 2 If the polynomial $P(x) \equiv p_0x^n + p_1x^{n-1} + \cdots + p_{n-1}x + p_n$ is equal to zero for more than n distinct values of x, then $P(x)$ is equal to zero for all values of x and each of the coefficients $p_0, p_1, \ldots, p_n$ is zero. In this case $P(x)$ is identically zero, that is

$$p_0x^n + p_1x^{n-1} + \cdots + p_{n-1}x + p_n \equiv 0.$$

We suppose that $P(x) = 0$ when $x = a_1, a_2, \ldots, a_n$. Then, by Property 1,

$$P(x) \equiv p_0(x - a_1)(x - a_2) \cdots (x - a_n).$$

Now suppose that $P(a) = 0$, where a is different from any of $a_1, a_2, \ldots, a_n$. Then

$$p_0(a - a_1)(a - a_2) \cdots (a - a_n) = 0,$$

and since none of the factors $(a - a_1), (a - a_2), \ldots, (a - a_n)$ is zero, p_0 must be zero. Hence

$$P(x) \equiv p_1x^{n-1} + p_2x^{n-2} + \cdots + p_n.$$

This polynomial of degree $n - 1$ vanishes for more than $n - 1$ values of x and so, applying the same argument, we see that $p_1 = 0$. Continuing in this way, we prove that

$$p_1 = p_2 = \cdots = p_n = 0.$$

As a corollary to Property 2, we have the important property that if, for all values of x,

$$\begin{aligned} p_0x^n + p_1x^{n-1} + \cdots + p_{n-1}x + p_n \\ = q_0x^n + q_1x^{n-1} + \cdots + q_{n-1}x + q_n, \end{aligned} \tag{1}$$

then $p_0 = q_0, p_1 = q_1, \ldots, p_n = q_n$.

To see this we consider the polynomial

$$(p_0 - q_0)x^n + (p_1 - q_1)x^{n-1} + \cdots + (p_{n-1} - q_{n-1})x + (p_n - q_n)$$

which, by hypothesis, is zero for all values of x. It follows that all its coefficients are zero, and so

$$p_0 = q_0, p_1 = q_1, \ldots, p_n = q_n. \tag{2}$$

The process of deducing from the identity (1) the results (2) is called *equating coefficients.*

4.2.3 Polynomials in Several Variables A polynomial in several variables $x, y, z, \ldots$ is a sum of terms of the form $kx^p y^q z^r \cdots$, where the indices $p, q, r, \ldots$ are positive integers or zero and k is a constant.

The degree of any term is the sum of its degrees with respect to the variables, so that a term $7xy^3z^2$ is of the 6th degree.

The degree of a polynomial is that of the term of the highest degree in the polynomial. For example $x^2y^2 + y^2 + 2x - 5$ is a polynomial of the 4th degree in x and y; $xyz - 2x^2 + 3z + 4$ is a polynomial of the 3rd degree in x, y and z.

4.2.4 The Method of Undetermined Coefficients The method of undetermined coefficients is based on the principle that if two polynomials in x are identically equal, coefficients of like powers of x must be equal. The principle is valid for polynomials in several variables $x, y, z, \ldots$ and the method of undetermined coefficients may be applied to such polynomials.

Example 6 Find the values of the constants a, b, c and d such that

$$r^3 \equiv ar(r-1)(r-2) + br(r-1) + cr + d. \quad (1)$$

Multiplying out, we have

$$r^3 \equiv a(r^3 - 3r^2 + 2r) + b(r^2 - r) + cr + d,$$

$$r^3 \equiv ar^3 + r^2(b - 3a) + r(c + 2a - b) + d. \quad (2)$$

We shall use the symbol $((r^n))$ to denote the coefficient(s) of r^n.

Equating $((r^3))$ on each side of (2) we have $1 = a$;

equating $((r^2))$, $0 = b - 3a$, i.e. $b = 3$
equating $((r))$, $0 = c + 2a - b$, i.e. $c = 1$;
equating $((r^0))$, $0 = d$.

Hence $r^3 \equiv r(r-1)(r-2) + 3r(r-1) + r$.

Alternatively, we may use the fact that identity (1) is true for all values of r. Substituting in turn the values $r = 0, 1, 2$ and 3, we obtain as before $d = 0$, $c = 1$, $b = 3$, $a = 1$.

Example 7 Factorise the expression

$$2x^2 - 3xy - 2y^2 + x + 13y - 15.$$

Since $$2x^2 - 3xy - 2y^2 \equiv (2x + y)(x - 2y)$$

we try

$$2x^2 - 3xy - 2y^2 + x + 13y - 15$$
$$\equiv \{(2x + y) + A\}\{(x - 2y) + B\} \quad (1)$$
$$\equiv 2x^2 - 3xy - 2y^2 + A(x - 2y) + B(2x + y) + AB,$$

A and B being constants.

Then $\quad x + 13y - 15 \equiv x(A + 2B) + y(B - 2A) + AB. \quad (2)$

Equating $((x))$ on each side we have $1 = A + 2B$;

equating $((y))$ on each side we have $13 = B - 2A$;

Hence $\quad A = -5,\ B = 3.$

By equating the constants on each side of (2) we obtain $AB = -15$, so that our original conjecture is justified. Hence

$$2x^2 - 3xy - 2y^2 + x + 13y - 15 \equiv (2x + y - 5)(x - 2y + 3).$$

Example 8 Prove that if $(a + b + c) = 0$ and $(bc + ca + ab) + 3m = 0$, then the expression E where $E = (x^2 + ax + m)(x^2 + bx + m)(x^2 + cx + m)$ contains no powers of x except those whose index is a multiple of three.

Given that the expression $x^6 + 16x^3 + 64$ has a factor of the form $x^2 - 2x + m$, resolve it into three quadratic factors of the form similar to E.

We have

$$E = (y + ax)(y + bx)(y + cx),$$

where

$$y = x^2 + m. \quad (1)$$

Hence

$$E = y^3 + (a + b + c)y^2x + (bc + ca + ab)yx^2 + abcx^3.$$

If $a + b + c = 0$ and $(bc + ca + ab) + 3m = 0$,

$$\begin{aligned} E &= y^3 - 3myx^2 + abcx^3 \\ &= y(y^2 - 3mx^2) + abcx^3 \\ &= (x^2 + m)(x^4 - mx^2 + m^2) + abcx^3 \quad \text{by (1)} \\ &= x^6 + abcx^3 + m^3 \quad (2) \end{aligned}$$

Thus E contains no powers of x except those whose index is a multiple of three.

If $x^6 + 16x^3 + 64$ is identically equal to E

$$64 = m^3, \quad \text{that is} \quad m = 4.$$

Hence if $x^6 + 16x^3 + 64$ has a factor of the form $x^2 - 2x + m$ and two other similar factors, as in E, we may assume that

$$x^6 + 16x^3 + 64 \equiv (x^2 - 2x + 4)(x^2 + bx + 4)(x^2 + cx + 4)$$
$$\equiv x^6 - 2bcx^3 + 64, \qquad \text{by (2).}$$

Therefore

$$-2bc = 16, \quad \text{that is} \quad bc = -8.$$
$$\text{But } -2 + b + c = 0, \quad \text{therefore} \quad b + c = 2.$$

Hence

$$b = 4, \quad c = -2 \text{ or } b = -2, \quad c = 4$$

and

$$x^6 + 16x^3 + 64 \equiv (x^2 - 2x + 4)^2(x^2 + 4x + 4).$$

The method of undetermined coefficients may be used to establish identities between expressions other than polynomials.

Example 9 Show that when a and b are positive constants, a positive constant R and a constant acute angle α may be found such that

$$a \sin\theta + b\cos\theta \equiv R\sin(\theta + \alpha).$$

We have, if possible, to choose R, α so that

$$a \sin\theta + b\cos\theta \equiv R\sin(\theta + \alpha)$$
$$\equiv R\sin\theta\cos\alpha + R\cos\theta\sin\alpha$$
$$\equiv (R\cos\alpha)\sin\theta + (R\sin\alpha)\cos\theta.$$

The identity is valid if

$$a = R\cos\alpha, \tag{1}$$

$$b = R\sin\alpha. \tag{2}$$

Squaring and adding corresponding sides of these equations we get

$$a^2 + b^2 = R^2$$
$$\therefore R = \sqrt{(a^2 + b^2)} \qquad (R > 0).$$

When a, b, R are positive, we see from (1) and (2) that $\sin\alpha$ and $\cos\alpha$ are positive. Hence α is an acute angle such that

$$\sin\alpha : \cos\alpha : 1 = b : a : R.$$

With this value of α,

$$a\sin\theta + b\cos\theta \equiv \sqrt{(a^2 + b^2)}\sin(\theta + \alpha).$$

Similarly, when a and b are positive,

$$a\sin\theta - b\cos\theta \equiv \sqrt{(a^2 + b^2)}\sin(\theta - \alpha),$$

and

$$a \cos \theta - b \sin \theta \equiv \sqrt{(a^2 + b^2)} \cos (\theta + \alpha),$$

where α is the same acute angle as above.

Exercises 4(b)

1. Express $n(n + 1)(2n + 1)$ in the form

$$An + Bn(n - 1) + Cn(n - 1)(n - 2),$$

where A, B, and C are constants independent of n.

2. Express $(2n - 1)(2n + 1)(2n + 3)$ in the form

$$A + B(2n) + C(2n)(2n - 1) + D(2n)(2n - 1)(2n - 2),$$

where A, B, C and D are constants independent of n.

3. Factorise the expressions

(a) $2x^2 - 3xy - 2y^2 + 2x + 11y - 12$,
(b) $6x^2 - 5xy - 6y^2 - 5x + 14y - 4$,
(c) $6x^2 + xy - y^2 - 3x + y$.

4. Write the following expressions in the form indicated:

(a) $\sin \theta + \cos \theta \equiv R \sin (\theta + \alpha)$,
(b) $\sin \theta - \sqrt{3} \cos \theta \equiv R \sin (\theta - \alpha)$,
(c) $3 \cos \theta - 4 \sin \theta \equiv R \cos (\theta + \alpha)$,

where in all cases R is a positive constant and α is a constant acute angle measured in degrees.

5. Determine the coefficients a, b, c in the polynomial $f(x)$, where

$$f(x) = ax^4 + bx^3 + cx^2$$

given that

$$f\{n(n + 1)\} - f\{n(n - 1)\} \equiv n^7.$$

Hence, or otherwise, find the sum of the seventh powers of the first n integers.

4.2.5 Factorisation of Cyclic Homogeneous Polynomials An expression in several variables is said to be *homogeneous* if all its terms are of the same degree.

An expression in two or more variables is said to be *symmetric* in these variables when its value is unaltered by the interchange of any two of the variables; it is said to be an *alternating* (or a *skew*) expression if the interchange multiplies the value of the expression by -1.

For example, $x + y + 2$ is a symmetric expression of the first degree in x and y, $x^2 - y^2$ is an alternating homogeneous expression of the second degree in x and y.

An expression in x, y and z which is unaltered when we write y for x, z for y and x for z is said to be *cyclic*, or to have *cyclic symmetry*. For example, $x + y + z$ is a cyclic homogeneous expression of the first degree; $x^2 + y^2 + z^2$ and $yz + zx + xy$ are cyclic homogeneous expressions of the second degree.

Every symmetric expression has cyclic symmetry.

The following examples illustrate methods of factorising cyclic homogeneous polynomials.

Example 10 Factorise

$$x^4(y^2 - z^2) + y^4(z^2 - x^2) + z^4(x^2 - y^2).$$

Let

$$E = x^4(y^2 - z^2) + y^4(z^2 - x^2) + z^4(x^2 - y^2). \quad (1)$$

When $x = y$, $E = 0$; therefore $x - y$ is a factor of E. When $x = -y$, $E = 0$; therefore $x + y$ is a factor of E. Similarly, $y - z$, $y + z$, $z - x$ and $z + x$ are factors of E.

Now E is of the sixth degree, so that in addition to the six factors already found there can be only a numerical factor k (say).

Hence

$$E = k(y - z)(z - x)(x - y)(y + z)(z + x)(x + y) \quad (2)$$

and comparing the coefficients of x^4y^2 in (1) and (2) we obtain $k = -1$. Therefore

$$E = -(y - z)(z - x)(x - y)(y + z)(z + x)(x + y).$$

Example 11 Factorise

$$x(y - z)^3 + y(z - x)^3 + z(x - y)^3.$$

Let

$$E = x(y - z)^3 + y(z - x)^3 + z(x - y)^3. \quad (1)$$

As in Example 10, $(y - z)(z - x)(x - y)$ is a factor of E. But E and this factor are both cyclic and E is a homogeneous expression of the fourth degree; hence the remaining factor must be a cyclic homogeneous expression of the first degree, and the only such expression is $k(x + y + z)$, where k is a numerical constant. Therefore

$$E = k(x + y + z)(y - z)(z - x)(x - y). \quad (2)$$

Comparing coefficients of xy^3 in (1) and (2), we obtain

$$E = (x + y + z)(y - z)(z - x)(x - y).$$

Example 12 Factorise

$$x(y^4 - z^4) + y(z^4 - x^4) + z(x^4 - y^4).$$

Let

$$E = x(y^4 - z^4) + y(z^4 - x^4) + z(x^4 - y^4). \tag{1}$$

As in the previous examples, $(y - z)(z - x)(x - y)$ is a factor of E. Since E and this factor are both cyclic and E is a homogeneous expression of the fifth degree, the remaining factor is a cyclic homogeneous expression of the second degree, and the most general expression of this type is

$$k_1(x^2 + y^2 + z^2) + k_2(yz + zx + xy),$$

where k_1 and k_2 are numerical constants. Therefore

$$E = (y - z)(z - x)(x - y)[k_1(x^2 + y^2 + z^2) + k_2(yz + zx + xy)]. \tag{2}$$

Comparing coefficients of x^4y in (1) and (2) we obtain $k_1 = 1$; comparing coefficients of x^3y^2 we have $k_2 = k_1$.

$$\therefore E = (y - z)(z - x)(x - y)(x^2 + y^2 + z^2 + yz + zx + xy).$$

Exercises 4(c)

Factorise the following expressions:

1. $(b + c)^3(b - c) + (c + a)^3(c - a) + (a + b)^3(a - b)$.
2. $a^3(b^2 - c^2) + b^3(c^2 - a^2) + c^3(a^2 - b^2)$.
3. $a^3 + b^3 + c^3 - 3abc$.
4. $a^4(b - c) + b^4(c - a) + c^4(a - b)$.
5. $a^2b^2(a - b) + b^2c^2(b - c) + c^2a^2(c - a)$.
6. $(b - c)(b + c)^4 + (c - a)(c + a)^4 + (a - b)(a + b)^4$.
7. $a(b - c)(b + c)^3 + b(c - a)(c + a)^3 + c(a - b)(a + b)^3$.
8. $(b - c)^5 + (c - a)^5 + (a - b)^5$.
9. $(bc + ca + ab)^3 - b^3c^3 - c^3a^3 - a^3b^3$.
10. $a^5(b - c) + b^5(c - a) + c^5(a - b)$.

4.3 Rational Functions

If

$$P(x) \equiv p_0x^n + p_1x^{n-1} + \cdots + p_{n-1}x + p_n$$

and

$$Q(x) \equiv q_0x^m + q_1x^{m-1} + \cdots + q_{m-1}x + q_m,$$

where m and n are positive integers or zero and $q_0 \neq 0$, then $P(x)/Q(x)$ is said to be a *rational function of* x. The term 'rational function' includes a polynomial, for if $m = 0$, $Q(x)$ reduces to the constant q_0 and $P(x)/Q(x)$ is a polynomial in x.

If $n < m$, $P(x)/Q(x)$ is said to be a *proper* fraction, but if $n \geqslant m$, it is an *improper* fraction. An improper fraction may be expressed, by division, as the sum of a polynomial and a proper fraction.

For example,

$$\frac{2x^2 + 6x + 1}{x + 2} = 2x + 2 - \frac{3}{x + 2}.$$

4.3.1 Partial Fractions The sum or difference of a number of proper fractions is itself a proper fraction. For example,

$$\frac{2}{x + 1} + \frac{1}{x + 2} = \frac{3x + 5}{x^2 + 3x + 2}$$

and

$$\frac{1}{x - 1} - \frac{x + 2}{x^2 + x + 1} = \frac{3}{x^3 - 1}.$$

The converse result is also true: a proper fraction $P(x)/Q(x)$ whose denominator $Q(x)$ breaks up into real factors may be expressed as the sum or difference of simpler fractions, known as *partial fractions*, each with one of the factors of $Q(x)$ as denominator. For example,

$$\frac{2x}{x^2 - 9} = \frac{1}{x - 3} + \frac{1}{x + 3}$$

and

$$\frac{x^3 + x^2 - x + 3}{x^4 - 1} = \frac{x - 1}{x^2 + 1} + \frac{1}{x - 1} - \frac{1}{x + 1}.$$

We now apply the method of undetermined coefficients to the problem of expressing as a sum or difference of partial fractions a rational function of x, say $N(x)/D(x)$ given in its lowest terms. Detailed discussion of this problem is to be found in text-books on algebra; we merely outline the rules by which the partial fractions may be found.

We assume that $N(x)$ is of lower degree than $D(x)$, i.e. that the fraction is proper. Should this not be the case the fraction must be expressed, by division, as the sum of a polynomial and a proper fraction (see Section 4.3).

Rule 1. To each non-repeated linear factor $(x - a)$ of $D(x)$, there corresponds a fraction of the form $A/(x - a)$, where A is a constant not equal to zero.

Example 13 Express $(x + 5)/(x - 3)(x + 1)$ in partial fractions.

Since the given fraction is proper, the partial fractions must be proper and so we assume that

$$\frac{x + 5}{(x - 3)(x + 1)} \equiv \frac{A}{x - 3} + \frac{B}{x + 1}.$$

This assumption is justified because it is equivalent to

$$x + 5 \equiv A(x + 1) + B(x - 3) \tag{1}$$

or

$$x + 5 \equiv x(A + B) + A - 3B;$$

and this identity is valid since unique values of A and B can be found to satisfy the equations $A + B = 1$, $A - 3B = 5$, obtained by equating coefficients.

In practice, after verifying that the original assumption is valid, it is shorter to use the method given below.

Substitute $x = 3$ in (1); then $8 = 4A$, i.e. $A = 2$. Substitute $x = -1$ in (1); then $4 = -4B$, i.e. $B = -1$. Therefore

$$\frac{x + 5}{(x - 3)(x + 1)} \equiv \frac{2}{x - 3} - \frac{1}{x + 1}.$$

From (1) we see that A can be immediately obtained by deleting the factor $(x - 3)$ in $(x + 5)/[(x - 3)(x + 1)]$ and substituting $x = 3$ in the resulting fraction. Similarly B can be immediately determined.

Partial fractions corresponding to all non-repeated linear factors of $D(x)$ can be found in this way.

Example 14 Express $(3x^3 - x^2 + 2)/x(x^2 - 1)$ in partial fractions.

By division the improper fraction

$$\frac{3x^3 - x^2 + 2}{x(x^2 - 1)} \equiv 3 + \frac{P(x)}{x(x + 1)(x - 1)},$$

where $P(x)$ is of degree less than three. We therefore assume that

$$\frac{3x^3 - x^2 + 2}{x(x^2 - 1)} \equiv 3 + \frac{A}{x} + \frac{B}{x + 1} + \frac{C}{x - 1}$$

and, clearing fractions, obtain

$$\begin{aligned} 3x^3 - x^2 + 2 \equiv 3x(x^2 - 1) &+ A(x^2 - 1) \\ &+ Bx(x - 1) + Cx(x + 1). \end{aligned} \tag{1}$$

Having verified that our first assumption is justified (since by equating coefficients in identity (1) we obtain three equations from which A, B and C may be uniquely determined), we substitute in turn the values $x = 0$, -1 and 1 in (1) and obtain $A = -2$, $B = -1$, $C = 2$. Therefore

$$\frac{3x^3 - x^2 + 2}{x(x^2 - 1)} = 3 - \frac{2}{x} - \frac{1}{x + 1} + \frac{2}{x - 1}.$$

From (1) we see that the value of A can be immediately determined by deleting the factor x in $(3x^3 - x^2 + 2)/[x(x - 1)(x + 1)]$ and substituting $x = 0$ in the resulting fraction. Similarly B and C may be immediately found.

It is useful to note that when $N(x)/D(x)$ is a proper fraction, the number of constants to be determined is equal to the degree of $D(x)$.

Rule 2. The quadratic factor $ax^2 + bx + c$ is said to be *irreducible* if it has no real linear factors. To such a factor in $D(x)$ there corresponds a partial fraction of the form

$$\frac{Ax + B}{ax^2 + bx + c}.$$

Example 15 Express $2/(x^4 + x^2 + 1)$ in partial fractions.

Since

$$x^4 + x^2 + 1 = (x^2 + x + 1)(x^2 - x + 1),$$

the given proper fraction must be expressed as the sum of two proper fractions with irreducible quadratic denominators. The numerators of such fractions may be of the first degree, and so we assume that

$$\frac{2}{x^4 + x^2 + 1} \equiv \frac{Ax + B}{x^2 + x + 1} + \frac{Cx + D}{x^2 - x + 1};$$

that is

$$2 \equiv (Ax + B)(x^2 - x + 1) + (Cx + D)(x^2 + x + 1).$$

Equating coefficients we have

$$\begin{aligned} ((x^3)):& \quad A + C = 0, \\ ((x^2)):& \quad B + D - A + C = 0, \\ ((x)):& \quad A - B + C + D = 0, \\ ((x^0)):& \quad B + D = 2. \end{aligned}$$

Solving these equations, we get $A = B = D = 1$, $C = -1$. Therefore

$$\frac{2}{x^4 + x^2 + 1} \equiv \frac{x + 1}{x^2 + x + 1} + \frac{1 - x}{x^2 - x + 1}.$$

Example 16 Express $(5x - 12)/(x + 2)(x^2 - 2x + 3)$ in partial fractions.

Assume that

$$\frac{5x - 12}{(x + 2)(x^2 - 2x + 3)} \equiv \frac{A}{x + 2} + \frac{Bx + C}{x^2 - 2x + 3}$$

so that

$$5x - 12 \equiv A(x^2 - 2x + 3) + (Bx + C)(x + 2). \qquad (1)$$

This assumption is justified since there are three coefficients to equate, and so the three unknowns can be found uniquely.

By substituting $x = -2$ we find that $A = -2$.

Equating coefficients we have

$$((x^2)): \quad 0 = A + B, \quad \therefore B = 2.$$
$$((x^0)): \quad -12 = 3A + 2C, \quad \therefore C = -3.$$

Hence

$$\frac{5x - 12}{(x + 2)(x^2 - 2x + 3)} \equiv \frac{2x - 3}{x^2 - 2x + 3} - \frac{2}{x - 2}.$$

Rule 3. If $D(x)$ contains a repeated linear factor, $(x - a)^2$ say, it would be correct to assume that the corresponding partial fraction is of the form $(A_1x + B_1)/(x - a)^2$, for the denominator is of the second degree and hence two undetermined coefficients are required.

But we may write

$$A_1x + B_1 \equiv A_1(x - a) + B_1 + aA_1;$$

that is

$$A_1x + B_1 \equiv A_1(x - a) + A_2, \qquad \text{where } A_2 = B_1 + aA_1.$$

Thus $(A_1x + B_1)/(x - a)^2$ is equivalent to

$$\frac{A_1(x - a) + A_2}{(x - a)^2} \quad \text{or} \quad \frac{A_1}{x - a} + \frac{A_2}{(x - a)^2},$$

and in general it is more convenient to have the partial fractions in this form.

Similarly, if $(x - a)^r$, but not $(x - a)^{r+1}$, is a factor of $D(x)$, we have corresponding to this factor r partial fractions of the form

$$\frac{A_1}{x - a} + \frac{A_2}{(x - a)^2} + \cdots + \frac{A_r}{(x - a)^r}.$$

Example 17 Express $(2x^2 + x - 2)/x^3(x - 1)$ in partial fractions.

Assume that

$$\frac{2x^2 + x - 2}{x^3(x - 1)} \equiv \frac{A}{x} + \frac{B}{x^2} + \frac{C}{x^3} + \frac{D}{x - 1},$$

so that

$$2x^2 + x - 2 \equiv Ax^2(x - 1) + Bx(x - 1) + C(x - 1) + Dx^3. \quad (1)$$

This assumption is justified since there are four coefficients to equate and so the four unknowns A, B, C and D can be uniquely determined.

Substituting in turn the values $x = 0$ and $x = 1$ in (1), we find that $C = 2$ and $D = 1$.

Equating coefficients, we have

$$((x^3)):\quad 0 = A + D, \quad \therefore A = -1.$$
$$((x)):\quad 1 = -B + C, \quad \therefore B = 1.$$

Therefore

$$\frac{2x^2 + x - 2}{x^3(x - 1)} \equiv \frac{1}{x - 1} - \frac{1}{x} + \frac{1}{x^2} + \frac{2}{x^3}.$$

It is useful to note that the value of C can be immediately determined by deleting x^3 in the denominator of the given fraction and substituting $x = 0$ in the resulting fraction. Similarly, D can be immediately found. We may then use the method given below.

From (1),

$$2x^2 + x - 2 \equiv Ax^2(x - 1) + Bx(x - 1) + 2(x - 1) + x^3;$$

that is

$$-(x^3 - 2x^2 + x) \equiv x(x - 1)(Ax + B),$$
$$-(x - 1) \equiv Ax + B.$$

Therefore

$$A = -1, B = 1.$$

Example 18 Express $(x^3 + 5x^2 + 4x + 5)/[(x - 1)(x^3 - 1)]$ in partial fractions.

The factors of the denominator are $(x - 1)^2(x^2 + x + 1)$ and so we assume that

$$\frac{x^3 + 5x^2 + 4x + 5}{(x - 1)^2(x^2 + x + 1)} \equiv \frac{A}{x - 1}$$
$$+ \frac{B}{(x - 1)^2} + \frac{Cx + D}{x^2 + x + 1};$$

that is

$$x^3 + 5x^2 + 4x + 5 \equiv A(x-1)(x^2+x+1) + B(x^2+x+1) + (Cx+D)(x-1)^2.$$

By putting $x = 1$, we get $B = 5$, and so

$$x^3 - x \equiv (x-1)\{A(x^2+x+1) + (Cx+D)(x-1)\}.$$

Therefore

$$x(x+1) \equiv A(x^2+x+1) + (Cx+D)(x-1).$$

By putting $x = 1$ we get $A = \frac{2}{3}$ and so

$$\tfrac{1}{3}(x^2+x-2) \equiv (Cx+D)(x-1).$$

Therefore

$$\tfrac{1}{3}(x+2) \equiv Cx + D.$$

Hence

$$\frac{x^3+5x^2+4x+5}{(x-1)(x^3-1)} \equiv \frac{1}{3}\left\{\frac{2}{x-1} + \frac{15}{(x-1)^2} + \frac{x+2}{x^2+x+1}\right\}.$$

Rule 4. If $D(x)$ contains a repeated factor $(ax^2 + bx + c)^r$, where $ax^2 + bx + c$ is irreducible, the corresponding partial fractions are

$$\frac{A_1x+B_1}{ax^2+bx+c} + \frac{A_2x+B_2}{(ax^2+bx+c)^2} + \frac{A_3x+B_3}{(ax^2+bx+c)^3} + \cdots + \frac{A_rx+B_r}{(ax^2+bx+c)^r}.$$

Example 19 Express $(2x^2 + x + 4)/x(x^2 + 2)^2$ in partial fractions.

Assume that

$$\frac{2x^2+x+4}{x(x^2+2)^2} \equiv \frac{A}{x} + \frac{B_1x+C_1}{x^2+2} + \frac{B_2x+C_2}{(x^2+2)^2}.$$

Therefore

$$2x^2 + x + 4 \equiv A(x^2+2)^2 + x(B_1x+C_1)(x^2+2) + x(B_2x+C_2).$$

By substituting $x = 0$, we find that $A = 1$.

Equating coefficients, we have

$$\begin{array}{lll} ((x^4)): & 0 = A + B_1, & \therefore B_1 = -1. \\ ((x^3)): & 0 = C_1. & \\ ((x^2)): & 2 = 4A + 2B_1 + B_2, & \therefore B_2 = 0. \\ ((x)): & 1 = 2C_1 + C_2, & \therefore C_2 = 1. \end{array}$$

Therefore

$$\frac{2x^2 + x + 4}{x(x^2 + 2)^2} \equiv \frac{1}{x} - \frac{x}{x^2 + 2} + \frac{1}{(x^2 + 2)^2} .$$

Exercises 4(d)

Express in terms of partial fractions:

1. $(x + 1)/(x^2 + 7x + 12)$.
2. $3x/(x - 4)(x + 2)$.
3. $(2x - 1)/(x^3 - x^2 - 2x)$.
4. $(3x^2 - 7x)/[(1 - x)(2 - x)(3 - x)]$.
5. $x^3/(16 - x^2)$.
6. $(20x - 5)/[(x - 1)(2x + 1)(2x + 3)]$.
7. $x^2/(x - 2)^3$.
8. $(3x^2 + 5x - 4)/[(x - 1)(x + 1)^2]$.
9. $(x^2 - 2)/(x^3 + 2x^2)$.
10. $(13 - x^2)/[(x - 2)(x + 1)^2]$.
11. $(x^2 + 6)/[(2x - 1)(x + 2)^2]$.
12. $(17x + 53)/[(x + 4)^2(2x - 7)]$.
13. $(1 - x)/(1 + x + x^2 + x^3)$.
14. $(9 - x)/[x(x^2 + 9)]$.
15. $(2x^2 - x)/(x^3 + 1)$.
16. $32/x(4 - x^2)^2$.
17. $81/x(9 + x^2)^2$.
18. $(x^5 - 1)/x^2(x^3 + 1)$.
19. $(x^3 - 1)/[(x^2 - 2x + 3)(x + 1)^2]$.
20. $(x^2 + 1)/(x - 1)^3$.
21. $(2x^3 + x + 3)/(x^2 + 1)^2$.
22. $x(1 + x + 5x^2 + x^3 + x^4)/(1 - x^3)^2$.
23. $x^3(x^2 + 4)/(x^2 + 2)^3$.
24. $16x/(1 - x^2)^3$.

4.4 Summation of Series

The expression $u_1 + u_2 + u_3 + \cdots + u_n$ formed by the addition of n terms each defined by some law $u_r = f(r)$, say, is called a (finite) series of n terms. The series could be written

$$f(1) + f(2) + f(3) + \cdots + f(n)$$

but the notation u_r, in which it is understood that r can take only positive integral values, is more convenient.

The Greek letter Σ (sigma) is used to denote summation and the series may be written briefly as $\Sigma_{r=1}^{n} u_r$.*

* For this and similar notations the subscripts and superscripts are often placed directly above or below the main symbol. The alternatives are used for convenience in printing.

The result of summing the series is usually denoted by S_n, i.e.

$$S_n = \sum_{r=1}^{n} u_r = u_1 + u_2 + u_3 + \cdots + u_n.$$

We assume that the student is familiar with elementary results:

(1) For the arithmetical progression (A.P.)

$$a,\ a + d,\ a + 2d,\ \ldots,\ a + (n - 1)d,$$

S_n, the sum to n terms, is given by

$$S_n = \tfrac{1}{2}n\{2a + (n - 1)d\}, \quad \text{or } \tfrac{1}{2}n(a + l),$$

where l is the nth term.

(2) For the geometrical progression (G.P.)

$$a,\ ax,\ ax^2,\ \ldots,\ ax^{n-1}, \quad \text{where } x \neq 1,$$

the sum $S_n = a(1 - x^n)/(1 - x)$.

Other useful results are the following:

(3) $$\sum_{r=1}^{n} r = 1 + 2 + 3 + \cdots + n = \tfrac{1}{2}n(n + 1)$$

(particular case of (1)),

$$\sum_{r=1}^{n} r^2 = 1^2 + 2^2 + 3^2 + \cdots + n^2 = \tfrac{1}{6}n(n + 1)(2n + 1),$$

$$\sum_{r=1}^{n} r^3 = 1^3 + 2^3 + 3^3 + \cdots + n^3 = \{\tfrac{1}{2}n(n + 1)\}^2,$$

which can readily be verified by the method of induction (Section 4.4.2).

(4) The arithmetico-geometrical progression is of the form

$$a,\ (a + d)x,\ (a + 2d)x^2,\ (a + 3d)x^3,\ \ldots$$

The nth term is $\{a + (n - 1)d\}x^{n-1}$, and the sum to n terms, S_n, is found as follows:

$$S_n = a + (a + d)x + (a + 2d)x^2 + \cdots + [a + (n - 1)d]x^{n-1},$$

$$xS_n = ax + (a + d)x^2 + \cdots + \{a + (n - 2)d\}x^{n-1} + \{a + (n - 1)d\}x^n.$$

Subtracting, we obtain

$$(1 - x)S_n = a + [dx + dx^2 + dx^3 + \cdots + dx^{n-1}] - \{a + (n - 1)d\}\, x^n.$$

The terms within the square brackets form a G.P. of $(n - 1)$ terms and common ratio x. Summing these, we have

$$(1 - x)S_n = a + \frac{dx(1 - x^{n-1})}{1 - x} - \{a + (n - 1)d\}x^n, \quad x \neq 1;$$

that is

$$S_n = \frac{a - x^n\{a + (n - 1)d\}}{1 - x} + \frac{dx(1 - x^{n-1})}{(1 - x)^2}.$$

The types of series whose summation is considered here may be classified as follows:

(a) series summed by using the results of (3),
(b) series summed by the *method of induction*,
(c) series summed by the *method of differences*.

4.4.1 Series Whose rth Term is a Polynomial in r

Example 20 Sum to n terms the series $2.3 + 3.4 + 4.5 + \cdots$.

In this case $u_r = (r + 1)(r + 2)$. Hence

$$\begin{aligned}\sum_1^n u_r &= \sum_1^n (r^2 + 3r + 2) \\ &= \sum_1^n r^2 + 3\sum_1^n r + 2n \\ &= \tfrac{1}{6}n(n + 1)(2n + 1) + \tfrac{3}{2}n(n + 1) + 2n, \quad \text{by (3) in Section 4.4} \\ &= \tfrac{1}{3}n(n^2 + 6n + 11).\end{aligned}$$

Example 21 Sum to n terms the series $1 + (1 + 2) + (1 + 2 + 3) + (1 + 2 + 3 + 4) + \cdots$

In this case $u_r = (1 + 2 + 3 + \cdots + r)$
$= \tfrac{1}{2}r(r + 1)$, by (3) in Section 4.4.

Thus

$$\begin{aligned}\sum_1^n u_r &= \tfrac{1}{2}\sum_1^n r^2 + \tfrac{1}{2}\sum_1^n r \\ &= \tfrac{1}{12}n(n + 1)(2n + 1) + \tfrac{1}{4}n(n + 1), \quad \text{by (3) in Section 4.4} \\ &= \tfrac{1}{6}n(n + 1)(n + 2).\end{aligned}$$

If u_r is a polynomial in r of degree higher than the third, this method of solution is not applicable without first calculating sums of higher powers of r. In some cases, however, it is possible to sum such a series by the method of differences (see Section 4.4.3) or, when the result to be proved is given, by the method of induction.

4.4.2 The Method of Induction This method is best illustrated by an example.

Example 22 Prove that the sum of n terms of the series $1.3.2^2 + 2.4.3^2 + 3.5.4^2 + \cdots$ is $\frac{1}{10}n(n+1)(n+2)(n+3)(2n+3)$.

First we assume that the result is true when $n = p$ and prove that on this assumption it is true when $n = p + 1$.

Suppose, then, that

$$\sum_1^p u_r = \tfrac{1}{10}p(p+1)(p+2)(p+3)(2p+3). \tag{1}$$

Since

$$u_r = r(r+2)(r+1)^2$$

$$u_{p+1} = (p+1)(p+3)(p+2)^2$$

$$\begin{aligned}\therefore \sum_1^{p+1} u_r &= \sum_1^p u_r + u_{p+1} \\ &= \tfrac{1}{10}p(p+1)(p+2)(p+3)(2p+3) \\ &\quad + (p+1)(p+3)(p+2)^2 \quad \text{by (1)} \\ &= \tfrac{1}{10}(p+1)(p+2)(p+3)[(2p^2+3p)+10(p+2)] \\ &= \tfrac{1}{10}(p+1)(p+2)(p+3)(p+4)(2p+5).\end{aligned}$$

But this is the result we should obtain by substituting $n = (p+1)$ in the given result. Hence if the given formula is true when $n = p$, it is true when $n = p + 1$. But the formula is true when $n = 1$ since $1.3.2^2 = \frac{1}{10}(1.2.3.4.5)$; hence it is true for $n = 2$. Since it is true for $n = 2$, it is true for $n = 3$, and proceeding in this way we may show that the given result is true for all positive integral values of n.

The reader should verify the results (3) in Section 4.4 in by this method.

4.4.3 The Method of Differences Consider the series $\Sigma_1^n u_r$. If u_r can be expressed in the form $v_{r+1} - v_r$, where v_r is a known function of r, then

$$\sum_1^n u_r = (v_2 - v_1) + (v_3 - v_2) + (v_4 - v_3)$$
$$+ \cdots + (v_n - v_{n-1}) + (v_{n+1} - v_n) = v_{n+1} - v_1.$$

The method of differences is particularly useful for summing a series each of whose terms is the reciprocal of a product of a constant number of factors in A.P., the first factors of successive terms being in the same A.P.

Example 23 Sum to n terms the series

$$\frac{1}{1.3.5}+\frac{1}{3.5.7}+\frac{1}{5.7.9}+\cdots.$$

Here

$$u_r=\frac{1}{(2r-1)(2r+1)(2r+3)}$$
$$=\frac{1}{4}\left\{\frac{1}{(2r-1)(2r+1)}-\frac{1}{(2r+1)(2r+3)}\right\}.$$

Summing as above, we obtain

$$S_n=\frac{1}{4}\left\{\frac{1}{3}-\frac{1}{(2n+1)(2n+3)}\right\}.$$

Example 24 Sum to n terms the series whose rth term is

$$\frac{1}{r(r+1)(r+3)}.$$

We write

$$u_r=\frac{r+2}{r(r+1)(r+2)(r+3)}$$
$$=\frac{1}{(r+1)(r+2)(r+3)}+\frac{2}{r(r+1)(r+2)(r+3)}$$
$$=\frac{1}{2}\left\{\frac{1}{(r+1)(r+2)}-\frac{1}{(r+2)(r+3)}\right\}$$
$$+\frac{2}{3}\left\{\frac{1}{r(r+1)(r+2)}-\frac{1}{(r+1)(r+2)(r+3)}\right\}$$

and summing we obtain

$$S_n=\frac{1}{2}\left\{\frac{1}{6}-\frac{1}{(n+2)(n+3)}\right\}+\frac{2}{3}\left\{\frac{1}{6}-\frac{1}{(n+1)(n+2)(n+3)}\right\}.$$

The method of differences may also be used to sum a series each of whose terms is the product of a constant number of factors in A.P., the first factors of successive terms being in the same A.P.

Example 25 Sum to n terms the series $3.5.7.9+5.7.9.11+\cdots$.

Here

$$u_r=(2r+1)(2r+3)(2r+5)(2r+7).$$

Let

$$v_{r+1}=(2r+1)(2r+3)(2r+5)(2r+7)(2r+9).$$

Then

$$v_r = (2r - 1)(2r + 1)(2r + 3)(2r + 5)(2r + 7),$$

and

$$v_{r+1} - v_r = 10(2r + 1)(2r + 3)(2r + 5)(2r + 7) = 10u_r.$$

Hence

$$u_r = \tfrac{1}{10}(v_{r+1} - v_r)$$

and

$$\sum_1^n u_r = \tfrac{1}{10}(v_{n+1} - v_1)$$

$$= \tfrac{1}{10}[(2n + 1)(2n + 3)(2n + 5)(2n + 7)(2n + 9) - 1.3.5.7.9].$$

The method of differences may be used to sum certain trigonometric series. For example, to sum the series

$$S_n = \sin A + \sin (A + B) + \sin (A + 2B) + \cdots + \sin [A + (n - 1)B],$$

when B is not a multiple of 2π, we multiply each term by $2 \sin \frac{1}{2}B$. Then

$$\begin{aligned} 2 \sin A \sin \tfrac{1}{2}B &= \cos (A - \tfrac{1}{2}B) - \cos (A + \tfrac{1}{2}B), \\ 2 \sin (A + B) \sin \tfrac{1}{2}B &= \cos (A + \tfrac{1}{2}B) - \cos (A + \tfrac{3}{2}B), \\ 2 \sin (A + 2B) \sin \tfrac{1}{2}B &= \cos (A + \tfrac{3}{2}B) - \cos (A + \tfrac{5}{2}B), \\ &\vdots \\ 2 \sin [A + (n - 1)B] \sin \tfrac{1}{2}B &= \cos [A + (n - \tfrac{3}{2})B] - \cos [A + (n - \tfrac{1}{2})B], \end{aligned}$$

and by addition

$$(2 \sin \tfrac{1}{2}B)S_n = \cos (A - \tfrac{1}{2}B) - \cos \{A + (n - \tfrac{1}{2})B\}.$$

Therefore

$$S_n = \frac{\sin [A + \frac{1}{2}(n - 1)B] \sin \frac{1}{2}nB}{\sin \frac{1}{2}B}. \tag{1}$$

The same method may be used to sum the series

$$C_n = \cos A + \cos (A + B) + \cos (A + 2B) + \cdots + \cos [A + (n - 1)B]$$

but we deduce its sum from (1) by substituting $(A + \frac{1}{2}\pi)$ for A. This gives

$$C_n = \frac{\cos [A + \frac{1}{2}(n - 1)B] \sin \frac{1}{2}nB}{\sin \frac{1}{2}B}.$$

If $\pi + B$ is written for B in the above two series, we obtain the series

$$\sin A - \sin (A + B) + \sin (A + 2B) + \cdots$$

$$\cos A - \cos (A + B) + \cos (A + 2B) + \cdots,$$

whose sum to n terms can also be found directly by using the multiplier $2 \sin \frac{1}{2}(\pi + B) = 2 \cos \frac{1}{2}B$.

Exercises 4(e)

Sum to n terms the following series:

1. $2.5 + 3.7 + 4.9 + \cdots$.
2. $1^2 + 4^2 + 7^2 + \cdots$.
3. $1.2.3 + 3.4.5 + 5.6.7 + \cdots$,
4. $1.3.4 + 2.5.7 + 3.7.10 + \cdots$.
5. $1^2.3 + 2^2.5 + 3^2.7 + \cdots$.
6. $1 + 2x + 3x^2 + 4x^3 + \cdots$.
7. $1 + 4x + 7x^2 + 10x^3 + \cdots$.
8. $\dfrac{1}{1.2.3} + \dfrac{1}{2.3.4} + \dfrac{1}{3.4.5} + \cdots$,
9. $1.3.5 + 3.5.7 + 5.7.9 + \cdots$.
10. $\cos A \cos 2A + \cos 2A \cos 3A + \cos 3A \cos 4A + \cdots$.

Use mathematical induction to establish the following results:

11. $1^2 + 2^2 + 3^2 + \cdots + n^2 = \frac{1}{6}n(n + 1)(2n + 1)$.
12. $1.2.4 + 2.3.5 + \cdots + n(n + 1)(n + 3) = \frac{1}{12}n(n + 1)(n + 2)(3n + 13)$.
13. $$\frac{1}{1.2.3.4} + \frac{1}{2.3.4.5} + \cdots + \frac{1}{n(n + 1)(n + 2)(n + 3)} = \frac{1}{18} - \frac{1}{3(n + 1)(n + 2)(n + 3)}.$$
14. $$1 - \frac{x}{1!} + \frac{x(x - 1)}{2!} - \cdots + (-1)^n \frac{x(x - 1) \cdots (x - n + 1)}{n!} = (-1)^n \frac{(x - 1)(x - 2) \cdots (x - n)}{n!}.$$
15. Prove by induction the formula
$$\sum_1^n r^3 = \frac{1}{4}n^2(n + 1)^2.$$
Use the formula to find $\Sigma_{n+1}^{2n} r^3$.
16. Show that if the statement
$$\cos x + \cos 2x + \cdots + \cos nx = \frac{\sin (n + \frac{1}{2})x}{2 \sin \frac{1}{2}x}$$
is true for $n = k$, then it is also true for $n = k + 1$. What further step would be needed to complete a proof by induction that the statement is true for all positive integral values of n? Can that step be taken?
17. Show that
$$\cos x + \cos 2x + \cdots + \cos nx = \frac{\sin (n + \frac{1}{2})x}{2\sin \frac{1}{2}x} - \frac{1}{2}$$

and deduce that

$$\sum_{r=1}^{n} \cos\frac{r\pi}{2n} = \tfrac{1}{2}\left(\cot\frac{\pi}{4n} - 1\right).$$

18. Find $\Sigma_{r=1}^{n}(r^3 + 3r^2 - r + 1)$.
19. Sum the series

(a) $\dfrac{1}{1.2} + \dfrac{1}{2.3} + \dfrac{1}{3.4} + \cdots + \dfrac{1}{n(n+1)}$.

(b) $1.3 + 2.4 + 3.5 + \cdots + 100\cdot 102$.

20. Sum the series

(a) $\dfrac{1}{1.3} + \dfrac{1}{3.5} + \dfrac{1}{5.7} + \cdots$ to n terms.

(b) $1.2.3 + 2.3.4 + 3.4.5 + \cdots$ to 15 terms.

(c) $x + 2x^2 + 3x^3 + \cdots$ to n terms.

21. Assuming that ${}^nC_r + {}^nC_{r-1} = {}^{n+1}C_r$, or otherwise, show that

$$(a + b)(a^n + {}^nC_1a^{n-1}b + \cdots + {}^nC_ra^{n-r}b^r + \cdots + b^n)$$
$$= a^{n+1} + {}^{n+1}C_1a^nb + \cdots + {}^{n+1}C_ra^{n-r+1}b^r + \cdots + b^{n+1}.$$

By induction deduce the binomial theorem for a positive integral index, namely:

$$(a + b)^n = a^n + {}^nC_1a^{n-1}b + \cdots + {}^nC_ra^{n-r}b^r + \cdots + b^n.$$

CHAPTER 5

ISOMORPHISM

In Chapter 1 we discussed the way in which two different systems can be analogous so that problems in one system can be solved by using the other system as a model. In this chapter we define more explicitly the kind of analogy between systems that is fundamental in mathematics.

5.1 Definitions

If two sets X, Y are in (1, 1) correspondence (Section 3.1) and each set has more than one element, then there is more than one way of associating elements of Y with elements of X that would demonstrate this (1, 1) correspondence. To take a very simple example, let $X = \{x_1, x_2, x_3\}$ and $Y = \{y_1, y_2, y_3\}$; then X and Y are in (1, 1) correspondence and we can demonstrate this in the following six ways:

$$\begin{array}{llllll}
(1) & x_1 \mapsto y_1 & (2) & x_1 \mapsto y_1 & (3) & x_1 \mapsto y_2 \\
 & x_2 \mapsto y_2 & & x_2 \mapsto y_3 & & x_2 \mapsto y_1 \\
 & x_3 \mapsto y_3 & & x_3 \mapsto y_2 & & x_3 \mapsto y_3 \\
(4) & x_1 \mapsto y_2 & (5) & x_1 \mapsto y_3 & (6) & x_1 \mapsto y_3 \\
 & x_2 \mapsto y_3 & & x_2 \mapsto y_2 & & x_2 \mapsto y_1 \\
 & x_3 \mapsto y_1 & & x_3 \mapsto y_1 & & x_3 \mapsto y_2.
\end{array}$$

Any particular mode of associating elements of Y with elements of X that sets up a (1, 1) correspondence between the two sets is called a *(1, 1) mapping of X onto Y* (mappings are discussed more fully in Chapter 12). If the element y in Y is associated with the element x in X, we say y is the *image* of x under the mapping.

Thus the above example shows that there are six different (1, 1) mappings of $\{x_1, x_2, x_3\}$ onto $\{y_1, y_2, y_3\}$; under mapping (1), the image of x_1 is y_1, whereas under mapping (3) the image of x_1 is y_2, and so on.

The idea of mapping is sometimes employed in situations where the sets X, Y have common elements or are the same set. For example the association

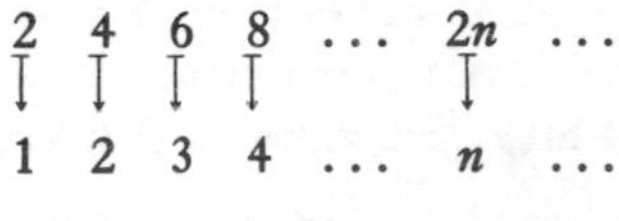

may be called a (1, 1) mapping of the set of positive even integers onto the set of positive integers; likewise we get a (1, 1) mapping of the set of positive real numbers onto the set of all real numbers by taking the image of x to be $\log_{10}x$, and a (1, 1) mapping of the set of real numbers onto itself by taking the image of x to be x^3.

A binary operation on a set X is any rule that associates with each pair of elements in X another element (normally also an element of X). Denoting such an operation by $*$ we denote by $x_1 * x_2$ the element that it associates with the pair of elements x_1, x_2. It is not necessary that $x_1 * x_2$ and $x_2 * x_1$ refer to the same element, but when an operation is such that the order of the two elements to which it is applied is immaterial then the operation is said to be *commutative.* The following are some examples of binary operations:

(a) addition and multiplication are both commutative operations on the set of real numbers;
(b) subtraction is a non-commutative operation on the set of real numbers;
(c) division is a non-commutative operation on the set of positive and negative real numbers (note that division is not strictly an operation on the set of all real numbers since division by zero is undefined);
(d) union and intersection are both commutative operations on the set of all subsets of a universal set S;
(e) 'series connection' and 'parallel connection' are both commutative operations on the set of switch arrangements (see Section 3.3.1).

The usual *equality sign* is used to mean 'is the same element as', e.g. if $*$ is a commutative operation we can write $x_1 * x_2 = x_2 * x_1$. Thus the same symbol may mean 'is the same number as', or 'is the same set as', or 'is the same switch arrangement as', depending on the context.

If $*$ is a commutative operation on the set X, and x, y are both variables on X, then we can say that $x * y = y * x$ for all x in X and all y in X, and we sometimes abbreviate this by writing $x * y \equiv y * x$ as in the case of an identity in real-number algebra.

The set X is said to be *closed* with respect to $*$ if $x * y$ is defined, and belongs to X, for all x in X and all y in X, i.e. $*$ is an operation on X that always produces elements within X. For example

(a) the set of real numbers is closed with respect to addition, multiplication, subtraction,
(b) the set of positive real numbers is closed with respect to addition, multiplication, division, but is not closed with respect to subtraction since this operation may give answers that are negative,

(c) the set of all subsets of a universal set S is closed with respect to union and intersection since these operations can produce only subsets of S.

5.2 Systems

A set, together with one or more operations, will be called a ***system.*** If X is a set on which an operation $*$ is defined, then the system given by X under the operation $*$ may be referred to as the system $X(*)$. If in addition another operation $\circledast$ is defined on X, then we also have a system $X(\circledast)$ given by X under the operation $\circledast$, and a system $X(*, \circledast)$ given by X under the two operations $*, \circledast$.

For example $R(+)$ denotes the system of real numbers under addition, and $R(+, \times)$ the system of real numbers under addition and multiplication, usually called simply the ***real-number system.***

The following are further examples of systems:

(a) the system of all subsets of a universal set S under the operations of union and intersection,
(b) the system of switch arrangements under the operations of series connection and parallel connection,
(c) the system of mechanical forces under the operation of 'taking the resultant'.

5.2.1 Isomorphic Systems Consider a system consisting of a set X under an operation $*$, and another system consisting of a set Y under an operation $\circledast$. Suppose that

(a) X and Y are closed with respect to $*$ and $\circledast$ respectively,
(b) there is a (1, 1) mapping of X onto Y such that if x_1, x_2 are elements of X with images y_1, y_2 in Y then the image of $x_1 * x_2$ is $y_1 \circledast y_2$.

The systems are then said to be ***isomorphic*** and the mapping is called an ***isomorphism.*** Systems that are isomorphic are analogous in the sense

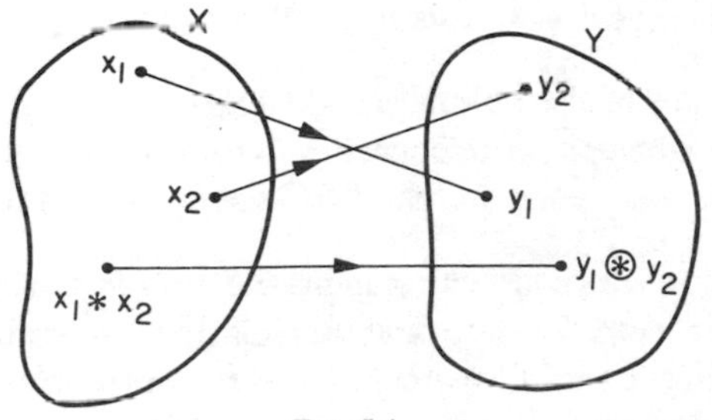

FIG. 5.1

that problems concerned with the operation $*$ in the first system can be translated into the second system and solved using the operation $\circledast$. Two systems $X(*, \square)$, $Y(\circledast, \triangle)$, each having *two* operations, are said to be isomorphic if there is a mapping that is an isomorphism of $X(*)$ onto $Y(\circledast)$ and also an isomorphism of $X(\square)$ onto $Y(\triangle)$.

We now look at some examples of isomorphic systems.

(a) *Isomorphism between two physical systems.* Corresponding to the system of switch arrangements under the operations of series connection and parallel connection (Section 3.3.1) we get another system if we think of electric wires as replaced by water-pipes, switches by control valves, and series and parallel connection by similar connections of these control-valve arrangements. Thus each switch arrangement has a corresponding control-valve arrangement, and vice versa, and it is easily seen that this correspondence is an isomorphism with respect to both types of connection.

(b) *Isomorphism between a physical system and a mathematical system.* We saw further in Section 3.3.1 that, corresponding to the arrangements of switches $A, B, C, \ldots$ we can introduce a mathematical system of elements $A, B, C, \ldots$ under operations $\cup, \cap$; this mathematical system is isomorphic to the system of switch arrangements under the operations of series connection and parallel connection. The problem of finding similar equivalent switching arrangements can be solved using the algebra (Boolean algebra) of the mathematical system.

(c) *Isomorphism between two mathematical systems.* Consider the system of positive real numbers under multiplication, the system of all real numbers under addition, and the mapping $x \mapsto \log_{10} x$. Clearly this mapping is an isomorphism of the first system onto the second; for it is a (1, 1) mapping and, if x_1, x_2 are any two positive numbers in the first system, the result of multiplying them is x_1x_2, whereas the result of adding their images in the second system is $\log_{10} x_1 + \log_{10} x_2$, i.e. $\log_{10} x_1x_2$, which is the image of x_1x_2. The familiar process of multiplication using log tables can be described as follows: to perform the operation (multiplication) in the first system it is sufficient to

(a) find the images of the numbers (take logs),
(b) perform the *corresponding* operation with the *images* (add the logs),
(c) interpret the final result for the first system (take antilog).

Further, general results concerning multiplication in the first system (e.g. multiplication is commutative and associative) must be equally true about addition in the second system and so the entire algebras of the two systems are analogous.

Example 1 A schoolboy does addition sums and multiplication sums using arithmetic tables written in special code. Due to a misunderstanding about the code he interprets all numbers in the tables at twice their true value. Show that he gets correct answers to addition sums and incorrect answers to multiplication sums.

To perform the addition $x_1 + x_2$ he reads from the table $\frac{1}{2}x_1 + \frac{1}{2}x_2$ to get half the correct answer; however he decodes this answer at twice its value to arrive at the correct answer.

To perform the multiplication x_1x_2 he reads from the table $\frac{1}{2}x_1\frac{1}{2}x_2$ to get $\frac{1}{4}x_1x_2$ and on decoding this becomes $\frac{1}{2}x_1x_2$—still incorrect.

In other words his misunderstanding may be thought of as a (1, 1) mapping $x \mapsto \frac{1}{2}x$ of the set of real numbers onto itself; this mapping is applied each time he attempts to express a number in code. As far as addition is concerned this mapping is an isomorphism, but not so for multiplication.

Exercises 5

1. Show that the mapping $x \mapsto kx$ (k a non-zero constant) is an isomorphism of the system $R(+)$ onto itself. Is it an isomorphism of the system $R(\times)$ onto itself?
2. The mapping $x \mapsto \log_{10} x$ is an isomorphism of the positive real numbers under division onto R under . . . . Suggest an operation to complete this statement and verify that the completed statement is correct.
3. Consider the system of non-zero real numbers under multiplication. Is the mapping $x \mapsto 1/x$ an isomorphism of this system onto itself? What is the answer if the operation is addition instead of multiplication?
4. Show that the mapping $x \mapsto b^x$ (b a positive constant) is an isomorphism of $R(+)$ onto the positive real numbers under multiplication, and also of $R(-)$ onto the positive real numbers under division.
5. Let operations $\oplus$, $\otimes$ on R be defined as follows:

$$x \oplus y = \text{the greater of } x \text{ and } y,$$

for example, $3 \oplus 2 = 3$, $3 \oplus 7 = 7$, $3 \oplus 3 = 3$;

$$x \otimes y = \text{the lesser of } x \text{ and } y,$$

for example, $3 \otimes 2 = 2$, $3 \otimes 7 = 3$, $3 \otimes 3 = 3$.

Show that

(a) the mapping $x \mapsto \log_{10} x$ is an isomorphism of $R(\oplus)$ onto the positive real numbers under $\oplus$, and of $R(\otimes)$ onto the positive real numbers under $\otimes$,

(b) the mapping $x \mapsto 1/x$ is an isomorphism of the non-zero real numbers under $\oplus$ onto the non-zero real numbers under $\otimes$,

(c) the mapping $x \mapsto x^2$ is an isomorphism of the positive real numbers under $\oplus$ onto the positive real numbers under $\oplus$, and of the negative real numbers under $\oplus$ onto the positive real numbers under $\otimes$.

5.3 Practical Considerations

Imagine that a civil engineer has constructed a model of a river estuary in order to study currents. There would be no objection to his inserting a few spots of dye into the water to help to detect a current; the dye simply reveals the presence of current, as smoke reveals wind.

It could be pointed out that this sort of trick introduces into the model elements that do not correspond to anything in the true set-up. This is so, but the presence of these elements and their interactions with the genuine elements do not affect the validity of the model. Without influencing the behaviour that is under investigation they simplify the investigation.

These considerations are relevant in connection with mathematical systems. It might appear that, ideally, we would like a strict isomorphism to exist between a physical system and the mathematical system used to represent it; this would require a (1, 1) correspondence between the elements of the two systems. In practice the mathematical system used may contain more elements than the physical system that it represents. This occurs, for example, when we use the real-number system to deal with problems where, strictly speaking, only positive numbers, or perhaps only positive integers, apply. In a system containing (say) only positive numbers a simple identity like

$$x^2 - y^2 \equiv (x - y)(x + y)$$

requires some qualification ($x > y$) to ensure that we are always talking about positive numbers; this leads to a very inconvenient kind of algebra. In problems concerned with real numbers, whatever limitation there may be on the numbers strictly applicable, we use the algebra of the entire real-number system to solve the problem, rejecting at the end any answers that are impermissible. In such cases the presence of extra elements in the mathematical system does not invalidate the representation but does simplify the algebra.

CHAPTER 6

ALGEBRAIC STRUCTURE

6.1 Comparison of Algebras

The complicated railway set-up at an important junction like Crewe can easily be modelled by using a suitable toy railway set. Such a toy set must include all the necessary types of elements (e.g. straight rails, curved rails, points, signals) and it must have enough elements of each type to provide the complete model of Crewe junction.

Two model railway sets may differ greatly in the number of elements they contain, but if they contain the same *types* of element then the same set of working instructions apply to both. In other words if we know how to fit straight rails together, straight rails to curved rails, and so on, then we can use a model set regardless of its size.

In the axioms (1.2)–(1.8) of Section 1.3.1 only the numbers 0, 1 are specifically mentioned. If S is any set of numbers which includes 0, 1 these axioms are valid in $S(+, \times)$: for example, S might be the set $\{0, \pm 1, \pm 2, \pm 3, \ldots\}$, or the set $\{0, 1, 2, 3, \ldots\}$, or the set $\{0, 1, 3, 5, \ldots\}$, or simply the set $\{0, 1\}$. Later we shall add other axioms which are *not* valid if S is any of the sets shown above, but we shall find that all the axioms of the system $R(+, \times)$ are valid for the system $Q(+, \times)$ where Q is the set of rational numbers. Hence $Q(+, \times)$ is a smaller version of $R(+, \times)$ in a sense similar to that in which one model railway set can be a smaller version of another.

We say that $Q(+, \times)$ and $R(+, \times)$ have the same *algebraic structure.* This means that the two systems are governed by the same set of axioms and so all algebraic results following from these axioms are valid in *both* systems. For example, the identity

$$x^2 - y^2 = (x - y)(x + y) \tag{1}$$

is valid whether x and y are variables on Q or on R. The results

$$x^2 - 1 = (x - 1)(x + 1) \tag{2}$$

$$x^2 - 2 = (x - \sqrt{2})(x + \sqrt{2}) \tag{3}$$

are particular cases of (1) and such results need not necessarily apply for both systems; clearly (3), which mentions the irrational number $\sqrt{2}$, does not apply to $Q(+, \times)$, but (2) applies to both $Q(+, \times)$ and $R(+, \times)$.

It is also possible to notice similarities between the algebras of two systems that have elements of a completely different nature, and consequently have different operations. For example the identities

$$A \cup (B \cup C) = (A \cup B) \cup C,$$
$$A \cap (B \cap C) = (A \cap B) \cap C,$$
$$A \cap (B \cup C) = (A \cap B) \cup (A \cap C)$$

in Boolean algebra are essentially the same as the identities

$$a + (b + c) = (a + b) + c, \; a(bc) = (ab)c, \; a(b + c) = ab + ac$$

in the algebra of real numbers, provided we think of union as the set operation corresponding to addition of numbers, intersection as the set operation corresponding to multiplication of numbers. The reader should be able to suggest other equivalent identities in the two algebras. We may also note that in Boolean algebra the universal set S and the empty set ϕ play roles which can be identified with the roles played by 1, 0 respectively in the algebra of real numbers; thus

$$A \cup \phi = A, \qquad A \cap S = A$$

correspond to

$$a + 0 = a, \qquad a \times 1 = a.$$

Much economy of thought and effort may be achieved if we fully exploit the common ground that can be found between various algebras. In this book we examine in some detail several mathematical systems (vectors, complex numbers, matrices), each different from the real-number system, and each having important applications. Before we can use these systems we must acquire familiarity with their algebras—a formidable task, were it not for the fact that there is so much similarity to the algebra of real numbers. In view of this a more abstract approach to algebra is very desirable.

6.2 Groups

We begin by looking at those axioms of the real-number system that refer solely to the operation of addition. It is usual to include in the axioms for real numbers that

(1) the system is closed with respect to addition,

since this property has a considerable streamlining effect on the algebra.

Other axioms already stated in Section 1.3.1 are now repeated for completeness:

(2) $x + 0 = x$,
(3) $x + y = y + x$,
(4) $(x + y) + z = x + (y + z)$.

The only other axiom necessary for addition is that

(5) corresponding to each number x in the system there is a number $(-x)$ such that $x + (-x) = 0$.

Note that (5) makes it unnecessary to regard subtraction as another operation with real numbers; instead we regard $x - y$ as being an alternative way of writing $x + (-y)$, e.g. subtraction of 3 from 5 is the same as addition of the two numbers 5, -3.

Example 1 Consider the set of numbers 1, 2, . . ., 12 appearing on the face of a clock. If x and y are two of these numbers, define 'clock addition' of x and y by the rule that $x + y$ means the number y places from x in the clockwise direction, e.g. $2 + 3 = 5$, $10 + 3 = 1$.

Next, define $(-x)$ to be the number opposite x across the diameter joining 6 and 12, e.g. $(-4) = 8$, $(-10) = 2$.

This set of 12 numbers, with the operation of addition as here defined, satisfies the axioms (1)–(5) if in (2) and (5) we read 12 for 0.

Example 2 Consider the set of all positive and negative real numbers, i.e. all real numbers except zero. This set under the operation of multiplication, satisfies the axioms (1)–(5) if in these axioms we replace addition by multiplication, replace 0 by 1, and replace $(-x)$ by $1/x$.

Example 3 Consider the system of all subsets of a universal set S under the operation of union. If in axioms (1)–(4) we replace addition by union, and replace 0 by ϕ (empty set), these axioms are satisfied. However, there is no way to interpret $(-x)$ and so axiom (5) is not satisfied by this system.

There are many examples of mathematical and physical systems with a single operation which satisfy axioms (1)–(5) suitably interpreted. We therefore state these axioms in abstract form as follows: denoting by $*$ a binary operation on a set G, we require that

(1′) G is closed with respect to $*$,
(2′) there is an element e of G such that $x * e = e * x = x$ for all x in G, (e is usually called the *identity* element, or *neutral* element, for $*$),
(3′) $x * y = y * x$ for all elements x, y in G ($*$ is commutative),

(4′) $(x * y) * z = x * (y * z)$ for all elements x, y, z in G ($*$ is associative),

(5′) corresponding to each element x in G there is an element x^{-1} in G such that $x * x^{-1} = x^{-1} * x = e$ (x^{-1} is usually called the *inverse* of x with respect to $*$).

Thus we may take G to be the set of real numbers, $*$ to be addition, e to be zero, and x^{-1} to be $(-x)$; also we may take G to be the set of all real numbers except zero, $*$ to be multiplication, e to be 1, and x^{-1} to be $1/x$. A system $G(*)$ satisfying the axioms (1′)–(5′) is called an *abelian group* (after the Norwegian mathematician Abel, 1802–29) or *commutative group*; if (3′) is omitted the system is called simply a *group*. The system $R(+)$, and the system of non-zero real numbers under multiplication, are both abelian groups.

The theory of groups has proved to be very significant for modern physics. A full study of the properties of groups is beyond the scope of this book but the following examples give three of the most important properties and illustrate the method of logical deduction from axioms.

Example 4 Show that the identity element is unique, i.e. there can be only one element e in G with the property (2′).

Suppose that f is in G and that $x * f = f * x = x$ for all x in G. In particular, putting $x = e$ in one of these equations, we have $e * f = e$. Likewise, putting $x = f$ in (2′), we have $e * f = f$. Hence $f = e$.

Note that it was only necessary to assume that f has the property $x * f = x$ for all x in G in order to prove that $f = e$. Similarly we can show that if g is an element of G with the property $g * x = x$ for all x in G, then $g = e$.

Example 5 Show that inverse elements are unique, i.e. to each element x there corresponds only one element with the property (5′).

Suppose that $\bar{x}$ is in G and that $x * \bar{x} = \bar{x} * x = e$. Now

$$\begin{aligned}
\bar{x} &= e * \bar{x} && \text{(by (2′)} \\
&= (x^{-1} * x) * \bar{x} && \text{(by (5′)} \\
&= x^{-1} * (x * \bar{x}) && \text{(by (4′)} \\
&= x^{-1} * e && \text{(by hypothesis)} \\
&= x^{-1} && \text{(by (2′).}
\end{aligned}$$

Note that only the part $x * \bar{x} = e$ of the hypothesis has been used in this proof. Similarly, the assumption $\bar{x} * x = e$ enables us to show that $\bar{x} = x^{-1}$.

Example 6 Show that $(x^{-1})^{-1} = x$, i.e. the elements x, x^{-1} are mutually inverse elements. [Particular cases of this result are the familiar results

$$-(-x) = x,$$

$$\frac{1}{1/x} = x \text{ for } x \neq 0,$$

in real-number algebra.]

From (5′) we have $x^{-1} * (x^{-1})^{-1} = e$.
Operate with x on both sides: $x * [x^{-1} * (x^{-1})^{-1}] = x * e$.
Use (4′) on l.h.s. (2′) on r.h.s.: $(x * x^{-1}) * (x^{-1})^{-1} = x$.
Use (5′) on l.h.s.: $e * (x^{-1})^{-1} = x$.
Use (3′) and (2′) on l.h.s.: $(x^{-1})^{-1} = x$.

Exercises 6(a)

1. If the operation $*$ is not associative how many different meanings can we attach to the expressions $a * b * c$, $a * b * c * d$, $a * b * c * d * e$ (keeping the order a, b, c, d, e)?
2. Verify that the rational numbers under addition form a commutative group, and under multiplication form a commutative group if zero is omitted.
3. Verify that the integers $0, \pm 2, \pm 4, \pm 6, \ldots$ under addition form a commutative group.
4. The integers $0, \pm 1, \pm 3, \pm 5, \ldots$ under addition do not form a group. Why?
5. Verify that the set of numbers 2^n, where $n = 0, \pm 1, \pm 2, \ldots$, under multiplication form a group.
6. Let $R(\alpha)$ denote a rotation through α and let the product of two such rotations $R(\alpha), R(\beta)$ be taken to mean a rotation through α followed by a rotation through β, i.e. $R(\alpha + \beta)$. Assume that α is measured positively for rotation in one sense, negatively for rotation in the other sense. Verify that the set of rotations $R(\alpha)$, where $-\infty < \alpha < \infty$, under the product operation here defined, is a commutative group.

   Assuming that $R(\alpha + 2k\pi) = R(\alpha)$ where $k = \pm 1, \pm 2, \ldots$, show that the set $R(\alpha)$, where $0 \leqslant \alpha < 2\pi$, is a commutative group.
7. Let R_0, R_1, R_2, R_3 be a subset of the rotations discussed in Exercise 6, with $R_0 = R(0)$, $R_1 = R(\pi/2)$, $R_2 = R(\pi)$, $R_3 = R(3\pi/2)$. This set of four rotations is a group, as can be verified from the table of products (Table 6.1).

TABLE 6.1

$\times$	R_0	R_1	R_2	R_3
R_0	R_0	R_1	R_2	R_3
R_1	R_1	R_2	R_3	R_0
R_2	R_2	R_3	R_0	R_1
R_3	R_3	R_0	R_1	R_2

What features of this table enable us to say that (a) the system is closed? (b) the operation is commutative? (c) there is an identity element? (d) each element has an inverse?

6.3 Fields

We began the study of groups by looking at the axioms of the real-number system that refer solely to the operation of addition. We now look at the axioms of the real-number system that refer solely to the operation of multiplication. It is usual to include the axiom that

(6) the system is closed with respect to multiplication,

together with those already stated in Section 1.3.1, namely

(7) $x1 = x$,
(8) $xy = yx$,
(9) $(xy)z = x(yz)$.

One further axiom is required:

(10) corresponding to each number x except $x = 0$ there is a number $1/x$ such that $x(1/x) = 1$.

Note that (10) makes it unnecessary to regard division as another operation since we can define $x \div y$ to mean $x(1/y)$.

There is only one small difference between these axioms and the axioms for an abelian group and, as we have already seen in Example 2, if the number 0 is omitted the real numbers under multiplication do in fact become an abelian group.

We look finally at an axiom for real numbers which refers to both addition and multiplication:

(11) $x(y + z) = xy + xz$.

The set of axioms (1)–(11) is a complete set of axioms for the system $R(+, \times)$.

More generally let $\oplus$, $\otimes$ denote two binary operations on a set F such that

(1) F under $\oplus$ is an abelian group,
(2) if the identity element with respect to $\oplus$ is omitted then F under $\otimes$ becomes an abelian group,
(3) $x \otimes (y \oplus z) = (x \otimes y) \oplus (x \otimes z)$. (i.e. $\otimes$ is distributive over $\oplus$).

We then call the system $F(\oplus, \otimes)$ a *field*. Thus $R(+, \times)$ is a field, and the set of rational numbers under addition and multiplication is also a field.

6.3.1 Properties of Fields From the properties of abelian groups we can immediately conclude that

(a) there is a unique identity element with respect to $\oplus$, i.e. there is an element e_0 in F such that

$$x \oplus e_0 = e_0 \oplus x = x$$

and no other elements in F have this property,

(b) there is a unique identity element with respect to $\otimes$, i.e. there is an element e_1 in F such that

$$x \otimes e_1 = e_1 \otimes x = x$$

and no other elements in F have this property,

(c) corresponding to each element x in F there is a unique element $\bar{x}$ such that $x \oplus \bar{x} = \bar{x} \oplus x = e_0$, i.e. $\bar{x}$ is the inverse of x with respect to $\oplus$, and it follows that x is the inverse of $\bar{x}$ with respect to $\oplus$,

(d) corresponding to each element x in F, except $x = e_0$, there is a unique element x^{-1} such that $x \otimes x^{-1} = x^{-1} \otimes x = e_1$, i.e. x^{-1} is the inverse of x with respect to $\otimes$, and it follows that x is the inverse of x^{-1} with respect to $\otimes$.

Example 7 Show that $x \otimes e_0 = e_0 \otimes x = e_0$ for any x in F. (In particular $x0 = 0x = 0$ for real numbers.)

Since $\otimes$ is commutative $x \otimes e_0 = e_0 \otimes x$ and it is sufficient to prove that either equals e_0.

Since $x \oplus e_0 = x$ for all x in F, we can substitute e_0 for x to get $e_0 \oplus e_0 = e_0$.

Do operation $\otimes$ with x on both sides:

$$x \otimes (e_0 \oplus e_0) = x \otimes e_0.$$

Use distributive law:

$$(x \otimes e_0) \oplus (x \otimes e_0) = x \otimes e_0.$$

Do operation $\oplus$ with $(\overline{x \otimes e_0})$ on both sides:

$$(\overline{x \otimes e_0}) \oplus (x \otimes e_0) \oplus (x \otimes e_0) = (\overline{x \otimes e_0}) \oplus (x \otimes e_0).$$

Use (c):

$$e_0 \oplus (x \otimes e_0) = e_0.$$

Use (a):

$$x \otimes e_0 = e_0.$$

Example 8 Show that if $x \otimes y = e_0$, then either $x = e_0$ or $y = e_0$. (For real numbers, if $xy = 0$ then $x = 0$ or $y = 0$. This result is of basic importance in solving equations.)

It is sufficient to suppose that $x \neq e_0$ and prove that $y = e_0$. If $x \neq e_0$, then x^{-1} exists and $x \otimes x^{-1} = e_1$.

Hence we can write

$$y = e_1 \otimes y = x^{-1} \otimes x \otimes y = x^{-1} \otimes e_0 = e_0$$

using the result of Example 7 in the final step.

In the above examples we have given proofs, valid for all fields, of elementary properties already well known for the field $R(+, \times)$. Clearly other properties of the real number field are capable of generalisation in this way. Let us first generalise the notation of real-number algebra by writing

$$x \oplus x \oplus x \oplus \cdots \text{ to } n \text{ terms} = nx,$$
$$x \otimes x \otimes x \otimes \cdots \text{ to } n \text{ terms} = x^n,$$
$$x \oplus \bar{y} = x \ominus y.$$

It is now easy to generalise results from real number algebra, e.g. the identities

$$x^2 - y^2 = (x - y)(x + y), \quad \frac{1}{x - y} + \frac{1}{x + y} = \frac{2x}{x^2 - y^2}$$

become respectively

$$x^2 \ominus y^2 = (x \ominus y) \otimes (x \oplus y), (x \ominus y)^{-1} \oplus (x \oplus y)^{-1} = (2x) \otimes (x^2 \ominus y^2)^{-1},$$

and the binomial theorem for a positive integral exponent becomes

$$(x \oplus y)^n = x^n \oplus n(x^{n-1} \otimes y) \oplus [n(n-1)/1.2](x^{n-2} \otimes y^2) \oplus \cdots \oplus y^n.$$

Thus we see that once we have established that a particular mathematical system is a field, we can regard its algebra as something with which we are already quite familiar. Only results that refer to particular elements (other than the two identity elements mentioned in the axioms) can differ from one field to another. It is especially important to realise that an equation may have more roots in one field than in another, e.g. the equation $x^2 - 2 = 0$ has no roots in the field $Q(+, \times)$, where Q is the set of rational numbers, whereas the same equation has the two roots $\pm\sqrt{2}$ in the field $R(+, \times)$.

It should also be noted that there need not be relations between the elements of a field corresponding to the relations $>$, $<$ between real numbers, and so there is no question of generalising inequalities of the real-number system in the way we have generalised identities.

Exercises 6(b)

1. An operation $\otimes$ on R is defined by the equation $a \otimes b = ab - a - b$. Determine if this operation is (a) commutative, (b) associative, (c) distributive over addition.
2. If the operation $\otimes$ is non-commutative, then it is said to be *left-distributive* over $\oplus$ if

$$a \otimes (b \oplus c) = (a \otimes b) \oplus (a \otimes c),$$

right-distributive over $\oplus$ if

$$(a \oplus b) \otimes c = (a \otimes c) \oplus (b \otimes c).$$

Show that the operation $\otimes$ defined by the equation

$$a \otimes b = ab/(2a + 1)$$

is non-commutative and non-associative, and determine to what extent it is distributive (a) over addition, (b) over the operation defined in Exercise 1.

3. Verify that the natural numbers 1, 2, 3, . . . under addition and multiplication do not form a field.
4. Verify that the set of numbers $a + b\sqrt{2}$, where a, b can be any pair of rational numbers, under addition and multiplication form a field.
5. If the system $F(\oplus, \otimes)$ is a field and A is a subset of F, which of the field axioms would you automatically assume must hold for $A(\oplus, \otimes)$ and which axioms would you say need to be tested?
6. Set subtraction is defined by writing $A - B = A \cap (A \cap B)'$. Use Venn diagrams to see if set subtraction is commutative. Determine if (a) intersection, (b) union is distributive over subtraction.
7. There are two 'cancellation laws' for a field:

$$a \oplus x = a \oplus y \Rightarrow x = y,$$

$$a \otimes x = a \otimes y \ (a \neq e_0) \Rightarrow x = y$$

(e_0 denotes the identity element with respect to $\oplus$). Derive these laws from the axioms for a field. If we interpret $\oplus$, $\otimes$ as union, intersection respectively, do these laws hold in the Boolean algebra of sets?

8. For the operations $\oplus$, $\otimes$ defined in Exercises 5, No. 5, what are the identity elements (a) if the set is restricted to the positive integers from 1 to 10, (b) if the set consists of all positive integers, (c) if the set consists of all real numbers?

Show that these operations are commutative and associative, and determine if $\otimes$ is distributive over $\oplus$. Do we get a field in (a), (b), or (c)?

CHAPTER 7

VECTORS

7.1 Introduction

Consider the following two problems from elementary physics:

(a) Two given forces act on a particle mass. What single force acting on the particle mass would have the same effect?

(b) A billiard ball is struck simultaneously by two cues. Knowing the speed and direction that each cue acting singly would have imparted to the ball, can we tell the speed and direction imparted by the double strike?

From (a) we are led to consider a physical system in which the elements are forces and an operation of 'addition' is defined thus: the 'sum' of two forces is the single force that has the same effect when acting on a particle mass.

Similarly from (b) we are led to consider a physical system in which the elements are velocities and an operation of 'addition' is defined thus: the 'sum' of two velocities is the velocity that results from their simultaneous application.

In both these physical systems the elements are *vector* quantities, i.e. quantities that involve the idea of direction as well as magnitude. In contrast, quantities (e.g. mass, temperature, length, area, volume) that can be completely specified by a real number (and hence by a point on a scale) are called *scalar* quantities.

The simplest example of a vector quantity is a straight-line *displacement*. Such a displacement can be represented by the line joining the initial and final positions. From Fig. 7.1 we see that if an object is displaced from A to B and then displaced from B to C, these two displacements have the same effect as a single displacement from A to C. The displacement AC is therefore called the sum of the displacements AB, BC and we may write

$$\text{disp } AB + \text{disp } BC = \text{disp } AC.$$

7.2 The Vector System

Any vector quantity can be represented by a directed line segment, i.e. the line AB may represent any vector quantity which has the same direction and (on some scale) the same magnitude as AB. It is a well-known physical law that this representation enables us to find the 'sum' of other vector quantities in the same way as for displacements.

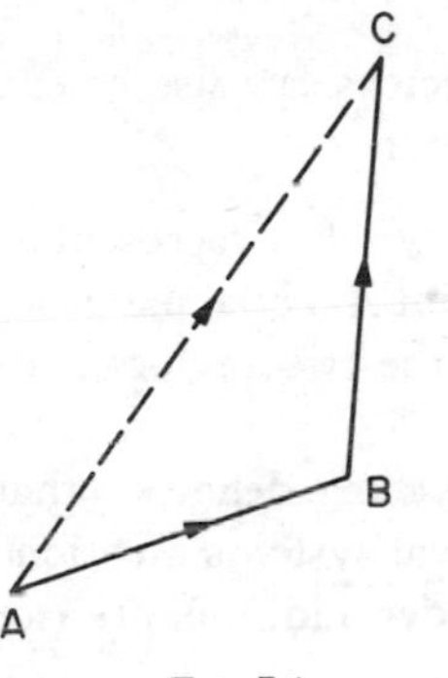

FIG. 7.1

The representation of a vector quantity by a line segment takes no account of the location of the line; if AB (see Fig. 7.2) represents a certain vector quantity, then so do CD, EF and any other line with the same magnitude (length) and direction as AB. We therefore consider a system $V(+)$ in which all line segments with the same magnitude and direction as AB are regarded as just one single element, called a *vector*. Any one of these line segments may be used to *represent* the vector, and any one of the symbols $\overrightarrow{AB}$, $\overrightarrow{CD}$, $\overrightarrow{EF}$, ... (see Fig. 7.2) may be used to *denote* the vector.

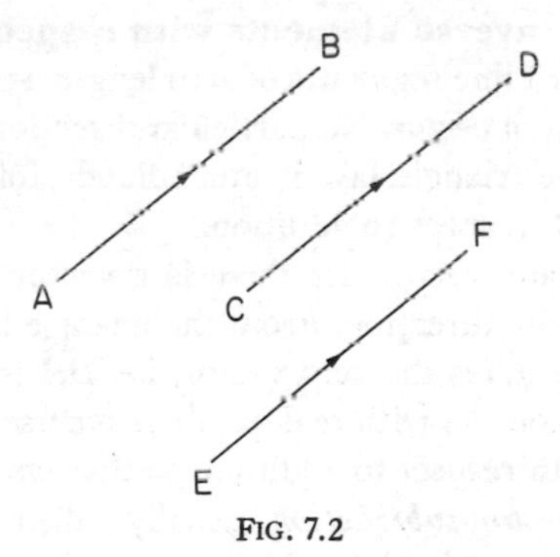

FIG. 7.2

7.2.1 Vector Addition An operation of *vector addition*, usually called simply *addition* and denoted by +, is defined by the following 'triangle' law:

Given two vectors we can select any representative line segment AB

for the first vector and a representative line segment BC, chosen with its starting point coinciding with the end-point of AB, for the second vector. The sum of the two vectors is then represented by AC, i.e. $\overrightarrow{AB} + \overrightarrow{BC} = \overrightarrow{AC}$ (see Fig. 7.1).

The rule for addition of vectors can also be expressed in the form of the following 'parallelogram' law:

Given two vectors we may select representative line segments AB, AC with a common starting point A. If the parallelogram $ABCD$ is completed (Fig. 7.3) then the sum of the two vectors is represented by AD, that is $\overrightarrow{AB} + \overrightarrow{AC} = \overrightarrow{AD}$.

It follows easily from the above definition that addition is commutative and associative. The physical systems mentioned in Section 7.1, and the system $V(+)$ of vectors under addition, are isomorphic.

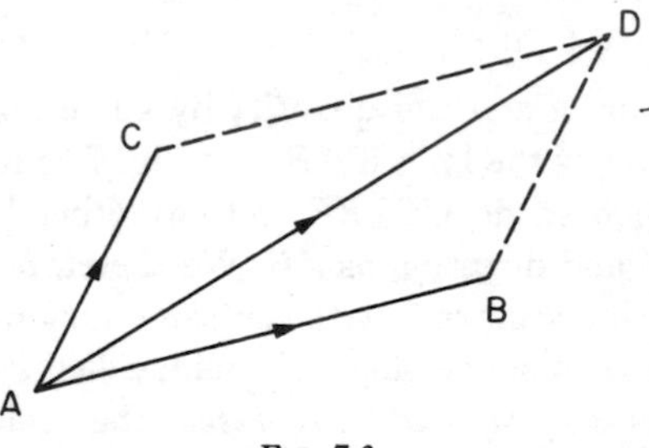

FIG. 7.3

7.2.2 Neutral and Inverse Elements with respect to Addition It is convenient to regard line segments of zero length as representing a vector, to be called the ***zero vector***. No particular direction is associated with this vector. From the triangle law it immediately follows that the zero vector is neutral with respect to addition.

Corresponding to any vector $\overrightarrow{AB}$ there is a vector $\overrightarrow{BA}$ with the same magnitude but opposite direction. From the triangle law we see that addition of $\overrightarrow{AB}$ and $\overrightarrow{BA}$ gives the zero vector, i.e. $\overrightarrow{BA}$ is the inverse of $\overrightarrow{AB}$ with respect to addition. As with real numbers we use a minus sign to indicate the inverse with respect to addition, so that we write $\overrightarrow{BA} = -\overrightarrow{AB}$.

An operation of ***vector subtraction***, usually called simply ***subtraction*** and denoted by $-$, can now be defined by the equation

$$\overrightarrow{AB} - \overrightarrow{CD} = \overrightarrow{AB} + (-\overrightarrow{CD}) = \overrightarrow{AB} + \overrightarrow{DC}.$$

Figure 7.4 illustrates the use of the triangle law for the operation of subtraction with vectors $\overrightarrow{AB}$, $\overrightarrow{CD}$; we draw a representative line segment BX for the vector $\overrightarrow{DC}$ so that

$$\overrightarrow{AX} = \overrightarrow{AB} + \overrightarrow{BX} = \overrightarrow{AB} + \overrightarrow{DC} = \overrightarrow{AB} - \overrightarrow{CD}.$$

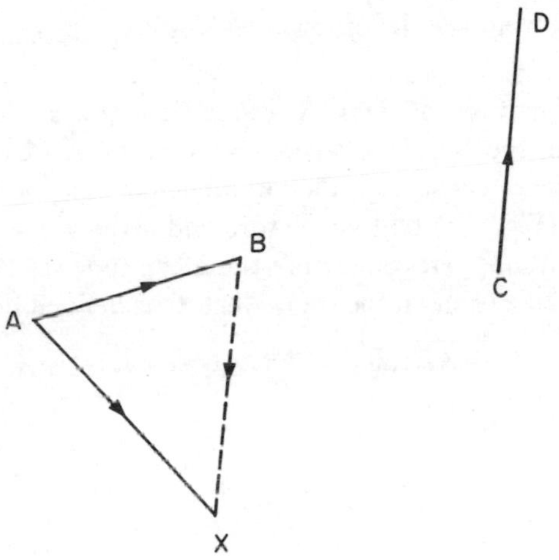

FIG. 7.4

7.2.3 Multiplication of a Vector by a Scalar It is natural to denote the sum $\overrightarrow{AB} + \overrightarrow{AB}$ by $2\overrightarrow{AB}$; from the triangle law we easily see that this vector is in the same direction as $\overrightarrow{AB}$ and has double the magnitude of $\overrightarrow{AB}$ (Fig. 7.5).

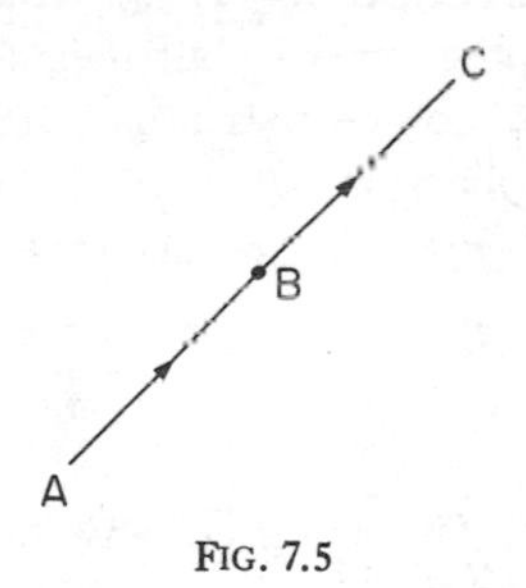

FIG. 7.5

In general we define $k\,\overrightarrow{AB}$, for any real number $k \geqslant 0$, to be the vector with the same direction as $\overrightarrow{AB}$ and with k times the magnitude of $\overrightarrow{AB}$, while $(-k)\overrightarrow{AB}$ is defined to be the vector $-(k\,\overrightarrow{AB})$.

The reader should check the associative law

$$k(l\overrightarrow{AB}) = kl\overrightarrow{AB} = l(k\overrightarrow{AB})$$

and the two distributive laws

$$(k + l)\overrightarrow{AB} = k\overrightarrow{AB} + l\overrightarrow{AB},$$

$$k(\overrightarrow{AB} + \overrightarrow{AC}) = k\overrightarrow{AB} + k\overrightarrow{AC},$$

which all follow from the definitions.

7.2.4 Scalar Product of Two Vectors By the angle between two vectors we mean the angle (between 0 and π) formed by two representative line segments chosen with a common starting point.

Let $\overrightarrow{AB}$, $\overrightarrow{AC}$ (Fig. 7.6) be two vectors and θ the angle between them. Let AB, AC (without arrows) denote the magnitudes of $\overrightarrow{AB}$, $\overrightarrow{AC}$ respectively. Another operation on the set of vectors is defined by the equation

$$\overrightarrow{AB}.\overrightarrow{AC} = AB.AC \cos \theta. \tag{7.1}$$

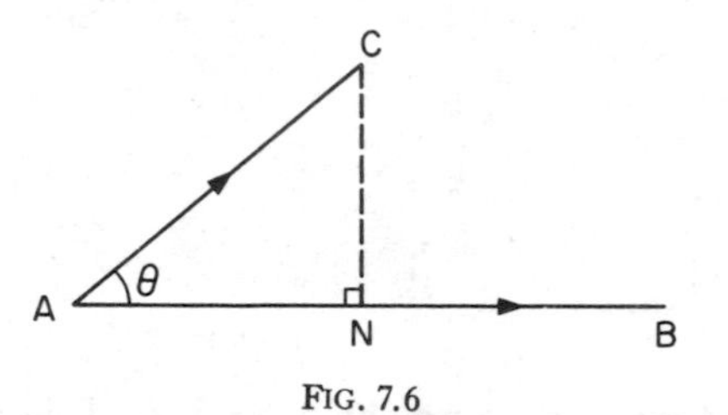

FIG. 7.6

This operation is called *scalar multiplication* (not to be confused with 'multiplication by a scalar' of Section 7.2.3). Note that the scalar product of two vectors is *not* another vector, but it is a *real number* (hence the term scalar). Thus the set of vectors is not closed with respect to the operation of scalar multiplication.

Since $AC \cos \theta = AN$ (Fig. 7.6), we see that

$$\overrightarrow{AB}.\overrightarrow{AC} = AB \times (\text{projection of } AC \text{ on } AB).$$

Likewise

$$\overrightarrow{AB}.\overrightarrow{AC} = AC \times (\text{projection of } AB \text{ on } AC).$$

In particular,

(a) if $\theta = 0$ (vectors in same direction), then $\cos \theta = 1$ and so the scalar product of the vectors becomes the ordinary product of their magnitudes, e.g. $\overrightarrow{AB}.\overrightarrow{AB} = AB^2$,

(b) if $\theta = \pi/2$ (perpendicular vectors) then $\cos \theta = 0$ and so the scalar product is zero,

(c) if $\theta = \pi$ (vectors in opposite directions) then $\cos\theta = -1$ and so the scalar product of the vectors is minus the product of their magnitudes.

The scalar product $\overrightarrow{AB}.\overrightarrow{AB}$ is sometimes denoted by $(\overrightarrow{AB})^2$. From (a) we see that $(\overrightarrow{AB})^2 = AB^2$.

As an example of a physical situation to which the idea of scalar multiplication can be applied, let $\overrightarrow{AB}$ represent a force acting on a particle that undergoes a displacement represented by $\overrightarrow{CD}$; the *work* done by the force in this displacement is $\overrightarrow{AB}.\overrightarrow{CD}$.

The reader should verify that the operation of scalar multiplication is commutative and is distributive over addition; also that

$$k(\overrightarrow{AB}.\overrightarrow{AC}) = (k\overrightarrow{AB}).\overrightarrow{AC} = \overrightarrow{AB}.(k\overrightarrow{AC}).$$

7.3 Subsets of V

The set V of vectors has two important subsets. Consider first the subset V_1 consisting of all vectors parallel to some fixed line, i.e. any two members of V_1 have either the same or opposite directions. It is easily seen that the set V_1 is closed with respect to the operation + and that the mapping

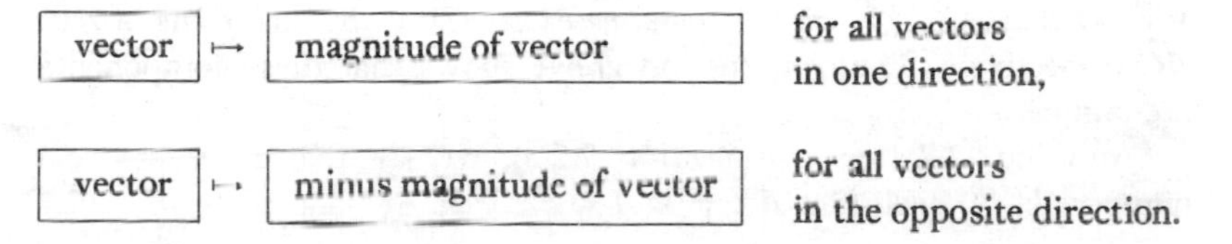

is an isomorphism of $V_1(+)$ onto $R(+)$; moreover the scalar product of two vectors in V_1 equals the ordinary product of their images in R. Hence the systems $V_1(+, .)$, $R(+, \times)$ are isomorphic; this means that it is not necessary to distinguish between the two systems and it is quite usual to refer to the system of one-dimensional vectors as the system $R(+, \times)$.

Several vectors are said to be *coplanar* if representative line segments, drawn from a common starting point, are coplanar. Consider next the subset V_2 consisting of all vectors with representative line segments lying in some fixed plane. In many applications of vectors the nature of the problem is such that the system $V_2(+,.)$ is adequate. When using this system we usually say that we are 'working with two-dimensional vectors'. Likewise, when using the full system $V(+,.)$ we say we are 'working with three-dimensional vectors', and we sometimes use V_3 rather than V to denote the set of all vectors.

7.4 Components of a Vector

Let $\overrightarrow{AB}$, $\overrightarrow{AC}$ be two non-parallel vectors in the set V_2 of vectors with representative line segments in a plane π. For any other vector in V_2, the representative line segment drawn from A must also lie in π. Thus if $\overrightarrow{AD}$ is in V_2, we can draw lines through D parallel to AB and AC and form a parallelogram $AXDY$ (see Fig. 7.7). It follows that

$$\overrightarrow{AD} = \overrightarrow{AX} + \overrightarrow{AY}.$$

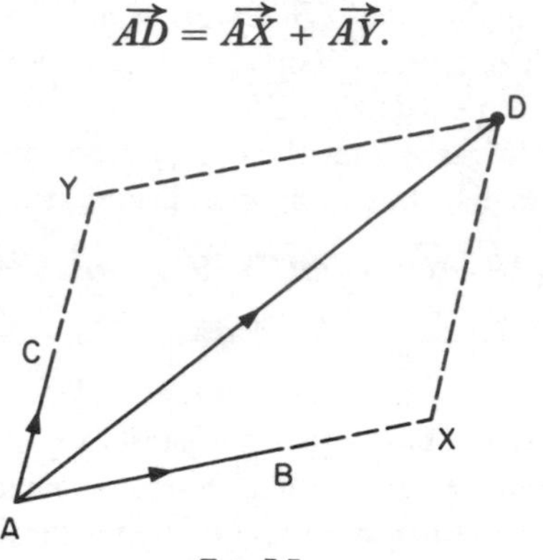

FIG. 7.7

We say that $\overrightarrow{AX}$, $\overrightarrow{AY}$ are the *components* of $\overrightarrow{AD}$ in the directions of $\overrightarrow{AB}$, $\overrightarrow{AC}$ respectively. The construction above shows that these components are unique.

From Fig. 7.7 (see also Section 7.2.3), we see that there are real numbers k_1, k_2 such that $\overrightarrow{AX} = k_1\overrightarrow{AB}$, $\overrightarrow{AY} = k_2\overrightarrow{AC}$, and so

$$\overrightarrow{AD} = k_1\overrightarrow{AB} + k_2\overrightarrow{AC}. \tag{7.2}$$

Any expression of the form $k_1\overrightarrow{AB} + k_2\overrightarrow{AC}$, where k_1 and k_2 are real numbers, is called a *linear combination* of the vectors $\overrightarrow{AB}$, $\overrightarrow{AC}$. We have thus proved an important property of the set V_2, viz. *any two non-parallel vectors in V_2 may be selected and all other vectors in the set can be expressed as linear combinations of these two.*

We say that the vectors $\overrightarrow{AB}$, $\overrightarrow{AC}$ are a *basis* for the set V_2. Any two non-parallel vectors in V_2 may be chosen as a basis for V_2.

A *unit vector* is a vector with unit magnitude. The basis most commonly used for V_2 is a pair of perpendicular unit vectors $\overrightarrow{OI}$, $\overrightarrow{OJ}$ say, and if co-ordinate axes are introduced as shown in Fig. 7.8 we have, for any vector $\overrightarrow{OP}$ in V_2,

$$\overrightarrow{OP} = \overrightarrow{OM} + \overrightarrow{ON} = x\overrightarrow{OI} + y\overrightarrow{OJ} \tag{7.3}$$

where (x, y) are the co-ordinates of P.

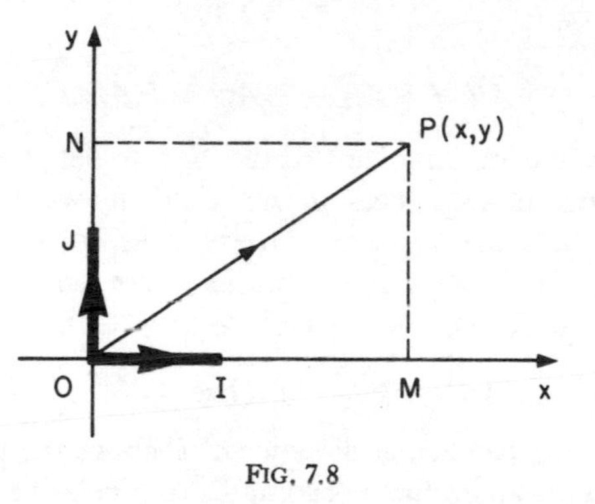

FIG. 7.8

The foregoing ideas may be extended to the set V_3 of all vectors.

Let $\overrightarrow{AB}$, $\overrightarrow{AC}$, $\overrightarrow{AD}$ be three vectors which are not coplanar, $\overrightarrow{AE}$ any other vector in V_3. Through E draw a line parallel to AD, meeting the plane containing AB and AC in F. Then

$$\overrightarrow{AE} = \overrightarrow{AF} + \overrightarrow{FE}.$$

Since AF lies in the same plane as AB and AC, we have

$$\overrightarrow{AF} = \overrightarrow{AG} + \overrightarrow{AH}$$

(see Fig. 7.9), where $\overrightarrow{AG}$, $\overrightarrow{AH}$ are parallel to $\overrightarrow{AB}$, $\overrightarrow{AC}$ respectively.

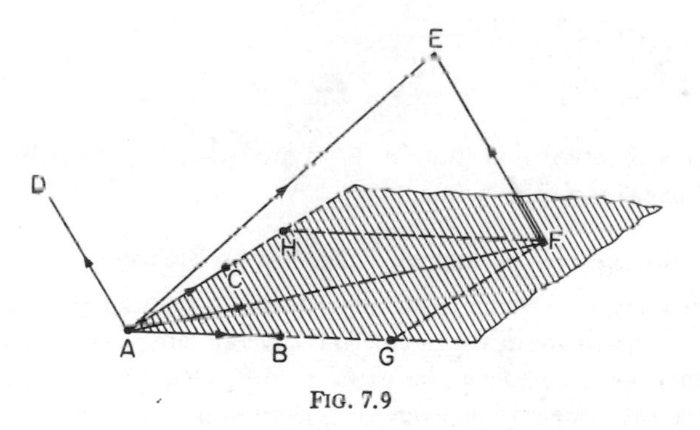

FIG. 7.9

Thus

$$\overrightarrow{AE} = \overrightarrow{AG} + \overrightarrow{AH} + \overrightarrow{FE},$$

where $\overrightarrow{AG}$, $\overrightarrow{AH}$, $\overrightarrow{FE}$ are called the *components* of $\overrightarrow{AE}$ in the directions of $\overrightarrow{AB}$, $\overrightarrow{AC}$, $\overrightarrow{AD}$ respectively. Real numbers k_1, k_2, k_3 can be found such that

$$\overrightarrow{AG} = k_1\overrightarrow{AB},\ \overrightarrow{AH} = k_2\overrightarrow{AC},\ \overrightarrow{FE} = k_3\overrightarrow{AD}$$

and so

$$\overrightarrow{AE} = k_1\overrightarrow{AB} + k_2\overrightarrow{AC} + k_3\overrightarrow{AD}. \tag{7.4}$$

Summarising, we say that any three non-coplanar vectors may be chosen as a *basis* for V_3. This means that any vector in V_3 can be expressed as a *linear combination* of these three.

In particular, if $\overrightarrow{OI}$, $\overrightarrow{OJ}$, $\overrightarrow{OK}$ are mutually perpendicular unit vectors then (see Fig. 7.10) for any vector $\overrightarrow{OP}$ we have

$$\overrightarrow{OP} = x\overrightarrow{OI} + y\overrightarrow{OJ} + z\overrightarrow{OK}, \tag{7.5}$$

where $z = NP$ = perpendicular distance of P above the plane containing OI and OJ, and (x, y) are the co-ordinates of N relative to axes OI, OJ.

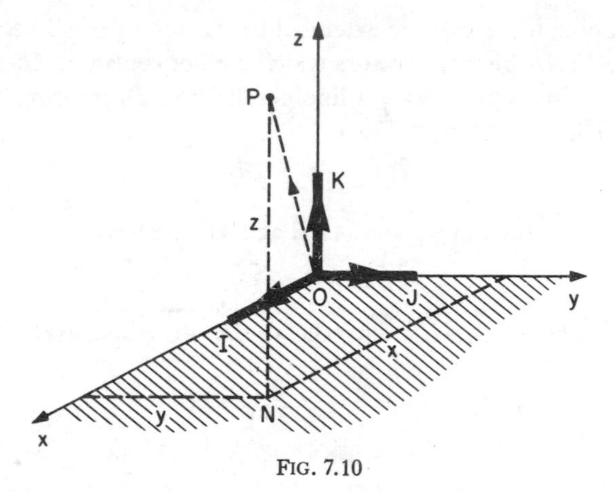

FIG. 7.10

We shall see (Section 8.1) that (x, y, z) are called the co-ordinates of P relative to axes OI, OJ, OK.

7.5 An Alternative Approach to the Vector System

It frequently requires some ingenuity to design enough symbols for the elements of a mathematical system. We have already seen (Section 1.2.1) how decimal notation solves the problem in the case of the real-number system. Vectors appear to be even more numerous than real numbers, but due to the discoveries of Section 7.4 a simple procedure is available. Let special symbols **i**, **j**, **k** be introduced for the unit vectors $\overrightarrow{OI}$, $\overrightarrow{OJ}$, $\overrightarrow{OK}$ respectively (Fig. 7.10); then by (7.5) any vector can be represented by a symbol of the form $x\mathbf{i} + y\mathbf{j} + z\mathbf{k}$ where x, y, z are real numbers, e.g. $2\mathbf{i} + 3\mathbf{j} - 7\mathbf{k}$, $-\mathbf{i} + 0\mathbf{j} + 5\mathbf{k}$ are typical vectors in this notation.

This is the notation that we shall mainly use, although a simpler notation is obtained if we replace $x\mathbf{i} + y\mathbf{j} + z\mathbf{k}$ by (x, y, z), e.g. $(2, 3, -7)$,

(−1, 0, 5) denote typical vectors. Following this line of thought the reader may appreciate why mathematicians, using hindsight, define a vector to be *an ordered triplet of real numbers*. We shall not discuss this simpler notation any further until Chapter 11.

For convenience single bold-face letters **a**, **b**, **c**, . . . are also used to denote vectors; thus we may declare $\mathbf{a} = 2\mathbf{i} + 3\mathbf{j} - 7\mathbf{k}$, $\mathbf{b} = -\mathbf{i} + 0\mathbf{j} + 5\mathbf{k}$, etc. In practice zeros are usually omitted, e.g. $-\mathbf{i} + 5\mathbf{k}$ is written for $-\mathbf{i} + 0\mathbf{j} + 5\mathbf{k}$. The zero vector is $0\mathbf{i} + 0\mathbf{j} + 0\mathbf{k}$ and is denoted by **0**.

Certain rules can now be established. Let

$$\mathbf{a} = x_1\mathbf{i} + y_1\mathbf{j} + z_1\mathbf{k}, \quad \mathbf{b} = x_2\mathbf{i} + y_2\mathbf{j} + z_2\mathbf{k}.$$

Since the components of a vector are unique we can say that $\mathbf{a} = \mathbf{b}$ if, and only if,

$$x_1 = x_2, \quad y_1 = y_2, \quad z_1 = z_2.$$

Since addition is associative we can write

$$\mathbf{a} + \mathbf{b} = (x_1\mathbf{i} + x_2\mathbf{i}) + (y_1\mathbf{j} + y_2\mathbf{j}) + (z_1\mathbf{k} + z_2\mathbf{k}),$$

and using the first distributive law of Section 7.2.3 we get

$$\mathbf{a} + \mathbf{b} = (x_1 + x_2)\mathbf{i} + (y_1 + y_2)\mathbf{j} + (z_1 + z_2)\mathbf{k}. \tag{7.6}$$

Likewise

$$\mathbf{a} - \mathbf{b} = (x_1 - x_2)\mathbf{i} + (y_1 - y_2)\mathbf{j} + (z_1 - z_2)\mathbf{k} \tag{7.7}$$

and

$$k\mathbf{a} = (kx_1)\mathbf{i} + (ky_1)\mathbf{j} + (kz_1)\mathbf{k} \quad (k \text{ real}). \tag{7.8}$$

Turning next to scalar multiplication, we first observe (see Section 7.2.4) that, since **i**, **j**, **k** are unit vectors,

$$\mathbf{i}^2 = \mathbf{j}^2 = \mathbf{k}^2 = 1, \tag{7.9}$$

and since they are mutually perpendicular,

$$\mathbf{i}.\mathbf{j} = \mathbf{j}.\mathbf{k} = \mathbf{k}.\mathbf{i} = 0. \tag{7.10}$$

Now

$$\mathbf{a}.\mathbf{b} = (x_1\mathbf{i} + y_1\mathbf{j} + z_1\mathbf{k}).(x_2\mathbf{i} + y_2\mathbf{j} + z_2\mathbf{k})$$

and using laws mentioned in the last sentence of Section 7.2.4 we may expand the r.h.s. to get

$$\mathbf{a}.\mathbf{b} = x_1x_2\mathbf{i}^2 + x_1y_2\mathbf{i}.\mathbf{j} + x_1z_2\mathbf{i}.\mathbf{k} + y_1x_2\mathbf{j}.\mathbf{i} + y_1y_2\mathbf{j}^2 + y_1z_2\mathbf{j}.\mathbf{k} + z_1x_2\mathbf{k}.\mathbf{i} + z_1y_2\mathbf{k}.\mathbf{j} + z_1z_2\mathbf{k}^2.$$

Using (7.9) and (7.10) we find that

$$\mathbf{a}.\mathbf{b} = x_1x_2 + y_1y_2 + z_1z_2. \tag{7.11}$$

The significance of these results is that we can now conveniently regard the vector system as a system of elements $x\mathbf{i} + y\mathbf{i} + z\mathbf{k}$, where x, y, z are real numbers, with an operation of addition defined by (7.6), an operation of 'multiplication by a scalar' defined by (7.8), and an operation of 'scalar multiplication' defined by (7.11).

Example 1 Given $\mathbf{a} = 2\mathbf{i} - \mathbf{j} + 3\mathbf{k}$, $\mathbf{b} = \mathbf{i} + 2\mathbf{j} - 4\mathbf{k}$, find

$$(\mathbf{a} + 2\mathbf{b}).(2\mathbf{a} - \mathbf{b}).$$

We have

$$2\mathbf{b} = 2\mathbf{i} + 4\mathbf{j} - 8\mathbf{k}, \quad \mathbf{a} + 2\mathbf{b} = 4\mathbf{i} + 3\mathbf{j} - 5\mathbf{k},$$

$$2\mathbf{a} = 4\mathbf{i} - 2\mathbf{j} + 6\mathbf{k}, \quad 2\mathbf{a} - \mathbf{b} = 3\mathbf{i} - 4\mathbf{j} + 10\mathbf{k}.$$

Hence

$$(\mathbf{a} + 2\mathbf{b}).(2\mathbf{a} - \mathbf{b}) = 4.3 + 3(-4) + (-5).10$$
$$= 12 - 12 - 50 = -50.$$

7.5.1 Magnitude of a Vector The magnitude of $\mathbf{a}$ is denoted by $|\mathbf{a}|$. It was shown in Section 7.2.4 that $\mathbf{a}^2 = |\mathbf{a}|^2$.

If $\mathbf{a} = x\mathbf{i} + y\mathbf{j} + z\mathbf{k}$ we have from (7.11) $\mathbf{a}^2 = x^2 + y^2 + z^2$.
Hence $|\mathbf{a}|^2 = x^2 + y^2 + z^2$, i.e.

$$|\mathbf{a}| = \sqrt{(x^2 + y^2 + z^2)}. \tag{7.12}$$

For example the magnitude of $2\mathbf{i} - 3\mathbf{j} + \mathbf{k}$ is $\sqrt{[2^2 + (-3)^2 + 1^2]}$, i.e. $\sqrt{14}$.

7.5.2 Unit Vector in a Given Direction The unit vector with the same direction as $\mathbf{a}$ is $k\mathbf{a}$ where $k = 1/|\mathbf{a}|$. Hence if $\mathbf{a} = x\mathbf{i} + y\mathbf{j} + z\mathbf{k}$, the unit vector in the direction of $\mathbf{a}$ is

$$\frac{1}{\sqrt{(x^2 + y^2 + z^2)}}(x\mathbf{i} + y\mathbf{j} + z\mathbf{k}).$$

For example the unit vector in the direction of $2\mathbf{i} - 3\mathbf{j} + \mathbf{k}$ is

$$(2\mathbf{i} - 3\mathbf{j} + \mathbf{k})/\sqrt{14}.$$

7.5.3 Angle Between Two Vectors If θ is the angle between $\mathbf{a}$ and $\mathbf{b}$, then by (7.1)

$$\mathbf{a}.\mathbf{b} = |\mathbf{a}||\mathbf{b}|\cos\theta;$$

that is

$$\cos\theta = \frac{\mathbf{a}.\mathbf{b}}{|\mathbf{a}||\mathbf{b}|}.$$

If $\mathbf{a} = x_1\mathbf{i} + y_1\mathbf{j} + z_1\mathbf{k}$ and $\mathbf{b} = x_2\mathbf{i} + y_2\mathbf{j} + z_2\mathbf{k}$, this result becomes

$$\cos\theta = \frac{x_1x_2 + y_1y_2 + z_1z_2}{\sqrt{(x_1^2 + y_1^2 + z_1^2)}\sqrt{(x_2^2 + y_2^2 + z_2^2)}}. \tag{7.13}$$

In particular the condition that $\mathbf{a}$ and $\mathbf{b}$ are perpendicular is that

$$x_1x_2 + y_1y_2 + z_1z_2 = 0. \tag{7.14}$$

Example 2 Show that the vectors $\mathbf{i} - 2\mathbf{j} + \mathbf{k}$, $2\mathbf{i} + 3\mathbf{j} + 4\mathbf{k}$ are perpendicular.

This follows since $1.2 + (-2).3 + 1.4 = 2 - 6 + 4 = 0$.

Example 3 Find the angle between the vectors $\mathbf{i} - 2\mathbf{j} + 2\mathbf{k}$, $4\mathbf{i} + 3\mathbf{k}$.

Let $\mathbf{a} = \mathbf{i} - 2\mathbf{j} + 2\mathbf{k}$, $\mathbf{b} = 4\mathbf{i} + 3\mathbf{k}$. Then

$$|\mathbf{a}| = \sqrt{(1 + 4 + 4)} = 3, \quad |\mathbf{b}| = \sqrt{(16 + 9)} = 5,$$

and $\mathbf{a}.\mathbf{b} = 1.4 + (-2).0 + 2.3 = 10$. Hence $\cos\theta = 10/3.5 = \frac{2}{3}$, giving $\theta = 48{\cdot}18^\circ$ approximately.

7.5.4 Two-dimensional Vectors Equation (7.3) shows that vectors in V_2 may be expressed in the form $x\mathbf{i} + y\mathbf{j}$, where x, y are real numbers, i.e. we may regard V_2 as the subset of V_3 that contains all vectors of the form $x\mathbf{i} + y\mathbf{j} + 0\mathbf{k}$. Continuing to omit the zero component we may write the rules for addition, multiplication by a scalar, and scalar multiplication as follows. If

$$\mathbf{a} = x_1\mathbf{i} + y_1\mathbf{j}, \quad \mathbf{b} = x_2\mathbf{i} + y_2\mathbf{j},$$

then

$$\begin{aligned} \mathbf{a} + \mathbf{b} &= (x_1 + x_2)\mathbf{i} + (y_1 + y_2)\mathbf{j}, \\ k\mathbf{a} &= kx_1\mathbf{i} + ky_1\mathbf{j}, \\ \mathbf{a}.\mathbf{b} &= x_1x_2 + y_1y_2. \end{aligned}$$

Furthermore we have $|\mathbf{a}| = \sqrt{(x_1^2 + y_1^2)}$, so that the unit vector in the direction of $\mathbf{a}$ is

$$\frac{1}{\sqrt{(x_1^2 + y_1^2)}}(x_1\mathbf{i} + y_1\mathbf{j}),$$

and the angle θ between $\mathbf{a}$ and $\mathbf{b}$ is given by

$$\cos\theta = \frac{x_1x_2 + y_1y_2}{\sqrt{(x_1^2 + y_1^2)}\sqrt{(x_2^2 + y_2^2)}}.$$

7.6 Vector Algebra

Bold-face letters $\mathbf{a}$, $\mathbf{b}$, $\mathbf{c}$, ... are also used as *variables* on the set of vectors. Thus we may state the commutative and associative laws for vector addition in the form of *identities*, viz.

$$\mathbf{a} + \mathbf{b} = \mathbf{b} + \mathbf{a},$$
$$\mathbf{a} + (\mathbf{b} + \mathbf{c}) = (\mathbf{a} + \mathbf{b}) + \mathbf{c}.$$

Likewise (Section 7.2.3) we have an associative law

$$k(l\mathbf{a}) = kl\mathbf{a} = l(k\mathbf{a})$$

and two distributive laws

$$(k + l)\mathbf{a} = k\mathbf{a} + l\mathbf{a},$$
$$k(\mathbf{a} + \mathbf{b}) = k\mathbf{a} + k\mathbf{b}.$$

From Section 7.2.4 we have

$$\mathbf{a}.\mathbf{b} = \mathbf{b}.\mathbf{a},$$
$$k(\mathbf{a}.\mathbf{b}) = (k\mathbf{a}).\mathbf{b} = \mathbf{a}.(k\mathbf{b}),$$
$$\mathbf{a}.(\mathbf{b} + \mathbf{c}) = \mathbf{a}.\mathbf{b} + \mathbf{a}.\mathbf{c}.$$

By taking the above identities as axioms we have the foundations for an algebra of vectors. Clearly this algebra has much in common with the algebra of the real-number system; for example

$$(\mathbf{a} + \mathbf{b}).(\mathbf{c} + \mathbf{d}) = \mathbf{a}.\mathbf{c} + \mathbf{a}.\mathbf{d} + \mathbf{b}.\mathbf{c} + \mathbf{b}.\mathbf{d},$$

and in particular

$$(\mathbf{a} + \mathbf{b})^2 = \mathbf{a}^2 + 2\mathbf{a}.\mathbf{b} + \mathbf{b}^2.$$

However the system $V(+, .)$ is not a field; it is not closed with respect to the operation . and the ideas of neutral element and inverse elements with respect to this operation just do not apply. We must therefore be careful not to copy the algebra of the real-number system in any respects where it relies on these ideas, e.g. the proof that $xy = 0$ implies $x = 0$ or $y = 0$ (Section 6.3.1, Example 8) makes use of multiplicative inverses, and in fact $\mathbf{a}.\mathbf{b} = 0$ does *not* imply $\mathbf{a} = \mathbf{0}$ or $\mathbf{b} = \mathbf{0}$ since it is true whenever $\mathbf{a}$ and $\mathbf{b}$ are perpendicular.

Example 4 Do Example 1 by an alternative method.

We can first write

$$(\mathbf{a} + 2\mathbf{b}).(2\mathbf{a} - \mathbf{b}) = 2\mathbf{a}^2 + 3\mathbf{a}.\mathbf{b} - 2\mathbf{b}^2.$$

Then since

$$\mathbf{a} = 2\mathbf{i} - \mathbf{j} + 3\mathbf{k}, \quad \mathbf{b} = \mathbf{i} + 2\mathbf{j} - 4\mathbf{k},$$

we have

$$\mathbf{a}^2 = 2^2 + (-1)^2 + 3^2 = 14, \quad \mathbf{b}^2 = 1^2 + 2^2 + (-4)^2 = 21,$$

$$\mathbf{a}.\mathbf{b} = 2.1 + (-1).2 + 3.(-4) = 2 - 2 - 12 = -12.$$

Finally

$$2\mathbf{a}^2 + 3\mathbf{a}.\mathbf{b} - 2\mathbf{b}^2 = 28 - 36 - 42 = -50.$$

7.6.1 Application of Vector Algebra to Geometry The examples in this section illustrate the use of vector algebra in connection with plane geometry. The ideas are equally applicable to solid geometry. However the full value of vectors in connection with geometry is realised when co-ordinates are used, and the subject of vectors applied to co-ordinate geometry is dealt with in Chapter 8.

Basically, the method used in this section is:

(a) A point O is selected as 'origin', and all lines joining O to other points P, Q, R, ... are used to define vectors $\mathbf{p}$, $\mathbf{q}$, $\mathbf{r}$, ... where

$$\mathbf{p} = \overrightarrow{OP}, \mathbf{q} = \overrightarrow{OQ}, \mathbf{r} = \overrightarrow{OR}, \ldots \qquad \text{(see Fig. 7.11).}$$

The vectors $\mathbf{p}$, $\mathbf{q}$, $\mathbf{r}$, ... are called the *position vectors* of the points P, Q, R, ... relative to the origin O.

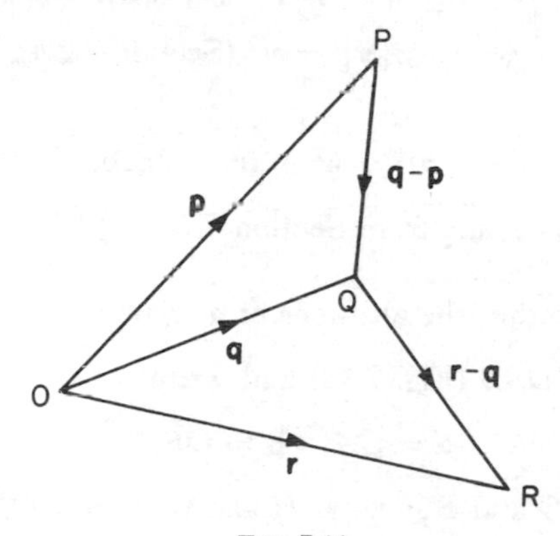

FIG. 7.11

(b) Lines joining two points not including O correspond to vectors as follows:

$$\overrightarrow{PQ} = \mathbf{q} - \mathbf{p}, \quad \overrightarrow{QR} = \mathbf{r} - \mathbf{q}, \quad \overrightarrow{PR} = \mathbf{r} - \mathbf{p}, \ldots.$$

(c) The problem is examined in terms of these vectors.

Example 5 Verify the cosine rule for a triangle by vector methods.

Let one of the vertices of the triangle be chosen as the origin O. Let the other vertices be A, B and define

$$\mathbf{a} = \overrightarrow{OA}, \quad \mathbf{b} = \overrightarrow{OB}.$$

It follows that $\overrightarrow{AB} = \mathbf{b} - \mathbf{a}$.

The cosine rule states that

$$AB^2 = OA^2 + OB^2 - 2OA.OB \cos\theta \qquad \text{(see Fig. 7.12).}$$

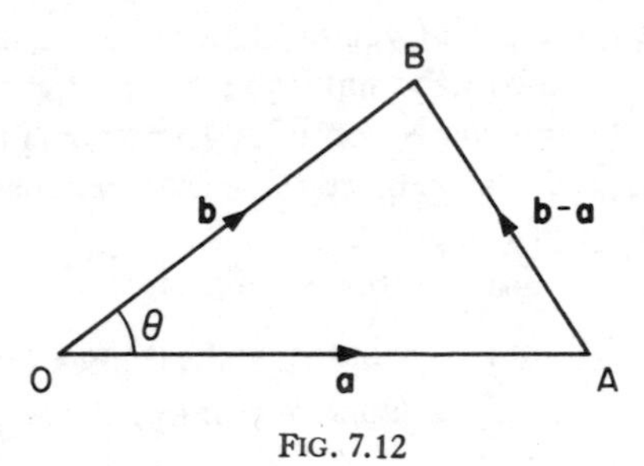

FIG. 7.12

In terms of vectors this becomes

$$|\mathbf{b} - \mathbf{a}|^2 = |\mathbf{a}|^2 + |\mathbf{b}|^2 - 2|\mathbf{a}||\mathbf{b}| \cos\theta.$$

Now for any vector $\mathbf{v}$ we have $|\mathbf{v}|^2 = \mathbf{v}^2$ (Section 7.2.4). Thus the cosine rule can be written

$$(\mathbf{b} - \mathbf{a})^2 = \mathbf{a}^2 + \mathbf{b}^2 - 2\mathbf{a}.\mathbf{b}.$$

This identity follows easily from Section 7.6.

Example 6 Prove that the altitudes of a triangle are concurrent.

Let the triangle be OAB (Fig. 7.13) and write

$$\mathbf{a} = \overrightarrow{OA}, \quad \mathbf{b} = \overrightarrow{OB}.$$

Let altitudes from O and B meet at H and write $\mathbf{h} = \overrightarrow{OH}$. We require to prove that the altitude from A passes through H. In terms of vectors, the

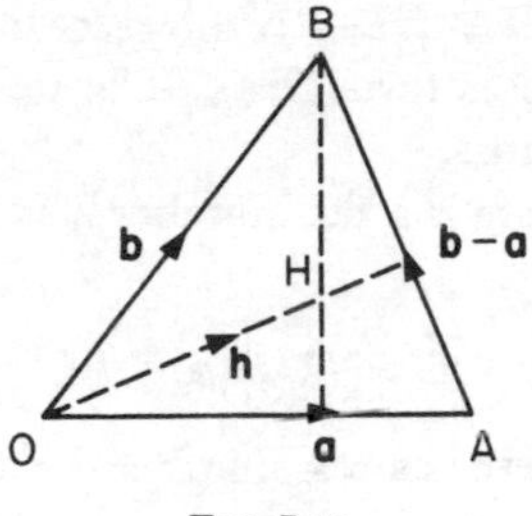

FIG. 7.13

available information is that $\overrightarrow{OH}$ is perpendicular to $\overrightarrow{AB}$ and $\overrightarrow{BH}$ is perpendicular to $\overrightarrow{OA}$, i.e.

$$\mathbf{h}.(\mathbf{b} - \mathbf{a}) = 0, \tag{a}$$

$$\mathbf{a}.(\mathbf{h} - \mathbf{b}) = 0. \tag{b}$$

We require to prove that $\overrightarrow{AH}$ is perpendicular to $\overrightarrow{OB}$, i.e.

$$\mathbf{b}.(\mathbf{h} - \mathbf{a}) = 0. \tag{c}$$

It is a straightforward matter to obtain (c) by addition of (a) and (b).

Example 7 Prove that the medians of a triangle are concurrent and that each median is divided in the ratio 2 : 1 by the point of intersection.

As before, let the triangle be OAB with $\mathbf{a} = \overrightarrow{OA}$, $\mathbf{b} = \overrightarrow{OB}$. Let L, M, N be the mid-points of the three sides (Fig. 7.14). Then

$$\overrightarrow{ON} = \tfrac{1}{2}\mathbf{a}, \quad \overrightarrow{OM} = \tfrac{1}{2}\mathbf{b}, \quad \overrightarrow{AL} = \tfrac{1}{2}(\mathbf{b} - \mathbf{a}),$$

$$\overrightarrow{OL} = \mathbf{a} + \tfrac{1}{2}(\mathbf{b} - \mathbf{a}) = \tfrac{1}{2}(\mathbf{a} + \mathbf{b}),$$

$$\overrightarrow{BN} = \tfrac{1}{2}\mathbf{a} - \mathbf{b}.$$

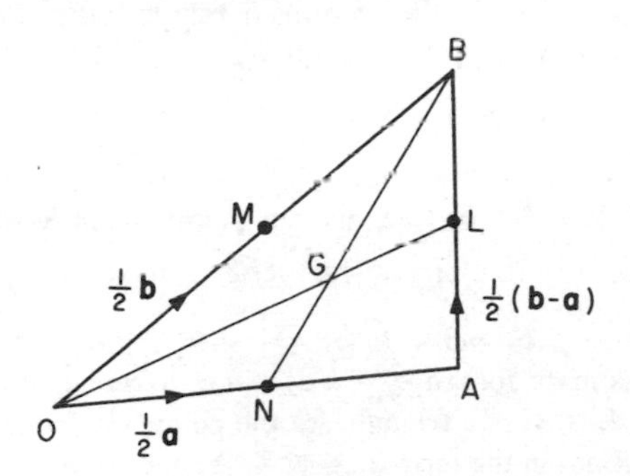

FIG. 7.14

Suppose that the medians OL, BN intersect at G; we require to prove that the median AM passes through G. It is sufficient to prove that $\overrightarrow{AG}$ and $\overrightarrow{AM}$ are parallel vectors.

Since G lies on OL there is a real number λ such that $\overrightarrow{OG} = \lambda\overrightarrow{OL}$; that is

$$\overrightarrow{OG} = (\lambda/2)(\mathbf{a} + \mathbf{b}).$$

Since G lies on BN there is a real number μ such that $\overrightarrow{BG} = \mu\overrightarrow{BN}$; that is

$$\overrightarrow{BG} = \mu(\tfrac{1}{2}\mathbf{a} - \mathbf{b}),$$

and so

$$\overrightarrow{OG} = \overrightarrow{OB} + \overrightarrow{BG} = \mathbf{b} + \mu(\tfrac{1}{2}\mathbf{a} - \mathbf{b}) = (\mu/2)\mathbf{a} + (1 - \mu)\mathbf{b}.$$

Equating the two expressions for $\overrightarrow{OG}$ we get

$$(\lambda/2)(\mathbf{a} + \mathbf{b}) = (\mu/2)\mathbf{a} + (1 - \mu)\mathbf{b}.$$

Hence

$$(\lambda/2 - \mu/2)\mathbf{a} = (1 - \mu - \lambda/2)\mathbf{b},$$

and since $\mathbf{a}$, $\mathbf{b}$ are not parallel vectors this requires

$$\lambda/2 - \mu/2 = 1 - \mu - \lambda/2 = 0.$$

Solving we get $\lambda = \mu = \tfrac{2}{3}$.

Thus

$$\overrightarrow{OG} = \tfrac{1}{3}(\mathbf{a} + \mathbf{b}) \text{ and so } \overrightarrow{AG} = \tfrac{1}{3}(\mathbf{a} + \mathbf{b}) - \mathbf{a} = \tfrac{1}{3}(\mathbf{b} - 2\mathbf{a}).$$

Finally

$$\overrightarrow{AM} = \tfrac{1}{2}\mathbf{b} - \mathbf{a} = \tfrac{1}{2}(\mathbf{b} - 2\mathbf{a}),$$

showing that $\overrightarrow{AG} = \tfrac{2}{3}\overrightarrow{AM}$. The required result follows since the above method applies equally to all three medians.

Exercises 7(a)

1. Given that A, B, C, D are four distinct points in space, show that
$$\overrightarrow{AD} = \overrightarrow{AB} + \overrightarrow{BC} + \overrightarrow{CD}.$$
2. Given $\mathbf{e}_1 = \mathbf{i} + \mathbf{j}$, $\mathbf{e}_2 = \mathbf{i} + \mathbf{k}$, $\mathbf{e}_3 = \mathbf{j} + \mathbf{k}$, express (a) $5\mathbf{i} + 2\mathbf{j} + \mathbf{k}$, and (b) $-\mathbf{i} + 4\mathbf{j} - \mathbf{k}$ in the form $k_1\mathbf{e}_1 + k_2\mathbf{e}_2 + k_3\mathbf{e}_3$, where k_1, k_2, k_3 are scalars.
3. The vertices A, B, C of a triangle are the points $(1, 2, -1)$, $(3, 4, 1)$, $(3, 5, 3)$ respectively. Find, in the form $x\mathbf{i} + y\mathbf{j} + z\mathbf{k}$, unit vectors in the directions of $\overrightarrow{AB}$, $\overrightarrow{BC}$, $\overrightarrow{CA}$. Verify that $\overrightarrow{AB} + \overrightarrow{BC} + \overrightarrow{CA} = \mathbf{0}$.

Evaluate the scalar products $\overrightarrow{AB}.\overrightarrow{BC}$, $\overrightarrow{BC}.\overrightarrow{CA}$, $\overrightarrow{CA}.\overrightarrow{AB}$.

Verify that $\overrightarrow{CA}.\overrightarrow{AB} + \overrightarrow{BC}.\overrightarrow{CA} + CA^2 = 0$, where $CA^2 = \overrightarrow{CA}.\overrightarrow{CA}$ is the square of the length of CA.

4. Given $\mathbf{u} = 3\mathbf{i} + \mathbf{j} + \mathbf{k}$ and $\mathbf{v} = 2\mathbf{i} - 2\mathbf{j} - 3\mathbf{k}$, find (a) $\mathbf{u} + 2\mathbf{v}$ and $\mathbf{u}.\mathbf{v}$, (b) the value of p for which $\mathbf{u} + p\mathbf{v}$ is orthogonal to $\mathbf{u}$.
5. Given $\mathbf{u} = 2\mathbf{i} - 3\mathbf{j} + \mathbf{k}$, $\mathbf{v} = 5\mathbf{i} + \mathbf{j} + 3\mathbf{k}$, find (a) $\mathbf{u} + 3\mathbf{v}$ and $\mathbf{u}.\mathbf{v}$, (b) the value of λ such that $\mathbf{u} + \lambda\mathbf{v}$ is orthogonal to $\mathbf{u}$.

   Find a formula for λ in (b) applicable if $\mathbf{u}$, $\mathbf{v}$ are general vectors.
6. Given $\mathbf{a} = 2\mathbf{i} + 3\mathbf{j} - 5\mathbf{k}$, $\mathbf{b} = 5\mathbf{i} + \mathbf{j} - 4\mathbf{k}$, find $\mathbf{a} + 2\mathbf{b}$ and $\mathbf{a} - 2\mathbf{b}$. Hence find $(\mathbf{a} + 2\mathbf{b}).(\mathbf{a} - 2\mathbf{b})$. Find an independent check of the value of the scalar product.
7. Given that $\mathbf{u}$, $\mathbf{v}$, $\mathbf{w}$ are the vectors

$$\mathbf{u} = 2\mathbf{i} - \mathbf{j} + \mathbf{k}, \quad \mathbf{v} = \mathbf{i} + \mathbf{j} + 2\mathbf{k}, \quad \mathbf{w} = 2\mathbf{i} - \mathbf{j} + 4\mathbf{k},$$

   determine scalars g and h such that $\mathbf{w} - g\mathbf{u} - h\mathbf{v}$ is perpendicular to both $\mathbf{u}$ and $\mathbf{v}$.
8. The sides AB, BC, CD of a quadrilateral represent the vectors $2\mathbf{i} - 3\mathbf{j} + \mathbf{k}$, $\mathbf{i} + \mathbf{j} + 2\mathbf{k}$, $\mathbf{i} + 2\mathbf{j} - 3\mathbf{k}$ respectively. Find the vectors represented by the diagonals of this quadrilateral and determine the angle between these diagonals.
9. Given $\mathbf{u} = \mathbf{i} + 2\mathbf{j} + \mathbf{k}$, $\mathbf{v} = \mathbf{i} - \mathbf{j} + \mathbf{k}$, find (a) the angle between $\mathbf{u}$ and $\mathbf{v}$, (b) the relation between a, b (scalars) if $\mathbf{r}$ is orthogonal to the vector $-\mathbf{i} + 3\mathbf{j} + \mathbf{k}$ where $\mathbf{r} = a\mathbf{u} + b\mathbf{v}$, and the values of a, b if $\mathbf{r}$ is a unit vector.
10. In the triangle OAB, C is the mid-point of AB and D is the mid-point of OB. Show that $\overrightarrow{DC} = \frac{1}{2}\overrightarrow{OA}$.
11. In the trapezium $OABC$ the sides OA and CB are parallel. Given that X is the mid-point of AB and Y is the mid-point of OC, show that

$$\overrightarrow{YX} = \tfrac{1}{2}(\overrightarrow{OA} + \overrightarrow{CB}).$$

12. Given that C, C' are the mid-points of any two straight lines AB, $A'B'$ show that

$$\overrightarrow{AA'} + \overrightarrow{BB'} = 2\overrightarrow{CC'}.$$

13. The points D, E, F respectively are the mid-points of the sides BC, CA, AB of the triangle ABC. Prove that

$$\overrightarrow{AD} + \overrightarrow{BE} + \overrightarrow{CF} = \mathbf{0}.$$

14. Illustrate geometrically the formula $\mathbf{a} = \frac{1}{2}(\mathbf{a} + \mathbf{b}) + \frac{1}{2}(\mathbf{a} - \mathbf{b})$.
15. Given $\mathbf{c} = b\mathbf{a} + a\mathbf{b}$, where $a = |\mathbf{a}|$, $b = |\mathbf{b}|$ show that $\mathbf{c}$ bisects the angle between $\mathbf{a}$ and $\mathbf{b}$.
16. The vertices A, B, C, D of a tetrahedron have position vectors $\mathbf{a}$, $\mathbf{b}$, $\mathbf{c}$, $\mathbf{d}$ respectively. Express the vectors $\overrightarrow{AB}$, $\overrightarrow{CD}$ in terms of $\mathbf{a}$, $\mathbf{b}$, $\mathbf{c}$, $\mathbf{d}$ and show that $\overrightarrow{AB}$ is perpendicular to $\overrightarrow{CD}$ if and only if

$$\mathbf{a}.\mathbf{c} + \mathbf{b}.\mathbf{d} = \mathbf{b}.\mathbf{c} + \mathbf{a}.\mathbf{d}.$$

Show that if AB is perpendicular to CD and AC perpendicular to BD, then AD is perpendicular to BC.

17. The points A, B, C, D have position vectors $\mathbf{a}, \mathbf{b}, \mathbf{c}, \mathbf{d}$ respectively. Given $\mathbf{a} + \mathbf{c} = \mathbf{b} + \mathbf{d}$, show that $ABCD$ is a parallelogram.

    Show also that (a) if the diagonals of the parallelogram are perpendicular then $\mathbf{a}.\mathbf{b} + \mathbf{c}.\mathbf{d} = \mathbf{a}.\mathbf{d} + \mathbf{b}.\mathbf{c}$, (b) if E is the mid-point of BC, then AE passes through a point of trisection of BD and this point is also a point of trisection of AE.

18. Show that if C is the mid-point of the segment AB and O is any other point then

$$2\overrightarrow{OC} = \overrightarrow{OA} + \overrightarrow{OB}.$$

$ABCD$ is a tetrahedron. P, Q are the mid-points of the edges AB, CD and R is the mid-point of PQ. Show that

$$\overrightarrow{OA} + \overrightarrow{OB} + \overrightarrow{OC} + \overrightarrow{OD} = 4\overrightarrow{OR}.$$

Deduce that in a tetrahedron the lines joining mid-points of opposite edges are concurrent. How many such lines are there?

19. Show that if $\mathbf{a}, \mathbf{b}, \mathbf{c}$ are position vectors of three collinear points then there exists scalars α, β, γ, such that

$$\alpha\mathbf{a} + \beta\mathbf{b} + \gamma\mathbf{c} = \mathbf{0}, \quad \alpha + \beta + \gamma = 0.$$

7.7 The Vector Product

Let OA, OB, OC be drawn from a point O and write $\mathbf{a} = \overrightarrow{OA}$, $\mathbf{b} = \overrightarrow{OB}$, $\mathbf{c} = \overrightarrow{OC}$. Suppose that $\mathbf{c}$ is perpendicular to both $\mathbf{a}$ and $\mathbf{b}$; then so also is $-\mathbf{c}$, as is $k\mathbf{c}$ for any real number k.

We say that the lines OA, OB, OC are a ***right-handed set***, or the vectors $\mathbf{a}, \mathbf{b}, \mathbf{c}$ are a right-handed set, if the direction of rotation from A to B is clockwise as seen by an observer looking in the direction of $\overrightarrow{OC}$ (see Fig. 7.15). If $\mathbf{a}, \mathbf{b}, \mathbf{c}$ form a right-handed set, then $\mathbf{a}, \mathbf{b}, -\mathbf{c}$ form a ***left-handed set***.

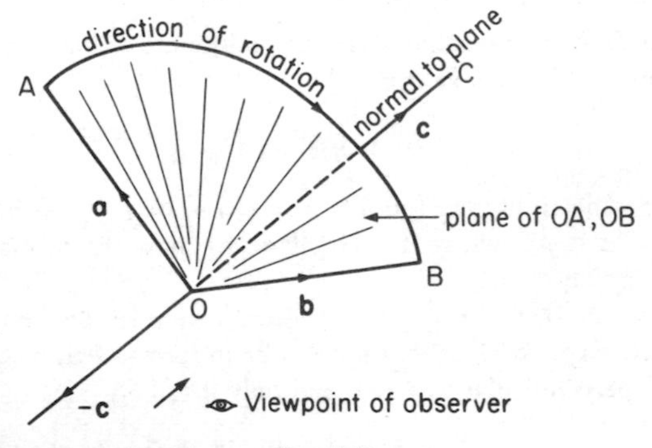

FIG. 7.15

It is conventional to choose the basis vectors $\mathbf{i}, \mathbf{j}, \mathbf{k}$ (see Section 7.4) so that they are a right-handed set; note that in this case $\mathbf{j}, \mathbf{k}, \mathbf{i}$ and $\mathbf{k}, \mathbf{i}, \mathbf{j}$ are also right-handed sets but that, for example, $\mathbf{i}, \mathbf{k}, \mathbf{j}$ (in which the cyclic order has been changed) is a left-handed set.

Corresponding to two non-parallel vectors $\mathbf{a}, \mathbf{b}$ there is a unique vector $\mathbf{c}$ such that

(a) $\mathbf{c}$ is perpendicular to both $\mathbf{a}$ and $\mathbf{b}$,
(b) $\mathbf{a}, \mathbf{b}, \mathbf{c}$ is a right-handed set,
(c) $|\mathbf{c}| = |\mathbf{a}||\mathbf{b}| \sin\theta$, where θ is the angle between $\mathbf{a}$ and $\mathbf{b}$.

This vector $\mathbf{c}$ is called the *vector product*, or *cross-product*, of $\mathbf{a}$ and $\mathbf{b}$, written $\mathbf{c} = \mathbf{a} \wedge \mathbf{b}$ (sometimes $\mathbf{a} \times \mathbf{b}$, or $[\mathbf{a}\ ;\ \mathbf{b}]$).

If $\mathbf{a}$ and $\mathbf{b}$ are parallel the above definition does not define a direction for $\mathbf{a} \wedge \mathbf{b}$ but (c) gives the value zero for $|\mathbf{a} \wedge \mathbf{b}|$; consequently we define $\mathbf{a} \wedge \mathbf{b} = \mathbf{0}$ in this case. In particular, $\mathbf{a} \wedge \mathbf{a} = \mathbf{0}$ for any vector $\mathbf{a}$, and so

$$\mathbf{i} \wedge \mathbf{i} = \mathbf{j} \wedge \mathbf{j} = \mathbf{k} \wedge \mathbf{k} = \mathbf{0}. \tag{7.15}$$

If $\mathbf{a}$ and $\mathbf{b}$ are perpendicular, then $\theta = \pi/2$, $\sin\theta = 1$ and so

$$|\mathbf{a} \wedge \mathbf{b}| = |\mathbf{a}|.|\mathbf{b}|;$$

in particular, it follows that

$$\mathbf{i} \wedge \mathbf{j} = \mathbf{k}, \quad \mathbf{j} \wedge \mathbf{k} = \mathbf{i}, \quad \mathbf{k} \wedge \mathbf{i} = \mathbf{j}. \tag{7.16}$$

Note that, unlike the scalar product $\mathbf{a}.\mathbf{b}$ (which is a real number), the vector product $\mathbf{a} \wedge \mathbf{b}$ is a *vector*, i.e. V is closed with respect to the operation $\wedge$.

The operation $\wedge$ is non-commutative, for

$$\mathbf{b} \wedge \mathbf{a} = -\mathbf{a} \wedge \mathbf{b} \tag{7.17}$$

since the definition requires that $\mathbf{b}, \mathbf{a}, \mathbf{b} \wedge \mathbf{a}$ must be a right-handed set. Important examples of (7.17), following from (7.16), are the results

$$\mathbf{j} \wedge \mathbf{i} = -\mathbf{k}, \quad \mathbf{k} \wedge \mathbf{j} = -\mathbf{i}, \quad \mathbf{i} \wedge \mathbf{k} = -\mathbf{j}. \tag{7.18}$$

Distributive law. From the definition of vector product we can verify by elementary geometry that

$$\mathbf{a} \wedge (\mathbf{b} + \mathbf{c}) = \mathbf{a} \wedge \mathbf{b} + \mathbf{a} \wedge \mathbf{c}.$$

Likewise

$$(\mathbf{a} + \mathbf{b}) \wedge \mathbf{c} = -\mathbf{c} \wedge (\mathbf{a} + \mathbf{b}) = -\mathbf{c} \wedge \mathbf{a} - \mathbf{c} \wedge \mathbf{b} = \mathbf{a} \wedge \mathbf{c} + \mathbf{b} \wedge \mathbf{c}.$$

Thus vector multiplication of brackets follows the usual pattern, e.g.

$$(\mathbf{a} + \mathbf{b}) \wedge (\mathbf{c} + \mathbf{d}) = \mathbf{a} \wedge \mathbf{c} + \mathbf{a} \wedge \mathbf{d} + \mathbf{b} \wedge \mathbf{c} + \mathbf{b} \wedge \mathbf{d},$$

except that each term on the r.h.s. must have the vectors in the correct order since this multiplication is non-commutative.

7.7.1 Vector Product in Component Form We may also verify geometrically that

$$k(\mathbf{a} \wedge \mathbf{b}) = (k\mathbf{a}) \wedge \mathbf{b} = \mathbf{a} \wedge (k\mathbf{b})$$

where k is a scalar. It now follows that if

$$\mathbf{a} = x_1\mathbf{i} + y_1\mathbf{j} + z_1\mathbf{k}, \quad \mathbf{b} = x_2\mathbf{i} + y_2\mathbf{j} + z_2\mathbf{k},$$

then

$$\begin{aligned}\mathbf{a} \wedge \mathbf{b} &= (x_1\mathbf{i} + y_1\mathbf{j} + z_1\mathbf{k}) \wedge (x_2\mathbf{i} + y_2\mathbf{j} + z_2\mathbf{k}) \\ &= x_1x_2\mathbf{i} \wedge \mathbf{i} + x_1y_2\mathbf{i} \wedge \mathbf{j} + x_1z_2\mathbf{i} \wedge \mathbf{k} + y_1x_2\mathbf{j} \wedge \mathbf{i} + y_1y_2\mathbf{j} \wedge \mathbf{j} \\ &\quad + y_1z_2\mathbf{j} \wedge \mathbf{k} + z_1x_2\mathbf{k} \wedge \mathbf{i} + z_1y_2\mathbf{k} \wedge \mathbf{j} + z_1z_2\mathbf{k} \wedge \mathbf{k}.\end{aligned}$$

Using (7.15), (7.16) and (7.18) we get

$$\mathbf{a} \wedge \mathbf{b} = (y_1z_2 - y_2z_1)\mathbf{i} + (z_1x_2 - z_2x_1)\mathbf{j} + (x_1y_2 - x_2y_1)\mathbf{k}. \quad (7.19)$$

This formula need not be remembered in detail since in practice we can find $\mathbf{a} \wedge \mathbf{b}$ by the following procedure:

(a) Set out the components of **a** and **b** in column form as shown below, where the first column has been repeated at the end because of the importance of the cyclic order of components throughout the procedure.

$$\begin{array}{cccc} \mathbf{i} & \mathbf{j} & \mathbf{k} & \mathbf{i} \\ x_1 & y_1 & z_1 & x_1 \\ x_2 & y_2 & z_2 & x_2. \end{array}$$

(b) From any two consecutive columns we get a component of $\mathbf{a} \wedge \mathbf{b}$ by cross-multiplication and subtraction, as indicated below.

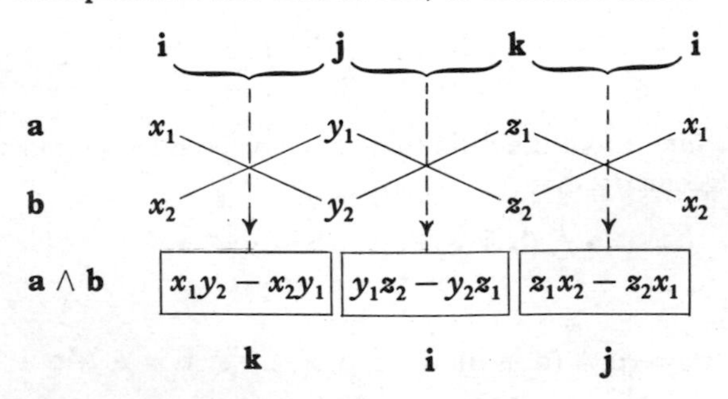

Note that each component of $\mathbf{a} \wedge \mathbf{b}$ comes from a cross-multiplication involving the *following* two columns. (See page 220 for the method of writing (7.19) in terms of a determinant.)

Example 8 Find $\mathbf{a} \wedge \mathbf{b}$ where $\mathbf{a} = 2\mathbf{i} - \mathbf{j} + 3\mathbf{k}$, $\mathbf{b} = \mathbf{i} + 2\mathbf{j} - 4\mathbf{k}$.

It is convenient to omit **i**, **j**, **k** and write the numbers only, viz.

$$\begin{array}{cccc} 2, & -1, & 3, & 2 \\ 1, & 2, & -4, & 1. \end{array}$$

Following the procedure outlined in (b) above, we have

$$\begin{aligned} (-1)(-4) - 3.2 &= 4 - 6 = -2, \\ 3.1 - 2(-4) &= 3 + 8 = 11, \\ 2.2 - (-1).1 &= 4 + 1 = 5, \end{aligned}$$

and so $\mathbf{a} \wedge \mathbf{b} = -2\mathbf{i} + 11\mathbf{j} + 5\mathbf{k}$.

7.7.2 Application of Vector Product to Area and Volume From Fig. 7.16 we see that the length of the perpendicular from B onto OA is h, where

$$h = OB \sin\theta = |\mathbf{b}| \sin\theta.$$

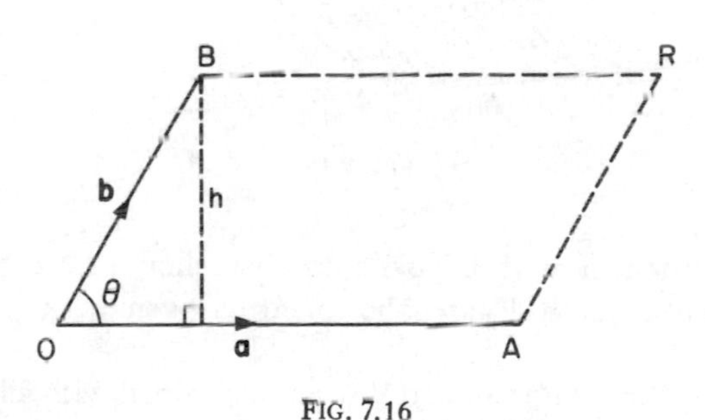

FIG. 7.16

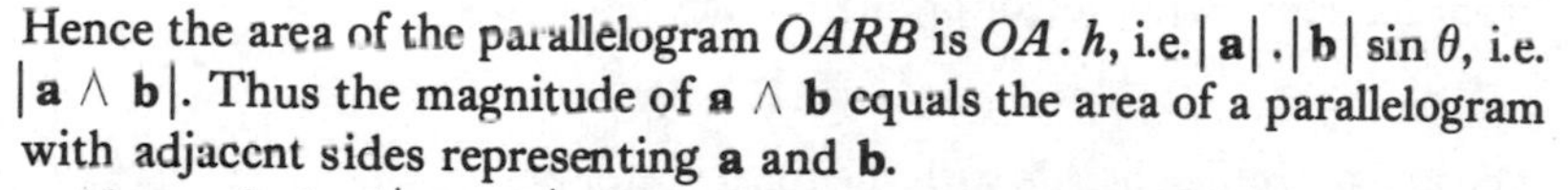

Hence the area of the parallelogram $OARB$ is $OA \,.\, h$, i.e. $|\mathbf{a}| \,.\, |\mathbf{b}| \sin\theta$, i.e. $|\mathbf{a} \wedge \mathbf{b}|$. Thus the magnitude of $\mathbf{a} \wedge \mathbf{b}$ equals the area of a parallelogram with adjacent sides representing **a** and **b**.

Alternatively, $\frac{1}{2}|\mathbf{a} \wedge \mathbf{b}|$ equals the area of the triangle OAB.

Example 9 Find the area of the triangle with adjacent sides representing the vectors $2\mathbf{i} - \mathbf{j} + 3\mathbf{k}$, $\mathbf{i} + 2\mathbf{j} - 4\mathbf{k}$.

From Example 8 the vector product of these two vectors is $-2\mathbf{i} + 11\mathbf{j} + 5\mathbf{k}$.

The magnitude of this product is

$$\sqrt{[(-2)^2 + 11^2 + 5^2]} = \sqrt{150} = 5\sqrt{6}.$$

Hence the area of the triangle is $(5\sqrt{6})/2$.

Volume of a parallelepiped. Let adjacent edges of a parallelepiped represent vectors **a**, **b**, **c** as shown in Fig. 7.17. The volume of the parallelepiped is given by

area of base × perpendicular height
= area of parallelogram $OARB \times h$,
$= |\mathbf{a} \wedge \mathbf{b}| \times$ projection of **c** on direction perpendicular to base,
$= |\mathbf{a} \wedge \mathbf{b}| \times$ projection of **c** on the direction of $\mathbf{a} \wedge \mathbf{b}$,
$= (\mathbf{a} \wedge \mathbf{b}).\mathbf{c}$.

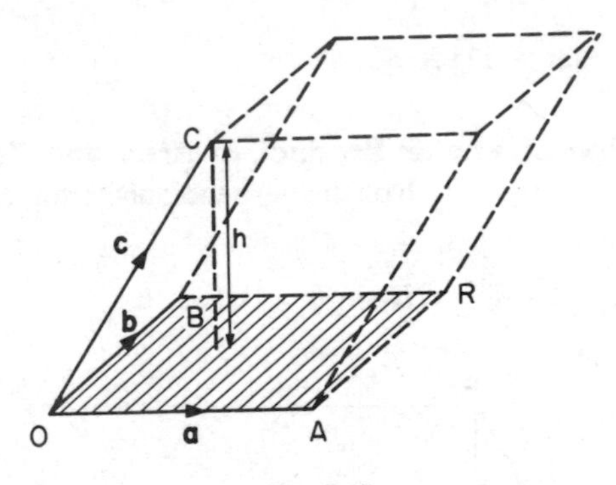

FIG. 7.17

If the direction of **c** is opposite to that shown in Fig. 7.17, then $(\mathbf{a} \wedge \mathbf{b}).\mathbf{c}$ is negative. In all cases the volume is given by $|(\mathbf{a} \wedge \mathbf{b}).\mathbf{c}|$.

Example 10 Find the volume of the parallelepiped with adjacent edges representing the vectors $2\mathbf{i} - \mathbf{j} + 3\mathbf{k}$, $\mathbf{i} + 2\mathbf{j} - 4\mathbf{k}$, $2\mathbf{i} - 2\mathbf{j} + \mathbf{k}$.

From Example 8, the vector product of the first two vectors is

$$-2\mathbf{i} + 11\mathbf{j} + 5\mathbf{k}.$$

The scalar product of this with the third vector is

$$(-2).2 + 11.(-2) + 5.1,$$

i.e. $-4 - 22 + 5$, i.e. -21. Hence the volume is 21.

The triple product $(\mathbf{a} \wedge \mathbf{b}).\mathbf{c}$ is called a *scalar triple product* since it represents a purely scalar quantity. The brackets may be omitted since the other interpretation $\mathbf{a} \wedge (\mathbf{b}.\mathbf{c})$ would imply the meaningless operation of vector-multiplication with the vector $\mathbf{a}$ and the scalar quantity $\mathbf{b}.\mathbf{c}$.

The connection between the scalar triple product and the volume of a parallelepiped shows easily that

$$\mathbf{a}.\mathbf{b} \wedge \mathbf{c} = \mathbf{b}.\mathbf{c} \wedge \mathbf{a} = \mathbf{c}.\mathbf{a} \wedge \mathbf{b} = \mathbf{a} \wedge \mathbf{b}.\mathbf{c} = \mathbf{b} \wedge \mathbf{c}.\mathbf{a} = \mathbf{c} \wedge \mathbf{a}.\mathbf{b}. \tag{7.20}$$

Thus the positions of the dot and $\wedge$ are unimportant, and the positions of the vectors may be changed so long as the cyclic order is unchanged; a change in cyclic order affects the sign so that, for example,

$$\mathbf{a}.\mathbf{c} \wedge \mathbf{b} = -\mathbf{a}.\mathbf{b} \wedge \mathbf{c}.$$

7.7.3 Condition that Three Vectors are Coplanar If $\mathbf{a}$, $\mathbf{b}$, $\mathbf{c}$ are coplanar, the parallelepiped with adjacent edges representing $\mathbf{a}$, $\mathbf{b}$, $\mathbf{c}$ has zero volume. Hence the condition that $\mathbf{a}$, $\mathbf{b}$, $\mathbf{c}$ should be coplanar is that $\mathbf{a} \wedge \mathbf{b}.\mathbf{c} = 0$.

Exercises 7(b)

1. A tetrahedron has vertices $ABCD$. Given $\mathbf{b} = \overrightarrow{AB}$, $\mathbf{c} = \overrightarrow{AC}$, $\mathbf{d} = \overrightarrow{AD}$, express the following in terms of $\mathbf{b}$, $\mathbf{c}$, $\mathbf{d}$: (a) $\overrightarrow{DB}$, $\overrightarrow{DC}$, (b) $\overrightarrow{BF}$, where F is the mid-point of DC, (c) the area of the triangle ACD.
2. Points A, B, C have position vectors

$$2\mathbf{i} + 2\mathbf{j} + 2\mathbf{k}, \quad \mathbf{i} - 2\mathbf{j} + \mathbf{k}, \quad 3\mathbf{i} - 3\mathbf{k},$$

respectively, where $\mathbf{i}$, $\mathbf{j}$, $\mathbf{k}$ are unit vectors along the axes. Find the angles BOC, COA, AOB. Find also the area of the triangle BOC and the volume of the tetrahedron $OABC$.
3. The position vectors, with respect to an origin O, of points A, B, C are $2\mathbf{i} + 2\mathbf{j} - \mathbf{k}$, $\mathbf{i} - \mathbf{j} + 3\mathbf{k}$, $-2\mathbf{i} + 2\mathbf{j} - 2\mathbf{k}$ respectively, where $\mathbf{i}, \mathbf{j}, \mathbf{k}$ are unit vectors in the directions of the axes Ox, Oy, Oz. Show that AB is perpendicular to AC and find a unit vector normal to the plane of the triangle ABC.
4. $\overrightarrow{OA}$, $\overrightarrow{OB}$, $\overrightarrow{OC}$ are, respectively, the vectors

$$2\mathbf{i} - 2\mathbf{j} + \mathbf{k}, \quad 4\mathbf{i} - \mathbf{j} - \mathbf{k}, \quad 3\mathbf{i} + 3\mathbf{k}.$$

Prove that in the tetrahedron $OABC$: (a) OA is perpendicular to AB and to AC, (b) the edge OB makes an angle of $60°$ with OC, (c) the area of the triangle ABC is $4\frac{1}{2}$ square units.
5. Two adjacent edges AB, AD of a parallelogram $ABCD$ represent respectively the vectors $\mathbf{a}$, $\mathbf{b}$. The mid-points of BC, CD are E, F respectively. Express the vectors $\overrightarrow{AE}$, $\overrightarrow{ED}$, $\overrightarrow{FE}$ in terms of $\mathbf{a}$ and $\mathbf{b}$.

   Given $\mathbf{a} = 2\mathbf{i} + 3\mathbf{j} + 4\mathbf{k}$, $\mathbf{b} = \mathbf{i} - 2\mathbf{j} + \mathbf{k}$, where $\mathbf{i}, \mathbf{j}, \mathbf{k}$ are unit vectors along the co-ordinate axes, find a unit vector normal to the plane of the parallelogram.

6. Two adjacent edges AB, AD of the rectangular base $ABCD$ of a prism represent the vectors $\mathbf{i} + 2\mathbf{j} + 4\mathbf{k}$, $2\mathbf{i} + \mathbf{j} - \mathbf{k}$. The upper face $EFGH$ of the prism is so placed that all the edges AE, BF, CG, DH are parallel and represent the vector $2\mathbf{i} + \mathbf{j}$. Find the vectors represented by the diagonal AC of the base and the diagonal AG of the prism and determine the angle between them. Evaluate the volume of the prism.
7. For any two vectors in V_2, $\mathbf{x} = x_1\mathbf{i} + x_2\mathbf{j}$, $\mathbf{y} = y_1\mathbf{i} + y_2\mathbf{j}$, let an operation $*$ be defined by the equation

$$\mathbf{x} * \mathbf{y} = x_1y_1\mathbf{i} + x_2y_2\mathbf{j}.$$

Show that the operation is commutative and associative, and that it is distributive over vector addition. Which vector is the identity element for the operation? Which vectors do not have inverses? Is it true that if $\mathbf{x} * \mathbf{y} = \mathbf{0}$ then either $\mathbf{x} = \mathbf{0}$ or $\mathbf{y} = \mathbf{0}$?
8. A *triple vector product* is a product of the form $\mathbf{a} \wedge (\mathbf{b} \wedge \mathbf{c})$ or $(\mathbf{a} \wedge \mathbf{b}) \wedge \mathbf{c}$. Use (7.19) to show that

$$\mathbf{a} \wedge (\mathbf{b} \wedge \mathbf{c}) = (\mathbf{a}.\mathbf{c})\mathbf{b} - (\mathbf{a}.\mathbf{b})\mathbf{c}$$

and verify that

$$\mathbf{a} \wedge (\mathbf{b} \wedge \mathbf{c}) \neq (\mathbf{a} \wedge \mathbf{b}) \wedge \mathbf{c}.$$

9. Use the result of Exercise 8 to show that

$$\mathbf{a} \wedge (\mathbf{b} \wedge \mathbf{c}) + \mathbf{b} \wedge (\mathbf{c} \wedge \mathbf{a}) + \mathbf{c} \wedge (\mathbf{a} \wedge \mathbf{b}) = \mathbf{0}.$$

CHAPTER 8

CO-ORDINATE GEOMETRY OF THREE DIMENSIONS

8.1 Co-ordinates of a Point in Space

In the rectangular cartesian system of co-ordinates, the position of a point is fixed by its perpendicular distances from three mutually perpendicular planes. Three such planes intersecting in three mutually perpendicular lines $x'Ox$, $y'Oy$, $z'Oz$ are shown in Fig. 8.1. Their point of intersection O is the *origin*, the lines $x'Ox$, $y'Oy$, $z'Oz$ (labelled so that Ox,

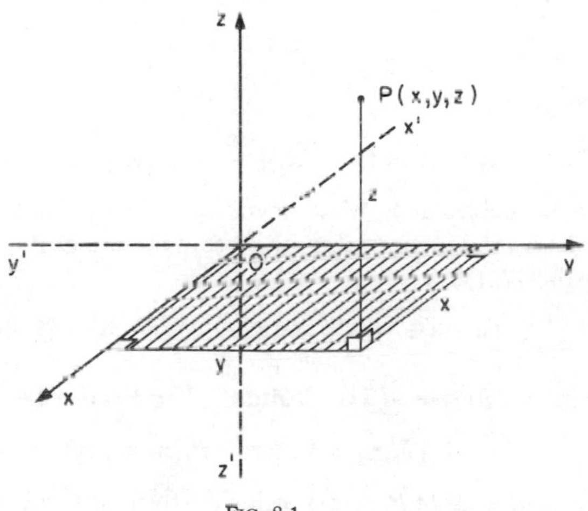

FIG. 8.1

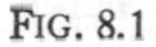

Oy, Oz are a right-handed set as defined in Section 7.7) are the *co-ordinate axes* and the planes yOz, zOx, xOy (known respectively as the yz-, zx-, xy-planes) are the *co-ordinate planes*. The point $P(x, y, z)$ lies at perpendicular distances x, y, z from the yz-, zx-, xy-planes respectively, x being positive when measured in the direction Ox, negative if measured in the direction Ox'. Similar sign conventions hold for y and z. The co-ordinate planes divide space into eight regions known as *octants*. The octant $Oxyz$, in which x, y, z are all positive, is called the *positive octant*. In the octant $Oxyz'$, x and y are positive and z is negative, and so on.

8.1.1 Vectors and Co-ordinates The vector $\overrightarrow{OP}$ drawn from the origin to the point P is called the ***position vector*** of P. If P has co-ordinates (x, y, z) then (see Section 7.4) we have $\overrightarrow{OP} = x\mathbf{i} + y\mathbf{j} + z\mathbf{k}$ where $\mathbf{i}, \mathbf{j}, \mathbf{k}$ denote unit vectors in the directions Ox, Oy, Oz respectively. It is usual to write $\mathbf{r} = \overrightarrow{OP}$, $r = |\mathbf{r}| = OP$.

Let P_1, P_2 have co-ordinates (x_1, y_1, z_1), (x_2, y_2, z_2) respectively so that

$$\mathbf{r}_1 = \overrightarrow{OP_1} = x_1\mathbf{i} + y_1\mathbf{j} + z_1\mathbf{k}$$

$$\mathbf{r}_2 = \overrightarrow{OP_2} = x_2\mathbf{i} + y_2\mathbf{j} + z_2\mathbf{k}.$$

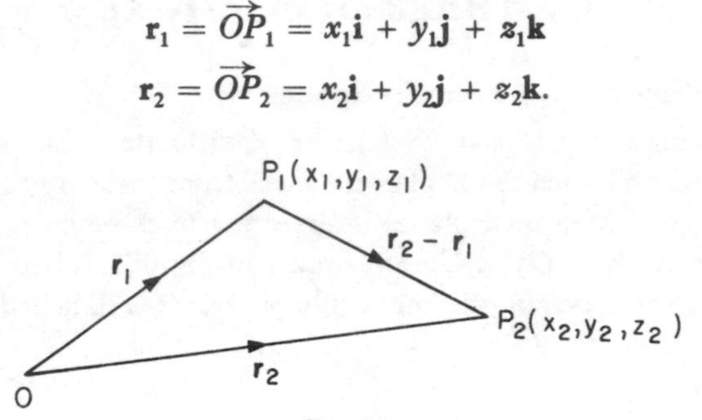

FIG. 8.2

Now

$$\overrightarrow{P_1P_2} = \mathbf{r}_2 - \mathbf{r}_1 = (x_2 - x_1)\mathbf{i} + (y_2 - y_1)\mathbf{j} + (z_2 - z_1)\mathbf{k}. \tag{8.1}$$

This formula enables us to write down the vector joining any two points when we know the co-ordinates of these points, e.g. the vector joining (1, 2, 3) to (4, −7, 11) is

$$(4 - 1)\mathbf{i} + (-7 - 2)\mathbf{j} + (11 - 3)\mathbf{k} = 3\mathbf{i} - 9\mathbf{j} + 8\mathbf{k}.$$

8.1.2 Distance Between Two Points The distance between

$$P_1(x_1, y_1, z_1) \quad \text{and} \quad P_2(x_2, y_2, z_2)$$

equals the magnitude of $\overrightarrow{P_1P_2}$, and using the formula (7.12) we get

$$P_1P_2 = \sqrt{[(x_2 - x_1)^2 + (y_2 - y_1)^2 + (z_2 - z_1)^2]}. \tag{8.2}$$

For example, the distance between the points (1, 2, 3) and (4, −7, 11) is $\sqrt{[3^2 + (-9)^2 + 8^2]} = \sqrt{154}$.

8.1.3 Section Formula Let $P(x, y, z)$ divide the join of $P_1(x_1, y_1, z_1)$ and $P_2(x_2, y_2, z_2)$ in the ratio $\lambda : \mu$ (Fig. 8.3). Then $\overrightarrow{P_1P}$ and $\overrightarrow{PP_2}$ are vectors in the same direction and with magnitudes in the ratio $\lambda : \mu$. Hence

$$\overrightarrow{P_1P} = \frac{\lambda}{\mu}\overrightarrow{PP_2};$$

that is

$$(x - x_1)\mathbf{i} + (y - y_1)\mathbf{j} + (z - z_1)\mathbf{k}$$
$$= \frac{\lambda}{\mu}(x_2 - x)\mathbf{i} + \frac{\lambda}{\mu}(y_2 - y)\mathbf{j} + \frac{\lambda}{\mu}(z_2 - z)\mathbf{k},$$

that is

$$x - x_1 = \frac{\lambda}{\mu}(x_2 - x), \quad y - y_1 = \frac{\lambda}{\mu}(y_2 - y), \quad z - z_1 = \frac{\lambda}{\mu}(z_2 - z).$$

Solving, we get

$$x = \frac{\lambda x_2 + \mu x_1}{\lambda + \mu}, \quad y = \frac{\lambda y_2 + \mu y_1}{\lambda + \mu}, \quad z = \frac{\lambda z_2 + \mu z_1}{\lambda + \mu}. \tag{8.3}$$

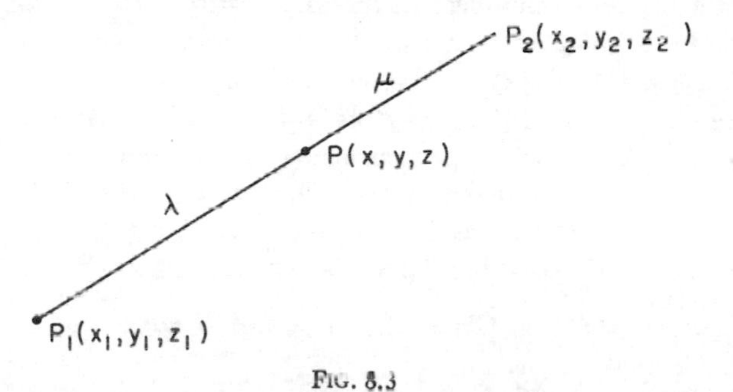

FIG. 8.3

In particular, the co-ordinates of the mid-point of P_1P_2 are $\frac{1}{2}(x_1 + x_2)$, $\frac{1}{2}(y_1 + y_2)$, $\frac{1}{2}(z_1 + z_2)$. The point P divides P_1P_2 internally or externally according as the ratio $\lambda : \mu$ is positive or negative; if the ratio is negative it is immaterial whether we assign a negative value to λ or to μ in (8.3).

Example 1 The line joining $A(1, 8, -1)$ and $B(4, -4, 2)$ meets the xz- and yz-planes at P and Q respectively. Find the co-ordinates of P and Q and the ratios in which they divide AB.

The co-ordinates of the point that divides AB in the ratio $k:1$ are

$$\frac{4k + 1}{k + 1}, \quad \frac{8 - 4k}{k + 1}, \quad \frac{2k - 1}{k + 1}. \tag{1}$$

If this point lies on the plane $y = 0$, its y-co-ordinate is zero and so $k = 2$. Hence $P \equiv (3, 0, 1)$, and since k is positive P divides AB internally in the ratio $2:1$.

If the point given by (1) lies on the plane $x = 0$, $k = -\frac{1}{4}$. Hence

$$Q \equiv (0, 12, -2),$$

and since k is negative Q divides AB externally in the ratio 1 : 4.

8.2 Direction of a Line

The direction of a line can be described by reference to a vector with the required direction. For example, the line through the points $A(1, 2, 3)$ and $B(4, -7, 11)$, taken in the sense from A to B, has the direction of $\overrightarrow{AB}$, i.e. of $3\mathbf{i} - 9\mathbf{j} + 8\mathbf{k}$; equally it has the direction of $6\mathbf{i} - 18\mathbf{j} + 16\mathbf{k}$, or of $3k\mathbf{i} - 9k\mathbf{j} + 8k\mathbf{k}$ for any positive k.

The line in the direction of the vector $\lambda\mathbf{i} + \mu\mathbf{j} + \nu\mathbf{k}$ is said to have *direction ratios* (D.R.'s) $[\lambda:\mu:\nu]$. Thus the line AB in the previous paragraph has D.R.'s $[3:-9:8]$; equally $[6:-18:16]$ are the D.R.'s of AB, as are any set of numbers in the same ratio.

It follows from (8.1) that the line P_1P_2 joining $P_1(x_1, y_1, z_1)$ and $P_2(x_2, y_2, z_2)$ has D.R.'s $[x_2 - x_1 : y_2 - y_1 : z_2 - z_1]$.

Let $\overrightarrow{OP}$ (Fig. 8.4) be the vector $\lambda\mathbf{i} + \mu\mathbf{j} + \nu\mathbf{k}$ so that P is the point (λ, μ, ν). Let PX, PY, PZ be perpendiculars from P onto the x-, y-, z-axes respectively, and let OP make angles α, β, γ respectively with the positive directions of these axes. From the triangle OXP, right-angled at X, we have $\lambda = OX = OP\cos\alpha$. Similarly from the triangles OYP, OZP we get

$$\mu = OY = OP\cos\beta, \quad \nu = OZ = OP\cos\gamma.$$

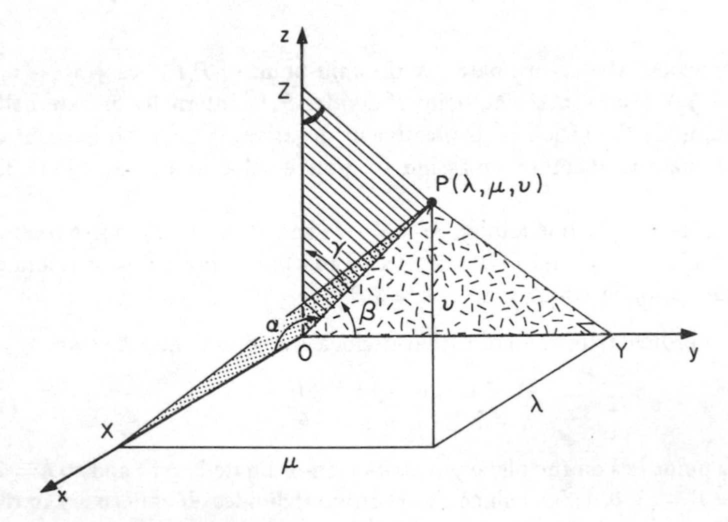

FIG. 8.4

Thus $\lambda : \mu : \nu = \cos\alpha : \cos\beta : \cos\gamma$.

In particular, if $l\mathbf{i} + m\mathbf{j} + n\mathbf{k}$ is the unit vector in the direction of $\lambda\mathbf{i} + \mu\mathbf{j} + \nu\mathbf{k}$ then l, m, n can be found in terms of α, β, γ in the same way as for λ, μ, ν with the added simplification that $OP = 1$. Thus

$$l = \cos\alpha, \quad m = \cos\beta, \quad n = \cos\gamma, \tag{8.4}$$

and from (7.12)

$$l^2 + m^2 + n^2 = 1, \tag{8.5}$$

which combines with (8.4) to give

$$\cos^2\alpha + \cos^2\beta + \cos^2\gamma = 1. \tag{8.6}$$

Summarising, the direction of a line may be specified by giving the special D.R.'s $[l : m : n]$ for which $l^2 + m^2 + n^2 = 1$; these are called the *direction cosines* (D.C.'s) of the line and are written $[l, m, n]$. The D.C.'s are, in fact, the cosines of the angles the line makes with the co-ordinate axes, but are usually found much more easily from their association with the unit vector in the direction of the line.

It has been shown (Section 7.5.2) that the unit vector in the direction of $\lambda\mathbf{i} + \mu\mathbf{j} + \nu\mathbf{k}$ is

$$\frac{1}{\sqrt{(\lambda^2 + \mu^2 + \nu^2)}}(\lambda\mathbf{i} + \mu\mathbf{j} + \nu\mathbf{k}).$$

It follows that if $[\lambda : \mu : \nu]$ are the D.R.'s of the line L, then the D.C.'s of L are

$$\left[\frac{\lambda}{\sqrt{(\lambda^2 + \mu^2 + \nu^2)}}, \frac{\mu}{\sqrt{(\lambda^2 + \mu^2 + \nu^2)}}, \frac{\nu}{\sqrt{(\lambda^2 + \mu^2 + \nu^2)}}\right].$$

The abbreviation

$$\left[\frac{\lambda, \mu, \nu}{\sqrt{(\lambda^2 + \mu^2 + \nu^2)}}\right]$$

is frequently used for these D.C.'s. For example, given that $[3 : -9 : 8]$ are the D.R.'s of a certain line, we may write the D.C.'s of the line as

$$\left[\frac{3, -9, 8}{\sqrt{154}}\right].$$

8.2.1 The Angle Between Two Lines The angle θ between two lines can be found by considering two vectors in the directions of the lines and using the result (7.13) for the angle between two vectors. If $[\lambda_1, \mu_1, \nu_1]$,

$[\lambda_2, \mu_2, \nu_2]$ are the D.R.'s of the two lines, $\lambda_1\mathbf{i} + \mu_1\mathbf{j} + \nu_1\mathbf{k}$, $\lambda_2\mathbf{i} + \mu_2\mathbf{j} + \nu_2\mathbf{k}$ are vectors in the directions of the lines and so

$$\cos\theta = \frac{\lambda_1\lambda_2 + \mu_1\mu_2 + \nu_1\nu_2}{\sqrt{(\lambda_1^2 + \mu_1^2 + \nu_1^2)}\sqrt{(\lambda_2^2 + \mu_2^2 + \nu_2^2)}}. \tag{8.7}$$

In particular, if $[l_1, m_1, n_1]$, $[l_2, m_2, n_2]$ are the D.C.'s of the two lines then since by (8.5) $l_1^2 + m_1^2 + n_1^2 = 1$, $l_2^2 + m_2^2 + n_2^2 = 1$, we have

$$\cos\theta = l_1l_2 + m_1m_2 + n_1n_2. \tag{8.8}$$

The condition that the lines be perpendicular is

$$\lambda_1\lambda_2 + \mu_1\mu_2 + \nu_1\nu_2 = 0. \tag{8.9}$$

Example 2 Show that the points $A(2, 4, 3)$, $B(4, 1, 9)$, $C(10, -1, 6)$ are the vertices of an isosceles right-angled triangle.

The D.R.'s of AB are $[4 - 2 : 1 - 4 : 9 - 3]$, i.e. $[2 : -3 : 6]$. Similarly, the D.R.'s of BC, AC are $[6 : -2 : -3]$, $[8 : -5 : 3]$ respectively.

By (8.9) the lines AB and BC are perpendicular; by (8.7) $\cos(\angle CAB) = 1/\sqrt{2}$ so that $\angle CAB = 45°$. Hence ABC is an isosceles right-angled triangle.

Example 3 If A, B, C, D are four points in space such that AB is perpendicular to CD and AC is perpendicular to BD, prove that AD is perpendicular to BC.

Let A, B, C, D be the points

$$(x_1, y_1, z_1), \quad (x_2, y_2, z_2), \quad (x_3, y_3, z_3), \quad (x_4, y_4, z_4)$$

respectively. The D.R.'s of AB, CD are

$$[x_2 - x_1 : y_2 - y_1 : z_2 - z_1], \quad [x_4 - x_3 : y_4 - y_3 : z_4 - z_3]$$

respectively and by (8.9)

$$(x_2 - x_1)(x_4 - x_3) + (y_2 - y_1)(y_4 - y_3) + (z_2 - z_1)(z_4 - z_3) = 0. \tag{1}$$

Similarly, the condition that AC be perpendicular to BD is that

$$(x_3 - x_1)(x_4 - x_2) + (y_3 - y_1)(y_4 - y_2) + (z_3 - z_1)(z_4 - z_2) = 0. \tag{2}$$

Subtracting corresponding sides of (1) and (2) we get

$$(x_4 - x_1)(x_3 - x_2) + (y_4 - y_1)(y_3 - y_2) + (z_4 - z_1)(z_3 - z_2) = 0,$$

which is the condition that AD and BC be perpendicular.

Note on Projection. Let d be the length of the line joining the points

$$P_1(x_1, y_1, z_1), \quad P_2(x_2, y_2, z_2),$$

that is

$$d = \sqrt{[(x_2 - x_1)^2 + (y_2 - y_1)^2 + (z_2 - z_1)^2]}.$$

Let $[l, m, n]$ be the D.C.'s of a line at angle θ to the line P_1P_2. The D.C.'s of P_1P_2 are

$$\left[\frac{x_2 - x_1}{d}, \frac{y_2 - y_1}{d}, \frac{z_2 - z_1}{d}\right]$$

and so the projection of P_1P_2 on the other line, viz. $d \cos \theta$, is given by

$$l(x_2 - x_1) + m(y_2 - y_1) + n(z_2 - z_1). \tag{8.10}$$

8.2.2 The Common Perpendicular to Two Given Directions Let $[\lambda_1 : \mu_1 : \nu_1]$, $[\lambda_2 : \mu_2 : \nu_2]$ be the D.R.'s of the given directions. Finding a direction perpendicular to both of these is equivalent to finding a vector perpendicular to the two vectors

$$\lambda_1\mathbf{i} + \mu_1\mathbf{j} + \nu_1\mathbf{k}, \quad \lambda_2\mathbf{i} + \mu_2\mathbf{j} + \nu_2\mathbf{k}.$$

Such a vector is

$$(\lambda_1\mathbf{i} + \mu_1\mathbf{j} + \nu_1\mathbf{k}) \wedge (\lambda_2\mathbf{i} + \mu_2\mathbf{j} + \nu_2\mathbf{k}) \qquad \text{(see Section 7.7)},$$

giving

$$[\mu_1\nu_2 - \mu_2\nu_1 : \nu_1\lambda_2 - \nu_2\lambda_1 : \lambda_1\mu_2 - \lambda_2\mu_1] \tag{8.11}$$

as D.R.'s for the common perpendicular; these D.R.'s of course treat the perpendicular as pointing in the direction that makes $[\lambda_1, \mu_1, \nu_1]$, $[\lambda_2, \mu_2, \nu_2]$ and the perpendicular a right-handed set.

In practice (8.11) can be found by writing

$$\lambda_1 : \mu_1 : \nu_1$$

$$\lambda_2 : \mu_2 : \nu_2$$

and following the procedure described in Section 7.7.1 for finding vector products.

Exercises 8(a)

1. Find the distance between the points $P(-2, 4, 3)$ and $Q(0, 1, -3)$. Given that PQ is produced to R so that $PQ = QR$, find the co-ordinates of R.
2. The point $P(x, y, z)$ moves so that its distance from $A(1, 2, 3)$ is equal to its distance from $B(-2, 3, 4)$. Find the equation of the locus of P.

If P moves so that its distance from A is twice its distance from B, what is the equation of the locus of P?

3. Find the co-ordinates of the centroid of the triangle whose vertices are $(1, 3, -4)$, $(-4, 2, -6)$, $(-3, 1, 1)$.
4. Show that the points $(4, 2, 3)$, $(1, 4, 9)$, $(-1, 10, 6)$ are the vertices of a right-angled isosceles triangle.
5. Find the co-ordinates of the point P in which the line joining the points $A(1, -2, 6)$ and $B(2, -4, 3)$ meets the xy-plane. In what ratio does P divide AB?
6. The projections of a straight line on the co-ordinate axes are 3, 6 and 2 respectively. Find the length of the line.
7. A straight line drawn through the point $P(-2, 1, 4)$ has D.R.'s $[6:-2:3]$. Find the D.C.'s of the line and also the co-ordinates of the points which lie on the line at a distance of 7 units from P.
8. If a straight line makes an angle of $60°$ with each of the x- and y-axes, what angle does it make with the z-axis?
9. Calculate the lengths of the sides and the sizes of the angles of the triangle whose vertices are $(1, 2, 3)$, $(3, 3, 5)$, $(3, 0, 5)$.
10. Given that A is the point $(3, 7, 5)$ and B is the point $(-3, 2, 6)$, find the length of the projection of AB on the straight line which joins the points $(7, 9, 4)$ and $(4, 5, -8)$.
11. Given that A, B, C, D are four points such that $AB^2 + CD^2 = AC^2 + BD^2$, show that BC is perpendicular to AD.
12. A, B, C are the points $(-4, -4, 7)$, $(-2, 6, 9)$, $(6, 6, 7)$ respectively, and O is the origin. A parallelepiped is drawn with OA, OB, OC as edges. Find (a) the co-ordinates of the other four vertices of the parallelepiped, (b) the length of its longest edge, (c) the scalar products $\overrightarrow{OB}.\overrightarrow{OC}$, $\overrightarrow{OC}.\overrightarrow{OA}$, $\overrightarrow{OA}.\overrightarrow{OB}$, (d) the magnitude of the greatest face angle of the parallelepiped.

8.3 Equation of a Plane

(a) *Given a Point on the Plane and the Direction of the Normal to the Plane.* Let $\mathbf{r}_1$ be the position vector of the point P_1 on the plane and let $\mathbf{a}$ be a vector normal to the plane. If $\mathbf{r}$ is the position vector of any other point P on the plane, then $\mathbf{r} - \mathbf{r}_1$ is the vector $\overrightarrow{P_1P}$. The basic property of

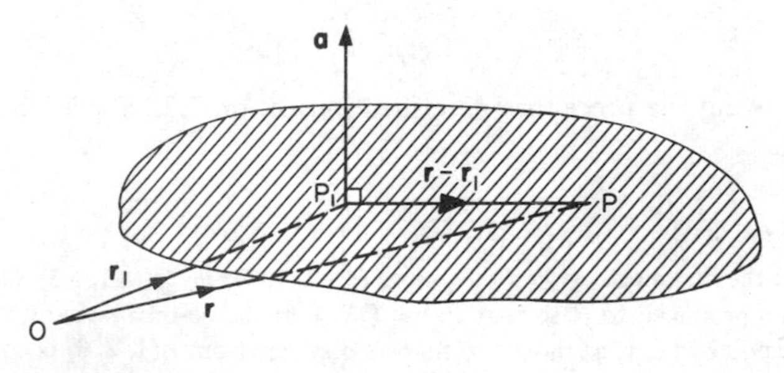

FIG. 8.5

the normal to a plane is that it is perpendicular to any line in the plane; hence

$$(\mathbf{r} - \mathbf{r}_1).\mathbf{a} = 0. \tag{8.12}$$

Since equation (8.12) is satisfied by the position vector $\mathbf{r}$ of *any* point P in the given plane, it is the *vector equation* of the plane.

To obtain the cartesian equation of the plane, let P_1 and P be the points (x_1, y_1, z_1) and (x, y, z) respectively, so that $\mathbf{r}_1 = x_1\mathbf{i} + y_1\mathbf{j} + z_1\mathbf{k}$ and $\mathbf{r} = x\mathbf{i} + y\mathbf{j} + z\mathbf{k}$. In addition put $\mathbf{a} = \lambda\mathbf{i} + \mu\mathbf{j} + \nu\mathbf{k}$ so that $[\lambda : \mu : \nu]$ are the D.R.'s of the normal to the plane. Equation (8.12) now becomes

$$\{(x - x_1)\mathbf{i} + (y - y_1)\mathbf{j} + (z - z_1)\mathbf{k}\}.\{\lambda\mathbf{i} + \mu\mathbf{j} + \nu\mathbf{k}\} = 0;$$

that is

$$\lambda(x - x_1) + \mu(y - y_1) + \nu(z - z_1) = 0. \tag{8.13}$$

This is the equation of the plane through (x_1, y_1, z_1) with normal in the direction $[\lambda : \mu : \nu]$.

(b) *Given the Direction of the Normal and the Perpendicular Distance from the Origin onto the Plane.* Let $\mathbf{n}$ be a unit vector normal to the plane and let $ON = p$ where N is the foot of the perpendicular from the origin O onto the plane.

If P is any point on the plane, PN lies in the plane and so is perpendicular to ON. Hence the projection of OP on ON is simply ON, i.e. p.

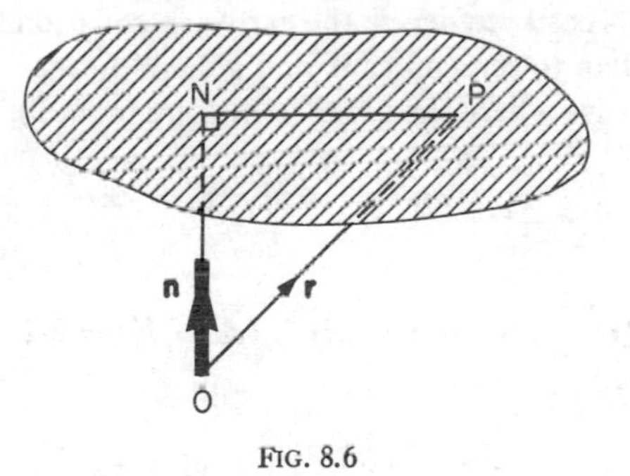

FIG. 8.6

From the definition of a scalar product we thus see that

$$\mathbf{r}.\mathbf{n} = p, \tag{8.14}$$

where $\mathbf{r}$ is the position vector of P. Equation (8.14) is the vector equation of the plane.

If we write $\mathbf{r} = x\mathbf{i} + y\mathbf{j} + z\mathbf{k}$, $\mathbf{n} = l\mathbf{i} + m\mathbf{j} + n\mathbf{k}$, (8.14) becomes

$$lx + my + nz = p. \tag{8.15}$$

This is the equation of a plane distant p from the origin when the direction $[l, m, n]$ is normal to the plane and pointing *away* from the origin. If the direction $[l, m, n]$ is normal to the plane and pointing *towards* the origin, the equation is

$$lx + my + nz = -p.$$

(c) *Given Three Points on the Plane.* Let the given points P_1, P_2, P_3 have position vectors $\mathbf{r}_1$, $\mathbf{r}_2$, $\mathbf{r}_3$ respectively. Since P_1P_2, P_2P_3 lie in the plane it follows that $\overrightarrow{P_1P_2} \wedge \overrightarrow{P_2P_3}$ is normal to the plane. Thus a vector normal to

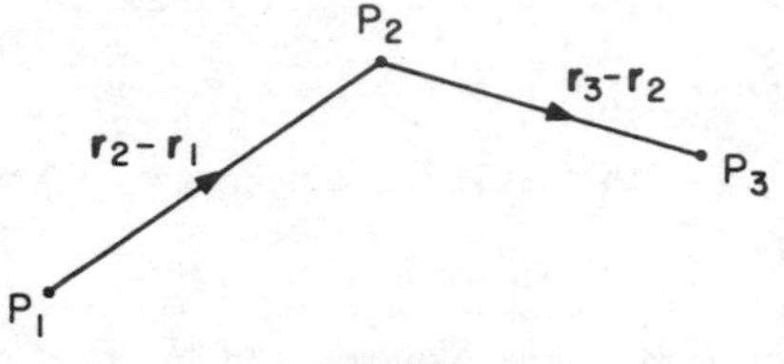

FIG. 8.7

the plane is found by obtaining the vector product $(\mathbf{r}_2 - \mathbf{r}_1) \wedge (\mathbf{r}_3 - \mathbf{r}_2)$. By (8.12) we now see that the vector equation of the plane is

$$(\mathbf{r} - \mathbf{r}_1).(\mathbf{r}_2 - \mathbf{r}_1) \wedge (\mathbf{r}_3 - \mathbf{r}_2) = 0. \tag{8.16}$$

Obviously $\mathbf{r}_2$ or $\mathbf{r}_3$ could replace $\mathbf{r}_1$ in the first bracket.

In practice it is easier to derive the cartesian equation from (8.16) in individual cases than to remember it in a general form.

Consider the case where the given points are $(a, 0, 0)$, $(0, b, 0)$, $(0, 0, c)$. Then

$$\mathbf{r}_1 = a\mathbf{i}, \quad \mathbf{r}_2 = b\mathbf{j}, \quad \mathbf{r}_3 = c\mathbf{k}.$$

Hence

$$(\mathbf{r}_2 - \mathbf{r}_1) \wedge (\mathbf{r}_3 - \mathbf{r}_2) = (b\mathbf{j} - a\mathbf{i}) \wedge (c\mathbf{k} - b\mathbf{j}) = bc\mathbf{i} + ac\mathbf{j} + ab\mathbf{k}.$$

Also

$$\mathbf{r} - \mathbf{r}_1 = (x - a)\mathbf{i} + y\mathbf{j} + z\mathbf{k},$$

and so (8.16) becomes

$$bc(x - a) + acy + abz = 0,$$

that is

$$\frac{x}{a} + \frac{y}{b} + \frac{z}{c} = 1. \tag{8.17}$$

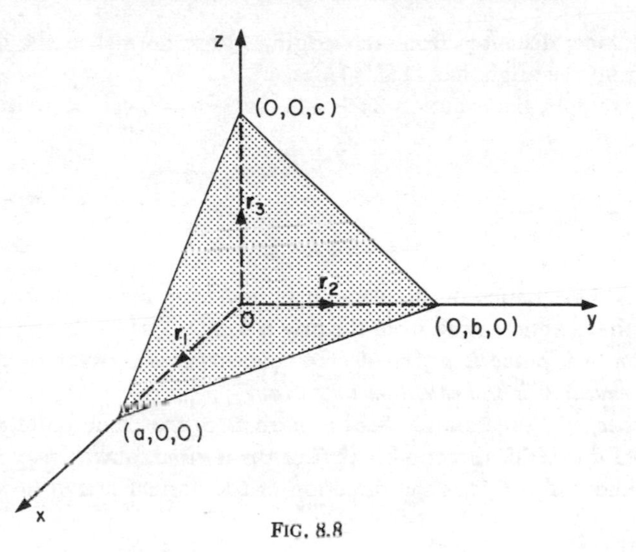

FIG. 8.8

This is the equation of the plane that makes intercepts of lengths a, b, c on Ox, Oy, Oz respectively.

8.3.1 The General Equation of the First Degree The general equation of the first degree can be written as

$$ax + by + cz + d = 0. \tag{8.18}$$

This may be written in the form

$$\frac{ax + by + cz}{\sqrt{(a^2 + b^2 + c^2)}} = \frac{-d}{\sqrt{(a^2 + b^2 + c^2)}};$$

that is

$$lx + my + nz = p, \tag{8.19}$$

where

$$l = \frac{\pm a}{\sqrt{(a^2 + b^2 + c^2)}}, \quad m = \frac{+b}{\sqrt{(a^2 + b^2 + c^2)}},$$

$$n = \frac{\pm c}{\sqrt{(a^2 + b^2 + c^2)}}, \quad p = \frac{\pm d}{\sqrt{(a^2 + b^2 + c^2)}},$$

and the upper or lower sign is chosen according to which is required to make $p \geqslant 0$.

Since $l^2 + m^2 + n^2 = 1$ it follows that $[l, m, n]$ are the D.C.'s of some line. Comparing (8.19) with (8.15) we conclude that (8.19) is the equation of the plane, distant p from the origin, whose normal in the direction away from the origin has D.C.'s $[l, m, n]$.

For example, the equation $2x - y + 2z + 6 = 0$ can be written

$$\tfrac{2}{3}x - \tfrac{1}{3}y + \tfrac{2}{3}z = -2;$$

that is

$$-\tfrac{2}{3}x + \tfrac{1}{3}y - \tfrac{2}{3}z = 2,$$

which is the equation of the plane, distant 2 from the origin, whose normal in the direction away from the origin has D.C.'s $[-\tfrac{2}{3}, \tfrac{1}{3}, -\tfrac{2}{3}]$. *Thus the equation of a plane is a first-degree equation and conversely any first-degree equation is the equation of a plane.*

Further, the direction $[a:b:c]$ is normal to the plane represented by (8.18); if $d < 0$ this direction is that of the normal drawn *away* from the origin and if $d > 0$ it is the direction of the normal drawn *towards* the origin.

Example 4 Find the equation of the plane through the point (1, 2, 3) and parallel to the plane $2x - 3y + 4z = 12$.

The direction $[2:-3:4]$ is normal to the plane and hence is normal to the required plane. It follows that the equation of the required plane is

$$2x - 3y + 4z = k$$

for some constant k. This passes through (1, 2, 3) if $2 - 6 + 12 = k$, i.e. $k = 8$. The required equation is $2x - 3y + 4z = 8$.

Example 5 Find the equation of the plane that passes through the points (3, 4, 1), (1, 1, −7) and (2, 2, −4).

First Method. We can use (8.16) with

$$\mathbf{r}_1 = 3\mathbf{i} + 4\mathbf{j} + \mathbf{k}, \quad \mathbf{r}_2 = \mathbf{i} + \mathbf{j} - 7\mathbf{k}, \quad \mathbf{r}_3 = 2\mathbf{i} + 2\mathbf{j} - 4\mathbf{k}.$$

Then

$$\begin{aligned}(\mathbf{r}_2 - \mathbf{r}_1) \wedge (\mathbf{r}_3 - \mathbf{r}_2) &= (-2\mathbf{i} - 3\mathbf{j} - 8\mathbf{k}) \wedge (\mathbf{i} + \mathbf{j} + 3\mathbf{k}) \\ &= -\mathbf{i} - 2\mathbf{j} + \mathbf{k},\end{aligned}$$

and

$$\mathbf{r} - \mathbf{r}_1 = (x - 3)\mathbf{i} + (y - 4)\mathbf{j} + (z - 1)\mathbf{k}.$$

Equation (8.16) now becomes $-(x - 3) - 2(y - 4) + (z - 1) = 0$, i.e. $x + 2y - z = 10$.

Second Method. The plane $ax + by + cz + d = 0$ passes through the given points if

$$3a + 4b + c + d = 0, \qquad (1)$$

$$a + b - 7c + d = 0, \qquad (2)$$

$$2a + 2b - 4c + d = 0. \qquad (3)$$

From (2) and (3)

$$c = \frac{1}{10}d,$$

and from (1) and (2)

$$b = -\frac{2}{10}d, \quad a = -\frac{1}{10}d.$$

Hence the equation of the required plane is $x + 2y - z = 10$.

8.3.2 The Angle Between Two Planes The angle θ between two planes is the angle between their respective normals. If the planes have equations

$$a_1x + b_1y + c_1z + d_1 = 0, \quad a_2x + b_2y + c_2z + d_2 = 0,$$

it follows from (8.7) that

$$\cos\theta = \frac{a_1a_2 + b_1b_2 + c_1c_2}{\sqrt{(a_1^2 + b_1^2 + c_1^2)}\sqrt{(a_2^2 + b_2^2 + c_2^2)}}.$$

Exercises 8(b)

1. Points A, B, C have respectively position vectors

   $$2\mathbf{j} + 3\mathbf{k}, \quad 2\mathbf{k} + 3\mathbf{j}, \quad 2\mathbf{i} + 3\mathbf{j}.$$

   Find the lengths and direction cosines of BC, CA, AB, and the equation of the plane ABC.
2. Find the equation of the plane through the point $(2, -3, 1)$ normal to the line joining $P(3, 4, -1)$ and $Q(2, -1, 5)$.
3. Points P, Q have co-ordinates $(1, 1, 1)$, $(-1, 3, 3)$ respectively. Find the equation of the plane that bisects the segment PQ at right angles.

   Determine whether the origin lies on the same side of the plane as P or as Q.
4. Find the equation of the plane that passes through the points $(1, 2, 0)$, $(3, 4, 2)$ and $(5, -3, 1)$.
5. Find the equations of the three planes that pass through the points $(2, 3, -4)$ and $(4, -1, 8)$ and are parallel to Ox, Oy and Oz respectively.

6. Find the equation of the plane that makes intercepts 2, 3 and 4 on the x-, y- and z-axes respectively. What is the perpendicular distance of the origin from this plane?
7. Find the equation of the plane that passes through the point (1, 0, 6) and is parallel to the plane $2x + 3y + 6z = 7$. What is the acute angle between this plane and the y-axis?
8. Find the equation of the plane that contains the x-axis and passes through the point (2, 1, 4).
9. Find the equation of the plane that bisects at right angles the straight line joining the points (3, 4, 8) and (5, −2, 4).
10. Show that the planes $2x + y - 3z + 5 = 0$, $5x - 7y + 2z + 3 = 0$ and $x + 10y - 11z + 12 = 0$ have a common line p which is equally inclined to the axes.
11. The plane $x/a + y/b + z/c = 1$ meets the axes of x, y, z in A, B, C respectively. Find the D.C.'s of the side AB of triangle ABC and prove that if angle A is equal to 60° then $3a^4 = b^2c^2 + c^2a^2 + a^2b^2$.

8.4 Equations of a Straight Line

Let $P_1(x_1, y_1, z_1)$ be a point on a given straight line and let $\mathbf{u}$ be a unit vector in the direction of the line. Let $P(x, y, z)$ be any other point on the line distant d from P_1 so that $\overrightarrow{P_1P} = d\mathbf{u}$. Then the position vectors $\mathbf{r}_1$, $\mathbf{r}$ of the points P_1, P respectively are related by the equation

$$\mathbf{r} = \mathbf{r}_1 + d\mathbf{u}. \tag{8.20}$$

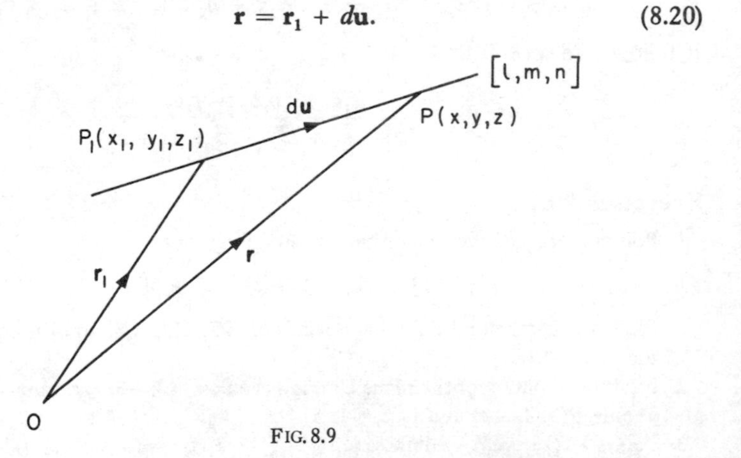

FIG. 8.9

Clearly if $d = 0$ the points P_1 and P coincide, while positive values of d give points P on one side of P_1 and negative values of d give points P on the other side of P_1.

If we regard d as a variable parameter, then eq. (8.20), in which $\mathbf{r}$ is now a *variable* position vector representing all points on the given line, is called the *vector equation of the line.*

Cartesian equations for the line can be deduced from (8.20). For if $[l, m, n]$ are the D.C.'s of the line we have $\mathbf{u} = l\mathbf{i} + m\mathbf{j} + n\mathbf{k}$ and by definition

$$\mathbf{r}_1 = x_1\mathbf{i} + y_1\mathbf{j} + z_1\mathbf{k}, \quad \mathbf{r} = x\mathbf{i} + y\mathbf{j} + z\mathbf{k}.$$

Equation (8.20) becomes

$$x\mathbf{i} + y\mathbf{j} + z\mathbf{k} = (x_1 + dl)\mathbf{i} + (y_1 + dm)\mathbf{j} + (z_1 + dn)\mathbf{k}.$$

Hence

$$\begin{aligned} x &= x_1 + dl, \\ y &= y_1 + dm, \\ z &= z_1 + dn, \end{aligned} \tag{8.21}$$

which are parametric equations for the line, again with d as parameter.

Elimination of d from (8.21) gives

$$\frac{x - x_1}{l} = \frac{y - y_1}{m} = \frac{z - z_1}{n}, \tag{8.22}$$

which gives the equations of the line in *symmetric* (or *standard*) form.

The equations (8.22) are equivalent to the equations

$$\frac{x - x_1}{\lambda} = \frac{y - y_1}{\mu} = \frac{z - z_1}{\nu}, \tag{8.22'}$$

where $[\lambda : \mu : \nu]$ are the D.R.'s of the line.

In particular, the equations of the line joining $P_1(x_1, y_1, z_1)$ and $P_2(x_2, y_2, z_2)$ are

$$\frac{x - x_1}{x_2 - x_1} = \frac{y - y_1}{y_2 - y_1} = \frac{z - z_1}{z_2 - z_1}. \tag{8.23}$$

Example 6 Find the equations of the line joining $A(1, 2, 4)$ and $B(2, 4, 2)$. Show that AB meets the xy-plane in a point of trisection of the line joining $C(6, 7, 1)$, $D(-3, 4, -2)$.

By (8.23) the equations of AB are

$$\frac{x - 1}{1} = \frac{y - 2}{2} = \frac{z - 4}{-2}.$$

AB meets the xy-plane at P, where $z = 0$ and

$$\frac{x - 1}{1} = \frac{y - 2}{2} = 2;$$

hence P is the point $(3, 6, 0)$.

By (8.3) the point that divides CD in the ratio $1:2$ is

$$(\tfrac{1}{3}(12 - 3), \tfrac{1}{3}(14 + 4), \tfrac{1}{3}(2 - 2)),$$

i.e. the point $(3, 6, 0)$.

Hence P is a point of trisection of CD.

8.4.1 The Line of Intersection of Two Planes Consider the planes whose equations are

$$a_1x + b_1y + c_1z + d_1 = 0, \tag{1}$$

$$a_2x + b_2y + c_2z + d_2 = 0. \tag{2}$$

These planes are parallel if $a_1:b_1:c_1 = a_2:b_2:c_2$. If they are not parallel, they intersect in a line $\mathscr{L}$. All points on $\mathscr{L}$ must satisfy both (1) and (2); conversely any point satisfying both (1) and (2) must lie on $\mathscr{L}$. It follows that the pair of simultaneous equations (1) and (2) are equations for $\mathscr{L}$.

For any line it is possible to draw two non-parallel planes passing through the line. Hence any line has equations in the form of a pair of simultaneous equations of first degree. The symmetric form (8.22), although it appears to give three simultaneous equations, amounts to two *independent* equations.

Given the equations of a line in the form of (1) and (2), we can determine the direction of the line as follows: since $\mathscr{L}$ lies in both planes it is perpendicular to both normal directions, i.e. $\mathscr{L}$ is the common perpendicular to the directions indicated by $[a_1:b_1:c_1]$, $[a_2:b_2:c_2]$ and so by (8.11) $\mathscr{L}$ has D.R.'s $[b_1c_2 - b_2c_1 : c_1a_2 - c_2a_1 : a_1b_2 - a_2b_1]$.

Example 7 Find in symmetric form the equations of the line of intersection of the planes $x + y + z = 5$, $4x + y + 2z = 15$.

First Method. The D.R.'s of the normals are $[1:1:1]$, $[4:1:2]$. Hence the line has direction $[1:2:-3]$.

To find one point on the line we can put $z = 0$ in both equations and solve $x + y = 5$, $4x + y = 15$ to get $x = \frac{10}{3}$, $y = \frac{5}{3}$. Thus $(\frac{10}{3}, \frac{5}{3}, 0)$ is on the line. We can now write down the equations for the line in the form

$$\frac{x - \frac{10}{3}}{1} = \frac{y - \frac{5}{3}}{2} = \frac{z}{-3}.$$

Second Method. Observing that in the symmetric form each equation has one variable missing, we may proceed by eliminating y and z in turn between the equations of the planes to obtain

$$3x + z = 10, \quad \text{that is} \quad x = \tfrac{1}{3}(10 - z);$$

$$2x - y = 5, \quad \text{that is} \quad x = \tfrac{1}{2}(5 + y).$$

Hence the equations may be written in the form

$$x = \frac{y+5}{2} = \frac{z-10}{-3}.$$

Check. The denominators 1, 2, −3 agree with the D.R.'s found in the first method. We can easily verify that (0, −5, 10) is a point on the line.

Example 8 Through the point (−1, 1, 2) a line is drawn parallel to the line of intersection of the planes $x - 2y + z = 3$ and $x + 6y - 5z = 0$. Find the equations of this line and the co-ordinates of the points where it meets the plane $x - 3y + 2z = 2$.

The direction of the line is perpendicular to [1:−2:1] and also to [1:6:−5]. Hence the line has D.R.'s [4:6:8], i.e. [2:3:4]. Hence the equations of the line can be written in the form

$$\frac{x+1}{2} = \frac{y-1}{3} = \frac{z-2}{4} = t \quad \text{(say)}$$

so that in terms of the parameter t

$$x = 2t - 1, \quad y = 3t + 1, \quad z = 4t + 2.$$

Substituting in the equation $x - 3y + 2z = 2$ we get

$$2t - 1 - 3(3t + 1) + 2(4t + 2) = 2, \quad \text{that is} \quad t = 2.$$

This gives (3, 7, 10) as the point of intersection.

8.4.2 The Equation of a Plane through the Line of Intersection of Two Given Planes Let the planes

$$a_1x + b_1y + c_1z + d_1 = 0, \tag{1}$$

$$a_2x + b_2y + c_2z + d_2 = 0, \tag{2}$$

intersect in a line $\mathscr{L}$. For any constant k, the equation

$$a_1x + b_1y + c_1z + d_1 + k(a_2x + b_2y + c_2z + d_2) = 0 \tag{8.24}$$

represents a plane since it is of the first degree. Points that satisfy (1) and (2) simultaneously must satisfy (8.24), and so (8.24) is the equation of a plane passing through $\mathscr{L}$.

Example 9 Find the equation of the plane that contains the line

$$(x-4)/2=(y-3)/5=(z+1)/(-2)$$

and passes through the point (2, −4, 2).

The line

$$(x-4)/2=(y-3)/5=(z+1)/(-2) \tag{1}$$

may be regarded as the line of intersection of the planes

$$(x-4)/2=(y-3)/5, \quad \text{that is} \quad 5x-2y=14$$

and

$$(y-3)/5=(z+1)/(-2), \quad \text{that is} \quad 2y+5z=1.$$

But by (8.24) the plane

$$5x-2y-14+k(2y+5z-1)=0 \tag{2}$$

passes through line (1) and will contain the point (2, −4, 2) if $k=-4$.

Hence the equation of the required plane is

$$x-2y-4z=2.$$

Example 10 The two planes $2x-y-z=3$ and $2x+y-2z=1$ meet in a line l. Find (a) the equation of the plane which contains the line $\frac{1}{2}x=y+1=1-z$ parallel to l; and (b) the equation of the plane through l that passes through the origin.

(a) As in Section 8.4.1, the D.R.'s of l are $[3:2:4]$.

The line $\frac{1}{2}x=y+1=1-z$ is the line of intersection of the planes

$$x-2y-2=0 \quad \text{and} \quad y+z=0,$$

and by (8.24) the equation of any plane through this line is of the form

$$x-2y-2+k(y+z)=0,$$

that is

$$x+(k-2)y+kz=2. \tag{1}$$

This plane is parallel to l if its normal is perpendicular to l, i.e. if the line with D.R.'s $[1:k-2:k]$ is perpendicular to l. This condition is fulfilled if, by (8.9),

$$3+2(k-2)+4k=0.$$

Hence $k=\frac{1}{6}$ and the equation of the required plane is

$$6x-11y+z=12.$$

(b) By (8.24), the plane

$$2x - y - z - 3 + k(2x + y - 2z - 1) = 0$$

passes through l. It will pass through the origin if $k = -3$.

Hence the equation of the required plane is

$$4x + 4y - 5z = 0.$$

8.4.3 Lines and Planes Parallel to Co-ordinate Planes or Axes If a line is drawn in the plane $z = c$ parallel to the line $y = mx$, $z = 0$ in the plane xOy, its equations are

$$y = mx, \quad z = c. \tag{1}$$

But this line passes through the point $(0, 0, c)$ and is perpendicular to Oz.

Hence by (8.22′) its equations are

$$\frac{x}{1} = \frac{y}{m} = \frac{z - c}{0}. \tag{2}$$

Equations (2) are taken as the symmetric form of equations (1).

In the same way, the symmetric form of the equations of the line drawn through (a, b, c) parallel to Oz is

$$\frac{x - a}{0} = \frac{y - b}{0} = \frac{z - c}{1};$$

that is $x = a$, $y = b$.

An equation of the form $by + cz + d = 0$ represents a plane parallel to Ox; normals to this plane have D.R.'s $[0:b:c]$ and hence are perpendicular to Ox.

An equation of the form $cz + d = 0$, i.e. $z =$ constant, represents a plane parallel to the xy-plane.

8.4.4 The Perpendicular from a Point onto a Plane Let the equation of the plane be $lx + my + nz = p$, $p \geqslant 0$, and let P_0 be the point (x_0, y_0, z_0). To find the distance of P_0 from the plane, change the origin to (x_0, y_0, z_0). Then the equation of the plane becomes

$$l(x + x_0) + m(y + y_0) + n(z + z_0) = p;$$

that is

$$lx + my + nz = p',$$

where $p' \equiv p - (lx_0 + my_0 + nz_0)$ is the distance of the plane from P_0, the new origin.

If P_0 is on the same side of the plane as the original origin O, $[l, m, n]$ are still the D.C.'s of the normal from the new origin P_0 to the plane; hence $p' \geqslant 0$. If P_0 and O are on opposite sides of the plane, $[l, m, n]$ are the D.C.'s of the normal from the plane to P_0; hence $p' \leqslant 0$.

It follows from Section 8.3.1 that the distance of (x_0, y_0, z_0) from the plane $ax + by + cz + d = 0$ is

$$\pm \frac{ax_0 + by_0 + cz_0 + d}{\sqrt{(a^2 + b^2 + c^2)}}. \tag{8.25}$$

If the positive sign is chosen when d is positive, and the negative sign when d is negative, this formula gives a positive result when (x_0, y_0, z_0) and the origin lie on the *same* side of the plane, a negative result when they are on *opposite* sides.

When the equations of the perpendicular, or the co-ordinates of the foot of the perpendicular, are required the method illustrated in Example 12 can be used.

Example 11 Find the perpendicular distance from the point (1, 3, 5) onto the plane $x - y + 2z + 4 = 0$.

By (8.25) the distance is

$$\frac{1 - 3 + 10 + 4}{\sqrt{(1 + 1 + 4)}} = \frac{12}{\sqrt{6}} = 2\sqrt{6}.$$

Example 12 Find the foot of the perpendicular from the point (1, 3, 5) onto the plane $x - y + 2z + 4 = 0$.

This perpendicular passes through the point (1, 3, 5) and has D.R.'s $[1:-1:2]$. Hence its equations are

$$\frac{x-1}{1} = \frac{y-3}{-1} = \frac{z-5}{2} = t \qquad \text{(say)}.$$

Thus any point on the line has co-ordinates $(t + 1, -t + 3, 2t + 5)$ for some t, and the point where the line meets the plane is given by

$$t + 1 + t - 3 + 4t + 10 + 4 = 0, \quad \text{that is} \quad t = -2.$$

It follows that the required point is $(-1, 5, 1)$.

8.4.5 The Perpendicular from a Point onto a Line In this case general results are cumbersome and it is simpler to illustrate a method that can be applied in all cases.

Example 13 Find the perpendicular distance of the point $(-1, 0, 7)$ from the line

$$\frac{x-1}{1}=\frac{y-2}{-1}=\frac{z-3}{-2},$$

the equations of the perpendicular, and the co-ordinates of the foot.

Putting $(x - 1)/1 = (y - 2)/-1 = (z - 3)/-2 = t$, we see that points on the line have co-ordinates $(t + 1, -t + 2, -2t + 3)$. The line joining the point $(-1, 0, 7)$ to a point on the given line has D.R.'s

$$[t + 2 : -t + 2 : -2t - 4]$$

and is perpendicular to the given line if

$$1(t + 2) - 1(-t + 2) - 2(-2t - 4) = 0,$$

i.e. $t = -\frac{4}{3}$.

It follows that the co-ordinates of the foot of the perpendicular are $(-\frac{1}{3}, \frac{10}{3}, \frac{17}{3})$ and by (8.2) the length of the perpendicular is $\sqrt{(40/3)}$.

The perpendicular has D.R.'s $[\frac{2}{3} : \frac{10}{3} : -\frac{4}{3}]$, i.e. $[1 : 5 : -2]$, and so its equations may be written as

$$\frac{x+1}{1}=\frac{y}{5}=\frac{z-7}{-2}.$$

8.4.6 The Shortest Distance between Two Skew Lines Two straight lines are said to be *skew* if they are not coplanar, i.e. if they neither intersect nor are parallel.

Figure 8.10 shows two skew lines AP, BQ which pass through the points $P(x_1, y_1, z_1)$, $Q(x_2, y_2, z_2)$. The shortest distance between them is the intercept HK that they make on their common perpendicular LM.

Since LM is perpendicular to AP and BQ, HK is the projection of PQ on LM; hence if $[l, m, n]$ are the D.C.'s of LM, by (8.10),

$$HK = |\, l(x_2 - x_1) + m(y_2 - y_1) + n(z_2 - z_1)|.$$

Example 14 illustrates how this can be used in finding the shortest distance. A different method is used in Example 15 in order to find the equations as well as the length of the line of shortest distance.

Example 14 Show that the length of the common perpendicular to the lines whose equations referred to rectangular axes are

$$(x - 5)/1 = y/2 = (z + 1)/(-1),$$

$$(x - 2)/1 = (y - 4)/(-1) = z/1,$$

is $\sqrt{14}$.

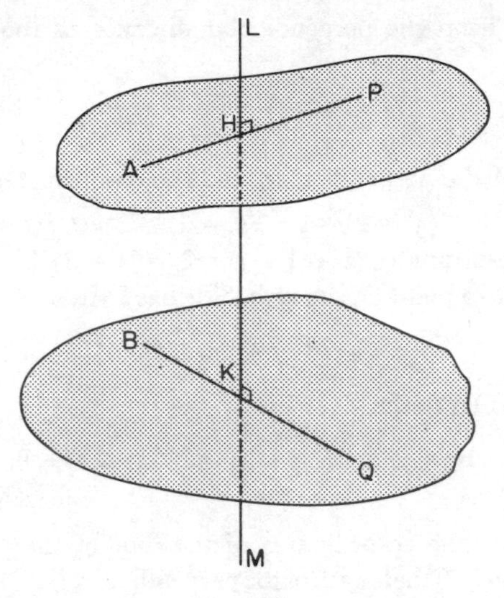

FIG. 8.10

The line $(x - 5)/1 = y/2 = (z + 1)/(-1)$ passes through $P(5, 0, -1)$; the line $(x - 2)/1 = (y - 4)/(-1) = z/1$ passes through $Q(2, 4, 0)$. The common perpendicular to the given lines has D.R.'s $[1:-2:-3]$, and so the D.C.'s of the common perpendicular are

$$[1/\sqrt{14}, -2/\sqrt{14}, -3/\sqrt{14}].$$

The projection of PQ on this line is

$$3(1/\sqrt{14}) + (-4)(-2/\sqrt{14}) + (-1)(-3/\sqrt{14}) = \sqrt{14},$$

and this is the shortest distance between the given lines.

Example 15 Find the length and the equations of the shortest distance between the lines $x = y - 1 = 4 - z$ and $x - 2y + 9 = 0$, $x + z - 10 = 0$.

Writing the lines in standard form we have

$$x/1 = (y - 1)/1 = (z - 4)/(-1) = \rho, \tag{1}$$

$$x/2 = (y - \tfrac{9}{2})/1 = (z - 10)/(-2) = \tau. \tag{2}$$

The co-ordinates of any point P on (1) may be taken as $(\rho, 1 + \rho, 4 - \rho)$ and similarly Q, any point on (2), is $(2\tau, \tau + \tfrac{9}{2}, 10 - 2\tau)$.

The D.R.'s of PQ are

$$[\rho - 2\tau : \rho - \tau - \tfrac{7}{2} : 2\tau - \rho - 6]$$

and by (8.11) the D.R.'s of the common perpendicular to (1) and (2) are found to be $[-1:0:-1]$. Hence PQ is the common perpendicular if

$$\frac{\rho - 2\tau}{-1} = \frac{\rho - \tau - 7/2}{0} = \frac{2\tau - \rho - 6}{-1};$$

that is if $\rho = 10$, $\tau = 13/2$.

With these values P is the point $(10, 11, -6)$ and Q is $(13, 11, -3)$, and by (8.2) $PQ = 3\sqrt{2}$.

By (8.22′), the equations of PQ are

$$\frac{x - 10}{-1} = \frac{y - 11}{0} = \frac{z + 6}{-1};$$

that is $x - z = 16, \quad y = 11.$

8.4.7 The Plane Containing Two Intersecting Lines The common perpendicular to the two lines is perpendicular to the plane.

Example 16 Verify that the lines

$$\frac{x-2}{1} = \frac{y-1}{2} = \frac{z}{3}, \quad \frac{x-1}{4} = \frac{y-2}{1} = \frac{z-3}{-2}$$

intersect, and find the equation of the plane containing them.

Writing

$$\frac{x - 2}{1} = \frac{y - 1}{2} = \frac{z}{3} = \rho, \tag{1}$$

$$\frac{x - 1}{4} = \frac{y - 2}{1} = \frac{z - 3}{-2} = \tau, \tag{2}$$

so that points on the first and second lines have co-ordinates

$$(\rho + 2, 2\rho + 1, 3\rho) \quad \text{and} \quad (4\tau + 1, \tau + 2, -2\tau + 3),$$

respectively, we see that there is a point of intersection if we can solve the equations

$$\rho + 2 = 4\tau + 1, \quad 2\rho + 1 = \tau + 2, \quad 3\rho = -2\tau + 3.$$

Solving the first two equations we get $\rho = \frac{5}{7}$, $\tau = \frac{3}{7}$, and this is found to satisfy the third equation also; hence there is a point of intersection, viz. $(\frac{19}{7}, \frac{17}{7}, \frac{15}{7})$.

The common perpendicular to the lines (1) and (2) has D.R.'s $[-7:14:-7]$, i.e. $[-1:2:-1]$ and so the equation of the plane can be written in the form

$$x - 2y + z + d = 0.$$

Since (2, 1, 0) is a point on the first line, it is a point on the plane, and we find that $2 - 2 + d = 0$, i.e. $d = 0$. Hence the equation of the plane is $x - 2y + z = 0$.

Another method is illustrated in the following example:

Example 17 Prove that the straight line

$$(x - 4)/3 = (y - 1)/2 = (z - 3)/1$$

intersects the line of intersection of the planes $x + y + 2z = 4$ and $3x - 2y - z = 3$, and find the equation of the plane that contains these two lines.

The straight line

$$(x - 4)/3 = (y - 1)/2 = (z - 3)/1 = \rho \tag{1}$$

meets the plane

$$x + y + 2z = 4 \tag{2}$$

where

$$(3\rho + 4) + (2\rho + 1) + 2(\rho + 3) = 4, \quad \text{that is} \quad \rho = -1;$$

that is at the point $(1, -1, 2)$.

This point satisfies the equation

$$3x - 2y - z = 3. \tag{3}$$

Hence (1) meets (2) and (3) in a point common to these planes, i.e. at a point on their line of intersection.

Any plane through the line of intersection of (2) and (3) is of the form

$$x + y + 2z - 4 + k(3x - 2y - z - 3) = 0. \tag{4}$$

It will contain (1) if the point (4, 1, 3) satisfies (4), i.e. if

$$k = -\tfrac{7}{4}.$$

Hence the equation of the required plane is

$$17x - 18y - 15z = 5.$$

Exercises 8(c)

1. A is the point (1, 3, 2), B the point (−1, 1, 1). Find (a) the length of the segment AB, (b) the direction cosines of AB, (c) the equations of the line AB. Find whether the point (3, 1, 3) lies on the line AB or not.
2. A is the point (2, −1, 2), B the point (3, 2, 1). Find (a) the length of the segment AB, (b) the direction cosines of AB, (c) the equations of the line AB, (d) the co-ordinates of the point P where the line AB meets the xOy-plane.

3. Find the co-ordinates of the point at which the line joining the points $(3, 1, 4)$, $(-2, 6, 1)$ meets the plane $2x + y - 3z = 3$.
4. A line passes through the point $P(1, 0, -1)$ and has direction cosines proportional to $(2, 2, 1)$. (a) Write down the equations of the line in parametric form. (b) What are its actual direction cosines? (c) Find the co-ordinates of the point Q where the line meets the plane

$$x + 4y - z + 1 - 0.$$

(d) Find the length PQ.
5. Find the D.C.'s of the line of intersection of the planes

$$x + y + z = 1, \quad 4x + y + 2z = 3,$$

and prove that it is perpendicular to the line $x = y = z$.
6. Find the equation of the line that passes through the point $(2, 3, 4)$ and is parallel to the line of intersection of the planes

$$x + y - 2z = 1 \quad \text{and} \quad 2x - 3y + z = -3.$$

7. Find the equation of the plane through the point $A(2, -3, 5)$ normal to the line joining $P(3, 2, -1)$ and $Q(2, -1, 1)$. Find also the D.R.'s of the perpendicular drawn from A to the line PQ.
8. Find the equation of the plane through the points $(2, 3, 1)$, $(1, 1, 3)$ and $(2, 2, 3)$. Find also the perpendicular distance from the point $(5, 6, 7)$ to this plane.
9. Find the equation of the plane that passes through the origin and contains the line of intersection of the planes

$$2x - y + 3z = 4 \quad \text{and} \quad 5x - 7y - 9z + 2 = 0.$$

10. Find the equation of the plane through the line

$$(x - 1)/2 = (y - 2)/1 = (z - 3)/2$$

that is parallel to the line $x/3 = y/1 = z/(-2)$.
11. Show that the two lines

$$(x - 1)/2 = (y + 2)/1 = z/3$$

and

$$(x - 2)/3 = (y - 2)/(-2) = (z - 4)/2$$

have a common point.

Find the perpendicular distance between the first line and the parallel line $(x - 1)/2 = (y - 7)/1 = (z - 1)/3$.
12. Show that the plane $2x + y + z = 3$ contains the line of intersection of the planes $x - y + 2z = 0$ and $3x + y + 2z = 4$.
13. Show that the line

$$(x - 1)/2 = (y - 2)/3 = (z - 3)/4$$

lies in the plane $x + 2y - 2z + 1 = 0$ and that the line

$$(x - 3)/3 = (y - 2)/2 = (z - 1)/4$$

lies in the plane $2x + y - 2z - 6 = 0$.

14. The straight line

$$(x - 6)/2 = (y - 4)/4 = (z + 6)/3$$

meets the yz-, zx-, xy-planes at P, Q and R respectively. Find the co-ordinates of the centroid of triangle PQR.

15. Find the direction cosines of the line $(2x - 1)/1 = y/2 = (2z + 9)/1$ and show that it is parallel to the plane $4x + y - 8z = 2$. Find the distance between the line and plane.

16. Find the distance of the point (1, 3, 5) from the plane

$$2x + y - 3z + 30 = 0$$

measured parallel to the straight line $x/3 = y/2 = z/6$.

17. Find the equation of the plane through the points (2, 3, 6), (3, 6, 2), (6, 2, 3) and determine the perpendicular distance from (5, 6, 7) to this plane.

18. A plane meets the co-ordinate axes at A, B, C, and the foot of the perpendicular from the origin O to the plane is P. If $OA = a$, $OB = b$, $OC = c$, find the co-ordinates of P.

    Prove that, if P is the centroid of the triangle ABC, then $|a| = |b| = |c|$.

19. Find the equation of the plane that passes through the point (5, 1, 2) and is perpendicular to the line $2x - 4 = y - 4 = z - 5$. Find also the co-ordinates of the point in which the line cuts the plane.

20. Find the D.C.'s of the line of intersection of the planes

$$x + y + z = 4, \quad 4x + y + 6z = 3,$$

and prove that it is perpendicular to the line $x = y = z$.

21. Find the ratio in which N, the foot of the perpendicular from the origin, divides the line joining the points $A(4, 6, 0)$ and $B(1, 2, -1)$ and prove that N does not lie between A and B.

22. The foot of the perpendicular from the point $P(4, 7, -9)$ to the line

$$(x - 2)/2 = (y + 1)/(-2) = (z + 3)/1$$

is Q. Find the length of PQ and the co-ordinates of Q.

23. Show that the lines

$$(x - 6)/3 = (y - 3)/2 = (z - 2)$$

and

$$5x + 4y + 7z - 26 = 2x + 3y + 2z - 11 = 0$$

are coplanar. Find the co-ordinates of the common point and the equation of the plane that contains these two lines.

24. A straight line is drawn through $P(1, 3, 4)$ to meet the line $x = -1$, $y = 1$ in Q and the line $x = 3$, $z = 3$ in R. Prove that the equations of the line PQR are $(x + 1)/2 = (y - 1)/2 = (z - 5)/(-1)$. Show that P is the mid-point of QR.

25. The plane $x/a + y/b + z/c = 1$ meets the axes Ox, Oy, Oz in A, B, C respectively, and L, M, N are mid-points of BC, CA, AB respectively. Prove that the three planes OAL, OBM, OCN, meet in the line $x/a = y/b = z/c$.

26. Find the equations of the line that passes through the origin and intersects each of the lines

$$(x - 1)/(-1) = (y + 2)/2 = (z - 3)/1$$

and

$$(x + 1)/3 = (y - 1)/2 = (z + 1)/(-1).$$

Find also the co-ordinates of its point of intersection with the first of the given lines.

27. Prove that any plane through the line L common to the two non-parallel planes $ax + by + cz + d = 0$, $a'x + b'y + c'z + d' = 0$ has the equation $\lambda(ax + by + cz + d) + \mu(a'x + b'y + c'z + d') = 0$, where λ, μ are finite constants and not both zero.

Find the equation of the plane through the line L that is parallel to the line $x/l = y/m = z/n$ and, hence or otherwise, show that the two lines are coplanar if, and only if,

$$d(la' + mb' + nc') = d'(la + mb + nc).$$

28. For each of the following pairs of lines find the equations and the magnitude of the shortest distance between the two lines:

(a) $(x - 5)/1 = (y - 4)/(-2) = (z - 4)/1$,
$(x - 1)/7 = (y + 2)/(-6) = (z + 4)/1$,
(b) $(x - 6)/2 = (y + 4)/5 = (z - 2)/1$, $(x + 1)/(-4) = (y - 9)/5 = (z - 5)/7$,
(c) $(x - 1)/4 = (y - 1)/3 = (z - 2)/(-2)$, $x/4 = (y - 5)/0 = (z - 15)/(-1)$,
(d) $(x - 2)/0 = (y - 5)/2 = (z - 1)/1$, $(x - 8)/2 = (y - 4)/2 = (z + 1)/(-1)$.

8.5 The Sphere

If the sphere has centre $P_0(x_0, y_0, z_0)$ and radius a, then any point on the sphere is distant a from P_0. If $\mathbf{r}_0$ is the position vector of P_0, it follows that the equation of the sphere is

$$|\mathbf{r} - \mathbf{r}_0| = a, \tag{8.26}$$

which becomes, in cartesian form

$$(x - x_0)^2 + (y - y_0)^2 + (z - z_0)^2 = a^2. \tag{8.27}$$

In particular, the equation of a sphere with centre the origin and radius a is

$$x^2 + y^2 + z^2 = a^2.$$

The general equation

$$x^2 + y^2 + z^2 + 2ux + 2vy + 2wz + k = 0 \tag{8.28}$$

can be written as

$$(x + u)^2 + (y + v)^2 + (z + w)^2 = u^2 + v^2 + w^2 - k$$

and so represents a sphere with centre $(-u, -v, -w)$ and radius $\sqrt{(u^2 + v^2 + w^2 - k)}$. The equation (8.28) is the most general form for the equation of a sphere.

Tangent Plane. If P_1 is a point on a sphere with centre P_0, all lines through P_1 perpendicular to P_0P_1 touch the sphere and all such tangent lines lie in the plane through P_1 perpendicular to P_0P_1. This plane is called the *tangent plane* at P_1.

If the position vectors of P_0, P_1 are $\mathbf{r}_0$, $\mathbf{r}_1$ respectively, then if $\mathbf{r}$ is the position vector of any point on the tangent plane it follows that $\mathbf{r} - \mathbf{r}_1$ is perpendicular to $\mathbf{r}_1 - \mathbf{r}_0$. Hence the equation of the tangent plane is

$$(\mathbf{r} - \mathbf{r}_1).(\mathbf{r}_1 - \mathbf{r}_0) = 0; \tag{8.29}$$

in other words the equation of the tangent plane at (x_1, y_1, z_1) on the sphere with centre (x_0, y_0, z_0) is

$$(x - x_1)(x_1 - x_0) + (y - y_1)(y_1 - y_0) + (z - z_1)(z_1 - z_0) = 0. \tag{8.30}$$

Example 18 Find the centre and radius of the sphere whose equation is

$$x^2 + y^2 + z^2 - 2x - 4y - 6z - 2 = 0.$$

Show that the intersection of this sphere and the plane $x + 2y + 2z - 20 = 0$ is a circle whose centre is the point (2, 4, 5), and find the radius of this circle.

The centre of the sphere is $C(1, 2, 3)$ and its radius is $\sqrt{(1 + 4 + 9 + 2)}$, i.e. 4. The normal through C to the given plane has equations

$$\frac{x-1}{1} = \frac{y-2}{2} = \frac{z-3}{2} = t$$

and meets the plane at A, where

$$(t + 1) + 2(2t + 2) + 2(2t + 3) - 20 = 0, \quad \text{that is} \quad t = 1.$$

Hence A, the centre of the circle of section, is the point (2, 4, 5).

By (8.2) $AC = 3$, and by Pythagoras' theorem the radius r of the circle and the radius R of the sphere are connected by the equation

$$R^2 = AC^2 + r^2, \text{ i.e. } 16 = 9 + r^2, \text{ i.e. } r = \sqrt{7}.$$

Example 19 Show that the line L:

$$\frac{x-7}{2} = \frac{y-4}{7} = \frac{z-13}{10}$$

touches the sphere S:

$$x^2 + y^2 + z^2 - 6x + 2y - 4z + 5 = 0.$$

Find the co-ordinates of P, the point of contact of L with S. Find also the equation of the sphere S' that touches S at P and passes through the centre of S.

The line

$$\frac{x-7}{2} = \frac{y-4}{7} = \frac{z-13}{10} = t$$

meets the sphere S where

$$(2t + 7)^2 + (7t + 4)^2 + (10t + 13)^2 - 6(2t + 7) + 2(7t + 4) \\ - 4(10t + 13) + 5 = 0.$$

This equation reduces to $(t + 1)^2 = 0$, and so L meets S in two coincident points at $P(5, -3, 3)$, i.e. L touches S at P.

The centre of S is $C(3, -1, 2)$. The line CP is a diameter of the sphere S' and so S' has radius $\frac{1}{2}CP$, i.e. $\frac{3}{2}$, and the centre of S' is the mid-point of CP, viz. $(4, -2, \frac{5}{2})$. Hence the equation of S' is

$$(x - 4)^2 + (y + 2)^2 + (z - \tfrac{5}{2})^2 = \tfrac{9}{4},$$

that is

$$x^2 + y^2 + z^2 - 8x + 4y - 5z + 24 = 0.$$

8.6 The Equation of a Surface

In co-ordinate geometry of two dimensions an equation represents a *curve*. Likewise, in co-ordinate geometry of three dimensions an equation represents a *surface*; examples we have met are the plane and the sphere. A *curve* is represented by a pair of simultaneous equations (intersection of two surfaces), a particular example being the straight line.

Consider first the equation

$$f(x, y) = 0 \tag{1}$$

in which z is absent. Points in the xy-plane that satisfy (1) lie on a curve C. If through any point $P(x_0, y_0, 0)$ on this curve a line PP' is drawn parallel to the z-axis, the co-ordinates of any point Q on the line PP' may be taken as (x_0, y_0, z_0). Then since P lies on (1), $f(x_0, y_0) = 0$ and it follows, since (1) is independent of z, that the co-ordinates of Q also satisfy (1) no matter what value z_0 may have. Hence the co-ordinates of any point on PP' satisfy (1). But P is any point on C and so (1) is the equation of the surface of the cylinder generated by straight lines drawn parallel to Oz through points on C. For example, the equation $x^2 + y^2 = a^2$ represents a right circular cylinder of radius a with axis along Oz. In the same way the equation $x^2/a^2 + z^2/c^2 = 1$ represents an elliptic cylinder with axis along Oy.

We now consider the more general equation

$$F(x, y, z) = 0. \tag{2}$$

Points in the plane $z = k$ whose co-ordinates satisfy (2) have x and y co-ordinates such that $F(x, y, k) = 0$. This means that these points lie on the curve C_k (say) in which the plane $z = k$ cuts the cylinder $F(x, y, k) = 0$. As k varies the curve C_k also varies and generates a surface of which (2) is the equation.

If $F(x, y, z) = 0$ is a homogeneous equation of degree n ($n > 0$) in x, y and z, the origin O lies on the surface and

$$F(kx, ky, kz) = k^n F(x, y, z)$$

for all values of k.

It follows that if the point $P(x, y, z)$ lies on the surface, $Q(kx, ky, kz)$, any point on the straight line OP, also lies on the surface. The surface is therefore a cone with vertex at the origin.

In particular $x^2 + y^2 - z^2 \tan^2 \alpha = 0$ is the equation of a circular cone, vertex at the origin, angle 2α at the vertex.

8.6.1 Quadric Surfaces All surfaces represented by equations of the second degree in x, y, and z are known as *quadric surfaces* or *quadrics*. It can be shown that any plane section of a quadric is a conic and so quadrics are also given the name *conicoids*.

If $P(\alpha, \beta, \gamma)$ lies on the surface

$$ax^2 + by^2 + cz^2 = 1, \tag{1}$$

$Q(-\alpha, -\beta, -\gamma)$ also lies on the surface, and the origin O is the mid-point of PQ. Hence all chords of (1) that pass through O are bisected at O. For this reason (1) is called a *central quadric*, O is called its *centre* and a chord through O is called a *diameter*.

Surface (1) is symmetrical about each of the co-ordinate planes, for if the point (α, β, γ) lies on (1) so do $(-\alpha, \beta, \gamma)$, $(\alpha, -\beta, \gamma)$ and $(\alpha, \beta, -\gamma)$. These three planes of symmetry are called the *principal planes* and the co-ordinate axes the *principal axes* of the quadric.

If in (1) we write $a = 1/A^2$, $b = 1/B^2$, $c = 1/C^2$ (i.e. a, b, c all positive) we obtain the equation

$$\frac{x^2}{A^2} + \frac{y^2}{B^2} + \frac{z^2}{C^2} = 1,$$

which is the standard form of the equation of an *ellipsoid* (Fig. 8.11(a)).

Sections of this surface parallel to the co-ordinate planes are ellipses. For example, the equations of the section of this surface by the plane $z = k$ are

$$\frac{x^2}{A^2} + \frac{y^2}{B^2} = 1 - \frac{k^2}{C^2}, \quad z = k$$

and these are the equations of an ellipse when $k^2 \leqslant C^2$. When $k^2 > C^2$ the plane does not cut the surface in real points. Similarly sections parallel to

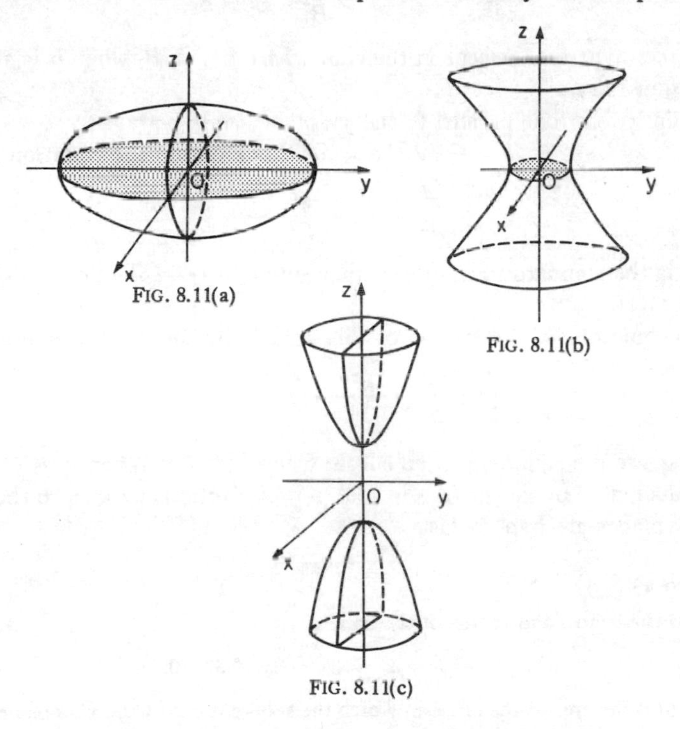

FIG. 8.11(a)

FIG. 8.11(b)

FIG. 8.11(c)

the other co-ordinate planes are ellipses. The surface meets the co-ordinate axes in the points $(\pm A, 0, 0)$, $(0, \pm B, 0)$, $(0, 0, \pm C)$.

If $A = B = C$, the surface is a sphere.

If in (1) we write $a = 1/A^2$, $b = 1/B^2$, $c = -1/C^2$, we obtain the equation

$$\frac{x^2}{A^2} + \frac{y^2}{B^2} - \frac{z^2}{C^2} = 1,$$

which is the standard form of the equation of a *hyperboloid of one sheet* (Fig. 8.11(b)).

Sections of the surface parallel to the yz- and zx-planes are hyperbolas; those parallel to the xy-plane are ellipses. For example, the section by the plane $z = k$ is given by

$$\frac{x^2}{A^2} + \frac{y^2}{B^2} = 1 + \frac{k^2}{C^2}, \quad z = k$$

and for all values of k, this is an ellipse.

The section by the plane $y = k$ is given by

$$\frac{x^2}{A^2} - \frac{z^2}{C^2} = 1 - \frac{k^2}{B^2}, \quad y = k,$$

which is a hyperbola except in the case where $k = \pm B$, when it is a pair of straight lines.

Similarly, sections parallel to the yz-plane are hyperbolas.

When $a = -1/A^2$, $b = -1/B^2$, $c = 1/C^2$, we obtain the equation

$$-\frac{x^2}{A^2} - \frac{y^2}{B^2} + \frac{z^2}{C^2} = 1,$$

which is the standard form of the equation of a *hyperboloid of two sheets* (Fig. 8.11(c)).

The equations of the section of this surface by the plane $z = k$ are

$$\frac{x^2}{A^2} + \frac{y^2}{B^2} = \frac{k^2}{C^2} - 1, \quad z = k.$$

These are the equations of an ellipse when $k^2 \geqslant C^2$. When $k^2 < C^2$ the plane does not cut the surface in real points. Sections parallel to the yz- and zx-planes are hyperbolas.

Exercises 8(d)

1. Find the centre and radius of the sphere

$$x^2 + y^2 + z^2 + 2x - 3y - 3 = 0.$$

Find (a) the area of the circle in which the sphere is cut by the xOz plane, and (b) the length of the chord of the sphere lying along the line $x = y = z$.

2. Find the centre and radius of the sphere

$$x^2 + y^2 + z^2 + 2x - 4y + 4z = 0.$$

Find (a) the direction cosines of the radius through the origin, (b) the equation of the plane α touching the sphere at the origin, and (c) the equation of the plane that touches the sphere and is parallel to α.

3. Determine the radius and the centre of the circle of intersection of the two spheres

$$x^2 + y^2 + z^2 - 2x + 4y - 164 = 0,$$
$$x^2 + y^2 + z^2 - 10x - 4y + 14z - 82 = 0.$$

4. Show that the plane

$$\text{(a)}\quad 3y + 4z - 37 = 0$$

touches the sphere

$$\text{(b)}\quad x^2 + y^2 + z^2 - 6x - 8y = 0,$$

and find the point of contact. Find also the equations of the planes that are parallel to (a) and cut (b) in a circle of radius 4 units.

5. Prove that the sphere

$$x^2 + y^2 + z^2 - 2x - 2y - 2z + 1 = 0$$

touches the co-ordinate axes and find the co-ordinates of the points of contact. Find also the centre and radius of the circle formed by the intersection of the sphere and the plane through these points of contact.

6. A line with direction ratios $[l : m : n]$ is drawn through the fixed point $(0, 0, a)$ to touch the sphere $x^2 + y^2 + z^2 = 2ax$. Prove that $m^2 + 2nl = 0$.

7. Show that the sphere

$$(x - \alpha - p)(x - \alpha) + (y - \beta - q)(y - \beta) + (z - \gamma - r)(z - \gamma) = a^2$$

intersects the sphere

$$(x - \alpha)^2 + (y - \beta)^2 + (z - \gamma)^2 = a^2$$

along a great circle of the second sphere.

8. The perpendicular distance of the centre of a sphere from a tangent plane equals the radius of the sphere. Use this fact to derive the condition that the plane $ax + by + cz + d = 0$ touches the sphere

$$x^2 + y^2 + z^2 + 2ux + 2vy + 2wz + k = 0.$$

CHAPTER 9

COMPLEX NUMBERS

9.1 Introduction

We saw in Section 6.3.1 that an equation may have more roots in one field than in another. With regard to the solution of equations the familiar field $R(+, \times)$ has certain disadvantages. Consider, for example, the two equations $x^2 - 1 = 0$, $x^2 + 1 = 0$. The first has two roots, viz. ± 1, whereas the second has no roots. Likewise the cubic equation $x^3 - 6x^2 + 11x - 6 = 0$ has three roots, viz. 1, 2, 3, whereas the cubic equation $x^3 - 1 = 0$ has only one root, viz. 1. Naturally it would be preferable to have an algebra in which equations of the same degree have the same number of roots.

It was pointed out in Section 5.3 that $R(+, \times)$ is sometimes used when a smaller system would suffice, because of its algebraic advantages. In this chapter we set out to find a still larger system than $R(+, \times)$ with still more algebraic advantages; more specifically, we want the larger system to be better from the point of view of solving equations.

Having learned to represent numbers by points on a straight line (real axis) we tend to see the introduction of negative, rational and irrational numbers as very natural extensions of the number system; equally, we tend to think of any further extension of the number system as being impossible. Rather than thinking about introducing *more* elements into the *same* system, it is more helpful to think in terms of building a *new system* with a completely new kind of element. Remember that two mathematical systems are equivalent for all practical purposes if the two systems are *isomorphic*; hence, for example, the system $V(+)$ of vectors is tantamount to an extension of $R(+)$ *because* $V(+)$ *includes the system* $V_1(+)$ *of one-dimensional vectors which is isomorphic to* $R(+)$ (see Section 7.3).

Thus what we need is a system part of which is isomorphic to $R(+, \times)$. Note that $V(+, \wedge)$ is *not* such a system since $V_1(\wedge)$ is not isomorphic to $R(\times)$; in fact the vector product of two vectors in V_1 is not in V_1.

Let us now state more precisely the aims of this chapter. We wish to find a mathematical system $C(+, \times)$ with two operations which can conveniently be called *addition* and *multiplication*, and with the following properties:

Property 1 There is a subset C_r of C such that the system $C_r(+, \times)$ is isomorphic to $R(+, \times)$.

Property 2 The presence of more elements in $C(+, \times)$ makes it a better system than $R(+, \times)$ from the point of view of solving equations.

Property 3 The system $C(+, \times)$ is a *field*. This will ensure that all the properties of a field, so familiar to us through the algebra of $R(+, \times)$, are valid in $C(+, \times)$.

Step 1. For the elements of C we require a set of written symbols, and for the operations of addition and multiplication we require some kind of addition and multiplication tables. By Property 1 we could begin by using the symbols 1, 2, 3, ... for those elements that correspond to the real numbers 1, 2, 3, ..., thereby re-using all real-number symbols for elements of C_r; operations with these elements must then obey the tables of ordinary arithmetic to give the required isomorphism, and the systems $C_r(+, \times)$, $R(+, \times)$ appear to be identical.

Step 2. Let us now replace Property 3 by the following more specific property:

Property 2a The system $C(+, \times)$ contains an element that satisfies the equation $x^2 + 1 = 0$.

Clearly, this element does not correspond to any real number and a new symbol must be used for it. Denoting it by i, we must then add to our tables the result $i^2 = -1$.

Step 3. By Property 3 it is necessary that C be closed w.r.t. multiplication. We now denote by $2i$ the element that results from multiplying 2 by i; in this way a set C_i of elements of the form bi (b any real number) is introduced. Likewise the symbol ib denotes the element that results from multiplying i by b, but of course we want multiplication to be *commutative* and so the symbols ib and bi must be alternative symbols for the same element.

Step 4. By Property 3 it is necessary that C be closed w.r.t. addition. If a, a member of C_r, is added to ib, a member of C_i, the result is denoted by $a + ib$. Thus a further set of elements of the form $a + ib$, where a and b are any real numbers, is introduced.

Step 5. So far we have introduced new elements to represent the results of operations when there seemed to be no reason to suppose these results could be found among the existing elements. In this respect our freedom is now at an end, and the results of all further operations are determined by Property 3.

For example, assuming Property 3 to hold, we would then have

$$(a + ib) + (c + id) = (a + c) + i(b + d), \tag{9.1}$$

and

$$\begin{aligned}(a + ib)(c + id) &= ac + ibc + iad + i^2bd \\ &= (ac - bd) + i(bc + ad),\end{aligned} \tag{9.2}$$

since $i^2 = -1$. Note that in both (9.1) and (9.2) the result is of the form $A + iB$ and hence is an element of the type introduced in Step 4.

Step 6. By Property 3 there must be elements e_0, e_1 which are identity elements w.r.t. addition and multiplication respectively. Obviously e_0, e_1 must be identity elements within the smaller system $C_r(+, \times)$; however we already know that 0, 1 are identity elements in $C_r(+, \times)$ and that these identity elements must be unique (see Section 6.3.1), so we must have $e_0 = 0$, $e_1 = 1$.

We can now see that some elements introduced in Step 4 were not new. The elements introduced in Step 4 included elements of the form $a + 0i$ and elements of the form $0 + bi$; but since 0, 1 are neutral elements we have

$$a + 0i = a + 0 = a \qquad \text{(which is an element of } C_r\text{)},$$
$$0 + bi = bi \qquad \text{(which is an element of } C_i\text{)}.$$

Thus for each element of C_r we have devised *two* symbols, e.g. 2 and $2 + 0i$, and likewise for each element of C_i, e.g. $3i$ and $0 + 3i$.

Step 7. In previous steps there has been the underlying assumption that it is possible to find a mathematical system with the Properties 1, 2a and 3. The final step must be to check that we have in fact produced such a system.

To describe the system and to check its properties it is best to make a completely fresh start as follows in Section 9.2.

9.2 The Complex Number System

Consider a set C of elements each of which is denoted by a symbol of the form $a + ib$, where a, b are real numbers, there being an element corresponding to each possible choice of a pair of real numbers a, b.

The elements of C will be called *complex numbers*, and eqs. (9.1) and (9.2) will be taken as *definitions* of the operations of addition and multiplication of complex numbers.

The real number a is called the *real part* of the complex number $a + ib$; likewise the real number b is called the *imaginary part* of $a + ib$.

Denote by C_r the subset of C consisting of elements of the form $a + i0$; these particular complex numbers are said to be *purely real.*

Denote by C_i the subset of C consisting of elements of the form $0 + ib$; these particular complex numbers are said to be *purely imaginary.* (The terms *real* and *imaginary* are somewhat unsuitable but unfortunately are well established in the literature of the subject.)

From (9.1) and (9.2) we easily see that $C_r(+, \times)$ is isomorphic to $R(+, \times)$. In practice the elements of C_r are usually written without the $i0$ as in Step 1 of the preliminary investigation. Thus, for example, we use the same symbol 1 to denote both the real number 1 and the complex number 1, meaning $1 + i0$.

Elements of C_i are usually written without the 0 as in Step 3 of the preliminary investigation, e.g. $2i$ is understood to mean the complex number $0 + 2i$.

Another kind of abbreviation is used when, for example, we denote the complex number $3 + (-2)i$ by $3 - 2i$.

9.2.1 Structure of the Complex Number System Examination of the definition of sum and product, given by (9.1) and (9.2), reveals that C is closed w.r.t. the two operations, that both operations are commutative and associative, and that multiplication is distributive w.r.t. addition. (This follows from the definitions together with the corresponding properties of the real number system.)

From (9.1) it follows that the complex number $0 + i0$ (for short, the complex number 0) is neutral in addition. Likewise from (9.2) we see that $1 + i0$ (for short, the complex number 1) is neutral in multiplication.

From (9.1) we have

$$(a + ib) + (-a - ib) = 0;$$

i.e. each element $a + ib$ has an additive inverse, viz. $-a - ib$.

From (9.2) we have

$$(a + ib)(a - ib) = a^2 + b^2, \tag{9.3}$$

where the r.h.s. is an element of C_r (purely real). This helps to show that any element $a + ib$, other than $0 + i0$, has a multiplicative inverse since we have

$$(a + ib)(c + id) = 1 \qquad \text{if } c = \frac{a}{a^2 + b^2},\ d = \frac{-b}{a^2 + b^2}.$$

Thus the complex number system is a field.

In addition we have from (9.2) that $(0 + i1)^2 = -1 + i0$ (for short, $i^2 = -1$), and so the complex number system has Properties 1, 2a and 3.

9.2.2 Practical Operations with Complex Numbers In practice we need not remember the formula (9.2) for the product of two complex numbers; instead we get the answer by multiplying as though $a + ib$ were an expression involving real numbers a, i, b, and then replacing i^2 by -1; for example,

$$(3 + 2i)(4 + 5i) = 12 + 8i + 15i + 10i^2 = 12 + 23i - 10 = 2 + 23i.$$

The complex numbers $a + ib$, $a - ib$ are said to be *conjugate* to each other; thus conjugate complex numbers differ only in the sign of their imaginary parts. The important result in connection with conjugate complex numbers is (9.3).

Subtraction. This follows from addition as for any field, and we get

$$(a + ib) - (c + id) = (a - c) + i(b - d). \qquad (9.4)$$

Division. The procedure is best illustrated by an example. To determine the quotient $(3 + 2i)/(4 + 5i)$, multiply numerator and denominator by the conjugate of $4 + 5i$, viz. $4 - 5i$. Thus

$$\frac{3 + 2i}{4 + 5i} = \frac{(3 + 2i)(4 - 5i)}{(4 + 5i)(4 - 5i)} = \frac{22 - 7i}{4^2 + 5^2} = \frac{22}{41} - \frac{7}{41}\,i.$$

In all cases the conjugate of the *denominator* is used.

Powers of i. Note that since $i^2 = -1$ it follows that $i^3 = i^2 . i = -i$, $i^4 = (i^2)^2 = (-1)^2 = 1$, etc.

Example 1 Express $(2 + 3i)^6$ in the form $a + ib$.

On using the Binomial theorem,

$$\begin{aligned}(2 + 3i)^6 &= 2^6 + 6.2^5.3i + 15.2^4.(3i)^2 \\ &\quad + 20.2^3.(3i)^3 + 15.2^2.(3i)^4 + 6.2.(3i)^5 + (3i)^6 \\ &= 64 + 576i + 2160i^2 + 4320i^3 + 4860i^4 + 2916i^5 + 729i^6 \\ &= 64 + 576i - 2160 - 4320i + 4860 + 2916i - 729 \\ &= 2035 - 828i.\end{aligned}$$

9.2.3 Algebra of the Complex Number System Single letters a, b, ... are frequently used to denote particular complex numbers; this is comparable to the practice with vectors (see Section 7.6), except that in this case no special precaution is taken to distinguish between these letters and similar letters which might be used to denote real numbers.

The letter z is the most common symbol for a *complex variable* (i.e. a variable on a set of complex numbers) just as x and y are the most common symbols for a *real variable* (i.e. a variable on a set of real numbers).

The conjugate of z is denoted by $\bar{z}$. Thus, if $z = x + iy$ then $\bar{z} = x - iy$ and (9.3) can be written as $z\bar{z} = x^2 + y^2$.

Equality. In the complex number system $a + ib$ and $c + id$ refer to different elements unless $a = c$ and $b = d$. Hence an equation involving complex numbers represents two simultaneous *real* equations (i.e. equations involving only real numbers).

For example the equation $z^2 + 1 = 0$ can be written as

$$(x + iy)^2 + 1 = 0 \quad \text{or} \quad x^2 + 2ixy + i^2y^2 + 1 = 0$$
$$\text{or} \quad x^2 - y^2 + 1 + i(2xy) = 0;$$

this amounts to the pair of simultaneous equations

$$x^2 - y^2 + 1 = 0, \quad 2xy = 0$$

for which the solutions are $x = 0, y = \pm 1$. Hence there are two solutions, $z = \pm i$.

Identities. Since the complex number system is a field, all the familiar identities of real-number algebra have their counterparts in the algebra of complex numbers. Thus for example, if z and a are complex numbers, then

$$z^2 - a^2 \equiv (z - a)(z + a),$$

$$\frac{1}{z - a} + \frac{1}{z + a} \equiv \frac{2z}{z^2 - a^2},$$

$$(z + a)^n \equiv z^n + nz^{n-1}a + \frac{n(n - 1)}{1.2} z^{n-2}a^2 + \cdots + a^n \quad (n = 1, 2, \ldots).$$

Exercises 9(a)

1. Express in the form $x + iy$

(a) $(2 - 3i)^2/(1 + 2i)$,
(b) $(2 + i)(2 + 3i)/(3 - 4i)$.

2. Find z in the form $x + iy$ given that

$$\frac{1}{1 + i} + \frac{1}{z} = \frac{1}{1 - i}.$$

3. Given $z = (2 + i)/(1 - i)$, find the real and imaginary parts of $z + 1/z$.

4. Solve for the real numbers x and y the equation

$$\left(\frac{1+i}{1-i}\right)^2 + \frac{1}{x+iy} = 1 + i.$$

5. Solve for the real numbers x and y the equation

$$\frac{1-i}{1+i} + \frac{5}{x+iy} = 1 - 3i.$$

6. Given that

$$(1+2i)z - 2iw = -i,$$
$$(1+i)z + w = 2i,$$

eliminate w to find z in the form $x + iy$. Then find w (in the same form).

7. Solve for z and w the simultaneous equations

$$(1+i)z + w = 2 - i,$$
$$z + (1+i)w = 6,$$

giving your answers in the form $x + iy$.

9.3 The Argand Diagram

Complex numbers may be represented geometrically in what is known as an *Argand diagram*, as follows.

Let rectangular axes be chosen and let P be the point with co-ordinates (x, y). Then P is taken to represent the complex number $x + iy$. Complex numbers that are 'purely real' are represented by points on the x-axis, which is called the *real axis*; 'purely imaginary' numbers are represented by points on the y-axis, which is called the *imaginary axis*. The origin represents the number 0.

The set C of all complex numbers corresponds to the entire plane of the diagram, sometimes called the *complex plane*.

If the length of OP is r and $\angle xOP = \theta$, then r is called the *modulus* of

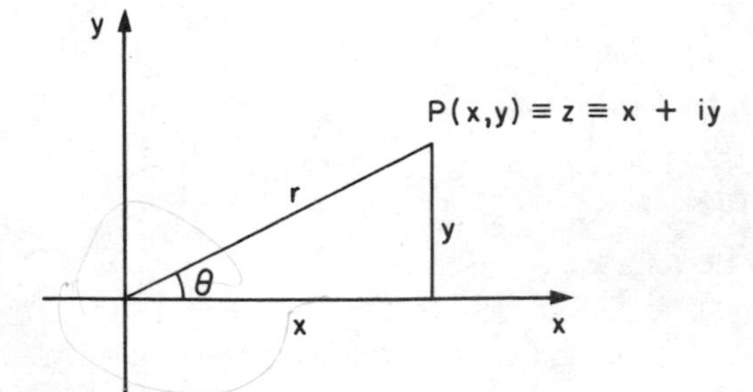

FIG. 9.1

z and written $|z|$, θ is called the ***argument*** or ***amplitude*** of z and written arg z or am z. We shall measure θ in radians unless otherwise stated.

Obviously $r = \sqrt{(x^2 + y^2)}$ and $\tan \theta = y/x$ (Fig. 9.1). There is a unique value of θ in the range $-\pi < \theta \leqslant \pi$; this is known as the ***principal value*** of arg z, other values being given by the formula $\theta + 2k\pi$ where k is any integer, not zero. In subsequent work, arg z will denote the principal value unless otherwise stated.

Since $x = r \cos \theta$ and $y = r \sin \theta$, z may be written in the cartesian form $x + iy$ or in the polar form $r(\cos \theta + i \sin \theta)$ which is often abbreviated as $r \angle \theta$.

If two numbers z and z' are equal, then $|z| = |z'|$ and arg z = arg z'.

Example 2 Represent the following complex numbers in an Argand diagram and express them in polar form: (a) $3 - 3i$, (b) -4, (c) $2i$, (d) $-3 + 4i$.

In Fig. 9.2, A, B, C and D represent the complex numbers $3 - 3i$, -4, $2i$ and $-3 + 4i$ respectively.

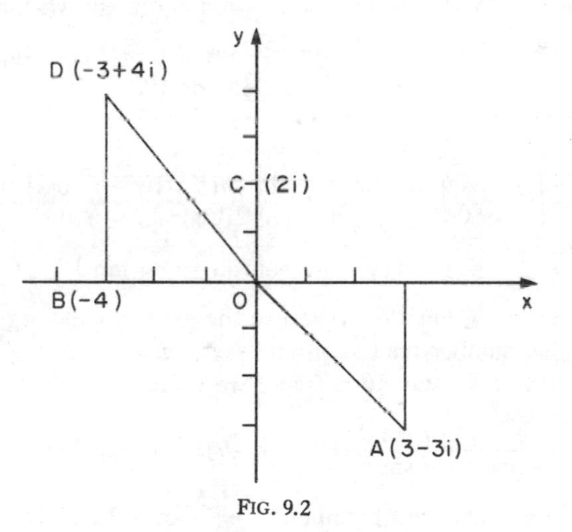

FIG. 9.2

(a) $OA = 3\sqrt{2}$ and $\angle xOA = -\frac{1}{4}\pi$; $3 - 3i = 3\sqrt{2} \angle (-\pi/4)$.
(b) $OB = 4$, $\angle xOB = \pi$; $-4 = 4 \angle \pi$.
(c) $OC = 2$, $\angle xOC = \frac{1}{2}\pi$; $2i = 2 \angle \frac{1}{2}\pi$.
(d) $OD = 5$, $\angle xOD = 180° - \angle BOD$ and $\tan \angle BOD = \frac{4}{3}$; that is $\angle xOD = 2{\cdot}214$ radians. Therefore $-3 + 4i = 5 \angle 2{\cdot}214$.

Alternatively, let $-3 + 4i = r(\cos \theta + i \sin \theta)$.

Then, equating real and imaginary parts on each side of this equation, we have

$$-3 = r\cos\theta \quad \text{and} \quad 4 = r\sin\theta,$$

whence

$$r^2 = 25, \quad r = 5.$$

Also,

$$\sin\theta = \tfrac{4}{5}, \quad \cos\theta = -\tfrac{3}{5}, \quad \theta = 2{\cdot}214 \text{ radians}.$$

Hence

$$-3 + 4i = 5\angle 2{\cdot}214.$$

Example 3 Find the modulus and argument of

$$\frac{1 - \cos\theta - i\sin\theta}{1 + \cos\theta + i\sin\theta}$$

when $0 < \theta < \pi$. What are the modulus and argument when $\pi < \theta < 2\pi$?

$$1 - \cos\theta = 2\sin^2\tfrac{1}{2}\theta, \quad 1 + \cos\theta = 2\cos^2\tfrac{1}{2}\theta,$$
$$\sin\theta = 2\sin\tfrac{1}{2}\theta\cos\tfrac{1}{2}\theta.$$

Hence

$$\frac{1 - \cos\theta - i\sin\theta}{1 + \cos\theta + i\sin\theta} = \frac{\sin\frac{1}{2}\theta(\sin\frac{1}{2}\theta - i\cos\frac{1}{2}\theta)}{\cos\frac{1}{2}\theta(\cos\frac{1}{2}\theta + i\sin\frac{1}{2}\theta)}$$
$$= -i\tan\tfrac{1}{2}\theta. = \tan\tfrac{1}{2}\theta\angle(-\tfrac{1}{2}\pi).$$

When $0 < \theta < \pi$, $\tan\frac{1}{2}\theta > 0$ and so the modulus and argument of the given complex number are $\tan\frac{1}{2}\theta$ and $-\frac{1}{2}\pi$ respectively.

When $\pi < \theta < 2\pi$, $\tan\frac{1}{2}\theta < 0$ and we write

$$\frac{1 - \cos\theta - i\sin\theta}{1 + \cos\theta + i\sin\theta} = (-\tan\tfrac{1}{2}\theta)i = (-\tan\tfrac{1}{2}\theta)\angle\tfrac{1}{2}\pi.$$

Hence the modulus and argument are $(-\tan\frac{1}{2}\theta)$ and $\frac{1}{2}\pi$ respectively.

Example 4 Find the modulus and argument of the complex number

$$(1 + i)(2 + i)/(3 - i).$$

Denoting the given complex number by z, we have

$$z = \frac{(1 + i)(2 + i)}{3 - i} = \frac{1 + 3i}{3 - i} = i = 1\angle\tfrac{1}{2}\pi.$$

Hence

$$|z| = 1 \quad \text{and} \quad \arg z = \tfrac{1}{2}\pi.$$

Example 5 Show that the point on the Argand diagram representing the complex number $a + b(1 + it)/(1 - it)$, where a and b are real constants and t is a real parameter, describes a circle as t varies.

Since

$$a + b\left(\frac{1 + it}{1 - it}\right) = a + b\,\frac{(1 + it)^2}{1 + t^2}$$

$$= a + b\left(\frac{1 - t^2}{1 + t^2}\right) + \frac{2ibt}{1 + t^2},$$

the co-ordinates (x, y) of the point on the Argand diagram are given by

$$x = a + b\left(\frac{1 - t^2}{1 + t^2}\right), \quad y = \frac{2bt}{1 + t^2}$$

and elimination of t gives $(x - a)^2 + y^2 = b^2$.

9.3.1 Vectorial Representation of Complex Numbers To each point on an Argand diagram there corresponds a position vector (Section 8.1.1), i.e. the vector represented by the line segment drawn from the origin to the point. Thus the complex number $x + iy$ (see Fig. 9.1) may be represented by the vector $\overrightarrow{OP}$. It should be noted that, although the point P representing a complex number is uniquely specified, in the vectorial representation the line segment OP can be replaced by any equal line segment in the same direction. Let A, B (Fig. 9.3) be the points

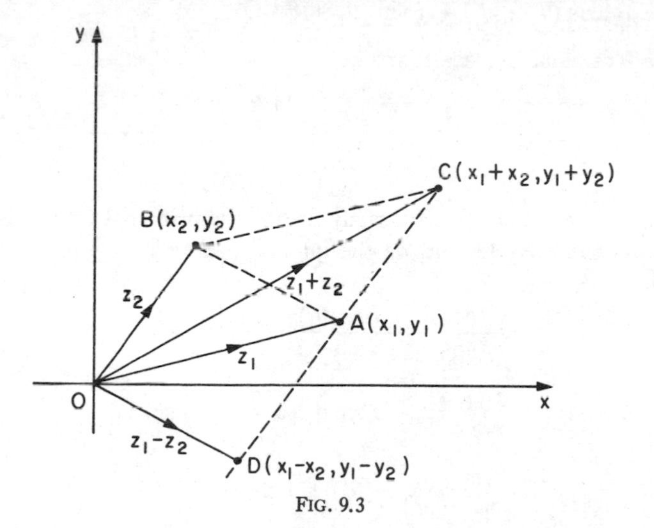

FIG. 9.3

representing the complex numbers z_1, z_2 respectively, where $z_1 = x_1 + iy_1$, $z_2 = x_2 + iy_2$. Then the point C which represents $z_1 + z_2$ has, from (9.1), co-ordinates $(x_1 + x_2, y_1 + y_2)$ and so $OACB$ is a parallelogram. In terms of the vectorial representation this means that the vector $\overrightarrow{OC}$ which represents $z_1 + z_2$ is the vector sum of the vectors $\overrightarrow{OA}$, $\overrightarrow{OB}$ representing z_1, z_2. Likewise the vector $\overrightarrow{OD}$ which represents $z_1 - z_2$ can be found by completing the parallelogram $ODAB$ and so we have $\overrightarrow{OD} = \overrightarrow{OA} - \overrightarrow{OB}$.

The system of complex numbers under addition as defined by (9.1) and the system of two-dimensional vectors under addition are isomorphic systems. Mathematicians usually define a complex number as an ordered pair of real numbers (x, y) which, although sometimes written in the form $x + iy$, may equally be left in the simple form (x, y). The basic difference between complex numbers and vectors lies in the definitions of multiplication for the two systems.

In the triangle OAC, $OC \leqslant OA + AC$, equality being possible only if O, A and C are collinear so that the triangle becomes a straight line. It follows that

$$|z_1 + z_2| \leqslant |z_1| + |z_2|. \tag{9.5}$$

9.3.2 Multiplication and Division of Complex Numbers in Polar Form Let $z_1 = r_1 \angle \theta_1$ and $z_2 = r_2 \angle \theta_2$. Then

$$\begin{aligned} z_1 z_2 &= r_1 r_2 (\cos\theta_1 + i \sin\theta_1)(\cos\theta_2 + i \sin\theta_2) \\ &= r_1 r_2 \{(\cos\theta_1 \cos\theta_2 - \sin\theta_1 \sin\theta_2) \\ &\quad + i(\sin\theta_1 \cos\theta_2 + \cos\theta_1 \sin\theta_2)\} \\ &= r_1 r_2 \{\cos(\theta_1 + \theta_2) + i \sin(\theta_1 + \theta_2)\}. \end{aligned} \tag{9.6}$$

From this result we see that

$$|z_1 z_2| = |z_1|.|z_2| \tag{9.7}$$

and

$$\arg(z_1 z_2) = \arg z_1 + \arg z_2. \tag{9.8}$$

The latter result is not necessarily true of the principal values since the right-hand side may lie outside the interval $(-\pi, \pi)$.

Again,

$$\begin{aligned} \frac{z_1}{z_2} &= \frac{r_1(\cos\theta_1 + i \sin\theta_1)}{r_2(\cos\theta_2 + i \sin\theta_2)} \\ &= \frac{r_1}{r_2}(\cos\theta_1 + i \sin\theta_1)(\cos\theta_2 - i \sin\theta_2) \\ &= \frac{r_1}{r_2}\{\cos(\theta_1 - \theta_2) + i \sin(\theta_1 - \theta_2)\}. \end{aligned} \tag{9.9}$$

Hence

$$|z_1/z_2| = r_1/r_2 = |z_1|/|z_2| \tag{9.10}$$

and

$$\arg (z_1/z_2) = \arg z_1 - \arg z_2. \tag{9.11}$$

The latter result is not necessarily true of the principal values.

If $z_1 = r \angle \theta$ and $z_2 = 1 \angle \phi$, we have by (9.6)

$$z_1 z_2 = r \angle (\theta + \phi).$$

Thus the effect of multiplying a complex number z_1 by a complex number with unit modulus and argument ϕ is to rotate the vector that represents z_1 counter-clockwise through an angle ϕ.

When $\phi = 90°$, $z_2 = i$, and so the vector that represents iz_1 is obtained by rotating counter-clockwise through 90° the vector that represents z_1.

More generally, if z_1 is represented by the vector $\overrightarrow{OP}$ and $z_2 = r_2 \angle \theta_2$, the product $z_1 z_2$ is represented by a vector $\overrightarrow{OQ}$ such that $\angle POQ = \theta_2$ and $OQ = r_2 . OP$.

Example 6 Find the modulus and argument of the complex number

$$\frac{(1 + \sqrt{3}i)^3(\sqrt{3} + i)}{(1 + i)^2}.$$

We have

$$\left|\frac{(1 + \sqrt{3}i)^3(\sqrt{3} + i)}{(1 + i)^2}\right| = \frac{|1 + \sqrt{3}i|^3|\sqrt{3} + i|}{|1 + i|^2} \quad \text{from (9.7) and (9.10)}$$

$$= \frac{2^3 . 2}{2} = 8.$$

Also

$$\arg \frac{(1 + \sqrt{3}i)^3(\sqrt{3} + i)}{(1 + i)^2} = 3 \arg (1 + \sqrt{3}i) + \arg (\sqrt{3} + i) - 2 \arg (1 + i)$$

from (9.8) and (9.11)

$$= 3 . \pi/3 + \pi/6 - 2 . \pi/4 = 2\pi/3.$$

9.3.3 Geometrical Construction for the Product of Two Complex Numbers Let P_1, P_2 and A represent the numbers z_1, z_2 and 1 respectively. Construct the triangle OP_2P directly similar to triangle OAP_1 (Fig. 9.4). Then

$$OP : OP_2 = OP_1 : OA; \quad \text{therefore} \quad OP = |z_1| . |z_2| = |z_1 z_2|.$$

Also

$$\angle xOP = \angle xOP_2 + \angle P_2OP = \angle xOP_2 + \angle xOP_1$$
$$= \arg z_2 + \arg z_1 = \arg (z_1z_2).$$

Hence P represents the number z_1z_2.

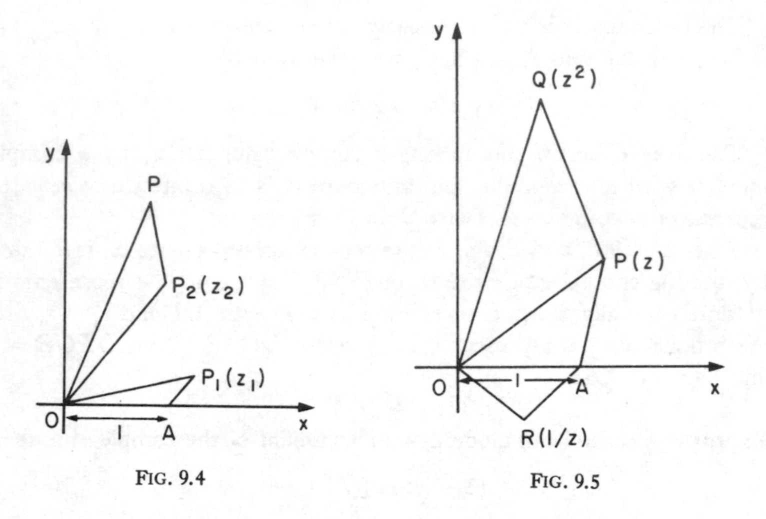

FIG. 9.4

FIG. 9.5

Example 7 Given that P represents the complex number z in an Argand diagram, show how to represent z^2 and $1/z$.

Join P to A, the point that represents the number 1 (Fig. 9.5).

Construct triangles ORA, OPQ directly similar to triangle OAP. Then, as above, it may be shown that Q represents the number z^2 and R represents $1/z$.

Example 8 In an Argand diagram, PQR is an equilateral triangle of which the circumcentre is at the origin. Given that P represents the number $2 + i$, find the numbers represented by Q and R.

In Fig. 9.6,

$$\angle POQ = \angle QOR = \angle ROP = \tfrac{2}{3}\pi \quad \text{and} \quad OP = OQ = OR.$$

Hence, if Q, R represent the complex numbers z_Q, z_R respectively,

$$z_Q = (2 + i) \times 1\angle\tfrac{2}{3}\pi \text{ and } z_R = (2 + i) \times 1\angle(-\tfrac{2}{3}\pi);$$

that is

$$z_Q = (2 + i) \times \tfrac{1}{2}(-1 + i\sqrt{3})$$
$$= -(1 + \tfrac{1}{2}\sqrt{3}) + i(\sqrt{3} - \tfrac{1}{2}).$$

Similarly

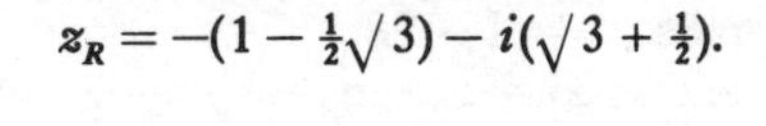

$$z_R = -(1 - \tfrac{1}{2}\sqrt{3}) - i(\sqrt{3} + \tfrac{1}{2}).$$

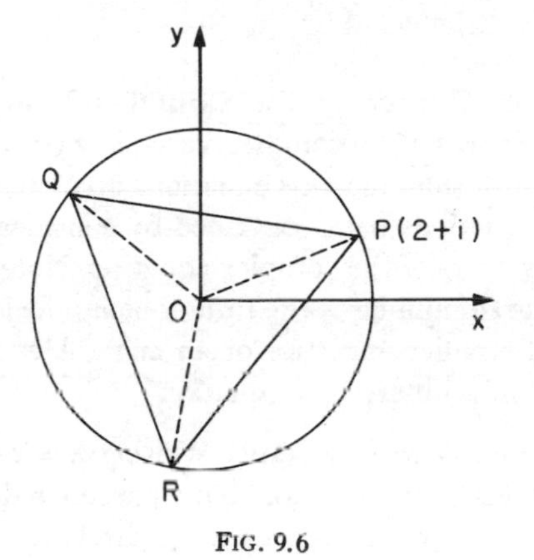

FIG. 9.6

Example 9 When the vertices A, B, C, D, of a square are taken anti-clockwise in that order, the points A, B represent the complex numbers $-1 + 4i$, $-3 + 0i$ in an Argand diagram. Find the complex numbers represented by the other vertices and by M, the centre of the square.

Let A, B, C, D and M (Fig. 9.7) represent the numbers z_A, z_B, z_C, z_D and z_M respectively.

Then $\overrightarrow{AB}$ represents $z_B - z_A$, that is $-2 - 4i$. Since $\angle CBA = \frac{1}{2}\pi$ and $BC = BA$, the complex number represented by $\overrightarrow{BA}$ is i times that

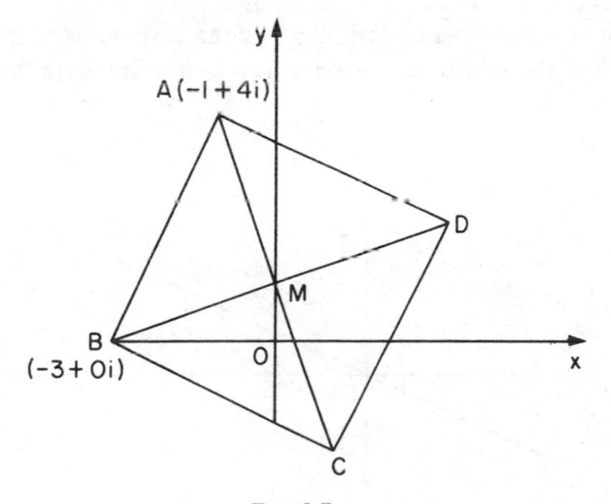

FIG. 9.7

represented by $\overrightarrow{BC}$, i.e. $\overrightarrow{BC}$ represents $(2 + 4i)/i$, i.e. $4 - 2i$. Hence $z_C - z_B = 4 - 2i$, and so $z_C = 1 - 2i$.

Now $\overrightarrow{DC} = \overrightarrow{AB}$ and so $z_C - z_D = z_B - z_A = -2 - 4i$. Hence $z_D = 3 + 2i$. Finally $z_M = \frac{1}{2}(z_B + z_D) = i$.

9.3.4 Equations of Curves in the Complex Plane We have seen that an equation involving the complex variable z $(= x + iy)$ is normally equivalent to a pair of simultaneous equations involving the real variables x and y, the real equations being obtained by considering separately the real and imaginary parts of the complex equation. Note, however, that $|z|$ and arg z are both *real* numbers, the former being the length of a line and the latter being the radian measure of an angle. Hence an equation involving z may be an ordinary real equation.

The equation $|z| = a$ (a a real constant). Writing $z = x + iy$ and using the definition of modulus (Section 9.3), this equation reduces to the single equation $x^2 + y^2 = a^2$ which represents a circle, centre at the origin, radius a, in the complex plane. This result could have been foreseen since $|z|$ is the distance from the origin to the point representing z, so all points satisfying the equation $|z| = a$ are at the same distance a from the origin. This equation can be written in the alternative form $z\bar{z} = a^2$.

The equation arg $z = \alpha$ (α a real constant). From the definition of arg z it follows that points satisfying this equation all lie on a 'semi-infinite' straight line starting from O and making an angle α radians with the real axis; points on the continuation of this line through O (the broken line in Fig. 9.8) do *not* satisfy the equation since these points have argument $\alpha + \pi$.

Note that the cartesian equation $y = x \tan \alpha$ represents the *entire* line, and we require the additional restriction $y \geqslant 0$ to indicate the solid line in Fig. 9.8.

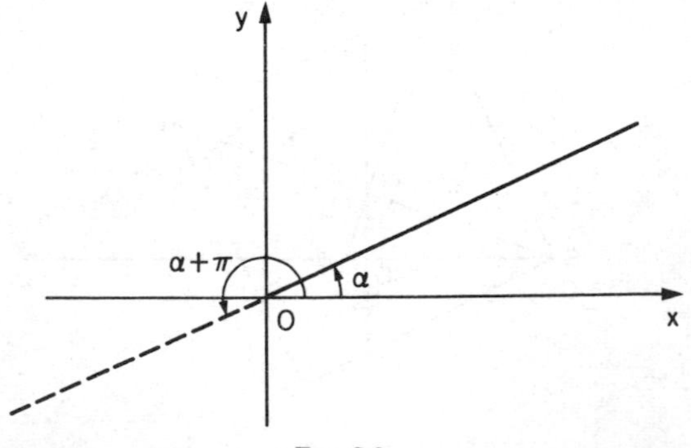

FIG. 9.8

To sum up, an equation involving z may be a real equation in which case it represents a curve in the complex plane. In this case the equation reduces to a single equation in x and y, and the curve may be found by

(a) putting $z = x + iy$ and deriving the x, y-equation of the curve, or
(b) using the geometrical significance of the terms in the z-equation to interpret this equation directly.

In connection with (b) the following observations are important:

(i) Let P_1, P_2 (Fig. 9.9) represent z_1, z_2 respectively. Then $\overrightarrow{P_1P_2}$ represents $z_2 - z_1$, $\overrightarrow{P_2P_1}$ represents $z_1 - z_2$. It follows that $\arg(z_2 - z_1)$, $\arg(z_1 - z_2)$ are the respective angles made by these vectors with the x-axis, and

$$\arg(z_2 - z_1) = \arg(z_1 - z_2) + \pi.$$

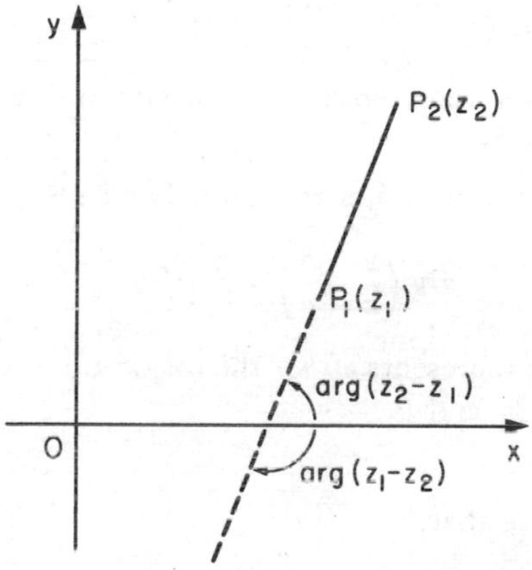

FIG. 9.9

Also, $|z_2 - z_1| = |z_1 - z_2| = P_1P_2$.

(ii) Let P_1, P_2, P (Fig. 9.10) represent complex numbers z_1, z_2, z respectively. From the triangle ABP we have

$$\begin{aligned} \angle APB &= \angle PBx - \angle PAx = \arg(z - z_2) - \arg(z - z_1) \\ &= \arg\{(z - z_2)/(z - z_1)\} \quad \text{by (9.11).} \end{aligned}$$

Example 10 Interpret geometrically the equations

(a) $|z - a| = k$, (b) $\arg\{(z - b)/(z - a)\} = k$,

where a and b are fixed complex numbers, z is a variable complex number, and k is a fixed real number.

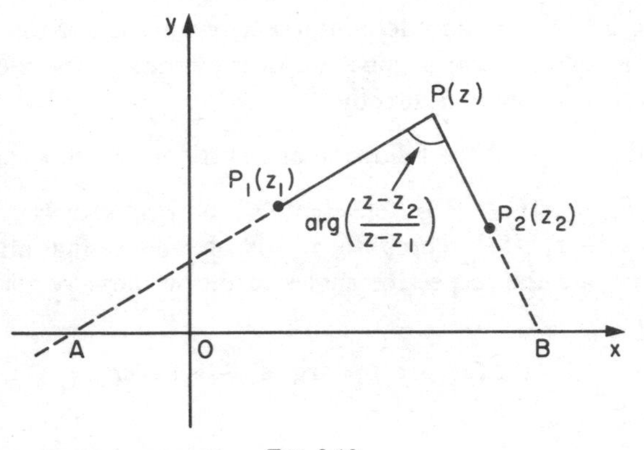

FIG. 9.10

(a) If A is the point representing a, the equation represents a circle, centre A, radius k.

(b) If P_1, P_2, P represent a, b, z respectively (see Fig. 9.10) then

$$\arg\left(\frac{z-b}{z-a}\right) = \angle P_1PP_2.$$

Hence the equation represents *either* the major arc *or* the minor arc P_1P_2 of a circle through P_1 and P_2.

Example 11 Prove that

(a) if $|z_1 + z_2| = |z_1 - z_2|$, then the difference between the arguments of z_1 and z_2 is $\frac{1}{2}\pi$;

(b) if $\arg\{(z_1 + z_2)/(z_1 - z_2)\} = \frac{1}{2}\pi$, then $|z_1| = |z_2|$.

Suppose that the points P and Q (Fig. 9.11) represent the complex numbers z_1 and z_2 respectively in an Argand diagram, and complete the parallelogram $OPRQ$.

Then the vectors $\overrightarrow{OR}$ and $\overrightarrow{QP}$ represent the complex numbers $z_1 + z_2$ and $z_1 - z_2$ respectively.

(a) If $|z_1 + z_2| = |z_1 - z_2|$, $OR = QP$ and so parallelogram $OPRQ$ has equal diagonals. Hence $OPRQ$ must be a rectangle, OP is perpendicular to OQ, and so $\arg z_1 - \arg z_2 = \pm\frac{1}{2}\pi$.

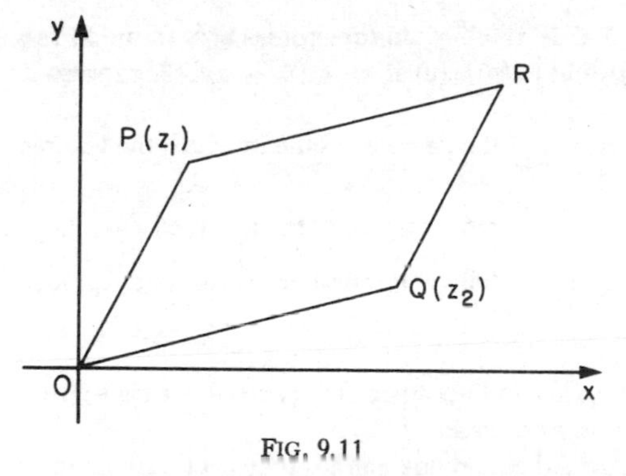

FIG. 9.11

(b) If

$$\arg\left(\frac{z_1 + z_2}{z_1 - z_2}\right) = \tfrac{1}{2}\pi,$$

the diagonals OR and QP of parallelogram $OPRQ$ are perpendicular. Hence $OPRQ$ is a rhombus, $OP = OQ$ and so $|z_1| = |z_2|$.

Example 12 Interpret geometrically the equations

(a) $|z + 3i|^2 - |z - 3i|^2 = 12$,
(b) $|z + ik|^2 + |z - ik|^2 = 10k^2, \quad k > 0$.

Putting $z = x + iy$ we get

(a) $|x + (y + 3)i|^2 - |x + (y - 3)i|^2 = 12$,

that is

$$x^2 + (y + 3)^2 - [x^2 + (y - 3)^2] = 12;$$

that is

$$y = 1 \text{ (straight line parallel to the } x\text{-axis).}$$

(b) $x^2 + (y + k)^2 + x^2 + (y - k)^2 = 10k^2$;

that is

$$x^2 + y^2 = 4k^2 \quad \text{(circle, centre the origin, radius } 2k\text{).}$$

Exercises 9(b)

1. Given $u = 1 - \sqrt{3}i$, $v = 1 + i$ show u, v, $u + v$ in an Argand diagram. Find $|u|$, $|v|$ and $\arg u$.
2. Find the modulus and argument of $(5 - i)/(2 - 3i)$.

3. Given $u = 3 + 2i$, $v = 2 + 3i$, (a) express u^2/v in the form $x + iy$ where x, y are real, (b) find $|u/v^2|$, (c) if $w = (u - v)/\sqrt{2}$, express w, w^2, w^3 in polar form.
4. Sketch the region of the complex plane in which the complex number z must lie if

$$\pi/6 < \arg z < 3\pi/4, \quad 1 < |z| < 2.$$

Which if any of the following numbers lie in that region:

$$u = 1 + i, \quad v = \sqrt{2} - \sqrt{3}i, \quad w = uv?$$

5. Find the modulus and argument of (a) $\cos\theta - i\sin\theta$, (b) $1 + i\tan\theta$, where $0 < \theta < \frac{1}{2}\pi$ in both cases.
6. (a) Write down the modulus and argument of each of the complex numbers $1 + i$, $-1 + i$, $1 - i$. (b) The points A, B, C, D in the Argand diagram correspond to the complex numbers $9 + i$, $4 + 13i$, $-8 + 8i$ and $-3 - 4i$. Prove that $ABCD$ is a square.
7. Find the modulus and argument of $(1 + i)/(1 - i)$, and of $\sqrt{2}/(1 - i)$. Show, without using tables, that the argument of $(1 + \sqrt{2} + i)/(1 - i)$ is $3\pi/8$.
8. Prove that if the real part of $(z + 1)/(z + i)$ is equal to 1, then the point z lies on a certain straight line in the complex plane.
9. The point representing z in the Argand diagram moves in such a way that $|(2z + 1)/(iz + 1)| = 2$. Show that it describes a straight line.
10. Show that if the ratio $(z - i)/(z - 1)$ is purely imaginary, then the point z lies on a circle whose centre is $\frac{1}{2}(1 + i)$ and radius $1/\sqrt{2}$.
11. Show that if the argument of $(z - 1)/(z + 1)$ is $\frac{1}{4}\pi$, then z lies on a fixed circle of radius $\sqrt{2}$ and centre $(0, 1)$.
12. The point P representing $z(= x + iy)$ in the Argand diagram lies on the line $6x + 8y = R$, where R is real. Q is the point representing R^2/z. Prove that the locus of Q is a circle, and find its centre and radius.
13. P is the point in the Argand diagram representing the complex number z, and Q is the point representing $1/(z - 3) + \frac{17}{3}$. Find the locus of Q as P describes the circle $|z - 3| = 3$.
14. Interpret the relation $|z - a| = |z - b|$ in the Argand diagram, where a, b, and z are complex numbers and $a \neq b$. Show that, when this relation holds, $\arg[(2z - a - b)/(a - b)] = \pm\frac{1}{2}\pi$.
15. (a) In the Argand diagram, find the locus of the point z if

$$\arg[(z - 2)/(z + 1)] = \tfrac{1}{2}\pi.$$

(b) Show that if the point z describes the circle $|z - 1| = 1$, then the point z^2 describes the curve $r = 2 + 2\cos\theta$.
16. (a) Show that if p is real, and the complex number

$$(1 + i)/(2 + pi) + (2 + 3i)/(3 + i)$$

is represented in the Argand diagram by a point on the line $x = y$, then $p = -5 \pm \sqrt{21}$.

(b) The complex numbers $z_1 = x_1 + iy_1$ and $z_2 = x_2 + iy_2$ are connected by the relation $z_1 = z_2 + 1/z_2$. If the point representing z_2 in the Argand diagram describes a circle of radius a with centre at the origin, show that the point representing z_1 describes the ellipse

$$x^2/(1 + a^2)^2 + y^2/(1 - a^2)^2 = 1/a^2.$$

17. Show that if z_1, z_2 are complex numbers,

$$|z_1 z_2| = |z_1|.|z_2| \quad \text{and} \quad \arg (z_1 z_2) = \arg z_1 + \arg z_2.$$

Given that the complex number $z = r (\cos \theta + i \sin \theta)$ is represented by the point P in an Argand diagram, find the complex number represented by the point P_1 which is the reflection of P in the y-axis. Show that if P_2 represents the complex number $-4/z$, then $OP_2 . OP_1 = 4$, where O is the origin, and that OP_1P_2 is a straight line.

By taking $z = x + iy$, and $4/z = u + iv$, where x, y, u, v are real, or otherwise, show that if P describes the line $x = c$ $(c > 0)$, in the Argand diagram the point P_2 describes a circle whose centre is on the real axis and which passes through the origin.

18. (a) Show that if z, z_0 are two complex numbers, $\bar{z}_0$ the conjugate of z_0 and $|z| = 1$, and if the numbers z, z_0, $z\bar{z}_0$, 1, 0 are represented in an Argand diagram by the points P, P_0, Q, A and the origin O respectively, then the triangles POP_0, AOQ are congruent. Hence, or otherwise, prove that $|z - z_0| = |z\bar{z}_0 - 1|$.

(b) Show that if the points representing two complex numbers z_1 and z_2 form with the origin an equilateral triangle, then $z_1^2 - z_1 z_2 + z_2^2 = 0$.

19. Prove that if z and a are complex numbers, then

$$|z + a|^2 + |z - a|^2 = 2(|z|^2 + |a|^2).$$

The complex number z is represented by a point on the circle whose centre is at the point $1 + 0i$, and whose radius is unity. Show in a diagram how $z - 2$ may be represented, and prove that

$$(z - 2)/z = i \tan (\arg z).$$

20. Given that

$$z \equiv \rho(\cos \theta + i \sin \theta) \quad \text{and} \quad a \equiv r(\cos \alpha + i \sin \alpha)$$

are two complex numbers so that $|z| = \rho$ and $|a| = r$, find the value of $|z - a|^2$ in terms of the real quantities r, ρ, θ, a. Deduce that if $\bar{a}$ is the conjugate of the number a,

$$|1 - \bar{a}z|^2 - |z - a|^2 = (1 - |z|^2)(1 - |a|^2).$$

21. (a) If A, B, P represent the numbers z_1, z_2, z respectively and $z = \lambda z_1 + \mu z_2$, where λ and μ are real numbers such that $\lambda + \mu = k$, a constant, prove that the locus of P is a straight line parallel to AB.

(b) Prove that if P represents the number z and $\arg [z/(z - i)] = \frac{1}{3}\pi$, then P lies on a circle.

22. $ABCD$ is a parallelogram whose diagonals AC and BD intersect at E. The angle AED is 45°, the length of AC is to the length of BD in the ratio 3 : 2, and the sense of the description of $ABCD$ is counter-clockwise. Given that the points A and C represent the complex numbers $-2 - 3i$ and $4 + i$ respectively in the Argand diagram, determine the numbers that are represented by the points B and D.
23. Given that the complex numbers z_1, z_2, z_3 are connected by the relation

$$2/z_1 = 1/z_2 + 1/z_3,$$

show that the points Z_1, Z_2, Z_3 representing them in an Argand diagram lie on a circle passing through the origin.

9.4 Indices and Demoivre's Theorem

Indices can be used in any field in the same way as they are used in the field of real numbers. Thus if z is a complex number and n is a positive integer, we write z^n to denote the complex number obtained by n multiplications (in accordance with (9.2)) of z by itself; we also write z^{-n} to denote the multiplicative inverse of z^n, so that $z^{-n}z^n = 1$ or $z^{-n} = 1/z^n$. If p and q are both integers we write $z^{p/q}$ to denote any complex number w with the property that $w^q = z^p$ and refer to such a number as a *qth root of* z^p. (Here we assume that the fraction p/q is in its lowest terms.)

Note that at this stage no meaning whatsoever is attached, even in the field of real numbers, to the idea of an *irrational* index, although this subject is taken up later (Chapters 17 and 22).

From (9.6) we have

$$(\cos\theta_1 + i\sin\theta_1)(\cos\theta_2 + i\sin\theta_2) = \cos(\theta_1 + \theta_2) + i\sin(\theta_1 + \theta_2),$$

and by repeated application of this result we get

$$(\cos\theta_1 + i\sin\theta_1)(\cos\theta_2 + i\sin\theta_2)\cdots(\cos\theta_n + i\sin\theta_n)$$
$$= \cos(\theta_1 + \theta_2 + \cdots + \theta_n) + i\sin(\theta_1 + \theta_2 + \cdots + \theta_n).$$

Putting $\theta_1 = \theta_2 = \cdots = \theta_n = \theta$, we now have

$$(\cos\theta + i\sin\theta)^n = \cos n\theta + i\sin n\theta. \qquad (1)$$

This result is known as *Demoivre's theorem for a positive integral index*. It is also true for negative indices, since

$$\begin{aligned}(\cos\theta + i\sin\theta)^{-n} &= \frac{1}{(\cos\theta + i\sin\theta)^n}\\ &= \frac{1}{\cos n\theta + i\sin n\theta} \qquad \text{by (1)}\\ &= \cos n\theta - i\sin n\theta\\ &= \cos(-n)\theta + i\sin(-n)\theta.\end{aligned}$$

Hence (1) can be extended to give

$$(\cos\theta + i\sin\theta)^p = \cos p\theta + i\sin p\theta \tag{2}$$

for any integer p, and this is *Demoivre's theorem for any integral index.*

The extension of (2) to include fractional indices raises the difficulty that $z^{p/q}$ does not denote a *single* complex number but rather denotes any qth root of z^p; hence the most we can ask is the question 'Is

$$\cos(p\theta/q) + i\sin(p\theta/q)$$

one of the values of

$$(\cos\theta + i\sin\theta)^{p/q}\text{?'}$$

Now

$$\begin{aligned}\{\cos(p\theta/q) + i\sin(p\theta/q)\}^q &= \cos p\theta + i\sin p\theta && \text{by (2)}\\ &= (\cos\theta + i\sin\theta)^p && \text{by (2)}\end{aligned}$$

Hence

$$\cos(p\theta/q) + i\sin(p\theta/q)$$

is a qth root of $(\cos\theta + i\sin\theta)^p$, and so is a value of $(\cos\theta + i\sin\theta)^{p/q}$. This is *Demoivre's theorem for a fractional index.*

Demoivre's theorem can be applied to any complex number by first putting the number in polar form. Consider for example the problem of evaluating $(1 + \sqrt{3}i)^5$ and $(1 + \sqrt{3}i)^{3/4}$; since

$$1 + \sqrt{3}i = 2[\cos(\pi/3) + i\sin(\pi/3)],$$

we have

$$\begin{aligned}(1 + \sqrt{3}i)^5 &= 2^5[\cos(\pi/3) + i\sin(\pi/3)]^5\\ &= 32[\cos(5\pi/3) + i\sin(5\pi/3)] = 16(1 - \sqrt{3}i),\end{aligned}$$

and

$$(1 + \sqrt{3}i)^{3/4} = 2^{3/4}[\cos(\pi/3) + i\sin(\pi/3)]^{3/4},$$

so that one of the values of $(1 + \sqrt{3}i)^{3/4}$ is

$$2^{3/4}[\cos(\pi/4) + i\sin(\pi/4)] = 2^{1/4}(1 + i).$$

Notation. Expressions of the form $\cos\theta + i\sin\theta$ occur so frequently that some kind of abbreviation is desirable. The notation $1\angle\theta$ has already been used. Later, in Chapter 22 (Section 22.7.1) we shall define e^z when the index z is a complex number and, in particular, if z is purely imaginary, say $z = i\theta$, we shall find that

$$e^{i\theta} = \cos\theta + i\sin\theta. \tag{3}$$

Although we shall not attempt to justify (3) at this stage, we shall nevertheless find immediate advantages in adopting $e^{i\theta}$ as an alternative symbol for the longer expression $\cos\theta + i\sin\theta$. By doing this we find that any complex number z can be expressed in *exponential form*, say

$$z = re^{i\theta},$$

since this is equivalent to the polar form introduced in Section 9.3. Note that

$$|re^{i\theta}| = r, \quad \arg(re^{i\theta}) = \theta,$$

and, in particular,

$$|e^{i\theta}| = 1, \quad \arg(e^{i\theta}) = \theta.$$

For example,

$$e^{i0} = 1, \quad e^{i\pi/2} = i, \quad e^{i\pi} = -1, \quad e^{i3\pi/2} = -i. \quad \text{(Fig. 9.12)}$$

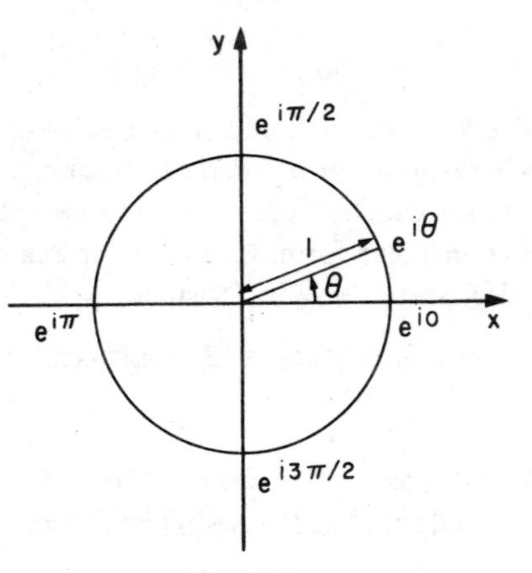

FIG. 9.12

Note also that

$$e^{i(\theta+2k\pi)} = e^{i\theta} \quad (k = \pm 1, \pm 2, \ldots),$$

since a change of $2k\pi$ in the argument of a complex number does not affect the number.

It follows from (9.6) and (9.9) that

$$(r_1e^{i\theta_1})(r_2e^{i\theta_2}) = r_1r_2e^{i(\theta_1+\theta_2)}, \quad r_1e^{i\theta_1}/r_2e^{i\theta_2} = (r_1/r_2)e^{i(\theta_1-\theta_2)},$$

and from Demoivre's theorem that

$$(e^{i\theta})^p = e^{ip\theta} \quad (p \text{ an integer}),$$

while $(e^{i\theta})^{p/q}$ has $e^{ip\theta/q}$ as *one* of its values. Thus the basic rules concerning operations with complex numbers in polar form are more easily remembered in this exponential form in which they follow the pattern of ordinary index laws.

9.4.1 Roots of a Complex Number We saw in Section 9.4 how to find *one* of the values of $z^{p/q}$. Two questions remain, viz. (a) how many other values does $z^{p/q}$ have? and (b) how do we find these other values?

Suppose that z has exponential form $re^{i\theta}$ and that $\rho e^{i\phi}$ is the exponential form of some value of $z^{p/q}$. Then

$$(\rho e^{i\phi})^q = (re^{i\theta})^p,$$

so that

$$\rho^q e^{iq\phi} = r^p e^{ip\theta};$$

that is

$$\rho^q(\cos q\phi + i \sin q\phi) = r^p(\cos p\theta + i \sin p\theta).$$

Equating real and imaginary parts we get

$$\rho^q \cos q\phi = r^p \cos p\theta, \quad \rho^q \sin q\phi = r^p \sin p\theta. \tag{1}$$

Squaring and adding we obtain $\rho^{2q} = r^{2p}$. Since ρ, the modulus of a complex number, is positive, it follows that ρ is the positive value of $r^{p/q}$; we shall write $\rho = r^{p/q}$ and in the remainder of this section interpret $r^{p/q}$ as *the positive value of* $r^{p/q}$. Equations (1) now become

$$\cos q\phi = \cos p\theta, \quad \sin q\phi = \sin p\theta,$$

giving

$$q\phi = p\theta + 2k\pi \quad (k = 0, \pm 1, \pm 2, \ldots);$$

that is

$$\phi = \frac{p\theta + 2k\pi}{q}.$$

Hence all values of $z^{p/q}$ are given by

$$r^{p/q}e^{i(p\theta + 2k\pi)/q} \quad (k = 0, \pm 1, \pm 2, \ldots).$$

When $k = 0$ we get the value already obtained in Section 9.4. When $k = 1, 2, \ldots, q - 1$ we get $(q - 1)$ other values. When $k = q$, since

$$e^{i(p\theta/q + 2\pi)} = e^{ip\theta/q},$$

we get the same value as when $k = 0$; similarly $k = q + 1, q + 2, \ldots$ and

$k = -1, -2, \ldots$ all give repetitions of previous values. This means that there are *exactly* q values of $z^{p/q}$ and they are found by putting $k = 0, 1, 2, \ldots, q - 1$ in the above formula. In an Argand diagram these roots all lie on the circumference of a circle, centre the origin, radius $r^{p/q}$, and are equally spaced at angular intervals of $2\pi/q$ around the circle.

Example 13 Find all the values of $(1 + \sqrt{3}i)^{3/4}$.

Since $1 + \sqrt{3}i = 2e^{i\pi/3}$, we have

$$(1 + \sqrt{3}i)^{3/4} = 2^{3/4}e^{i(\pi+2k\pi)/4} \quad (k = 0, 1, 2, 3).$$

When $k = 0$ we get $2^{3/4}e^{i\pi/4} = 2^{1/4}(1 + i)$.
When $k = 1$ we get $2^{3/4}e^{i3\pi/4} = 2^{1/4}(-1 + i)$.
When $k = 2$ we get $2^{3/4}e^{i5\pi/4} = -\ 2^{1/4}(1 + i)$.
When $k = 3$ we get $2^{3/4}e^{i7\pi/4} = 2^{1/4}(1 - i)$.

These values are represented by the points P_0, P_1, P_2, P_3 respectively in Fig. 9.13.

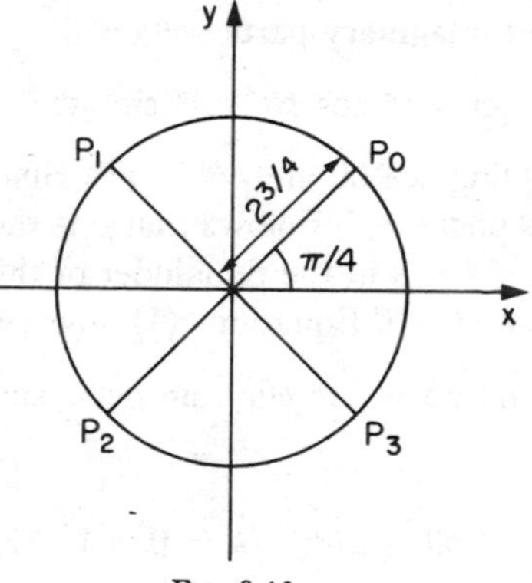

FIG. 9.13

Example 14 Find the fifth roots of -1.

Since $-1 = e^{i\pi}$, we have

$$(-1)^{1/5} = e^{i(2k+1)\pi/5} \ (k = 0, 1, 2, 3, 4).$$

The values are $e^{i\pi/5}$, $e^{i3\pi/5}$, -1, $e^{i7\pi/5}$, $e^{i9\pi/5}$.

Example 15 Find the square roots of i.

Since $i = e^{i\pi/2}$, we have

$$i^{1/2} = e^{i(\pi/2+2k\pi)/2} \quad (k = 0, 1).$$

The values are $e^{i\pi/4}$, $e^{i5\pi/4}$, that is $(1/\sqrt{2})(1+i)$, $-(1/\sqrt{2})(1+i)$.

These results are in marked contrast to the situation with real numbers. If q is even, a positive real number has 2 qth roots, while a negative real number has none, and if q is odd all real numbers have just one qth root.

9.4.2 Use of Demoivre's Theorem in the Derivation of Trigonometric Formulae If n is a positive integer we have, by Demoivre's theorem,

$$(\cos\theta + i\sin\theta)^n = \cos n\theta + i\sin n\theta,$$

and we also have, by the binomial theorem,

$$(\cos\theta + i\sin\theta)^n = \cos^n\theta + n\cos^{n-1}\theta(i\sin\theta) + \cdots.$$

It will be convenient to write $c = \cos\theta$, $s = \sin\theta$ and then

$$\begin{aligned}\cos n\theta + i\sin n\theta &= (c+is)^n\\ &= c^n + {}^nC_1c^{n-1}(is) + {}^nC_2c^{n-2}(is)^2\\ &\quad + {}^nC_3c^{n-3}(is)^3 + \cdots + (is)^n\\ &= c^n + {}^nC_1c^{n-1}is - {}^nC_2c^{n-2}s^2 - {}^nC_3c^{n-3}is^3 + \cdots,\end{aligned}$$

where

$${}^nC_r = \frac{n(n-1)(n-2)\cdots(n-r+1)}{1.2.3.\cdots.r} \quad (r = 1, 2, \ldots).$$

Equating real and imaginary parts we get

$$\cos n\theta = \cos^n\theta - {}^nC_2\cos^{n-2}\theta\sin^2\theta + \cdots, \tag{9.12}$$

$$\sin n\theta = {}^nC_1\cos^{n-1}\theta\sin\theta - {}^nC_3\cos^{n-3}\theta\sin^3\theta + \cdots. \tag{9.13}$$

Also

$$\tan n\theta = \frac{\sin n\theta}{\cos n\theta} = \frac{{}^nC_1c^{n-1}s - {}^nC_3c^{n-3}s^3 + \cdots}{c^n - {}^nC_2c^{n-2}s^2 + \cdots},$$

and dividing numerator and denominator by c^n we get

$$\tan n\theta = \frac{{}^nC_1\tan\theta - {}^nC_3\tan^3\theta + \cdots}{1 - {}^nC_2\tan^2\theta + \cdots}. \tag{9.14}$$

Since only even powers of $\sin\theta$ occur in (9.12) we can use the result $\sin^2\theta = 1 - \cos^2\theta$ to express $\cos n\theta$ entirely in terms of powers of $\cos\theta$. Likewise from (9.13) we can express $\sin n\theta$ entirely in terms of powers of $\sin\theta$. It is not necessary to remember the formulae (9.12)–(9.14) since it is easier to derive the results in any particular case, as is shown in the following example.

Example 16 Express $\cos 6\theta$ in terms of powers of $\cos\theta$.

We have

$$\begin{aligned}\cos 6\theta + i\sin 6\theta &= (c+is)^6\\ &= c^6 + 6c^5(is) + 15c^4(is)^2\\ &\quad + 20c^3(is)^3 + 15c^2(is)^4 + 6c(is)^5 + (is)^6.\\ &= c^6 + i6c^5s - 15c^4s^2 - i20c^3s^3 + 15c^2s^4 + i6cs^5 - s^6.\end{aligned}$$

Equating real parts we get

$$\begin{aligned}\cos 6\theta &= \cos^6\theta - 15\cos^4\theta\sin^2\theta + 15\cos^2\theta\sin^4\theta - \sin^6\theta\\ &= \cos^6\theta - 15\cos^4\theta\,(1-\cos^2\theta) + 15\cos^2\theta\,(1-\cos^2\theta)^2\\ &\quad - (1-\cos^2\theta)^3\\ &= 32\cos^6\theta - 48\cos^4\theta + 18\cos^2\theta - 1.\end{aligned}$$

Consider next the problem of expressing powers of $\cos\theta$ and $\sin\theta$ in terms of cosines and sines of multiples of θ. We first establish formulae (9.15) and (9.16) on which the method is based; the method is then illustrated by Examples 17 and 18.

From the equations

$$e^{i\theta} = \cos\theta + i\sin\theta, \quad e^{-i\theta} = \cos\theta - i\sin\theta,$$

we get

$$\cos\theta = \tfrac{1}{2}(e^{i\theta} + e^{-i\theta}), \tag{9.15}$$

$$\sin\theta = \frac{1}{2i}(e^{i\theta} - e^{-i\theta}). \tag{9.16}$$

Example 17 Express $\sin^5\theta$ in terms of sines of multiples of θ.

From (9.16) we have

$$\begin{aligned}\sin^5\theta &= [1/(2i)^5]\,(e^{i\theta} - e^{-i\theta})^5\\ &= (1/32i)\,(e^{i5\theta} - 5e^{i3\theta} + 10e^{i\theta} - 10e^{-i\theta} + 5e^{-i3\theta} - e^{-i5\theta})\\ &= (1/32i)\,[e^{i5\theta} - e^{-i5\theta}) - 5(e^{i3\theta} - e^{-i3\theta}) + 10(e^{i\theta} - e^{-i\theta})]\\ &= (1/16)\,(\sin 5\theta - 5\sin 3\theta + 10\sin\theta) \quad \text{by (9.16)}.\end{aligned}$$

Example 18 Express $\cos^3\theta\sin^4\theta$ in terms of cosines of multiples of θ.

From (9.15) and (9.16) we have

$$\begin{aligned}\cos^3\theta\sin^4\theta &= (1/2^3)\,(e^{i\theta} + e^{-i\theta})^3[1/(2i)^4](e^{i\theta} - e^{-i\theta})^4\\ &= (1/2^7)(e^{i2\theta} - e^{-i2\theta})^3\,(e^{i\theta} - e^{-i\theta})\\ &= (1/2^7)(e^{i6\theta} - 3e^{i2\theta} + 3e^{-i2\theta} - e^{-i6\theta})(e^{i\theta} - e^{-i\theta})\\ &= (1/2^7)[(e^{i7\theta} + e^{-i7\theta}) - (e^{i5\theta} + e^{-i5\theta}) - 3(e^{i3\theta} + e^{-i3\theta})\\ &\quad + 3(e^{i\theta} + e^{-i\theta})]\\ &= (1/64)(\cos 7\theta - \cos 5\theta - 3\cos 3\theta + 3\cos\theta) \quad \text{by (9.15)}.\end{aligned}$$

Exercises 9(c)

1. Evaluate in the form $x + iy$, where x, y are real,

   (a) $[\cos(\pi/6) + i\sin(\pi/6)]^9$,
   (b) $[\cos(\pi/12) + i\sin(\pi/12)]^5/[\cos(\pi/6) + i\sin(\pi/6)]^2$.

2. Show that if

$$[(\cos\beta + i\sin\beta)/(\cos\beta - i\sin\beta)]^6 = -1,$$

   then $\beta = (2n + 1)\pi/12$, where n is any integer.
3. Find the three roots of the equation $z^3 = 1$. Show that if z_1, z_2 are the two roots with non-zero imaginary parts, then $z_1^2 = z_2$ and $z_2^2 = z_1$. Show that the points representing the three roots on an Argand diagram are the vertices of an equilateral triangle.
4. Find the five roots of the equation $z^5 = i$.
5. Show that the nth roots of 1 form a geometric progression $1, w, w^2, \ldots, w^{n-1}$ where $w = \cos(2\pi/n) + i\sin(2\pi/n)$. Deduce that the sum of the roots is zero.
6. Find the cube roots of $1 + i$.
7. Find the cube roots of $2i - 2$.
8. Show that $(\sqrt{2} + 1 + i)/(\sqrt{2} + 1 - i) = \cos(\pi/4) + i\sin(\pi/4)$. Obtain all the values of

$$[(\sqrt{2} + 1 + i)/(\sqrt{2} + 1 - i)]^{1/4}$$

   in the form $x + iy$, where x, y are real.
9. Express the complex number $1 + i$ in the form $r(\cos\theta + i\sin\theta)$. Hence, or otherwise, prove that, n being any positive integer,

$$(1 + i)^n + (1 - i)^n = 2[2^{n/2}\cos(n\pi/4)].$$

   Expand $(1 + i)^n$ using the binomial theorem and deduce that

$$1 - {}^nC_2 + {}^nC_4 - \cdots = 2^{n/2}\cos(n\pi/4),$$

$${}^nC_1 - {}^nC_3 + {}^nC_5 - \cdots = 2^{n/2}\sin(n\pi/4).$$

10. Show that

   (a) $\sin 7\theta = 7\sin\theta - 56\sin^3\theta + 112\sin^5\theta - 64\sin^7\theta$,
   (b) $\cos 8\theta = \cos^8\theta(1 - 28\tan^2\theta + 70\tan^4\theta - 28\tan^6\theta + \tan^8\theta)$,
   (c) $(\sin 6\theta)/\sin\theta = 32\cos^5\theta - 32\cos^3\theta + 6\cos\theta$.

11. Show that

   (a) $16\cos^4\theta\sin\theta = \sin 5\theta + 3\sin 3\theta + 2\sin\theta$,
   (b) $2^6\sin^5\theta\cos^2\theta = \sin 7\theta - 3\sin 5\theta + \sin 3\theta + 5\sin\theta$,
   (c) $64\sin^7\theta = 35\sin\theta - 21\sin 3\theta + 7\sin 5\theta - \sin 7\theta$.

CHAPTER 10

FACTORISATION OF POLYNOMIALS AND THEORY OF ALGEBRAIC EQUATIONS

10.1 Complex Polynomials

In Section 9.1 we mentioned the disadvantages of the field $R(+, \times)$ from the point of view of solving equations. The discussion of equations in this chapter is based on the field $C(+, \times)$ and naturally this leads to simpler, more satisfactory results. This is not to say that we intend to ignore equations in $R(+, \times)$ in favour of some more elegant, but useless, theory. Remember that $C(+, \times)$ includes $C_r(+, \times)$ which is isomorphic to $R(+, \times)$, and we show how results for equations in $R(+, \times)$ can be extracted from results for equations in the larger field $C(+, \times)$.

Solving algebraic equations is closely related to the subject of factorising polynomials. For example, in $R(+, \times)$ we solve the equation $x^2 - x - 6 = 0$ by factorising the l.h.s. so that we get $(x - 3)(x + 2) = 0$, giving that *either* $x - 3 = 0$ *or* $x + 2 = 0$. In Example 8 of Section 6.3.1 we have shown that this link between solving equations and factorising exists in *every* field. The two topics are therefore treated together throughout this chapter.

We shall denote by $P(z)$ the polynomial

$$p_0z^n + p_1z^{n-1} + \cdots + p_{n-1}z + p_n, \tag{10.1}$$

in which z is a complex variable and $p_0, p_1, \ldots, p_n$ are given complex numbers, $p_0 \neq 0$. The remainder and factor theorems, proved for 'real' polynomials in Section 4.2.1, extend immediately to polynomials in any field and hence, in particular, to $P(z)$. Thus we can say, if a is any complex number,

(a) when $P(z)$ is divided by $z - a$, the remainder is $P(a)$,
(b) $z - a$ is a factor of $P(z)$ if, and only if, $P(a) = 0$,
(c) if $P(z) = 0$ when z has any one of the n distinct values $a_1, a_2, \ldots, a_n$, then $P(z) = p_0(z - a_1)(z - a_2) \cdots (z - a_n)$,
(d) if $P(z) = 0$ for more than n distinct values of z, then $P(z) = 0$ for all z and $p_0 = p_1 = \cdots = p_n = 0$.

The equation $P(z) = 0$ is the general algebraic equation of degree n in the complex number field. We see from (d) that this equation, like the corresponding equation for the real number field, has *at most n* roots, but in the complex number field it is possible to show that this equation has *exactly* n roots. In this connection a fundamental result is as follows:

Gauss' theorem The equation $P(z) = 0$ always has at least one root.

A proof of this theorem would be beyond the scope of this book. A direct consequence of the theorem is the following result:

Theorem 1 The equation $P(z) = 0$ has exactly n roots.

Proof. If z_1 is a root, then from (b) we see that $(z - z_1)$ is a factor of $P(z)$, i.e. $P(z) = (z - z_1)P_1(z)$, where $P_1(z)$ is a polynomial of degree $n - 1$.

Applying Gauss' theorem again, this time to the equation $P_1(z) = 0$, we see that $P_1(z)$ must have a factor $(z - z_2)$ so that

$$P(z) = (z - z_1)(z - z_2)P_2(z),$$

where $P_2(z)$ is a polynomial of degree $n - 2$. Continuing in this way we find that

$$P(z) = p_0(z - z_1)(z - z_2) \cdots (z - z_n).$$

It now follows from (b) that $z_1, z_2, \ldots, z_n$ are roots of the equation $P(z) = 0$.

Theorem 1 can be stated in the alternative form of

Theorem 2 The polynomial $P(z)$ reduces to n linear factors.

Note that in the proof of Theorem 1 it cannot be assumed that $z_1, z_2, \ldots, z_n$ are all distinct numbers. If the same number z_i occurs r times, then $P(z)$ has a *repeated* factor $(z - z_i)^r$ which must be counted as r factors for the purposes of Theorem 2, and the equation $P(z) = 0$ has a multiple root z_i which must be counted as r roots for the purpose of Theorem 1.

10.1.1 Related Equations If z in the equation $P(z) = 0$ is replaced by az, the resulting equation may be written as $P(az) = 0$. Denoting the roots of $P(z) = 0$ by z_i $(i = 1, 2, \ldots, n)$, the roots of $P(az) = 0$ are z_i/a; for substitution of $z = z_i/a$ in $P(az)$ gives $P(z_i)$. Likewise the roots of the equation $P(z + b) = 0$ are $z_i - b$, and the roots of the equation $P(az + b) = 0$ are $(z_i - b)/a$.

For example, it is known that the equation $z^2 + 1$ has roots $\pm i$. We can deduce that the roots of $4z^2 + 1 = 0$, i.e. $(2z)^2 + 1 = 0$ are $\pm i/2$, and that the roots of $z^2 - 6z + 10 = 0$, i.e. $(z - 3)^2 + 1 = 0$ are $3 \pm i$.

The equation $P(1/z) = 0$ may be written as

$$p_n z^n + p_{n-1} z^{n-1} + \cdots + p_1 z + p_0 = 0,$$

and the roots of this equation are $1/z_i$ for all $z_i \neq 0$.

10.1.2 Relations between the Roots and the Coefficients We have seen that

$$P(z) = p_0(z - z_1)(z - z_2) \cdots (z - z_n),$$

and multiplying together the factors on the r.h.s. this becomes

$$P(z) = p_0(z^n - S_1 z^{n-1} + S_2 z^{n-2} - \cdots + (-1)^n S_n), \qquad (1)$$

where

$S_1 = z_1 + z_2 + \cdots + z_n$ (sum of roots),

$S_2 = z_1 z_2 + z_1 z_3 + \cdots$ (sum of all possible products of two roots),

$S_3 = z_1 z_2 z_3 + z_1 z_2 z_4 + \cdots$ (sum of all possible products of three roots),

$\vdots$

$S_n = z_1 z_2 \cdots z_n$ (product of roots).

Now

$$P(z) = p_0\{z^n + (p_1/p_0)z^{n-1} + \cdots + (p_{n-1}/p_0)z + (p_n/p_0)\} \qquad (2)$$

and equating coefficients in the alternative forms (1) and (2) of $P(z)$ we get

$$S_1 = -p_1/p_0,\ S_2 = p_2/p_0,\ \ldots,\ S_n = (-1)^n\, p_n/p_0. \qquad (10.2)$$

In particular, if the roots of the quadratic equation $az^2 + bz + c = 0$ are α, β, then $\alpha + \beta = -b/a$, $\alpha\beta = c/a$; if the roots of the cubic equation $az^3 + bz^2 + cz + d = 0$ are α, β, γ, then

$$\alpha + \beta + \gamma = -b/a, \quad \alpha\beta + \beta\gamma + \alpha\gamma = c/a, \quad \alpha\beta\gamma = -d/a.$$

By means of the above relations any symmetric function of the roots of an algebraic equation may be expressed in terms of the coefficients.

Example 1 Given that α, β, γ are the roots of the equation

$$z^3 + pz^2 + qz + r = 0,$$

find

$$\text{(a) } \alpha^2 + \beta^2 + \gamma^2 \quad \text{and} \quad \text{(b) } \alpha^3 + \beta^3 + \gamma^3.$$

We have

$$\left.\begin{aligned} \alpha + \beta + \gamma &= -p, \\ \alpha\beta + \beta\gamma + \gamma\alpha &= \quad q, \\ \alpha\beta\gamma &= -r. \end{aligned}\right\} \qquad (1)$$

(a) Using the identity

$$\alpha^2 + \beta^2 + \gamma^2 = (\alpha + \beta + \gamma)^2 - 2(\alpha\beta + \beta\gamma + \gamma\alpha)$$

we find from (1) that

$$\alpha^2 + \beta^2 + \gamma^2 = (-p)^2 - 2q = p^2 - 2q.$$

(b) Using the identity

$$\begin{aligned} \alpha^3 + \beta^3 + \gamma^3 = (\alpha + \beta + \gamma)^3 \\ - 3(\alpha + \beta + \gamma)(\alpha\beta + \beta\gamma + \gamma\alpha) + 3\alpha\beta\gamma \end{aligned}$$

we find from (1) that

$$\alpha^3 + \beta^3 + \gamma^3 = (-p)^3 - 3(-p)q + 3(-r) = -p^3 + 3pq - 3r.$$

Alternatively, since α satisfies the given equation we have

$$\alpha^3 = -(p\alpha^2 + q\alpha + r),$$

with similar equations for β^3 and γ^3. Adding these three equations we get

$$\begin{aligned} \alpha^3 + \beta^3 + \gamma^3 &= -p(\alpha^2 + \beta^2 + \gamma^2) - q(\alpha + \beta + \gamma) - 3r \\ &= -p(p^2 - 2q) - q(-p) - 3r \qquad \text{from (1) and the result of (a)} \\ &= -p^3 + 3pq - 3r. \end{aligned}$$

Example 2 Given that α, β, γ are the roots of the equation

$$z^3 + (2 + i)z + 3i = 0,$$

evaluate $\alpha^3 + \beta^3 + \gamma^3$.

We have

$$\left.\begin{aligned} \alpha + \beta + \gamma &= 0, \\ \alpha\beta + \beta\gamma + \gamma\alpha &= 2 + i, \\ \alpha\beta\gamma &= -3i. \end{aligned}\right\} \qquad (1)$$

From the given equation

$$\alpha^3 = -(2 + i)\alpha - 3i,$$

and similarly for β^3 and γ^3. Adding, we get

$$\begin{aligned} \alpha^3 + \beta^3 + \gamma^3 &= -(2 + i)(\alpha + \beta + \gamma) - 9i \\ &= -9i \qquad \text{from (1).} \end{aligned}$$

10.1.3 Methods of Solving Algebraic Equations Since the complex number system is a field we easily derive that

(a) the linear equation $az + b = 0$ has the single root $-b/a$, and
(b) the quadratic equation $az^2 + bz + c = 0$ has two roots

$$\frac{-b \pm \sqrt{(b^2 - 4ac)}}{2a}.$$

Example 3 To find the roots of the equation

$$3z^2 - (2 + 11i)z + 3 - 5i = 0.$$

From (b) we have

$$z = \tfrac{1}{6}\{2 + 11i \pm \sqrt{[(2 + 11i)^2 - 12(3 - 5i)]}\}$$
$$= \tfrac{1}{6}[(2 + 11i \pm \sqrt{(-153 + 104i)}].$$

The square root may be found by the method of Section 9.4.1 or by solving for x and y the equation $(x + iy)^2 = -153 + 104i$, that is

$$x^2 - y^2 = -153, \quad 2xy = 104.$$

This gives $y = 52/x$ and $x^2 - 52^2/x^2 = -153$; thus

$$x^4 + 153x^2 - 52^2 = 0,$$

that is

$$(x^2 - 16)(x^2 + 169) = 0,$$

giving $x = \pm 4$, $y = \pm 13$, like signs being taken together. Hence

$$z = \tfrac{1}{6}\{2 + 11i \pm (4 + 13i)\}$$
$$= 1 + 4i \quad \text{or} \quad -\tfrac{1}{3}(1 + i).$$

Formulae have been found for the roots of algebraic equations of degree three and degree four, but these formulae are seldom used since other methods for solving these equations are more convenient in practice.

(c) The method of Section 9.4.1 may be used to solve an equation of the form $az^n + b = 0$. Other types of equation that can be solved by this method are illustrated in the following examples:

Example 4 Solve the equation

$$z^6 + z^5 + z^4 + z^3 + z^2 + z + 1 = 0,$$

and deduce that

$$\cos(2\pi/7) + \cos(4\pi/7) + \cos(6\pi/7) = -\tfrac{1}{2}.$$

Since

$$z^6 + z^5 + z^4 + z^3 + z^2 + z + 1 = (z^7 - 1)/(z - 1) \text{ if } z \neq 1,$$

we consider the equation $z^7 - 1 = 0$, for which the roots are $e^{i2k\pi/7}$ ($k = 0, 1, 2, 3, 4, 5, 6$). When $k = 0$ we get 1, which is not a root of the given equation. When $k = 1, 2, 3, 4, 5, 6$ we get the roots of the given equation; these roots can be written alternatively as $e^{i2k\pi/7}$ ($k = \pm 1, \pm 2, \pm 3$) and since their sum must be -1, we have

$$\sum_{k=1}^{3} \left(e^{i2k\pi/7} + e^{-i2k\pi/7}\right) = -1,$$

that is

$$2\{\cos(2\pi/7) + \cos(4\pi/7) + \cos(6\pi/7)\} = -1.$$

Example 5 Solve the equation $(z + 1)^3 + (z - 1)^3 = 0$.

From the given equation we get

$$\left(\frac{z+1}{z-1}\right)^3 = -1 = e^{i(2k+1)\pi};$$

that is

$$\frac{z+1}{z-1} = e^{i(2k+1)\pi/3} \quad (k = 0, 1, 2);$$

hence

$$z = \frac{e^{i(2k+1)\pi/3} + 1}{e^{i(2k+1)\pi/3} - 1}.$$

Writing $\theta_k = (2k + 1)\pi/3$, we have

$$\begin{aligned} z &= \frac{1 + \cos\theta_k + i\sin\theta_k}{-(1 - \cos\theta_k) + i\sin\theta_k} \\ &= \frac{2\cos^2\frac{1}{2}\theta_k + 2i\sin\frac{1}{2}\theta_k\cos\frac{1}{2}\theta_k}{-2\sin^2\frac{1}{2}\theta_k + 2i\sin\frac{1}{2}\theta_k\cos\frac{1}{2}\theta_k} \\ &= \frac{2\cos\frac{1}{2}\theta_k(\cos\frac{1}{2}\theta_k + i\sin\frac{1}{2}\theta_k)}{2\sin\frac{1}{2}\theta_k(-\sin\frac{1}{2}\theta_k + i\cos\frac{1}{2}\theta_k)} = -i\cot\tfrac{1}{2}\theta_k. \end{aligned}$$

Hence the roots are

$$-i\cot(2k + 1)\pi/6 \quad (k = 0, 1, 2).$$

10.2 Polynomials with Real Coefficients

The coefficients $p_0, p_1, p_2, \ldots, p_n$ in (10.1) are complex numbers, so that $p_r = a_r + ib_r$ (say) where a_r, b_r are real numbers ($r = 0, 1, 2, \ldots, n$). If all these coefficients are purely real, i.e. $b_r = 0$ $(r = 0, 1, 2, \ldots, n)$, then, omitting the zero imaginary parts in the usual way, we may write

$$P(z) = a_0 z^n + a_1 z^{n-1} + \cdots + a_{n-1} z + a_n, \tag{10.3}$$

and we say that $P(z)$ has ***real coefficients.***

We now look at some results which apply to this particular kind of polynomial.

Theorem 3 If z_1 is a root of the equation $P(z) = 0$ and $P(z)$ has real coefficients, then $\bar{z}_1$ is also a root.

Proof. Writing $z_1 = x_1 + iy_1$, we have

$$P(z_1) = a_0(x_1 + iy_1)^n + a_1(x_1 + iy_1)^{n-1} + \cdots + a_n.$$

Expanding on the r.h.s. and using the results $i^2 = i^6 = i^{10} \cdots = -1$, $i^4 = i^8 = \cdots = 1$, we get

$$P(z_1) = X + iY,$$

where X contains only even powers of y_1 and Y contains only odd powers of y_1. It follows that

$$a_0(x_1 - iy_1)^n + a_1(x_1 - iy_1)^{n-1} + \cdots + a_n = X - iY,$$

since y_1 has been replaced by $-y_1$. Now $P(z_1) = 0$, since z_1 is a root of the equation $P(z) = 0$, and this implies that $X + iY = 0$, i.e. $X = 0$, $Y = 0$. Hence $X - iY = 0$, i.e. $P(\bar{z}_1) = 0$, so that $\bar{z}_1$ is also a root.

Theorem 3 shows that roots of the equation $P(z) = 0$ that are not purely real occur in conjugate pairs.

Example 6 Solve the equation $z^4 - 2z^3 + 3z^2 - 2z + 2 = 0$ given that $1 + i$ is a root.

By Theorem 3, $1 - i$ is also a root. It follows that $z - (1 + i)$ and $z - (1 - i)$ are both factors of the l.h.s., i.e.

$$[z - (1 + i)][z - (1 - i)]$$

divides the l.h.s. Multiplying, we get that $z^2 - 2z + 2$ divides the l.h.s., and dividing out we find that the other factor is $z^2 + 1$, i.e. $(z - i)(z + i)$. Hence the roots are $1 \pm i$, $\pm i$.

Corresponding to a purely real root z_i of the equation $P(z) = 0$ there is a linear factor $z - z_i$ of $P(z)$; clearly the two coefficients, viz. $1, -z_i$, in this factor are purely real.

Corresponding to a pair of conjugate roots z_i, $\bar{z}_i$ there are two linear factors $z - z_i$, $z - \bar{z}_i$ which when multiplied together give the quadratic factors $z^2 - (z_i + \bar{z}_i)z + z_i\bar{z}_i$; writing $z_i = x_i + iy_i$ so that $\bar{z}_i = x_i - iy_i$ and hence $z_i + \bar{z}_i = 2x_i$, $z_i\bar{z}_i = x_i^2 + y_i^2$, we see that this quadratic factor may be written as $z^2 - 2x_iz + x_i^2 + y_i^2$ and so has purely real coefficients.

We are led to the following conclusion:

Theorem 4 A polynomial $P(z)$ with real coefficients may be reduced to linear and/or quadratic factors with all factors having real coefficients.

Example 7 To find factors with purely real coefficients for $z^4 + 1$.

Solving the equation $z^4 + 1 = 0$ we get

$$z = e^{i(2k+1)\pi/4} \quad (k = 0, 1, 2, 3).$$

That is,

$$z = \frac{1}{\sqrt{2}}(1 + i), \quad \frac{1}{\sqrt{2}}(-1 + i),$$

$$\frac{1}{\sqrt{2}}(-1 - i), \quad \frac{1}{\sqrt{2}}(1 - i).$$

The first and fourth roots as written are a conjugate pair; the two linear factors corresponding to these two roots combine to give the quadratic factor $z^2 - \sqrt{2}z + 1$ (putting $x_i = y_i = 1/\sqrt{2}$ in the general formula above). Likewise the second and third roots are a conjugate pair and give rise to a quadratic factor $z^2 + \sqrt{2}z + 1$. Hence

$$z^4 + 1 = (z^2 - \sqrt{2}z + 1)(z^2 + \sqrt{2}z + 1).$$

Example 8 Find factors with purely real coefficients for

$$z^6 + z^5 + z^4 + z^3 + z^2 + z + 1.$$

From Example 4 the roots of the equation

$$z^6 + z^5 + z^4 + z^3 + z^2 + z + 1 = 0$$

are

$$\cos(2k\pi/7) \pm i \sin(2k\pi/7) \quad (k = 1, 2, 3).$$

Corresponding to these six roots we get six linear factors which, grouped in pairs, give three quadratic factors

$$[z^2 - 2z\cos(2\pi/7) + 1][z^2 - 2z\cos(4\pi/7) + 1] \times [z^2 - 2z\cos(6\pi/7) + 1]$$

with purely real coefficients.

10.3 Real Polynomials

We shall use the term *real polynomial* for a polynomial $P(x)$ in the field of real numbers, i.e. the variable x is a real variable and the coefficients are real numbers.

Writing

$$P(x) = a_0x^n + a_1x^{n-1} + \cdots + a_{n-1}x + a_n,$$

we see that the real polynomial $P(x)$ can be identified with the complex polynomial $P(z)$ (with real coefficients) shown in (10.3); furthermore, if the variable z in (10.3) is restricted to take values in C_r, i.e. if $z = x + iy$ and we put $y = 0$, then $P(z)$ and $P(x)$ are for all practical purposes equivalent. Thus the problem of solving the equation $P(x) = 0$ in $R(+, \times)$ is equivalent to the problem of finding purely real roots of the equation $P(z) = 0$; likewise, the problem of factorising $P(x)$ is equivalent to the problem of factorising $P(z)$ into factors with real coefficients.

Example 9 To find the factors of $x^6 - 1$.

Consider the polynomial $z^6 - 1$ where z is a complex variable. The equation $z^6 - 1 = 0$ has roots

$$e^{i2k\pi/6} \quad (k = 0, 1, 2, 3, 4, 5),$$

that is

$$1, \ \tfrac{1}{2}(1 + \sqrt{3}i), \ \tfrac{1}{2}(-1 + \sqrt{3}i), \ -1,$$
$$-\tfrac{1}{2}(1 + \sqrt{3}i), \ \tfrac{1}{2}(1 - \sqrt{3}i).$$

The second and sixth roots are conjugate and the corresponding linear factors combine to give the quadratic factor $z^2 - z + 1$.

Likewise, corresponding to the third and fifth roots there is a quadratic factor $z^2 + z + 1$.

We can now write

$$z^6 - 1 = (z - 1)(z + 1)(z^2 - z + 1)(z^2 + z + 1).$$

It follows that $x^6 - 1 = (x - 1)(x + 1)(x^2 - x + 1)(x^2 + x + 1)$.

In a procedure such as that used in Example 9, one seldom troubles to use the symbol z for a complex variable different from the symbol x used for the real variable. Instead one uses the same variable x throughout, interpreting it as a real variable or as a complex variable as appropriate at the time.

In fact it is common practice in most mathematical literature not to distinguish between real polynomials and complex polynomials with real coefficients, or between real numbers and purely real complex numbers.

As a result there is an apparent tendency to switch from real numbers to complex numbers and vice versa whenever it suits our purposes; we are, of course, using the isomorphism between the two systems $C_r(+, \times)$ and $R(+, \times)$, and the lack of repeated explanations need cause no confusion once the point has been properly grasped.

Exercises 10

1. Find all the roots of the following equations:

   (a) $z^2 - (3 + 5i)z + 8i - 4 = 0$,
   (b) $z^2 + (4 - 6i)z = 9 + 15i$,
   (c) $z^6 - 2z^3 + 2 = 0$.

2. By substituting $t = x^2 + x$, or otherwise, solve the equation
$$(x^2 + x - 10)(x^2 + x - 20) = 80.$$

3. By substituting $t = x + 1/x$, or otherwise, solve the equations

   (a) $2x^4 - 13x^3 + 24x^2 - 13x + 2 = 0$,
   (b) $6x^4 - 25x^3 + 37x^2 - 25x + 6 = 0$.

4. Write down the solutions of the equation $w^4 = 16$ and deduce the solutions of the equation $(z + 1)^4 = 16(z - 1)^4$.

5. Find all solutions of the equations

   (a) $z^6 + z^4 + z^2 + 1 = 0$, (b) $z^4 = (z + 1)^4$.

6. Verify that $2 + i$ is a root of the equation $4z^3 - z^2 - 40z + 75 = 0$. Write down the other complex root, and hence deduce the real root.

7. Verify that $z = -\frac{1}{2} + i\sqrt{3}/2$ is one solution of the equation
$$z^4 + 5z^2 + 4z + 5 = 0.$$
A second solution can now be written down. State that solution and the theorem that justifies your statement.
   Use these two solutions to find one quadratic factor of $z^4 + 5z^2 + 4z + 5$. Then find the other.

8. Verify that $x = 1 + 2i$ is a solution of
$$2x^3 - x^2 + 4x + 15 = 0.$$
State the other complex solution and hence obtain the third solution. Factorise the cubic into factors with real coefficients.

9. Find the five solutions of $z^5 = 1$. Then express $z^5 - 1$ as the product of linear factors, and hence as the product of one linear and two quadratic factors with real coefficients.

10. Solve the equation $z^3 = i(z - 1)^3$, and show that the points in the Argand diagram that represent the roots are collinear.

11. Express the four roots of the equation $(z - 2)^4 + (z + 1)^4 = 0$ in the form
$$a_\nu + ib_\nu \quad (\nu = 1, 2, 3, 4),$$
where a_ν, b_ν are real numbers.

12. Write down the five roots of $z^5 - 1 = 0$. Show that the roots of the equation $(5 + z)^5 - (5 - z)^5 = 0$ can be written in the form

$$5i \tan (r\pi/5), \quad r = 0, \pm 1, \pm 2.$$

13. Find in the form $p + iq$, where p and q are real, all the solutions of the equations

(a) $z^2 - 4iz - 4 - 2i = 0$, (b) $z^6 + 8i = 0$.

14. Determine the roots of the equation $z^5 = 1$ and describe their positions in the Argand diagram.

Let ω be the root, other than 1, that lies in the first quadrant. Given that $u = \omega + \omega^4$ and $v = \omega^2 + \omega^3$, prove that

$$u + v = uv = -1 \quad \text{and} \quad u - v = +\sqrt{5}.$$

Deduce that $\cos 72° = (\sqrt{5} - 1)/4$.

15. Prove that, with the exception of one zero root, the roots of the equation $(1 + z)^n = (1 - z)^n$ are all imaginary.

16. Given that α, β, γ are the roots of the equation

$$x^3 + px^2 + r = 0,$$

express $\alpha^2 + \beta^2 + \gamma^2$ and $\beta^2\gamma^2 + \gamma^2\alpha^2 + \alpha^2\beta^2$ in terms of p and r.

17. Given that α, β, γ are the roots of the equation

$$x^3 + px^2 + qx + r = 0,$$

express $\alpha^3 + \beta^3 + \gamma^3$ in terms of p, q, r, and show that

$$1/\alpha^3 + 1/\beta^3 + 1/\gamma^3 = (3pqr - q^3 - 3r^2)/r^3.$$

18. Find the condition for the roots of the equation

$$x^3 + px^2 + qx + r = 0$$

to be in geometric progression, and solve the equation

$$3x^3 - 26x^2 + 52x - 24 = 0.$$

19. If α, β, γ are the roots of the equation $x^3 + px^2 + qx + r = 0$, express $\alpha\beta/\gamma + \beta\gamma/\alpha + \gamma\alpha/\beta$ in terms of p, q and r.

20. If the roots of the equation

$$x^4 - ax^3 + bx^2 - abx + 1 = 0$$

are $\alpha, \beta, \gamma, \delta$, show that

$$(\alpha + \beta + \gamma)(\alpha + \beta + \delta)(\alpha + \gamma + \delta)(\beta + \gamma + \delta) = 1.$$

21. Find the nth roots of unity and prove that their sum is zero. If ω is a complex fifth root of unity, prove that $\omega + 1/\omega$ is real and satisfies the equation $x^2 + x - 1 = 0$.

Hence show that

$$\cos (2\pi/5) = \tfrac{1}{4}(-1 + \sqrt{5}), \quad \cos (\pi/5) = \tfrac{1}{4}(1 + \sqrt{5}).$$

CHAPTER 11

LINEAR DEPENDENCE, DETERMINANTS AND SIMULTANEOUS LINEAR EQUATIONS

11.1 Vectors in R^3

In three-dimensional co-ordinate geometry (x, y, z) denotes a *point* and the set of all such points is denoted by R^3—this set is usually referred to as *three-dimensional space*. As was mentioned in Section 7.5, (x, y, z) may also denote the vector $x\mathbf{i} + y\mathbf{j} + z\mathbf{k}$, i.e. the same symbol may be used for the *point* and for the *position vector* of the point. Moreover when this notation for vectors is being used it is usual to denote the set of all three-dimensional vectors (previously denoted by V_3) by R^3. Likewise two-dimensional vectors are sometimes written in the form (x, y) and the set of all such vectors is denoted by V_2 or R^2. We refer to R^2 and R^3 as *vector spaces*.

In effect the system $R^2(+, .)$ may be regarded as a system of elements (x, y) (i.e. each element is an ordered pair of real numbers) subject to the rules (see Section 7.5.4)

$$(x_1, y_1) + (x_2, y_2) = (x_1 + x_2, y_1 + y_2),$$
$$(x_1, y_1) . (x_2, y_2) = x_1x_2 + y_1y_2,$$

and an operation of multiplication by a scalar may be included by adding the rule

$$k(x, y) = (kx, ky).$$

Likewise the system $R^3(+, .)$ may be regarded as a system of elements (x, y, z) subject to the rules (from (7.6), (7.11) and (7.8))

$$(x_1, y_1, z_1) + (x_2, y_2, z_2) = (x_1 + x_2, y_1 + y_2, z_1 + z_2),$$
$$(x_1, y_1, z_1) . (x_2, y_2, z_2) = x_1x_2 + y_1y_2 + z_1z_2$$
$$k(x, y, z) = (kx, ky, kz).$$

11.1.1 Linear Dependence

The vector $\mathbf{r}$ is said to be a *linear combination* of the vectors $\mathbf{r}_1, \mathbf{r}_2, \ldots, \mathbf{r}_m$ if scalars $k_1, k_2, \ldots, k_m$ exist such that

$$\mathbf{r} = k_1\mathbf{r}_1 + k_2\mathbf{r}_2 + \cdots + k_m\mathbf{r}_m. \qquad (11.1)$$

For example, if

$$\mathbf{r}_1 = (1, 2, 3), \quad \mathbf{r}_2 = (4, 5, 6), \quad \mathbf{r}_3 = (7, 8, 9), \quad \mathbf{r}_4 = (6, 9, 12), \quad (1)$$

then $\mathbf{r}_4$ is a linear combination of $\mathbf{r}_1$ and $\mathbf{r}_2$ since $\mathbf{r}_4 = 2\mathbf{r}_1 + \mathbf{r}_2$; also $\mathbf{r}_4$ is a linear combination of $\mathbf{r}_1$, $\mathbf{r}_2$, $\mathbf{r}_3$ since $\mathbf{r}_4 = 2\mathbf{r}_1 + \mathbf{r}_2 + 0\mathbf{r}_3$.

A set of non-zero vectors $\{\mathbf{r}_1, \mathbf{r}_2, \ldots, \mathbf{r}_m\}$ is said to be *linearly dependent* if scalars $k_1, k_2, \ldots, k_m$ exist, not all zero, such that

$$k_1\mathbf{r}_1 + k_2\mathbf{r}_2 + \cdots + k_m\mathbf{r}_m = \mathbf{0}. \qquad (11.2)$$

Thus in (1) above we see that $\{\mathbf{r}_1, \mathbf{r}_2, \mathbf{r}_4\}$ is a linearly dependent set since $2\mathbf{r}_1 + \mathbf{r}_2 - \mathbf{r}_4 = \mathbf{0}$; also $\{\mathbf{r}_1, \mathbf{r}_2, \mathbf{r}_3, \mathbf{r}_4\}$ is linearly dependent since $2\mathbf{r}_1 + \mathbf{r}_2 + 0\mathbf{r}_3 - \mathbf{r}_4 = \mathbf{0}$.

From (11.2) we see that if $k_1 \neq 0$ we may divide throughout the equation by k_1 and thereby express $\mathbf{r}_1$ as a linear combination of the other vectors; likewise if $k_2 \neq 0$ we can express $\mathbf{r}_2$ as a linear combination of the other vectors, and so on.

A set of vectors that is not linearly dependent is said to be *linearly independent*.

For example the set $\{(1, 0, 0), (0, 1, 0), (0, 0, 1)\}$, i.e. $\{\mathbf{i}, \mathbf{j}, \mathbf{k}\}$, is linearly independent since

$$k_1\mathbf{i} + k_2\mathbf{j} + k_3\mathbf{k} = (k_1, k_2, k_3),$$

and this is $\mathbf{0}$ only if $k_1 = k_2 = k_3 = 0$.

Suppose two vectors $\mathbf{r}_1$, $\mathbf{r}_2$ are related by the equation $k_1\mathbf{r}_1 + k_2\mathbf{r}_2 = \mathbf{0}$ and k_1, k_2 are not both zero. If $k_2 \neq 0$ we may write $\mathbf{r}_2 = -(k_1/k_2)\mathbf{r}_1$ and it follows that $\mathbf{r}_1$, $\mathbf{r}_2$ are parallel (see Section 7.2.3). Conversely if $\mathbf{r}_1$, $\mathbf{r}_2$ are parallel it follows immediately that $\{\mathbf{r}_1, \mathbf{r}_2\}$ is linearly dependent. Hence

$$\{\mathbf{r}_1, \mathbf{r}_2\} \text{ linearly dependent} \Leftrightarrow \mathbf{r}_1, \mathbf{r}_2 \text{ are parallel.} \qquad (11.3)$$

(The zero vector is regarded as being parallel to every vector.)

Next consider three vectors $\mathbf{r}_1$, $\mathbf{r}_2$, $\mathbf{r}_3$ related by the equation

$$k_1\mathbf{r}_1 + k_2\mathbf{r}_2 + k_3\mathbf{r}_3 = \mathbf{0},$$

and suppose that one of the scalars, k_3 say, is non-zero. Then

$$\mathbf{r}_3 = -(k_1/k_3)\mathbf{r}_1 - (k_2/k_3)\mathbf{r}_2,$$

say $\mathbf{r}_3 = \alpha\mathbf{r}_1 + \beta\mathbf{r}_2$, and it follows (see Fig. 11.1) that $\mathbf{r}_3$ lies in the plane of $\mathbf{r}_1$ and $\mathbf{r}_2$. Conversely, it follows from (7.2) (Section 7.4) that if $\mathbf{r}_1$, $\mathbf{r}_2$, $\mathbf{r}_3$ are coplanar then $\{\mathbf{r}_1, \mathbf{r}_2, \mathbf{r}_3\}$ is linearly dependent. Hence

$$\{\mathbf{r}_1, \mathbf{r}_2, \mathbf{r}_3\} \text{ linearly dependent} \Leftrightarrow \mathbf{r}_1, \mathbf{r}_2, \mathbf{r}_3 \text{ are coplanar.} \qquad (11.4)$$

(The zero vector is regarded as being coplanar with every pair of vectors.)

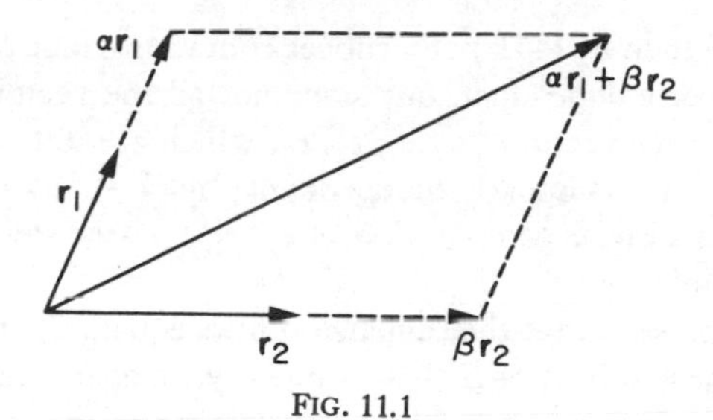

FIG. 11.1

Let $\{\mathbf{r}_1, \mathbf{r}_2, \mathbf{r}_3, \mathbf{r}_4\}$ be any set of four vectors. Suppose three of them, say $\mathbf{r}_1, \mathbf{r}_2, \mathbf{r}_3$, are not coplanar; then by (7.4) we can find scalars k_1, k_2, k_3 so that

$$\mathbf{r}_4 = k_1\mathbf{r}_1 + k_2\mathbf{r}_2 + k_3\mathbf{r}_3$$

and we see that $\{\mathbf{r}_1, \mathbf{r}_2, \mathbf{r}_3, \mathbf{r}_4\}$ is linearly dependent. If $\mathbf{r}_1, \mathbf{r}_2, \mathbf{r}_3$ are coplanar it follows trivially from (11.4) that $\{\mathbf{r}_1, \mathbf{r}_2, \mathbf{r}_3, \mathbf{r}_4\}$ is linearly dependent. Hence

$$\{\mathbf{r}_1, \mathbf{r}_2, \ldots, \mathbf{r}_m\} \text{ is necessarily linearly dependent if } m > 3. \quad (11.5)$$

11.1.2 The Dimension of a Set of Vectors A set S of vectors is said to have *dimension* 1 (dim $S = 1$) if all the vectors in S are parallel, e.g. $\{(1,2,3), (-2,-4,-6), (3,6,9)\}$ has dimension 1, as also has the set V_1 (or R) of *all* vectors parallel to some fixed direction (Section 7.3).

We say dim $S = 2$ if not all the vectors in S are parallel but all are coplanar, e.g. $\{(1,2,3), (4,5,6), (6,9,12)\}$ has dimension 2, as has the set R of *all* vectors in some fixed plane.

We say dim $S = 3$ if not all the vectors in S are coplanar, e.g. $\{(1,0,0), (0,1,0), (0,0,1)\}$ has dimension 3, as has the set R^3 of *all* vectors.

Let $S = \{\mathbf{r}_1, \mathbf{r}_2, \ldots, \mathbf{r}_m\}$.

If dim $S = 1$ it follows from (11.3) that any subset of S that contains two or more vectors is linearly dependent; *any* vector in S can be expressed as a scalar multiple of (say) $\mathbf{r}_1$—we say $\{\mathbf{r}_1\}$ is a *basis* for the set, but obviously any non-zero vector in S can be chosen to form a basis in this way.

If dim $S = 2$ it follows from (11.4) that any subset of S containing three or more vectors is linearly dependent; since not all vectors in S are parallel there must be (at least) two non-parallel vectors, say $\mathbf{r}_1, \mathbf{r}_2$, so that $\{\mathbf{r}_1, \mathbf{r}_2\}$ is linearly independent and by (7.2) any vector in S can be expressed as a linear combination of $\mathbf{r}_1, \mathbf{r}_2$—we say $\{\mathbf{r}_1, \mathbf{r}_2\}$ is a *basis* for S in this case.

If dim $S = 3$ then by (11.5) any subset containing four or more vectors is certainly linearly dependent, but since not all the vectors are coplanar there must be three vectors, say $\mathbf{r}_1, \mathbf{r}_2, \mathbf{r}_3$, which are not coplanar and so by (11.4) $\{\mathbf{r}_1, \mathbf{r}_2, \mathbf{r}_3\}$ is linearly independent; by (7.4) any vector in S can be expressed as a linear combination of $\mathbf{r}_1, \mathbf{r}_2, \mathbf{r}_3$—we say $\{\mathbf{r}_1, \mathbf{r}_2, \mathbf{r}_3\}$ is a *basis* for S in this case.

It can now be seen that the dimension of S equals the number of vectors in the largest subset of S that is linearly independent. If dim $S = s$ ($s = 1, 2, 3$), then one or more subsets containing s vectors are linearly independent and any such subset can be chosen as a basis for S, i.e. any vector in S can be expressed as a linear combination of the s vectors in the subset. In particular, dim $R^3 = 3$ and the subset $\{\mathbf{i}, \mathbf{j}, \mathbf{k}\}$ can be chosen as a basis for R^3 since it is linearly independent and contains three vectors.

11.1.3 Determinants and Linear Dependence If $\mathbf{r}_1 = (a_1, b_1, c_1)$, $\mathbf{r}_2 = (a_2, b_2, c_2)$, then

$$\mathbf{r}_1, \mathbf{r}_2 \text{ parallel} \Leftrightarrow a_1/a_2 = b_1/b_2 = c_1/c_2,$$

assuming that none of a_2, b_2, c_2 is zero. In other words

$$\{\mathbf{r}_1, \mathbf{r}_2\} \text{ linearly dependent}$$
$$\Leftrightarrow a_1b_2 - a_2b_1 = b_1c_2 - b_2c_1 = a_1c_2 - a_2c_1 = 0,$$

and in this latter form the statement is true even if some of a_2, b_2, c_2 are zero. In terms of *determinants*, this may be written as

$$\{\mathbf{r}_1, \mathbf{r}_2\} \text{ linearly dependent} \Leftrightarrow \begin{vmatrix} a_1 & b_1 \\ a_2 & b_2 \end{vmatrix} = \begin{vmatrix} b_1 & c_1 \\ b_2 & c_2 \end{vmatrix} = \begin{vmatrix} a_1 & c_1 \\ a_2 & c_2 \end{vmatrix} = 0, \quad (1)$$

each *2 × 2 determinant* being defined according to the rule

$$\begin{vmatrix} p & q \\ r & s \end{vmatrix} = ps - qr.$$

Consider the array of numbers

$$\begin{matrix} a_{11} & a_{12} & \cdots & a_{1n} \\ a_{21} & a_{22} & \cdots & a_{2n} \\ \vdots & & & \\ a_{m1} & a_{m2} & \cdots & a_{mn} \end{matrix} \qquad \text{(A)}$$

with *m rows* and *n columns*, where $m \geqslant 2$ and $n \geqslant 2$. If two rows and two columns are selected and all other rows and columns are omitted, we are left with a 2×2 array; this can be done in ${}^mC_2 \times {}^nC_2$ ways, and the corresponding 2×2 determinants are called the *2×2 determinants of* (A). For example

$$\begin{matrix} 1 & 2 & 3 & 4 \\ 5 & 6 & 7 & 8 \\ 9 & 10 & 11 & 12 \end{matrix}$$

has ${}^3C_2 \times {}^4C_2$ (i.e. 18) 2×2 determinants, including

$$\begin{vmatrix} 1 & 2 \\ 5 & 6 \end{vmatrix} = -4, \quad \begin{vmatrix} 1 & 3 \\ 5 & 7 \end{vmatrix} = -8, \quad \begin{vmatrix} 1 & 2 \\ 9 & 10 \end{vmatrix} = -8, \quad \begin{vmatrix} 2 & 3 \\ 6 & 7 \end{vmatrix} = -4,$$

and so on.

The array

$$\begin{matrix} a_1 & b_1 & c_1 \\ a_2 & b_2 & c_2 \end{matrix} \tag{B}$$

is formed from the components of $\mathbf{r}_1$, $\mathbf{r}_2$ and (1) can be written as

$$\{\mathbf{r}_1, \mathbf{r}_2\} \text{ linearly dependent} \Leftrightarrow \text{all the } 2 \times 2 \text{ determinants of (B) are zero} \tag{11.6}$$

It should be noted that the formula (7.19) for a vector product can be expressed in terms of determinants; thus

$$(x_1, y_1, z_1) \wedge (x_2, y_2, z_2) = \left(\begin{vmatrix} y_1 & z_1 \\ y_2 & z_2 \end{vmatrix}, - \begin{vmatrix} x_1 & z_1 \\ x_2 & z_2 \end{vmatrix}, \begin{vmatrix} x_1 & y_1 \\ x_2 & y_2 \end{vmatrix} \right).$$

If $\{\mathbf{r}_1, \mathbf{r}_2, \mathbf{r}_3\}$ is linearly dependent, so that $\mathbf{r}_1$, $\mathbf{r}_2$, $\mathbf{r}_3$ are coplanar, then $\mathbf{r}_1 . \mathbf{r}_2 \wedge \mathbf{r}_3 = 0$, and conversely. Writing

$$\mathbf{r}_1 = (a_1, b_1, c_1), \quad \mathbf{r}_2 = (a_2, b_2, c_2), \quad \mathbf{r}_3 = (a_3, b_3, c_3)$$

we have

$$\mathbf{r}_2 \wedge \mathbf{r}_3 = \left(\begin{vmatrix} b_2 & c_2 \\ b_3 & c_3 \end{vmatrix}, - \begin{vmatrix} a_2 & c_2 \\ a_3 & c_3 \end{vmatrix}, \begin{vmatrix} a_2 & b_2 \\ a_3 & b_3 \end{vmatrix} \right),$$

$$\mathbf{r}_1 . \mathbf{r}_2 \wedge \mathbf{r}_3 = a_1 \begin{vmatrix} b_2 & c_2 \\ b_3 & c_3 \end{vmatrix} - b_1 \begin{vmatrix} a_2 & c_2 \\ a_3 & c_3 \end{vmatrix} + c_1 \begin{vmatrix} a_2 & b_2 \\ a_3 & b_3 \end{vmatrix}.$$

Definition.

$$\begin{vmatrix} a_1 & b_1 & c_1 \\ a_2 & b_2 & c_2 \\ a_3 & b_3 & c_3 \end{vmatrix} = a_1 \begin{vmatrix} b_2 & c_2 \\ b_3 & c_3 \end{vmatrix} - b_1 \begin{vmatrix} a_2 & c_2 \\ a_3 & c_3 \end{vmatrix} + c_1 \begin{vmatrix} a_2 & b_2 \\ a_3 & b_3 \end{vmatrix}$$

the l.h.s. being called a *3×3 determinant*, or *third-order determinant*.

The *3×3 determinants of* (A) are the ${}^mC_3 \times {}^nC_3$ third-order determinants that can be found by selecting three rows and three columns from (A) and omitting all the other rows and columns.

It is a direct consequence of this definition that

$$\mathbf{r}_1 . \mathbf{r}_2 \wedge \mathbf{r}_3 = \begin{vmatrix} a_1 & b_1 & c_1 \\ a_2 & b_2 & c_2 \\ a_3 & b_3 & c_3 \end{vmatrix} \tag{11.7}$$

and that

$$\{\mathbf{r}_1, \mathbf{r}_2, \mathbf{r}_3\} \text{ linearly dependent} \Leftrightarrow \begin{vmatrix} a_1 & b_1 & c_1 \\ a_2 & b_2 & c_2 \\ a_3 & b_3 & c_3 \end{vmatrix} = 0. \tag{11.8}$$

Note also that the formula (7.19) for a vector product can now be expressed in the compact form

$$(x_1, y_1, z_1) \wedge (x_2, y_2, z_2) = \begin{vmatrix} \mathbf{i} & \mathbf{j} & \mathbf{k} \\ x_1 & y_1 & z_1 \\ x_2 & y_2 & z_2 \end{vmatrix} .$$

Let S denote a set of three or more vectors, $S = \{\mathbf{r}_1, \mathbf{r}_2, \ldots, \mathbf{r}_m\}$ and let

$$\begin{matrix} a_1 & b_1 & c_1 \\ a_2 & b_2 & c_2 \\ \vdots & \vdots & \vdots \\ a_m & b_m & c_m \end{matrix} \tag{C}$$

be the array formed from the components of these vectors. It now follows from Section 11.1.2 that

$\dim S = 3 \Leftrightarrow$ at least one 3×3 determinant of (C) is non-zero, (11.9)

$\dim S = 2 \Leftrightarrow$ all 3×3 determinants of (C) are zero, and at least one 2×2 determinant of (C) is non-zero, (11.10)

$\dim S = 1 \Leftrightarrow$ all 2×2 determinants of (C) are zero. (11.11)

Example 1 Determine the dimension of the set of vectors

$$\{(1, 4, 0), \quad (1, 4, 3), \quad (2, 8, 5), \quad (0, 1, -1)\}$$

and select a basis for the set.

We begin by examining the 3×3 determinants of the array

$$\begin{matrix} 1 & 4 & 0 \\ 1 & 4 & 3 \\ 2 & 8 & 5 \\ 0 & 1 & -1 \end{matrix}$$

For example,

$$\begin{vmatrix} 1 & 4 & 0 \\ 1 & 4 & 3 \\ 2 & 8 & 5 \end{vmatrix} = \begin{vmatrix} 4 & 3 \\ 8 & 5 \end{vmatrix} - 4 \begin{vmatrix} 1 & 3 \\ 2 & 5 \end{vmatrix} + 0 \begin{vmatrix} 1 & 4 \\ 2 & 8 \end{vmatrix}$$
$$= -4 - 4(-1) + 0 = 0,$$

$$\begin{vmatrix} 1 & 4 & 0 \\ 1 & 4 & 3 \\ 0 & 1 & -1 \end{vmatrix} = \begin{vmatrix} 4 & 3 \\ 1 & -1 \end{vmatrix} - 4 \begin{vmatrix} 1 & 3 \\ 0 & -1 \end{vmatrix} + 0 \begin{vmatrix} 1 & 4 \\ 0 & 1 \end{vmatrix}$$
$$= -7 - 4(-1) + 0 = -3.$$

Since we have already found a non-zero 3×3 determinant we can conclude that the dimension of the set is 3 without further evaluation of determinants.

The non-zero determinant shows that $\{(1, 4, 0)\ (1, 4, 3)\ (0, 1, -1)\}$ is linearly independent, so this subset can be chosen as a basis for the set.

11.1.4 Determinants and Linear Combinations Let $\mathbf{r}_1 = (a_1, b_1)$, $\mathbf{r}_2 = (a_2, b_2)$ be two non-parallel vectors in R^2. Then if $\mathbf{r} = (a, b)$ is *any* vector in R^2 we can find scalars k_1, k_2 such that $\mathbf{r} = k_1\mathbf{r}_1 + k_2\mathbf{r}_2$. In terms of components, this means that

$$k_1a_1 + k_2a_2 = a,$$
$$k_1b_1 + k_2b_2 = b,$$

from which we can easily find that

$$k_1 = \frac{ab_2 - a_2b}{a_1b_2 - a_2b_1}, \quad k_2 = \frac{a_1b - ab_1}{a_1b_2 - a_2b_1},$$

that is

$$k_1 = \frac{\begin{vmatrix} a & b \\ a_2 & b_2 \end{vmatrix}}{\begin{vmatrix} a_1 & b_1 \\ a_2 & b_2 \end{vmatrix}}, \quad k_2 = \frac{\begin{vmatrix} a_1 & b_1 \\ a & b \end{vmatrix}}{\begin{vmatrix} a_1 & b_1 \\ a_2 & b_2 \end{vmatrix}}.$$

Similarly, if

$$\mathbf{r}_1 = (a_1, b_1, c_1), \quad \mathbf{r}_2 = (a_2, b_2, c_2), \quad \mathbf{r}_3 = (a_3, b_3, c_3)$$

are three non-coplanar vectors in R^3, and $\mathbf{r} = (a, b, c)$ is *any* vector in R^3, we can find scalars k_1, k_2, k_3 such that

$$\mathbf{r} = k_1\mathbf{r}_1 + k_2\mathbf{r}_2 + k_3\mathbf{r}_3. \tag{1}$$

Expressing (1) in terms of components we get three simultaneous equations which can be solved for k_1, k_2, k_3; the method that follows is more convenient.

Eliminate k_3 from (1) by forming the vector product of both sides with $\mathbf{r}_3$; thus

$$\mathbf{r}_3 \wedge \mathbf{r} = k_1\mathbf{r}_3 \wedge \mathbf{r}_1 + k_2\mathbf{r}_3 \wedge \mathbf{r}_2 + \mathbf{0}. \tag{2}$$

Eliminate k_2 from (2) by forming the scalar product of both sides with $\mathbf{r}_2$; thus

$$\mathbf{r}_2 \cdot \mathbf{r}_3 \wedge \mathbf{r} = k_1\mathbf{r}_2 \cdot \mathbf{r}_3 \wedge \mathbf{r}_1 + 0 + 0.$$

We now get

$$k_1 = \frac{\mathbf{r}_2 \cdot \mathbf{r}_3 \wedge \mathbf{r}}{\mathbf{r}_2 \cdot \mathbf{r}_3 \wedge \mathbf{r}_1} = \frac{\mathbf{r} \cdot \mathbf{r}_2 \wedge \mathbf{r}_3}{\mathbf{r}_1 \cdot \mathbf{r}_2 \wedge \mathbf{r}_3}$$

and so from (11.7) we have

$$k_1 = \frac{\begin{vmatrix} a & b & c \\ a_2 & b_2 & c_2 \\ a_3 & b_3 & c_3 \end{vmatrix}}{\Delta},$$

where

$$\Delta = \begin{vmatrix} a_1 & b_1 & c_1 \\ a_2 & b_2 & c_2 \\ a_3 & b_3 & c_3 \end{vmatrix}$$

Similarly it can be shown that

$$k_2 = \frac{\begin{vmatrix} a_1 & b_1 & c_1 \\ a & b & c \\ a_3 & b_3 & c_3 \end{vmatrix}}{\Delta}, \quad k_3 = \frac{\begin{vmatrix} a_1 & b_1 & c_1 \\ a_2 & b_2 & c_2 \\ a & b & c \end{vmatrix}}{\Delta},$$

and it should be noted that the three determinants appearing in the numerators of k_1, k_2, k_3 can be obtained from Δ by replacing the components of $\mathbf{r}_1$, $\mathbf{r}_2$, $\mathbf{r}_3$ respectively by those of $\mathbf{r}$.

Exercises 11(a)

1. Given

$$S = \{(2, 5), (4, 10), (3, 9), (5, 14), (4, 7)\},$$

verify that $\dim S = 2$ and that $\{(2, 5), (3, 9)\}$ is a basis for S. Express $(4, 7)$ as a linear combination of the two vectors in this basis.

2. Given $\mathbf{r}_1 = (1, 2, 3)$, $\mathbf{r}_2 = (2, 3, 4)$, $\mathbf{r}_3 = (3, 4, 5)$, show that $\{\mathbf{r}_1, \mathbf{r}_2, \mathbf{r}_3\}$ is linearly dependent, and find a linear relation between these three vectors.
3. Given $\mathbf{r}_1 = (1, 2, 3)$, $\mathbf{r}_2 = (2, 3, 4)$, $\mathbf{r}_3 = (3, 4, 0)$, show that $\{\mathbf{r}_1, \mathbf{r}_2, \mathbf{r}_3\}$ is linearly independent and express $\mathbf{i}$, $\mathbf{j}$, $\mathbf{k}$ as linear combinations of $\mathbf{r}_1$, $\mathbf{r}_2$, $\mathbf{r}_3$, where $\mathbf{i} = (1, 0, 0)$, $\mathbf{j} = (0, 1, 0)$, $\mathbf{k} = (0, 0, 1)$.
4. Given $\mathbf{r}_1 = (1, 2, 3)$, $\mathbf{r}_2 = (3, 2, 1)$, $\mathbf{r}_3 = (2, 3, 4)$, show that $\{\mathbf{r}_1, \mathbf{r}_2, \mathbf{r}_3\}$ is linearly dependent, and find a linear relation between these three vectors. What is the dimension of the set $\{\mathbf{r}_1, \mathbf{r}_2, \mathbf{r}_3\}$?
5. Show that the set $\{(0, 0, 3), (0, 2, 3), (1, 2, 3)\}$ has dimension 3. Alter the value of the first component of the third vector so that the dimension of the set is reduced to 2, and alter the value of the second component of the first vector so that the dimension of the set is further reduced to 1.
6. Given

$$S = \{(1, -1, 3), (2, 0, 1), (0, 2, 1), (1, 2, 0), (3, -3, 4)\},$$

verify that $\dim S = 3$ and that

$$\{(1, -1, 3), (2, 0, 1), (0, 2, 1)\}$$

is a basis for S. Express $(1, 2, 0)$ as a linear combination of the vectors in this basis.

7. Given

$$S = \{(1, 2, 1), (2, 1, 2), (7, 8, 7), (-1, 4, -1), (1, 11, 1)\},$$

verify that $\dim S = 2$ and that $\{(1, 2, 1), (2, 1, 2)\}$ is a basis for S. Express $(7, 8, 7)$, $(-1, 4, -1)$, and $(1, 11, 1)$ as linear combinations of the vectors in this basis.

11.2 Properties of Determinants

We have defined 2×2 determinants as follows:

$$\begin{vmatrix} a_1 & b_1 \\ a_2 & b_2 \end{vmatrix} = a_1b_2 - a_2b_1. \tag{1}$$

We then used 2×2 determinants to define 3×3 determinants thus:

$$\begin{vmatrix} a_1 & b_1 & c_1 \\ a_2 & b_2 & c_2 \\ a_3 & b_3 & c_3 \end{vmatrix} = a_1 \begin{vmatrix} b_2 & c_2 \\ b_3 & c_3 \end{vmatrix} - b_1 \begin{vmatrix} a_2 & c_2 \\ a_3 & c_3 \end{vmatrix} + c_1 \begin{vmatrix} a_2 & b_2 \\ a_3 & b_3 \end{vmatrix} \tag{2}$$

From (1) and (2) we have

$$\begin{aligned} \begin{vmatrix} a_1 & b_1 & c_1 \\ a_2 & b_2 & c_2 \\ a_3 & b_3 & c_3 \end{vmatrix} &= a_1(b_2c_3 - b_3c_2) - b_1(a_2c_3 - a_3c_2) + c_1(a_2b_3 - a_3b_2) \\ &= a_1b_2c_3 + a_2b_3c_1 + a_3b_1c_2 - a_1b_3c_2 - a_2b_1c_3 - a_3b_2c_1. \end{aligned} \tag{3}$$

This can be represented diagrammatically by repeating the first two rows immediately below the determinant and drawing diagonals as shown:

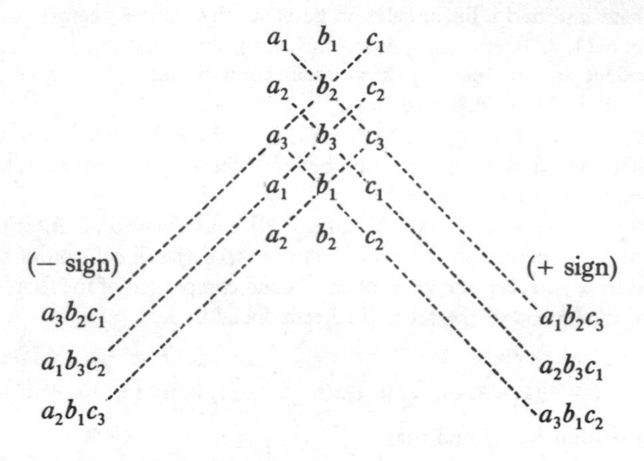

The three diagonals sloping down to the right give the terms with plus signs and the three diagonals sloping down to the left give the terms with minus signs.

For example, to evaluate

$$\begin{vmatrix} 1 & 4 & 0 \\ 1 & 4 & 3 \\ 0 & 1 & -1 \end{vmatrix},$$

we may write

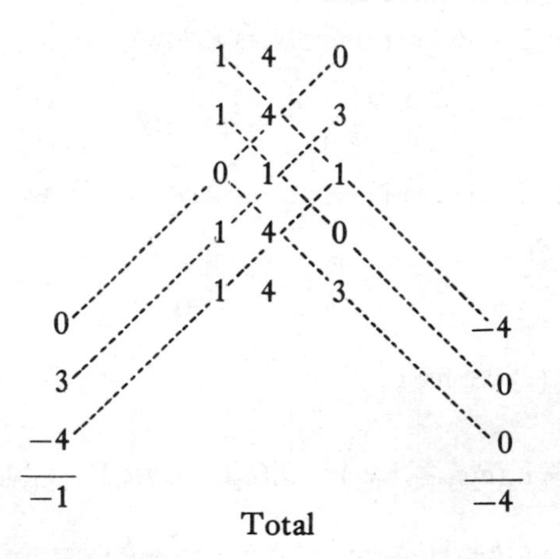

Hence the value of the determinant is $-4-(-1)=-3$.

In this section we shall be considering properties of 3×3 determinants but it is a very simple matter to verify that these properties also hold for 2×2 determinants.

Writing

$$\Delta = \begin{vmatrix} a_1 & b_1 & c_1 \\ a_2 & b_2 & c_2 \\ a_3 & b_3 & c_3 \end{vmatrix} \tag{4}$$

we have from (11.7)

$$\Delta = \mathbf{r}_1 . \mathbf{r}_2 \wedge \mathbf{r}_3 \tag{5}$$

where $\mathbf{r}_1 = (a_1, b_1, c_1)$, $\mathbf{r}_2 = (a_2, b_2, c_2)$, $\mathbf{r}_3 = (a_3, b_3, c_3)$.

It follows (Section 7.7.2) that $\Delta = \pm V$, where V is the volume of the parallelepiped with adjacent edges $\mathbf{r}_1$, $\mathbf{r}_2$, $\mathbf{r}_3$. From (5) several important properties of third-order determinants follow directly:

Property 1 If two rows are interchanged, the determinant retains its numerical value but changes sign. For example,

$$\begin{vmatrix} a_3 & b_3 & c_3 \\ a_2 & b_2 & c_2 \\ a_1 & b_1 & c_1 \end{vmatrix} = - \begin{vmatrix} a_1 & b_1 & c_1 \\ a_2 & b_2 & c_2 \\ a_3 & b_3 & c_3 \end{vmatrix} = \begin{vmatrix} a_2 & b_2 & c_2 \\ a_1 & b_1 & c_1 \\ a_3 & b_3 & c_3 \end{vmatrix} .$$

Property 2 If one row of a determinant is proportional to another, the value of the determinant is zero.

For example,

$$\begin{vmatrix} 1 & -1 & 2 \\ 2 & -2 & 4 \\ 3 & 7 & -5 \end{vmatrix} = 0$$

since in this case $\mathbf{r}_2 = 2\mathbf{r}_1$ and hence $\mathbf{r}_1 . \mathbf{r}_2 \wedge \mathbf{r}_3 = 0$.

Property 3 If the numbers in any row of a determinant are multiplied by the same factor, the determinant is multiplied by that factor.

For example,

$$\begin{vmatrix} a_1 & b_1 & c_1 \\ ka_2 & kb_2 & kc_2 \\ a_3 & b_3 & c_3 \end{vmatrix} = k \begin{vmatrix} a_1 & b_1 & c_1 \\ a_2 & b_2 & c_2 \\ a_3 & b_3 & c_3 \end{vmatrix}$$

since

$$\mathbf{r}_1 . (k\mathbf{r}_2) \wedge \mathbf{r}_3 = k(\mathbf{r}_1 . \mathbf{r}_2 \wedge \mathbf{r}_3).$$

Property 4 We have

$$\begin{vmatrix} (a_1+\alpha_1) & (b_1+\beta_1) & (c_1+\gamma_1) \\ a_2 & b_2 & c_2 \\ a_3 & b_3 & c_3 \end{vmatrix} = \begin{vmatrix} a_1 & b_1 & c_1 \\ a_2 & b_2 & c_2 \\ a_3 & b_3 & c_3 \end{vmatrix} + \begin{vmatrix} \alpha_1 & \beta_1 & \gamma_1 \\ a_2 & b_2 & c_2 \\ a_3 & b_3 & c_3 \end{vmatrix}.$$

For, writing $\rho_1 = (\alpha_1, \beta_1, \gamma_1)$, we have

$$(\mathbf{r}_1 + \rho_1).(\mathbf{r}_2 \wedge \mathbf{r}_3) = (\mathbf{r}_1 . \mathbf{r}_2 \wedge \mathbf{r}_3) + (\rho_1 . \mathbf{r}_2 \wedge \mathbf{r}_3)$$

by elementary vector algebra. Corresponding results are obtained from the identities

$$\mathbf{r}_1 . (\mathbf{r}_2 + \rho_2) \wedge \mathbf{r}_3 = (\mathbf{r}_1 . \mathbf{r}_2 \wedge \mathbf{r}_3) + (\mathbf{r}_1 . \rho_2 \wedge \mathbf{r}_3),$$
$$\mathbf{r}_1 . \mathbf{r}_2 \wedge (\mathbf{r}_3 + \rho_3) = (\mathbf{r}_1 . \mathbf{r}_2 \wedge \mathbf{r}_3) + (\mathbf{r}_1 . \mathbf{r}_2 \wedge \rho_3).$$

Property 5 If to the numbers in one row are added constant multiples of the numbers in another row the value of the determinant is unaltered. For example

$$\begin{vmatrix} a_1 & b_1 & c_1 \\ a_2 & b_2 & c_2 \\ a_3 & b_3 & c_3 \end{vmatrix} = \begin{vmatrix} (a_1+ka_2) & (b_1+kb_2) & (c_1+kc_2) \\ a_2 & b_2 & c_2 \\ a_3 & b_3 & c_3 \end{vmatrix}$$

$$= \begin{vmatrix} (a_1+ka_2+la_3) & (b_1+kb_2+lb_3) & (c_1+kc_2+lc_3) \\ (a_2+ma_3) & (b_2+mb_3) & (c_2+mc_3) \\ a_3 & b_3 & c_3 \end{vmatrix},$$

the first equation being true since

$$(\mathbf{r}_1 + k\mathbf{r}_2).(\mathbf{r}_2 \wedge \mathbf{r}_3) = \mathbf{r}_1 . (\mathbf{r}_2 \wedge \mathbf{r}_3) + k\mathbf{r}_2 . (\mathbf{r}_2 \wedge \mathbf{r}_3)$$

and

$$\mathbf{r}_2 . \mathbf{r}_2 \wedge \mathbf{r}_3 = 0,$$

while the second equation is obtained by two further applications of the same principle.

Property 6 If a determinant is rewritten so that its first, second and third rows become respectively the first, second and third columns, the value of the determinant is unaltered, i.e.

$$\begin{vmatrix} a_1 & b_1 & c_1 \\ a_2 & b_2 & c_2 \\ a_3 & b_3 & c_3 \end{vmatrix} = \begin{vmatrix} a_1 & a_2 & a_3 \\ b_1 & b_2 & b_3 \\ c_1 & c_2 & c_3 \end{vmatrix}.$$

This is easily verified by direct use of (3). As a consequence of this property, *each of the Properties 1–5 gives rise to a dual property in which 'row' is everywhere replaced by 'column'*, e.g. from Property 5 we get

$$\begin{vmatrix} a_1 & b_1 & c_1 \\ a_2 & b_2 & c_2 \\ a_3 & b_3 & c_3 \end{vmatrix} = \begin{vmatrix} (a_1 + kb_1) & b_1 & c_1 \\ (a_2 + kb_2) & b_2 & c_2 \\ (a_3 + kb_3) & b_3 & c_3 \end{vmatrix}$$

$$= \begin{vmatrix} (a_1 + kb_1 + lc_1) & (b_1 + mc_1) & c_1 \\ (a_2 + kb_2 + lc_2) & (b_2 + mc_2) & c_2 \\ (a_3 + kb_3 + lc_3) & (b_3 + mc_3) & c_3 \end{vmatrix}.$$

Example 2 Evaluate the determinant

$$\begin{vmatrix} 100 & 101 & 102 \\ 101 & 102 & 103 \\ 102 & 103 & 104 \end{vmatrix}.$$

Denoting the given determinant by Δ, we can subtract the first row from the second to obtain

$$\Delta = \begin{vmatrix} 100 & 101 & 102 \\ 1 & 1 & 1 \\ 102 & 103 & 104 \end{vmatrix}.$$

Again, subtracting the first row from the third, we get

$$\Delta = \begin{vmatrix} 100 & 101 & 102 \\ 1 & 1 & 1 \\ 2 & 2 & 2 \end{vmatrix}$$

$$= 0 \quad \text{by Property 2.}$$

The first two steps of the above simplification may be done simultaneously; however with simultaneous steps we should always check that the same result could be achieved in successive steps, and until some experience has been gained it is advisable to proceed one step at a time.

11.2.1 Minors of a Determinant Each of the nine 'positions' for a number in a 3 × 3 determinant can best be described by stating the row and column to which the number belongs, e.g. in Δ (see eq. (4), Section 11.2) we describe c_2 as being in the second row and third column.

Corresponding to each position there is a 2 × 2 determinant, viz. the determinant obtained when the entire row and entire column defining that position are both deleted. The nine such 2 × 2 determinants are called the *minors* of Δ. For example, the minor corresponding to c_2 is obtained by deleting the second row and third column and is

$$\begin{vmatrix} a_1 & b_1 \\ a_3 & b_3 \end{vmatrix}.$$

Denoting by $A_1, B_1, \ldots$ the minors corresponding to $a_1, b_1, \ldots$ we now see that eq. (2) of Section 11.2 can be expressed in the form

$$\Delta = a_1A_1 - b_1B_1 + c_1C_1.$$

Using Property 1 we can deduce similar results in which the determinant is expanded by the second row or by the third row, i.e.

$$\begin{aligned} \Delta &= -a_2A_2 + b_2B_2 - c_2C_2 \\ &= a_3A_3 - b_3B_3 + c_3C_3. \end{aligned}$$

From Property 6 we get similar expansions by *columns*, e.g.

$$\Delta = a_1A_1 - a_2A_2 + a_3A_3.$$

In all these expansions the sign to be taken with any particular minor is given as follows:

$$\begin{vmatrix} + & - & + \\ - & + & - \\ + & - & + \end{vmatrix}.$$

In the 2×2 determinant

$$\begin{vmatrix} a_1 & b_1 \\ c_1 & d_1 \end{vmatrix}$$

the minors of a_1, b_1, c_1, d_1 are simply the numbers d_1, c_1, b_1, a_1 respectively.

A determinant can be expanded in a number of ways since we may choose any row, or any column, for the expansion. Obviously a row or a column that has some zeros is a good choice.

Example 3 Evaluate the determinant

$$\begin{vmatrix} 1 & 5 & 7 \\ 2 & -3 & 1 \\ 1 & 3 & -1 \end{vmatrix}$$

Adding the second row to the third row the given determinant becomes

$$\begin{vmatrix} 1 & 5 & 7 \\ 2 & -3 & 1 \\ 3 & 0 & 0 \end{vmatrix}$$

and expanding by the bottom row this equals

$$3\begin{vmatrix} 5 & 7 \\ -3 & 1 \end{vmatrix} = 3 \times 26 = 78.$$

11.2.2 Determinants of Any Order The expansion of a third-order determinant in terms of second-order determinants suggests the following definition of a *fourth-order determinant*:

$$\begin{vmatrix} a_1 & b_1 & c_1 & d_1 \\ a_2 & b_2 & c_2 & d_2 \\ a_3 & b_3 & c_3 & d_3 \\ a_4 & b_4 & c_4 & d_4 \end{vmatrix} = a_1 \begin{vmatrix} b_2 & c_2 & d_2 \\ b_3 & c_3 & d_3 \\ b_4 & c_4 & d_4 \end{vmatrix} - b_1 \begin{vmatrix} a_2 & c_2 & d_2 \\ a_3 & c_3 & d_3 \\ a_4 & c_4 & d_4 \end{vmatrix}$$

$$+ c_1 \begin{vmatrix} a_2 & b_2 & d_2 \\ a_3 & b_3 & d_3 \\ a_4 & b_4 & d_4 \end{vmatrix} - d_1 \begin{vmatrix} a_2 & b_2 & c_2 \\ a_3 & b_3 & c_3 \\ a_4 & b_4 & c_4 \end{vmatrix}$$

$$= a_1A_1 - b_1B_1 + c_1C_1 - d_1D_1$$

where A_1, B_1, ... denote minors corresponding to a_1, b_1, ..., minors being 3×3 determinants in this case. Proceeding in this way we can define determinants of *any* order. All the properties discussed in Section 11.2 extend to determinants of higher order. Evaluation of such determinants is basically a matter of reduction to determinants of small order.

Example 4 Evaluate the determinant

$$\begin{vmatrix} 1 & -2 & 3 & -4 \\ -2 & 3 & -4 & 1 \\ 3 & -4 & 1 & -2 \\ -4 & 1 & -2 & 3 \end{vmatrix}.$$

To the second, third and fourth rows add respectively 2, −3 and 4 times the first row, giving

$$\begin{vmatrix} 1 & -2 & 3 & -4 \\ 0 & -1 & 2 & -7 \\ 0 & 2 & -8 & 10 \\ 0 & -7 & 10 & -13 \end{vmatrix}.$$

Expanding by the first column we now get

$$\begin{vmatrix} -1 & 2 & -7 \\ 2 & -8 & 10 \\ -7 & 10 & -13 \end{vmatrix}$$

which, expanded by the first row, gives

$$-(104 - 100) - 2(70 - 26) - 7(20 - 56) = 160.$$

Example 5 Evaluate the determinant

$$\begin{vmatrix} 1 & 1 & 1 & 1 \\ 1 & (1+a) & 1 & 1 \\ 1 & 1 & (1+b) & 1 \\ 1 & 1 & 1 & (1+c) \end{vmatrix}$$

Subtract the first column from each of the three other columns, giving

$$\begin{vmatrix} 1 & 0 & 0 & 0 \\ 1 & a & 0 & 0 \\ 1 & 0 & b & 0 \\ 1 & 0 & 0 & c \end{vmatrix}.$$

Expanding by the first row, this gives

$$\begin{vmatrix} a & 0 & 0 \\ 0 & b & 0 \\ 0 & 0 & c \end{vmatrix} = abc.$$

Exercises 11(b)

1. Evaluate

(a) $\begin{vmatrix} 2 & 4 & 16 \\ 3 & 9 & 81 \\ 5 & 25 & 625 \end{vmatrix}$, (b) $\begin{vmatrix} 1! & 2! & 3! \\ 2! & 3! & 4! \\ 3! & 4! & 5! \end{vmatrix}$.

2. Given

$$\Delta = \begin{vmatrix} a & h & g \\ h & b & f \\ g & f & c \end{vmatrix},$$

prove that $\Delta = abc + 2fgh - af^2 - bg^2 - ch^2$.

3. By expanding directly, show that

(a) $$\begin{vmatrix} 0 & c & b \\ c & 0 & a \\ b & a & 0 \end{vmatrix} = 2abc,$$

(b) $$\begin{vmatrix} a & b & c \\ c & a & b \\ b & c & a \end{vmatrix} = a^3 + b^3 + c^3 - 3abc,$$

(c) $$\begin{vmatrix} 1 & a & -b \\ -a & 1 & c \\ b & -c & 1 \end{vmatrix} = 1 + a^2 + b^2 + c^2,$$

(d) $$\begin{vmatrix} 1 & 1 & 1 \\ z & x & y \\ y & z & x \end{vmatrix} = x^2 + y^2 + z^2 - yz - zx - xy,$$

(e) $$\begin{vmatrix} \cos(x+y) & \sin(x+y) & -\cos(x+y) \\ \sin(x-y) & \cos(x-y) & \sin(x-y) \\ \sin 2x & 0 & \sin 2y \end{vmatrix} = \sin 2(x+y).$$

4. If

$$\Delta = \begin{vmatrix} 1 & 1 & 1 \\ \sin A & \sin B & \sin C \\ \cos A & \cos B & \cos C \end{vmatrix},$$

prove, by expanding from the first row, that

$$\Delta = \sin(A - B) + \sin(B - C) + \sin(C - A).$$

By subtracting columns before expansion and by converting differences into products, show that

$$\Delta = -4 \sin \tfrac{1}{2}(A - B) \sin \tfrac{1}{2}(B - C) \sin \tfrac{1}{2}(C - A).$$

5. Show that

$$\begin{vmatrix} 1 & 1 & 1 \\ 1 & 1+a & 1 \\ 1 & 1 & 1+b \end{vmatrix} = ab,$$

and that

$$\begin{vmatrix} 1+a & 1 & 1 \\ 1 & 1+b & 1 \\ 1 & 1 & 1+c \end{vmatrix} = abc(1 + 1/a + 1/b + 1/c).$$

6. Evaluate the determinant

$$\begin{vmatrix} b+c & b-c & c-b \\ a-c & c+a & c-a \\ a-b & b-a & a+b \end{vmatrix}.$$

7. Solve for x the equation

$$\begin{vmatrix} x & a & b \\ a & x & b \\ a & b & x \end{vmatrix} = 0.$$

8. Find the roots of the equation

$$\begin{vmatrix} x-3 & 1 & -1 \\ 1 & x-5 & 1 \\ -1 & 1 & x-3 \end{vmatrix} = 0.$$

9. Show that $x = 3$ is a root of the equation

$$\begin{vmatrix} x & -6 & -1 \\ 3 & -2x & x-4 \\ -2 & 3x & x-2 \end{vmatrix} = 0,$$

and solve it completely.

10. Find the roots of the following equations:

(a) $\begin{vmatrix} x^3 & 3 & 8 \\ x & 2 & 2 \\ 1 & 3 & 1 \end{vmatrix} = 0,$ (b) $\begin{vmatrix} 1 & 1 & 1 \\ x & 2 & 1 \\ x^3 & 8 & 1 \end{vmatrix} = 0.$

11. Solve the equation

$$\begin{vmatrix} x^2+x+2 & x^2 & 0 \\ x+4 & 2 & x^2 \\ 1 & 1 & 1 \end{vmatrix} = 0.$$

12. Solve the following equations in x:

(a) $$\begin{vmatrix} -1 & 1 & 1 \\ x^2+2ax & -ax & ax+2a^2 \\ x^2+2ax & ax+2a^2 & -ax \end{vmatrix} = 0,$$

(b) $$\begin{vmatrix} -1 & 2 & 2 \\ x^2+ax & -ax & ax+a^2 \\ x^2+ax & ax+a^2 & -ax \end{vmatrix} = 0.$$

13. Find the value of the determinant

$$\begin{vmatrix} a_1+\lambda a_2 & b_1+\lambda b_2 & c_1+\lambda c_2 \\ a_2+\mu a_3 & b_2+\mu b_3 & c_2+\mu c_3 \\ a_3+\nu a_1 & b_3+\nu b_1 & c_3+\nu c_1 \end{vmatrix}$$

in terms of λ, μ, ν and D, where D is the value of the determinant when $\lambda = \mu = \nu = 0$.

14. Evaluate the determinants

(a) $\begin{vmatrix} 8 & 3 & 4 & 7 \\ 4 & 1 & 2 & 3 \\ 11 & 3 & 7 & 6 \\ 9 & 2 & 3 & 5 \end{vmatrix}$, (b) $\begin{vmatrix} 7 & 13 & 10 & 6 \\ 5 & 9 & 7 & 4 \\ 8 & 12 & 11 & 7 \\ 4 & 10 & 6 & 3 \end{vmatrix}$, (c) $\begin{vmatrix} 3 & 13 & 11 & 9 \\ 1 & 4 & 17 & 8 \\ 1 & 8 & 0 & -2 \\ 2 & 6 & 24 & 11 \end{vmatrix}$.

15. Show that

$$\begin{vmatrix} a+1 & 1 & 1 & 1 \\ 1 & b+1 & 1 & 1 \\ 1 & 1 & c+1 & 1 \\ 1 & 1 & 1 & d+1 \end{vmatrix} = abcd(1/a + 1/b + 1/c + 1/d + 1).$$

16. Show that

$$\begin{vmatrix} a & 1 & 1 & 1 \\ 1 & a & 1 & 1 \\ 1 & 1 & a & 1 \\ 1 & 1 & 1 & a \end{vmatrix} = (a+3)(a-1)^3.$$

17. Prove that, if $a^2 + b^2 + c^2 + d^2 = 1$, then

$$\begin{vmatrix} a^2-1 & ab & ac & ad \\ ba & b^2-1 & bc & bd \\ ca & cb & c^2-1 & cd \\ da & db & dc & d^2-1 \end{vmatrix} = 0.$$

11.3 Vectors in R^n

The approach to the vector systems $R^2(+, .)$ and $R^3(+, .)$ used in Section 11.1 can be generalised to define vector systems $R^4(+, .)$, $R^5(+, .)$, and so on. For any positive integer n we consider the system $R^n(+, .)$ with elements $(x_1, x_2, \ldots, x_n)$ (i.e. each element consists of n real number components) and with operations $+$, . defined by the rules

$$(x_1, x_2, \ldots, x_n) + (y_1, y_2, \ldots, y_n) = (x_1 + y_1, x_2 + y_2, \ldots, x_n + y_n) \tag{11.12}$$

$$(x_1, x_2, \ldots, x_n).(y_1, y_2, \ldots, y_n) = x_1y_1 + x_2y_2 + \cdots + x_ny_n, \tag{11.13}$$

together with an operation of 'multiplication by a scalar' defined by the rule

$$k(x_1, x_2, \ldots, x_n) = (kx_1, kx_2, \ldots, kx_n). \tag{11.14}$$

The elements in this system are called *n-dimensional vectors*, or simply vectors in R^n, and R^n is called an *n-dimensional vector space*.

The results established in Sections 11.1.2–11.1.4 for vectors in R^3 extend to vectors in R^n, although the idea of a geometrical representation of vectors (upon which much of our thinking about vectors in R^3 was based) does not apply for $n > 3$. In this section we state the main results without proof.

As usual, bold-face letters will be used to denote vectors and the zero-vector $(0, 0, \ldots, 0)$ will be denoted by $\mathbf{0}$ regardless of dimension. *Linear combination* and *linear dependence* can be defined exactly as before (see (11.1) and (11.2)); thus if

$$\mathbf{r}_1 = (1, 2, 3, 4, 5), \qquad \mathbf{r}_2 = (6, 7, 8, 9, 10),$$

$$\mathbf{r}_3 = (11, 12, 13, 14, 15), \quad \mathbf{r}_4 = (8, 11, 14, 17, 20),$$

then $\mathbf{r}_4$ is a linear combination of $\mathbf{r}_1, \mathbf{r}_2$ since $\mathbf{r}_4 = 2\mathbf{r}_1 + \mathbf{r}_2$ and $\mathbf{r}_4$ is also a linear combination of $\mathbf{r}_1, \mathbf{r}_2, \mathbf{r}_3$ since $\mathbf{r}_4 = 2\mathbf{r}_1 + \mathbf{r}_2 + 0\mathbf{r}_3$.

Let $S = \{\mathbf{r}_1, \mathbf{r}_2, \ldots, \mathbf{r}_m\}$ be a set of m vectors in R^n and let

$$\begin{aligned} \mathbf{r}_1 &= (a_{11}, \quad a_{12}, \quad \ldots, \quad a_{1n}), \\ \mathbf{r}_2 &= (a_{21}, \quad a_{22}, \quad \ldots, \quad a_{2n}), \\ &\vdots \\ \mathbf{r}_m &= (a_{m1}, \quad a_{m2}, \quad \ldots, \quad a_{mn}) \end{aligned}$$

so that the component array is

$$\begin{matrix} a_{11} & a_{12} & \ldots & a_{1n} \\ a_{21} & a_{22} & \ldots & a_{2n} \\ \vdots & \vdots & & \vdots \\ a_{m1} & a_{m2} & \ldots & a_{mn} \end{matrix} \tag{A}$$

Theorem 1 If $m > n$, then S is necessarily linearly dependent.

If $m = n$ the array (A) is 'square' and there is a corresponding $n \times n$ determinant $\Delta(S)$, where

$$\Delta(S) = \begin{vmatrix} a_{11} & a_{12} & \ldots & a_{1n} \\ a_{21} & a_{22} & \ldots & a_{2n} \\ \vdots & \vdots & & \\ a_{n1} & a_{n2} & \ldots & a_{nn} \end{vmatrix}.$$

Theorem 2 If S is a set of n vectors in R^n then

$$S \text{ linearly dependent} \Leftrightarrow \Delta(S) = 0.$$

Definition. Suppose that no subset of S with more than s vectors in it is linearly independent, but that S has at least one linearly independent subset with s vectors in it. We then say S has *dimension* s, and write $\dim S = s$.

It follows from Theorem 1 that the largest possible linearly independent set of vectors in R^n has n vectors in it. Hence $\dim S \leqslant n$.

Theorem 3 $\text{Dim } S = s \Leftrightarrow$ all $(s + 1) \times (s + 1)$ determinants of (A) are zero, and at least one $s \times s$ determinant of (A) is non-zero.

Example 6 Find the dimension of the set of 4-dimensional vectors

$$\{(1, 2, -1, 4), \quad (2, -2, 0, 1), \quad (3, 0, -1, 5), \\ (8, -2, -2, 11), \quad (6, -6, 0, 3)\}.$$

Considerable labour is saved if we observe that the third vector in the set is the sum of the first two, and the fifth is three times the second. Hence the dimension of the set is at most 3; we can ignore the third and fifth vectors and consider the component-array

$$\begin{matrix} 1 & 2 & -1 & 4 \\ 2 & -2 & 0 & 1 \\ 8 & -2 & -2 & 11. \end{matrix}$$

The 3×3 determinants are

$$\begin{vmatrix} 1 & 2 & -1 \\ 2 & 2 & 0 \\ 8 & -2 & -2 \end{vmatrix}, \quad \begin{vmatrix} 1 & 2 & 4 \\ 2 & -2 & 1 \\ 8 & -2 & 11 \end{vmatrix}, \quad \begin{vmatrix} 1 & -1 & 4 \\ 2 & 0 & 1 \\ 8 & -2 & 11 \end{vmatrix}, \quad \begin{vmatrix} 2 & 1 & 1 \\ -2 & 0 & 1 \\ -2 & -2 & 11 \end{vmatrix}$$

and on evaluation are all found to be zero.

Proceeding to evaluate the 2×2 determinants we readily find a non-zero one, for example

$$\begin{vmatrix} 1 & 2 \\ 2 & -2 \end{vmatrix} = -6.$$

Hence the dimension of the set is 2.

Theorem 4 If dim $S = s$ and B is a linearly independent subset of S with s vectors in it, then any vector in S can be expressed as a linear combination of the s vectors in B.

The set B in Theorem 4 is called a *basis* for S.

In particular, dim $R^n = n$ and we have

Theorem 5 If B is a linearly independent set of n vectors in R^n, say

$$B = \{\mathbf{r}_1, \mathbf{r}_2, \ldots, \mathbf{r}_n\},$$

then any vector $\mathbf{r}$ in R^n can be expressed in the form

$$\mathbf{r} = k_1\mathbf{r}_1 + k_2\mathbf{r}_2 + \cdots k_n\mathbf{r}_n$$

and the scalars $k_1, k_2, \ldots, k_n$ are given by

$$k_1 = \frac{\Delta(B_1)}{\Delta(B)}, \quad k_2 = \frac{\Delta(B_2)}{\Delta(B)}, \quad \ldots, \quad k_n = \frac{\Delta(B_n)}{\Delta(B)},$$

where B_1 is obtained from B by replacing $\mathbf{r}_1$ by $\mathbf{r}$, B_2 is obtained from B by replacing $\mathbf{r}_2$ by $\mathbf{r}$, and so on.

Writing

$$\Delta(B) = \begin{vmatrix} a_{11} & a_{12} & \dots & a_{1n} \\ a_{21} & a_{22} & \dots & a_{2n} \\ \vdots & \vdots & & \vdots \\ a_{n1} & a_{n2} & \dots & a_{nn} \end{vmatrix},$$

then if $\mathbf{r} = (b_1, b_2, \dots, b_n)$ Theorem 5 gives

$$k_1 = \frac{\begin{vmatrix} b_1 & b_2 & \dots & b_n \\ a_{21} & a_{22} & \dots & a_{2n} \\ \vdots & \vdots & & \vdots \\ a_{n1} & a_{n2} & \dots & a_{nn} \end{vmatrix}}{\Delta(B)}, \qquad k_2 = \frac{\begin{vmatrix} a_{11} & a_{12} & \dots & a_{1n} \\ b_1 & b_2 & \dots & b_n \\ a_{31} & a_{32} & \dots & a_{3n} \\ \vdots & \vdots & & \vdots \\ a_{n1} & a_{n2} & \dots & a_{nn} \end{vmatrix}}{\Delta(B)},$$

and so on.

Exercises 11(c)

1. Show that $\{(1, 0, 0, 0), (0, 1, 0, 0), (0, 0, 1, 0), (0, 0, 0, 1)\}$ is a basis for R^4.
2. Show that $\{(1, 0, 0, \dots, 0), (0, 1, 0, \dots, 0), \dots, (0, 0, 0, \dots, 1)\}$ is a basis for R^n.
3. Given $\mathbf{r}_1 = (1, 1, 1, 1)$, $\mathbf{r}_2 = (0, 1, 1, 1)$, $\mathbf{r}_3 = (0, 0, 1, 1)$, $\mathbf{r}_4 = (0, 0, 0, 1)$, show that $\{\mathbf{r}_1, \mathbf{r}_2, \mathbf{r}_3, \mathbf{r}_4\}$ is a basis for R^4. Express the vectors $\mathbf{i}, \mathbf{j}, \mathbf{k}, \mathbf{l}$ in terms of $\mathbf{r}_1$, $\mathbf{r}_2$, $\mathbf{r}_3$, $\mathbf{r}_4$, where $\mathbf{i} = (1, 0, 0, 0)$, $\mathbf{j} = (0, 1, 0, 0)$, $\mathbf{k} = (0, 0, 1, 0)$, $\mathbf{l} = (0, 0, 0, 1)$.
4. Given $\mathbf{r}_1 = (1, 1, 1, 0)$, $\mathbf{r}_2 = (1, 1, 0, 1)$, $\mathbf{r}_3 = (1, 0, 1, 1)$, $\mathbf{r}_4 = (0, 1, 1, 1)$, show that $\{\mathbf{r}_1, \mathbf{r}_2, \mathbf{r}_3, \mathbf{r}_4\}$ is a basis for R^4. Express the vectors $\mathbf{i}, \mathbf{j}, \mathbf{k}, \mathbf{l}$ (see Question 3) in terms of $\mathbf{r}_1$, $\mathbf{r}_2$, $\mathbf{r}_3$, $\mathbf{r}_4$.
5. Given $\mathbf{r}_1 = (1, 1, 0, 0)$, $\mathbf{r}_2 = (0, 1, 1, 0)$, $\mathbf{r}_3 = (0, 0, 1, 1)$, $\mathbf{r}_4 = (1, 0, 0, 1)$, show that $\{\mathbf{r}_1, \mathbf{r}_2, \mathbf{r}_3, \mathbf{r}_4\}$ is linearly dependent and find a linear relation between these four vectors.
6. Find the dimension of the set
$$\{(1, 1, 0, 0), \quad (1, 0, 1, 0), \quad (1, 0, 0, 1), \quad (0, 1, 1, 0) \quad (0, 1, 0, 1), \quad (0, 0, 1, 1)\}.$$
7. Find the dimension of the set
$$\{(0, 1, 2, 3), \quad (3, 2, 1, 0), \quad (1, 1, 1, 1), \quad (3, 1, -1, -3), \quad (2, 1, 0, -1)\}.$$

11.4 Linear Simultaneous Equations

The equations

$$x + 2y + 3z = 4,$$

$$x + 2y + 3z = 8,$$

are obviously inconsistent since they require that $x + 2y + 3z$ should have two different values. Clearly no solution can be found satisfying both equations.

The set of equations

$$\left.\begin{aligned} x + 2y + 3z &= 4, \\ 5x + 6y + 7z &= 8, \\ 7x + 10y + 13z &= 1 \end{aligned}\right\} \tag{1}$$

is also inconsistent, but this is less obvious. To see this we first note that the l.h.s. of the third equation is a linear combination of the l.h. sides of the first two equations; more precisely

$$(7x + 10y + 13z) = 2(x + 2y + 3z) + (5x + 6y + 7z). \tag{2}$$

Hence if x_0, y_0, z_0 are values of x, y, z satisfying the first two equations, i.e.

$$x_0 + 2y_0 + 3z_0 = 4, \quad 5x_0 + 6y_0 + 7z_0 = 8,$$

we see from (2) that

$$7x_0 + 10y_0 + 13z_0 = 2 \times 4 + 8 = 16. \tag{3}$$

For x_0, y_0, z_0 to satisfy the third equation in (1) we require that $7x_0 + 10y_0 + 13z_0 = 1$ which is inconsistent with (3). Thus a solution satisfying all three equations simultaneously is impossible.

Returning to (3) we see that the set of equations (1) is consistent if in the r.h.s. of the third equation the number 1 is changed to 16. But in this case the derivation of (3) should convince us that if x_0, y_0, z_0 satisfies the first two equations then x_0, y_0, z_0 *must* satisfy the third equation. In other words the third equation is redundant, and the proposed set of equations, viz.

$$\left.\begin{aligned} x + 2y + 3z &= 4, \\ 5x + 6y + 7z &= 8, \\ 7x + 10y + 13z &= 16 \end{aligned}\right\} \tag{4}$$

is effectively the same as the set

$$\left.\begin{aligned} x + 2y + 3z &= 4, \\ 5x + 6y + 7z &= 8. \end{aligned}\right\} \tag{5}$$

We say the set (4) *reduces* to the set (5). We can say that the third equation in (4) is a linear combination of the other two in the sense that

$$\text{third equation} = \text{twice first equation} + \text{second equation}. \tag{6}$$

The reader should construct other examples, if necessary, to establish convincingly that a linear relationship like (6) between equations always implies redundancy, whereas a relationship like (2), which is true for the l.h. sides of the equations but does not extend to the full equations, implies inconsistency.

We now identify the l.h. sides of eqs. (1) with the set of vectors $\{\mathbf{r}_1, \mathbf{r}_2, \mathbf{r}_3\}$, where

$$\mathbf{r}_1 = (1, 2, 3), \quad \mathbf{r}_2 = (5, 6, 7), \quad \mathbf{r}_3 = (7, 10, 13), \tag{7}$$

i.e. the components of $\mathbf{r}_1$ are the coefficients of x, y, z in the first equation, and so on. We can then identify the relation (2) with the relation

$$\mathbf{r}_3 = 2\mathbf{r}_1 + \mathbf{r}_2, \tag{8}$$

which is satisfied by these vectors and which shows that $\{\mathbf{r}_1, \mathbf{r}_2, \mathbf{r}_3\}$ is a linearly dependent set.

The full equations in (1) can likewise be identified with the set of vectors $\{\mathbf{r}_1', \mathbf{r}_2', \mathbf{r}_3'\}$, where

$$\mathbf{r}_1' = (1, 2, 3, 4), \quad \mathbf{r}_2' = (5, 6, 7, 8), \quad \mathbf{r}_3' = (7, 10, 13, 1), \tag{9}$$

the fourth component being the numbers on the r.h. sides of the equations. These fourth components are such that the relation (8) does not extend to the vectors (9) and $\{\mathbf{r}_1', \mathbf{r}_2', \mathbf{r}_3'\}$ is a linearly independent set. The set of vectors corresponding to the full equations in (4) is $\{\mathbf{r}_1', \mathbf{r}_2', \mathbf{r}_3'\}$, where $\mathbf{r}_1', \mathbf{r}_2'$ are defined by (9) and $\mathbf{r}_3' = \{7, 10, 13, 16\}$. For these vectors we have the relation $\mathbf{r}_3' = 2\mathbf{r}_1' + \mathbf{r}_2'$ which can be identified with the relation (6) and which shows that $\{\mathbf{r}_1', \mathbf{r}_2', \mathbf{r}_3'\}$ is a linearly dependent set.

More generally, consider the set of equations

$$\left.\begin{aligned} a_{11}x_1 + a_{12}x_2 + \cdots + a_{1n}x_n &= b_1, \\ a_{21}x_1 + a_{22}x_2 + \cdots + a_{2n}x_n &= b_2, \\ &\vdots \\ a_{m1}x_1 + a_{m2}x_2 + \cdots + a_{mn}x_n &= b_m, \end{aligned}\right\} \tag{11.15}$$

which is a set of m equations for the n unknowns $x_1, x_2, \ldots, x_n$. We shall write

$$\mathbf{r}_1 = (a_{11}, a_{12}, \ldots, a_{1n}), \quad \mathbf{r}_2 = (a_{21}, a_{22}, \ldots, a_{2n}), \\ \ldots, \mathbf{r}_m = (a_{m1}, a_{m2}, \ldots, a_{mn})$$

and call these the *primary row vectors* of the equations.

We shall write

$$\mathbf{r}_1' = (a_{11}, a_{12}, \ldots, a_{1n}, b_1), \quad \mathbf{r}_2' = (a_{21}, a_{22}, \ldots, a_{2n}, b_2), \\ \ldots, \mathbf{r}_m' = (a_{m1}, a_{m2}, \ldots, a_{mn}, b_m).$$

and call these the *complete row vectors* of the equations.

We shall denote by $\mathscr{R}$ the set of primary row vectors and by $\mathscr{R}'$ the set of complete row vectors, i.e.

$$\mathscr{R} = \{\mathbf{r}_1, \mathbf{r}_2, \ldots, \mathbf{r}_m\}, \quad \mathscr{R}' = \{\mathbf{r}_1', \mathbf{r}_2', \ldots, \mathbf{r}_m'\}.$$

Each vector of $\mathscr{R}'$ represents one equation from (11.15). If one vector of $\mathscr{R}'$ is a linear combination of the others then the corresponding equation is redundant. Several equations may be redundant in this way, and we have

Lemma 1 If dim $\mathscr{R}' = s$ where $s < m$, then $(m - s)$ equations in (11.15) are redundant and the set reduces to a set of s equations. We may take as the reduced set any set of s equations from (11.15) for which the corresponding vectors form a basis for $\mathscr{R}'$.

For example, any of the three equations in (4) may be declared redundant, the other two being taken as the reduced set, since for these equations any pair of complete row vectors is a basis for $\mathscr{R}'$.

Another result, suggested by the discussion of (1), is as follows:

Lemma 2 If $\mathscr{R}$ is linearly dependent while $\mathscr{R}'$ is linearly independent, then eqs. (11.15) are inconsistent.

Note that if any set of vectors in $\mathscr{R}'$ is linearly dependent the corresponding set of vectors in $\mathscr{R}$ must also be linearly dependent; consequently dim $\mathscr{R} \leqslant$ dim $\mathscr{R}'$. Suppose that dim $\mathscr{R}' = s$. By Lemma 1 eqs. (11.15) can be replaced by a reduced set containing s equations. The reduced set of complete row vectors is linearly independent, and applying Lemma 2 to the reduced set of equations we see that if the reduced set of primary row vectors is linearly dependent the equations are inconsistent, i.e. the equations are inconsistent if dim $\mathscr{R} < s$. We conclude that

Lemma 3 If dim $\mathscr{R} \neq$ dim $\mathscr{R}'$ then eqs. (11.15) are inconsistent.

Thus we need not attempt to solve any set of equations for which dim $\mathscr{R} \neq$ dim $\mathscr{R}'$. Since the vectors in $\mathscr{R}$ are n-dimensional it follows from Theorem 1 that dim $\mathscr{R} \leqslant n$. We consider separately the two cases dim $\mathscr{R}$ = dim $\mathscr{R}' = n$, dim $\mathscr{R}$ = dim $\mathscr{R}' = s$, where $s < n$.

11.4.1 dim $\mathscr{R}$ = dim $\mathscr{R}' = n$

By Lemma 1 the equations reduce to a set of n equations, which we may suppose are

$$\left.\begin{aligned} a_{11}x_1 + a_{12}x_2 + \cdots + a_{1n}x_n &= b_1, \\ a_{21}x_1 + a_{22}x_2 + \cdots + a_{2n}x_n &= b_2, \\ &\vdots \\ a_{n1}x_1 + a_{n2}x_2 + \cdots + a_{nn}x_n &= b_n. \end{aligned}\right\} \tag{11.16}$$

The determinant $\Delta(\mathscr{R})$ formed by the components of the vectors in $\mathscr{R}$ is given by

$$\Delta(\mathscr{R}) = \begin{vmatrix} a_{11} & a_{12} & \dots & a_{1n} \\ a_{21} & a_{22} & \dots & a_{2n} \\ \vdots & \vdots & & \vdots \\ a_{n1} & a_{n2} & \dots & a_{nn} \end{vmatrix},$$

and since $\mathscr{R}$ is linearly independent we have $\Delta(\mathscr{R}) \neq 0$.

Let us write

$$\mathbf{c}_1 = (a_{11}, a_{21}, \dots, a_{n1}), \quad \mathbf{c}_2 = (a_{12}, a_{22}, \dots, a_{n2}),$$
$$\dots, \mathbf{c}_n = (a_{1n}, a_{2n}, \dots, a_{nn}), \quad \mathbf{b} = (b_1, b_2, \dots, b_n);$$

we shall call these vectors the *column vectors* of eqs. (11.16) and write $C = \{\mathbf{c}_1, \mathbf{c}_2, \dots, \mathbf{c}_n\}$. Equations (11.16) may be written in the form

$$x_1\mathbf{c}_1 + x_2\mathbf{c}_2 + \cdots + x_n\mathbf{c}_n = \mathbf{b},$$

so that the problem of solving eqs. (11.16) is equivalent to the problem of expressing $\mathbf{b}$ as a linear combination of the vectors in C. Since $\Delta(C) = \Delta(\mathscr{R}) \neq 0$, C is linearly independent; by Theorem 5 a solution exists and (in the notation used in Theorem 5) the solution is

$$x_1 = \frac{\Delta(C_1)}{\Delta(C)}, \quad x_2 = \frac{\Delta(C_2)}{\Delta(C)}, \quad \dots, \quad x_n = \frac{\Delta(C_n)}{\Delta(C)},$$

where $C_1, C_2, \dots, C_n$ are obtained from C by replacing $\mathbf{c}_1, \mathbf{c}_2, \dots, \mathbf{c}_n$ respectively by $\mathbf{b}$.

Using Property 6 of determinants we may express this result as follows:

Theorem 6 (Cramer's Rule) If $\Delta(\mathscr{R}) \neq 0$, then eqs. (11.16) have the unique solution

$$x_1 = \frac{\Delta_1}{\Delta(\mathscr{R})}, \quad x_2 = \frac{\Delta_2}{\Delta(\mathscr{R})}, \quad \dots, \quad x_n = \frac{\Delta_n}{\Delta(\mathscr{R})},$$

where Δ_1 is the determinant obtained from $\Delta(\mathscr{R})$ by replacing the first column by the components of $\mathbf{b}$, Δ_2 is the determinant obtained from $\Delta(\mathscr{R})$ by replacing the second column by the components of $\mathbf{b}$, and so on.

Written more fully, Theorem 6 gives

$$x_1 = \frac{\begin{vmatrix} b_1 & a_{12} & \dots & a_{1n} \\ b_2 & a_{22} & \dots & a_{2n} \\ \vdots & \vdots & & \vdots \\ b_n & a_{n2} & \dots & a_{nn} \end{vmatrix}}{\Delta(\mathscr{R})}, \qquad x_2 = \frac{\begin{vmatrix} a_{11} & b_1 & \dots & a_{1n} \\ a_{21} & b_2 & \dots & a_{2n} \\ \vdots & \vdots & & \vdots \\ a_{n1} & b_n & \dots & a_{nn} \end{vmatrix}}{\Delta(\mathscr{R})},$$

and so on.

If $\mathbf{b} = \mathbf{0}$, i.e. $b_1 = b_2 = \cdots = b_n = 0$, then eqs. (11.16) are said to be *homogeneous*. In this case the solution is simply $x_1 = x_2 = \cdots = x_n = 0$.

With large sets of equations the calculations involved may be very extensive and it is usual to employ a desk calculator or a computer. In these circumstances, however, Cramer's rule is not the most practical method for finding the solution. Another approach is discussed in Section 13.3.

11.4.2 dim $\mathscr{R}$ = dim $\mathscr{R}'$ = s, where $s < n$ Since $\mathscr{R}$ is linearly dependent, we have $\Delta(\mathscr{R}) = 0$. By Theorem 3, $\Delta(\mathscr{R})$ must have at least one non-zero $s \times s$ minor; we may suppose that the equations and unknowns in (11.16) are so ordered that

$$\begin{vmatrix} a_{11} & a_{12} & \cdots & a_{1s} \\ a_{21} & a_{22} & \cdots & a_{2s} \\ \vdots & \vdots & & \vdots \\ a_{s1} & a_{s2} & \cdots & a_{ss} \end{vmatrix} \neq 0,$$

and eqs. (11.16) then reduce to

$$\begin{aligned} a_{11}x_1 + a_{12}x_2 + \cdots + a_{1s}x_s + \cdots + a_{1n}x_n &= b_1, \\ a_{21}x_1 + a_{22}x_2 + \cdots + a_{2s}x_s + \cdots + a_{2n}x_n &= b_2, \\ &\vdots \\ a_{s1}x_1 + a_{s2}x_2 + \cdots + a_{ss}x_s + \cdots + a_{sn}x_n &= b_s. \end{aligned} \tag{11.17}$$

We may assign values arbitrarily to the unknowns $x_{s+1}, \ldots, x_n$ in eqs. (11.17), giving a set of s equations in the s unknowns $x_1, x_2, \ldots, x_s$. This last set of equations may be solved by Theorem 6; since values have been assigned arbitrarily to the other unknowns, there are infinitely many solutions of eqs. (11.16).

Theorem 7 If dim $\mathscr{R}$ = dim $\mathscr{R}'$ = s, where $s < n$, then values can be assigned arbitrarily to $(n - s)$ unknowns and the resulting equations solved for the other s unknowns.

Note that if the equations in $x_1, x_2, \ldots, x_n$ are homogeneous, values can be assigned to $x_{s+1}, \ldots, x_n$ so that solutions other than the trivial solution $x_1 = x_2 = \cdots = x_n = 0$ are obtained.

Theorem 8 Non-trivial solutions for a set of homogeneous equations can be found if, and only if, $\Delta(\mathscr{R}) = 0$.

Example 7 Solve the equations

$$\begin{aligned} x + 2y + 3z &= 1, \\ 2x - y + 2z &= -5, \\ x + 2y - z &= 9. \end{aligned}$$

In the notation used above, we have

$$\Delta(\mathscr{R}) = \begin{vmatrix} 1 & 2 & 3 \\ 2 & -1 & 2 \\ 1 & 2 & -1 \end{vmatrix} = \begin{vmatrix} 1 & 2 & 3 \\ 2 & -1 & 2 \\ 0 & 0 & -4 \end{vmatrix} = -4(-5) = 20,$$

$$\Delta_1 = \begin{vmatrix} 1 & 2 & 3 \\ -5 & -1 & 2 \\ 9 & 2 & -1 \end{vmatrix} = \begin{vmatrix} 1 & 2 & 3 \\ -5 & -1 & 2 \\ -1 & 0 & 3 \end{vmatrix} = -1(7) + 3(9) = 20,$$

$$\Delta_2 = \begin{vmatrix} 1 & 1 & 3 \\ 2 & -5 & 2 \\ 1 & 9 & -1 \end{vmatrix} = \begin{vmatrix} 1 & 1 & 4 \\ 2 & -5 & 4 \\ 1 & 9 & 0 \end{vmatrix} = 4(23) - 4(8) = 60,$$

$$\Delta_3 = \begin{vmatrix} 1 & 2 & 1 \\ 2 & -1 & -5 \\ 1 & 2 & 9 \end{vmatrix} = \begin{vmatrix} 1 & 2 & 1 \\ 2 & -1 & -5 \\ 0 & 0 & 8 \end{vmatrix} = 8(-5) = -40.$$

Hence by Theorem 6, $x = 1$, $y = 3$, $z = -2$.

Example 8 Find the values of λ for which the equations

$$\begin{aligned} 2\lambda x - 3y + (\lambda - 3)z &= 0, \\ 3x - 2y + z &= 0, \\ 4x - \lambda y + 2z &= 0, \end{aligned}$$

have non-trivial solutions, and find the solutions corresponding to these values of λ.

By Theorem 8 there are non-trivial solutions provided that

$$\begin{vmatrix} 2\lambda & -3 & (\lambda - 3) \\ 3 & -2 & 1 \\ 4 & -\lambda & 2 \end{vmatrix} = 0,$$

that is

$$\begin{vmatrix} 2\lambda & -3 & (\lambda - 3) \\ 3 & -2 & 1 \\ -2 & (4 - \lambda) & 0 \end{vmatrix} = 0.$$

Expanding by the third column, we get

$$(\lambda - 3)(8 - 3\lambda) + 2\lambda^2 - 8\lambda + 6 = 0,$$

giving

$$(\lambda - 3)(6 - \lambda) = 0,$$

that is

$$\lambda = 3 \quad \text{or} \quad 6.$$

When $\lambda = 3$ the equations become

$$\begin{aligned} 6x - 3y \phantom{{}+ 2z} &= 0, \\ 3x - 2y + z &= 0, \\ 4x - 3y + 2z &= 0. \end{aligned}$$

It is not difficult to find a non-zero 2×2 determinant in the array of coefficients; conveniently there is one in the top-left corner, viz.

$$\begin{vmatrix} 6 & -3 \\ 3 & -2 \end{vmatrix}.$$

Thus these equations reduce to a set of two equations which can be written as

$$\begin{aligned} 6x - 3y &= 0, \\ 3x - 2y &= -z, \end{aligned}$$

where we suppose that a value has been assigned to z. The solution for x, y is $x = z$, $y = 2z$.

Hence any values of x, y, z such that $x:y:z = 1:2:1$ are solutions of the given equations when $\lambda = 3$.

When $\lambda = 6$ the equations become

$$\begin{aligned} 12x - 3y + 3z &= 0, \\ 3x - 2y + z &= 0, \\ 4x - 6y + 2z &= 0. \end{aligned}$$

These equations reduce to the two equations

$$\begin{aligned} 12x - 3y &= -3z, \\ 3x - 2y &= -z, \end{aligned}$$

where again we suppose that a value has been assigned to z. Solving for x, y we get $x = -\frac{1}{5}z$, $y = \frac{1}{5}z$. Hence any values of x, y, z such that $x:y:z = -1:1:5$ are solutions of the given equations when $\lambda = 6$.

Example 9 Find integer values of λ for which the equations

$$\begin{aligned} 3\lambda x - 4y \phantom{{}+\lambda z} &= 0, \\ 3x - \lambda y + z &= 0, \\ 9x - 6y + \lambda z &= 0, \end{aligned}$$

have non-trivial solutions, and find the solutions corresponding to these values of λ.

By Theorem 8 there are non-trivial solutions provided that

$$\begin{vmatrix} 3\lambda & -4 & 0 \\ 3 & -\lambda & 1 \\ 9 & -6 & \lambda \end{vmatrix} = 0.$$

Expanding by the third column we get

$$\lambda(-3\lambda^2 + 12) - (-18\lambda + 36) = 0,$$

giving

$$3\lambda^3 - 30\lambda + 36 = 0,$$

that is

$$(\lambda - 2)(\lambda^2 + 2\lambda - 6) = 0.$$

It follows that $\lambda = 2$ is the only integer value. When $\lambda = 2$ the equations become

$$\begin{aligned} 6x - 4y \qquad &= 0, \\ 3x - 2y + z &= 0, \\ 9x - 6y + 2z &= 0. \end{aligned}$$

The 2×2 determinants occurring in the coefficients of x and y are zero but a non-zero 2×2 determinant can be found in the coefficients of y and z, e.g.

$$\begin{vmatrix} -4 & 0 \\ -2 & 1 \end{vmatrix},$$

which occurs in the top-right corner. Thus the equations reduce to the two equations

$$\begin{aligned} 4y \qquad &= 6x, \\ 2y - z &= 3x, \end{aligned}$$

where we may suppose that a value has been assigned to x. Solving for y, z we get $y = \frac{3}{2}x$, $z = 0$. Hence the solutions corresponding to $\lambda = 2$ are values of x, y, z such that $x:y:z = 2:3:0$.

Example 10 Solve the equations

$$\begin{aligned} 2x - 3y + 4z + t &= 3, \\ x + 2y - z + 3t &= 5, \\ 7x \qquad\quad + 5z + 11t &= 21, \\ 3x - y + 3z + 4t &= 8. \end{aligned}$$

Note that the fourth equation may be obtained by adding the first two equations, also that the third equation may be obtained by adding twice the first equation to three times the second equation. Without evaluation of determinants we thus know that dim $\mathscr{R}' \leqslant 2$. Since

$$\begin{vmatrix} 2 & -3 \\ 1 & 2 \end{vmatrix} = 7 \neq 0,$$

we see that dim $\mathscr{R}'$ = dim $\mathscr{R} = 2$.

By Theorem 7 we can assign values arbitrarily to two of the four unknowns, e.g. we may solve for x and y the equations

$$\begin{aligned} 2x - 3y &= 3 - 4z - \ t, \\ x + 2y &= 5 + \ z - 3t, \end{aligned}$$

where z and t are regarded as known. This gives

$$x = \frac{\begin{vmatrix} (3 - 4z - \ t) & -3 \\ (5 + \ z - 3t) & 2 \end{vmatrix}}{\begin{vmatrix} 2 & -3 \\ 1 & 2 \end{vmatrix}} = 3 - \tfrac{1}{7}(5z + 11t),$$

$$y = \frac{\begin{vmatrix} 2 & (3 - 4z - \ t) \\ 1 & (5 + \ z - 3t) \end{vmatrix}}{\begin{vmatrix} 2 & -3 \\ 1 & 2 \end{vmatrix}} = 1 + \tfrac{1}{7}(6z - 5t).$$

With any values substituted for z and t, this gives a solution of the given equations.

11.4.3 Consistency It has been shown in Sections 11.4.1 and 11.4.2 that equations (11.16) can be solved in all cases where dim $\mathscr{R}$ = dim $\mathscr{R}'$. This combines with Lemma 3 to give

Theorem 9 Equations (11.16) are consistent or inconsistent according as the dimensions of $\mathscr{R}$ and $\mathscr{R}'$ are equal or unequal.

A set of homogeneous equations is always consistent. This is obvious, either from Theorem 9 or from the fact that the equations can be seen to have the trivial solution $x_1 = x_2 = \cdots = x_n = 0$.

Example 11 Examine for consistency the equations

$$\begin{aligned} x + 2y - z &= 4, \\ 2x \qquad + 3z &= -6, \\ 13x + 10y + 7z &= 1. \end{aligned}$$

The component-array for $\mathscr{R}'$ is

$$\begin{matrix} 1 & 2 & -1 & 4 \\ 2 & 0 & 3 & -6 \\ 13 & 10 & 7 & 1 \end{matrix}$$

One 3×3 determinant of this array is

$$\begin{vmatrix} 2 & -1 & 4 \\ 0 & 3 & -6 \\ 10 & 7 & 1 \end{vmatrix} = 2\begin{vmatrix} 3 & -6 \\ 7 & 1 \end{vmatrix} + 10\begin{vmatrix} -1 & 4 \\ 3 & -6 \end{vmatrix} = 90 - 60 = 30.$$

Since we have found a non-zero 3×3 determinant it follows that $\dim \mathscr{R}' = 3$. However,

$$\Delta(\mathscr{R}) = \begin{vmatrix} 1 & 2 & -1 \\ 2 & 0 & 3 \\ 13 & 10 & 7 \end{vmatrix} = 2\begin{vmatrix} 2 & -1 \\ 10 & 7 \end{vmatrix} + 3\begin{vmatrix} 1 & 2 \\ 13 & 10 \end{vmatrix} = 48 - 48 = 0$$

and so $\dim \mathscr{R} < 3$.

It follows that the equations are inconsistent.

11.4.4 Geometrical Interpretation of Linear Equations When the number of unknowns is two, a simple interpretation is possible. The set of equations

$$\begin{aligned} a_1x + b_1y &= c_1, \\ a_2x + b_2y &= c_2, \\ a_3x + b_3y &= c_3, \\ &\vdots \\ a_mx + b_my &= c_m, \end{aligned}$$

represents a set of straight lines in the xy-plane. A solution $x = x_0$, $y = y_0$ of the equations exists if there is a point (x_0, y_0) lying on all the lines.

The possible situations which may arise are as follows:

(a) If $\dim \mathscr{R}' = 3$, the lines are not concurrent, i.e. there are no points common to all the lines (Fig. 11.2).

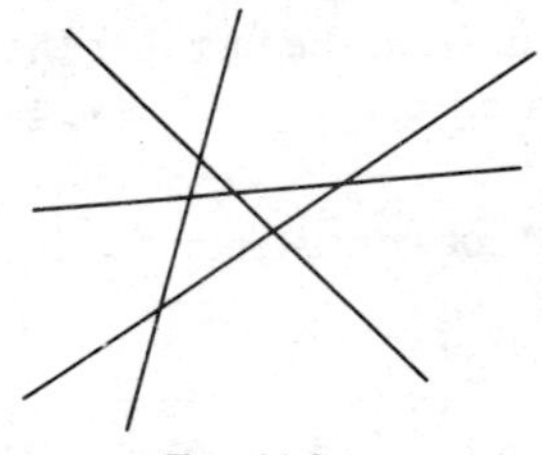

FIG. 11.2

(b) If $\dim \mathscr{R}' = \dim \mathscr{R} = 2$, the lines are concurrent, i.e. there is a single common point (Fig. 11.3).

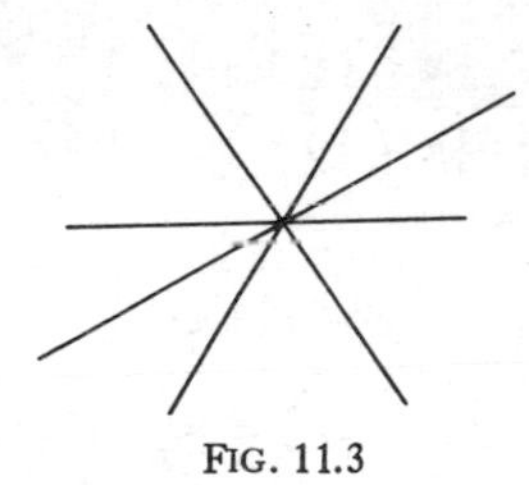

FIG. 11.3

(c) If $\dim \mathscr{R}' = 2$ and $\dim \mathscr{R} = 1$, the lines are all parallel, at least two of them being distinct, so there can be no common point (Fig. 11.4).

FIG. 11.4

(d) If $\dim \mathscr{R}' = \dim \mathscr{R} = 1$, the lines all coincide, i.e. there is a complete line of common points.

In a similar way the solutions of linear equations in three unknowns can be interpreted in terms of intersections of planes. However, a geometrical interpretation is not possible when the number of unknowns is more than three.

Exercises 11(d)

1. Solve by determinants the following sets of simultaneous equations:

(a) $$\begin{aligned} 5x + 3y + 3z &= 48, \\ 2x + 6y - 3z &= 18, \\ 8x - 3y + 2z &= 21, \end{aligned}$$

(b) $$\begin{aligned} x - y + z &= 1, \\ x - 2y + 4z &= 8, \\ x + 3y + 9z &= 27, \end{aligned}$$

(c) $$\begin{aligned} x + y + z &= 5, \\ x + 2y + 3z &= 11, \\ 3x + y + 4z &= 13, \end{aligned}$$

(d) $$\begin{aligned} x + y - z &= 1, \\ 8x + 3y - 6z &= 1, \\ -4x - y + 3z &= 1, \end{aligned}$$

(e) $$\begin{aligned} 4x + 3y + 5z &= 11, \\ 9x + 4y + 15z &= 13, \\ 12x + 10y - 3z &= 4. \end{aligned}$$

2. Solve the equations

$$\tfrac{1}{2}x + \tfrac{1}{3}y + \tfrac{1}{4}z = 1,$$
$$\tfrac{1}{3}x + \tfrac{1}{4}y + \tfrac{1}{5}z = 1,$$
$$\tfrac{1}{4}x + \tfrac{1}{5}y + \tfrac{1}{6}z = 1$$

by means of determinants.

If

$$\begin{aligned} y - z &= ax, \\ z - x &= by, \\ x - y &= cz, \end{aligned}$$

and x, y, z are not all zero, show, without solving any equations, that

$$abc + a + b + c = 0.$$

3. Determine which, if any, of the following two sets of equations are consistent and, when possible, solve them,

$$\text{(a)}\ \begin{aligned} x + y + z &= 1, \\ 2x + 4y - 3z &= 9, \\ 3x + 5y - 2z &= 11. \end{aligned} \qquad \text{(b)}\ \begin{aligned} 2x - y + z &= 7, \\ 3x + y - 5z &= 13, \\ x + y + z &= 5. \end{aligned}$$

4. Find the values of λ for which the following equations are consistent:

$$\begin{aligned} 3x + \lambda y &= 5, \\ \lambda x - 3y &= -4, \\ 3x - y &= -1. \end{aligned}$$

Solve the equations for these values of λ.

5. Find the value of λ for which the following equations are consistent:

$$\begin{aligned} 4x + \lambda y &= 10, \\ x - 2y &= 8, \\ 5x + 7y &= 6. \end{aligned}$$

Find the values of x and y corresponding to this value of λ.

6. Find the values of λ for which the equations

$$\begin{aligned} (2 - \lambda)x + 2y + 3 &= 0, \\ 2x + (4 - \lambda)y + 7 &= 0, \\ 2x + 5y + 6 - \lambda &= 0, \end{aligned}$$

are consistent, and find the values of x and y corresponding to each of these values of λ.

7. Find the condition for the equations

$$\begin{aligned} (a - b - c)x + 2ay + 2a &= 0, \\ 2bx + (b - c - a)y + 2b &= 0, \\ 2cx + 2cy + (c - a - b) &= 0, \end{aligned}$$

to have a common solution, and show that when this condition is satisfied the equations have infinitely many common solutions.

8. Show that the equations

$$\begin{aligned} 2x + 3y &= 4, \\ 3x + \lambda y &= -1, \\ \lambda x - 2y &= c, \end{aligned}$$

are consistent for real values of λ if

$$4c^2 - 156c - 439 \geqslant 0.$$

9. Find the values of λ for which the equations

$$\begin{aligned} \lambda x + y + \sqrt{2}z &= 0, \\ x + \lambda y + \sqrt{2}z &= 0, \\ \sqrt{2}x + \sqrt{2}y + (\lambda - 2)z &= 0, \end{aligned}$$

have a solution other than $x = y = z = 0$. Find also the ratios $x:y:z$ which correspond to each of these values of λ.

10. If no two of the numbers a, b, c are equal, find the condition that the equations

$$\begin{aligned} x + y + z &= 0, \\ x/a + y/b + z/c &= 0, \\ a^2x + b^2y + c^2z &= 0, \end{aligned}$$

may have non-trivial solutions. Find the ratios $x:y:z$ when $a = 1$, $b = -3$, $c = 2$.

11. Find all the values of t for which the equations

$$\begin{aligned} (t - 1)x + (3t + 1)y + 2tz &= 0, \\ (t - 1)x + (4t - 2)y + (t + 3)z &= 0, \\ 2x + (3t + 1)y + 3(t - 1)z &= 0, \end{aligned}$$

have non-trivial solutions and find the ratios of $x:y:z$ when t has the smallest of these values. What happens when t has the greatest of these values?

12. Eliminate x, y and z from the equations

$$\begin{aligned} ax + hy + gz &= 0, \\ hx + by + fz &= 0, \\ gx + fy + cz &= 0. \end{aligned}$$

13. Prove that

$$\Delta \equiv \begin{vmatrix} a & b & c \\ b & c & a \\ c & a & b \end{vmatrix} = -(a + b + c)(a^2 + b^2 + c^2 - bc - ca - ab).$$

Show that there are three real values of λ for which the equations

$$\begin{aligned} (a - \lambda)x + by + cz &= 0, \\ bx + (c - \lambda)y + az &= 0, \\ cx + ay + (b - \lambda)z &= 0, \end{aligned}$$

are simultaneously true, and that the product of these values of λ is Δ.

CHAPTER 12

MAPPINGS

12.1 Relations and Mappings

We saw in Section 5.1 that if two sets X, Y are in (1, 1)-correspondence there are different ways of associating elements of Y with elements of X—i.e. different (1, 1)-mappings of X onto Y—that demonstrate this (1, 1)-correspondence.

Let us now consider two sets X, Y that are not necessarily in (1, 1)-correspondence. Any way of associating elements of Y with elements of X is called a *relation from X to Y* provided that with *each* element of X is associated *one or more* elements of Y. It is not required that each element of Y be the associate of one or more elements of X and so it may happen that only a proper subset Y' of Y is involved in the relation.

The main ideas in connection with relations can be adequately illustrated by a very simple representation in which sets are represented (as in Venn diagrams) by sets of points bounded by closed curves, an arrow being drawn from a point P in X towards each point in Y associated with P (Fig. 12.1).

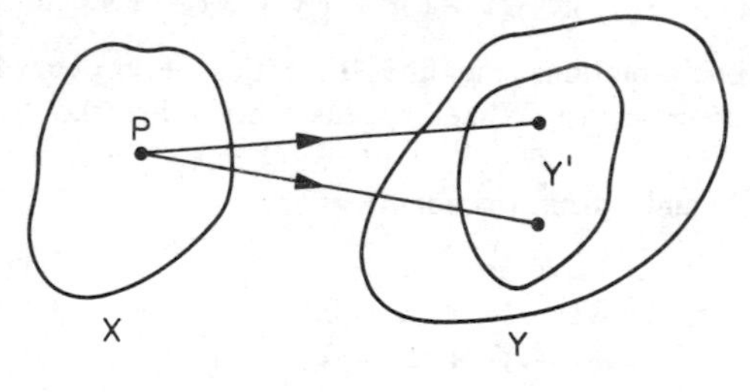

FIG. 12.1

Consider the following examples in each of which X is the set of first-year students at present attending a certain university:

(1) Let Y be the set of lecture courses at present being given in the university.

If we associate with each student in X the lecture courses taken by that student, we get a relation from X to Y. An individual student normally takes several courses and so in Fig. 12.2 several arrows run from the same student in X; a particular course is taken by a number of students so a number of arrows run towards the same course in Y. Clearly only the subset Y' consisting of courses taken by first-year students is involved in the relation.

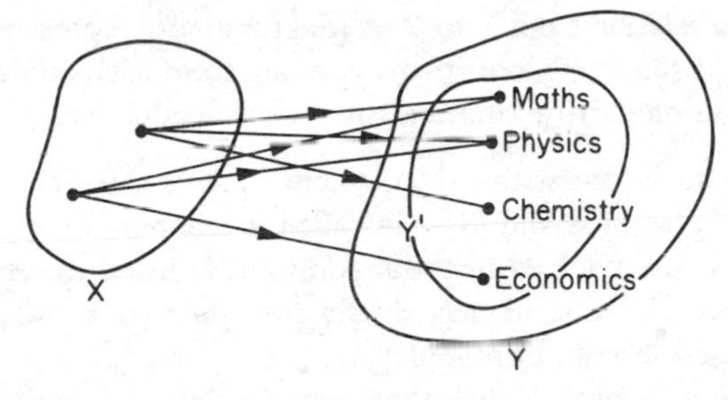

FIG. 12.2

(2) Let Y be the set of text-books owned by students attending the university.

For simplicity we shall assume that each student in X necessarily owns several text-books and that each text-book is the property of only one student. If we associate with each student the text-books owned by that student we get a relation from X to Y. Note that in this case (Fig. 12.3) several arrows run from each student in X but only one arrow can run towards a particular book in Y.

Only the set Y' of books owned by first-year students is involved in the relation.

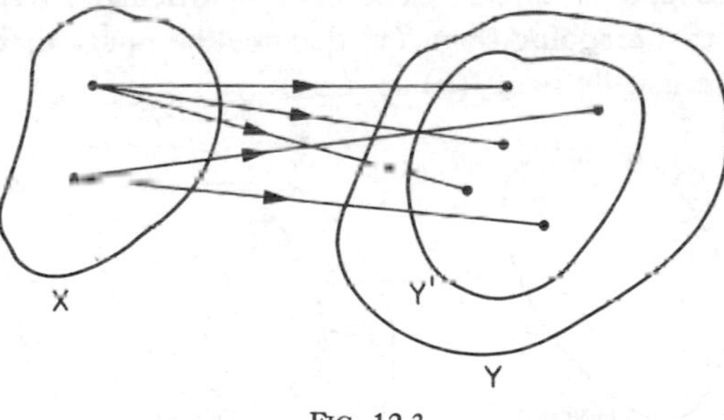

FIG. 12.3

(3) Let Y be the set of residences where students attending the university live.

If we associate with each student in X the residence where that student lives we get a relation from X to Y. The reader should illustrate this relation diagrammatically and note that only one arrow runs from each student in X, but a number of arrows may run towards the same residence in Y.

(4) Assume that each student is allocated a locker in the university, and let Y be the set of lockers.

If we associate with each student in X the locker allocated to that student we get a relation from X to Y. A diagrammatic representation of this relation would show only one arrow running from each student in X, and not more than one arrow running towards the same locker in Y.

A relation such that with *each* element in X is associated *exactly one* element in Y—e.g. (3) and (4)—is called a *mapping of X into Y*; if all elements in Y are involved in the mapping it is usual to refer to a mapping of X *onto* Y. Other names, chiefly *transformation* and *function*, are sometimes used instead of mapping.

A mapping of X onto Y such that each element of Y is the associate of exactly one element of X is called a *(1, 1)-mapping of X onto Y*. For example, in (4) we have a (1, 1)-mapping of X onto Y', where Y' is the set of lockers allocated to first-year students.

In terms of the diagrammatic representation a mapping is a relation in which *exactly one arrow runs from each element in X*; if, in addition, exactly one arrow runs towards each element in Y then the mapping is called a (1, 1)-mapping. The set X is called the *domain* of the mapping. If x is an element of X and if the mapping associates with x the element y in Y, then we say y is the *image* of x under the mapping; the set Y' of images of all elements x in the domain is called the *codomain*, or *range*, of the mapping.

Particular mappings may be denoted by particular letters f, g, F, G, ...; if f denotes a mapping then $f(x)$ denotes the image of x under f (see Fig. 12.4). (We usually read $f(x)$ as 'f of x'.)

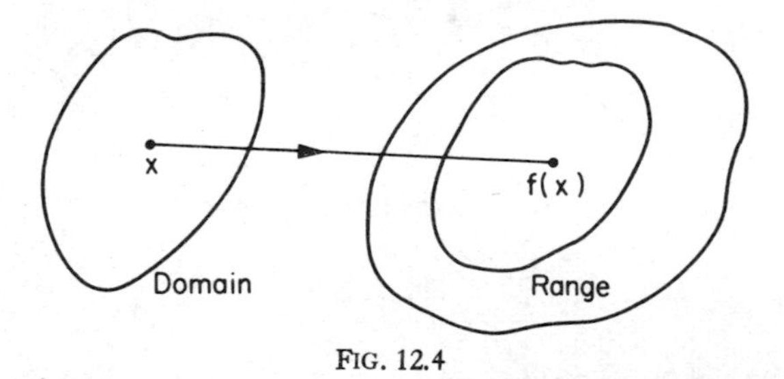

FIG. 12.4

Suppose x is a *variable* on X and we write $y = f(x)$; then y can refer to *any* element within the range Y' of f, but, whereas we regard ourselves free to specify x arbitrarily within the elements of X, the element to which y refers is determined once x has been specified. We say y is a *dependent variable* on Y' and in contrast x may be called an *independent variable* on X.

At this stage we should note carefully that in order to describe a mapping f we must do the following two things:

(a) state the domain of f,
(b) indicate the image $f(x)$ of each x in the domain.

The range of f is determined by (a) and (b), since it is the set of images of all elements in the domain. For a domain consisting of a finite number of elements we can do (b) simply by listing these elements together with their images; however the mappings (3) and (4) above are typical examples where we have an *image rule*, i.e. we can state the image in the form of a general rule, without the need to specify images individually. Mostly we shall be concerned with mappings where the domains are infinite sets and where, fortunately, it is possible to do (b) in terms of an image rule.

12.1.1 Some Basic Mappings We shall now describe some very simple mappings which have been of fundamental importance in our earlier work.

(1) Let f denote the mapping defined as follows:
 (a) The domain of f is R, the set of real numbers.
 (b) Let a straight line be drawn and an origin O chosen on this line. For any x in R let $f(x) = P$ where P is the point on the line distant $|x|$ units from O, on one side of O if $x > 0$ and on the opposite side if $x < 0$.

 Clearly f is the familiar mapping of R onto the points of the real axis.

(2) Let the mapping g be defined thus:
 (a) The domain of g is the set of points on a given plane.
 (b) For any point P on the plane, $g(P) = (x, y)$ where x and y are the co-ordinates of P relative to a pair of axes chosen in the plane.

 Then g is the familiar mapping of the points on a plane onto the set R^2 of number pairs (x, y).

(3) Let the mapping h be defined thus:
 (a) The domain of h is the set of all points in space.
 (b) For any point P, $h(P) = (x, y, z)$ where x, y, z are the co-ordinates of P relative to a chosen set of axes.

 Then h is the familiar mapping of the points in space onto the set R^3 of number triads (x, y, z).

The above three mappings are so well known that we normally use them without any acknowledgment of the fact that we are doing so, as for example when we refer to the set of points on a line, on a plane, in space as the set R, R^2, R^3 respectively.

(4) Let the mapping F be defined thus:
   (a) The domain of F is the set of points in space.
   (b) For any point P, $F(P) = \mathbf{r}$ where $\mathbf{r}$ is the position vector of P relative to some origin O.
   Then F is the familiar mapping of points in space onto the set V_3 of vectors.

(5) Let the mapping G be defined thus:
   (a) The domain of G is the set C of complex numbers.
   (b) For any complex number $x + iy$, $G(x + iy) = P$ where P is the point with co-ordinates (x, y) in some plane.
   Then G is the familiar mapping of C onto the points of a plane (Argand diagram).

(6) Let the mapping H be defined thus:
   (a) The domain of H is the set C of complex numbers.
   (b) For any complex number $x + iy$, $H(x + iy) = x\mathbf{i} + y\mathbf{j}$.
   Then H is the familiar mapping of C onto the set V_2 of vectors (alternative use of Argand diagram).

(7) Many physical quantities are 'measured' so as to map them into the set R of real numbers. For example, we may regard a thermometer as a device which maps a set of physical states into R. Some physical quantities (e.g. force, velocity) are mapped into the set V_3 of vectors.

12.1.2 The Identity Mapping The mappings we have considered above are mappings from elements of one kind onto elements of a different kind, e.g. numbers onto points, points onto vectors, physical entities onto numbers, etc. However we frequently have mappings in which the elements in the domain and range are of the same kind, and in some cases the domain and range are the same set. For example, consider the mapping f defined thus:

(a) The domain of f is the set C of complex numbers.
(b) For any complex number $x + iy$, $f(x + iy) = x - iy$; i.e. the image of each complex number is its conjugate.

The range of f is C. In general a complex number is not the same as its image under f, although certain complex numbers (ones which are purely real) are their own images.

For any set X there is an *identity mapping*, usually denoted by I, and defined thus:

(a) The domain of I is X.

(b) For any element x in X, $I(x) = x$, i.e. every element is its own image under I.

Such a mapping may seem too trivial to serve any useful purpose, but it will be found to have value in the development of algebra.

Exercises 12(a)

1. Assume that for each word in language A there is a word in language B with exactly the same meaning. Under what further assumptions does this give

   (a) a mapping from language A *into* language B,
   (b) a mapping from language A *onto* language B,
   (c) a (1, 1)-mapping from language A onto language B?

2. For any pair of real numbers m, c let the image of the line $y = mx + c$ be the point (m, c). Describe the domain and range of this mapping.
3. For any pair of real numbers a, b where $a \neq 0$, $b \neq 0$ let the image of the line $(x/a) + (y/b) = 1$ be the point (a, b). Describe the domain and range of this mapping.
4. For any pair of real numbers a, b where $a \neq 0$, $b \neq 0$ associate with the ellipse $(x^2/a^2) + (y^2/b^2) = 1$ the point (a, b). Suggest a further condition so that this association becomes a mapping, and describe the domain and range of your mapping.
5. Denote by L the set of lines that can be drawn from the origin O in all possible directions. If a line in L has D.C.'s $[l, m, n]$ let the image of that line be the point (l, m, n). Describe the range of this mapping. State the images of the three positive co-ordinate axes.
6. Denote by S the set of points on the surface of a sphere, centre at the origin. Let the image of a point in S be the position vector of that point. Describe the range of this mapping.
7. Denote by S the set of points, other than the origin O, inside the sphere $x^2 + y^2 + z^2 = a^2$. Let the image of a point P in S be the point P' where P' lies on the line OP and $OP \cdot OP' = a^2$. Describe the range of this mapping.

12.2 Inverse Relations and Mappings

Corresponding to any relation from the set X onto the set Y there is an *inverse relation* defined as follows:

(a) The inverse relation is a relation from Y to X.
(b) If in the original relation the element y in Y is associated with the element x in X, then in the inverse relation x is associated with y.

In the diagrammatic representation the inverse relation is obtained from the original relation simply by *reversing the directions of all the arrows.*

As an example consider the relation (1) in Section 12.1; the inverse relation is a relation from the set of lecture courses to the set of first-year students, each lecture course having associated with it any student who takes that course.

It is easily seen that if the inverse relation is similarly 'inverted' the result is the original relation, i.e. if one relation is the inverse of another relation, then the two relations are *mutually* inverse.

Consider a relation that is a mapping f, i.e. each element x in X has a *unique* image $f(x)$ in Y. It may happen that two or more elements in X have the *same* image in Y, say $f(x_1) = f(x_2)$ where $x_1 \neq x_2$. Then the inverse relation is *not* a mapping since it associates *two* elements x_1, x_2 in X with an element in Y (see Fig. 12.5). However if f is a (1, 1)-mapping then

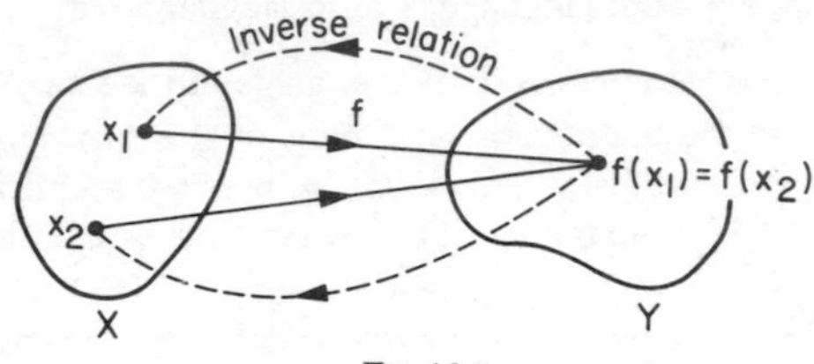

FIG 12.5

the inverse relation is also a (1, 1)-mapping, called the ***inverse mapping*** and denoted by f^{-1}. Again we can say that f and f^{-1} are *mutually* inverse since, in symbols, $(f^{-1})^{-1} = f$. The domain of f^{-1} is the range of f and the range of f^{-1} is the domain of f. Furthermore, writing $y = f(x)$, we have $x = f^{-1}(y)$.

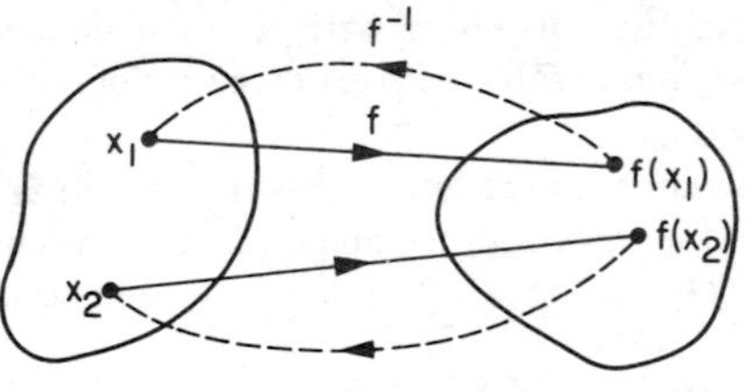

FIG. 12.6

The mappings in Section 12.1.1 are (1, 1)-mappings and so have inverses; e.g. in (6), H^{-1} is a mapping of V_2 onto C, and

$$H^{-1}(x\mathbf{i} + y\mathbf{j}) = x + iy.$$

Consider the mapping f defined thus: f maps V_2 into R and

$$f(x\mathbf{i} + y\mathbf{j}) = \sqrt{(x^2 + y^2)},$$

i.e. the image of each vector is its magnitude. Clearly f is not a (1, 1)-mapping since, for example, $3\mathbf{i} + 4\mathbf{j}$, $4\mathbf{i} + 3\mathbf{j}$, $5\mathbf{i}$, $5\mathbf{j}$, $(5/2)\,\mathbf{i} + 5\sqrt{3}/2\mathbf{j}$ are all vectors with the same image (magnitude) 5. Hence f does not have an inverse.

Consider also the mapping p that has as domain the set of possible events when a die is cast and, for any event E, $p(E)$ is defined to be the probability that E occurs. The mapping p does not have an inverse since it is not a (1, 1)-mapping.

Exercises 12(b)

1. Let f be a mapping in which a line of length l is mapped onto the real number l, g a mapping in which a rectangle of area A is mapped onto the real number A. If we assume that lines and rectangles are to be distinguished only by their dimensions, and not by their positions in space or any other properties, do f^{-1}, g^{-1} exist?
2. Given a fixed line l in a fixed plane, let the image of any line in the plane be the real number α where α is the angle the line makes with l such that $0 < \alpha < \pi/2$. Does this mapping have an inverse?
3. Given a fixed line l in a fixed plane π, then if O is a fixed point on l let the image of any point P on π be the real number α where α is the angle, measured in the anti-clockwise direction, that OP makes with l. Does this mapping have an inverse?
4. The domain of the mapping f is the set of points $\{(x, y) : y > 0\}$. The image of the point P in the domain is the point P' where $OP' = OP$, O being the origin, and the angle that OP' makes with the positive x-axis is double that made by OP, both angles being measured in the anti-clockwise direction. What is the range of f? Describe f^{-1}. Explain why f^{-1} would not exist if the word *double* above were replaced by *treble*.
5. Let f denote the mapping in Question 7 of Exercises 12(a). State the image-rule for f^{-1}.

12.3 Composite Mappings

Let f be a mapping of X onto Y, g a mapping of Y onto Z. Each element x in X has (under f) a unique image y in Y, and y in turn has (under g) a unique image z in Z; this associates with each element x of X a unique element z in Z giving a *composite mapping* of X onto Z. Since we write $y = f(x)$ and $z = g(y)$, we may combine these equations and write $z = g(f(x))$ (in words, 'g of f of x'). We refer to the composite mapping as the mapping 'g of f', denoted by $g \circ f$; thus

$$g \circ f(x) = g(f(x)).$$

As examples, consider the following composite mappings where g, G, H are the mappings defined in (2), (5), (6) of Section 12.1.1:

(a) the composite mapping $g \circ G$ maps C onto R^2 and

$$g \circ G(x + iy) = (x, y).$$

(b) the composite mapping $G \circ H^{-1}$ maps V_2 onto the set of points on a plane and $G \circ H^{-1}(x\mathbf{i} + y\mathbf{j}) = P$, where P has co-ordinates (x, y).

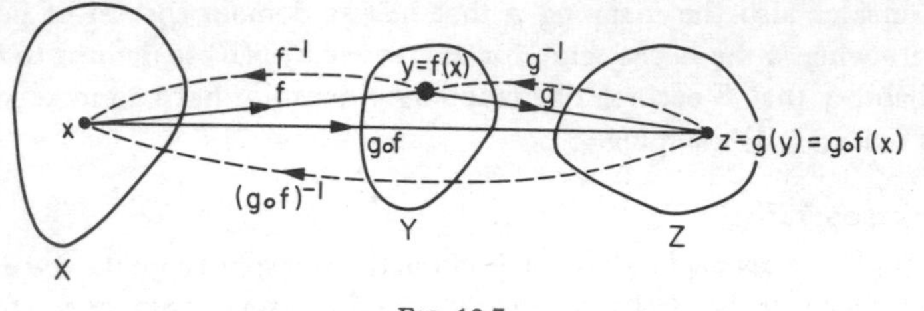

FIG. 12.7

If $Z = X$, i.e. f maps X onto Y and g maps Y onto X, then two composite mappings are possible, viz. $g \circ f, f \circ g$; $g \circ f$ maps X onto itself whereas $f \circ g$ maps Y onto itself, and so the two mappings are different. In particular, if $g = f^{-1}$ we find that $f^{-1} \circ f = I$ where I is the identity mapping from X onto itself, and $f \circ f^{-1} = I$ where I is the identity mapping of Y onto itself; an alternative way of saying this is that for any element x in X we have $f^{-1} \circ f(x) = x$, and for any element y in Y we have $f \circ f^{-1}(y) = y$.

If $X = Y = Z$ then the two composite mappings $g \circ f, f \circ g$ are both mappings of X onto itself but are not necessarily the same mapping, as the following example shows: let f, g be mappings of C onto itself with

$$f(x + iy) = x - iy \quad \text{and} \quad g(x + iy) = y + ix,$$

and then

$$g \circ f(x + iy) = g(x - iy) = -y + ix,$$

whereas

$$f \circ g(x + iy) = f(y + ix) = y - ix.$$

Reversal rule. Assume that the mappings f, g have inverses f^{-1}, g^{-1}. Then $g \circ f$ also has an inverse and it can be seen from Fig. 12.7 that

$$(g \circ f)^{-1} = f^{-1} \circ g^{-1}.$$

Note that the order of f, g is different on the two sides of this equation.

Again using g, G, H to denote the mappings defined in (2), (5), (6) of Section 12.1.1, the reader should satisfy himself that, if P is the point with co-ordinates (x, y),

$$g \circ G(x + iy) = g(P) = (x, y),$$

$$G^{-1} \circ g^{-1}(x, y) = G^{-1}(P) = x + iy,$$

i.e.
$$G^{-1} \circ g^{-1} = (g \circ G)^{-1}.$$

Exercises 12(c)

1. The two lines l_1, l_2 are at right-angles to each other. Under the mapping f the image of a point P in the plane of the two lines is the reflection of P in l_1 (i.e. P and its image P' are equidistant from l_1 but on opposite sides, PP' being perpendicular to l_1). Similarly, under g the image of P is its reflection in l_2.
   Indicate on a diagram how to obtain the image of P under $f \circ g$ and under $g \circ f$, and say whether or not you think $f \circ g = g \circ f$. Do the same with the lines l_1, l_2 not at right-angles.
2. Show that if the mappings f, g are reflections in two lines (see previous question) then $(g \circ f)^{-1} = f \circ g$.
3. Let the mapping f_θ be a rotation through θ about the point O, i.e. O is a point on a given plane and the image of any other point P on the plane is the point P', where $OP' = OP$ and OP' makes an angle θ, measured anti-clockwise, with OP. Find the value of θ such that the equation $f_\alpha \circ f_\beta = f_\beta \circ f_\alpha = f_\theta$ is true.
   Is $f_\alpha \circ (f_\beta \circ f_\gamma) = (f_\alpha \circ f_\beta) \circ f_\gamma$?
4. The two lines l_1, l_2 intersect at O at angle α, the direction of rotation from l_1 to l_2 being anti-clockwise. Let the mappings f, g be reflections in these two lines, and let the mapping f_θ be a rotation through θ about O (see previous question). Verify that
$$g \circ f = f_{2\alpha}, \quad f \circ g = f_{-2\alpha}.$$
5. For general mappings f, g, h is $f \circ (g \circ h) = (f \circ g) \circ h$?

12.4 Linear Transformations

The term *transformation* is frequently used in preference to *mapping* when the domain and range are both sets of points. We now look at some important examples of transformations from the set R^2 onto itself.

(a) *Translation of axes.* If new axes $O'x'$, $O'y'$ are chosen as shown in Fig. 12.8, then the co-ordinates of a point P change from (x, y) to (x', y') where

$$x' = x - a, \quad y' = y - b. \tag{1}$$

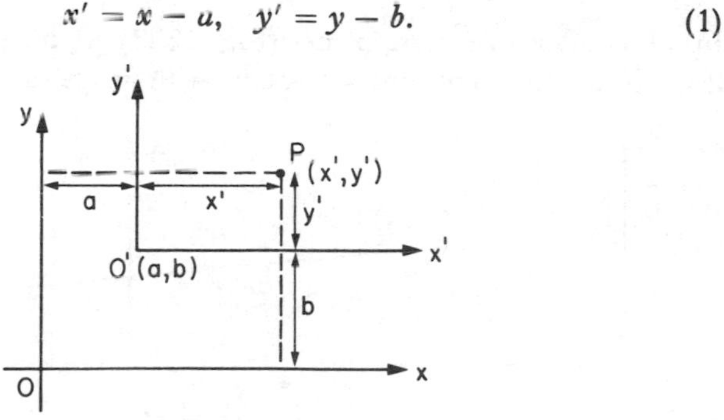

FIG. 12.8

It is sometimes convenient to think in terms of two separate planes, an xy-plane and an $x'y'$-plane (Fig. 12.9), eqs. (1) being the image rule of a transformation from one plane to the other. In particular the point (a, b) on the xy-plane is mapped onto the point (0, 0) on the $x'y'$-plane.

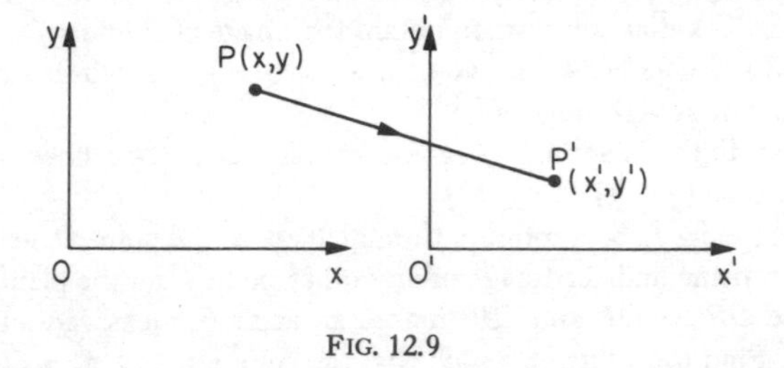

FIG. 12.9

(b) *Rotation of axes.* When the axes are rotated through an angle α as shown in Fig. 12.10 the co-ordinates of a point P change from (x, y) to (x', y') where (see Section 2.6)

$$x' = x \cos \alpha + y \sin \alpha, \quad y' = -x \sin \alpha + y \cos \alpha. \tag{2}$$

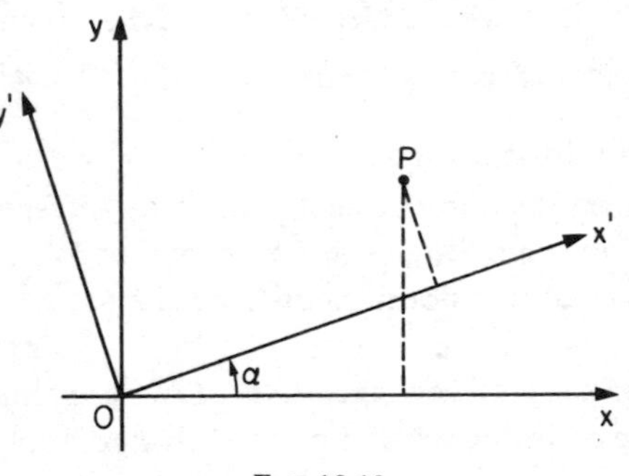

FIG. 12.10

In terms of two separate planes (Fig. 12.11) we have a transformation from the xy-plane onto the $x'y'$-plane with image rule given by eqs. (2).

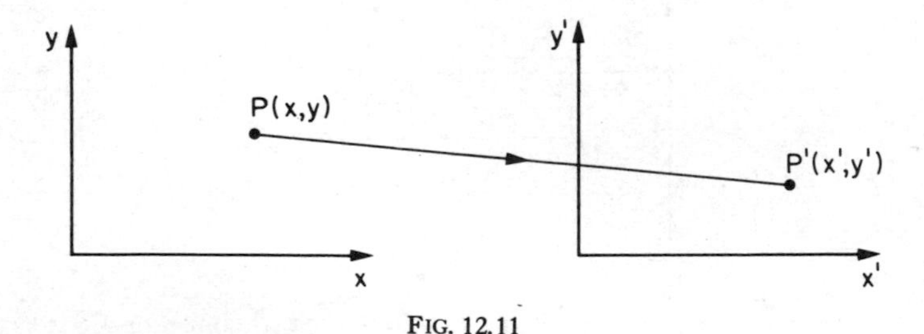

FIG. 12.11

(c) *Interchange of axes.* Let new axes Ox', Oy' be chosen so that Ox' coincides with Oy and Oy' coincides with Ox. Then the co-ordinates of a point P change from (x, y) to (x', y') where

$$x' = y, \quad y' = x. \tag{3}$$

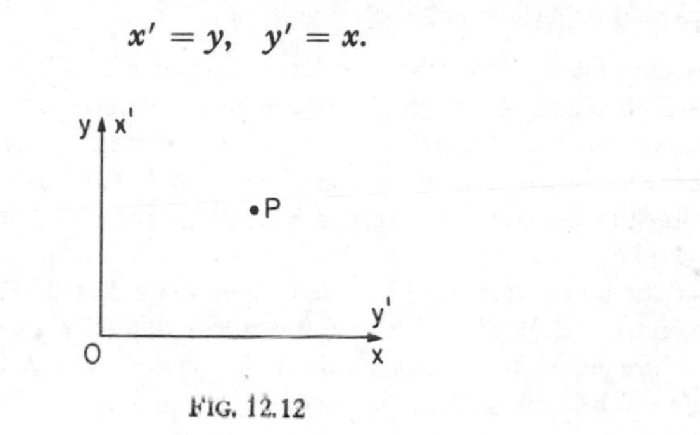

FIG. 12.12

Again the situation can be interpreted as a transformation from an xy-plane onto an $x'y'$-plane, the image rule being eqs. (3).

At this stage the reader should take note that the image rule of a transformation need not necessarily take the form of algebraic equations like (1)–(3). For example, we get a perfectly valid image rule by defining x' to be the number of men presently aged between x years and y years and living in Britain, and defining y' to be the number of women with the same qualifications. However, in all the cases of practical interest in this course the image rule can be expressed in the form of algebraic equations; moreover, in many cases these equations are *linear*, i.e. in the case of transformations from R^2 onto itself the equations have the form

$$x' = a_1x + b_1y, \quad y' = a_2x + b_2y, \tag{4}$$

where a_1, b_1, a_2, b_2 are constants.

Any pair of equations of the form (4) defines a transformation from R^2 onto (or into) itself, and consequently the name *linear transformation* is frequently applied to such a pair of equations, regardless of the context in which the equations have arisen.

More generally the name *linear transformation* is applied to any set of equations of the form

$$\begin{aligned}
y_1 &= a_{11}x_1 + a_{12}x_2 + \cdots + a_{1n}x_n,\\
y_2 &= a_{21}x_1 + a_{22}x_2 + \cdots + a_{2n}x_n,\\
&\vdots\\
y_m &= a_{m1}x_1 + a_{m2}x_2 + \cdots + a_{mn}x_n,
\end{aligned}$$

where $a_{11}, a_{12}, \ldots, a_{mn}$ are all constants; such a set of equations defines a transformation from R^n onto (or into) R^m, the image of $(x_1, x_2, \ldots, x_n)$ being $(y_1, y_2, \ldots, y_m)$.

Exercises 12(d)

1. Given a fixed vector **t** let *translation by* **t** be a transformation from R^2 onto itself in which the point P has image P', where $\overrightarrow{PP'} = \mathbf{t}$. Show that the transformation (a) above is of this type, and state the appropriate vector **t**.
2. For what value of θ is the transformation f_θ (rotation through θ about O, defined in Question 3 of Exercises 12(c)) equivalent to the transformation (b) above?
3. Let the transformation f be a reflection in the line $y = x$ (see Question 1 of Exercises 12(c)). Show that the transformation (c) above is equivalent to f.

   More generally let f be a reflection in the line $y = mx$. Show that the image rule can be expressed in the form of the equations

$$x' = \left(\frac{1-m^2}{1+m^2}\right)x + \left(\frac{2m}{1+m^2}\right)y,$$

$$y' = \left(\frac{2m}{1+m^2}\right)x - \left(\frac{1-m^2}{1+m^2}\right)y.$$

4. A *uniform stretch* in the y-direction is a transformation from R^2 onto itself given by the image rule $x' = x$, $y' = ky$, where k is a positive constant. (If $k < 1$ the transformation is really a *contraction* in the y-direction.)

   Show that points on a straight line map onto points that are also on a straight line, and that points on a circle map onto points that are on an ellipse. [*Hint*: Replace x by x', y by $(1/k)y'$ in the equation of a curve to get the equation of the *image* curve.]
5. A *dilatation* is a transformation from R^2 onto itself given by the image rule $x' = kx$, $y' = ky$ where k is a positive constant. Show that points on a straight line map onto points that are also on a straight line, and that points on a circle of radius a map onto points that are on a circle of radius ka.
6. The image of a point P inside a circle, centre O, radius a, is defined to be the point P' where P' lies on the line OP and $OP'.OP = a^2$. Show that this image rule can be expressed in the form

$$x' = \frac{a^2x}{x^2+y^2}, \quad y' = \frac{a^2y}{x^2+y^2}.$$

   Show also that the image rule of the inverse transformation is

$$x = \frac{a^2x'}{x'^2+y'^2}, \quad y = \frac{a^2y'}{x'^2+y'^2}.$$

   Deduce that points on a straight line not passing through O map onto points that lie on a circle.
7. Taking a dilatation to mean the transformation from R^3 onto itself given by the image rule $x' = kx$, $y' = ky$, $z' = kz$ where k is a positive constant, show that under a dilatation a sphere of radius a is transformed into a sphere of radius ka.

8. For the transformation described in Question 7 of Exercises 12(a), show that the image rule may be expressed in the form

$$x' = \frac{a^2x}{x^2 + y^2 + z^2}, \quad y' = \frac{a^2y}{x^2 + y^2 + z^2}, \quad z' = \frac{a^2z}{x^2 + y^2 + z^2},$$

or in the inverse form

$$x = \frac{a^2x'}{x'^2 + y'^2 + z'^2}, \quad y = \frac{a^2y'}{x'^2 + y'^2 + z'^2}, \quad z = \frac{a^2z'}{x'^2 + y'^2 + z'^2}.$$

Deduce that points on a plane not passing through O map onto points on a sphere.

12.5 Functions of a Complex Variable

The term *function* is normally used in preference to *mapping* when the domain and range are both sets of numbers (real or complex). Let f be a function in which the domain and range are sets of complex numbers; then we write

$$z' = f(z)$$

where z is an *independent complex variable* on the domain and z' is a *dependent complex variable* on the range. Of main interest are the cases where the image rule can be expressed as an equation; for example,

$$z' = z^2, \quad z' = 2z^2 + (2 + 3i)z - i.$$

Values of z and z' can be represented on an Argand diagram; normally two separate Argand diagrams, one called the z-plane and the other called the z'-plane, are used (Fig. 12.13). Writing

$$z = x + iy, \quad z' = x' + iy',$$

we have

$$x' + iy' = f(x + iy)$$

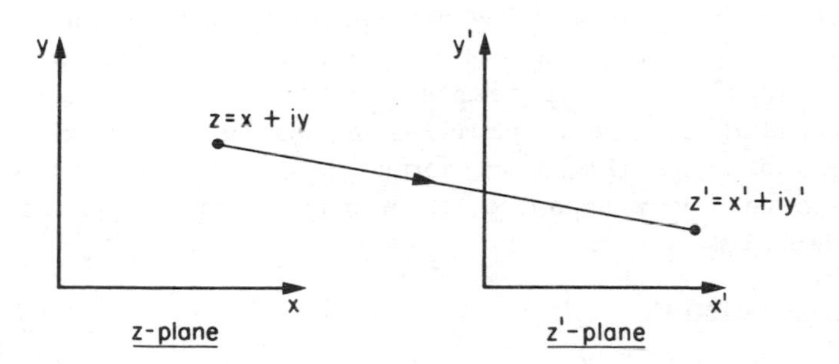

FIG. 12.13

and so

$$x' = \text{real part of } f(x + iy),$$
$$y' = \text{imaginary part of } f(x + iy),$$

giving the image rule in the form of two equations expressing the co-ordinates (x', y') of the image point in terms of the co-ordinates (x, y) of the point with which it is associated.

Example 1 Express the equation $z' = z^2$ in the form of two real equations.

Write $z = x + iy$, $z' = x' + iy'$, so that

$$x' + iy' = (x + iy)^2 = x^2 - y^2 + 2ixy.$$

Hence

$$x' = x^2 - y^2, \quad y' = 2xy.$$

Example 2 Express the equation $z' = (z - i)/(z + i)$ in the form of two real equations.

Write $z = x + iy$, $z' = x' + iy'$, so that

$$x' + iy' = \frac{x + i(y - 1)}{x + i(y + 1)} = \frac{[x + i(y - 1)][x - i(y + 1)]}{x^2 + (y + 1)^2}$$
$$= \frac{x^2 + y^2 - 1 - 2ix}{x^2 + (y + 1)^2}.$$

Hence

$$x' = \frac{x^2 + y^2 - 1}{x^2 + (y + 1)^2}, \quad y' = \frac{-2x}{x^2 + (y + 1)^2}.$$

We see that a single equation in complex variables can be used to describe a transformation from one plane onto another; this same transformation is described by *two* equations when real variables are used. However, given a pair of equations in real variables, we cannot necessarily find a corresponding equation in complex variables. Thus functions in which the image rule is an equation in complex variables form a rather special set of transformations from R^2 into R^2; this turns out to be a very important set in more advanced applications of mathematics.

Exercises 12(e)

[For brevity the mapping $f: z \mapsto z'$, $z' = f(z)$ is frequently referred to as the mapping $z' = f(z)$.]

1. The mapping $z' = z - a - ib$, where a, b are real constants, is a well-known linear transformation. Which one is it?
2. Show that for any complex number a, the mapping $z' = az$ is a linear transformation.
3. From your knowledge about multiplication of complex numbers in polar form give a suitable complex number a such that the mapping $z' = az$ is equivalent to a rotation about O through α (Question 3 of Exercises 12(c)). Likewise give a complex number b such that the mapping $z' = bz$ is equivalent to a rotation of axes through α (in the usual anti-clockwise direction). Use this to derive the equations
$$x' = x \cos \alpha + y \sin \alpha, \quad y' = -x \sin \alpha + y \cos \alpha$$
for rotation of axes.
4. Interpret each of the mappings $z' = \bar{z}$, $z' = -\bar{z}$, $z' = i\bar{z}$ as a reflection in a line.
5. Show that the mapping $z' = \bar{z}(1 - m^2 + 2im)/(1 + m^2)$ is equivalent to a reflection in the line $y = mx$.
6. Show that the mapping $z' = a^2/\bar{z}$ is equivalent to the mapping in Question 6 of Exercises 12(d).
7. If z is a complex variable and f is the mapping $z' = z^2$ with domain
$$\{z : Im\,(z) > 0\},$$
what is the range of f? Does f have an inverse? If g is the mapping $z' = z^2$ with domain C, the set of all complex numbers, does g have an inverse?

12.6 Functions of a Real Variable

Much of the remainder of this book is concerned with mappings in which the domain X and the range Y are both sets of real numbers. In this situation, as remarked at the beginning of Section 12.5, the word *function* rather than *mapping* is normally used. If $y = f(x)$, where f denotes a function, x is an independent variable on X and y is a dependent variable on Y, then we sometimes express the fact that the value of y depends on the value of x by saying that *y is a function of x*. In order to emphasise that the values of x are real numbers we may say that f is a *function of a real variable*; to emphasise that the values of y are real numbers we may say that f is a *real-valued function*. In this context the definition of mapping (Section 12.1) implies that with each (real-number) value of x in X is associated *exactly one* (real-number) value of y in Y.

We shall hereafter assume that, unless otherwise stated, the word *function* refers to a mapping of the above type. Such functions arise naturally in connection with physical problems. From (a), (b), (c), (d) in Section 1.2.3 we get the following simple examples:

(a) Let h m be the height of the stone t s after release from a height of 100 m; then h is a function of t.

Let us be more precise about the function that we have in mind. Since presumably we are interested in what happens *after* the stone is released, we may as well ignore negative values of t and specify the domain as the set $\{t: t \geqslant 0\}$. If we assume the elementary law that the distance fallen is

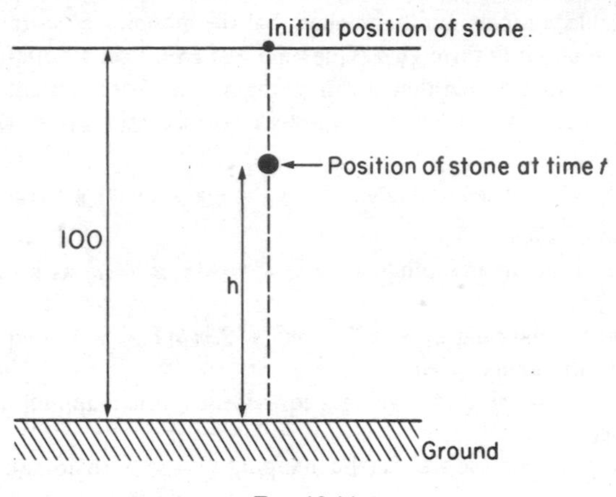

FIG. 12.14

$\frac{1}{2}gt^2$ m, where g m/s^2 is the (supposed constant) acceleration due to gravity, then we have the image rule

$$h = \begin{cases} 100 - \frac{1}{2}gt^2 & \text{before the stone hits the ground,} \\ 0 & \text{after the stone hits the ground.} \end{cases}$$

The stone hits the ground at the instant when $100 - \frac{1}{2}gt^2 = 0$, i.e. $t = \sqrt{(200/g)}$, so the image rule may be written in the form

$$h = \begin{cases} 100 - \frac{1}{2}gt^2 & \text{for } 0 \leqslant t \leqslant \sqrt{(200/g)}, \\ 0 & \text{for } t > \sqrt{(200/g)}. \end{cases}$$

The above domain and image rule define a function that is highly relevant to this particular physical situation. However it is most probable that we are only interested in what happens *before* the stone hits the ground, in which case the function we would consider is the simpler function with domain $\{t: 0 \leqslant t \leqslant \sqrt{(200/g)}\}$ and image rule $h = 100 - \frac{1}{2}gt^2$ for all t in this domain. In either case the range of the function is the set $\{h: 0 \leqslant h \leqslant 100\}$.

(b) Let r cm be the radius of the cross-section at distance x cm from the vertex of the cone; then r is a function of x.

Clearly it is meaningless to talk about the cross-section at distance (say) 13 cm from the vertex, and we are naturally led to consider a function with domain $\{x: 0 < x \leqslant 12\}$. By simple geometry we get the image rule $r = \frac{1}{6}x$, and the range of the function is $\{r: 0 < r \leqslant 2\}$.

Is it meaningful to talk about the cross-section when $x = 0$? It is best to regard this as a limiting case, i.e. since the values of r get closer and closer to zero as x gets closer and closer to zero, we *define* the value of r corresponding to $x = 0$ to be $r = 0$. We now have in mind a function with

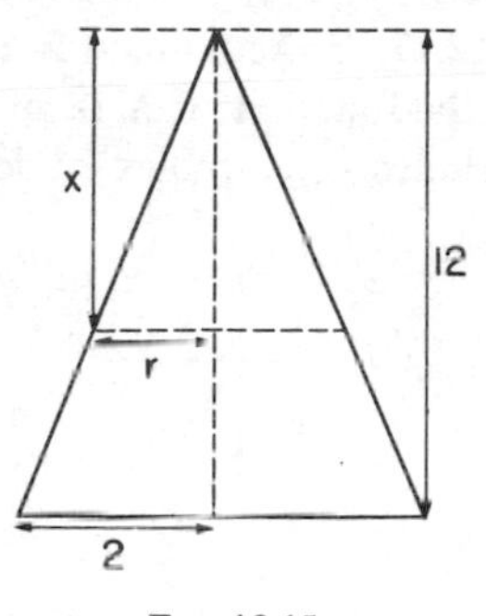

FIG. 12.15

domain $\{x: 0 \leqslant x \leqslant 12\}$, image rule $r = \frac{1}{6}x$, and range $\{r: 0 \leqslant r \leqslant 2\}$; the only difference from the previous function is that we have included $x = 0$ in the domain by declaring, a little artificially perhaps, that the image of 0 is 0. Surprisingly we shall discover later that there are advantages in considering functions with closed rather than open intervals as domains.

(c) Let A cm^2 be the area of the rectangle of side x cm; since the total perimeter must be 400 cm it follows that A is a function of x. It is physically meaningful to bend the wire into such a rectangle only if

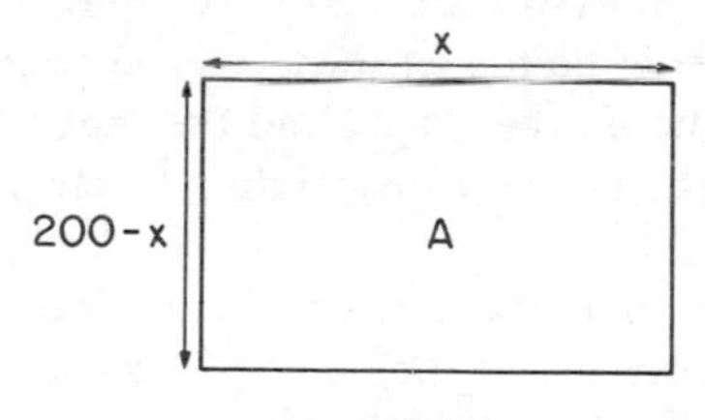

FIG. 12.16

$0 < x < 200$, but we can consider $x = 0$ and $x = 200$ as limiting cases where the rectangle has zero area. This leads us to consider the function with domain $\{x: 0 \leqslant x \leqslant 200\}$ and image rule $A = x(200 - x)$ for all x in this domain.

(d) Let p be the probability of getting the result x when the die is thrown; then p is a function of x. The obvious function to consider in this case has domain $\{1, 2, 3, 4, 5, 6\}$ and image rule $p = \frac{1}{6}$ for all x in this domain; the range of the function is the single number $\frac{1}{6}$.

12.6.1 Graph of a Function Arrow diagrams of the type generally used for mappings (see Section 12.1) are rarely used in connection with functions of a real variable. Instead, a function may be illustrated by a table in which the values $y_1, y_2, \ldots, y_n$ associated with $x_1, x_2, \ldots, x_n$ are listed as shown in Fig. 12.17. If the domain X is a finite set such a table can completely describe the function; if X is an infinite set, only selected values of x can be included, e.g. tables of logarithms, trigonometric tables.

x	y
x_1	y_1
x_2	y_2
.	.
.	.
.	.
.	.
x_n	y_n

FIG. 12.17

Alternatively we may choose axes for x and y, and plot the points $(x_1, y_1), (x_2, y_2), \ldots, (x_n, y_n)$. The set of points corresponding to all associated pairs in the function is called the *graph* of the function. The domain X is represented by the set of points on the x-axis corresponding to the x-co-ordinates of the points on the graph, and the range Y is represented by the set of points on the y-axis corresponding to the y-co-ordinates of the points on the graph.

If, as is usually the case, the domain X is an interval $a \leqslant x \leqslant b$, then the graph of f is the curve $y = f(x)$, $a \leqslant x \leqslant b$ (see Fig. 12.18(b)).

Let L be the point on the x-axis corresponding to a value $x = x_1$ in the domain of f (Fig. 12.19(a)); a line through L parallel to the y-axis must meet the graph of f *exactly once*, showing that x_1 has exactly one image. A curve such as that shown in Fig. 12.19(b) cannot be the graph of a function since it associates more than one image with x_1.

Figure 12.20 shows the curve $y^2 = x$. Since lines parallel to the y-axis meet this curve *twice*, the curve cannot be the graph of a function. However, the curve may be divided into two parts as shown, and if we

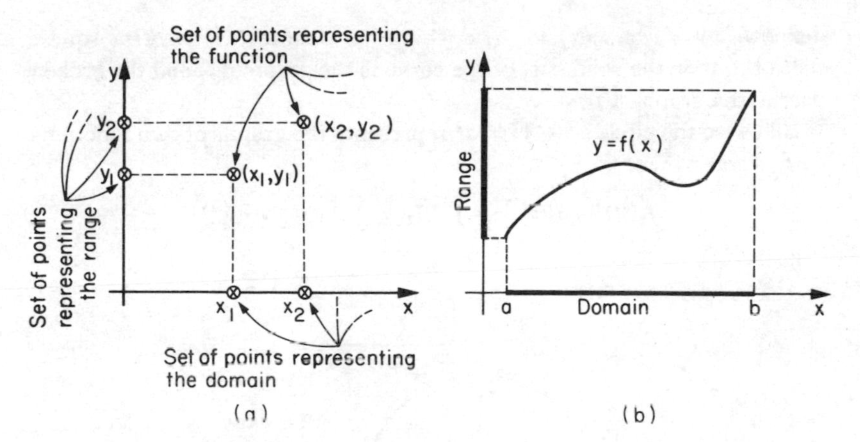

FIG. 12.18

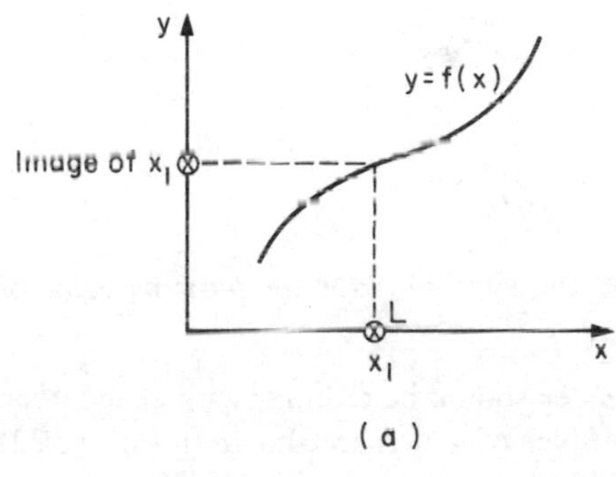

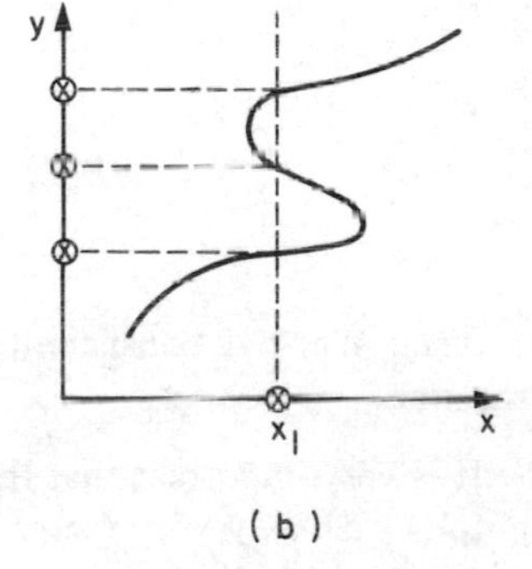

FIG. 12.19

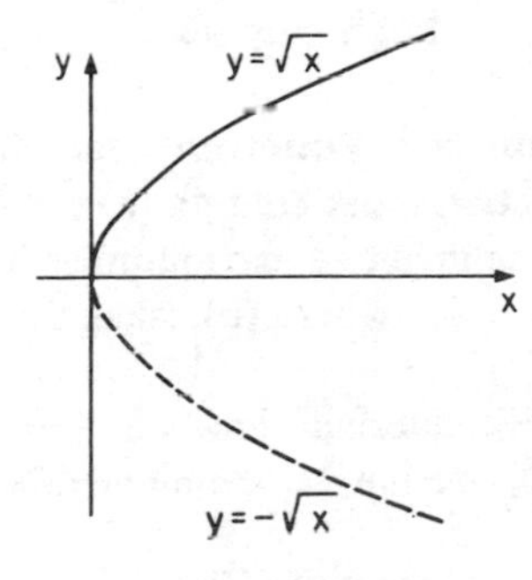

FIG. 12.20

define $f_1(x) = \sqrt{x}$, $f_2(x) = -\sqrt{x}$ where $\sqrt{x}$ denotes the *positive* square root of x, then the solid part of the curve is the graph of f_1 and the broken part is the graph of f_2.

Likewise the circle $x^2 + y^2 = a^2$ represents the graphs of two functions f_1, f_2 where

$$f_1(x) = \sqrt{(a^2 - x^2)}, \quad f_2(x) = -\sqrt{(a^2 - x^2)}$$

(Fig. 12.21).

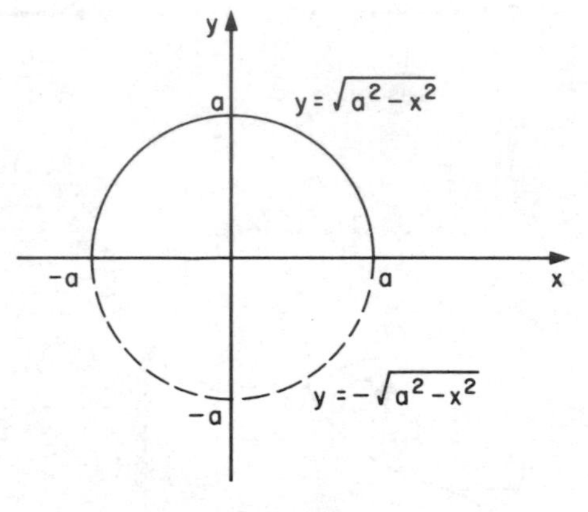

FIG. 12.21

Note. We shall consistently use the symbol $\sqrt{\ }$ for the *positive* value of the root.

It is very important that the reader should be familiar with elementary graphs such as $y = x^n$ (some examples of which are shown in Fig. 12.22) and graphs of trigonometric (or circular) functions (Fig. 12.23).

Note. In defining the functions sin, cos, tan, we define the image of x to be the trigonometric sine, cosine, tangent, respectively of an angle of x *radians*, where 1 radian $= 180/\pi$ degrees.

12.6.2 The Description of a Function As with any other mapping, the description of a function must contain (a) a statement of the domain, and (b) a statement of the image of each number in the domain. We shall be concerned only with cases where (b) takes the form of some kind of image rule.

As an indication of the abbreviations used in such descriptions, the function discussed in (c), Section 12.6, may be described as *the function f where*

$$f(x) = x(200 - x), \quad 0 \leqslant x \leqslant 200;$$

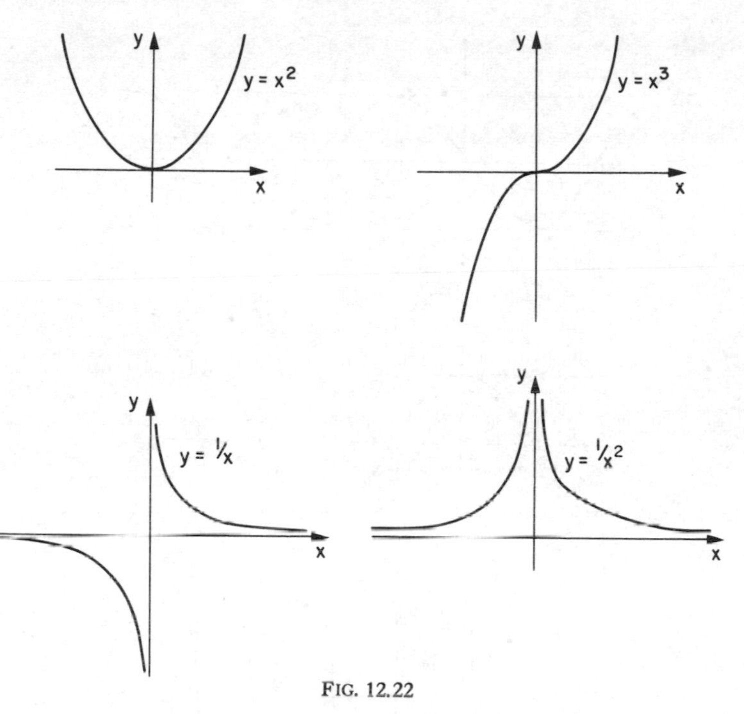

FIG. 12.22

alternatively we may refer to

$$f : x \mapsto x(200 - x), \quad 0 \leqslant x \leqslant 200,$$

or to

$$f : x \mapsto A, \quad A = x(200 - x), \quad 0 \leqslant x \leqslant 200.$$

Frequently there is no explicit mention of domain, and this is meant to imply that *the domain is the set of all numbers for which the image rule defines an image.* For example, if $f(x) = 1/x$, the implied domain is $\{x : x \neq 0\}$, and if $f(x) = \sqrt{x}$ the implied domain is $\{x : x \geqslant 0\}$. In other words it is assumed, unless otherwise stated, that the domain of f is the set of all numbers x for which $f(x)$ can be evaluated.

Example 3 State the domain of the function f, given that

(a) $f(x) = \dfrac{1}{x^2 + x - 6}$, (b) $f(x) = \sqrt{(x^2 + x - 6)}$,

(c) $f(x) = \dfrac{1}{\sqrt{(x^2 + x - 6)}}$.

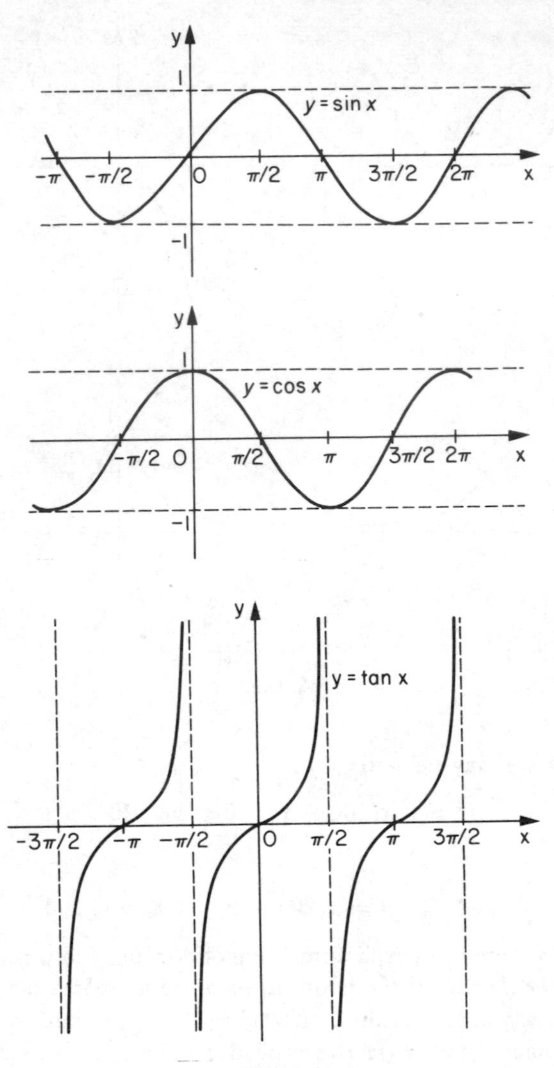

FIG. 12.23

(a) Since $f(x) = 1/[(x - 2)(x + 3)]$, we see that $f(x)$ can be evaluated except when $x = 2$ or $x = -3$. Hence the implied domain is

$$\{x : x \neq 2 \text{ or } -3\}.$$

(b) Since $f(x) = \sqrt{[(x - 2)(x + 3)]}$, we see that $f(x)$ can be evaluated provided that $(x - 2)(x + 3) \geqslant 0$. Hence the implied domain is

$$\{x : x \geqslant 2 \text{ or } x \leqslant -3\}.$$

(c) In (b) we saw that the square root could be evaluated if $x \geqslant 2$ or $x \geqslant -3$. In the present case $f(x)$ cannot be evaluated if the value of the square root is zero. Hence the implied domain is

$$\{x : x > 2 \quad \text{or} \quad x < -3\}.$$

For the functions sin, cos the domain is the set of all real numbers, and for the function tan the domain is

$$\{x : x \neq \pm\pi/2, \pm\frac{3\pi}{2}, \pm\frac{5\pi}{2}, \ldots\}$$

(see Fig. 12.23).

Example 4 State the domain of the function f, given that

(a) $f(x) = \dfrac{1}{\sin x}$, (b) $f(x) = \tan\left(\dfrac{x}{x+1}\right)$.

(a) We can evaluate $f(x)$ except when $\sin x = 0$. Hence the implied domain is $\{x : x \neq 0, \pm\pi, \pm 2\pi, \pm 3\pi \ldots\}$.

(b) We can evaluate $x/(x + 1)$ except when $x = -1$; otherwise we can evaluate $f(x)$ except when $x/(x + 1) = \pm\pi/2, \pm 3\pi/2, \ldots$. Writing $x/(x + 1) = \theta$, we may solve for x to get $x = \theta/(1 - \theta)$. Hence the implied domain of f is

$$\{x : x \neq -1, x \neq \theta/(1 - \theta) \text{ where } \theta = \pm\pi/2, \pm 3\pi/2, \ldots\}.$$

Exercises 12(f)

1. From each corner of a square sheet of metal of side 18 cm small squares of side x cm are removed, and the remainder of the sheet is made to form an open box by turning up the edges. The volume V cm^3 of this box is a function of x. What function do we have in mind?
2. A piece of wire of length 48 cm is to be cut and each piece bent to form the largest possible square. The total area A cm^2 of the two squares is a function of x, where x cm is the distance of the cut from one end of the original piece of wire. What function do we have in mind?
3. A rectangle is inscribed in a circle of radius 3 cm by choosing two points, not diametrically opposite, on the circumference of the circle, joining these points to form the first side, and then completing the rectangle. If the first side has length x cm and the rectangle has area A cm^2, then A is a function of x. What function do we have in mind?
4. If

$$f(x) = \frac{x^2 + 3x + 2}{x^2 + 5x + 4},$$

what are the values of $f(1)$, $f(-1)$?

5. Given $f(x) = x^2 + x + 1$, find $f(3) + f(2)$, $f(3 + 2)$, $f(3) \times f(2)$, $f(3 \times 2)$, $f(3)/f(2)$, $f(3/2)$.
6. What is the domain and range of the function f, given that

   (a) $f(x) = \sqrt{(9 - x^2)}\sqrt{(4 - x^2)}$? (b) $f(x) = \sqrt{[(9 - x^2)(4 - x^2)]}$?

7. Give the domains of the following functions:

   (a) $f: x \mapsto (x^2 + 3x + 3)/(x^2 - x - 6)$,
   (b) $f: x \mapsto \sqrt{(x^2 - x - 6)}$,
   (c) $f: x \mapsto \sqrt{(x^2 + 3x + 3)}$, (d) $f: x \mapsto \log_{10} x$,
   (e) $f: x \mapsto \log_{10} |x|$.

8. State the domains of the trigonometric functions sec, cosec, cot.
9. State the domain of the function f, given that

   (a) $f(x) = \sin 3x$, (b) $f(x) = \tan (x/2)$.

12.6.3 Extension and Restriction of a Function In (c), Section 12.6, we considered the function

$$f: x \mapsto A, \quad A = x(200 - x), \quad 0 \leqslant x \leqslant 200.$$

Here $f(x) = x(200 - x)$, and obviously $f(x)$ can be evaluated for *any* value of x, not just for values in the domain of f. Hence we may define other functions such as

$$\begin{aligned} g: x &\mapsto x(200 - x), \quad 0 \leqslant x \leqslant 1000, \\ h: x &\mapsto x(200 - x), \quad -1000 \leqslant x \leqslant 1000, \\ i: x &\mapsto x(200 - x), \quad x \text{ in } R, \end{aligned}$$

that share the same image rule as f, but have larger domains than f. Such functions are called *extensions* of f; thus g, h, i are extensions of f, and furthermore h, i are extensions of g, i is an extension of h.

Alternatively, we may say that f, g, h are *restrictions* of i, and that f, g are restrictions of h, and so on. In particular, we say that f is the *restriction* of i (or g, or h) *to the domain* $\{x: 0 \leqslant x \leqslant 200\}$.

The above is the most natural and commonly occurring type of extension. We shall say that the function g is an *extension* of f if the domain of g includes the domain of f and if $g(x) = f(x)$ for all x in the domain of f; this means that the image rule applied to numbers outside the domain of f need not necessarily be the same as that applied to numbers within the domain of f. Consider, for example, the function $f: x \mapsto 1/x$, $x \neq 0$; if we wish to extend this function to the domain R, it is simply a matter of defining an image for $x = 0$ but the image rule of f cannot be used to do this—we may arbitrarily define the image of $x = 0$ to be (say) 0, or 1, or any other number, and in each case we get an extension of f.

12.6.4 Elementary Functions For the moment we may regard the *elementary functions* as consisting of

(1) powers of x, i.e. $f(x) = x^n$, where n is a rational number,
(2) the circular functions.

To this list we must add the following functions which are introduced later:

(3) inverse circular functions,
(4) logarithmic and exponential functions,
(5) hyperbolic and inverse hyperbolic functions.

In this book we are mainly concerned with functions for which the image rule takes the form of an equation involving various combinations of these elementary functions, e.g.

$$y = x^2/\sin x, \quad y = \sin(x^2), \quad \tan y = \sin x.$$

It will be assumed in future that *equation* means *equation in terms of elementary functions.*

The reader should not assume that, in the context of functions of a real variable, the words *image rule* and *equation* are synonymous, as the following two examples illustrate:

(a) Let p kg/cm² denote the atmospheric pressure as measured by the barometer at Prestwick airport t s after midnight on a certain date. Clearly p is a function of t since there is a single value of p corresponding to each value of t, but we do not know any *equation* relating p with t. Using a barograph—an instrument that records atmospheric pressure graphically on a slowly rotating cylinder—we can get a graph of the appropriate function, but this graph is not a simple curve with known equation.

(b) Let r denote the value of x rounded off to the nearest integer, the nearest *even* integer being chosen if x is half-way between two consecutive integers. In this way a single value of r is associated with each value of x and so r is a function of x. However, we cannot relate r and x by an *equation.* The graph of this function is shown in Fig. 12.24.

In (a), Section 12.6, we have seen an example of an image rule that cannot be expressed as a *single* equation applicable to the whole domain, but is expressed as *two* equations each of which applies to *part* of the domain. A typical example of a similar function occurring in engineering is the following:

(c) A beam of length 12 m is simply supported at two points each 4 m

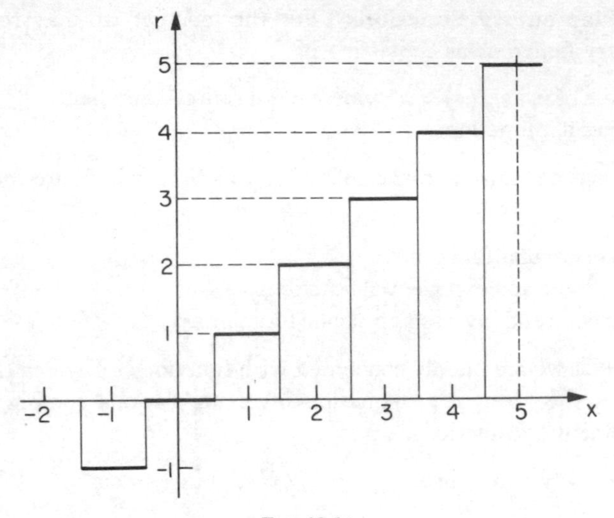

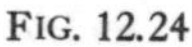

FIG. 12.24

from the centre, and the beam is under a uniform load distribution of 1 N/m (Fig. 12.25). Let S N be the shearing force at distance x m from the centre; then it can be shown that

$$S = \begin{cases} -6 - x & \text{for } -6 \leqslant x < -4, \\ -x & \text{for } -4 < x < 4, \\ 6 - x & \text{for } \quad 4 < x \leqslant 6. \end{cases}$$

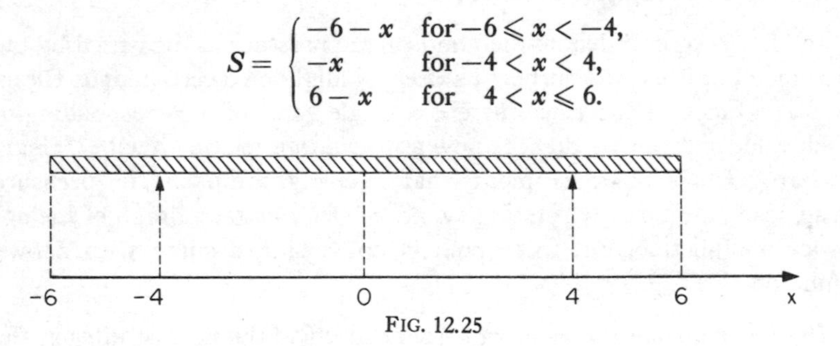

FIG. 12.25

Further, if M N m is the bending-moment at distance x m from the centre, then it can be shown that

$$M = \begin{cases} \frac{1}{2}(6 + x)^2 & \text{for } -6 \leqslant x < -4, \\ \frac{1}{2}\, x^2 - 6 & \text{for } -4 \leqslant x < 4, \\ \frac{1}{2}(6 - x)^2 & \text{for } \quad 4 \leqslant x \leqslant 6. \end{cases}$$

12.6.5 Explicit and Implicit Equations Equations like

$$y = x^2 + 2x + 1, \quad y = \sin^2 x,$$

where the l.h.s. contains the single term y and the r.h.s. is free from y, are said to express y *explicitly* in terms of x.

An *image formula* for the function f is a formula for $f(x)$ in terms of elementary functions; clearly an image rule in the form of an explicit equation is equivalent to an image formula, since we may replace y by $f(x)$, e.g. $f(x) = x^2 + 2x + 1$, $f(x) = \sin^2 x$, etc.

Equations like

$$\text{(a)}\ \frac{x-y}{x+y} = x, \quad \text{(b)}\ y^2 + 2xy - 1 = 0, \quad \text{(c)}\ y^5 + x^3y^3 + xy^2 - 3 = 0,$$

where y occurs other than in the manner described above for the explicit equation, are said to express y ***implicitly*** in terms of x.

From (a) we get

$$x - y = x^2 + xy,$$

so that

$$y = (x - x^2)/(x + 1),$$

and we see that (a) can be expressed in explicit form.

From (b) we get y in terms of x by solving a quadratic equation; thus $y = -x \pm \sqrt{(x^2 + 1)}$, so that *two* values of y are associated with each value of x. Equation (b) combines the image rules of two functions f_1, f_2 where

$$f_1(x) = -x + \sqrt{(x^2 + 1)}, \quad f_2(x) = -x - \sqrt{(x^2 + 1)}.$$

Note that (c) is a fifth degree equation for y and we have no general method for solving this. It is therefore impossible to express y explicitly in terms of x in this case; values of y corresponding to particular values of x must be individually computed using approximation methods, and of course the equation may have several roots.

It is important to remember that an implicit equation may not by itself be an adequate image rule. If the equation associates more than one value of y with a value of x, then some other condition to single out *one* value of y must be included in the image rule. For example, returning to (b) we see that the image rule $y^2 + 2xy - 1 = 0, y > -x$ would be appropriate for f_1, and the image rule $y^2 + 2xy - 1 = 0, y < -x$ would be appropriate for f_2.

12.6.6 Classification of Functions The simplest functions are those for which the image of x can be computed from the value of x by performing a finite number of basic arithmetic operations (addition, subtraction, multiplication, division, root extraction).

If the image rule has the form

$$y = p_0x^n + p_1x^{n-1} + \cdots + p_{n-1}x + p_n$$

where n is a non-negative integer and $p_0, p_1, \ldots, p_n$ are constants, the function is a ***polynomial function.*** A constant function is a special case of a polynomial function.

If the image rule has the form

$$y = P(x)/Q(x)$$

where P, Q are both polynomial functions, the function is a ***rational function.*** Rational functions include polynomial functions as special cases.

If the image rule has the form

$$y^n + R_1(x)y^{n-1} + \cdots + R_{n-1}(x)y + R_n(x) = 0$$

where n is a non-negative integer and $R_1, R_2, \ldots, R_n$ are rational functions, the function is ***algebraic.*** Algebraic functions include rational functions and polynomial functions as special cases.

Examples of such image rules are

$$y = 3x + 4, \quad y = 2x^4 - 3x^3 + 7x^2 - 1 \quad \text{(polynomial functions)},$$

$$y = \frac{x^3 + 2x - 7}{x - 1}, \quad y = \frac{x^2 - 1}{x^2 + x} \quad \text{(rational functions)},$$

$$y^2 + \left(\frac{x^2 - 1}{x^2 + 1}\right)y - \left(\frac{2x - 3}{x^3 - 7}\right) = 0, \quad y^3 + \left(\frac{x}{x + 3}\right)y^2 + \left(\frac{3x}{x + 2}\right) = 0$$

(algebraic functions).

In the case of algebraic functions the equation is not usually the complete image rule, but must be accompanied by a further condition (see Section 12.6.5).

A function that is not algebraic is called ***transcendental.*** This includes functions for which the image rule is an equation involving elementary functions other than (1) (Section 12.6.4); it also includes functions for which the image rule cannot be expressed in the form of an equation.

Exercises 12(g)

1. Show that the equation $x = (y + 1)/(y - 1)$ determines y as a function of x with domain $\{x: x \neq 1\}$.
2. Find the domain of the function

$$f: x \mapsto y, \quad x^2 + xy + y^2 = 3, \quad x + 2y \geqslant 0.$$

3. Show that the function

$$f: x \mapsto y, \quad xy = (x + 2y)/(x^2 + 1)$$

has domain $\{x: x \neq 1\}$.

4. Show that the function

$$f: x \mapsto y, \quad y^4 - 2xy^2 - 1 = 0, \quad y > 0$$

has domain R.

5. Find the domain of the function

$$f: x \mapsto y, \quad y \geqslant x, \quad \text{and} \quad \begin{cases} y^2 - 2xy - 2x^2 + 2 = 0 \text{ if } x > 1, \\ y^2 - 2xy + 2x^2 - 2 = 0 \text{ if } x < 1. \end{cases}$$

6. Show that the functions f_1 and f_2 are identical, where

$$f_1(x) = 1/(\sqrt{(x+a)} + \sqrt{a}), \quad f_2(x) = \begin{cases} (\sqrt{(x+a)} - \sqrt{a})/x & \text{if } x \neq 0 \\ 1/2\sqrt{a} & \text{if } x = 0. \end{cases}$$

7. Show that the function

$$f: x \mapsto \frac{x - 1/(x-1)}{x^2 + 1/(x+1)}$$

is a rational function.

8. Show that the function $f: x \mapsto \sqrt{(\sqrt{x} + x)}$ is an algebraic function. [*Hint*: Write $y = f(x)$ and show that $y^4 - 2xy^2 + x^2 - x = 0$.]
9. If f_1, f_2 are (a) both polynomials, (b) both rational functions, (c) one a polynomial and one a rational function, and $f(x) = f_1(x) + f_2(x)$, what kind of function is f?
10. Same as Question 9 but with $f(x) = f_1(x)\,f_2(x)$.

12.6.7 Inverse Functions It was pointed out in Section 12.6.1 that if f is a function and x_1 is a number in the domain of f, then the line $x = x_1$ should meet the graph of f exactly once. If, for any number c in the range of f, the line $y = c$ meets the graph of f exactly once, the function is a (1, 1) function.

For example the function $f: x \mapsto x^2$ is not (1, 1) since for any positive c the line $y = c$ meets the graph of f twice (see Fig. 12.22). However the restriction of f to the domain $\{x: x \geqslant 0\}$, i.e. the function $f_+: x \mapsto x^2$, $x \geqslant 0$, is a (1, 1) function. Likewise the restriction of f to the domain $\{x: x \leqslant 0\}$, i.e. the function $f_-: x \mapsto x^2$, $x \leqslant 0$, is a (1, 1) function.

In Fig. 12.26 the graph of f_+ is the solid part of the curve and the graph of f_- is the broken part of the curve.

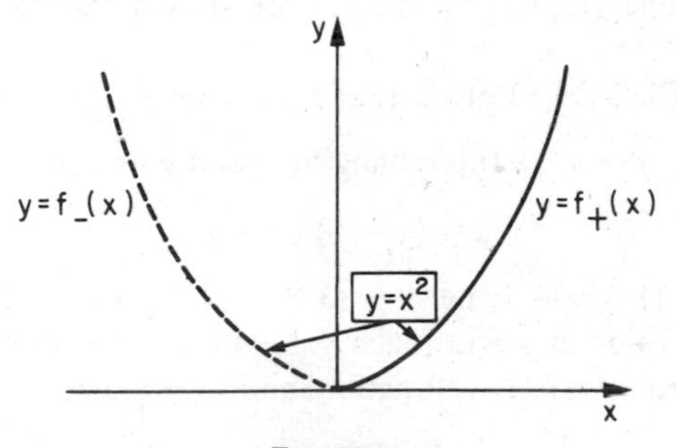

FIG. 12.26

Corresponding to a (1, 1) function f there is an inverse f^{-1} (see Section 12.2).

If the image rule for f has the form of an equation relating the dependent variable y to the independent variable x, this same equation is the image rule for f^{-1} if we simply regard y as independent variable and x as dependent variable. Likewise the graph of f is also the graph of f^{-1} if we imagine the roles of the two axes to interchange so that the independent variable is plotted on the y-axis (see Fig. 12.27).

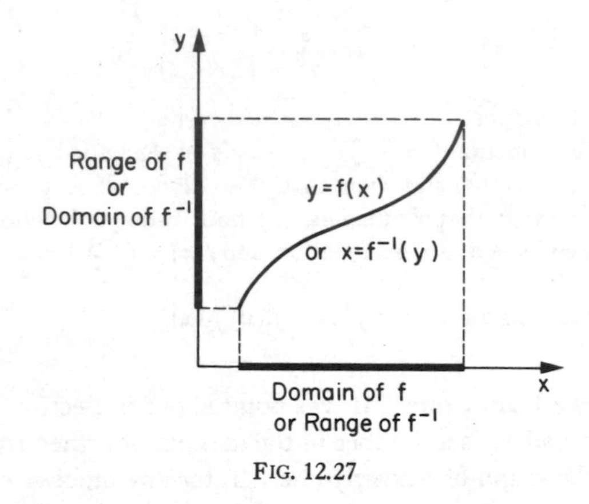

FIG. 12.27

In practice it is much preferable to retain x as the symbol for the independent variable in both f and f^{-1}. We then get the image rule for f^{-1} by simply interchanging x and y in the image rule for f, and the graph of f^{-1} is the reflection in the line $y = x$ of the graph of f (see transformation (c), Section 12.4).

Example 5 Find $f^{-1}(x)$ given that $f(x) = x^3$.

Write $y = x^3$. Interchanging x and y we get $x = y^3$, i.e. $y = x^{1/3}$. Hence $f^{-1}(x) = x^{1/3}$. The graphs of f and f^{-1} are shown together in Fig. 12.28.

Example 6 Find $f^{-1}(x)$ given that $f(x) = (x - 1)/(x + 1)$.

Write $y = (x - 1)/(x + 1)$. Interchanging x and y we get

$$x = (y - 1)/(y + 1),$$

so that $x(y + 1) = y - 1$, i.e. $y = (1 + x)/(1 - x)$. Again this equation associates *one* value of y with each value of x in the domain $\{x : x \neq 1\}$. This confirms that f is a (1, 1) function, and shows that

$$f^{-1}(x) = (1 + x)/(1 - x).$$

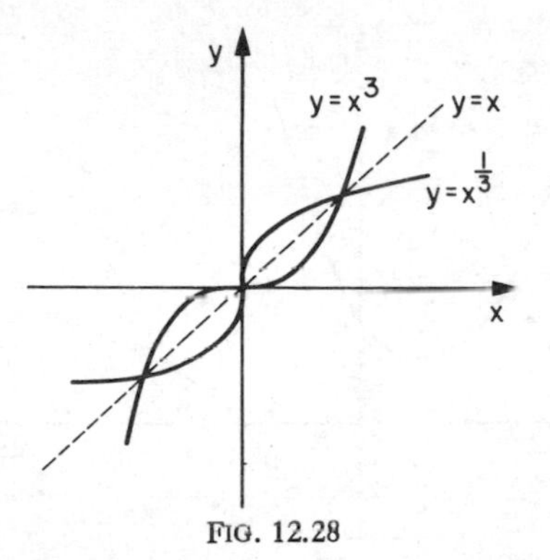

FIG. 12.28

Note that the domain of f is $\{x : x \neq -1\}$, whereas the domain of f^{-1} is $\{x : x \neq 1\}$.

If the function f is not (1, 1), then it is a simple matter to select a function f_r that is a restriction of f and is a (1, 1) function. The graph of f_r is just *part* of the curve $y = f(x)$, and if this curve is reflected in the line $y = x$ only the part that corresponds to the graph of f_r is included in the graph of f_r^{-1}.

For example, if the curve $y = x^2$ is reflected in the line $y = x$, the reflected curve $x = y^2$ includes the graphs of two functions f_1, f_2 where $f_1(x) = \sqrt{x}, f_2(x) = -\sqrt{x}$. Given $f(x) = x^2$ we can define restrictions f_+, f_- as shown in Fig. 12.26, and from Fig. 12.29 we now see that $y = \sqrt{x}$ is

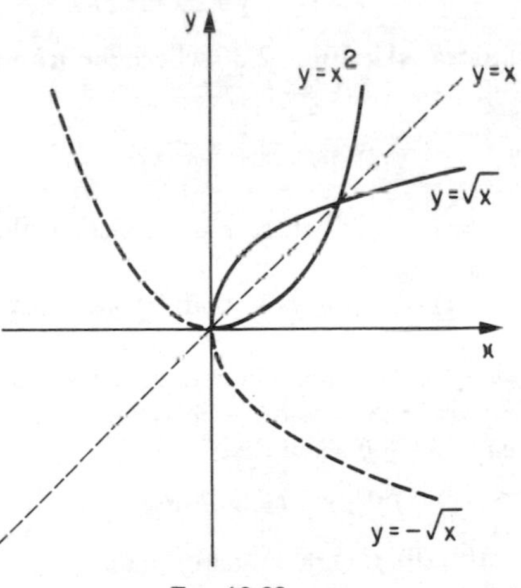

FIG. 12.29

the reflection of the graph of f_+, $y = -\sqrt{x}$ is the reflection of the graph f_-. Thus $f_+^{-1}(x) = \sqrt{x}, f_-^{-1}(x) = -\sqrt{x}$.

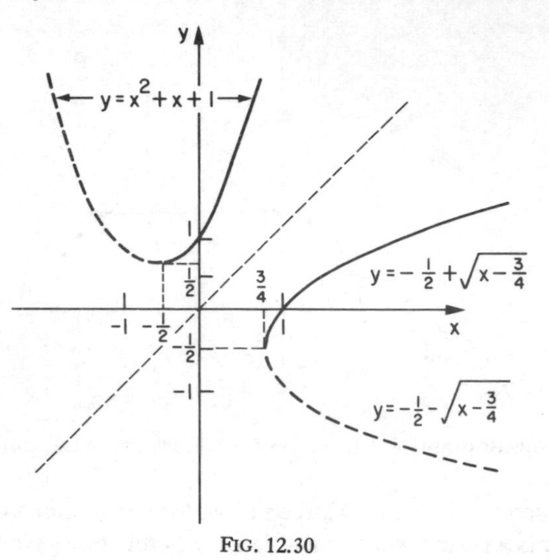

FIG. 12.30

Example 7 Find $f^{-1}(x)$ when $f(x) = x^2 + x + 1$ and the domain of f is $\{x: x \geqslant -\frac{1}{2}\}$.

From the equation $y = x^2 + x + 1$ we get $x = -\frac{1}{2} \pm \sqrt{(y - \frac{3}{4})}$; for x to be in the given domain we require $x = -\frac{1}{2} + \sqrt{(y - \frac{3}{4})}$, ignoring the negative square root. Hence, interchanging x and y we get

$$f^{-1}(x) = -\tfrac{1}{2} + \sqrt{(x - \tfrac{3}{4})}.$$

The situation is illustrated in Fig. 12.30 where the graphs of f and f^{-1} are the solid curves.

Exercises 12(h)

1. Show that if $f(x) = -x + c$, where c is a constant, then $f^{-1} = f$. Illustrate graphically.
2. Given $f(x) = (1 - x)/(2 + x)$, show that if $f(x_2) = f(x_1)$ then $x_2 = x_1$. Find $f^{-1}(x)$.
3. Given $f(x) = (ax + b)/(cx + d)$, where a, b, c, d are constants, show that if $x_1 \neq x_2$ then $f(x_1) \neq f(x_2)$ unless $ad - bc = 0$ in which case $f(x)$ is constant. Assuming that $ad - bc \neq 0$ show that

$$f^{-1}(x) = (b - dx)/(cx - a).$$

4. Given (a) $f(x) = 10^x$, (b) $f(x) = 2^x$, find $f^{-1}(x)$.
5. Sketch the parabola $y = x^2 + 2x + 1$. Show that the function $f: x \mapsto x^2 + 2x + 1$ is not a (1, 1) function. Find the inverse of the restriction of f to the domain $\{x: x \geqslant -1\}$ and also the inverse of the restriction of f to the domain $\{x: x \leqslant -1\}$.

6. Show that the function $f: x \mapsto |x - 1|$ is not a (1, 1) function. If f_1 is the restriction of f to the domain $\{x: x \geqslant 1\}$, and f_2 is the restriction of f to the domain $\{x: x \leqslant 1\}$ show that

$$f_1^{-1}(x) = x + 1, \quad f_2^{-1}(x) = 1 - x.$$

7. Sketch the parabola $y^2 + 4y - x + 7 = 0$. Indicate the part of the parabola that is the graph of the function

$$f: x \mapsto y, \quad y^2 + 4y - x + 7 = 0, \quad y \geqslant -2.$$

Show that f^{-1} is the restriction to the domain $\{x : x \geqslant -2\}$ of the function $g : x \mapsto x^2 + 4x + 7$.

8. Sketch the ellipse $x^2 + xy + y^2 = 4$ and show that the line $x = c$ meets the ellipse twice if $-4/\sqrt{3} < c < 4/\sqrt{3}$. Indicate the part of the ellipse that is the graph of the function $f: x \mapsto y$, $x^2 + xy + y^2 = 4$, $y \geqslant -\frac{1}{2}x$. Find f^{-1}.

12.6.8 Inverse Circular Functions The equations $y = \sin x$, $x = \sin y$ describe inverse relations, the symbol x being used as independent variable in each case. The curve $x = \sin y$ (Fig. 12.31) shows that to each value of x between -1 and 1 there correspond infinitely many values of y. Instead of the usual sine function let us now consider its restriction to the domain $\{x: -\pi/2 \leqslant x \leqslant \pi/2\}$. This restriction is a (1, 1) function and so has an inverse; it is to this inverse function that the name $\sin^{-1}$, or arcsin, is applied.

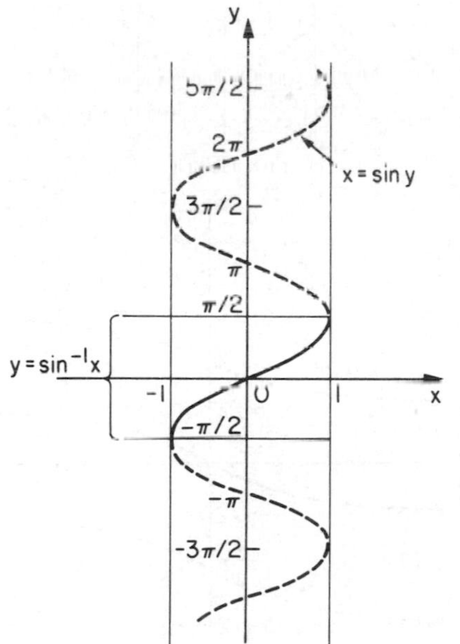

FIG. 12.31

Hence the equation $y = \sin^{-1} x$ means that $-\pi/2 \leqslant y \leqslant \pi/2$ and that $x = \sin y$.

The graph of $\sin^{-1}$ is the solid part of the curve in Fig. 12.31.

Similarly we denote by $\cos^{-1}$, or arccos, the inverse of the restriction of the cosine function to the domain $\{x: 0 \leqslant x \leqslant \pi\}$ and by $\tan^{-1}$, or arctan, the inverse of the restriction of the tangent function to the domain $\{x: -\pi/2 < x < \pi/2\}$.

Thus the equation $y = \cos^{-1} x$ means that $0 \leqslant y \leqslant \pi$ and that $x = \cos y$,

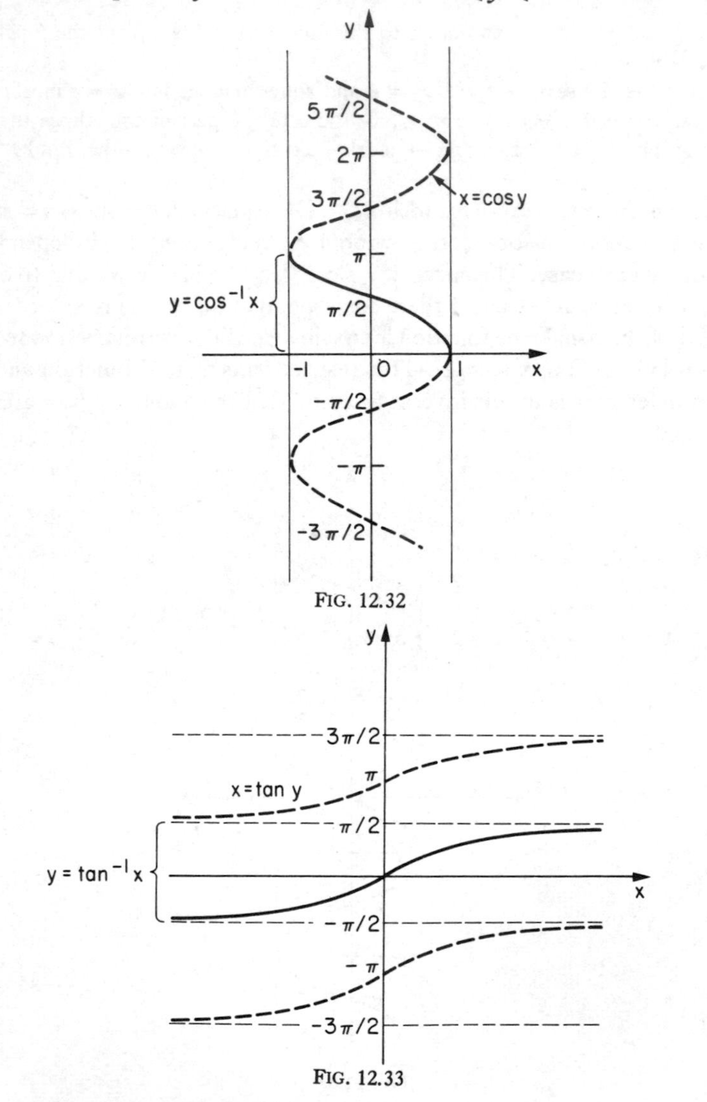

FIG. 12.32

FIG. 12.33

and the equation $y = \tan^{-1}x$ means that $-\pi/2 < y < \pi/2$ and that $x = \tan y$.

The graphs of $\cos^{-1}$, $\tan^{-1}$ are the solid parts of the curves in Figs. 12.32, 12.33.

Note. It is important to distinguish between *inverse* and *reciprocal.* The reciprocal of $\sin x$ is $1/\sin x$, which may be written as $(\sin x)^{-1}$, and *this should not be confused with* $\sin^{-1} x$.

Identities for trigonometric functions have their counterparts for inverse trigonometric functions. From the identity

$$\cos(\pi/2 - \theta) = \sin\theta, \tag{1}$$

we get the corresponding identity

$$\cos^{-1} x = \pi/2 - \sin^{-1} x. \tag{12.1}$$

To prove (12.1) we put $y = \sin^{-1} x$; this means that

$$-\pi/2 \leqslant y \leqslant \pi/2 \tag{2}$$

and that

$$\sin y = x. \tag{3}$$

Now from (1) we have

$$\cos(\pi/2 - y) = \sin y = x, \quad \text{from (3).}$$

Also from (2) we have $0 \leqslant \pi/2 - y \leqslant \pi$, and so

$$\pi/2 - y = \cos^{-1} x,$$

that is

$$\pi/2 - \sin^{-1} x = \cos^{-1} x.$$

From the identity

$$\tan(\alpha + \beta) = \frac{\tan\alpha + \tan\beta}{1 - \tan\alpha\tan\beta} \tag{4}$$

we get

$$\tan^{-1} a + \tan^{-1} b = \tan^{-1}\left(\frac{a+b}{1-ab}\right) \tag{12.2}$$

provided, of course, that the l.h.s. is between $-\pi/2$ and $\pi/2$ (if not, then the l.h.s. and r.h.s. differ by π).

To prove (12.2) we write $\tan^{-1} a = \alpha$, $\tan^{-1} b = \beta$ so that

$$a = \tan\alpha, \quad b = \tan\beta, \qquad -\pi/2 < \alpha < \pi/2, \quad \pi/2 < \beta < \pi/2.$$

From (4) we have

$$\tan(\alpha + \beta) = \frac{a+b}{1-ab}$$

and so, if $-\pi/2 < \alpha + \beta < \pi/2$, we have

$$\alpha + \beta = \tan^{-1}\left(\frac{a+b}{1-ab}\right),$$

that is

$$\tan^{-1} a + \tan^{-1} b = \tan^{-1}\left(\frac{a+b}{1-ab}\right).$$

(Since $-\pi < \alpha + \beta < \pi$ it follows that either $\alpha + \beta$, or $\alpha + \beta + \pi$, or $\alpha + \beta - \pi$, must lie between $-\pi/2$ and $\pi/2$; now

$$\tan(\alpha + \beta \pm \pi) = \tan(\alpha + \beta) = \frac{a+b}{1-ab},$$

so that either $\alpha + \beta$, or $\alpha + \beta + \pi$, or $\alpha + \beta - \pi$, equals

$$\tan^{-1}(a+b)/(1-ab).)$$

Exercises 12(i)

1. State the values of the following:

$$\sin^{-1}(1/2),\quad \sin^{-1}(1/\sqrt{2}),\quad \sin^{-1}(\sqrt{3}/2),\quad \sin^{-1}(0),\quad \sin^{-1}(1);$$
$$\tan^{-1}(1/\sqrt{3}),\quad \tan^{-1}(\sqrt{3}),\quad \tan^{-1}(1),\quad \tan^{-1}(0).$$

2. Given $x = \sin^{-1}(3/5)$ and $y = \cos^{-1}(12/13)$, find, without using tables, the value of $\tan(x + y)$.
3. Prove that $\tan^{-1}(\frac{1}{2}) + \tan^{-1}(\frac{1}{3}) = \pi/4$.
4. Solve the equation $\tan^{-1}(1 - x) + \tan^{-1}(1 + x) = \tan^{-1}(1/2)$.
5. Solve the equation $\tan^{-1}(x + 1) + \tan^{-1}(x - 1) = \pi/6$.
6. Show that

$$\cos^{-1} a + \cos^{-1} b = \cos^{-1}\{ab - \sqrt{[(1-a^2)(1-b^2)]}\}$$

if the l.h.s. is between 0 and π. Check the formula when $a = 1/2$, $b = \sqrt{3}/2$.

7. Show that $2\tan^{-1} t = \sin^{-1}[2t/(1 + t^2)]$ if the l.h.s. is between $-\pi/2$ and $\pi/2$.
8. Prove that $\cos[2\tan^{-1}\sqrt{(y/x)}] = (x - y)/(x + y)$.
9. Each of the equations

$$\text{(a) } \cot^{-1} x = \tan^{-1}(1/x), \quad \text{(b) } \cot^{-1} x = \pi/2 - \tan^{-1} x$$

is sometimes used to define the function $\cot^{-1}$. In each case give a reason why the equation is a reasonable definition. By giving the range of $\cot^{-1}$ in each case show that the two definitions do not give the same function. Sketch the graph of $\cot^{-1}$ in each case and verify that $\cot^{-1}$ is the inverse of the restriction of the cotangent function to

(a) $\{x: -\pi/2 < x < \frac{\pi}{2}, x \neq 0\}$, (b) $\{x: 0 < x < \pi\}$.

10. Given that $\sec^{-1}$ is defined by the equation $\sec^{-1} x = \cos^{-1}(1/x)$, show that $\sec^{-1}$ is the inverse of the restriction of the secant function to

$$\{x: 0 \leqslant x \leqslant \pi,\ x \neq \pi/2\}.$$

Sketch the graph of $\sec^{-1}$.

11. Given that $\operatorname{cosec}^{-1}$ is defined by the equation $\operatorname{cosec}^{-1} x = \sin^{-1}(1/x)$, show that $\operatorname{cosec}^{-1}$ is the inverse of the restriction of the cosecant function to

$$\{x: -\pi/2 \leqslant x \leqslant \pi/2,\ x \neq 0\}.$$

Sketch the graph of $\operatorname{cosec}^{-1}$.

12.6.9 Composite Functions Consider the function

$$f: x \mapsto y,\ y = \sin(3x^2 + 2x - 5).$$

The function f may be regarded as a composite function, say $f = g \circ h$, by introducing the function $h: x \mapsto u,\ u = 3x^2 + 2x - 5$ and the function $g: u \mapsto y,\ y = \sin u$. We have $g(x) = \sin x$, $h(x) = 3x^2 + 2x - 5$ and $g(h(x)) = \sin(3x^2 + 2x - 5) = f(x)$.

Given formulae for $g(x)$, $h(x)$ a formula for $g(h(x))$ is obtained by replacing x by $h(x)$ in the formula for $g(x)$, e.g.

(a) if $g(x) = x^3$, $h(x) = x^2 + 1$, then $g(h(x)) = (x^2 + 1)^3$.
(b) if $g(x) = 1/\sqrt{x}$, $h(x) = 2x + 3$, then $g(h(x)) = 1/\sqrt{(2x + 3)}$,
(c) if $g(x) = \sin^{-1} x$, $h(x) = 3x^2$, then $g(h(x)) = \sin^{-1}(3x^2)$.

Conversely, given a formula for $f(x)$ we can sometimes spot suitable formulae $g(x)$, $h(x)$ such that $f(x) = g(h(x))$, e.g.

(a) given $f(x) = (x^2 + 1)^3$, put $x^2 + 1 = h(x)$, $x^3 = g(x)$,
(b) given $f(x) = 1/\sqrt{(2x + 3)}$, put $2x + 3 = h(x)$, $1/\sqrt{x} = g(x)$,
(c) given $f(x) = \sin^{-1}(3x^2)$, put $3x^2 = h(x)$, $\sin^{-1} x = g(x)$.

Example 8 If $f(x) = 3x^2 - 2x + 1/x$, what is meant by $f(x + 1)$, $f(2x)$, $f(-x)$, $f(1/x)$, $f(\sin x)$?

By direct substitution for x in the formula for $f(x)$ we get

$$f(x + 1) = 3(x + 1)^2 - 2(x + 1) + \frac{1}{x + 1} = 3x^2 + 4x + 1 + \frac{1}{x + 1},$$

$$f(2x) = 3(2x)^2 - 2(2x) + \frac{1}{2x} = 12x^2 - 4x + \frac{1}{2x},$$

$$f(-x) = 3(-x)^2 - 2(-x) + \frac{1}{-x} = 3x^2 + 2x - \frac{1}{x},$$

$$f\left(\frac{1}{x}\right) = 3\left(\frac{1}{x}\right)^2 - 2\left(\frac{1}{x}\right) + \frac{1}{1/x} = \frac{3}{x^2} - \frac{2}{x} + x,$$

$$f(\sin x) = 3(\sin x)^2 - 2(\sin x) + \frac{1}{\sin x} = 3\sin^2 x - 2\sin x + \frac{1}{\sin x}.$$

Example 9 Suggest alternative ways of writing $f(x)$ in the form $g(h(x))$ when $f(x) = 1/\sqrt{(x+1)}$.

(a) Take $h(x) = x + 1$, $g(x) = 1/\sqrt{x}$.
(b) Take $h(x) = \sqrt{(x+1)}$, $g(x) = 1/x$.

Exercises 12(j)

1. Given $f(x) = (x^2 - 5)/(2x + 1)$, find $f(1/x)$, $f(-x)$, $f(x - 1)$, $f(\sin x)$.
2. Given $f(x) = 2^x$, show that $f(x + 3) - f(x - 1) = (15/2) f(x)$.
3. Given $f(x) = 3^x$, show that $f(x + 2) - f(x - 1) = (26/3) f(x)$.
4. Given $f(x) = 4x - 5 - x^2$, show that $f(2 + k) = f(2 - k)$.
5. Given $f(x) = x^2 - 1$, $g(x) = 2x + 4$, show that

$$g((x^2 - 5)/2) = f(x), \quad f(\sqrt{(2x + 5)}) = g(x).$$

6. If $f(x)$ is linear, i.e. $f(x) = ax + b$, where a, b are constants, what more can you say about the form of $f(x)$ if for all values of x

   (a) $f(x + 1) = f(x) + 1$?, (b) $f(2x) = 2f(x)$?,
   (c) $f(2x + 1) = 2f(x) + 1$?

7. Given $f(x) = ax + b$, $g(x) = cx + d$, where a, b, c, d are constants, show that $f \circ g$ and $g \circ f$ are both linear functions and that

$$f \circ g = g \circ f \text{ if } ad - bc = d - b.$$

8. The function f is called *bilinear* if $f(x) = (ax + b)/(cx + d)$ where a, b, c, d are constants. Show that if f and g are both bilinear then $f \circ g$ and $g \circ f$ are also bilinear.
9. Given $g(x) = 1 + f(x)$, show that $g^{-1}(x) = f^{-1}(x - 1)$, $g^{-1}(x + 1) = f^{-1}(x)$.
10. Suppose that $f(x) < g(x)$ for all values of x. Is it true that for any other function h, and for all values of x, we can say

    (a) $h \circ f(x) < h \circ g(x)$?, (b) $f \circ h(x) < g \circ h(x)$?,

    [*Hint*: Try some examples.]
11. If f_1, f_2 are two functions we denote by $f_1 + f_2$, $f_1 f_2$, f_1/f_2 the function f such that $f(x) = f_1(x) + f_2(x)$, $f(x) = f_1(x) f_2(x)$, $f(x) = f_1(x)/f_2(x)$ respectively. State which of the following are true and which are false:

    (a) $f \circ (g + h) = f \circ g + f \circ h$, (b) $(g + h) \circ f = g \circ f + h \circ f$,
    (c) $f \circ (gh) = (f \circ g)(f \circ h)$, (d) $(gh) \circ f = (g \circ f)(h \circ f)$,
    (e) $f \circ (g/h) = (f \circ g)/(f \circ h)$, (f) $(g/h) \circ f = (g \circ f)/(h \circ f)$.

    [*Hint*: Try some examples.]

CHAPTER 13

MATRICES

13.1 Introduction

Consider the equations

$$\begin{aligned} x' &= a_1x + b_1y, \\ y' &= a_2x + b_2y, \end{aligned} \tag{13.1}$$

where a_1, b_1, a_2, b_2 are constants. Such a pair of equations may arise in a variety of situations; for example:

(a) if (x', y') are the co-ordinates of the point (x, y) after a rotation of axes through 45° in the anti-clockwise direction, then (see Section 2.6)

$$x' = \frac{1}{\sqrt{2}}(x + y),$$

$$y' = \frac{1}{\sqrt{2}}(-x + y).$$

(b) suppose A works x days using materials costing £12 per day and is paid £15 per day for his labour, while B works y days using materials costing £10 per day and is paid £14 per day for his labour, then if £x' is the total labour cost and £y' is the total cost of materials we have

$$\begin{aligned} x' &= 15x + 14y, \\ y' &= 12x + 10y. \end{aligned}$$

In particular if A works 7 days and B works 5 days then $x' = 175$, $y' = 134$.

In (a) we are concerned with a mapping from R^2 onto R^2. However we may likewise interpret the equations in (b) as a mapping from R^2 onto R^2 in which, for example, the point (7, 5) has image (175, 134). It is convenient to regard eqs. (13.1) as representing in all cases a mapping from R^2 onto R^2 and a diagram of the type shown in Fig. 13.1 may be used. From the form of eqs. (13.1) we see that the mapping is a linear transformation (Section 12.4).

Obviously eqs. (13.1) describe exactly the same linear transformation

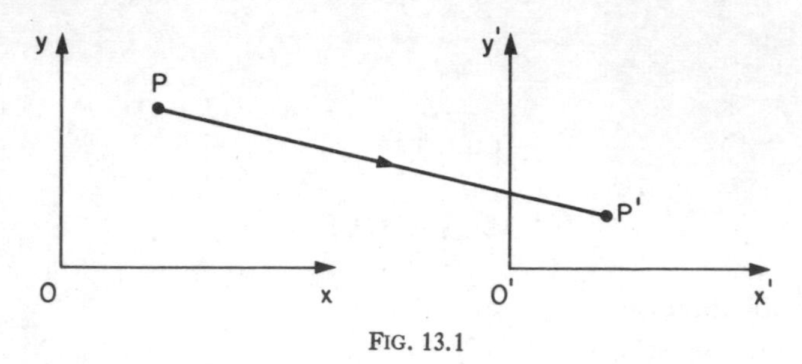

FIG. 13.1

as is described by the equations

$$u = a_1 s + b_1 t,$$
$$v = a_2 s + b_2 t,$$

where different letters have been used as variables. Provided we are consistent about the lay-out of the equations, i.e. about the positions of the variables in the equations, then the transformation is adequately described by the bracketed array of numbers

$$\begin{bmatrix} a_1 & b_1 \\ a_2 & b_2 \end{bmatrix}. \tag{13.2}$$

We refer to (13.2) as the *matrix* of eqs. (13.1), or the matrix of the linear transformation (13.1).

We see that the equations in (a) and (b) above have the respective matrices

$$\begin{bmatrix} 1/\sqrt{2} & 1/\sqrt{2} \\ -1/\sqrt{2} & 1/\sqrt{2} \end{bmatrix}, \quad \begin{bmatrix} 15 & 14 \\ 12 & 10 \end{bmatrix}.$$

Obviously a change in any one of the numbers, or a change in the positioning of the numbers, must be regarded as giving a different matrix. We express this by saying that two matrices

$$\begin{bmatrix} a_1 & b_1 \\ a_2 & b_2 \end{bmatrix}, \quad \begin{bmatrix} \alpha_1 & \beta_1 \\ \alpha_2 & \beta_2 \end{bmatrix}$$

are equal if, and only if, $a_1 = \alpha_1$, $b_1 = \beta_1$, $a_2 = \alpha_2$, $b_2 = \beta_2$ in which case we write

$$\begin{bmatrix} a_1 & b_1 \\ a_2 & b_2 \end{bmatrix} = \begin{bmatrix} \alpha_1 & \beta_1 \\ \alpha_2 & \beta_2 \end{bmatrix}.$$

Assume that

$$\begin{vmatrix} a_1 & b_1 \\ a_2 & b_2 \end{vmatrix} = \Delta \neq 0.$$

Then (Section 11.4.1) we may solve equations (13.1) to get

$$x = \frac{b_2}{\Delta} x' \quad - \frac{b_1}{\Delta} y',$$

$$y = - \frac{a_2}{\Delta} x' + \frac{a_1}{\Delta} y'.$$

This shows that the inverse mapping is also a linear transformation and that it has the matrix

$$\begin{bmatrix} \dfrac{b_2}{\Delta} & -\dfrac{b_1}{\Delta} \\ -\dfrac{a_2}{\Delta} & \dfrac{a_1}{\Delta} \end{bmatrix}.$$

Consequently this last-mentioned matrix is called the *inverse* of the matrix (13.2) and we write

$$\begin{bmatrix} a_1 & b_1 \\ a_2 & b_2 \end{bmatrix}^{-1} = \begin{bmatrix} \dfrac{b_2}{\Delta} & -\dfrac{b_1}{\Delta} \\ -\dfrac{a_2}{\Delta} & \dfrac{a_1}{\Delta} \end{bmatrix}.$$

For example,

$$\begin{bmatrix} 1/\sqrt{2} & 1/\sqrt{2} \\ -1/\sqrt{2} & 1/\sqrt{2} \end{bmatrix}^{-1} = \begin{bmatrix} 1/\sqrt{2} & -1/\sqrt{2} \\ 1/\sqrt{2} & 1/\sqrt{2} \end{bmatrix},$$

since in this case

$$\Delta = \begin{vmatrix} 1/\sqrt{2} & 1/\sqrt{2} \\ -1/\sqrt{2} & 1/\sqrt{2} \end{vmatrix} = 1;$$

of course this result may alternatively be deduced from the fact that the inverse of a rotation through 45° is a rotation through −45°.

As an example of a transformation with zero determinant, consider the transformation

$$x' = x + y,$$
$$y' = 2x + 2y.$$

Here

$$\Delta = \begin{vmatrix} 1 & 1 \\ 2 & 2 \end{vmatrix} = 0.$$

Note that R^2 is mapped onto the set of points on the line $y' = 2x'$. This is not a (1, 1) mapping and so does not have an inverse.

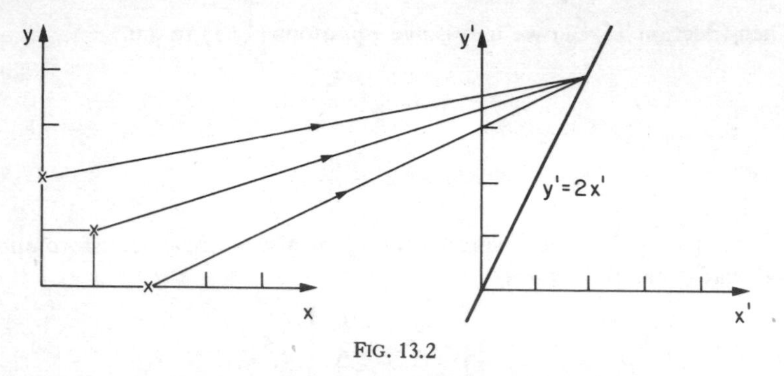

FIG. 13.2

Next suppose that f denotes the transformation (13.1) and g denotes the transformation

$$\begin{aligned} x'' &= \alpha_1 x' + \beta_1 y', \\ y'' &= \alpha_2 x' + \beta_2 y'. \end{aligned} \tag{13.3}$$

Then by direct substitution from (13.1) into (13.3) we get

$$\begin{aligned} x'' &= (\alpha_1 a_1 + \beta_1 a_2)x + (\alpha_1 b_1 + \beta_1 b_2)y, \\ y'' &= (\alpha_2 a_1 + \beta_2 a_2)x + (\alpha_2 b_1 + \beta_2 b_2)y. \end{aligned} \tag{13.4}$$

Equations (13.4) represent the composite mapping $g \circ f$ and show that it also is a linear transformation. The matrix of this composite transformation is called the *product* of the matrices of g and f and we write

$$\begin{bmatrix} \alpha_1 & \beta_1 \\ \alpha_2 & \beta_2 \end{bmatrix} \begin{bmatrix} a_1 & b_1 \\ a_2 & b_2 \end{bmatrix} = \begin{bmatrix} \alpha_1 a_1 + \beta_1 a_2 & \alpha_1 b_1 + \beta_1 b_2 \\ \alpha_2 a_1 + \beta_2 a_2 & \alpha_2 b_1 + \beta_2 b_2 \end{bmatrix}.$$

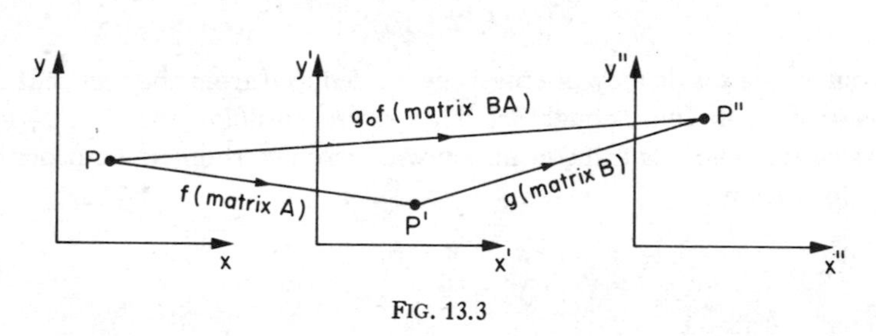

FIG. 13.3

Note that the *order* of the matrices in multiplication is important; for $f \circ g$ is not necessarily the same as $g \circ f$ (see Section 12.3) and in general

$$\begin{bmatrix} a_1 & b_1 \\ a_2 & b_2 \end{bmatrix} \begin{bmatrix} \alpha_1 & \beta_1 \\ \alpha_2 & \beta_2 \end{bmatrix} \neq \begin{bmatrix} \alpha_1 & \beta_1 \\ \alpha_2 & \beta_2 \end{bmatrix} \begin{bmatrix} a_1 & b_1 \\ a_2 & b_2 \end{bmatrix}.$$

If (x, y), (x', y') are regarded as vectors rather than points, then eqs. (13.1) represent a mapping from V_2 onto V_2. In Fig. 13.1 we would then regard $\overrightarrow{O'P'}$ as the image of $\overrightarrow{OP}$. Now suppose that in Fig. 13.4 $\overrightarrow{OP}$ has image $\overrightarrow{O'P'}$ under the transformation f given by eqs. (13.1), and has image $\overrightarrow{O'Q'}$ under the transformation g given by the equations

$$x' = \alpha_1 x + \beta_1 y,$$
$$y' = \alpha_2 x + \beta_2 y.$$

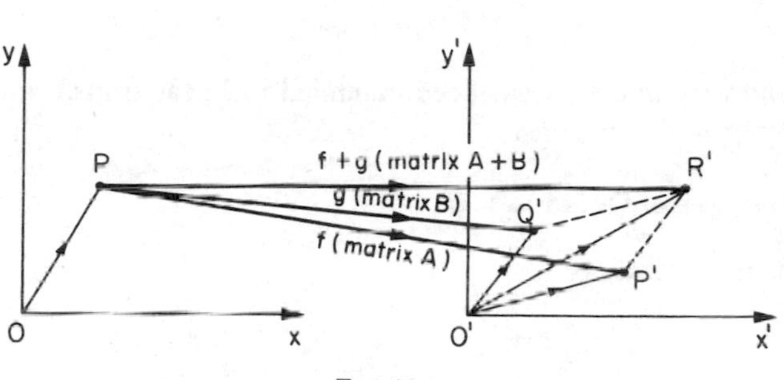

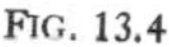
Fig. 13.4

The transformation given by the equations

$$x' = (a_1 + \alpha_1)x + (b_1 + \beta_1)y,$$
$$y' = (a_2 + \alpha_2)x + (b_2 + \beta_2)y$$

is denoted by $f + g$ and under this transformation $\overrightarrow{OP}$ has image $\overrightarrow{O'R'}$ where $\overrightarrow{O'R'} = \overrightarrow{O'P'} + \overrightarrow{O'Q'}$. The matrix of $f + g$ is said to be the *sum* of the matrices of f and g and we write

$$\begin{bmatrix} a_1 & b_1 \\ a_2 & b_2 \end{bmatrix} + \begin{bmatrix} \alpha_1 & \beta_1 \\ \alpha_2 & \beta_2 \end{bmatrix} = \begin{bmatrix} a_1 + \alpha_1 & b_1 + \beta_1 \\ a_2 + \alpha_2 & b_2 + \beta_2 \end{bmatrix}.$$

Finally, if k is a constant, the transformation given by the equations

$$x' = ka_1 x + kb_1 y,$$
$$y' = ka_2 x + kb_2 y$$

is denoted by kf and under this transformation $\overrightarrow{OP}$ has image $\overrightarrow{O'P''}$ where $O'P'' = kO'P'$. The matrix of kf is said to be k times the matrix of f (product of a scalar and a matrix) and we write

$$k\begin{bmatrix} a_1 & b_1 \\ a_2 & b_2 \end{bmatrix} = \begin{bmatrix} ka_1 & kb_1 \\ ka_2 & kb_2 \end{bmatrix}.$$

These are the basic ideas underlying the definitions used in the remainder of this chapter in which the concept of a matrix and of

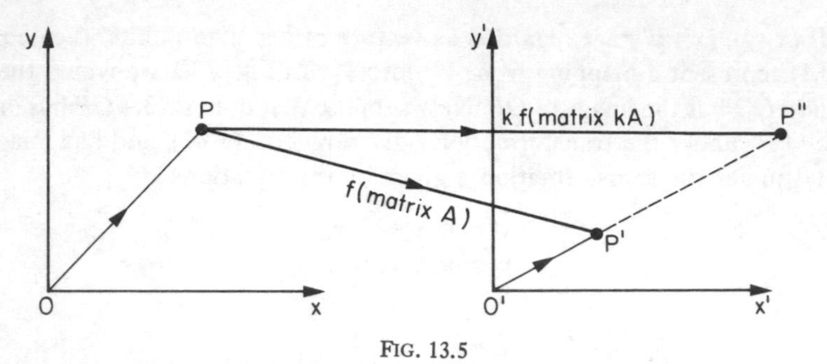

FIG. 13.5

operations with matrices has been extended to linear transformations in any number of dimensions.

13.2 Matrices of Order $m \times n$

Consider the equations

$$\left.\begin{aligned} y_1 &= a_{11}x_1 + a_{12}x_2 + \cdots + a_{1n}x_n, \\ y_2 &= a_{21}x_1 + a_{22}x_2 + \cdots + a_{2n}x_n, \\ &\vdots \\ y_m &= a_{m1}x_1 + a_{m2}x_2 + \cdots + a_{mn}x_n \end{aligned}\right\} \tag{13.5}$$

relating the variables $y_1, y_2, \ldots, y_m$ to the variables $x_1, x_2, \ldots, x_n$. This is a generalisation of eqs. (13.1) but note that, when a large number of variables is involved, it is usual to adopt a suffix notation for variables and a double-suffix notation for coefficients. Equations (13.5) may be regarded as representing a linear transformation from R^n to R^m and the bracketed array of coefficients

$$\begin{bmatrix} a_{11} & a_{12} & \cdots & a_{1n} \\ a_{21} & a_{22} & \cdots & a_{2n} \\ \vdots & \vdots & & \vdots \\ a_{m1} & a_{m2} & \cdots & a_{mn} \end{bmatrix} \tag{13.6}$$

is called the *matrix* of the transformation, or simply the matrix of eqs. (13.5). For the purposes of this book it is assumed that $a_{11}, a_{12}, \ldots$ are real numbers; however, matrices with complex numbers are quite common in more advanced work.

Many authors prefer parentheses to the square brackets used here to denote a matrix; the important thing is to avoid any confusion between the brackets used for a matrix and the straight lines which denote a determinant.

The matrix (13.6) has *m rows* and *n columns*. For this reason it is called a matrix of *order* $m \times n$, or simply an $m \times n$ *matrix*. If $m = n$ we say the matrix is *square*, otherwise *rectangular*. The matrices discussed in Section 13.1 were all 2×2 square matrices; examples of other matrices are as follows:

(a) Corresponding to a rotation of axes in 3-dimensional co-ordinate geometry we have

$$x' = l_1x + m_1y + n_1z,$$
$$y' = l_2x + m_2y + n_2z,$$
$$z' = l_3x + m_3y + n_3z,$$

where $[l_1, m_1, n_1]$, $[l_2, m_2, n_2]$, $[l_3, m_3, n_3]$ are the direction cosines of the x'-axis, y'-axis, z'-axis respectively.

Thus rotation of axes in three dimensions is a transformation from R^3 onto R^3 and the corresponding 3×3 matrix is

$$\begin{bmatrix} l_1 & m_1 & n_1 \\ l_2 & m_2 & n_2 \\ l_3 & m_3 & n_3 \end{bmatrix}.$$

(b) Let l be a line in the direction of the unit vector $\mathbf{u} = (a, b, c)$. For any point $P(x, y, z)$ the projection p of OP on this line is given by $p = ax + by + cz$.

All points P are thus mapped onto points Q on the line l, i.e. we have a mapping from R^3 onto R, and the corresponding 1×3 matrix is $[a, b, c]$.

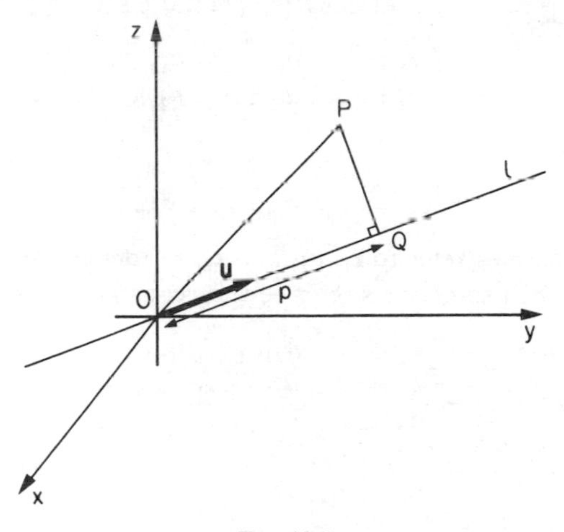

FIG. 13.6

(c) Let π denote the plane $l_3x + m_3y + n_3z = c$ and let (l_1, m_1, n_1), (l_2, m_2, n_2) be unit vectors in the plane π which are perpendicular and may be used as directions for axes of x', y' respectively. Projecting vectors (x, y, z) onto the plane π we get a mapping from R^3 onto R^2 given by the equations

$$x' = l_1x + m_1y + n_1z,$$
$$y' = l_2x + m_2y + n_2z,$$

and the corresponding 2×3 matrix is

$$\begin{bmatrix} l_1 & m_1 & n_1 \\ l_2 & m_2 & n_2 \end{bmatrix}.$$

A matrix with a single row (e.g. example (b) above) is frequently called a *row vector* since for most purposes it may be regarded as equivalent to a vector. Likewise a matrix with a single column is called a *column vector*.

13.2.1 Abbreviated Notations In the matrix (13.6) the number appearing in the ith row and jth column is denoted by a_{ij}. This statement is true for all integers i, j subject to $1 \leqslant i \leqslant m$, $1 \leqslant j \leqslant n$. Consequently the entire matrix is adequately described by writing $[a_{ij}]_{m \times n}$ and, except when we wish to draw special attention to the order of the matrix, we write it as simply $[a_{ij}]$.

More briefly still, the entire matrix may be denoted by a single letter, normally a capital. For example A, or $A_{m \times n}$, may denote the matrix (13.6).

The rows of the matrix (13.6) are closely associated with the primary row vectors $\mathbf{r}_1, \mathbf{r}_2, \ldots, \mathbf{r}_m$ of equations (13.5) since

$$\mathbf{r}_1 = (a_{11}, a_{12}, \ldots, a_{1n}),$$
$$\mathbf{r}_2 = (a_{21}, a_{22}, \ldots, a_{2n}),$$
$$\vdots$$
$$\mathbf{r}_m = (a_{m1}, a_{m2}, \ldots, a_{mn}).$$

We shall sometimes refer to $\mathbf{r}_1, \mathbf{r}_2, \ldots, \mathbf{r}_m$ as *row vectors of the matrix*. Likewise we shall sometimes refer to the vectors

$$\mathbf{c}_1 = (a_{11}, a_{21}, \ldots, a_{m1}),$$
$$\mathbf{c}_2 = (a_{12}, a_{22}, \ldots, a_{m2}),$$
$$\vdots$$
$$\mathbf{c}_n = (a_{1n}, a_{2n}, \ldots, a_{mn})$$

as *column vectors of the matrix*.

13.2.2 Addition of Matrices Two matrices may be added provided they are of the same order. If A and B are both $m \times n$ matrices, say

$$A = [a_{ij}], \quad B = [b_{ij}]$$

then we define their sum $A + B$ as follows:

$$A + B = [c_{ij}], \quad \text{where } c_{ij} = a_{ij} + b_{ij}.$$

For example,

$$\begin{bmatrix} 1 & -2 & 6 \\ 3 & 0 & -4 \end{bmatrix} + \begin{bmatrix} 2 & 0 & -5 \\ 1 & -1 & 7 \end{bmatrix} = \begin{bmatrix} 3 & -2 & 1 \\ 4 & -1 & 3 \end{bmatrix}.$$

13.2.3 Zero Matrices The $m \times n$ matrix with zero in each position is called the $m \times n$ *zero matrix*, e.g. the 3×3 zero matrix and 2×3 zero matrix are

$$\begin{bmatrix} 0 & 0 & 0 \\ 0 & 0 & 0 \\ 0 & 0 & 0 \end{bmatrix}, \quad \begin{bmatrix} 0 & 0 & 0 \\ 0 & 0 & 0 \end{bmatrix}.$$

If O denotes the $m \times n$ zero matrix, clearly we have $A + O = A$ for any $m \times n$ matrix A. Zero matrices of all orders are denoted by O as the order is usually clear from the context.

13.2.4 Subtraction of Matrices If $A = [a_{ij}]$ then we write

$$-A = [-a_{ij}],$$

e.g.

$$-\begin{bmatrix} 1 & -2 & 3 \\ -4 & -5 & 6 \end{bmatrix} = \begin{bmatrix} -1 & 2 & -3 \\ 4 & 5 & -6 \end{bmatrix}.$$

If A and B are matrices of the same order we further write $A - B$ to denote the sum $A + (-B)$. Thus

$$\begin{bmatrix} 1 & 5 & -2 \\ 2 & 4 & 3 \end{bmatrix} - \begin{bmatrix} 1 & -2 & 3 \\ -4 & -5 & 6 \end{bmatrix} = \begin{bmatrix} 0 & 7 & -5 \\ 6 & 9 & -3 \end{bmatrix}.$$

13.2.5 Multiplication of a Matrix by a Scalar If $A = [a_{ij}]$ and k is a constant, then we write kA to denote the matrix $[c_{ij}]$ where $c_{ij} = ka_{ij}$. For example,

$$3\begin{bmatrix} 1 & 2 & -3 \\ 4 & -5 & 6 \end{bmatrix} = \begin{bmatrix} 3 & 6 & -9 \\ 12 & -15 & 18 \end{bmatrix}.$$

13.2.6 Multiplication of Matrices If f denotes the transformation represented by eqs. (13.5) and g denotes the transformation represented by the equations

$$\left.\begin{aligned} z_1 &= b_{11}y_1 + b_{12}y_2 + \cdots + b_{1m}y_m, \\ z_2 &= b_{21}y_1 + b_{22}y_2 + \cdots + b_{2m}y_m, \\ &\vdots \\ z_p &= b_{p1}y_1 + b_{p2}y_2 + \cdots + b_{pm}y_m, \end{aligned}\right\} \tag{13.7}$$

then, if A is the matrix of f and B is the matrix of g, we write BA to denote the matrix of $g \circ f$. Note that f is a transformation from R^n onto R^m and g is a transformation from R^m onto R^p and clearly the idea of a composite transformation $g \circ f$ does not make sense unless the domain of g (R^m) is the same as the range of f. Hence we say two matrices A, B are *conformable for multiplication* in the order BA if, and only if, the number of columns in B equals the number of rows in A. Writing

$$B_{p\times m} \; A_{m\times n}$$

we see that the two *inner* suffices must agree, e.g. a 5×9 matrix multiplied by a 9×11 matrix, or a 3×8 matrix multiplied by an 8×2 matrix, are both permissible.

By substituting from eqs. (13.5) into eqs. (13.7) we find that BA is a $p \times n$ matrix and that if $\mathbf{r}_1, \mathbf{r}_2, \ldots, \mathbf{r}_p$ denote the row vectors of B while $\mathbf{c}_1, \mathbf{c}_2, \ldots, \mathbf{c}_n$ denote the column vectors of A, then

$$BA = \begin{bmatrix} \mathbf{r}_1.\mathbf{c}_1 & \mathbf{r}_1.\mathbf{c}_2 & \cdots & \mathbf{r}_1.\mathbf{c}_n \\ \mathbf{r}_2.\mathbf{c}_1 & \mathbf{r}_2.\mathbf{c}_2 & \cdots & \mathbf{r}_2.\mathbf{c}_n \\ \vdots & \vdots & & \vdots \\ \mathbf{r}_p.\mathbf{c}_1 & \mathbf{r}_p.\mathbf{c}_2 & \cdots & \mathbf{r}_p.\mathbf{c}_n \end{bmatrix}$$

where dots denote the usual scalar products with vectors. Thus

$$\begin{bmatrix} 3 & 7 & -2 \\ 2 & 1 & -1 \\ 4 & 0 & 3 \end{bmatrix} \begin{bmatrix} 1 & 2 \\ -1 & 3 \\ 0 & 5 \end{bmatrix} = \begin{bmatrix} -4 & 17 \\ 1 & 2 \\ 4 & 23 \end{bmatrix}$$

where, for example, the number 1 in the second row and first column of the product is found by taking the scalar product of the second row vector and first column vector of the respective matrices on the l.h.s.

An $m \times n$ matrix A and an $n \times m$ matrix B are conformable for multiplication in either order, BA or AB; since BA is an $n \times n$ matrix whereas AB is an $m \times m$ matrix, the order in which the matrices are multiplied affects the result. As a special case, if $m = n$ we see that any two $n \times n$ square matrices can be multiplied in either order and the result is another $n \times n$ square matrix; but even in this case we usually find that AB and BA are different, e.g.

$$\begin{bmatrix} 1 & 2 \\ 3 & 4 \end{bmatrix}\begin{bmatrix} 1 & -1 \\ 0 & 2 \end{bmatrix} = \begin{bmatrix} 1 & 3 \\ 3 & 5 \end{bmatrix}, \qquad \begin{bmatrix} 1 & -1 \\ 0 & 2 \end{bmatrix}\begin{bmatrix} 1 & 2 \\ 3 & 4 \end{bmatrix} = \begin{bmatrix} -2 & -2 \\ 6 & 8 \end{bmatrix}.$$

Thus matrix multiplication is ***not*** a commutative operation.

A square matrix A can be multiplied by itself and in this case the question of the order in which the matrices are multiplied does not arise—the unique result is denoted by A^2. Likewise the product $A(A^2)$ is denoted by A^3, and so on.

Note that if $AB = O$ *it does not necessarily follow that either* $A = O$ *or* $B = O$, e.g.

$$\begin{bmatrix} 1 & 1 \\ 1 & 1 \end{bmatrix}\begin{bmatrix} 1 & 1 \\ -1 & -1 \end{bmatrix} = \begin{bmatrix} 0 & 0 \\ 0 & 0 \end{bmatrix}.$$

13.2.7 Unit Matrices The leading diagonal of a matrix is the diagonal running from top left to bottom right. The $n \times n$ square matrix in which all the numbers on the leading diagonal are 1 and all other numbers are zero, is called the $n \times n$ ***unit matrix***. Thus the matrices

$$\begin{bmatrix} 1 & 0 & 0 \\ 0 & 1 & 0 \\ 0 & 0 & 1 \end{bmatrix}, \quad \begin{bmatrix} 1 & 0 \\ 0 & 1 \end{bmatrix}$$

are the 3×3 unit matrix and 2×2 unit matrix respectively.

Unit matrices of all orders are usually denoted by I. For any $m \times n$ matrix A we have

$$IA = A,$$

where I is the $m \times m$ unit matrix; also

$$AI = A,$$

where I is the $n \times n$ unit matrix.

13.2.8 Transpose of a Matrix The matrix A' is called the ***transpose*** of the matrix A if the first, second, etc. ... row vectors of A are respectively the first, second, etc. ... column vectors of A'. It follows that the column vectors of A are the row vectors of A'; also that the number in the ith row and jth column of A is the same as the number in the jth row and the ith column of A'. For example the transpose of

$$\begin{bmatrix} 1 & 3 & -2 \\ 5 & 6 & 4 \end{bmatrix}$$

is

$$\begin{bmatrix} 1 & 5 \\ 3 & 6 \\ -2 & 4 \end{bmatrix}.$$

Clearly, if A' is the transpose of A then A is the transpose of A'. Writing down the transpose A' of a given matrix A may be regarded as an operation on A—we say we are *transposing* A, and the prime may be regarded as a symbol denoting this operation. Performing the operation of transposition twice in succession brings us back to the matrix we started with, i.e. $(A')' = A$.

Exercises 13(a)

1. Given that

$$A = \begin{bmatrix} 1 & 2 & -1 \\ 3 & 0 & 5 \end{bmatrix}, \quad B = \begin{bmatrix} 2 & -1 & 1 \\ 2 & 1 & 0 \end{bmatrix}, \quad C = \begin{bmatrix} 3 & 1 \\ 1 & 3 \end{bmatrix}, \quad D = \begin{bmatrix} 1 & 2 \\ 0 & 1 \end{bmatrix},$$

find, where possible, $A + B$, $A + C$, $C + D$, CA, AB, AC, $A'C'$.

2. Given that

$$A = \begin{bmatrix} 1 & 2 \\ 3 & 4 \end{bmatrix},$$

find $A + 2I$, $A + 3I$, $(A + 2I)(A + 3I)$, where I is the appropriate unit matrix. Verify that

$$(A + 2I)(A + 3I) = A^2 + 5A + 6I.$$

3. Given that

$$A = \begin{bmatrix} 1 & 2 \\ 3 & 4 \end{bmatrix}, \qquad B = \begin{bmatrix} 4 & 3 \\ 1 & 2 \end{bmatrix},$$

find $A + 2B$, $A + 3B$, $(A + 2B)(A + 3B)$. Verify that

$$(A + 2B)(A + 3B) \neq A^2 + 5AB + 6B^2,$$

but that

$$(A + 2B)(A + 3B) = A^2 + 2BA + 3AB + 6B^2.$$

4. B is the matrix

$$\begin{bmatrix} 0 & 1 & 1 \\ 1 & 0 & 1 \\ 1 & 1 & 0 \end{bmatrix}.$$

Find λ if $B^2 - \lambda B - \lambda^2 I = I$.

5. Given

$$A = \begin{bmatrix} a & b & b \\ b & a & b \\ b & b & a \end{bmatrix},$$

show that A^2 has a similar form and find values for a, b such that $A^2 = I$.

6. Verify that the operation of matrix addition is commutative and associative, i.e. $A + B = B + A$, $(A + B) + C = A + (B + C)$, assuming that A, B, C are conformable for addition.

7. Verify that the operation of multiplication by a scalar obeys the following laws:

$$(k_1 + k_2)A = k_1A + k_2A, \quad (k_1k_2)A = k_1(k_2A) = k_2(k_1A),$$
$$k(A + B) = kA + kB, \quad k(AB) = (kA)B = A(kB),$$

where A, B are matrices such that these operations are possible, and k_1, k_2, k are scalars.

8. The operation of matrix multiplication is associative, i.e. if A, B, C are matrices such that the products AB, $(AB)C$ are possible, then the products BC, $A(BC)$ are also possible and $(AB)C = A(BC)$. Verify this in the case where A, B, C are 2×2 matrices; also when A is a 3×2 matrix, B a 2×2 matrix, C a 2×3 matrix.

9. Since matrix multiplication is non-commutative there are two possible distributive laws, viz.

$$A(B + C) = AB + AC, \quad (B + C)A = BA + CA.$$

Both these laws hold when the matrices are conformable for the appropriate operations. Verify this in the case where A, B, C are 2×2 matrices; also when A is a 3×2 matrix and B, C are 2×3 matrices.

10. The system of matrices under the operations of matrix addition and matrix multiplication is *not* a field. Support this statement by identifying at least one field axiom that is not satisfied by this system.

11. A square matrix in which all the numbers except those in the leading diagonal are zero is called a *diagonal* matrix. Show that if D_1, D_2 are two diagonal matrices of the same order, then D_1D_2 is also a diagonal matrix and that $D_1D_2 = D_2D_1$.

    Show that if A denotes the matrix

$$\begin{bmatrix} -1 & 0 & 0 \\ 2 & 1 & 0 \\ 1 & 1 & -1 \end{bmatrix},$$

then $A^2 = I$ and that if $P = AD_1A$, $Q = AD_2A$, where D_1, D_2 are both 3×3 diagonal matrices, then $PQ = QP$.

12. Assuming that A and B are conformable for multiplication, prove that $(AB)' = B'A'$. (This is the ***reversal rule for transposition.***)

13.2.9 Inverse Matrix Given a set of values for $x_1, x_2, \ldots, x_n$ eqs. (13.5) determine a corresponding set of values for $y_1, y_2, \ldots, y_m$. Regarding (13.5) as a linear transformation, T say, from R^n onto R^m, we say that under T the vector $(x_1, x_2, \ldots, x_n)$ has image $(y_1, y_2, \ldots, y_m)$. Given a vector $(y_1, y_2, \ldots, y_m)$ in R^m can we find a vector $(x_1, x_2, \ldots, x_n)$ that has the given vector as image? This amounts to asking if we can solve equations (13.5) for $x_1, x_2, \ldots, x_n$, and clearly T is not a (1, 1) transformation unless

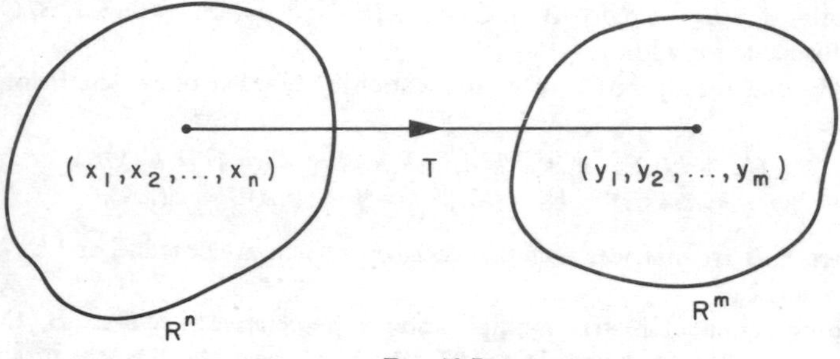

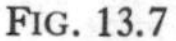
FIG. 13.7

this solution is unique. From the study of linear equations in Section 11.4 it follows that the solution exists and is unique only if $m = n$ (i.e. the matrix of (13.5) is square) and the determinant of the coefficients is non-zero. Examples of transformations with non-square matrices are given in (b) and (c) of Section 13.2, and it is easily seen that these are not (1, 1) transformations.

The linear transformation

$$\left.\begin{aligned} y_1 &= a_{11}x_1 + a_{12}x_2 + \cdots + a_{1n}x_n, \\ y_2 &= a_{21}x_1 + a_{22}x_2 + \cdots + a_{2n}x_n, \\ &\vdots \\ y_n &= a_{n1}x_1 + a_{n2}x_2 + \cdots + a_{nn}x_n, \end{aligned}\right\} \tag{13.8}$$

with square matrix A, where

$$A = \begin{bmatrix} a_{11} & a_{12} & \cdots & a_{1n} \\ a_{21} & a_{22} & \cdots & a_{2n} \\ \vdots & \vdots & & \vdots \\ a_{n1} & a_{n2} & \cdots & a_{nn} \end{bmatrix},$$

is a (1, 1) transformation provided that

$$\begin{vmatrix} a_{11} & a_{12} & \cdots & a_{1n} \\ a_{21} & a_{22} & \cdots & a_{2n} \\ \vdots & \vdots & & \vdots \\ a_{n1} & a_{n2} & \cdots & a_{nn} \end{vmatrix} \neq 0.$$

This determinant is denoted by $|A|$ and called the *determinant of the matrix A*. When $|A| \neq 0$ (13.8) is a (1, 1) transformation and so has an inverse; the inverse transformation may be found by solving (13.8) to get (Theorem 6, Section 11.4.1)

$$x_1 = \frac{\begin{vmatrix} y_1 & a_{12} & \cdots & a_{1n} \\ y_2 & a_{22} & \cdots & a_{2n} \\ \vdots & \vdots & & \vdots \\ y_n & a_{n2} & \cdots & a_{nn} \end{vmatrix}}{|A|}, \qquad x_2 = \frac{\begin{vmatrix} a_{11} & y_1 & \cdots & a_{1n} \\ a_{21} & y_2 & \cdots & a_{2n} \\ \vdots & \vdots & & \vdots \\ a_{n1} & y_n & \cdots & a_{nn} \end{vmatrix}}{|A|}, \ldots.$$

Let us denote by $A_{11}, A_{12}, \ldots$ the minors of A corresponding to $a_{11}, a_{12}, \ldots$. Then in the above solution the numerator may be expanded in each case by the column containing $y_1, y_2, \ldots, y_n$ to give

$$x_1 = \frac{1}{|A|}\{A_{11}y_1 - A_{21}y_2 + \cdots + (-1)^{n+1} A_{n1}y_n\},$$

$$x_2 = \frac{1}{|A|}\{-A_{12}y_1 + A_{22}y_2 - \cdots + (-1)^{n} A_{n2}y_n\},$$

and so on.

This shows that the inverse of the transformation (13.8) is also a linear transformation and its matrix, denoted by A^{-1}, is given by

$$A^{-1} = \frac{1}{|A|}\begin{bmatrix} A_{11} & -A_{21} & \cdots & (-1)^{n+1}A_{n1} \\ -A_{12} & A_{22} & \cdots & (-1)^{n}A_{n2} \\ \vdots & \vdots & & \vdots \\ (-1)^{n+1}A_{1n} & (-1)^{n}A_{2n} & \cdots & A_{nn} \end{bmatrix}.$$

If we denote the transformation (13.8) by T, then A is the matrix of T and A^{-1} the matrix of the inverse transformation T^{-1}. The product matrix $A^{-1}A$ is the matrix of the composite transformation $T^{-1} \circ T$ (see Section 13.2.6); but $T^{-1} \circ T$ is the identity transformation, which has matrix I. Thus

$$A^{-1}A = I,$$

and we can similarly deduce that

$$AA^{-1} = I.$$

In practice we may find A^{-1} by the following procedure:

(a) Evaluate $|A|$. If $|A| = 0$ the matrix A is said to be *singular* and A^{-1} does not exist. If $|A| \neq 0$ proceed to the next step.

(b) Replace each number in A by the minor of $|A|$ corresponding to that number, or by minus the minor, depending on the position of the number, e.g. replace a_{11} by A_{11}, a_{12} by $-A_{12}$, a_{ij} by $(-1)^{i+j}A_{ij}$. (Note that the sign to take with each minor is given by the same rule as in Section 11.2.1 for the expansion of a determinant in terms of minors).

(c) Transpose the matrix obtained in (b). This gives a matrix usually called the *adjugate* of A.

(d) Multiply the adjugate matrix obtained in (c) by $1/|A|$. This gives A^{-1}.

(e) Check the result by verifying that $A^{-1}A = I$.

Note: An alternative method for finding A^{-1} is given in Section 13.3.4.

Example 1 Find the inverse of the matrix

$$\begin{bmatrix} 13 & -3 & -2 \\ -3 & 5 & 6 \\ -1 & 3 & 14 \end{bmatrix}.$$

Following the procedure outlined above, we have:

(a) $$\begin{vmatrix} 13 & -3 & -2 \\ -3 & 5 & 6 \\ -1 & 3 & 14 \end{vmatrix} = \begin{vmatrix} 12 & 0 & 12 \\ -3 & 5 & 6 \\ -1 & 3 & 14 \end{vmatrix} = 12\begin{vmatrix} 1 & 0 & 0 \\ -3 & 5 & 9 \\ -1 & 3 & 15 \end{vmatrix}$$

$$= 12\begin{vmatrix} 5 & 9 \\ 3 & 15 \end{vmatrix} = 576.$$

(b) The minor corresponding to 13 is

$$\begin{vmatrix} 5 & 6 \\ 3 & 14 \end{vmatrix} = 52,$$

and continuing to evaluate minors we construct the matrix

$$\begin{bmatrix} 52 & 36 & -4 \\ 36 & 180 & -36 \\ -8 & -72 & 56 \end{bmatrix}, \quad \text{that is} \quad 4\begin{bmatrix} 13 & 9 & -1 \\ 9 & 45 & -9 \\ -2 & -18 & 14 \end{bmatrix},$$

in which, of course, the signs of some minors have been altered in accordance with the usual rule.

(c) The transpose of the matrix constructed in (c) is

$$4\begin{bmatrix} 13 & 9 & -2 \\ 9 & 45 & -18 \\ -1 & -9 & 14 \end{bmatrix}.$$

(d) Finally, the inverse matrix is

$$\frac{1}{144}\begin{bmatrix} 13 & 9 & -2 \\ 9 & 45 & -18 \\ -1 & -9 & 14 \end{bmatrix}.$$

(e) Multiplying out, we get

$$\frac{1}{144}\begin{bmatrix} 13 & 9 & -2 \\ 9 & 45 & -18 \\ -1 & -9 & 14 \end{bmatrix}\begin{bmatrix} 13 & -3 & -2 \\ -3 & 5 & 6 \\ -1 & 3 & 14 \end{bmatrix} = \begin{bmatrix} 1 & 0 & 0 \\ 0 & 1 & 0 \\ 0 & 0 & 1 \end{bmatrix}.$$

Exercises 13(b)

1. In a certain linear transformation the point (x_1, x_2, x_3) has image (y_1, y_2, y_3), where

$$y_1 = x_1 + x_2 + x_3, \quad y_2 = -x_1 + x_2 - 2x_3, \quad y_3 = x_2 + x_3.$$

Write down the matrix of the transformation and find the matrix of the inverse transformation.

2. Given

$$A = \begin{bmatrix} 1 & -1 & -1 \\ -1 & 1 & -1 \\ -1 & -1 & 1 \end{bmatrix}, \quad B = \begin{bmatrix} 1 & 1 & -1 \\ -1 & 1 & -1 \\ -1 & -1 & 1 \end{bmatrix}$$

find A^{-1} and show that B is singular.

3. Given that

$$A = \begin{bmatrix} 2 & 1 & 4 \\ 3 & 2 & 5 \\ 0 & -1 & 1 \end{bmatrix}$$

find A^{-1} and $A^{-1}A'$, where A' denotes the transpose of A.

4. Given that

$$A = \begin{bmatrix} 1 & 2 & 3 \\ 2 & 4 & 5 \\ 3 & 5 & 6 \end{bmatrix}, \quad B = \begin{bmatrix} 1 & -3 & 2 \\ -3 & 3 & -1 \\ 2 & a & b \end{bmatrix},$$

find AB and hence determine a, b so that $B = A^{-1}$.

5. Show that if a square matrix A satisfies the matrix equation

$$A^2 - \mu A - \lambda I = O,$$

where μ and λ are scalars ($\lambda \neq 0$) and I is the appropriate unit matrix, then $A^{-1} = (1/\lambda)(A - \mu I)$. [*Hint*: Multiply the given equation by A^{-1}.]

Verify that if

$$A = \begin{bmatrix} 2 & -3 \\ -2 & 1 \end{bmatrix},$$

then $A^2 - 3A - 4I = O$ and hence find A^{-1}. Verify also that if

$$B = \begin{bmatrix} 2 & 1 & 1 \\ 1 & 2 & 1 \\ 1 & 1 & 2 \end{bmatrix},$$

then $B^2 - 5B + 4I = O$ and hence find B^{-1}.

6. Given that X is a row vector and X' its transpose, show that the matrix $X'X$ does not have an inverse.
7. Given

$$A = \begin{bmatrix} 1 & c & c \\ c & 1 & c \\ c & c & 1 \end{bmatrix},$$

find the non-zero value of c such that A^2 is a diagonal matrix. Deduce the inverse matrix A^{-1} when c has this value. Find A^{-1} when $c = -1$.

8. Prove that $(AB)^{-1} = B^{-1}A^{-1}$ (reversal rule for inverses). [*Hint*: Start from the equation $(AB)(AB)^{-1} = I$, multiply both sides by $(B^{-1}A^{-1})$ and use the fact that multiplication is associative.]
9. The matrix A is defined as

$$A = \begin{bmatrix} 1 & 1 & \mu \\ 2 & -\mu & 3 \\ 1 & -2 & 2 \end{bmatrix}.$$

(a) Given $\mu = -1$, find A^{-1}. (b) Find the two values of μ for which A is singular. (c) Show that if μ takes the smaller of the values for which A is singular, then the planes represented by the equations

$$\begin{aligned} x + y + \mu z &= 1, \\ 2x - \mu y + 3z &= 3, \\ x - 2y + 2z &= 2, \end{aligned}$$

meet on a line. [*Hint*: show that the plane $x - 2y + 2z = 2$ passes through the line of intersection of the other two planes.]

13.3 Matrix Equations

Suppose that A is a non-singular $n \times n$ square matrix and that B is a matrix with n rows. Then it is reasonable to ask the question 'what matrix X satisfies the equation

$$AX = B\text{?'} \tag{1}$$

Conformability requires that X must have n rows, and if B has p columns then X must be an $n \times p$ matrix. Multiplying both sides of eq. (1) by A^{-1} we get

$$A^{-1}(AX) = A^{-1}B.$$

(Note that multiplication has been done in the same order on both sides.) Since multiplication is associative (Question 8 of Exercises 13(a)), we can now write

$$(A^{-1}A)X = A^{-1}B,$$

and since $A^{-1}A = I$, $IX = X$ we get

$$X = A^{-1}B. \qquad (2)$$

Thus (2) is the solution of eq. (1). We immediately see the similarity to the algebra of the real number system where $ax = b \Leftrightarrow x = a^{-1}b$ if $a \neq 0$; however, because matrix multiplication is non-commutative, the order of A^{-1} and B in (2) is important, unlike the situation with real numbers.

Example 2 Given that

$$A = \begin{bmatrix} 1 & 2 & 3 \\ 2 & 1 & 0 \\ 0 & 1 & -1 \end{bmatrix}, \quad B = \begin{bmatrix} 4 & 15 \\ 2 & 6 \\ -1 & -1 \end{bmatrix},$$

solve the equation $AX = B$.

The solution is

$$X = A^{-1}B = \frac{1}{9}\begin{bmatrix} -1 & 5 & -3 \\ 2 & -1 & 6 \\ 2 & -1 & -3 \end{bmatrix}\begin{bmatrix} 4 & 15 \\ 2 & 6 \\ -1 & -1 \end{bmatrix} = \begin{bmatrix} 1 & 2 \\ 0 & 2 \\ 1 & 3 \end{bmatrix}.$$

Likewise if A is a non-singular $n \times n$ square matrix and B has n columns, the equation

$$XA = B$$

may be solved by forming the product of each side with A^{-1} in the reverse order to that used with eq. (1). Thus

$$(XA)A^{-1} = BA^{-1},$$

and this leads to the solution $X = BA^{-1}$.

13.3.1 Linear Transformation in Matrix Form Let us define two column vectors X, Y as follows:

$$X = \begin{bmatrix} x_1 \\ x_2 \\ \vdots \\ x_n \end{bmatrix}, \quad Y = \begin{bmatrix} y_1 \\ y_2 \\ \vdots \\ y_m \end{bmatrix}.$$

Equations (13.5) can now be written in the form of the single matrix equation

$$Y = AX, \tag{1}$$

where

$$A = \begin{bmatrix} a_{11} & a_{12} & \cdots & a_{1n} \\ a_{21} & a_{22} & \cdots & a_{2n} \\ \vdots & \vdots & & \vdots \\ a_{m1} & a_{m2} & \cdots & a_{mn} \end{bmatrix}.$$

(The reader should perform the multiplication AX to confirm that the matrix equation (1) is equivalent to eqs. (13.5)).

Likewise eqs. (13.7) can be written in the form

$$Z = BY, \tag{2}$$

where

$$Z = \begin{bmatrix} z_1 \\ z_2 \\ \vdots \\ z_p \end{bmatrix}, \quad B = \begin{bmatrix} b_{11} & b_{12} & \cdots & b_{1m} \\ b_{21} & b_{22} & \cdots & b_{2m} \\ \vdots & \vdots & & \vdots \\ b_{p1} & b_{p2} & \cdots & b_{pm} \end{bmatrix}.$$

As an example of the economy achieved by this notation, consider the lengthy algebraic procedure of substituting for $y_1, y_2, \ldots, y_m$ in eqs. (13.7) from eqs. (13.5). This can be done by substituting for Y in eq. (2) from eq. (1); thus

$$Z = B(AX) = (BA)X.$$

Note that this brings out the advantage of defining the product matrix as the matrix of the composite transformation (Section 13.2.6).

The linear transformation (13.8) can also be expressed as the matrix equation (1); in this case A is an $n \times n$ square matrix and X, Y both have n components. Assuming that $|A| \neq 0$ there is an inverse matrix A^{-1}; this was defined in Section 13.2.9 as the matrix of the inverse transformation, and from this definition it was deduced that $A^{-1}A = AA^{-1} = I$. It is worth noting that we may alternatively define A^{-1} by the property $A^{-1}A = I$, and then, solving eq. (1) by the method of Section 13.3, we get

$$X = A^{-1}Y,$$

showing that A^{-1} is the matrix of the inverse transformation. The latter approach emphasises the similarity of matrix algebra to the algebra of the real number system.

13.3.2 Linear Simultaneous Equations in Matrix Form The equations

$$\begin{aligned} a_{11}x_1 + a_{12}x_2 + \cdots + a_{1n}x_n &= b_1, \\ a_{21}x_1 + a_{22}x_2 + \cdots + a_{2n}x_n &= b_2, \\ &\vdots \\ a_{m1}x_1 + a_{m2}x_2 + \cdots + a_{mn}x_n &= b_m, \end{aligned}$$

may be written as the matrix equation

$$AX = B,$$

where

$$A = \begin{bmatrix} a_{11} & a_{12} & \cdots & a_{1n} \\ a_{21} & a_{22} & \cdots & a_{2n} \\ \vdots & \vdots & & \vdots \\ a_{m1} & a_{m2} & \cdots & a_{mn} \end{bmatrix}, \quad X = \begin{bmatrix} x_1 \\ x_2 \\ \vdots \\ x_n \end{bmatrix}, \quad B = \begin{bmatrix} b_1 \\ b_2 \\ \vdots \\ b_m \end{bmatrix}.$$

With these equations we also associate the *augmented matrix* $\tilde{A}$, defined by

$$\tilde{A} = \begin{bmatrix} a_{11} & a_{12} & \cdots & a_{1n} b_1 \\ a_{21} & a_{22} & \cdots & a_{2n} b_2 \\ \vdots & \vdots & & \vdots \\ a_{m1} & a_{m2} & \cdots & a_{mn} b_m \end{bmatrix}.$$

The *rank* of a matrix is the dimension (see Section 11.3) of the set of row vectors of the matrix.

Theorem 9 of Section 11.4.3 implies that the equations are consistent or inconsistent according as the ranks of A and $\tilde{A}$ are equal or unequal. Theorem 7 of Section 11.4.2 implies that if rank A = rank $\tilde{A} = s$, where $s < n$, then values can be assigned arbitrarily to $(n - s)$ variables and the resulting equations solved for the other s variables; thus the solution is not unique.

In the case where A is a non-singular $n \times n$ square matrix, the equations have the unique solution

$$X = A^{-1}B;$$

this follows from Section 13.3, but in this case rank A = rank $\tilde{A} = n$ and the result is essentially equivalent to Theorem 6 of Section 11.4.1—the method we have been using to find A^{-1} (Section 13.2.9) does, of course, depend on this theorem.

Example 3 Solve the equations

$$\begin{aligned} 13x_1 - 3x_2 - 2x_3 &= 19 \\ -3x_1 + 5x_2 + 6x_3 &= 19 \\ -x_1 + 3x_2 + 14x_3 &= 65. \end{aligned}$$

The equations may be written as

$$\begin{bmatrix} 13 & -3 & -2 \\ -3 & 5 & 6 \\ -1 & 3 & 14 \end{bmatrix} \begin{bmatrix} x_1 \\ x_2 \\ x_3 \end{bmatrix} = \begin{bmatrix} 19 \\ 19 \\ 65 \end{bmatrix}$$

and the solution is thus

$$\begin{aligned} \begin{bmatrix} x_1 \\ x_2 \\ x_3 \end{bmatrix} &= \begin{bmatrix} 13 & -3 & -2 \\ -3 & 5 & 6 \\ -1 & 3 & 14 \end{bmatrix}^{-1} \begin{bmatrix} 19 \\ 19 \\ 65 \end{bmatrix} \\ &= \frac{1}{144} \begin{bmatrix} 13 & 9 & -2 \\ 9 & 45 & -18 \\ -1 & -9 & 14 \end{bmatrix} \begin{bmatrix} 19 \\ 19 \\ 65 \end{bmatrix} \quad \text{(see Example 1)} \\ &= \begin{bmatrix} 2 \\ -1 \\ 5 \end{bmatrix}. \end{aligned}$$

Hence $x_1 = 2$, $x_2 = -1$, $x_3 = 5$.

The matrix method of solving linear simultaneous equations is particularly advantageous when we have to solve a number of sets of equations all having the same matrix A but having different matrices B for their right-hand sides; once A^{-1} has been found all the solutions follow quickly by forming the product $A^{-1}B$ in each case.

13.3.3 Elementary Row Operations A set of simultaneous equations may be written in many equivalent forms as follows:

(a) both sides of an equation may be multiplied by some constant,
(b) to one equation may be added some multiple of another equation,
(c) the order of the equations may be changed.

Taking for example the equations solved in Example 3, the first equation might be written alternatively as $26x_1 - 6x_2 - 4x_3 = 38$, or it might be replaced by the equation $7x_1 + 7x_2 + 10x_3 = 57$ obtained by adding twice the second equation to it; and of course it makes no difference if we change the order of the equations.

This means that, in terms of the matrix form $AX = B$ of the equations, equivalent forms are obtained by the following operations on the matrix A, provided that the same operations are performed on B:

(a) each element of a row may be multiplied by a constant,
(b) to one row vector may be added a multiple of another row vector,
(c) the order of the rows may be changed.

Such operations are called *elementary row operations*. In practice they may be carried out by first doing the operations to the unit matrix and then multiplying the given matrix by the amended unit matrix. For example consider the matrix

$$\begin{bmatrix} a_1 & b_1 & c_1 \\ a_2 & b_2 & c_2 \\ a_3 & b_3 & c_3 \end{bmatrix}$$

and the equivalent matrix

$$\begin{bmatrix} 2a_1 & 2b_1 & 2c_1 \\ a_2 + 3a_3 & b_2 + 3b_3 & c_2 + 3c_3 \\ a_3 & b_3 & c_3 \end{bmatrix}$$

obtained by doubling the first row and adding three times the third row to the second row. The same row operations applied to the unit 3×3 matrix give

$$\begin{bmatrix} 2 & 0 & 0 \\ 0 & 1 & 3 \\ 0 & 0 & 1 \end{bmatrix}$$

and multiplication of the original matrix by this matrix has the same effect as the operations, i.e.

$$\begin{bmatrix} 2 & 0 & 0 \\ 0 & 1 & 3 \\ 0 & 0 & 1 \end{bmatrix} \begin{bmatrix} a_1 & b_1 & c_1 \\ a_2 & b_2 & c_2 \\ a_3 & b_3 & c_3 \end{bmatrix} = \begin{bmatrix} 2a_1 & 2b_1 & 2c_1 \\ a_2 + 3a_3 & b_2 + 3b_3 & c_2 + 3c_3 \\ a_3 & b_3 & c_3 \end{bmatrix}.$$

13.3.4 Alternative Method of Finding the Inverse Matrix In solving simultaneous equations by the traditional method of 'elimination' we are in fact using elementary row operations. In Example 3 we might begin by eliminating x_1 from the first equation by adding thirteen times the third equation to it; we might also eliminate x_1 from the second equation by subtracting three times the third equation from it; we might now multiply through the third equation by -1 and relabel it as the first equation, and so on. A complete solution following these lines is shown in Table 13.1, where on the left we give the matrix A of the equations and its equivalent forms after suitable row operations, and on the right we do the same row operations to the unit matrix, thereby obtaining a matrix multiplier which would have the same effect as these operations.

TABLE 13.1

Matrix A			Unit matrix			
13	−3	−2	1	0	0	
−3	5	6	0	1	0	
−1	3	14	0	0	1	
1	−3	−14	0	0	−1	3rd row multiplied by −1 and
−3	5	6	0	1	0	interchanged with 1st row.
13	−3	−2	1	0	0	
1	−3	−14	0	0	−1	
0	−4	−36	0	1	−3	3 times 1st row added.
0	36	180	1	0	13	13 times 1st row subtracted.
1	−3	−14	0	0	−1	
0	1	9	0	−1/4	3/4	Multiplied by −1/4.
0	0	−144	1	9	−14	36 times second row (in its present form) added.
1	−3	0	−7/72	−7/8	13/36	14 times third row (as amended below) added.
0	1	0	1/16	5/16	−1/8	9 times third row (as amended below) subtracted.
0	0	1	−1/144	−1/16	7/72	Multiplied by −1/144.
1	0	0	13/144	1/16	−1/72	3 times second row added.
0	1	0	1/16	5/16	−1/8	
0	0	1	−1/144	−1/16	7/72	

At this stage we have only one variable in each equation and so, had we done the same operations on the right-hand sides of the equations we would now have arrived at the solution. To perform these operations on the right-hand sides we may apply the above matrix multiplier to the column vector B formed by these right-hand sides, i.e. multiply B by R, where

$$R=\begin{bmatrix} 13/144 & 1/16 & -1/72 \\ 1/16 & 5/16 & -1/8 \\ -1/144 & -1/16 & 7/72 \end{bmatrix}=\frac{1}{144}\begin{bmatrix} 13 & 9 & -2 \\ 9 & 45 & -18 \\ -1 & -9 & 14 \end{bmatrix}.$$

Thus the solution of the equations is

$$X=\frac{1}{144}\begin{bmatrix} 13 & 9 & -2 \\ 9 & 45 & -18 \\ -1 & -9 & 14 \end{bmatrix}\begin{bmatrix} 19 \\ 19 \\ 65 \end{bmatrix}=\begin{bmatrix} 2 \\ -1 \\ 5 \end{bmatrix}.$$

Comparing with the previous method of solving these equations we can now identify the matrix R with the inverse A^{-1}. More generally, from the elementary row operations which reduce a matrix A to the unit matrix, we can find a matrix R by performing these same operations on the unit matrix, and $R = A^{-1}$. The method of setting out the work is as shown in the above example; errors can arise from trying to do several steps at once and, if in doubt, one should proceed one step at a time so that each equivalent form follows from its predecessor by a simple elementary row operation. We start with the matrix A and a unit matrix on its right, and we finish with a unit matrix on the left and A^{-1} on the right. A systematic procedure should be used and the following, illustrated by the example, is recommended:

(1) Get 1 in the top left corner.
(1a) Use the first row to get zeros in the first column below this 1.
(2) Get 1 in the second row, second column.
(2a) Use the second row to get zeros in the second column below this 1.
(3) Get 1 in the third row, third column so that the matrix A is now reduced to the 'triangular' form

$$\begin{bmatrix} 1 & \times & \times \\ 0 & 1 & \times \\ 0 & 0 & 1 \end{bmatrix}$$

(3a) Use the third row to get zeros in the third column above this 1.
(4) Use the second row to get zero in the position above the 1 in the second column. The matrix A is now reduced to the unit matrix.

These rules, described here for a 3×3 matrix, are readily adapted to deal with any square matrix.

The method of elementary row operations for finding an inverse matrix may seem less convenient than the previous method when dealing with a 3×3 matrix. However, for larger matrices, particularly when a computer is used to do the calculations, it is certainly the more practical approach.

Exercises 13(c)

1. Given

$$A = \begin{bmatrix} 2 & 1 & 2 \\ 2 & -4 & 1 \\ 1 & 1 & 1 \end{bmatrix},$$

find A^{-1} and hence solve the matrix equations

$$\text{(a) } AX = \begin{bmatrix} 1 \\ 2 \\ 1 \end{bmatrix}, \qquad \text{(b) } AY = \begin{bmatrix} 1 & 1 \\ 2 & 0 \\ 1 & 1 \end{bmatrix}.$$

2. Given that

$$A = \begin{bmatrix} 1 & 1 & 1 \\ 3 & 2 & 1 \\ 2 & 2 & 1 \end{bmatrix}, \quad B = \begin{bmatrix} 1 & 1 \\ 1 & 0 \\ 1 & 1 \end{bmatrix},$$

find A^{-1} and use your result to solve the matrix equation $AX = B$.

3. Find A^{-1}, where

$$A = \begin{bmatrix} 1 & 3 & 3 \\ 1 & 4 & 3 \\ 1 & 3 & 4 \end{bmatrix}.$$

Hence solve the matrix equation $AX = B$, where

$$B = \begin{bmatrix} 2 & 3 \\ 3 & 1 \\ 1 & 0 \end{bmatrix},$$

and the equations

$$\begin{aligned} x + 3y + 3z &= 4, \\ x + 4y + 3z &= 3, \\ x + 3y + 4z &= 6. \end{aligned}$$

4. Given

$$A = \begin{bmatrix} 1 & 1 & 0 \\ 0 & 1 & 1 \\ 1 & 0 & 1 \end{bmatrix},$$

find A^{-1} and hence solve for x, y, z the equations

$$x + y = a,\ y + z = b,\ z + x = c.$$

5. Express the equations

$$\begin{aligned} x_1 + x_2 + \ \ x_3 + \ \ x_4 &= a, \\ x_2 + \ \ x_3 + \ \ x_4 &= b, \\ x_3 + 2x_4 &= c, \\ 2x_3 + \ \ x_4 &= d, \end{aligned}$$

in matrix form and hence, by finding an inverse matrix, express x_1, x_2, x_3, x_4 in terms of a, b, c, d.

6. (a) Use matrix methods to solve the equations

$$\begin{aligned} 3x - \ \ y - 2z &= \ \ 2, \\ 2y - \ \ z &= -1, \\ 6x - 5y \qquad &= \ \ 3. \end{aligned}$$

(b) Given that A and B are matrices such that $AB = O$ prove that if B is non-singular then $A = O$. If neither A nor B is a zero matrix, what conclusions can you draw?

7. Given

$$A = \begin{bmatrix} 1 & 1 & 1 \\ 0 & 1 & 1 \\ 0 & 0 & 1 \end{bmatrix},$$

show by induction, or otherwise, that

$$A^n = \begin{bmatrix} 1 & n & \frac{1}{2}n(n+1) \\ 0 & 1 & n \\ 0 & 0 & 1 \end{bmatrix}.$$

Find A^{-1} and $(A^n)^{-1}$, and verify that $(A^{-1})^2 = (A^2)^{-1}$.

CHAPTER 14

DIFFERENTIAL CALCULUS

14.1 Limits

The tangent to a curve at a point P is usually first introduced as a line that

(a) touches the curve at P, or
(b) has two coincident points of intersection with the curve at P, or
(c) is the limiting position of the chord PQ as Q approaches P.

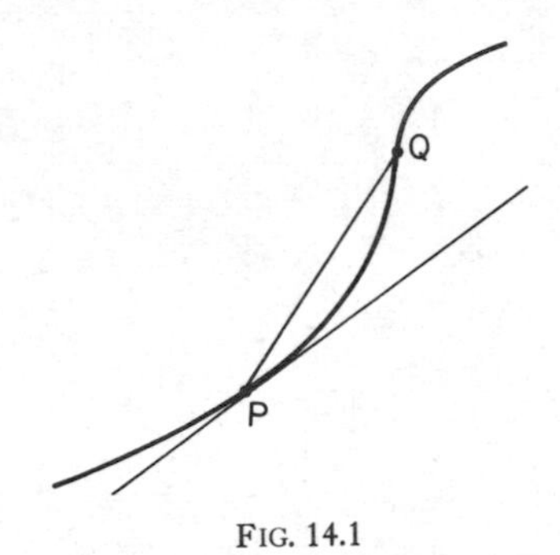

FIG. 14.1

In analytic geometry a point is represented by numerical co-ordinates and a curve by an equation, so the question arises—what is the equation of the tangent to the curve $y = f(x)$ at a point (x_0, y_0) on the curve? Although the above definitions enable us to visualise the tangent and to draw it fairly accurately on a diagram, they do not indicate very clearly how to obtain such precise information as its equation.

As an example, suppose we want the equation of the tangent to the curve $y = x^2$ at the point (1, 1). One method is to interpret (b) as meaning that the equation for the points of intersection must have equal roots—a method which could be used in this example, since the equation for the points of intersection is quadratic, but is not generally practicable. A more useful approach, suggested by (c), is as follows.

Let the co-ordinates of Q be (x, y) so that the slope of the line PQ is $(y - 1)/(x - 1)$; that is,

$$\text{slope of } PQ = \frac{x^2 - 1}{x - 1}$$

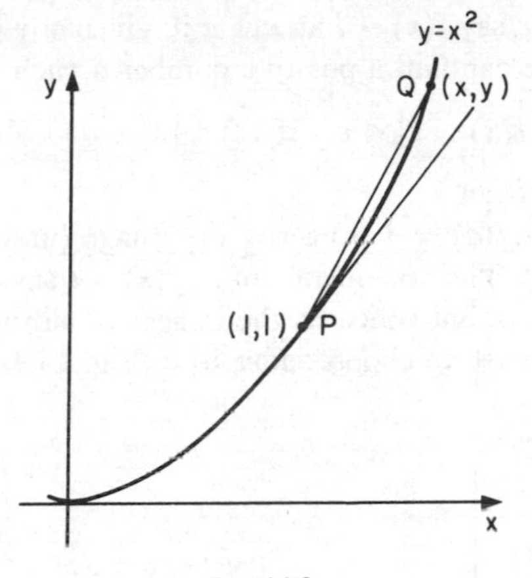

FIG. 14.2

since the co-ordinates of Q must satisfy the equation of the curve, namely $y = x^2$. If we give x some values that get closer and closer to 1 we find that the slope gets closer and closer to 2, e.g. for $x = 1{\cdot}1, 1{\cdot}01, 1{\cdot}001, 1{\cdot}0001, \ldots$ the slopes are $2{\cdot}1, 2{\cdot}01, 2{\cdot}001, 2{\cdot}0001, \ldots$. In fact if we nominate any slope, however close to 2 but different from 2, there is some chord that has this slope, e.g. for slope of $2{\cdot}0013$ take $x = 1{\cdot}0013$, for slope of $1{\cdot}9997$ take $x = 0{\cdot}9997$, and so on, so we are led to conclude that the slope of the tangent is 2. It follows that the equation of the tangent is $y - 1 = 2(x - 1)$. We regard 2 as a 'limiting value' of the formula $(x^2 - 1)/(x - 1)$.

Thus the idea of 'limiting position' has led to the idea of 'limiting value'. Statements about numbers are usually fairly unambiguous, so with hindsight we now see that more satisfactory definitions are possible if we begin by defining 'limiting value'. From this new starting point we can define the tangent to a curve and, as we shall see later, many other important concepts, e.g. area bounded by a curve.

The abbreviation 'limit' for 'limiting value' will be used from now on, and the symbol $\rightarrow$ will be used for the word 'approaches'.

Definition 1 We say $f(x) \rightarrow l$ as $x \rightarrow a$, or $f(x)$ has limit l as $x \rightarrow a$ (usually written $\lim_{x \rightarrow a} f(x) = l$), if the values of $f(x)$ can be made to differ from l by as little as we please by choosing x sufficiently close to a.

This definition is not entirely free from ambiguity and a better definition is given below for the sake of completeness. A full comprehension of this more difficult definition is not necessary in what follows.

Definition 1a We say $f(x) \to l$ as $x \to a$ if, given any positive number ε, however small, we can find a positive number δ such that

$$|f(x) - l| < \varepsilon \quad \text{if} \quad 0 < |x - a| < \delta.$$

Notes on the definition

(1) The statement $f(a) = l$ concerns the image (under f) of the single point a (Fig. 14.3). The statement $\lim_{x \to a} f(x) = l$ says *nothing* about the image of the point a, but concerns the images of all the points (infinitely many) one might wish to choose close to a (Fig. 14.4).

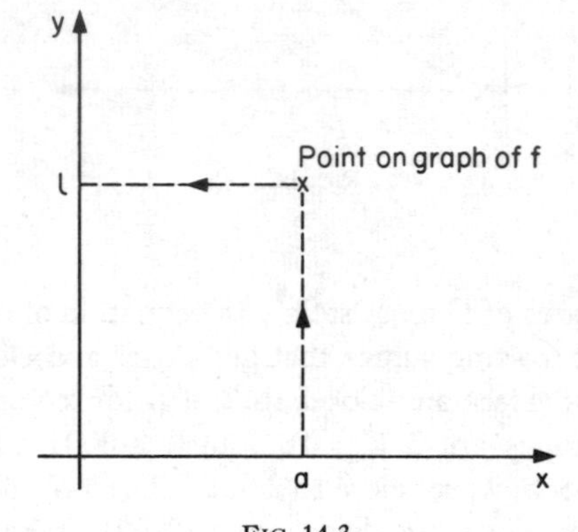

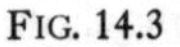
FIG. 14.3

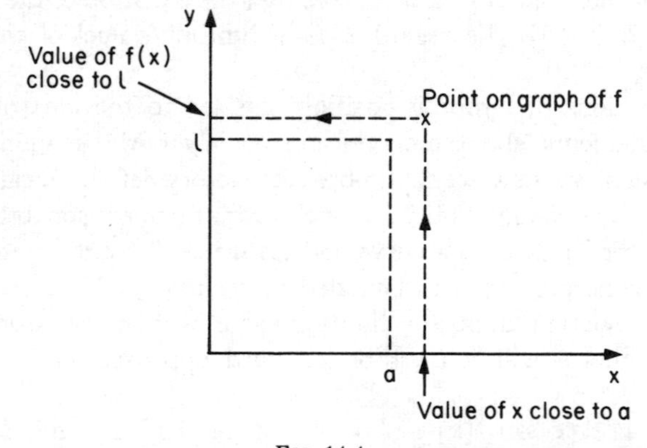

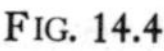
FIG. 14.4

If $f(x) = x^2$ then *both* the statements

$$f(2) = 4, \quad \lim_{x \to 2} f(x) = 4$$

are true (Fig. 14.5) and the distinction between them is not important.

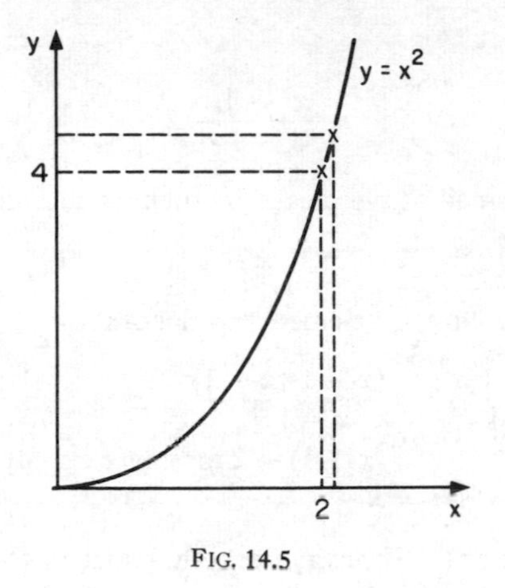

FIG. 14.5

If $f(x) = (x^2 - 1)/(x - 1)$, then as we have already seen

$$\lim_{x \to 1} f(x) = 2,$$

whereas $f(1)$ is undefined (when $x = 1$ the formula gives the meaningless result 0/0) so the distinction between the 'value at $x = 1$' and the 'limit as $x \to 1$' is very important.

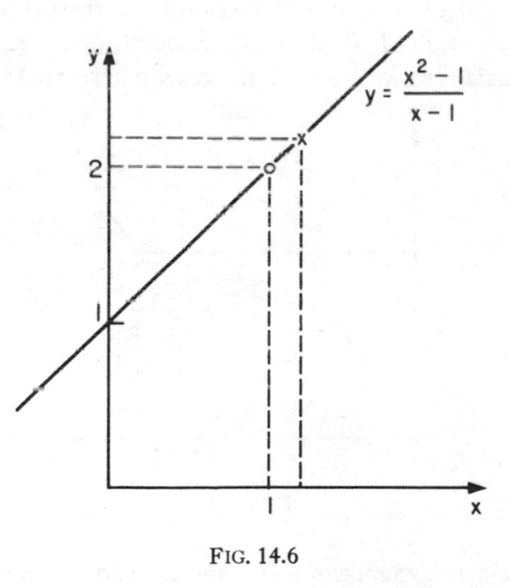

FIG. 14.6

(2) The definition is not a basis for *finding* limits but rather a criterion for determining whether or not a suspected limit is correct. For example to show that $\lim_{x \to 1} (x^2 - 1)/(x - 1) = 2$ we must show that the

difference

$$\frac{x^2-1}{x-1}-2 \tag{1}$$

can be made as small as we please by making the difference

$$x-1 \tag{2}$$

sufficiently small. Now by elementary elgebra

$$\begin{aligned}\frac{x^2-1}{x-1}-2 &= \frac{(x-1)(x+1)}{x-1}-2\\ &= (x+1)-2 \text{ for any } x \text{ except } x=1,\\ &= x-1.\end{aligned}$$

Thus the difference (1) is always exactly equal to the difference (2) and the conclusion is now obvious.

Fortunately as we familiarise ourselves with the most important limit situations we can avoid the tedious business of repeatedly testing limits in terms of the definition.

14.1.1 Methods for Finding Limits For most of the functions we have encountered it would be true to say that if P is the point on the graph corresponding to $x = a$ then all points corresponding to values of x close to a are close to P (Fig. 14.7). From this we might expect that $\lim_{x\to a} f(x)$ equals $f(a)$, at least in all cases where $f(a)$ is defined.

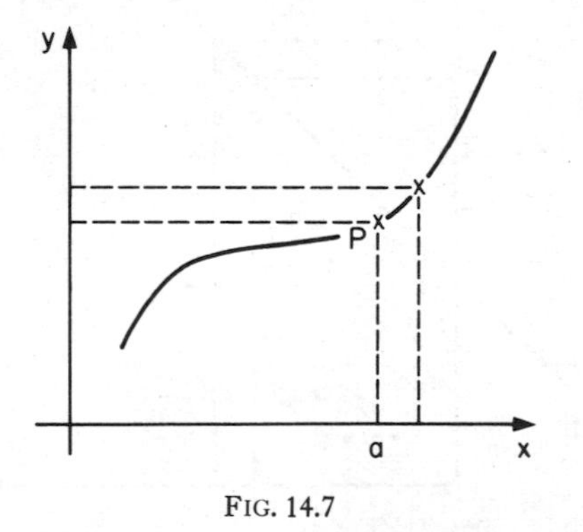

FIG. 14.7

However there are exceptions to this. Consider the function r, where $r(x)$ denotes the value of x rounded off to the nearest integer—this function was introduced in Section 12.6.4 and its graph is shown there (Fig. 12.24). Since half-integers are rounded off to the nearest *even* integer, we have, for example, $r(2{\cdot}5) = 2$; but from the graph we see that if x is

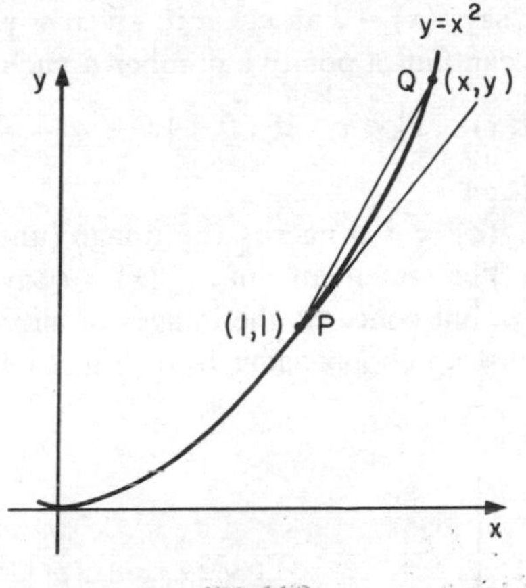

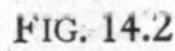
FIG. 14.2

since the co-ordinates of Q must satisfy the equation of the curve, namely $y = x^2$. If we give x some values that get closer and closer to 1 we find that the slope gets closer and closer to 2, e.g. for $x = 1{\cdot}1$, $1{\cdot}01$, $1{\cdot}001$, $1{\cdot}0001$, ... the slopes are $2{\cdot}1$, $2{\cdot}01$, $2{\cdot}001$, $2{\cdot}0001$, ... In fact if we nominate any slope, however close to 2 but different from 2, there is some chord that has this slope, e.g. for slope of $2{\cdot}0013$ take $x = 1{\cdot}0013$, for slope of $1{\cdot}9997$ take $x = 0{\cdot}9997$, and so on, so we are led to conclude that the slope of the tangent is 2. It follows that the equation of the tangent is $y - 1 = 2(x - 1)$. We regard 2 as a 'limiting value' of the formula $(x^2 - 1)/(x - 1)$.

Thus the idea of 'limiting position' has led to the idea of 'limiting value'. Statements about numbers are usually fairly unambiguous, so with hindsight we now see that more satisfactory definitions are possible if we begin by defining 'limiting value'. From this new starting point we can define the tangent to a curve and, as we shall see later, many other important concepts, e.g. area bounded by a curve.

The abbreviation 'limit' for 'limiting value' will be used from now on, and the symbol $\to$ will be used for the word 'approaches'.

Definition 1 We say $f(x) \to l$ as $x \to a$, or $f(x)$ has limit l as $x \to a$ (usually written $\lim_{x \to a} f(x) = l$), if the values of $f(x)$ can be made to differ from l by as little as we please by choosing x sufficiently close to a.

This definition is not entirely free from ambiguity and a better definition is given below for the sake of completeness. A full comprehension of this more difficult definition is not necessary in what follows.

Definition 1a We say $f(x) \to l$ as $x \to a$ if, given any positive number ε, however small, we can find a positive number δ such that

$$|f(x) - l| < \varepsilon \quad \text{if} \quad 0 < |x - a| < \delta.$$

Notes on the definition

(1) The statement $f(a) = l$ concerns the image (under f) of the single point a (Fig. 14.3). The statement $\lim_{x \to a} f(x) = l$ says *nothing* about the image of the point a, but concerns the images of all the points (infinitely many) one might wish to choose close to a (Fig. 14.4).

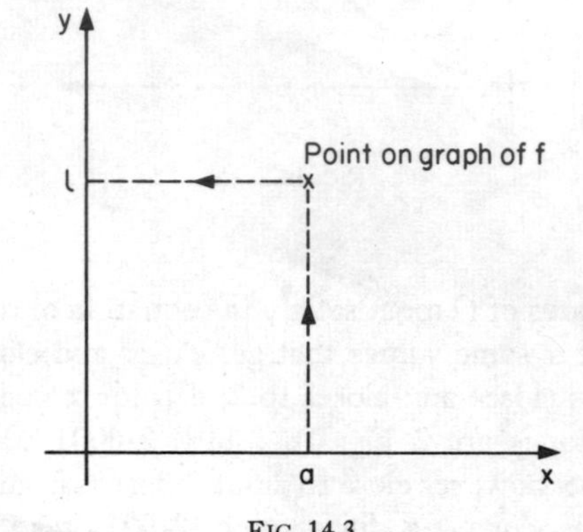

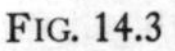
FIG. 14.3

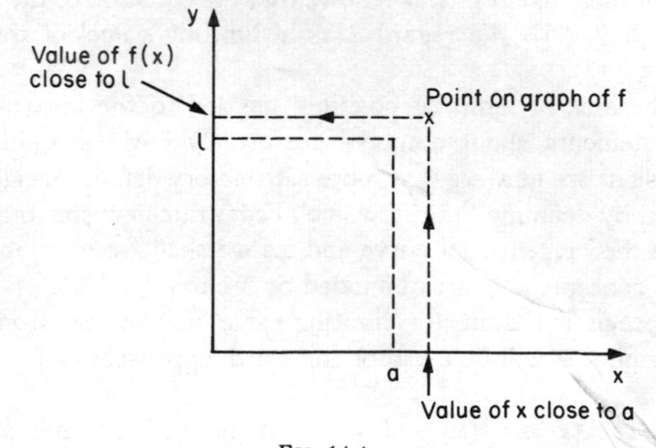

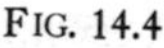
FIG. 14.4

If $f(x) = x^2$ then *both* the statements

$$f(2) = 4, \quad \lim_{x \to 2} f(x) = 4$$

are true (Fig. 14.5) and the distinction between them is not important.

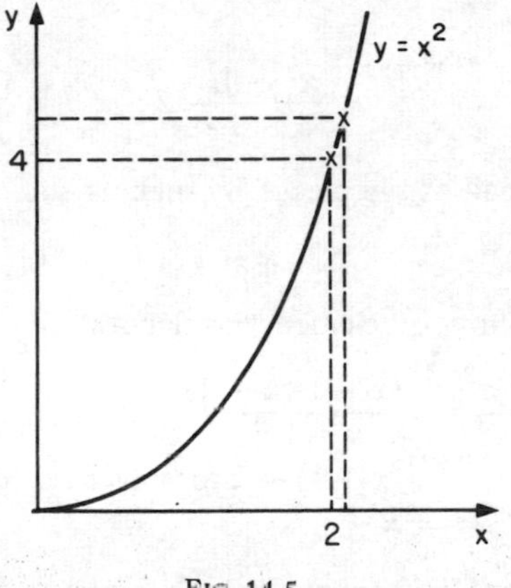

FIG. 14.5

If $f(x) = (x^2 - 1)/(x - 1)$, then as we have already seen

$$\lim_{x \to 1} f(x) = 2,$$

whereas $f(1)$ is undefined (when $x = 1$ the formula gives the meaningless result 0/0) so the distinction between the 'value at $x = 1$' and the 'limit as $x \to 1$' is very important.

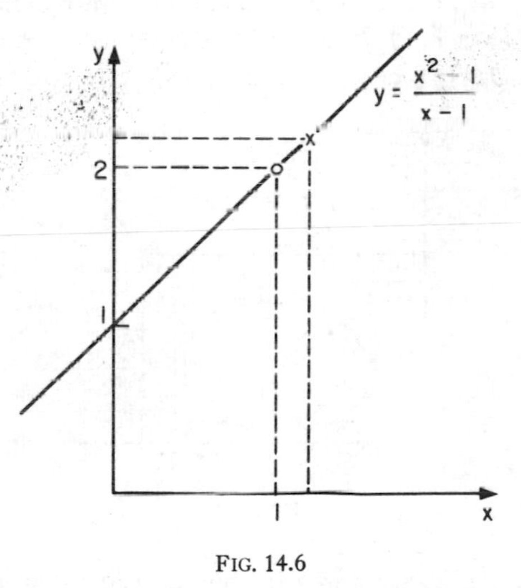

FIG. 14.6

(2) The definition is not a basis for *finding* limits but rather a criterion for determining whether or not a suspected limit is correct. For example to show that $\lim_{x \to 1} (x^2 - 1)/(x - 1) = 2$ we must show that the

difference

$$\frac{x^2 - 1}{x - 1} - 2 \qquad (1)$$

can be made as small as we please by making the difference

$$x - 1 \qquad (2)$$

sufficiently small. Now by elementary elgebra

$$\begin{aligned} \frac{x^2 - 1}{x - 1} - 2 &= \frac{(x - 1)(x + 1)}{x - 1} - 2 \\ &= (x + 1) - 2 \text{ for any } x \text{ except } x = 1, \\ &= x - 1. \end{aligned}$$

Thus the difference (1) is always exactly equal to the difference (2) and the conclusion is now obvious.

Fortunately as we familiarise ourselves with the most important limit situations we can avoid the tedious business of repeatedly testing limits in terms of the definition.

14.1.1 Methods for Finding Limits For most of the functions we have encountered it would be true to say that if P is the point on the graph corresponding to $x = a$ then all points corresponding to values of x close to a are close to P (Fig. 14.7). From this we might expect that $\lim_{x \to a} f(x)$ equals $f(a)$, at least in all cases where $f(a)$ is defined.

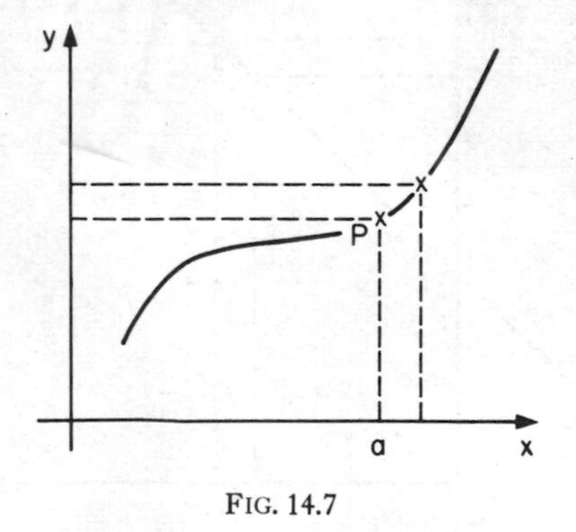

FIG. 14.7

However there are exceptions to this. Consider the function r, where $r(x)$ denotes the value of x rounded off to the nearest integer—this function was introduced in Section 12.6.4 and its graph is shown there (Fig. 12.24). Since half-integers are rounded off to the nearest *even* integer, we have, for example, $r(2 \cdot 5) = 2$; but from the graph we see that if x is taken close to $2\cdot 5$, this does not guarantee that $r(x)$ is close to 2, e.g. $r(2\cdot 51) = 3$. Hence $\lim_{x\to 2\cdot 5} r(x) \neq 2$, and likewise $\lim_{x\to 2\cdot 5} r(x) \neq 3$. Since no number can be found to satisfy the requirements for being a limit, we say that $\lim_{x\to 2\cdot 5} r(x)$ does not exist.

Rule 1. If there is a formula for $f(x)$ in terms of elementary functions (see Section 12.6.4) and $f(a)$ is defined, then $\lim_{x\to a} f(x) = f(a)$.

For example

$$\lim_{x\to 3} \frac{x+4}{2x+1} = 1, \quad \lim_{x\to \pi/4} \sin x = 1/\sqrt{2}, \quad \lim_{x\to 1} \tan^{-1} x = \pi/4,$$

where the limits may be found by simply giving x the values 3, $\pi/4$, 1 respectively.

Note. There is no formula for $r(x)$ in terms of elementary functions.

Rule 2. If

$$f(x) = \frac{(x-a)^a\, g(x)}{(x-a)^a\, h(x)},$$

then $f(a)$ is undefined, but

$$f(x) = \frac{g(x)}{h(x)} \qquad \text{for all } x \neq a$$

and so

$$\lim_{x\to a} f(x) = \lim_{x\to a} \frac{g(x)}{h(x)}.$$

(Remember that $\lim_{x\to a} f(x)$ is concerned with values of x close to a but *different* from a.)

As an example of the application of this rule consider again

$$\lim_{x\to 1} \frac{x^2-1}{x-1}.$$

Now

$$\frac{x^2-1}{x-1} = \frac{(x-1)(x-1)}{x-1}$$
$$= x+1 \qquad \text{for all } x \neq 1.$$

Hence

$$\lim_{x\to 1} \frac{x^2-1}{x-1} = \lim_{x\to 1} (x+1) = 2.$$

Example 1 Find

$$\lim_{x \to 3} \frac{x^2 - 9}{x^2 - x - 6}.$$

The given expression is undefined when $x = 3$. Factorising, we get

$$\frac{x^2 - 9}{x^2 - x - 6} = \frac{(x - 3)(x + 3)}{(x - 3)(x + 2)}$$
$$= \frac{x + 3}{x + 2} \qquad \text{for all } x \neq 3.$$

Hence

$$\lim_{x \to 3} \frac{x^2 - 9}{x^2 - x - 6} = \lim_{x \to 3} \frac{x + 3}{x + 2} = \frac{6}{5}.$$

Example 2 Find

$$\lim_{x \to 0} \frac{1 + 1/x}{1 - 1/x}.$$

The given expression is undefined when $x = 0$. We may write

$$\frac{1 + 1/x}{1 - 1/x} = \frac{x(x + 1)}{x(x - 1)}$$
$$= \frac{x + 1}{x - 1} \qquad \text{for all } x \neq 0.$$

Hence

$$\lim_{x \to 0} \frac{1 + 1/x}{1 - 1/x} = \lim_{x \to 0} \frac{x + 1}{x - 1} = -1.$$

Example 3 Find

$$\lim_{x \to 0} \frac{1 - \sqrt{(1 - x)}}{x}.$$

The given expression is undefined when $x = 0$. To put it into a form where Rule 2 can be applied we first rationalise the numerator. Thus

$$\frac{1 - \sqrt{(1 - x)}}{x} = \frac{[1 - \sqrt{(1 - x)}][1 + \sqrt{(1 - x)}]}{x[1 + \sqrt{(1 - x)}]}$$
$$= \frac{x}{x[1 + \sqrt{(1 - x)}]}$$
$$= \frac{1}{1 + \sqrt{(1 - x)}} \qquad \text{for all } x \neq 0.$$

Hence

$$\lim_{x \to 0} \frac{1-\sqrt{(1-x)}}{x} = \frac{1}{2}.$$

Example 4 Find

$$\lim_{x \to 2} \frac{x^3 - 6x^2 + 12x - 8}{x^3 - 3x^2 + 4}.$$

The given expression is undefined when $x = 2$. Factorising, we get

$$\frac{x^3 - 6x^2 + 12x - 8}{x^3 - 3x^2 + 4} = \frac{(x-2)^3}{(x-2)^2(x+1)}$$

$$= \frac{x-2}{x+1} \quad \text{for all } x \neq 2.$$

Hence

$$\lim_{x \to 2} \frac{x^3 - 6x^2 + 12x - 8}{x^3 - 3x^2 + 4} = 0.$$

The above technique may not be suitable if the formula for $f(x)$ is non-algebraic. For example, the important result

$$\lim_{x \to 0} \frac{\sin x}{x} = 1 \tag{14.1}$$

cannot be derived in this way, or at least not until we have learned how to express $\sin x$ as an infinite series. We prove it here by an alternative method. Figure 14.8 (unit circle, angle of x radians at the centre) shows that

$$\frac{\sin x}{x} = \frac{\text{area of triangle } OXP}{\text{area of sector } OXP}$$

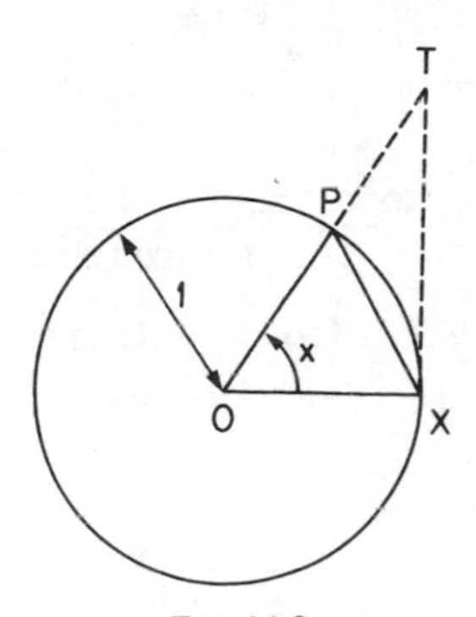

FIG. 14.8

and we would guess that this tends to 1 as $x \to 0$. If XT is the tangent to the circle at X, then comparing the areas of the triangles OXP, OXT and the sector OXP we see that

$$\tfrac{1}{2}\sin x < \tfrac{1}{2}x < \tfrac{1}{2}\tan x,$$

that is

$$1 < \frac{x}{\sin x} < \frac{1}{\cos x} \quad \text{or} \quad 1 > \frac{\sin x}{x} > \cos x.$$

Hence the difference between 1 and $(\sin x)/x$ is always less than the difference between 1 and $\cos x$. Since $\cos x \to 1$ as $x \to 0$ it follows that $(\sin x)/x \to 1$ as $x \to 0$.

Theorems about limits If $\lim_{x \to a} f_1(x) = l_1$ and $\lim_{x \to a} f_2(x) = l_2$, the following theorems are true.

Theorem 1 $\lim_{x \to a} \{f_1(x) \pm f_2(x)\} = l_1 \pm l_2$.

Theorem 2 $\lim_{x \to a} f_1(x) f_2(x) = l_1 l_2$.

Theorem 3 $\lim_{x \to a} f_1(x)/f_2(x) = l_1/l_2$ provided that $l_2 \neq 0$.

As a special case of Theorem 2 we see that if $\lim_{x \to a} f(x) = l$ and k is a constant, then $\lim_{x \to a} kf(x) = kl$.

Although these theorems may appear to be a statement of the obvious, nevertheless formal proofs are rather difficult and are not given here. The following examples show some applications of these theorems in finding limits.

Example 5 Find

$$\lim_{x \to 0} \frac{\tan x}{x}.$$

Write

$$\frac{\tan x}{x} = \frac{\sin x}{x} \cdot \frac{1}{\cos x}.$$

We know that $(\sin x)/x \to 1$, $1/(\cos x) \to 1$, as $x \to 0$. Hence by Theorem 2, $\lim_{x \to 0} (\tan x)/x = 1$.

Example 6 Find

$$\lim_{x \to 0} \frac{x}{\sin 2x - \sin x}.$$

Write

$$\frac{x}{\sin 2x - \sin x} = \frac{x}{\sin x} \frac{1}{(2 \cos x - 1)}.$$

Since $(\sin x)/x \to 1$, it follows from Theorem 3 that $x/(\sin x) \to 1$ as $x \to 0$. Since $1/(2 \cos x - 1) \to 1$ as $x \to 0$ it now follows from Theorem 2 that the required limit is 1.

Example 7 Find

$$\lim_{x \to 0} \frac{1 - \cos x}{x}.$$

Write

$$\frac{1 - \cos x}{x} = \frac{2 \sin^2 (x/2)}{x} = \frac{\sin (x/2)}{x/2} \sin x/2.$$

If $t = x/2$ then $t \to 0$ as $x \to 0$, and

$$\frac{\sin (x/2)}{x/2} = \frac{\sin t}{t} \to 1.$$

Since $\sin (x/2) \to 0$ as $x \to 0$ it follows from Theorem 2 that the required limit is 0.

14.1.2 One-sided Limits Sometimes $f(x)$ is given not by one but by several formulae, e.g. shearing force S and bending-moment M, described in Section 12.6.4, graphs of which are shown in Figs. 14.9, 14.10. The notion of *left-hand* and *right-hand limits* is useful in such cases. The definition of a left-hand limit is identical with that of a limit except that we interpret 'x close to a' as meaning 'x close to a but less than a' and we say that x tends to *a from the left*, writing $x \to a-$; thus the left-hand limit is indicated by $\lim_{x \to a-} f(x)$. Similarly the right-hand limit is indicated by $\lim_{x \to a+} f(x)$. It follows that $\lim_{x \to a} f(x) = l$ if, and only if,

$$\lim_{x \to a-} f(x) = \lim_{x \to a+} f(x) = l.$$

From Fig. 14.9 we see that

$$\lim_{x \to 4-} S = -4, \quad \lim_{x \to 4+} S = 2,$$

and since these two limits are unequal it follows that $\lim_{x \to 4} S$ does not exist. Likewise

$$\lim_{x \to -4-} S = -2, \quad \lim_{x \to -4+} S = 4,$$

and $\lim_{x \to -4} S$ does not exist.

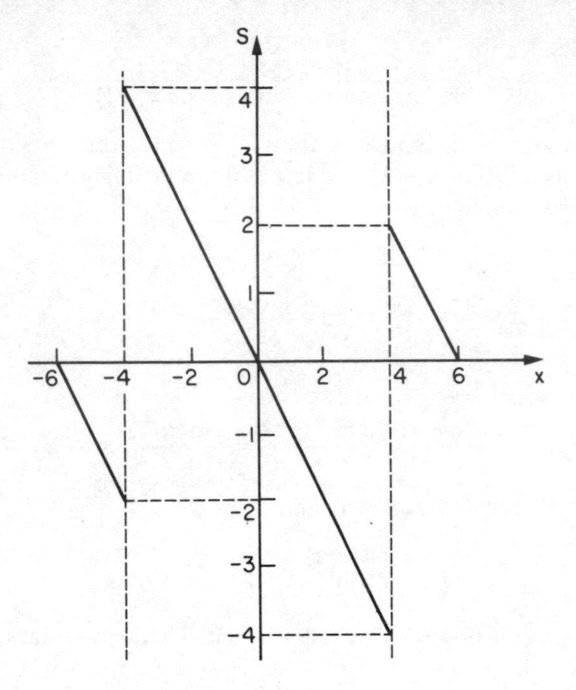

FIG. 14.9

From Fig. 14.10 we see that

$$\lim_{x \to 4-} M = \lim_{x \to 4+} M = 2$$

and so $\lim_{x \to 4} M = 2$; likewise

$$\lim_{x \to -4-} M = \lim_{x \to -4+} M = 2$$

and so $\lim_{x \to -4} M = 2$.

Example 8 Discuss $\lim_{x \to 0} f(x)$, where

$$f(x) = \begin{cases} x^2 + 1 & \text{if } x < 0, \\ x^2 - 1 & \text{if } x > 0. \end{cases}$$

We have

$$\lim_{x \to 0-} f(x) = \lim_{x \to 0} (x^2 + 1) = 1,$$

and

$$\lim_{x \to 0+} f(x) = \lim_{x \to 0} (x^2 - 1) = -1.$$

Hence $\lim_{x \to 0} f(x)$ does not exist.

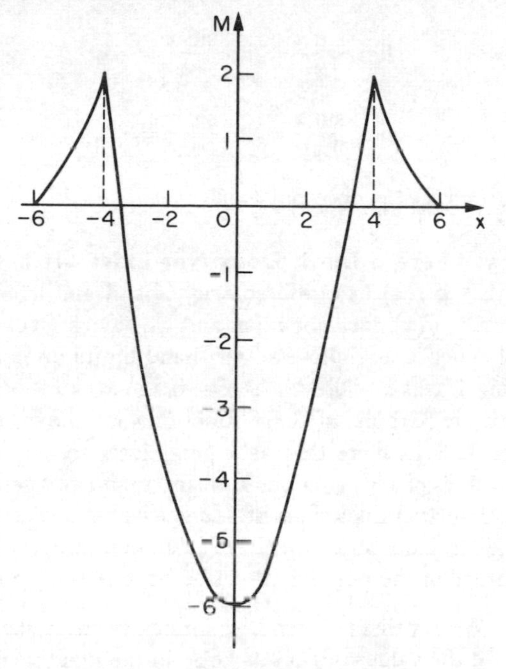

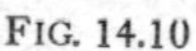
FIG. 14.10

Example 9 Discuss $\lim_{x\to 0} f(x)$, where

$$f(x)=\begin{cases} x+1 & \text{if } x<0,\\ 0 & \text{if } x=0,\\ \dfrac{\sin x}{x} & \text{if } x>0. \end{cases}$$

We have

$$\lim_{x\to 0-} f(x) = \lim_{x\to 0}(x+1) = 1,$$

$$\lim_{x\to 0+} f(x) = \lim_{x\to 0}\frac{\sin x}{x} = 1.$$

Hence $\lim_{x\to 0} f(x) = 1$. On the other hand $f(0) = 0$ by prescription, which emphasises the fact that it is the values of $f(x)$ for x close to 0 but *different* from 0 that determine the *limit*.

Example 10 Discuss

$$\lim_{x\to 0}\frac{\sin x}{|x|}.$$

We have

$$\lim_{x\to 0-}\frac{\sin x}{|x|}=\lim_{x\to 0}\frac{\sin x}{-x}=-1,$$

$$\lim_{x\to 0+}\frac{\sin x}{|x|}=\lim_{x\to 0}\frac{\sin x}{x}=\quad 1.$$

Hence $\lim_{x\to 0}(\sin x)/|x|$ does not exist.

14.1.3 Cases Where a Limit Does Not Exist It has already been pointed out that if $f(x)$ has different right-hand and left-hand limits at $x = a$, then $\lim_{x\to a} f(x)$ does not exist, and we have seen examples of this. It may also happen that right- and left-hand limits do not exist, typical examples being $1/x$ as $x \to 0$, $1/x^2$ as $x \to 0$, $\tan x$ as $x \to n(\pi/2)$ where n is an odd integer. (Graphs of these functions are shown in Figs. 12.22, 12.23, Section 12.6.1.) Note that as x gets closer to zero from the right ($x \to 0+$) the values of $1/x$ get larger and larger, and as x gets closer to zero from the left ($x \to 0-$) the values of $1/x$ get larger in the negative direction. Similarly for $\tan x$ as $x \to n(\pi/2)$ (n an odd integer), but the values of $1/x^2$ get larger in the positive direction as $x \to 0$ from *either* side.

Definition 2 We say that $f(x)$ tends to infinity (written $f(x) \to \infty$) as $x \to a$ if we can make the values of $f(x)$ as large in the positive direction as we wish by choosing x sufficiently close to a.

Definition 2a We say that $f(x) \to \infty$ as $x \to a$ if, given a positive number N, however large, we can find a positive number δ such that $f(x) > N$ if $0 < |x - a| < \delta$.

The reader can easily construct similar definitions to cover the cases $f(x) \to -\infty$ as $x \to a$, $f(x) \to \infty$ as $x \to a-$, $f(x) \to \infty$ as $x \to a+$, $f(x) \to -\infty$ as $x \to a-$, $f(x) \to -\infty$ as $x \to a+$.

Thus $1/x \to -\infty$ as $x \to 0-$, $1/x \to \infty$ as $x \to 0+$, $1/x^2 \to \infty$ as $x \to 0$, $\tan x \to \infty$ as $x \to \pi/2-$, $\tan x \to -\infty$ as $x \to \pi/2+$.

Rule 3. If $f(x) = g(x)/h(x)$, where $g(x) \to l \neq 0$ and $h(x) \to 0$ as $x \to a$, then $|f(x)| \to \infty$ as $x \to a$, and if, moreover, $l/h(x)$ is positive (negative) for all values of x close to a then $f(x) \to \infty(-\infty)$; if $g(x)$ and $h(x)$ both $\to 0$ as $x \to a$ no general conclusion can be drawn.

The following three examples concerning limits as $x \to 0$ illustrate the situation where $g(x)$ and $h(x)$ both $\to 0$, and the reader should note that a different conclusion is reached in each case:

(a)
$$\frac{x^2}{x^3+5x^2}=\frac{1}{x+5} \quad \text{for all } x \neq 0$$
$$\to \tfrac{1}{5} \text{ as } x \to 0.$$

(b) $$\frac{x}{x^3+5x^2}=\frac{1}{x^2+5x} \quad \text{for all } x \neq 0$$
$$\to -\infty \text{ as } x \to 0-, \quad \to \infty \text{ as } x \to 0+.$$

(c) $$\frac{x^2}{\sin x}=\frac{x}{(\sin x)/x} \quad \text{for all } x \neq 0$$
$$\to 0 \text{ as } x \to 0.$$

14.1.4 An Important Limit We shall now prove that

$$\lim_{x \to a}\frac{x^n-a^n}{x-a}=na^{n-1} \qquad (n \text{ any rational number}) \qquad (14.2)$$

Suppose $a \neq 0$. Then

(a) if n is a positive integer

$$\frac{x^n-a^n}{x-a}=x^{n-1}+x^{n-2}a+x^{n-3}a^2+\cdots$$
$$+xa^{n-2}+a^{n-1} \text{ for all } x \neq a$$

and the result follows immediately.

(b) if n is a negative integer, say $n=-m$ where m is a positive integer,

$$\frac{x^n-a^n}{x-a}=\frac{-1}{x^m a^m}\frac{x^m-a^m}{x-a}$$
$$\to -\frac{1}{a^{2m}}ma^{m-1} \text{ as } x \to a \qquad \text{using (a);}$$

that is

$$\lim_{x \to a}\frac{x^n-a^n}{x-a}=-ma^{-m-1}=na^{n-1}.$$

(c) if n is fractional, say $n=p/q$ where p, q are both integers, write $y=x^{1/q}$, $b=a^{1/q}$ so that

$$\frac{x^n-a^n}{x-a}=\frac{y^p-b^p}{y^q-b^q}$$
$$=\frac{y^p-b^p}{y-b}\,\frac{y-b}{y^q-b^q} \qquad \text{for all } y \neq b$$
$$\to pb^{p-1}\frac{1}{qb^{q-1}} \qquad \text{as } y \to b \quad (\text{i.e. as } x \to a);$$

that is

$$\lim_{x \to a}\frac{x^n-a^n}{x-a}=\frac{p}{q}b^{p-q}=\frac{p}{q}a^{(p/q)-1}=na^{n-1}.$$

Note that if $a = 0$ the result is true only if $n \geqslant 1$; for if $a = 0$ then

$$\frac{x^n - a^n}{x - a} = x^{n-1}$$

$$\to 0 \quad \text{as } x \to 0 \text{ if } n - 1 > 0.$$

$$\to \infty \text{ as } x \to 0+ \text{ if } n - 1 < 0.$$

Exercises 14(a)

Evaluate the following limits:

1. $\lim_{x \to 0} \frac{2x^2 + 3x^3 + x^7}{3x^2 + 4x^5 + x^6}$

2. $\lim_{x \to -1} \frac{x + 1}{x^2 - 1}$

3. $\lim_{x \to 2} \frac{x^2 - 4}{x^2 + x - 6}$

4. $\lim_{x \to 1} \frac{x^2 - 4x + 3}{x^2 - 3x + 2}$

5. $\lim_{x \to a} \frac{x^2 - a^2}{x^2 + 2ax - 3a^2}$

6. $\lim_{x \to 0} \frac{a - \sqrt{(a^2 - x^2)}}{x^2}$

7. $\lim_{x \to 0} \frac{x^3 - 2x + 5}{2x^3 - 7}$

8. $\lim_{x \to 3} \frac{27 - x^3}{x^2 - 3x}$

9. $\lim_{x \to 2-} \frac{\sqrt{(4 - x^2)}}{\sqrt{(6 - 5x + x^2)}}$

10. $\lim_{x \to 0} \frac{\sqrt{(1 - x)} - \sqrt{(1 + x)}}{x}$ (rationalise numerator)

11. $\lim_{x \to 0} \frac{1 - \cos^2 x}{x^2}$

12. $\lim_{x \to 0} \frac{\sin 2x}{\sqrt{x}}$

13. $\lim_{x \to 0} \frac{\sin kx}{x}$ (put $t = kx$ and let $t \to 0$)

14. $\lim_{x \to 0} \frac{\sin kx}{\sin x}$

15. $\lim_{x \to \pi/2} (\sec x - \tan x)$ (put $x = \pi/2 - t$ and let $t \to 0$)

16. $\lim_{x \to 0} \frac{\sin^{-1} x}{x}$ (put $x = \sin t$ and let $t \to 0$)

17. $\lim_{x \to 2} \frac{x^5 - 32}{x - 2}$ (use the result (14.2))

18. Find the limit of

    (a) $\frac{\tan x}{|x|}$ as $x \to 0-$ and as $x \to 0+$

    (b) $\frac{x - 5}{|x - 5|}$ as $x \to 5-$ and as $x \to 5+$.

19. Discuss $\lim_{x\to 0} f(x)$ where

$$f(x) = \begin{cases} x - 1 & \text{if } x < 0, \\ 1 & \text{if } x = 0, \\ (\sin x)/x & \text{if } x > 0. \end{cases}$$

14.2 Continuity

A glance at the graphs in Figs. 12.22, 12.23 (Section 12.6.1) will show that in some cases (x^2, x^3, $\sin x$, $\cos x$) the graph is a single continuous curve and in other cases ($1/x$, $1/x^2$, $\tan x$) the graph consists of two or more unconnected curves. We say the graphs of $1/x$ and $1/x^2$ are broken, or *discontinuous*, at $x = 0$ and likewise the graph of $\tan x$ at $x = \pm\pi/2$, $\pm 3\pi/2, \ldots$. Similarly the graph of r, where $r(x)$ is the value of x rounded off to the nearest integer (see Fig. 12.24, Section 12.6.4) is discontinuous at $x = \pm\frac{1}{2}, \pm\frac{3}{2}, \ldots$.

The graphs of shearing force S and bending moment M shown in Figs. 14.9, 14.10 both consist of three distinct parts, but in the case of M these parts connect up whereas the graph of S has three unconnected parts; hence we say the graph of S is discontinuous at $x = -4$ and at $x = 4$, but the graph of M is continuous throughout its domain.

Unfortunately a quick glance at the graph is not always sufficient. For example, the graph of $(\sin x)/x$ is strictly speaking a curve with a point missing, viz. the point corresponding to $x = 0$, where $(\sin x)/x$ is undefined, and so this graph is regarded as discontinuous at $x = 0$. Consider next the function f where

$$f(x) = \begin{cases} (\sin x)/x & \text{if } x \neq 0 \\ \frac{1}{2} & \text{if } x = 0 \end{cases}.$$

The graph of f (see Fig. 14.11) consists of a curve with a point missing,

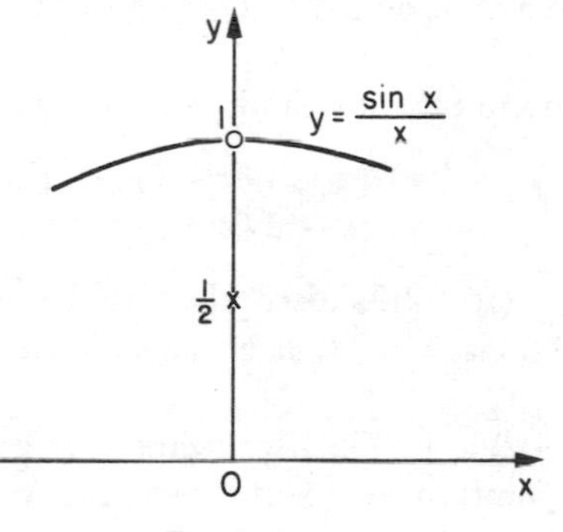

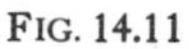
FIG. 14.11

together with the isolated point $(0, \frac{1}{2})$; this graph is also regarded as discontinuous at $x = 0$. If instead of $f(0) = \frac{1}{2}$ we have $f(0) = 1$ (the value of $\lim_{x \to 0} (\sin x)/x$), the missing point on the curve is now filled in and there is no isolated point; in this case the graph is regarded as continuous.

Having outlined in a fairly commonsense sort of way how the idea of continuity applies to the graph of a function, we now give a formal definition. At the same time we apply the idea of continuity to the function rather than to its graph, but the reader should continue to visualise the situation in terms of a graph.

Definition 3 We say the function f is *continuous* at the point a (or $f(x)$ is continuous at $x = a$) if $f(a)$ is defined and if, in addition, $\lim_{x \to a} f(x) = f(a)$; otherwise we say f is *discontinuous* at the point a.

We say f is *continuous on the interval I* if f is continuous at each point of I.

With elementary functions and combinations thereof, discontinuities occur at isolated points where the image is not defined (division by zero occurs) e.g.

$1/x^n$ $(n > 0)$ is discontinuous at $x = 0$,
$1/(x - a)^n$ $(n > 0)$ is discontinuous at $x = a$,
$1/(x - a)^n(x - b)^m$ $(n > 0, m > 0)$ is discontinuous at $x = a$ and at $x = b$,
$\tan x$ is discontinuous at $x = \pm\pi/2, \pm 3\pi/2, \pm 5\pi/2, \ldots$,
$x/\sin x$ is discontinuous at $x = 0, \pm\pi, \pm 2\pi, \pm 3\pi, \ldots$,
$(\sin x)/x$ is discontinuous at $x = 0$.

If $f(x)$ is given by several formulae, then discontinuities may occur at points of transition from one formula to another (e.g. the shearing force S, Fig. 14.9, is discontinuous at $x = -4$ and at $x = 4$), although this need not necessarily happen (e.g. the bending moment M, Fig. 14.10, is continuous at all points of its domain including the points $x = -4$ and $x = 4$).

Example 11 Discuss the continuity of the function f, where

$$f(x) = \begin{cases} 1/(x-3)(x+2) & \text{if } x < 0, \\ 1/(x-2)(x+3) & \text{if } x \geqslant 0. \end{cases}$$

Note that $1/(x - 3)(x + 2)$ is discontinuous at $x = 3$ and at $x = -2$ and one of these points, viz. $x = -2$, is in the region $x < 0$ where this formula applies.

Likewise $1/(x - 2)(x + 3)$ is discontinuous at $x = 2$ and at $x = -3$, the former of these points being in the region $x \geqslant 0$ where this formula applies.

At the transition point $x = 0$ we have

$$1/[(x-3)(x+2)] = 1/[(x-2)(x+3)] = -1/6.$$

Hence $\lim_{x\to 0} f(x) = -1/6 = f(0)$, and so f is continuous at $x = 0$.
Thus the only points of discontinuity are $x = -2$ and $x = 2$.

Theorem 4 If $f(x)$ and $g(x)$ are both continuous at $x = a$, then

$$f(x) \pm g(x),\ f(x)g(x),\ f(x)/g(x)$$

are continuous at $x = a$, the last mentioned only if $g(a) \neq 0$.

Proof. Since $f(a)$ and $g(a)$ are both defined, the above expressions are all defined when $x = a$ except for the last expression in the case where $g(a) = 0$.

Since

$$\lim_{x\to a} f(x) = f(a) \text{ and } \lim_{x\to a} g(x) = g(a),$$

we also have

$$\lim_{x\to a}[f(x) \pm g(x)] = f(a) \pm g(a),$$

$$\lim_{x\to a} f(x)g(x) = f(a)g(a)$$

$$\lim_{x\to a}\left[\frac{f(x)}{g(x)}\right] = \frac{f(a)}{g(a)} \quad \text{provided that } g(a) \neq 0,$$

from the corresponding theorems about limits. This proves the theorem.

Now it is easily seen that if $f(x) = \text{constant}$ or $f(x) = x$, then $f(x)$ is continuous for all x. It follows from the above theorem that any polynomial is continuous everywhere, and any rational function is continuous except at points where its denominator has the value zero.

14.2.1 A Fundamental Property of a Continuous Function It is frequently of interest to know if $f(x)$ has a *largest* value, or *smallest* value, for all values of x in some interval $a \leqslant x \leqslant b$. The idea of continuity plays an important role in this connection.

Consider values of x in the interval $-1 \leqslant x \leqslant 1$. Then if $f(x) = 1/x$, there is clearly no largest value of $f(x)$ since $f(x) \to \infty$ as $x \to 0+$. Furthermore if $f(x) = (\sin x)/x$ (see Fig. 14.11) again $f(x)$ has no largest value; for although all values of $f(x)$ are less than 1 there is no point where $f(x) = 1$, and any value less than 1 is certainly not the largest value. In both these examples $f(x)$ is discontinuous at $x = 0$.

Theorem 5 If $f(x)$ is continuous on the interval $a \leqslant x \leqslant b$, there is (at least) one point, x_1 say, in that interval where $f(x)$ has its largest value and (at least) one point, x_2 say, where $f(x)$ has its smallest value, i.e.

$$f(x_2) \leqslant f(x) \leqslant f(x_1)$$

for all x in the interval. Moreover, if k is any number between $f(x_1)$ and $f(x_2)$, there is (at least) one point, x_3 say, in the interval such that $f(x_3) = k$.

A formal proof of Theorem 5 is beyond the scope of this book but the graph in Fig. 14.12 shows how plausible the result is; a continuous curve has a highest point, at $x = x_1$ say, a lowest point, at $x = x_2$ say, and if the line $y = k$ lies between the highest and lowest points then somewhere, at $x = x_3$ say, the curve must cross the line. Of course such a line may be crossed several times by the curve, e.g. the line $y = k'$ in Fig. 14.12.

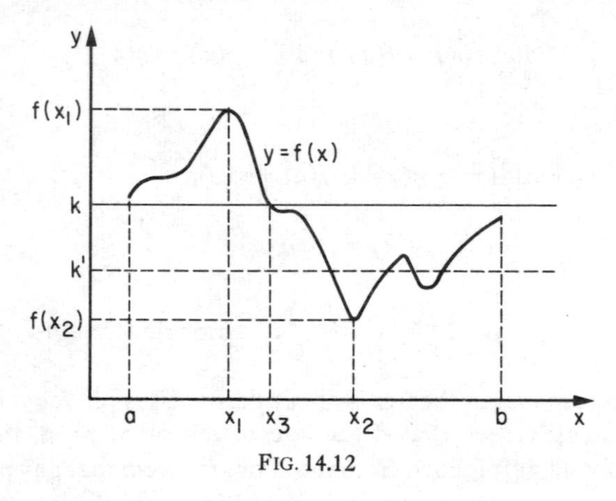

FIG. 14.12

It is important that the interval be closed. For example if $f(x) = x^2$, then on the closed interval $0 \leqslant x \leqslant 2$, $f(x)$ has a smallest value zero and a largest value 4, whereas on the open interval $0 < x < 2$, $f(x)$ has no smallest value and no largest value.

Exercises 14(b)

Find the values of x for which the following functions of x are discontinuous:

1. $\dfrac{x+4}{x-2}$
2. $\dfrac{x^4-1}{x^2}$
3. $\dfrac{x^2-4}{x^2-x-2}$
4. $\tan 2x$
5. $\tan (x + \pi/4)$
6. $\operatorname{cosec} 3x$
7. $\dfrac{\sin^2 x}{1-\cos x}$
8. $\dfrac{x-5}{|x-5|}$

9. $f(x)$ where $f(x) = \begin{cases} (\sin x)/x & \text{if } x < 0, \\ 1 & \text{if } x = 0, \\ x/\sin x & \text{if } x > 0. \end{cases}$

10. $f(x)$ where $f(x) = \begin{cases} (1/x)\tan x & \text{if } x < 0, \\ 1/(x \tan x) & \text{if } x > 0. \end{cases}$

11. Sketch the graph of the function f where $f(x)$ is the number of pence charged for a taxi ride of x km, the fare being a basic charge of 15p, plus 5p per $\frac{1}{4}$ km or part thereof. State the values of x for which f is discontinuous.

14.3 Rate of Change

The word 'rate' has its origins in the idea of 'ratio', or 'proportion'. Thus we refer to a death rate of 17 *per thousand*, a tax-rate of 30p *in the pound*, a flow rate of 25 litres *per second*, a rate of charge (for gas) of 10p *per therm*, and so on. The speed of a motorist, measured in (say) kilometres *per hour*, illustrates the commonest kind of rate which is called a rate *with respect to time*, or *time rate*; on the other hand a motorist may calculate his rate *with respect to fuel*, measuring the rate in (say) kilometres *per litre*.

Frequently the idea of rate is applied to the *change* in some quantity that may be increasing or decreasing, e.g. the rate of spread of an epidemic, the rate of increase in the population, the rate of decline in the purchasing power of the pound, etc. It is this idea of 'rate of change' that concerns us primarily in this chapter.

14.3.1 Uniform Rate and Variable Rate With a uniform charge of 10p per article a purchaser pays £1 for 10 articles, £2 for 20 articles, etc., so that in all cases the ratio

$$\frac{\text{total cost}}{\text{number of articles purchased}} \tag{1}$$

equals the rate of charge.

Likewise a motorist travelling at a uniform speed of 50 km/h travels 50 km in 1 h, 100 km in 2 h, etc., so that in all cases the ratio

$$\frac{\text{distance travelled}}{\text{time taken}} \tag{2}$$

equals the speed.

With a non-uniform charge, e.g. 12p per article for the first 10 articles, 8p per article for the next 50, and 6p per article thereafter, the ratio (1) is called the *average rate of charge*, just as the ratio (2) is called the *average speed* for a journey to which the idea of uniform speed may not apply.

In an accelerating (or decelerating) car the speedometer needle may move, without stopping, through the 50 km/h position. In that case there is one *instant* of time, say $t = t_0$, when the reading is exactly 50 km/h but there is no *period* of time, from t_0 to t_1 (say), however small the difference between t_0 and t_1, when the speed is uniform or the average speed equal to 50 km/h. The real significance of the number 50 in this case is that it represents the *limit* of this average speed (in km/h) as $t_1 \to t_0$; it is natural, however, to refer to 50 km/h as the speed *at the instant* t_0. This idea of instantaneous rate applies in a similar way to any time rate, and the idea can easily be extended to other rates also, as discussed below.

Figures 14.13, 14.14 show cross-sections of two buildings, one with a traditional slant roof and one with a curved roof. At the point x_0 on the floor the headroom is denoted by h_0, and at x_1 the headroom is denoted by h_1, so that the gain in headroom due to a move from x_0 to x_1 is $(h_1 - h_0)$, i.e. an average rate of gain of $(h_1 - h_0)/(x_1 - x_0)$ with respect to distance moved. In Fig. 14.13 *any* position x_1 in the left half of the diagram gives the same value for this average rate—the word *average* is thus unnecessary since the rate is uniform. In Fig. 14.14 different positions x_1 give different average rates and we take the *limit* of the average rate as $x_1 \to x_0$ to be the rate of gain of headroom *at the point* x_0.

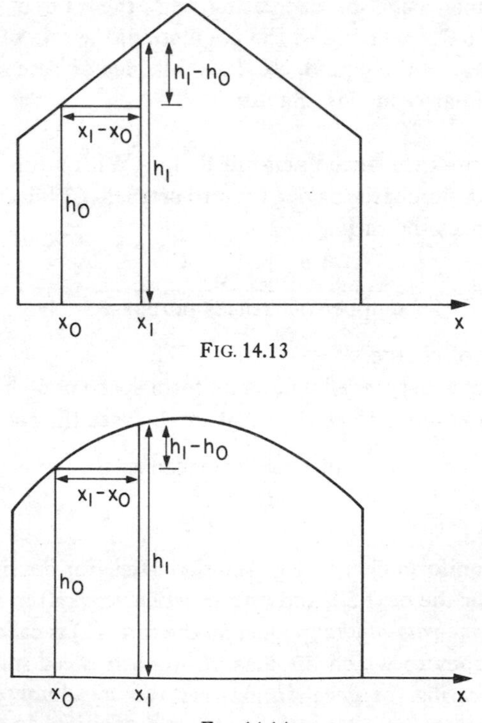

FIG. 14.13

FIG. 14.14

We may summarise as follows, using the abbreviation 'w.r.t.' for 'with respect to': if x changes from the value x_0 to the value x_1, then the change in $f(x)$ is $f(x_1) - f(x_0)$, the average rate of change of $f(x)$ w.r.t. x is $[f(x_1) - f(x_0)]/(x_1 - x_0)$, and the rate of change of $f(x)$ w.r.t. x at the point $x = x_0$ is defined to be

$$\lim_{x_1 \to x_0} \frac{f(x_1) - f(x_0)}{x_1 - x_0}.$$

14.3.2 Rate of Change and Slope of Graph; Singular Points We see in Fig. 14.15 that the average rate of change of $f(x)$ w.r.t. x,

$$\frac{f(x_1) - f(x_0)}{x_1 - x_0},$$

equals the slope of the chord P_0P_1. As $x_1 \to x_0$, i.e. as P_1 gets closer to P_0 on the curve, this chord gets closer to the tangent at P_0 and so we can identify the rate of change of $f(x)$ w.r.t. x at $x = x_0$ with the slope of the tangent at $x = x_0$. In particular we can identify the question of the existence of a rate of change at $x = x_0$ with the question of the existence of a tangent at $x = x_0$, and the following example illustrates the importance of this question.

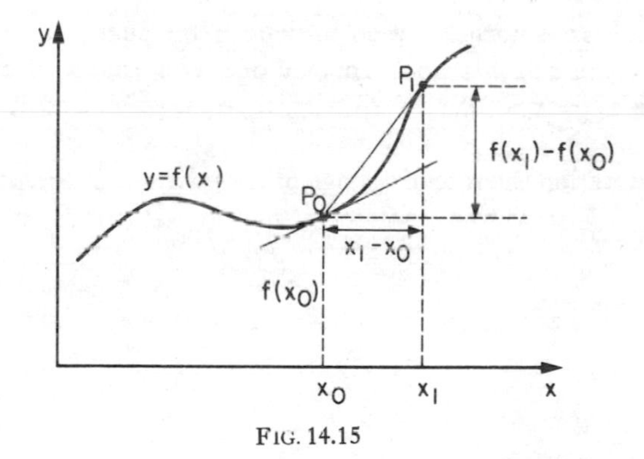

FIG. 14.15

Figure 14.16 shows a cross-section of an upper room with a dormer window, the solid lines representing the floor and ceiling of the room. The point x_0 on the floor is so placed that $h_1 - h_0 \to d$ as $x_1 \to x_0+$ and so $(h_1 - h_0)/(x_1 - x_0) \to \infty$ as $x_1 \to x_0+$ (see Rule 3). Hence the rate of change of headroom is not defined at the point x_0 since the appropriate limit does not exist; this is equivalent to saying that there is no tangent to the ceiling (Fig. 14.16) at the point $x = x_0$ because the ceiling is discontinuous there.

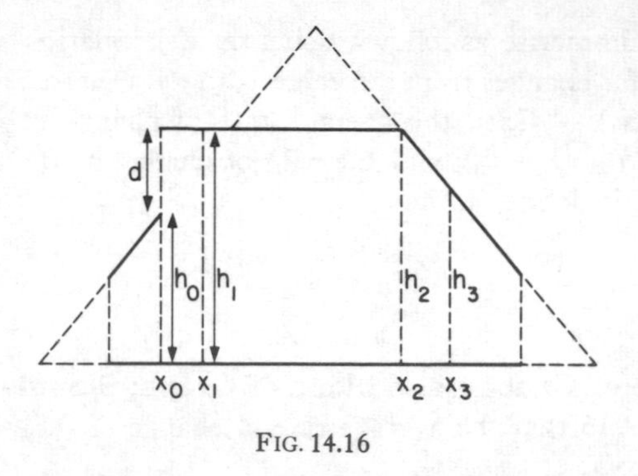

FIG. 14.16

Continuing with the same illustration, consider next the rate of change of headroom at the point x_2. The reader can easily verify that $(h_3 - h_2)/(x_3 - x_2)$ tends to a limit as $x_3 \to x_2+$, but tends to a different limit, viz. zero, as $x_3 \to x_2-$. Thus the rate of change of headroom is not defined at $x = x_2$ since here again the appropriate limit does not exist; this is equivalent to saying that there is no tangent to the ceiling at the point $x = x_2$ because the ceiling has a sharp angle there.

14.4 The Derivative

The symbol Δx is normally used to denote 'the change in x', particularly when the change in x is small. Instead of saying that x_1 is close to x_0, we may write $x_1 = x_0 + \Delta x$; if $x_1 > x_0$ then $\Delta x > 0$, and if $x_1 < x_0$ then $\Delta x < 0$.

In this notation the rate of change of $f(x)$ w.r.t. x at the point $x = x_0$ is

$$\lim_{\Delta x \to 0} \frac{f(x_0 + \Delta x) - f(x_0)}{\Delta x}.$$

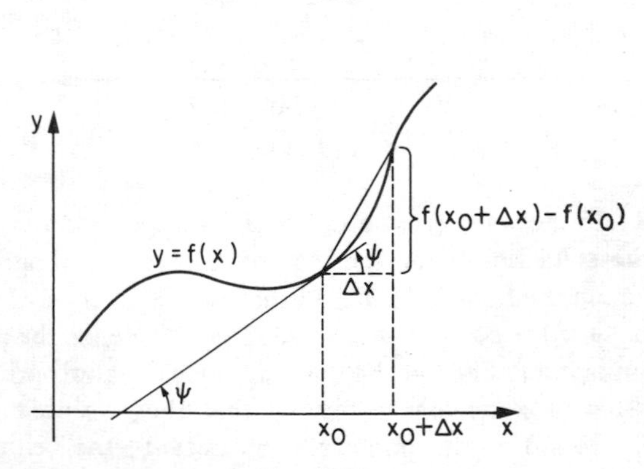

FIG. 14.17

This rate of change may be calculated for any point x_0 in the domain of f where the limit exists. Hence we may drop the suffix 0 and write

$$f'(x) = \lim_{\Delta x \to 0} \frac{f(x + \Delta x) - f(x)}{\Delta x} \tag{14.3}$$

thereby defining a function f' such that, for any point x_0 in its domain, $f'(x_0)$ equals the rate of change of $f(x)$ w.r.t. x at the point $x = x_0$. In other words, for any point x_0 in the domain of f', $f'(x_0)$ equals the slope of the tangent to the curve $y = f(x)$ at the point $x = x_0$. The function f' is called the *derivative* of the function f.

Alternatively we may write $y = f(x)$ and then denote $f(x + \Delta x) - f(x)$ by Δy (see Fig. 14.18) so that

$$f'(x) = \lim_{\Delta x \to 0} \frac{\Delta y}{\Delta x}.$$

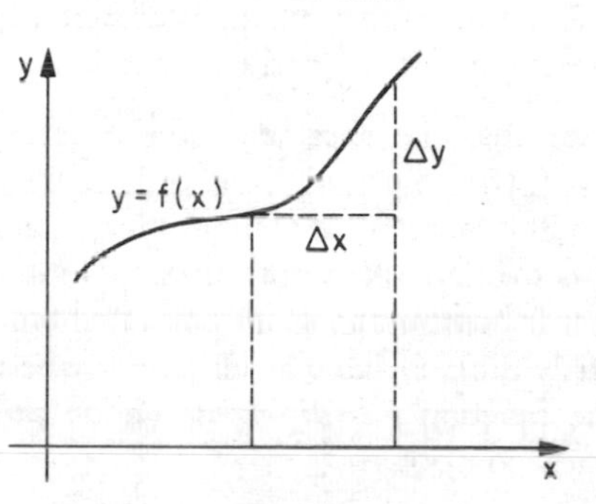

FIG. 14.18

This leads to the notation dy/dx for $f'(x)$, dy/dx being an abbreviated form of $\lim_{\Delta x \to 0} \Delta y/\Delta x$, and we shall refer to dy/dx as the *derivative of y with respect to x*. Other symbols, e.g. y', $y^{(1)}$, y_1 may be used for this derivative when there is no risk of confusion over the variable with respect to which the derivative is to be taken.

If the limit in (14.3) exists at the point $x = x_0$ we say f is *differentiable* at $x = x_0$. If the limit does not exist at $x = x_0$ then we say f is *non-differentiable* at $x = x_0$. It follows that the domain of f' consists of all points in the domain of f where f is differentiable.

Example 12 Find $f'(0)$ when (a) $f(x) = 2 - \cos x$, (b) $f(x) = 2 + \sqrt{x}$.

By definition,

$$f'(0) = \lim_{\Delta x \to 0} \frac{f(\Delta x) - f(0)}{\Delta x}.$$

In (a) this becomes

$$\lim_{\Delta x \to 0} \frac{1 - \cos(\Delta x)}{\Delta x},$$

that is 0 (see Example 10), so that $f'(0) = 0$. In (b) $f'(0)$ becomes

$$\lim_{\Delta x \to 0} \frac{\sqrt{\Delta x}}{\Delta x} = \lim_{\Delta x \to 0} \frac{1}{\sqrt{\Delta x}},$$

which does not exist; in this case f is not differentiable at $x = 0$.

The following abnormalities in the graph of f indicate points where f is non-differentiable:

(a) Tangent parallel to the y-axis. (Slope of tangent is then undefined—sometimes described as *infinite slope*.)
(b) Point of discontinuity.
(c) Sharp angle. (Slope of tangent changes abruptly at such a point but the slope *at* the point is not clearly defined.)

In Fig. 14.9 we see that the shearing force S is non-differentiable at $x = -4$ and at $x = 4$, the points where it is discontinuous. In Fig. 14.10 we see that the bending moment M is non-differentiable at $x = -4$ and at $x = 4$ although it is continuous at both these points. This illustrates the fact that a function is continuous at all points where it is differentiable, but is not necessarily differentiable at all points where it is continuous.

We say that the function *f is differentiable on the interval I* if it is differentiable at each point of I.

14.4.1 Differentiation The process of finding $f'(x)$ from a given $f(x)$ is called *differentiation*, or, more explicitly, *differentiation with respect to x*. We may regard differentiation as a mapping D from a set of functions $\{f_1, f_2, \ldots\}$ onto another set of functions $\{f_1', f_2', \ldots\}$ in which the image of any function f is its derivative f', i.e. $D(f) = f'$. In practice we usually omit the brackets and write $Df = f'$, the symbol D being called an *operator*. Alternatively we may write

$$\frac{\mathrm{d}}{\mathrm{d}x} f(x) = f'(x),$$

regarding the symbol $\mathrm{d}/\mathrm{d}x$ as meaning the operation of differentiation w.r.t. x.

14.4.2 Fundamental Results

(1) $\dfrac{\mathrm{d}}{\mathrm{d}x}(x^n) = nx^{n-1}$ for any rational number n.

Proof. Write $f(x) = x^n$, so that

$$\frac{f(x+\Delta x)-f(x)}{\Delta x} = \frac{(x+\Delta x)^n - x^n}{\Delta x}$$

$$= \frac{t^n - x^n}{t - x} \qquad \text{where } t = x + \Delta x.$$

$$= nx^{n-1} \text{ as } t \to x \qquad \text{by (14.2).}$$

Obviously if $t \to x$ we have $\Delta x \to 0$ and so $f'(x) = nx^{n-1}$. Note that $f'(0)$ is not defined if $n < 1$, i.e. if $n < 1$ then x^n is not differentiable at $x = 0$.

Particular cases of this result are the following:

$$\frac{d}{dx}(x^{53}) = 53x^{52}, \quad \frac{d}{dx}\left(\frac{1}{x^{29}}\right) = \frac{d}{dx}(x^{-29}) = -29x^{-30} = -\frac{29}{x^{30}},$$

$$\frac{d}{dx}(\sqrt{x}) = \frac{d}{dx}(x^{1/2}) = \tfrac{1}{2}x^{-1/2} = \frac{1}{2\sqrt{x}},$$

$$\frac{d}{dx}(\sqrt[3]{x^{26}}) = \frac{d}{dx}(x^{26/3}) = \frac{26}{3}x^{23/3},$$

$$\frac{d}{dx}\left(\frac{1}{\sqrt[4]{x^5}}\right) = \frac{d}{dx}(x^{-5/4}) = -\frac{5}{4}x^{-9/4}.$$

(2) $\dfrac{d}{dx}(\sin x) = \cos x.$

Proof. Write $f(x) = \sin x$, so that

$$\frac{f(x+\Delta x)-f(x)}{\Delta x} = \frac{\sin(x+\Delta x) - \sin x}{\Delta x}$$

$$= \frac{2\cos(x + \frac{1}{2}\Delta x)\sin(\frac{1}{2}\Delta x)}{\Delta x}$$

$$= \cos(x + \tfrac{1}{2}\Delta x)\ \frac{\sin(\frac{1}{2}\Delta x)}{(\frac{1}{2}\Delta x)}$$

$$\to \cos x \text{ as } \Delta x \to 0.$$

Note that $\sin x$, $\cos x$ are interpreted as usual to mean the trigonometric sine, cosine of x *radians*; the result

$$\lim_{x\to 0} \frac{\sin x}{x} = 1$$

requires this interpretation.

(3) $\dfrac{d}{dx}(\cos x) = -\sin x.$

Proof. Write $f(x) = \cos x$ so that

$$\begin{aligned}\frac{f(x+\Delta x)-f(x)}{\Delta x} &= \frac{\cos(x+\Delta x)-\cos x}{\Delta x}\\ &= \frac{-2\sin(x+\frac{1}{2}\Delta x)\sin(\frac{1}{2}\Delta x)}{\Delta x}\\ &= -\sin(x+\tfrac{1}{2}\Delta x)\frac{\sin(\frac{1}{2}\Delta x)}{(\frac{1}{2}\Delta x)}\\ &\to -\sin x \text{ as } \Delta x \to 0.\end{aligned}$$

Exercises 14(c)

Find $f'(x)$ by evaluating

$$\lim_{\Delta x\to 0}\frac{f(x+\Delta x)-f(x)}{\Delta x}$$

when $f(x)$ is given as follows:

1. $f(x) = x^2 - 4x + 7$
2. $f(x) = 3/\sqrt{x}$
3. $f(x) = 1/(3x^2 + 5)$
4. $f(x) = \sqrt{(2x + 7)}$
5. $f(x) = (4x + 3)/(x + 1)$
6. $f(x) = x \sin x$
7. $f(x) = x \cos x$
8. $f(x) = \sin(x^2)$
9. Verify that if $f(x) = \begin{cases}\sqrt{(2-x)} & \text{if } x < 0,\\ \sqrt{(2+x)} & \text{if } x > 0,\end{cases}$
   then the function f is not differentiable at $x = 0$.
10. Find

$$\lim_{h\to 0}\frac{1}{h}\left(\frac{1}{\sqrt{(1+h)}}-1\right)$$

and use your result to determine ***from first principles*** the value of $f'(1)$ given that $f(x) = 1/\sqrt{x}$. What can you say about $f(0)$ and $f'(0)$?

14.4.3 Rules of Differentiation From the definition (14.3) the following rules may be established, using the theorems about limits given in Section 14.1.1. We shall express these rules in terms of the functions f, g and also in terms of variables u, v where $u = f(x)$, $v = g(x)$.

Addition rule

$$\frac{d}{dx}[f(x) \pm g(x)] = f'(x) \pm g'(x),$$

that is

$$\text{if } y = u \pm v, \qquad \text{then } \frac{dy}{dx} = \frac{du}{dx} \pm \frac{dv}{dx}.$$

Proof

$$\frac{d}{dx}[f(x)+g(x)] = \lim_{\Delta x \to 0} \frac{[f(x+\Delta x)+g(x+\Delta x)]-[f(x)+g(x)]}{\Delta x}$$

$$= \lim_{\Delta x \to 0} \frac{[f(x+\Delta x)-f(x)]+[g(x+\Delta x)-g(x)]}{\Delta x}$$

$$= \lim_{\Delta x \to 0} \frac{f(x+\Delta x)-f(x)}{\Delta x} + \lim_{\Delta x \to 0} \frac{g(x+\Delta x)-g(x)}{\Delta x}$$

by theorem 1 of Section 14.1.1

$$= f'(x)+g'(x),$$

and likewise for

$$\frac{d}{dx}[f(x)-g(x)].$$

Thus, for example,

$$\frac{d}{dx}(\sin x + \cos x) = \cos x - \sin x,$$

$$\frac{d}{dx}\left(x^{53}+\frac{1}{x^{29}}\right) = 53x^{52} - \frac{29}{x^{30}}.$$

Chain rule

$$\frac{d}{dx} f \circ g(x) = f' \circ g(x) \times g'(x),$$

that is if y is a function of v and v is a function of x, then y is a function of x and

$$\frac{dy}{dx} = \frac{dy}{dv}\frac{dv}{dx}.$$

Proof

$$\frac{d}{dx} f \circ g(x) = \lim_{\Delta x \to 0} \frac{f \circ g(x+\Delta x) - f \circ g(x)}{\Delta x}$$

$$= \lim_{\Delta x \to 0} \frac{f \circ g(x+\Delta x) - f \circ g(x)}{g(x+\Delta x)-g(x)} \frac{g(x+\Delta x)-g(x)}{\Delta x}$$

$$= \lim_{\Delta x \to 0} \frac{f(v+\Delta v)-f(v)}{\Delta v} \frac{g(x+\Delta x)-g(x)}{\Delta x}$$

(putting $v = g(x)$)

$$= f'(v)g'(x)$$ by Theorem 2 of Section 14.1.1 (Note: $\Delta v \to 0$ as $\Delta x \to 0$)

$$= f' \circ g(x)g'(x).$$

For example:

(a) If $y = (2x^2 + 3)^2$, then putting $v = 2x^2 + 3$ we have $y = v^2$, $dy/dv = 2v$ and so

$$\frac{dy}{dx} = 2v\frac{dv}{dx} = 2v \, . \, 4x = 8x(2x^2 + 3).$$

(b) If $y = (3x^5 + 5x^2 - 1)^{27}$, then putting $v = 3x^5 + 5x^2 - 1$ we have $y = v^{27}$, $dy/dv = 27v^{26}$ and so

$$\frac{dy}{dx} = 27v^{26}\frac{dv}{dx} = 27(3x^5 + 5x^2 - 1)^{26}(15x^4 + 10x)$$
$$= 135x(3x^3 + 2)(3x^5 + 5x^2 - 1)^{26}.$$

(c) If $y = 1/\sqrt{(x^2 + 1)}$, then putting $v = x^2 + 1$ we have $y = v^{-1/2}$, $dy/dv = -\frac{1}{2}v^{-3/2}$, and so

$$\frac{dy}{dx} = -\tfrac{1}{2}v^{-3/2}\frac{dv}{dx} = -\frac{1}{2(x^2 + 1)^{3/2}}2x = -\frac{x}{(x^2 + 1)^{3/2}}.$$

(d) If $y = \sin^{27} x$, then putting $v = \sin x$ we have $y = v^{27}$, $dy/dv = 27v^{26}$, and so

$$\frac{dy}{dx} = 27v^{26}\frac{dv}{dx} = 27 \sin^{26} x \cos x.$$

(e) If $y = \sin(x^{27})$, then putting $v = x^{27}$ we have $y = \sin v$, $dy/dv = \cos v$, and so

$$\frac{dy}{dx} = \cos v \frac{dv}{dx} = 27x^{26} \cos(x^{27}).$$

(f) If $y = \sqrt{(\cos x)}$, then

$$\frac{dy}{dx} = \frac{-\sin x}{2\sqrt{(\cos x)}} \qquad (\text{put } v = \cos x).$$

(g) If $y = \cos(\sqrt{x})$, then

$$\frac{dy}{dx} = -\frac{1}{2\sqrt{x}}\sin(\sqrt{x}) \qquad (\text{put } v = \sqrt{x}).$$

Extended Chain Rule: We can easily extend the chain rule to show that if y is a function of v_1, v_1 a function of v_2, and v_2 a function of x, then

$$\frac{dy}{dx} = \frac{dy}{dv_1}\frac{dv_1}{dv_2}\frac{dv_2}{dx},$$

and more generally if y is a function of v_1, v_1 a function of v_2, . . ., v_n a function of x, then

$$\frac{dy}{dx} = \frac{dy}{dv_1}\frac{dv_1}{dv_2} \cdots \frac{dv_n}{dx}.$$

For example:

(a) If $y = \sin\sqrt{(x^2+1)}$, then putting $v_1 = \sqrt{(x^2+1)}$, $v_2 = x^2+1$, we have $y = \sin v_1$, $v_1 = v_2^{1/2}$, giving $dy/dv_1 = \cos v_1$, $dv_1/dv_2 = \frac{1}{2}v_2^{-1/2}$, and so

$$\frac{dy}{dx} = \cos v_1\left(\tfrac{1}{2}v_2^{-1/2}\right)\frac{dv_2}{dx} = \frac{x}{\sqrt{(x^2+1)}}\cos\sqrt{(x^2+1)}.$$

(b) If $y = \sqrt{[\sin(x^2+1)]}$, then putting $v_1 = \sin(x^2+1)$, $v_2 = x^2+1$, we have $y = v_1^{1/2}$, $v_1 = \sin v_2$, giving $dy/dv_1 = \frac{1}{2}v_1^{-1/2}$, $dv_1/dv_2 = \cos v_2$, and so

$$\frac{dy}{dx} = \tfrac{1}{2}v_1^{-1/2}\cos v_2\frac{dv_2}{dx} = \frac{x\cos(x^2+1)}{\sqrt{[\sin(x^2+1)]}}.$$

(c) If $y = 1/\sin^2(2x+3)$, then putting $v_1 = \sin(2x+3)$, $v_2 = 2x+3$, we have $y = v_1^{-2}$, $v_1 = \sin v_2$, giving $dy/dv_1 = -2v_1^{-3}$, $dv_1/dv_2 = \cos v_2$, and so

$$\frac{dy}{dx} = -\frac{2\cos v_2}{v_1^3}\frac{dv_2}{dx} = -\frac{4\cos(2x+3)}{\sin^3(2x+3)}.$$

With a little practice the procedure can easily be carried out without explicit mention of the intermediate variables $v_1, v_2, \ldots, v_n$.

Product Rule

$$\frac{d}{dx}[f(x)\,g(x)] = f(x)\,g'(x) + f'(x)\,g(x),$$

that is if $y = uv$, then

$$\frac{dy}{dx} = u\frac{dv}{dx} + v\frac{du}{dx}.$$

Proof

$$\frac{d}{dx}[f(x)g(x)] = \lim_{\Delta x\to 0}\frac{f(x+\Delta x)\,g(x+\Delta x) - f(x)\,g(x)}{\Delta x}$$

$$= \lim_{\Delta x\to 0}\frac{[f(x+\Delta x) - f(x)]\,g(x+\Delta x) + f(x)\,[g(x+\Delta x) - g(x)]}{\Delta x}$$

$$= f'(x)g(x) + f(x)g'(x)$$

by Theorems 1 and 2 of Section 14.1.1.

For example:

(a) If $y = x^2 \sin x$, then

$$\frac{dy}{dx} = x^2 \cos x + 2x \sin x.$$

(b) If $y = \sqrt{x} \cos x$, then

$$\frac{dy}{dx} = -\sqrt{x} \sin x + \frac{1}{2\sqrt{x}} \cos x.$$

Quotient Rule

$$\frac{d}{dx}\left[\frac{f(x)}{g(x)}\right] = \frac{g(x)f'(x) - f(x)g'(x)}{[g(x)]^2},$$

that is if $y = u/v$, then

$$\frac{dy}{dx} = \frac{v\, du/dx - u\, dv/dx}{v^2}.$$

The proof is left as an exercise for the reader.

For example:

(a) If $y = (x^2 + 1)/(x^2 - 1)$, then

$$\frac{dy}{dx} = \frac{(x^2 - 1)2x - (x^2 + 1)2x}{(x^2 - 1)^2} = \frac{-4x}{(x^2 - 1)^2}.$$

(b) If $y = \sin x/(1 + \cos x)$, then

$$\frac{dy}{dx} = \frac{(1 + \cos x) \cos x - \sin x(-\sin x)}{(1 + \cos x)^2} = \frac{1}{1 + \cos x}.$$

The quotient rule may be used to derive another fundamental result, viz.

$$\frac{d}{dx}(\tan x) = \sec^2 x.$$

For writing $y = \tan x = \sin x/\cos x$, we get

$$\frac{dy}{dx} = \frac{\cos x \cos x - \sin x(-\sin x)}{\cos^2 x} = \frac{1}{\cos^2 x} = \sec^2 x.$$

Inverse Rule. If f has an inverse f^{-1}, so that $y = f(x)$ and $x = f^{-1}(y)$, then

$$\frac{d}{dy} f^{-1}(y) = 1/f'(x),$$

that is

$$\frac{dx}{dy}=\frac{1}{dy/dx}.$$

Proof. Since $y = f(x)$, we see, on using the chain rule to differentiate with respect to y that

$$1 = f'(x)\frac{dx}{dy}, \quad \text{that is} \quad \frac{dx}{dy} = 1/f'(x).$$

For example, if $y = x^2$, then $dy/dx = 2x$, while $x = \sqrt{y}$ and so

$$\frac{dx}{dy}=\frac{1}{2\sqrt{y}}=\frac{1}{2x}=\frac{1}{dy/dx}.$$

This rule enables us to derive the following fundamental results:

$$\frac{d}{dx}(\sin^{-1}x)=\frac{1}{\sqrt{(1-x^2)}}, \quad \frac{d}{dx}(\cos^{-1}x)=-\frac{1}{\sqrt{(1-x^2)}},$$
$$\frac{d}{dx}(\tan^{-1}x)=\frac{1}{1+x^2}.$$

Proof. Write $y = \sin^{-1} x$, so that $x = \sin y$, where $-\pi/2 \leqslant y \leqslant \pi/2$. From $x = \sin y$ we get $dx/dy = \cos y$ and hence

$$\frac{dy}{dx}=\frac{1}{\cos y}=\frac{1}{\pm\sqrt{(1-x^2)}}.$$

From the inequalities $-\pi/2 \leqslant y \leqslant \pi/2$ we see that only the positive value of the root should be taken. Hence

$$\frac{dy}{dx}=\frac{1}{\sqrt{(1-x^2)}}.$$

Note that dy/dx is not defined for $x = \pm 1$, i.e. $\sin^{-1} x$ is non-differentiable at the points $x = \pm 1$. (See Fig. 12.31 which shows that the tangent is parallel to the y-axis at these points.)

Proof of the other two results is similar.

Example 13 Differentiate the following with respect to x:

(a) $\sin^{-1}(1/x)$, (b) $\tan^{-1}(1 + x^2)$.

(a) If $y = \sin^{-1}(1/x)$, then

$$\frac{dy}{dx}=\frac{1}{\sqrt{[1-(1/x)^2]}}\left(-\frac{1}{x^2}\right) \quad \text{by the chain rule,}$$

that is

$$\frac{dy}{dx} = -\frac{1}{x\sqrt{(x^2-1)}}.$$

(b) If $y = \tan^{-1}(1 + x^2)$, then

$$\frac{dy}{dx} = \frac{1}{1+(1+x^2)^2}2x \quad \text{by the chain rule,}$$

that is

$$\frac{dy}{dx} = \frac{2x}{2+2x^2+x^4}.$$

Exercises 14(d)

1. Find the points where the curve $y = 6x - 10x^2 + 3x^3$ has slope 2.
2. Show that the curve $y = 6x + 3x^2 + x^3$ has positive slope everywhere.
3. Find the angle at which the curve $y = (1 - x)(x^2 + 2)$ crosses (a) the x-axis, (b) the y-axis. (It is required to find the angle between the axis and the tangent to the curve where it crosses the axis.)

Differentiate the following w.r.t. x, simplifying your answers where possible:

4. $2x^3(x + 2)^2$
5. $(3x^2 - 2x + 1)\sqrt{x}$
6. $1/\sqrt{(1 - 3x)}$
7. $3/(1 - 2x)^2$
8. $(x + 5)/(2x - 1)$
9. $[x(1 - x)]^{1/2}$
10. $(x - 1/x^4)^{1/2}$
11. $[x + \sqrt{(1 + x^2)}]^n$
12. $(1 - x)/(1 - x^3)^{1/2}$
13. $(x^2 + 2x + 7)/(3x - 1)^{1/2}$
14. $(3x + 1)^5/(2 - x)^{10}$
15. $\tan^3 x$
16. $\sin^5 3x$
17. $\sqrt{(\sin x)}$
18. $\cos x/(1 + \sin x)^2$
19. $(1 - \sin x)/(1 - \cos x)$
20. $x \sin x/(1 + \cos x)$
21. $[(2 + \sin^2 x)/(1 - \sin x)]^{1/2}$
22. $\cot(1/x)$
23. $\cot[1/(x^2 + 1)]$
24. $\sin^{-1}\sqrt{(x - 1)}$
25. $\sin^{-1}\sqrt{(1 - x^2)}$
26. $\sin^{-1}(\cos x)$
27. $\sin^{-1}[2ax\sqrt{(1 - a^2x^2)}]$
28. $\sin^{-1}[x^2/(a^2 + x^2)]$
29. $\cos^{-1}[2x\sqrt{(1 - x^2)}]$
30. $\tan^{-1}[1/(1 - x^2)]$
31. $\tan^{-1}[(a + bx)/(a - bx)]$
32. $\tan^{-1}(m \tan x)$
33. $\tan^{-1}[2\sqrt{x}/(1 - x)]$
34. $\tan^{-1}[(1 - \sqrt{x})/(1 + \sqrt{x})]$
35. $\tan^{-1}[\frac{1}{2}x^2/\sqrt{(1 + x^2)}]$
36. $\tan^{-1}[4\sqrt{x}/(1 - 4x)]$
37. $\tan^{-1}[(3 - x)\sqrt{x}/(1 - 3x)]$
38. $\tan^{-1}[(\cos x - \sin x)/(\cos x + \sin x)]$
39. $(x^2 - 2)\sin^{-1}\frac{1}{2}x + \frac{1}{2}x\sqrt{(4 - x^2)}$
40. Obtain and simplify the derivatives of

$$\cos^{-1}\frac{a\cos x + b}{a + b\cos x} \quad \text{and} \quad \tan^{-1}\left[\left(\frac{a-b}{a+b}\right)^{1/2}\tan\tfrac{1}{2}x\right],$$

and explain the significance of your results.

41. Prove that, if $\sin\theta = 2\sin\phi$ and $x = \cos\theta - 2\cos\phi$, then $dx/d\theta = x\tan\phi$.

14.4.4 Implicit Differentiation From the chain rule it follows that

$$\frac{d}{dx}f(y)=f'(y)\frac{dy}{dx}.$$

For example

$$\frac{d}{dx}(y^2)=2y\frac{dy}{dx}, \quad \frac{d}{dx}(\sin y)=\cos y\frac{dy}{dx}.$$

From an implicit equation (see Section 12.6.5) we can thus find dy/dx by differentiating both sides of the equation with respect to x. For example

(a) if $x^2+y^2=1$, then

$$2x+2y\frac{dy}{dx}=0 \quad \text{and so} \quad \frac{dy}{dx}=-\frac{x}{y},$$

(b) if $\sqrt{x}+\sin y=x^2$, then

$$\frac{1}{2\sqrt{x}}+\cos y\frac{dy}{dx}=2x \quad \text{and so} \quad \frac{dy}{dx}=\frac{2x-1/2\sqrt{x}}{\cos y}.$$

Special care is needed with terms that contain both x and y; for example

$$\frac{d}{dx}(xy)=x\frac{dy}{dx}+y, \quad \frac{d}{dx}(x^2y^3)=3x^2y^2\frac{dy}{dx}+2xy^3$$

(using the product rule),

$$\frac{d}{dx}\left(\frac{x^2}{\sin y}\right)=\frac{2x\sin y-x^2\cos y\, dy/dx}{\sin^2 y}$$

(using the quotient rule)

$$\frac{d}{dx}[\sin(xy)]=\cos(xy)(x\frac{dy}{dx}+y)$$

(chain rule and product rule).

Example 14 Find dy/dx given that

(a) $x^2+xy+y^2=1$, (b) $\sin(x/y)=xy$.

Equation (a) gives

$$2x+x\frac{dy}{dx}+y+2y\frac{dy}{dx}=0,$$

that is

$$\frac{dy}{dx}=-\frac{2x+y}{x+2y}.$$

Equation (b) gives

$$\cos(x/y)\frac{y - x\,dy/dx}{y^2} = x\frac{dy}{dx} + y,$$

that is

$$\frac{dy}{dx} = \frac{y[\cos(x/y) - y^2]}{x[\cos(x/y) + y^2]}.$$

Exercises 14(e)

Find dy/dx when x and y are related by the following equations:

1. $x^3 + y^3 = a^3$
2. $\sqrt{x} + \sqrt{y} = \sqrt{a}$
3. $1 + x^2y + xy^2 = 0$
4. $x^m y^n = a^{m+n}$
5. $a^m x^n = b^n y^m$
6. $(x + y)^2 = ax^2 + by^2$
7. $(x + y)^3 = 3axy$
8. $ax^2 + 2hxy + by^2 + 2gx + 2fy + c = 0$
9. $y = \sin(x + y)^2$.
10. Show that the point (1, 1) lies on the curve $x^3y + x^2y^2 + xy^3 = 3$ and that the tangent to the curve at this point makes an angle of 135° with the positive direction of the x-axis.
11. Show that the line $x = 1$ and the curve $x^3y + x^2y^2 + xy^3 = 3$ meet once only, and that the acute angle between the line and the curve at the point of intersection is 45°.
12. Find dy/dx in terms of x and y given that x and y are related by the equation

$$x^4y^2 + x^2y^4 = 26x^6 - 6.$$

Hence determine the slope of the curve $x^4y^2 + x^2y^4 = 26x^6 - 6$ at each of the points where it meets the line $x = 1$.

13. Given that $y = x + \sin(xy)$, find dy/dx in terms of x and y. Verify that the curve $y = x + \sin(xy)$ passes through the origin O and find the equation of its tangent at O. Show that this tangent meets the curve again at an infinite set of points $P_1, P_2, P_3, \ldots$ in the positive quadrant and at an infinite set of points $Q_1, Q_2, Q_3, \ldots$ in the negative quadrant.
14. Show that the equation of the tangent to the curve $x^{2/3} + y^{2/3} = a^{2/3}$ at the point (x_1, y_1) on it, is

$$x/x_1^{1/3} + y/y_1^{1/3} = a^{2/3}.$$

Deduce that the length of the tangent intercepted between the co-ordinate axes is the same for all tangents to this curve.

15. Show that the tangent to the curve $x^3 + y^3 = 3axy$ at the point (x_1, y_1) is

$$x(x_1^2 - ay_1) + y(y_1^2 - ax_1) = ax_1y_1.$$

Write down the equation of the tangent at the point $(6a/7, -12a/7)$ and verify that it meets the curve again at the point $(-16a/21, 4a/21)$.

14.4.5 Parametric Differentiation A relation between x and y is sometimes given by expressing both x and y in terms of another variable t (say) called a *parameter*. If $x = f(t)$, $y = g(t)$, then by the chain rule we have

$$\frac{dy}{dx} = g'(t)\frac{dt}{dx},$$

and since $dt/dx = 1/(dx/dt)$ by the inverse rule, it follows that

$$\frac{dy}{dx} = \frac{g'(t)}{f'(t)}, \quad \text{that is} \quad \frac{dy}{dx} = \frac{dy/dt}{dx/dt}.$$

For example:

(a) If $x = \cos t$, $y = \sin t$, then

$$\frac{dy}{dx} = \frac{\cos t}{-\sin t} = -\cot t.$$

(b) If $x = t + 1/t$, $y = 3t^{1/2} + t^{3/2}$, then

$$\frac{dy}{dx} = \frac{\frac{3}{2}t^{-1/2} + \frac{3}{2}t^{1/2}}{1 - 1/t^2} = \frac{3t^{3/2}}{2(t-1)}.$$

Exercises 14(f)

Find dy/dx in terms of the parameter t in the following cases:

1. $x = t^2$, $y = t + 1/t$
2. $x = \cos 2t$, $y = 4 \sin t + 3$
3. $x = \sqrt{(1 - t^2)}$, $y = \sin^{-1} t$
4. $x = \sec t$, $y = \tan t$
5. $x = \sin^3 t$, $y = \cos^3 t$
6. $x = \sin t - t \cos t$, $y = t \sin t + \cos t$
7. $x = \cos 2t$, $y = t + \sin 2t$.
8. Show that if $x = t + 1/t$, $y = t^m + 1/t^m$ then

$$(x^2 - 4)(dy/dx)^2 = m^2(y^2 - 4).$$

9. Show that if $x = at^2$, $y = at/(1 - t)^2$ then the tangent to the curve at the point $t = 2$ has the equation $3x + 4y = 20a$.
10. Show that the equation of the tangent at the point whose parameter is t on the curve $x = a \cos^3 t$, $y = a \sin^3 t$ is

$$x \sin t + y \cos t - a \sin t \cos t = 0.$$

11. A curve has the parametric equations

$$x = 2 \cos t + \cos 2t, \quad y = 2 \sin t - \sin 2t.$$

Show that the slope of the curve, at the point whose parameter is t, is $-\tan \frac{1}{2}t$. Show also that the equation of the tangent to the curve at this point is

$$x \sin \tfrac{1}{2}t + y \cos \tfrac{1}{2}t = \sin \tfrac{3}{2}t.$$

12. Given that the tangent at the point $P(at^2, at^3)$ on the curve $ay^2 = x^3$ meets the curve again at Q, find the co-ordinates of Q.

    Prove that if N is the foot of the perpendicular from P to the x-axis, R is the point where the tangent at P cuts the y-axis, and O is the origin, then OQ and RN are equally inclined to the x-axis.

14.4.6 Significance of the Sign of the Derivative We say that $f(x)$ is *increasing in the interval* $a \leqslant x \leqslant b$ if $f(x)$ increases as x increases from a to b inclusive. We say that $f(x)$ is *increasing at the point* x_0 if there is some neighbourhood of x_0 (i.e. some interval with the point x_0 inside it) in which $f(x)$ is increasing.

We say that $f(x)$ is *decreasing in the interval* $a \leqslant x \leqslant b$ if $f(x)$ decreases as x increases from a to b inclusive, and that $f(x)$ is *decreasing at the point* x_0 if there is some neighbourhood of x_0 in which $f(x)$ is decreasing.

Figure 14.19 shows the graph of a differentiable function f such that $f(x)$ is increasing at all points in the intervals $p \leqslant x < q$ and $s < x \leqslant v$, and decreasing at all points in the interval $q < x < s$.

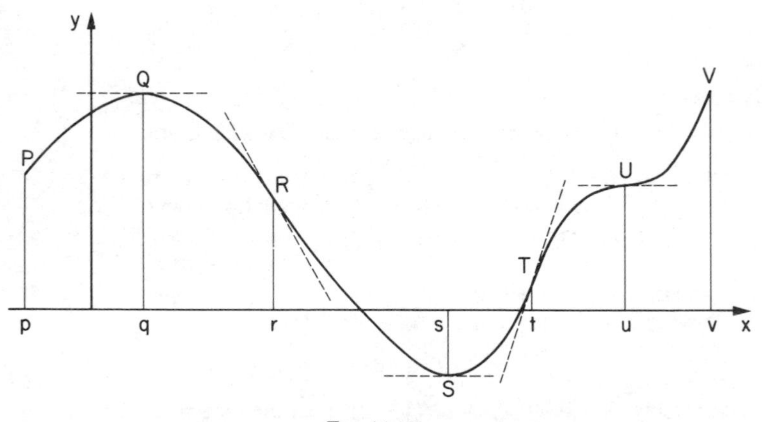

FIG. 14.19

The *gradient*, or *slope*, of a curve at a point on the curve is defined to be the slope of the tangent to the curve at that point. The slope at the point R (frequently referred to as 'the slope at $x = r$'), for example, is the slope of the tangent at R and is shown in Fig. 14.19 by a broken line as are the slopes at the points Q, S, T and U.

Thus the slope of the curve $y = f(x)$ at any point *equals the value of* $f'(x)$ *at that point*. It follows that $f(x)$ is increasing at a point where $f'(x) > 0$ and $f(x)$ is decreasing at a point where $f'(x) < 0$. A point where $f'(x) = 0$ is called a *stationary point*; at a stationary point the tangent is parallel to the x-axis, e.g. Q, S, U (Fig. 14.19) are stationary points.

In other words,

$$\frac{dy}{dx} > 0 \Rightarrow y \text{ increasing,}$$

$$\frac{dy}{dx} < 0 \Rightarrow y \text{ decreasing,}$$

$$\frac{dy}{dx} = 0 \Rightarrow y \text{ stationary.}$$

Stationary points like Q and S, where $f'(x)$ changes sign on passing through the point, are called *turning points*; such points must be distinguished from other stationary points like U where the sign of $f'(x)$ does not change.

At Q we see that the sign of $f'(x)$ is positive on the left ($x < q$) and negative on the right ($x > q$); we say Q is a *maximum turning point* (max. T.P.). At S the sign of $f'(x)$ changes from negative on the left to positive on the right; we say S is a *minimum turning point* (min. T.P.).

14.4.7 Higher Derivatives The derivative of the function f' is denoted by f'', the derivative of f'' is denoted by f''', and so on, that is

$$\frac{d}{dx}f'(x) = f''(x), \quad \frac{d}{dx}f''(x) = f'''(x), \ldots.$$

Writing $y = f(x)$, $dy/dx = f'(x)$, we may continue as follows:

$$\frac{d^2y}{dx^2} = f''(x), \quad \frac{d^3y}{dx^3} = f'''(x), \ldots.$$

Other notations, e.g. y', y'', y''', ... or y_1, y_2, y_3, ..., may be used for these derivatives from time to time.

For example:

(a) If $y = (ax + b)^m$, then

$$\frac{dy}{dx} = am(ax + b)^{m-1},$$

$$\frac{d^2y}{dx^2} = a^2m(m - 1)(ax + b)^{m-2},$$

$$\vdots$$

$$\frac{d^ny}{dx^n} = a^nm(m - 1)(m - 2) \cdots (m - n + 1)(ax + b)^{m-n}.$$

(b) If $y = \sin(ax + b)$, then

$$\frac{dy}{dx} = a\cos(ax + b) = a\sin(ax + b + \tfrac{1}{2}\pi),$$

$$\frac{d^2y}{dx^2} = -a^2\sin(ax + b) = a^2\sin(ax + b + \pi),$$

$$\vdots$$

$$\frac{d^ny}{dx^n} = a^n\sin(ax + b + \tfrac{1}{2}n\pi).$$

Usually it is not possible to find a general formula for the nth derivative as has been done in the above examples (but see also the theorem of Leibniz below). The following example illustrates the use of implicit differentiation to find a second derivative.

Example 15 Find d^2y/dx^2 given that $x^2 + y^2 = 1$.

We have

$$x^2 + y^2 = 1, \tag{1}$$

and hence

$$2x + 2y\frac{dy}{dx} = 0, \tag{2}$$

giving

$$\frac{dy}{dx} = -\frac{x}{y}.$$

Differentiating (2) with respect to x (after dividing through by 2) we get

$$1 + y\frac{d^2y}{dx^2} + \left(\frac{dy}{dx}\right)^2 = 0. \tag{3}$$

(Note the use of the product rule to differentiate $y\,dy/dx$.)

From (3) we have

$$1 + y\frac{d^2y}{dx^2} + \frac{x^2}{y^2} = 0,$$

that is

$$\frac{d^2y}{dx^2} = -\frac{x^2 + y^2}{y^3} = -\frac{1}{y^3} \qquad \text{by (1).}$$

Note. Although $dx/dy = 1/(dy/dx)$, in general $d^2x/dy^2 \neq 1/(d^2y/dx^2)$. In Example 15 the same procedure, only differentiating with respect to y, gives $d^2x/dy^2 = -1/x^3$, and this is not equal to $1/(d^2y/dx^2)$.

If x and y are given in terms of a parameter t, then it is usual to denote derivatives with respect to t by dots, i.e.

$$\dot{x} = \frac{dx}{dt}, \quad \dot{y} = \frac{dy}{dt}, \quad \ddot{x} = \frac{d^2x}{dt^2}, \quad \ddot{y} = \frac{d^2y}{dt^2}, \ldots.$$

Then

$$\frac{dy}{dx} = \frac{\dot{y}}{\dot{x}} \quad \text{(see Section 14.4.5) and so}$$

$$\frac{d^2y}{dx^2} = \frac{d}{dx}\left(\frac{\dot{y}}{\dot{x}}\right) = \frac{d}{dt}\left(\frac{\dot{y}}{\dot{x}}\right) \times \frac{dt}{dx} \quad \text{by the chain rule}$$

$$= \frac{\dot{x}\ddot{y} - \ddot{x}\dot{y}}{\dot{x}^2} \times \frac{1}{\dot{x}} \quad \text{using the quotient rule and the}$$

inverse rule.

Hence

$$\frac{d^2y}{dx^2} = \frac{\dot{x}\ddot{y} - \ddot{x}\dot{y}}{\dot{x}^3}. \tag{14.4}$$

Example 16 Find d^2y/dx^2 when $x = \sin t$, $y = \cos t$.

We have

$$\dot{x} = \cos t, \quad \ddot{x} = -\sin t, \quad \dot{y} = -\sin t, \quad \ddot{y} = -\cos t.$$

Hence, from (14.4),

$$\frac{d^2y}{dx^2} = \frac{-\cos^2 t - \sin^2 t}{\cos^3 t} = -\frac{1}{\cos^3 t} = -\sec^3 t.$$

In the following theorem, which is an extension of the product rule to higher derivatives, we use the notation

$$u_r = \frac{d^ru}{dx^r}, \quad v_r = \frac{d^rv}{dx^r}.$$

Theorem of Leibniz If $y = uv$, then

$$\frac{d^ny}{dx^n} = u_n v + {}^nC_1 u_{n-1} v_1 + {}^nC_2 u_{n-2} v_2 + \cdots + {}^nC_r u_{n-r} v_r + \cdots + u v_n,$$

where the coefficients ${}^nC_1$, ${}^nC_2$, ... are those which occur in the binomial expansion of $(1 + x)^n$.

This theorem is easily verified for $n = 1, 2, 3, \ldots$ using the product rule. For if $y = uv$, then

$$\frac{dy}{dx} = u_1v + uv_1,$$

and so

$$\frac{d^2y}{dx^2} = (u_2v + u_1v_1) + (u_1v_1 + uv_2) = u_2v + 2u_1v_1 + uv_2,$$
$$\frac{d^3y}{dx^3} = u_3v + 3u_2v_1 + 3u_1v_2 + uv_3,$$
$$\vdots$$

A general proof by induction may be given; this is left as an exercise for the reader.

Example 17 Find d^6y/dx^6 when $y = x^3 \sin x$.

Put $u = \sin x$, $v = x^3$ so that

$$\begin{array}{ll} u_1 = \cos x, & v_1 = 3x^2, \\ u_2 = -\sin x, & v_2 = 6x, \\ u_3 = -\cos x, & v_3 = 6, \\ u_4 = \sin x, & v_4 = 0, \\ u_5 = \cos x, & v_r = 0 \quad \text{for } r \geqslant 4. \\ u_6 = -\sin x. & \end{array}$$

Now

$$\begin{aligned} \frac{d^6y}{dx^6} &= u_6v + 6u_5v_1 + 15u_4v_2 + 20u_3v_3 + 15u_2v_4 + 6u_1v_5 + uv_6 \\ &= -x^3 \sin x + 18x^2 \cos x + 90x \sin x - 120 \cos x \\ &= 6(3x^2 - 20) \cos x - x(x^2 - 90) \sin x. \end{aligned}$$

Exercises 14(g)

For brevity y_r is written for d^ry/dx^r.

1. Given $y = 2x/(1 - x^2)$, show that $y_2 = 4x(x^2 + 3)/(1 - x^2)^3$.
2. Given $y = \sec x$, show that $y_2 = 2 \sec^3 x - \sec x$.
3. Given $y = \tan x$, show that $y_3 = 2(1 + 3 \tan^2 x)(1 + \tan^2 x)$.
4. Given $y = \sqrt{(a^2 - x^2)}$, show that $x(a^2 - x^2)y_2 = a^2y_1$.
5. Given $y = x \sin (1/x)$, show that $x^4y_2 + y = 0$.
6. Given $y = \tan (m \tan^{-1} x)$, show that $(1 + x^2)y_2 = 2(my - x)y_1$.
7. Given $y = \sin (m \sin^{-1} x)$, show that $(1 - x^2)y_2 - xy_1 + m^2y = 0$.
8. Given $y = 2x - \tan^{-1} x$ show that $d^2x/dy^2 = -2x(1 + x^2)/(1 + 2x^2)^3$.
9. If $y = A \tan (\frac{1}{2}x) + B[2 + x \tan (\frac{1}{2}x)]$, where A and B are any constants, show that $(1 + \cos x)y_2 = y$.
10. Find y_2 when (a) $x^2 + xy + y^2 = 1$, (b) $x = \sec t$, $y = \tan t$.

11. Given $y(2x - y) = 4$ show that $y_2 = 4/(x - y)^3$.
12. Given $x = at + 2at^2$, $y = 2at^3 + 6at^4$, show that $y_2 = 12t/a(1 + 4t)$.
13. Given $x = \tan t$, $y = \tan pt$, where p is a constant, show that

$$(1 + x^2)y_2 = 2(py - x)y_1.$$

14. Given $x = \cos t$, $y = \cos 2pt$, show that $(1 - x^2)y_2 - xy_1 + 4p^2y = 0$.
15. Given $x = \lambda^3 + 1$, $y = \lambda^2 + 1$, where λ is a variable, show that

$$\frac{d^2y/dx^2}{(dy/dx)^4}$$

is constant.

16. Find the value of d^5/dx^5 $(16 \sin^4 x \cos x)$ when $x = \pi/10$.
17. Given $x = t + 1/t$, $y = t^m + 1/t^m$ show that $(x^2 - 4)y_2 + xy_1 - m^2y = 0$.
18. Find y_2 if $x = 3 \cos \theta - \cos 3\theta$, $y = 3 \sin \theta - \sin 3\theta$.
19. Given $x = 4b \cos \theta - b \cos 4\theta$, $y = 4b \sin \theta - b \sin 4\theta$, find y_1 in terms of θ and show that

$$y_2 = \frac{5}{16b} \sec^3 \tfrac{5}{2}\theta \operatorname{cosec} \tfrac{3}{2}\theta.$$

20. Given $x = \cos \theta$, $y = \sin n\theta \operatorname{cosec} \theta$, show that

$$(1 - x^2)y_1 - xy + n \cos n\theta = 0,$$
$$(1 - x^2)y_2 - 3xy_1 + (n^2 - 1)y = 0.$$

Show that, if $n = 7$, the latter equation is satisfied by a polynomial of the form $x^6 + bx^4 + cx^2 + d$, and find the values of b, c and d.

21. Given $y = 1/x$, show that $y_n = (-1)^n n!/x^{n+1}$.
22. Given $y = (1 + x)/(1 - x)$, show that $y_n = 2(n!)/(1 - x)^{n+1}$.
23. Given $y = 1/x(1 - x)$, show that $y_n = n! [(-1)^n/x^{n+1} + 1/(1 - x)^{n+1}]$.
24. Given $y = \tan^{-1} x$, show that $(1 + x^2)y_2 + 2xy_1 = 0$ and then use the theorem of Leibniz to show that

$$(1 + x^2)y_{n+2} + 2(n + 1)xy_{n+1} + n(n + 1)y_n = 0.$$

25. Given $y = (x^2 - 1)^n$, show that $(1 - x^2)y_1 + 2nxy = 0$, and hence that

$$(1 - x^2)y_{n+2} - 2xy_{n+1} + n(n + 1)y_n = 0.$$

14.4.8 Significance of the Sign of the Second Derivative It follows from Section 14.4.6 that $f'(x)$ is increasing at any point where $f''(x) > 0$, i.e. the slope of the curve $y = f(x)$ is increasing, and so the curve is bending upwards. We say the curve is *concave upwards* (or *convex downwards*) at such a point; in the neighbourhood of the point the curve lies *above* the tangent at that point.

Likewise at any point where $f''(x) < 0$ the curve $y = f(x)$ is *convex upwards* (or *concave downwards*); in the neighbourhood of the point the curve lies *below* the tangent at that point.

Thus the curve in Fig. 14.19 is convex upwards between P and R, concave upwards between R and T, convex upwards between T and U, and concave upwards between U and V. At the points R, T and U the curve changes its direction of bending; thus the sign of $f''(x)$ changes from $-$ to $+$ at R, from $+$ to $-$ at T, and from $-$ to $+$ at U. These points are called *points of inflexion* and clearly we must have $f''(x) = 0$ at these points, i.e. $f''(r) = f''(t) = f''(u) = 0$. The curve *crosses* its tangent at a point of inflexion.

Summarising,

$$\frac{d^2y}{dx^2} > 0 \Leftrightarrow \text{curve is concave upwards,}$$

$$\frac{d^2y}{dx^2} < 0 \Leftrightarrow \text{curve is convex upwards,}$$

$$\frac{d^2y}{dx^2} = 0 \text{ and changes sign} \Leftrightarrow \text{curve has a point of inflexion.}$$

Example 18 Find the points of inflexion on the curves

(a) $y = x^3 + 6x^2 + 7x + 1$, (b) $y = x^4 + 4x^3 + 6x^2 - 2$.

From (a) we get

$$\frac{dy}{dx} = 3x^2 + 12x + 7, \quad \frac{d^2y}{dx^2} = 6x + 12.$$

Clearly $d^2y/dx^2 = 0$ and changes sign at the point $x = -2$, i.e. the point $(-2, 3)$ is a point of inflexion.

From (b) we get

$$\frac{dy}{dx} = 4x^3 + 12x^2 + 12x, \quad \frac{d^2y}{dx^2} = 12x^2 + 24x + 12 = 12(x + 1)^2.$$

The only point where we have $d^2y/dx^2 = 0$ is at $x = -1$, but d^2y/dx^2 does not change sign there (in fact $d^2y/dx^2 \geqslant 0$ for all x); hence there are no points of inflexion on this curve.

Note that R (Fig. 14.19) is a point of inflexion where the tangent is backward sloping, T is a point of inflexion where the tangent is forward sloping, i.e. $f'(r) < 0, f'(t) > 0$. However U is a rather special point of inflexion which is also a stationary point, i.e. $f'(u) = 0$.

14.4.9 Criteria for Maximum and Minimum Turning Points The curve $y = f(x)$ has a max. T.P. at $x = a$ if $f'(a) = 0$ and $f'(x)$ changes sign from $+$ to $-$ as x increases through the value a.

Alternatively, there is a max. T.P. at $x = a$ if $f'(a) = 0$ and $f''(a) < 0$.

The curve $y = f(x)$ has a min. T.P. at $x = a$ if $f'(a) = 0$ and $f'(x)$ changes sign from $-$ to $+$ as x increases through the value a.

Alternatively, there is a min. T.P. at $x = a$ if $f'(a) = 0$ and $f''(a) > 0$.

The alternative form does not enable us to decide the nature of the turning point if $f''(a) = 0$; it may also be inconvenient if $f''(x)$ is difficult to find.

Example 19 Find the turning points on the following curves and determine their nature:

(a) $y = 2x^3 - 15x^2 + 36x - 28$, (b) $y = (x-1)^4$, (c) $y = x/\sqrt{(x-1)}$.

(a) We have

$$\frac{dy}{dx} = 6x^2 - 30x + 36 = 6(x-2)(x-3), \quad \frac{d^2y}{dx^2} = 12x - 30.$$

Hence $dy/dx = 0$ when $x = 2$ or $x = 3$.
When $x = 2$, $d^2y/dx^2 = -6 < 0$ so that this point is a max. T.P.
When $x = 3$, $d^2y/dx^2 = 6 > 0$ so that this point is a min. T.P.

(b) We have

$$\frac{dy}{dx} = 4(x-1)^3, \quad \frac{d^2y}{dx^2} = 12(x-1)^2.$$

Hence $dy/dx = 0$ only when $x = 1$.
When $x = 1$, $d^2y/dx^2 = 0$ so this test fails to determine the nature of the point. However it is easy to see that dy/dx changes sign from $-$ to $+$ as x increases through the value 1, so the point is a min. T.P.

(c) We have

$$\frac{dy}{dx} = \frac{\sqrt{(x-1)} - x/[2\sqrt{(x-1)}]}{x-1} = \frac{x-2}{2(x-1)^{3/2}}.$$

Hence $dy/dx = 0$ only when $x = 2$.
In this case it is more laborious to find d^2y/dx^2 than to check that dy/dx changes sign from $-$ to $+$ as x increases through the value 2. The point is a min. T.P.

Exercises 14(h)

1. Find the position and nature of the stationary points on the following curves:

(a) $y = x^2 + 4x + 3$, (b) $y = 5 + 4x - x^2$,
(c) $y = (x+5)^3$, (d) $y^3 = x^3/(x-1)$,
(e) $x = t^3 + 3t^2, y = t^4 - 8t^2$.

2. Find a and b if the curve $y = x^2(x + a)(x + b)$ has stationary points when $x = 1$ and $x = 2$.
3. Show that the curve $y = (3/x) - x^3$ has no stationary points but that it has two points of inflexion.
4. For the curve $y = (2x + 1)^6$ show that

   (a) $\frac{dy}{dx} = 0$ and changes sign at $x = -\frac{1}{2}$,

   (b) $\frac{d^2y}{dx^2} = 0$ and does not change sign at $x = -\frac{1}{2}$.

   What is the nature of the stationary point at $x = -\frac{1}{2}$?
5. For the curve $y = (2x + 1)^5$ show that

   (a) $\frac{dy}{dx} = 0$ and does not change sign at $x = -\frac{1}{2}$,

   (b) $\frac{d^2y}{dx^2} = 0$ and changes sign at $x = -\frac{1}{2}$.

   What is the nature of the stationary point at $x = -\frac{1}{2}$?
6. Obtain the values of x for which $x^3(x - 1)^2$ is stationary, determining which give maxima and which give minima.
7. Given $f(x) = (x - a)(x^2 - 2bx + c)$, where a, b, c are constants, find $f'(x)$ and $f''(x)$ and show that $f'(b)f''(b) = 2f(b)$. Show also that the curve $y = f(x)$ has a point of inflexion, and that if $b = a$ the point of inflexion lies on the x-axis.
8. Given that $y^4 + xy + 1 = 0$, find dy/dx and d^2y/dx^2 in terms of x and y. Verify that the curve $y^4 + xy + 1 = 0$ has no stationary points and no points of inflexion.
9. The equation of a plane curve is $y^3 + x^3 - 9xy + 1 = 0$, and (x_1, y_1) is a point on the curve at which the tangent is parallel to the x-axis. Show that, at (x_1, y_1), $d^2y/dx^2 = 18/(27 - x_1^3)$. Show also that the stationary values of y occur at the points for which $x = (27 \pm 3\sqrt{78})^{1/3}$, and determine which of these gives a maximum and which a minimum.
10. Find the equations of the tangents to the curve $y = x^3 - 3/x$ at the points of inflexion on the curve.
11. Prove that $(0, b)$, $(b, 0)$ are the only points of inflexion on the curve $x^3 + 3axy + y^3 = b^3$ $(b \neq a)$.

14.4.10 Fundamental Properties of Differentiable Functions In the following theorems we shall denote by I_c the closed interval $a \leqslant x \leqslant b$, and by I_o the open interval $a < x < b$. A point in the interval other than an end-point is referred to as an *interior point*.

Theorem 6 If f is continuous on I_c and differentiable on I_o, then the largest value of $f(x)$ in I_c occurs either at a max. T.P. or else at an end-point, and the smallest value of $f(x)$ in I_c occurs either at a min. T.P. or else at an end-point.

Proof. By Theorem 5, $f(x)$ has a largest value attained at some point x_1 (say) of I_c. Suppose x_1 is an interior point of the interval. Then x_1 is in I_o and f is differentiable at x_1. Since $f(x_1) \geqslant f(x)$ for any x in I_c it follows that

$$f(x_1 + \Delta x) - f(x_1) \leqslant 0,$$

and so

$$\frac{f(x_1 + \Delta x) - f(x_1)}{\Delta x} \leqslant 0 \text{ if } \Delta x > 0,$$
$$\geqslant 0 \text{ if } \Delta x < 0.$$

But

$$\lim_{\Delta x \to 0} \frac{f(x_1 + \Delta x) - f(x_1)}{\Delta x}$$

exists since f is differentiable at x_1, and so this limit must be zero, i.e. $f'(x_1) = 0$. It follows that there is a max. T.P. at $x = x_1$.

Similarly $f(x)$ has a smallest value, $f(x_2)$ say, where x_2 is an end-point, or else there is a min. T.P. at $x = x_2$.

In Fig. 14.20 two graphs are shown to illustrate how the extreme values may occur at turning points and how they may occur at end-points. In Fig. 14.21 the graph of $x^{2/3}$ shows that its smallest value in the interval $-1 \leqslant x \leqslant 1$ occurs when $x = 0$ but this is *not* a min. T.P.; $x^{2/3}$ is not differentiable at $x = 0$ so the conditions of Theorem 6 are not satisfied.

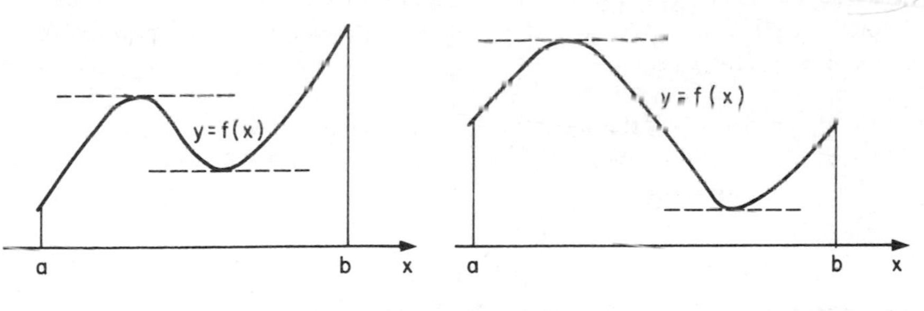

FIG. 14.20

The importance of Theorem 6 is that for a function f satisfying the conditions of the theorem we can find the largest and smallest values of $f(x)$ in I_c by considering ***only the two end-values and values at turning points.***

Example 20 Find the maximum and minimum values of $2x^3 - 3x^2$ for values of x in the interval $-1 \leqslant x \leqslant 2$.

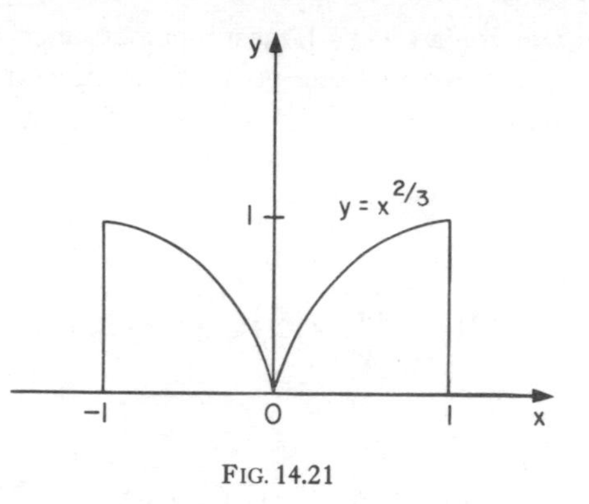

FIG. 14.21

Write $f(x) = 2x^3 - 3x^2$, so that

$$f'(x) = 6x^2 - 6x = 6x(x - 1).$$

Thus stationary values occur when $x = 0$ and when $x = 1$. We could easily check that both stationary points are turning points, but for the purposes of the problem it is sufficient to evaluate $f(x)$. Thus $f(0) = 0$, $f(1) = -1$. For the end-values we have $f(-1) = -5$, $f(2) = 4$. From the values we have found the maximum is 4 and the minimum is -5; these are the maximum and minimum values of $f(x)$ for all x in the interval.

Theorem 7 (Rolle's theorem) If f is continuous on I_c and differentiable on I_o and if $f(a) = f(b) = 0$, then there is (at least) one interior point x_0 (say) such that $f'(x_0) = 0$.

Proof. If $f(x) = 0$ for all x in I_c then the theorem obviously holds. Otherwise $f(x)$ has either a largest or a smallest value different from zero, and this value must be attained at an interior point x_0 (say). From Theorem 6 we would then have $f'(x_0) = 0$.

Theorem 8 (Mean value theorem) If f is continuous on I_c and differentiable on I_o then there is (at least) one interior point x_0 (say) such that

$$f'(x_0) = \frac{f(b) - f(a)}{b - a}.$$

Geometrically the theorem states that there is at least one point of the curve between A and B where the tangent is parallel to the chord AB (Fig. 14.22). Rolle's theorem is the particular case where A and B lie on the x-axis.

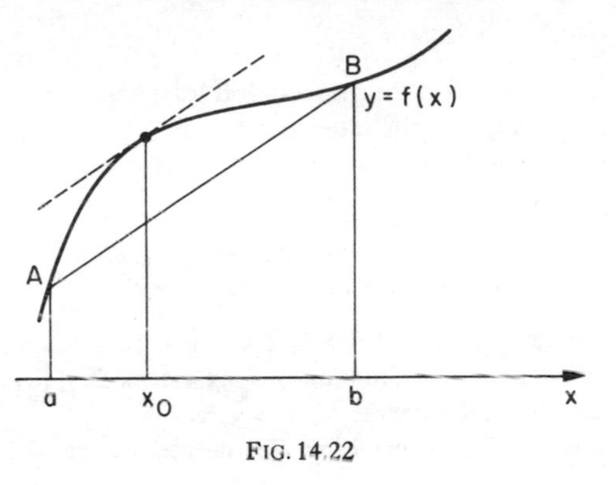

FIG. 14.22

Proof. Consider the function F where

$$F(x) = f(x) - f(a) - \frac{f(b) - f(a)}{b - a}(x - a).$$

Clearly $F(b) = F(a) = 0$, and F is continuous on I_c and differentiable on I_o with

$$F'(x) = f'(x) - \frac{f(b) - f(a)}{b - a}.$$

By Rolle's theorem there is an interior point x_0 (say) such that $F'(x_0) = 0$, that is

$$f'(x_0) = \frac{f(b) - f(a)}{b - a}.$$

Theorem 9 If f is continuous on I_c and differentiable on I_o and if in I_c $f(x)$ has only one stationary value which is a maximum turning value (or minimum turning value), then this value is the maximum (or minimum) for the whole interval.

Proof. Suppose there is a max. T.P. at $x = x_1$ and no other stationary points in I_c. By Theorem 6 we require to show that the value of $f(x)$ at an end-point cannot exceed $f(x_1)$. Suppose $f(a) > f(x_1)$.

There must be some neighbourhood of x_1 in which $f(x_1)$ is the largest value. Select a point x_2 in this neighbourhood, left of x_1, so that $x_2 < x_1$ and $f(x_2) < f(x_1)$. This means that $f(x_1)$ lies between $f(x_2)$ and $f(a)$, so by Theorem 5 (Section 14.2.1) there must be a point, x_3 say, between a and x_2 such that $f(x_3) = f(x_1)$. Now consider the interval $x_3 \leqslant x \leqslant x_1$; by the mean value theorem there must be a point, x_4 say, in this interval such that $f'(x_4) = 0$ which violates the supposition that $f(x)$ has no stationary values in I_c other than the stationary value at x_1.

Exercises 14(i)

1. Given $f(x) = x^3 + x^2$, find the greatest and least values of $f(x)$ (a) on the interval $-2 \leqslant x \leqslant 0$, (b) on the interval $-1 \leqslant x \leqslant 1$.
2. Find the greatest and least values of

   (a) $4x - 3x^2 - 6$ on the interval $0 \leqslant x \leqslant 1$,
   (b) $2x^3 - 15x^2 + 36x - 28$ on the interval $1 \leqslant x \leqslant 3$,
   (c) $5x^{1/2} + x^{-5/2}$ on the interval $1 \leqslant x \leqslant 4$.

3. Given $y = 2 + 2x - x^2$ for $-1 \leqslant x \leqslant 2$, and $y = 16/x + x - 8$ for $2 < x \leqslant 6$, find (a) the maximum and minimum turning values of y, (b) the greatest and least values of y, on the interval $-1 \leqslant x \leqslant 6$.
4. Find the greatest and least values of $f(x)$ on the interval $-1 \leqslant x \leqslant 1$, given

$$f(x) = (3x^2 - 5)/(2x^2 - x - 6).$$

   What is the answer if the interval $-2 \leqslant x \leqslant 2$ is taken?
5. Find the least value of each of the following expressions, x and θ being real: (a) $\sqrt{(x^2 + 4x + 6)}$, (b) $\cos^2 \theta + 4 \cos \theta + 6$.

CHAPTER 15

APPLICATIONS OF DIFFERENTIAL CALCULUS

15.1 Introduction

We recall that if $y = f(x)$, then the value of dy/dx at the point $x = x_0$, i.e. $f'(x_0)$, measures

(a) the rate of change of y with respect to x at the point $x = x_0$, or, in other words,

(b) the slope of the curve $y = f(x)$ at the point $x = x_0$.

We recall also that y is increasing, decreasing or stationary as x increases through the value x_0 according as $f'(x_0)$ is positive, negative or zero.

These ideas are fundamental in all applications of the derivative.

15.1.1 One-dimensional Motion Suppose the point P moves in a straight line and let the position of P at any time t be given by its s-coordinate, where s is the distance of P, measured along the line, from a fixed point O on the line. If s is measured positively on one side of O, negatively on the other side, then s is a function of t, $s = f(t)$ say, and ds/dt measures the velocity of the moving point. On writing $v = ds/dt$, it follows that when $v > 0$, s is increasing and so P is moving along the curve in the direction left to right if s is measured positively on the right-hand side of O as in Fig. 15.1 (or right to left if s is measured positively

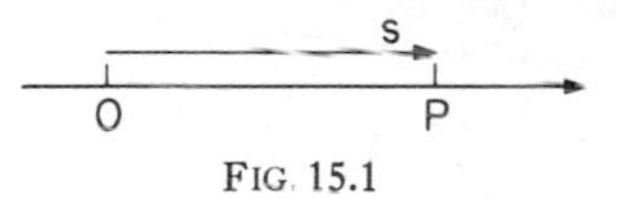

FIG. 15.1

on the left-hand side), and when $v < 0$, s is decreasing, so that P is moving in the opposite direction, i.e.

$$v > 0 \Leftrightarrow \text{motion is in the positive } s\text{-direction},$$
$$v < 0 \Leftrightarrow \text{motion is in the negative } s\text{-direction}.$$

The term *speed* is used for the quantity $|v|$, i.e. $|ds/dt|$; thus speed is never negative and does not distinguish between the two possible directions of motion.

The rate of change of v is called the acceleration, i.e.

$$\text{acceleration} = \frac{dv}{dt} = \frac{d^2s}{dt^2} = a \quad \text{(say).}$$

Thus a can take both negative and positive values; when a is negative v is decreasing, and the word ***deceleration*** is sometimes used for negative acceleration.

The *distance–time curve*, i.e. the curve $s = f(t)$, is frequently used in problems of one-dimensional motion. Figure 15.2 shows a distance–time curve along with the associated velocity–time and acceleration–time curves.

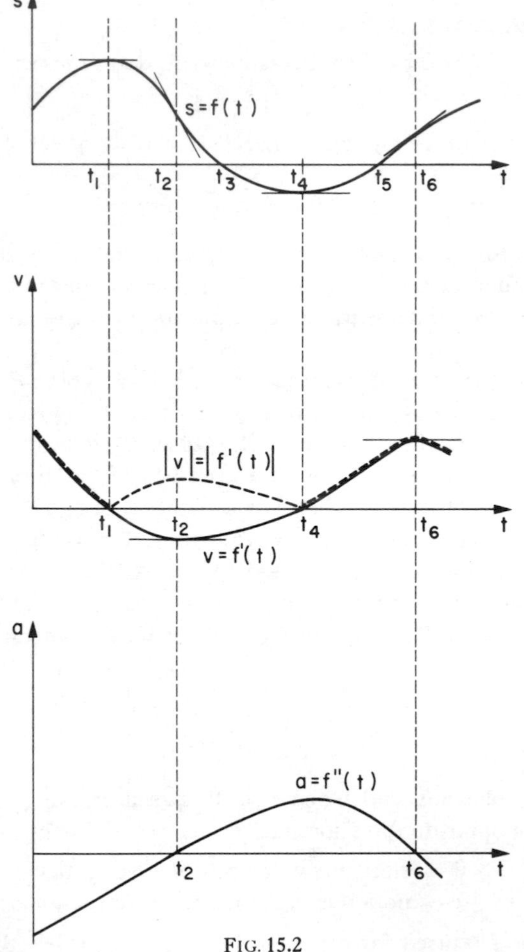

FIG. 15.2

The following observations about the distance–time curve should be noted:

(a) Points (e.g. t_3, t_5) where the curve meets the t-axis represent times when the moving point P is at O (Fig. 15.1).

(b) The slope of the curve is v; when the curve is rising we have $v > 0$, and when the curve is falling we have $v < 0$.
(c) T.P.s (e.g. at t_1, t_4) represent times when the direction of motion is reversed; at such times $v = 0$.
(d) When the curve is convex upwards we have $a < 0$ and when the curve is concave upwards we have $a > 0$.
(e) Points of inflexion (e.g. at t_2, t_6) represent times when $a = 0$.

Along with the velocity–time curve in Fig. 15.2 is shown the speed–time curve (broken curve). Note that when the velocity is negative an increase in velocity implies a decrease in speed. It will be seen from the graphs that

increasing speed $\Leftrightarrow$ v and a have the same sign,
decreasing speed $\Leftrightarrow$ v and a have opposite signs.

Example 1 Given that $s = t^3 - 6t^2 + 9t$ determine

(a) when the velocity is increasing and when it is decreasing,
(b) when the speed is increasing and when it is decreasing,
(c) the distance between the positions at $t = 0$ and $t = 4$,
(d) the distance travelled between $t = 0$ and $t = 4$.

We have

$$v = \frac{ds}{dt} = 3t^2 - 12t + 9 = 3(t-1)(t-3).$$

Hence

$$v > 0 \text{ if } t < 1 \text{ or } t > 3 \quad \text{and} \quad v < 0 \text{ if } 1 < t < 3.$$

Also

$$a = \frac{d^2s}{dt^2} = 6t - 12.$$

Hence

$$a > 0 \text{ if } t > 2 \quad \text{and} \quad a < 0 \text{ if } t < 2,$$

i.e. v is increasing if $t > 2$ and decreasing if $t < 2$.

However $|v|$ is increasing if $1 < t < 2$ or $t > 3$ and decreasing if $t < 1$ or $2 < t < 3$.

When $t = 0$ we have $s = 0$, when $t = 4$ we have $s = 4$; hence the distance between these positions is 4 units.

Since the direction of motion reverses at $t = 1$ and $t = 3$, we find the corresponding positions which are $s = 4$ and $s = 0$ respectively. It follows that the total distance travelled is 4 units forwards, 4 units backwards and 4 units forwards again, i.e. 12 units.

15.1.2 Related Rates If y is a function of x and both x and y are changing with respect to time, then there must be some connection between their two rates of change. This connection is given by the chain rule, since we must have

$$\frac{dy}{dt} = \frac{dy}{dx}\frac{dx}{dt}.$$

Example 2 Water pours into a right circular cylindrical tank at the rate of 4 cm^3/s. At what rate is the level rising in the tank?

The volume (V cm^3) of water in the tank at any time t is given by

$$V = \pi r^2 h \quad (r \text{ a constant})$$

where r cm is the radius of the tank and h cm is the depth of water in the tank at time t. Now

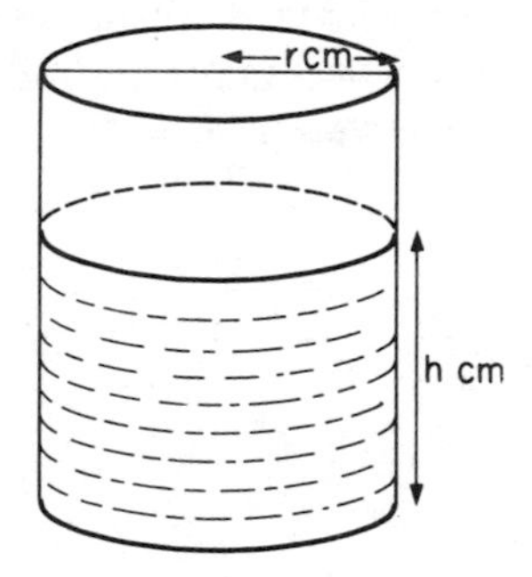

FIG. 15.3

$$\frac{dV}{dt} = \frac{dV}{dh}\frac{dh}{dt}$$

and so

$$\frac{dV}{dt} = \pi r^2 \frac{dh}{dt}.$$

But $dV/dt = 4$ and so $dh/dt = 4/\pi r^2$. Hence the level is rising at the rate of $4/\pi r^2$ cm/s.

Example 3 Solve the problem of Example 2 for the case in which the tank is conical, as shown in Fig. 15.4.

In this case $V = \frac{1}{3}\pi r^2 h$ (r, h both variables).

Since the semi-vertical angle of the cone is 30° we have

$$r = (1/\sqrt{3})\, h.$$

Hence

$$V = (\pi/9)\, h^3.$$

Thus

$$\frac{\mathrm{d}V}{\mathrm{d}t} = \frac{\mathrm{d}V}{\mathrm{d}h}\frac{\mathrm{d}h}{\mathrm{d}t} = (\pi/3)\, h^2\, \mathrm{d}h/\mathrm{d}t.$$

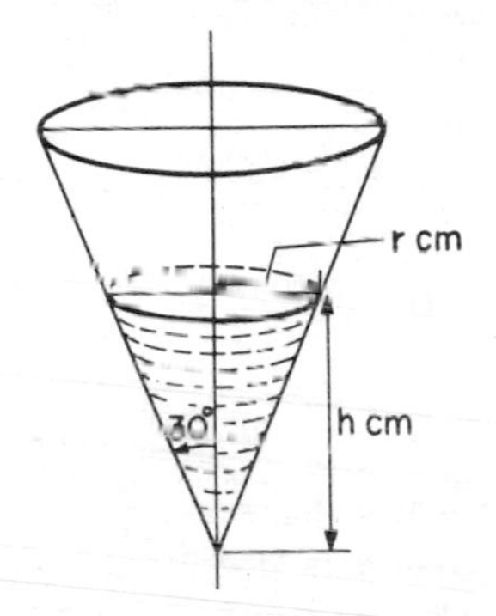

FIG. 15.4

That is

$$4 = (\pi/3)\, h^2\, \mathrm{d}h/\mathrm{d}t$$

giving

$$\frac{\mathrm{d}h}{\mathrm{d}t} = \frac{12}{\pi h^2}\ \mathrm{cm/s}.$$

Hence the level is rising at the rate of $12/\pi h^2$ cm/s.

Example 4 An aeroplane flying in a straight path at a constant height of 3 km passes directly over a searchlight on the ground and is subsequently caught in the beam of the searchlight with the beam at an angle of 60° to the level ground. In order to keep the aeroplane in the beam it is found necessary to start lowering the beam at 4 degrees/s. What is the speed of the aeroplane (in km/h) at that moment?

Let S be the position of the searchlight, O the point on the flight path 3 km above S. Let P be the position of the aeroplane at time t, and suppose the distance OP is x km while SP makes an angle of θ radians with the ground.

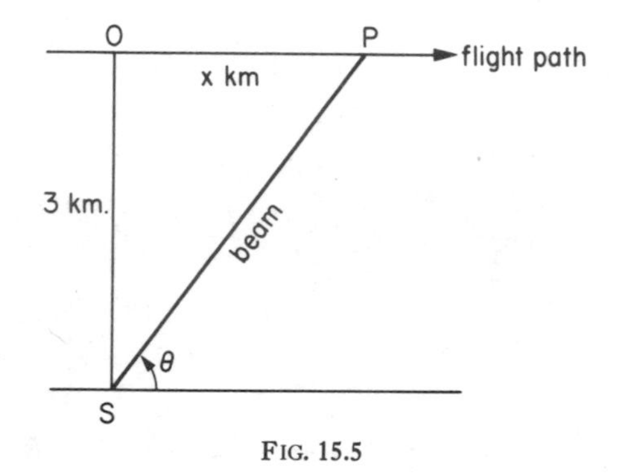

FIG. 15.5

We have $x = 3 \cot \theta$. Hence

$$\frac{dx}{dt} = -3 \operatorname{cosec}^2 \theta . \frac{d\theta}{dt}.$$

When the line SP rotates at 4 degrees/s, θ is changing at $4\pi/180$ rad/s. Since SP rotates clockwise θ is decreasing so that $d\theta/dt$ is negative. Hence, when $\theta = \pi/3$ we have $d\theta/dt = -4\pi/180$ (assuming t measured in seconds) and so

$$\frac{dx}{dt} = (3 \operatorname{cosec}^2 \pi/3) \frac{4\pi}{180} = \frac{16\pi}{180}.$$

It follows that x is increasing at the rate of $16\pi/180$ km/s, i.e. 320π km/h, and so the plane is flying at 320π km/h.

Example 5 The radius of a sphere increases uniformly from R to $3R$ in time T. Initially the sphere contains gas at pressure P and at all times the pressure p and volume v of the gas satisfy the relation $pv^N = K$, where N and K are constants. Show that when the radius of the sphere is $2R$ the pressure of the gas is decreasing at the rate $3NP/2^{3N}T$.

It is necessary in this problem to keep clear the distinction between a variable such as p, which is a function of time, and a particular value of that variable, e.g. P, which is a constant. To assist the reader all constants have been denoted by capital letters.

Let r be the radius of the sphere at time t, so that

$$p(\tfrac{4}{3}\pi r^3)^N = K.$$

In particular, $P(\frac{4}{3}\pi R^3)^N = K$, and so, replacing K in the first equation,

$$p(\tfrac{4}{3}\pi r^3)^N = P(\tfrac{4}{3}\pi R^3)^N,$$

that is

$$p = PR^{3N} r^{-3N}.$$

Hence

$$\frac{dp}{dr} = -3NPR^{3N} r^{-3N-1}$$

and so

$$\frac{dp}{dt} = -3NPR^{3N} r^{-3N-1} \frac{dr}{dt}.$$

Since r increases uniformly by $2R$ in time T, it follows that dr/dt has the constant value $2R/T$. Hence

$$\frac{dp}{dt} = -3NPR^{3N} r^{-3N-1} \frac{2R}{T},$$

and putting $r = 2R$ this gives

$$\frac{dp}{dt} = -\frac{3NP}{2^{3N}T}.$$

Exercises 15(a)

1. A particle moves along a straight line so that at time t s its distance s m from the origin is given by $s = (2t - 3)(2 + t^2/6)$. Find the distance of the particle from the origin when its velocity is 6 m/s.
2. A particle moves along a straight line according to the law

$$s = t^3 - 9t^2 + 24t.$$

   When is the speed of the particle increasing? Find the total distance travelled in the period from $t = 0$ to $t = 5$.
3. In a triangle ABC the sides AB, AC are 7 cm, 9 cm respectively. Given that A increases at the steady rate of 1° per minute, find the rate at which the area of the triangle is increasing when $A = 30°$.
4. A street light is 3·6 m above the ground and a man of height 1·8 m is walking away from it at the rate of 1·5 m/s. Find the rate of growth of his shadow.
5. At a certain port, t h after high water, the height h m of the tide above a fixed datum is given by $h = 10 + 7\cos(\pi t/6)$. At what rate in cm/min is the tide falling one hour after high water?

6. A ship leaves port at noon steaming due north at 12 knots. Another ship leaves the same port one hour later, steaming due east at 16 knots. How fast are they separating at 4 p.m. the same day?
7. A spherical block of ice is melting. At any instant the rate at which the volume of the block is decreasing is k times the surface area of the block. Show that the rate of change of the radius is constant.
8. A particle moves on a straight line so that its velocity at any instant is proportional to the square of the distance from a fixed point on the line. Show that the acceleration is proportional to the cube of that distance.
9. The depth of water in a hemispherical tank of radius 120 cm is increasing steadily at the rate of 4 cm/minute. Find the rate of increase of the area of the free surface of the water when the depth is 48 cm.
10. The distances u and v of a small object and its image from a lens of focal length f are connected by the formula $1/u + 1/v = 1/f$. When an object is 4 cm from the lens of focal length 2·5 cm, it is moved towards the lens at 9 cm/s. Find the rate at which the image recedes from the lens.
11. A tank is in the shape of a frustum of a right circular cone, with upper radius 120 cm and lower radius 60 cm and depth 144 cm. When the depth of water is 60 cm the level is rising at the rate of 4 cm/min. How fast is the water pouring in?
12. A ship steams at a speed of 20 km/h along a course parallel to a straight shore, always maintaining a distance of 4 km from the shore line. A radar station is situated ahead of the ship on the shore line. At the instant when the ship is 5 km from the radar station
    (a) how fast is the distance between the radar station and the ship decreasing?
    (b) how fast is the line from the radar station to the ship rotating?
13. An aeroplane flying at an altitude of 3000 m passes directly over an observer on the ground. Subsequently the observer notices that his line of sight to the plane makes an angle of $\pi/3$ rad with the horizontal and that this angle is decreasing at the rate of 0·09 rad/s. Find the speed of the aeroplane and also the rate at which the distance between the plane and the observer is increasing at that moment.
14. A cameraman televising a 100 m sprint is located 10 m from the track and in line with the finishing tape. When the runners are 10 metres from the finishing tape his camera is turning at the rate of 27° per second. What is the speed (in m/s) of the runners at that instant?

    Assuming the runners maintain a constant speed over the last 10 m, at what rate (in degrees per second) will the camera be turning when they reach the finishing tape?
15. A conical funnel has a radius of 10 cm at the top and a height of 20 cm. Oil is poured through the funnel and is coming out at a rate of 10 cm^3/s when the depth of oil in the funnel is 12 cm and is rising at a rate of 1 cm/s. At what rate is oil being poured into the top of the funnel at that moment, assuming the funnel sits upright and the radius of the opening at the bottom is negligible.

16. For a body projected upwards from the earth's surface, the velocity v and distance s from the centre of the earth are related by the equation

$$v = \sqrt{\left(v_0^2 - 2gR + \frac{2gR^2}{s}\right)},$$

where v_0, g, R are constants. Deduce from this equation that the acceleration is towards the earth and is inversely proportional to s^2.

17. A point P moves along the curve $y = 8/x^2$ so that its x-co-ordinate increases at the steady rate of 3 units per second. Q is the fixed point (0, 4). At what rate is the distance between P and Q increasing when $x = 2$, and what is the rate of change of the slope of the line PQ when $x = 2$?

15.1.3 Problems Concerning Maximum and Minimum Values

Many problems reduce to the fundamental problem of finding the maximum or minimum value of $f(x)$ in some closed interval, where the function f is continuous on that interval and differentiable at all points of the interval except (possibly) the end-points. The necessary mathematical tools have been developed in Section 14.4.10 (Theorem 6 and Theorem 9).

Example 6 How should a wire of length 48 cm be bent to produce the rectangle of maximum area?

Suppose the wire is bent at a point x cm from one end to form one side of length x cm. If the other side of the rectangle is y cm we require that $2x + 2y = 48$, i.e. $y = 24 - x$. Hence the area A of the rectangle is given by $A = x(24 - x) = 24x - x^2$.

If we allow the values $x = 0$ and $x = 24$ as limiting cases where one side of the rectangle has shrunk to zero, we can now regard the problem as that of finding the maximum value of $24x - x^2$ for values of x in the interval $0 \leqslant x \leqslant 24$. A is continuous and differentiable on this interval, and

$$\frac{dA}{dx} = 24 - 2x.$$

Clearly there is a max. T.P. when $x = 12$ and this gives the maximum value for the interval. Thus the wire should be bent to form a square of side 12 cm.

Example 7 A circular cylinder is cut from a cone as shown in Fig. 15.6. Given that the radius of the base of the cone is 2 cm and its height is 12 cm, find the maximum volume that the cylinder can have.

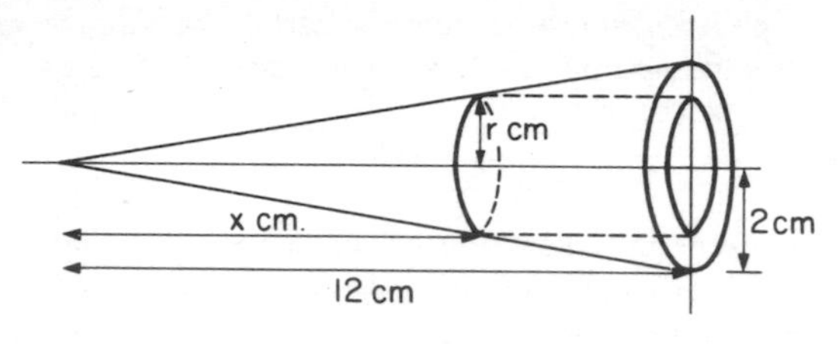

FIG. 15.6

Suppose the top of the cylinder is x cm from the vertex of the cone (see Fig. 15.6). Then the height of the cylinder is $(12 - x)$ cm. Let the radius of the cylinder be r cm. From similar triangles we have

$$\frac{r}{x} = \frac{2}{12}, \quad \text{that is} \quad r = \frac{1}{6}x.$$

Hence the volume (V cm^3) of the cylinder is given by

$$V = \pi r^2(12 - x) = (\pi/36)(12x^2 - x^3).$$

We wish to find the maximum value of V on the interval $0 \leqslant x \leqslant 12$. Clearly V is continuous and differentiable on this interval, and

$$\frac{dV}{dx} = \frac{\pi}{36}(24x - 3x^2) = \frac{\pi}{12}x(8 - x).$$

The stationary values of V occur when $x = 0$ and $x = 8$ and are $V = 0$, $V = \frac{64}{9}\pi$ respectively. Comparing with the value at $x = 12$, viz. $V = 0$, we see that $\frac{64}{9}\pi$ is the maximum value for the interval.

Note 1. It is essential to the method that the quantity being maximised or minimised be expressed in terms of only *one* variable. The choice of this one variable may affect the amount of labour involved, as shown in the next example.

Example 8 Find the rectangle of greatest area that can be inscribed in a circle of radius a.

Using variables x, y as shown in Fig. 15.7, the area A of the rectangle is given by $A = 4xy$.

We can express A in terms of x only, since $x^2 + y^2 = a^2$. Thus $A = 4x\sqrt{(a^2 - x^2)}$, and we require the maximum value of A in the interval $0 \leqslant x \leqslant a$.

Alternatively we may use the variable θ. Since $x = a \cos\theta, y = a \sin\theta$ we have

$$A = 4a^2 \sin\theta \cos\theta = 2a^2 \sin 2\theta.$$

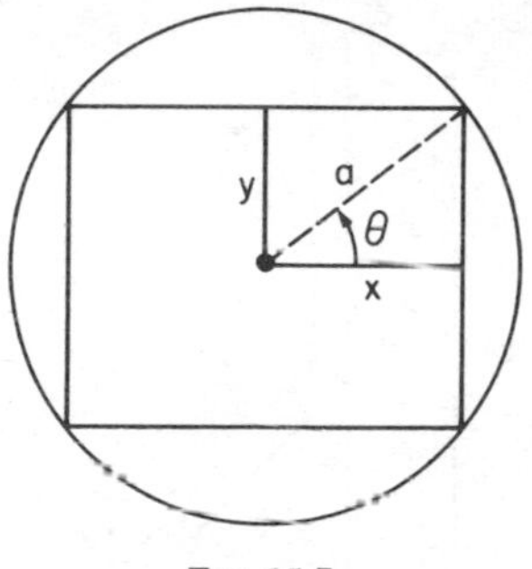

FIG. 15.7

We require the maximum value of A in the interval $0 \leqslant \theta \leqslant \pi/2$. Now

$$\frac{dA}{d\theta} = 4a^2 \cos 2\theta,$$

so that A has a single stationary value in the interval, occurring at $\theta = \pi/4$.

Further,

$$\frac{d^2A}{d\theta^2} = -8a^2 \sin 2\theta$$

and so when $\theta = \pi/4$ we have $d^2A/d\theta^2 = -8a^2 < 0$. Thus the stationary point is a max. T.P. and gives the maximum value for the whole interval.

When $\theta = \pi/4$ the rectangle is a square of side $\sqrt{2}a$.

Note 2. Implicit differentiation is sometimes a useful labour-saving device, as in the next example.

Example 9 A ship is sailing due south at 20 km/h and a second ship is sailing due east at 15 km/h. At a certain time the second ship is 100 km due south of the first ship. When will the ships come closest to each other?

Let the points A, B denote the positions of the ships at the time, $t = 0$ say, when the second ship is 100 km due south of the first ship (Fig. 15.8). Let P_1, P_2 denote the positions of the ships t h later; then $AP_1 = 20t$ km, $BP_2 = 15t$ km and so if $P_1P_2 = s$ km we have $s^2 = (100 - 20t)^2 + 225t^2$. Hence

$$2s\frac{ds}{dt} = -40(100 - 20t) + 450t = 1250t - 4000.$$

Since $s > 0$ for all t this last equation is sufficient to indicate that s has a minimum turning value when $t = 4000/1250 = 3\frac{1}{5}$, and no other

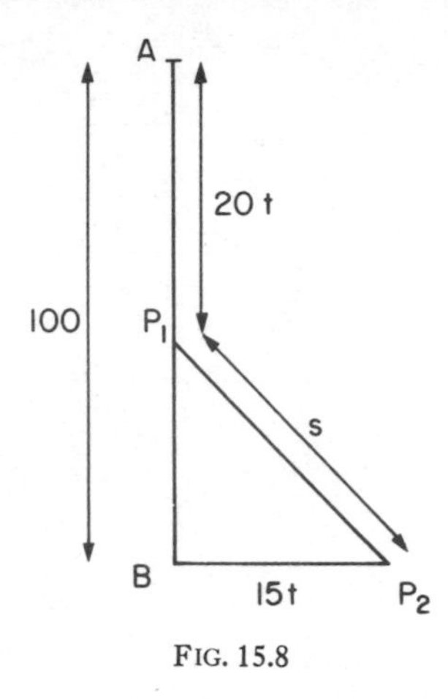

FIG. 15.8

stationary values whatever time interval we consider. Hence the ships are closest $3\frac{1}{3}$ h after being in the positions A, B.

Note 3. The method being used in this section can easily be extended to deal with the case where it is necessary to consider values of x lying in several intervals. This is illustrated in the following example.

Example 10 The right-angled triangle shown in Fig. 15.9 has sides $a = 8$ cm, $b = 6$ cm, $c = 10$ cm. A person knowing only that $c = 10$ cm finds a and b by measuring one of them and then calculating the length of the other by use of Pythagoras' theorem. If the area A of the triangle is now calculated using $A = \frac{1}{2}ab$, what is the maximum error that can occur in the area, assuming that the measurement of the side was accurate to within 5%?

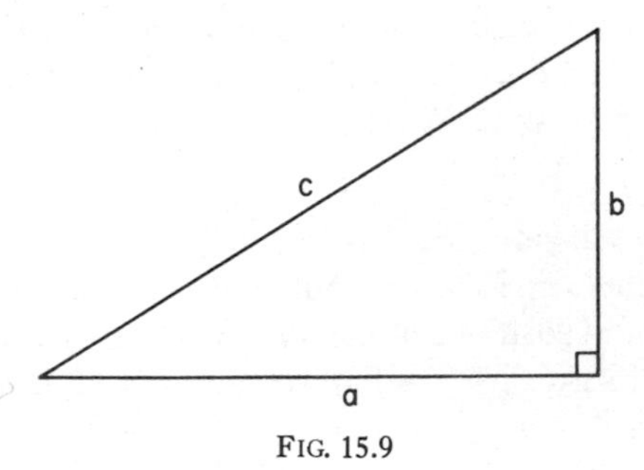

FIG. 15.9

Suppose x cm is the measurement taken. Since this is accurate to within 5% we have $5\cdot7 \leqslant x \leqslant 6\cdot3$ or else $7\cdot6 \leqslant x \leqslant 8\cdot4$, depending on whether a or b is measured. The length of the side not measured is taken as

$\sqrt{(100 - x^2)}$ and so

$$A = \tfrac{1}{2} x \sqrt{(100 - x^2)},$$

$$\frac{dA}{dx} = \tfrac{1}{2}\sqrt{(100 - x^2)} - \tfrac{1}{2}\frac{x^2}{\sqrt{(100 - x^2)}} = \frac{50 - x^2}{\sqrt{(100 - x^2)}}$$

Although A has a stationary value when $x = \sqrt{50}$, this is outside both the intervals under consideration. Hence for each interval the extreme values occur at the end-points. For the interval $5 \cdot 7 \leqslant x \leqslant 6 \cdot 3$ we find that the largest value of A is approximately 24·46, occurring when $x = 6 \cdot 3$, and the smallest value of A is approximately 23·42, occurring when $x = 5 \cdot 7$.

For the interval $7 \cdot 6 \leqslant x \leqslant 8 \cdot 4$, we find that the largest value of A is approximately 24·70, occurring when $x = 7 \cdot 6$, and the smallest value of A is approximately 22·79, occurring when $x = 8 \cdot 4$.

For the combined intervals we see that the largest value of A is 24·70 and the smallest value is 22·79.

The correct value for A is 24, so at one extreme A is too large by 0·70 and at the other extreme A is too small by 1·21. Hence the maximum error that can occur is an error of 1·21.

Note 4. Even if the values of x under consideration lie in a single interval, in order to satisfy the requirements about continuity and differentiability it may be necessary to split the interval into several smaller intervals and proceed as in Example 10. This is illustrated in Example 11.

Example 11 From the bending-moment formulae given at the end of Section 12.6.4, determine the maximum value of M, and also the maximum value of $|M|$.

We require the maximum value of M for values of x in the interval $-6 \leqslant x \leqslant 6$. Since we have given a complete graph of M (Fig. 14.10, Section 14.1.2), we have already solved this problem. However, working solely from the formulae we must note that M is not differentiable on the interval $-6 < x < 6$ (it is not differentiable at $x = \pm 4$) and it happens that the single stationary value (at $x = 0$) and the end-values (at $x = -6$ and $x = 6$) do not include the maximum value for the interval.

On the closed intervals $-6 \leqslant x \leqslant -4$, $-4 \leqslant x \leqslant 4$, $4 \leqslant x \leqslant 6$, M is continuous, and on the corresponding open intervals $-6 < x < -4$, $-4 < x < 4$, $4 < x < 6$, M is differentiable. Hence it is sufficient to consider stationary values and end-values for these intervals; when $x = -6$, -4, 4, 6 we find $M = 0, 2, 2, 0$ respectively and it is easily shown that the only stationary point is at $x = 0$, when $M = -6$. Thus the maximum value of M on the interval $-6 \leqslant x \leqslant 6$ is 2, and the maximum value of $|M|$ is 6.

Note 5. Some care may be needed in interpreting the problem so that the correct quantity is maximised, or minimised. The following example illustrates this point.

Example 12 Right circular tin cans have to be made with a specified volume. No waste is involved in cutting the tin for the curved surface of the can, but the top and bottom are each cut from a square piece of tin and the corners of these squares are wasted. Find the dimensions of the most economical cans.

We must be careful to interpret the most economical cans as meaning the cans that use the least tin *including the waste.* Let the radius of the can be r and the height h. If the specified volume of the can is V, then we must have $\pi r^2 h = V$, that is $h = V/\pi r^2$. To make the circular top of the can we require a square of side $2r$, and similarly for the circular bottom.

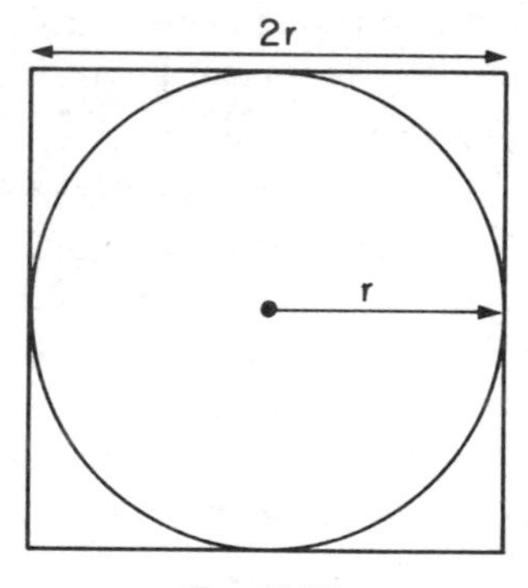

FIG. 15.10

Since the curved surface area is $2\pi rh$, it follows that the amount A of tin required is given by

$$\begin{aligned} A &= 2\pi rh + 8r^2 \\ &= \frac{2V}{r} + 8r^2 \qquad \text{(expressing } A \text{ in terms of one variable).} \end{aligned}$$

Hence

$$\frac{dA}{dr} = -\frac{2V}{r^2} + 16r,$$

and it is easily seen that the only stationary point is a min. T.P. when $r = \frac{1}{2}V^{1/3}$.

Since A is differentiable with respect to r for all $r > 0$, it follows that the minimum value of A occurs when $r = \frac{1}{2}V^{1/3}$; this gives $h = (4/\pi)V^{1/3}$.

Exercises 15(b)

1. Show that in Exercises 12(f) (a) the maximum volume of the box in Question 1 is 432 cm^3, (b) the minimum total area in Question 2 is 72 cm^2, (c) the maximum area of the rectangle in Question 3 is 18 cm^2.
2. The amplification of a wireless valve in a certain circuit is proportional to $x/(x^2 + a^2)$, where x is a variable and a is constant. Show that the amplification is greatest when $x = a$.
3. If y m is the height attained by a projectile when it has travelled x m horizontally, then (neglecting air resistance)

$$y = x \tan \alpha - \frac{16x^2}{(u \cos \alpha)^2}$$

where u m/s is the initial velocity and α is the angle between the direction of projection and the horizontal. Find the maximum value of y if $u = 800$ and $\alpha = 45°$.
4. A long rectangular sheet of tin is $2a$ cm wide and is to be bent along its central longitudinal axis to form a V-shaped trough. What is the depth of the trough if its cross-sectional area is the largest possible?
5. The sum of the squares of two positive real numbers is unity. Show that the product of the two numbers is greatest if the numbers are equal.
6. In the previous problem the word 'squares' can be replaced by 'cubes' or more generally by 'nth powers' and the conclusion is the same.
7. A man is in a boat 3 km from the nearest point A of a straight coast. He wishes to reach a point B on the coast, 10 km from A, in the shortest possible time. If he can row at 4 km/h and run at 5 km/h, how far from A should he land?
8. A, B and C are three points on the edge of a circular lake, A and C being diametrically opposite. A man swims in a straight line from A to B at a steady speed of 3 knots, climbs into a boat at B and from there rows in a straight line to C at a steady speed of 4 knots. Given that the diameter of the lake is 3 nautical miles show that, for all possible positions of B on the edge, the longest time taken for the complete journey from A to C is $1\frac{1}{4}$ h. (1 knot = 1 nautical mile/hour.)
9. Towns A, B, C are located 6 km west, 6 km east, and 10 km south, respectively, of a point D. A road is to run north from C to a point P and from P a branch road is to run to A and another branch road to B. Show that the minimum total length $PA + PB + PC$ for the three roads is $(10 + 6\sqrt{3})$ km.
10. From the vertex A of a triangle ABC a line is drawn to meet the opposite side BC at P, thus dividing ABC into two triangles ABP, APC. Find the position of P that gives the maximum value for the product of the areas of these two triangles.
11. Find the dimensions of the rectangle of greatest area that can be inscribed in an ellipse with semi-axes a and b, the sides of the rectangle being parallel to these axes.
12. Show that the rectangle of greatest area that can be inscribed in a given triangle has half the area of the triangle.

13. A rectangle is inscribed in a right-angled triangle so as to have one angle coincident with the right angle. Show that its area is maximum when the opposite corner bisects the hypotenuse.
14. For the rectangle in the previous problem, show that there is neither a maximum nor a minimum value for the perimeter.
15. Assuming that for beams with rectangular cross-section the flexural rigidity is proportional to the breadth and to the cube of the depth, find the breadth and depth of the rectangular beam of maximum flexural rigidity that can be cut from a cylindrical log of radius a.
16. A light mounted on a vertical pole is to illuminate a small object floating in a pool at horizontal distance d from the light. Assume that the illumination at the object varies inversely as the square of the distance between the object and the light and directly as $\cos\theta$, where θ is the angle that the light rays make with the normal to the water surface at the object. Express the illumination at the object in terms of d and θ. Hence determine the height at which the light should be mounted to give maximum illumination at the object.
17. A wall h m high stands d m from a tall building. Find the shortest ladder that can reach the building from the ground outside the wall.
18. How far from the wall of a house must a man stand in order that a window 2 m high, whose sill is 6 m above his eye-level, may subtend the greatest vertical angle?
19. A cone of vertical angle 2θ is inscribed in a sphere of radius a. Show that the volume of the cone is $\frac{8}{3}\pi a^3 \sin^2\theta \cos^4\theta$. Deduce that the volume of the cone cannot exceed $\frac{8}{27}$ times the volume of the sphere.
20. It can be shown that when a body of weight W is drawn slowly along a rough horizontal plane, the pull on the rope is equal to $\mu W/(\cos\theta + \mu \sin\theta)$, where μ is the coefficient of friction (constant). Show that the pull required is least when $\tan\theta = \mu$.
21. A cylinder, diameter 2 cm, height 1 cm, stands on a table with its axis vertical. A hollow cone is placed over it so that its base lies on the table and its curved surface just touches the upper rim of the cylinder all the way around. Show that the least volume the cone can have is $\frac{9}{4}\pi$ cm^3.
22. A right circular cone with a flat base has its vertex at the centre of a sphere of radius a, and lies entirely within the sphere. Show that the cone has maximum volume if its height is $a/\sqrt{3}$.
23. If a small object is placed at a distance u in front of a thin lens of focal length f, the distance of the image behind the lens is v where $1/u + 1/v = 1/f$. Show that the minimum distance between the image and the object is $4f$.
24. A straight line is drawn through a fixed point (h, k) and meets the coordinate axes in A and B. Show that the area of the triangle AOB is least when the line AB is bisected at the fixed point, O being the origin.
25. A sector is cut out of a circular piece of canvas and the bounding radii of the part that remains are drawn together to form a conical tent. Show that, if the tent is to have maximum volume, the angle of the sector used for the tent must be $(2\sqrt{6}/3)\pi$ radians.

26. The intensity of heat varies inversely as the square of the distance from the centre of heat. Two centres of heat are distance c apart and one is T times as hot as the other. Show that the coolest point on the line between the two centres is distant $cT^{1/3}/(1 + T^{1/3})$ from the hotter centre.
27. A beam 16 m long, weight 10 kg/m and supported freely at one end A and at a point D 4 m from the other end B, has a load of 40 kg hung from a point C 10 m from A. Under these conditions the bending moment M is given by

$$M = \begin{cases} 5x^2 - 60x & \text{along } AC, \\ 5x^2 - 20x - 400 & \text{along } CD, \\ 5x^2 - 160x - 1280 & \text{along } DB, \end{cases}$$

where x m is the distance from A. Find the greatest value of M and where it occurs. Find also the greatest value of $|M|$ and where it occurs.
28. A uniform beam $AOCB$ consists of two portions smoothly jointed at C. It is supported at the ends A, B and at the mid-point O. If

$$AO = OB = a, \; OC = \tfrac{1}{2}a,$$

then the bending moment M is given by

$$M = \begin{cases} \tfrac{1}{4}w\,(x + a)(2x + a) & \text{along the portion } AO, \\ \tfrac{1}{4}w\,(x - a)(2x - a) & \text{along the portion } OCB, \end{cases}$$

where x is the distance from O measured positively in the direction OCB. Show that the greatest value of M is $\frac{1}{4}wa^2$ and that this is also the greatest value of $|M|$ (w = weight per unit length).
29. From a fixed point A on the circumference of a circle of radius a the perpendicular AQ is drawn on to the tangent at a variable point P. If AP makes an angle θ with the diameter through A, prove that the area of the triangle APQ is $2a^2 \sin\theta \cos^3\theta$.

    Find the maximum area of the triangle.
30. Prove that the weight of the heaviest right circular cylinder that can be cut from a given sphere of uniform material is $\frac{1}{3}\sqrt{3}$ times the weight of the sphere.
31. A curve is given by the parametric equations

$$x = 1/(1 + t^2), \quad y = t^3/(1 + t^2).$$

Find the equation of the tangent to the curve at the point whose parameter is t, and show that the area of the triangle formed by the tangent and the co-ordinate axes is not greater than $3\sqrt{3}/8$ units.
32. A canister, of total length l, is made of metal which is thin compared with the dimensions of the canister. It consists of a cylinder of radius r closed at its ends by cones of vertical angle 2θ. The weight of metal per unit area for the cones is n times that for the cylinder. Prove that if only θ varies, the weight of the canister cannot have a true minimum unless n is greater than 2 and r is less than $\frac{1}{4}l\sqrt{(n^2 - 4)}$.
33. A and B are two points on either side of a straight line which separates two different types of country. M and N are the feet of the perpendiculars from A

and B respectively on this line. $MA = a$, $NB = b$, $MN = c$, and P is a point on the line between M and N distant x from M. If, in the type of country containing A, a man can walk with velocity u and, in the type containing B, with velocity v, find the time taken along the path APB, and show that when this time is a minimum $\sin\angle MAP/\sin\angle NBP = u/v$.

34. A right circular cylinder is inscribed in a given right circular cone so that one circular end is on the base of the cone and the circumference of the other end on the surface of the cone. Prove that the maximum volume of such a cylinder is $\frac{4}{9}$ that of the cone.

35. A right circular cylinder is inscribed in a right circular cone of height h and with base of radius a, one plane end of the cylinder being in contact with the base of the cone. Show that there is always a cylinder of maximum volume, but that there is no proper cylinder of maximum total superficial area (that is, the sum of the areas of the curved surface and the two plane ends) unless $2a$ is less than h.

36. A right circular cone is circumscribed to a sphere of radius a, with the base of the cone touching the sphere. Find an expression for the volume of the cone in terms of a and the semi-vertical angle of the cone, and show that when the volume of the cone is a minimum it is double the volume of the sphere.

37. A tree trunk, in the form of a frustum of a cone, is h m long, and the greatest and least diameters are a and b m respectively. A beam of square cross-section is cut from the tree. Show that if $2a > 3b$, then the beam has maximum volume when its length is $\frac{1}{3}\,ha/(a-b)$. What is the length of the beam for maximum volume when $2a < 3b$?

38. The brightness of a small surface varies inversely as the square of its distance r from the source of light and directly as the cosine of the angle between r and the normal to the surface. Two equal light sources are situated at the points A and B, not in the same vertical line, at heights a and $2a$ above a horizontal plane. The verticals through A and B meet the plane at M and N, where $MN = 3a$. If a small surface parallel to the plane is moved along the line MN, show that its brightness is a minimum when it is situated at a point of trisection of MN.

39. The illumination of an area by a source of light is proportional to

$$\frac{x}{(x^2+a^2)^{1/2}} - \frac{x}{(x^2+b^2)^{1/2}},$$

where a and b $(>a)$ are constants, and x can be varied. Find the value of x which gives maximum illumination.

40. The vertices of a quadrilateral are the centres of the circles

$$x^2 + y^2 + 2\lambda x = 0, \quad x^2 + y^2 + 2y/\lambda = 0,$$

and the points of intersection of these circles. Prove that the area of the quadrilateral is the same for all values of λ.

Find the length of the common chord of the circles, and show that it has a stationary value when the circles are of equal radius.

41. Given that r_1, r_2 are the focal distances of a point on an ellipse whose major axis is $2a$, find the maximum and minimum values of $r_1r_2(r_1 - r_2)$, $(r_1 > r_2)$, distinguishing between the cases where the eccentricity is greater than or less than $1/\sqrt{3}$.

    Illustrate by sketching the graph of the function $x(x - a)(2a - x)$.
42. Prove that the length of the tangent to the ellipse $b^2x^2 + a^2y^2 = a^2b^2$ intercepted between the axes has one finite stationary value. Prove analytically that it is a minimum and find this value.

15.1.4 Approximations and Small Errors From Definition 14.3 of the derivative $f'(x)$ it follows immediately that

$$\frac{f(x + \Delta x) - f(x)}{\Delta x} \simeq f'(x) \qquad \text{for small } \Delta x,$$

where the symbol $\simeq$ denotes *approximate* equality.

Hence

$$f(x + \Delta x) - f(x) \simeq f'(x)\,\Delta x$$

for small Δx or, in other words, $\Delta y \simeq (dy/dx)\,\Delta x$ for small Δx.

This approximation is shown diagrammatically in Fig. 15.11, P and Q are two points on the curve $y = f(x)$, PT is the tangent to the curve at P, and we are saying that the length RQ is approximately equal to the length RT provided that Q is close to P.

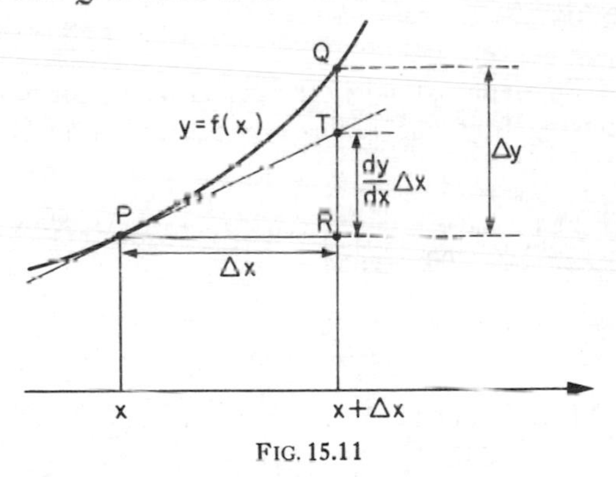

FIG. 15.11

Consequently, if the values of $f(x)$ and $f'(x)$ are both known at one point, the value of $f(x)$ anywhere in the neighbourhood of that point can be calculated approximately.

Example 13 Calculate $\sqrt{3{\cdot}987}$ approximately.

Write $y = \sqrt{x}$. Then $y = 2$ when $x = 4$. Also $dy/dx = 1/2\sqrt{x} = \frac{1}{4}$ when $x = 4$. Put $3{\cdot}987 = 4 + \Delta x$, i.e. $\Delta x = -0{\cdot}013$.

Now $\Delta y \simeq (\mathrm{d}y/\mathrm{d}x)\,\Delta x$ and so when $x = 4$, $\Delta x = -0{\cdot}013$, we get

$$\Delta y \simeq \tfrac{1}{4}\,(-0{\cdot}013) = -0{\cdot}00325.$$

Hence $\sqrt{3{\cdot}987} \simeq 2 - 0{\cdot}00325 = 1{\cdot}99675 = 1{\cdot}997$ say.

Example 14 Find approximately the increase in volume of a sphere due to an increase in radius from 9 to 9·13 cm.

Denoting the volume by V cm^3 and the radius by r cm, we have

$$V = \tfrac{4}{3}\pi r^3, \quad \frac{\mathrm{d}V}{\mathrm{d}r} = 4\pi r^2,$$

so that when $r = 9$ we have $\mathrm{d}V/\mathrm{d}r = 324\pi$.

Now $\Delta V \simeq 4\pi r^2\,\Delta r$ and so when $r = 9$, $\Delta r = 0{\cdot}13$ we have

$$\Delta V \simeq 324\pi(0{\cdot}13) = 42{\cdot}12\pi.$$

Suppose that $y = f(x)$ and that this relationship is used to obtain the value of y for some particular value of x. If, however, an erroneous value of x is used then a resulting error in y must be expected and the question arises—what is the relation between the two errors? Denoting the errors in x, y by Δx, Δy respectively this problem is equivalent to that of finding the change Δy in y due to a change Δx in x and the foregoing approximation $\Delta y \simeq (\mathrm{d}y/\mathrm{d}x)\,\Delta x$ is again useful.

Example 15 Find approximately the percentage error in the volume of a sphere calculated from the radius when there is a 2% error in the radius.

As in Example 14 we have $V = \tfrac{4}{3}\pi r^3$, $\Delta V \simeq 4\pi r^2\,\Delta r$. Now $\Delta r = 2r/100$, and so $\Delta V \simeq 4\pi r^2 \,.\, 2r/100$. Hence

$$\frac{\Delta V}{V} \simeq \frac{4\pi r^2}{\tfrac{4}{3}\pi r^3} \cdot \frac{2r}{100} = \frac{6}{100},$$

showing that ΔV is 6% of V.

Example 16 Solve Example 10 approximately.

Regard a, b as variables. Suppose a is measured and b calculated. Then

$$A = \tfrac{1}{2}a\sqrt{(100 - a^2)}, \quad \frac{\mathrm{d}A}{\mathrm{d}a} = \frac{50 - a^2}{\sqrt{(100 - a^2)}},$$

and so

$$\Delta A \simeq \frac{50 - a^2}{\sqrt{(100 - a^2)}}\,\Delta a.$$

Put $a = 8$ and consider values of Δa between $-0\cdot4$ and $+0\cdot4$; we find that ΔA lies between $-0\cdot93$ and $+0\cdot93$ approximately. If b is measured, we write

$$A = \tfrac{1}{2}b\sqrt{(100 - b^2)}, \quad \Delta A \simeq \frac{50 - b^2}{\sqrt{(100 - b^2)}}\,\Delta b.$$

Put $b = 6$ and consider values of Δb lying between $-0\cdot3$ and $+0\cdot3$; we find that ΔA lies between $-0\cdot52$ and $+0\cdot52$ approximately.

Altogether the maximum possible error is approximately $0\cdot93$. In this case the error of 5% in the measurement is too large to give a very good result by the approximate method.

15.1.5 Newton's Approximation to a Root of an Equation Suppose we wish to solve the equation $f(x) = 0$, which we know to have a root near x_1. Denote the correct root by $x_1 + \Delta x$; then $f(x_1 + \Delta x) = 0$. By the approximation of Section 15.1.4 we have

$$f(x_1 + \Delta x) \simeq f(x_1) + f'(x_1)\,\Delta x.$$

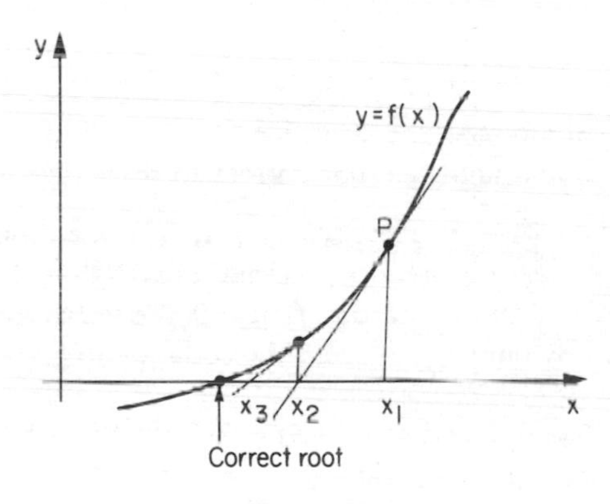

FIG. 15.12

Hence $f(x_1) + f'(x_1)\,\Delta x \simeq 0$, and so $\Delta x \simeq -f(x_1)/f'(x_1)$. Thus from the first approximate root x_1 we get a second approximate root, x_2 say, where

$$x_2 = x_1 - \frac{f(x_1)}{f'(x_1)}.$$

Proceeding in this way we can get a third approximate root, x_3 say, by taking

$$x_3 = x_2 - \frac{f(x_2)}{f'(x_2)}$$

and so on.

Figure 15.13 shows how a first approximate root x_1 to the right of the correct root may lead to a second approximate root x_2 on the left, but this need not matter. If a sufficiently close first approximation is taken the process will give successive approximations tending towards the correct root.

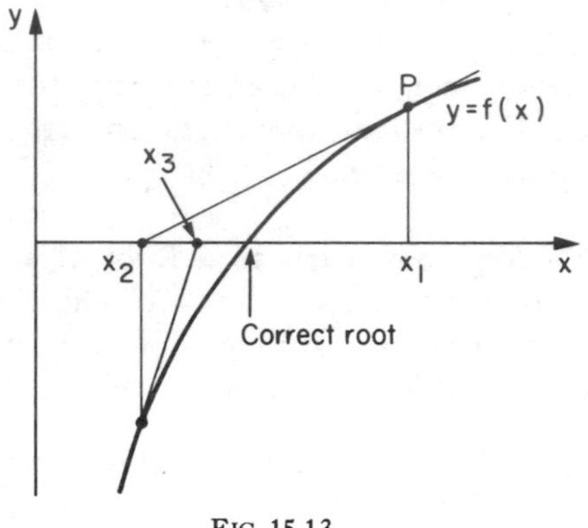

FIG. 15.13

Example 17 Show that the equation $x^3 + 2x - 1 = 0$ has a root between 0 and 1, and find this root correct to three decimal places.

Write $f(x) = x^3 + 2x - 1$. Then $f(0) = -1$, $f(1) = 2$. Since $f(x)$ is continuous it follows by Theorem 5 (Section 14.2.1) that there must be a value of x between 0 and 1 such that $f(x) = 0$. A good first approximation may be obtained by omitting x^3 from the equation to get $2x - 1 = 0$, i.e. $x_1 = 0{\cdot}5$.

Now $f'(x) = 3x^2 + 2$ and so $f'(0{\cdot}5) = 2{\cdot}7500$. Since $f(0{\cdot}5) = 0{\cdot}1250$, we get as a second approximation

$$x_2 = 0{\cdot}5000 - \frac{0{\cdot}1250}{2{\cdot}7500} = 0{\cdot}4545.$$

We now evaluate $f(0{\cdot}4545)$, $f'(0{\cdot}4545)$ to get $0{\cdot}0029$, $2{\cdot}6198$ respectively. The third approximation x_3 is now given by

$$x_3 = 0{\cdot}4545 - \frac{0{\cdot}0029}{2{\cdot}6198} = 0{\cdot}4534.$$

Finally $f(0{\cdot}4534)$ is found to have zeros in the first four decimal places and since $f'(0{\cdot}4534)$ is roughly $2{\cdot}6$ it is clear that the fourth approximation will not differ from the third in the first three decimal places. Hence the root is $0{\cdot}453$ correct to three decimal places.

Example 18 Show that the equation $\cos x = x$ has only one positive root and find it correct to two decimal places.

We see from the graphs (Fig. 15.14) that there is only one positive root and it is near $\pi/4$.

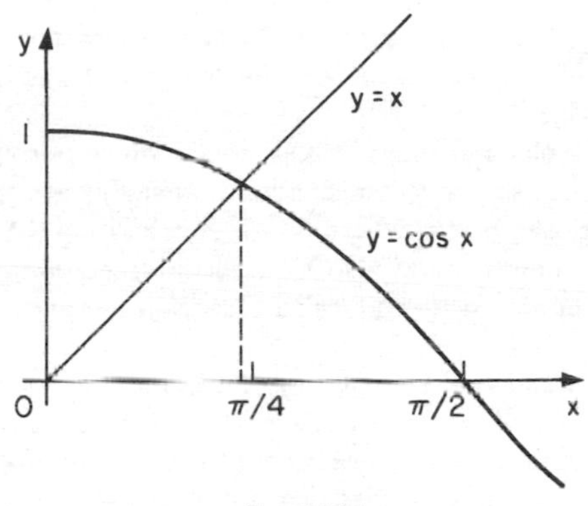

FIG. 15.14

Write

$$f(x) = x - \cos x, \quad f'(x) = 1 + \sin x.$$

The results are set out in Table 15.1. The last column gives the correction needed for the next approximation, and so the process stops when this correction is sufficiently small. We see that the root is 0·74 correct to two decimal places.

TABLE 15.1

	x(rad)	$\sin x$	$\cos x$	$f(x)$	$f'(x)$	$-\dfrac{f(x)}{f'(x)}$
1st approximation	0·785	0·7069	0·7073	0·0777	1·7069	−0·046
2nd approximation	0·739	0·6736	0·7392	−0·0002	1·6736	0·0000

Exercises 15(c)

1. Evaluate $\sqrt[3]{101}$ approximately, given that $\sqrt[3]{100} = 4 \cdot 6416$.
2. Find approximations to $\sqrt[3]{125 \cdot 06}$ and tan 46°.
3. If $pv^{1 \cdot 4}$ = constant and the value of v is increased by 5%, find approximately the percentage change in the value of p.
4. Calculate approximately the difference in $\sin x$ corresponding to a difference of one minute in x when $x = \pi/4$, and check your answer using tables.
5. The bending moment M in a beam of length l is, under certain conditions, given by the formula $M = \frac{1}{2}w(lx - x^2)$, w being the weight per unit length of

the beam. Given that $w = 2\cdot 5$ and $l = 50$ cm and the length x, which should have been 20 cm, is measured as $20\cdot 1$ cm, find the approximate percentage error in M.

6. The side a of a triangle ABC is calculated from the formula
$$a^2 = b^2 + c^2 - 2bc \cos A.$$
If c is exactly 10 cm, A is exactly 60° and b is measured as 30 cm with a possible error of $0\cdot 3$ cm, calculate the value of a and estimate the possible error in this value.

7. A battery of E volts and internal resistance r ohms is connected in circuit with a variable resistance R ohms. Assume that E, r are constants and that the power developed is RI^2, where $I = E/(R + r)$. Find the value of R which gives maximum power. Show also by a calculus method that when $R = 2r$ a small increase of $a\%$ in R produces a reduction in power of approximately $\frac{1}{3}a\%$.

8. The distances u and v of a small object and its image from a lens of focal length f are connected by the law $1/u + 1/v = 1/f$. Use a calculus method to show that for a small object placed 9 cm from a lens of focal length 3 cm, a 2% increase in the distance between the object and the lens produces approximately a 1% decrease in the distance between the image and the lens. Show also that if the lens is replaced by another with 2% greater focal length, the object again being a distance of 9 cm from the lens, the result is an increase of approximately 3% in the distance between the image and the lens.

9. The current i passing through a tangent galvanometer is such that
$$i = k \tan \theta,$$
where θ is the angle of deflection and k is a constant. Show that the proportional change in the current corresponding to a small change α rad in θ is given approximately by $2\alpha \operatorname{cosec} 2\theta$.

10. Show that the equation $x^3 + 3x^2 + 6x - 3 = 0$ has only one real root. Prove that this lies between 0 and 1, and find it correct to one place of decimals.

11. Prove that the equation $x^3 - 4x + 1 = 0$ has a root lying between 1 and 2 and find it correct to two decimal places.

12. Find to three places of decimals the root of the equation $x = \tan x$ that lies between π and $3\pi/2$. (*Hint*: To facilitate the use of tables, put $x = \pi + z$ and find the root of the resulting equation that lies between $z = 0$ and $z = \pi/2$.)

13. Show by means of a rough graph that the least positive root of the equation $\tan x = \frac{1}{2}x$ lies between π and $3\pi/2$. Find this least root correct to four significant figures.

14. Show that the equation $x^4 - 5x + 2 = 0$ has two real roots, both of which are positive. (Note that the roots are given by the intersections of the line $y = 5x - 2$ and the curve $y = x^4$.)

    Find the larger root correct to four significant figures.

15. Show that the curve $y = x(x - 4)/(x^2 + 3)^{1/2}$ has one turning point, and find the corresponding value of x correct to two decimal places.

15.2 Curvature

Let P, Q be two points on the curve $y = f(x)$ at distances s, $s + \Delta s$ respectively measured along the curve from a fixed point A on the curve. Let the tangents at P and Q make angles ψ and $\psi + \Delta\psi$ respectively with Ox. Then $\Delta\psi$ measures the change in direction along the curve between

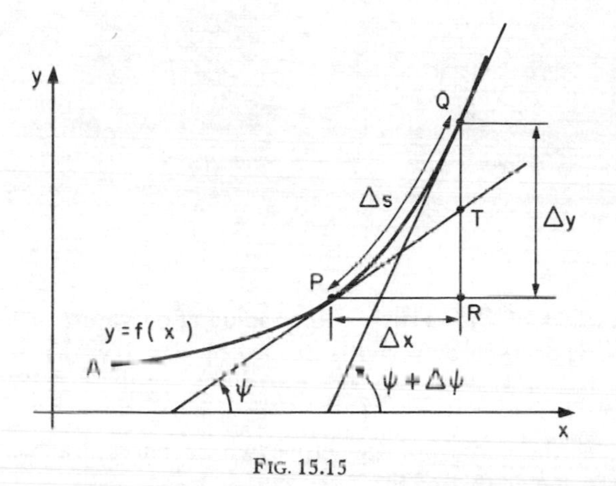

FIG. 15.15

P and Q; we refer to $\Delta\psi/\Delta s$ as the average curvature of the arc PQ and we define the curvature at P (denoted by $\varkappa$) as the limit of this average as $Q \to P$ along the curve, i.e. as $\Delta s \to 0$. Hence

$$\varkappa = \mathrm{d}\psi/\mathrm{d}s.$$

If ψ increases as s increases then $\varkappa$ is positive, otherwise $\varkappa$ is negative. This means that $\varkappa$ is positive or negative according as the tangent turns anticlockwise or clockwise when we move along the curve in the direction of increasing s.

If, as is usually done, we measure s so as to increase when x increases, then $\varkappa$ is positive or negative according as the curve is concave upwards or convex upwards.

15.2.1 Curvature of a Circle In the case of a circle of radius a (Fig. 15.16) the angle $\Delta\psi$ between the tangents at P and Q equals the angle subtended by PQ at the centre of the circle. Hence $\Delta s = a\,\Delta\psi$ and so $\Delta\psi/\Delta s = 1/a =$ constant. Hence $\mathrm{d}\psi/\mathrm{d}s = 1/a$. Likewise, at a point P on the upper semi-circle $\mathrm{d}\psi/\mathrm{d}s = -1/a$.

This means that the curvature of a circle equals (apart from sign) the reciprocal of the radius.

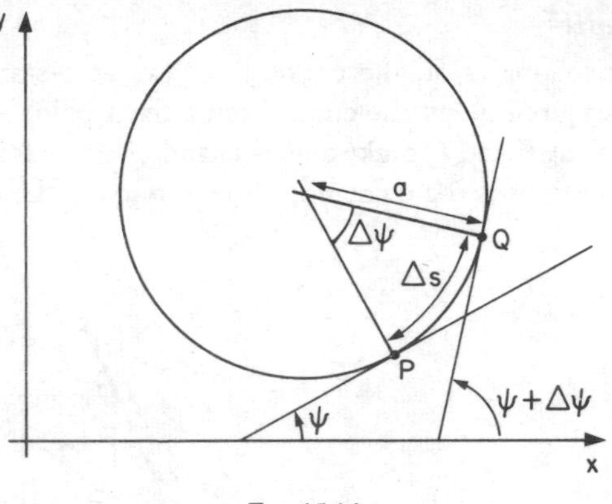

FIG. 15.16

15.2.2 Radius of Curvature The radius of curvature at any point P on a curve is denoted by ρ and is defined by $\rho = 1/\varkappa$, that is

$$\rho = \mathrm{d}s/\mathrm{d}\psi.$$

Thus $|\rho|$ is the radius of the circle whose curvature has the same numerical value as that of the given curve at P.

15.2.3 Basic Property of Arc-length The difficulty about calculating curvature for curves other than the circle is that in general we do not know the arc-length. However, even without a formula for arc-length, one basic property of arc-length can be derived.

Whatever length Δs we attribute to the arc PQ (Fig. 15.15) it must exceed the length of the *chord* PQ and (at least for sufficiently small Δs) it must be less than the sum of the lengths PT and TQ, that is

$$PQ < \Delta s < PT + TQ.$$

Hence

$$\frac{PQ}{\Delta x} < \frac{\Delta s}{\Delta x} < \frac{PT}{\Delta x} + \frac{TQ}{\Delta x}. \tag{1}$$

Now

$$PQ = \sqrt{(PR^2 + RQ^2)} = \sqrt{[(\Delta x)^2 + (\Delta y)^2]},$$

and so

$$\frac{PQ}{\Delta x} = \sqrt{\left[1 + \left(\frac{\Delta y}{\Delta x}\right)^2\right]} \to \sqrt{\left[1 + \left(\frac{dy}{dx}\right)^2\right]} \quad \text{as } \Delta x \to 0. \tag{2}$$

Remembering that $RT = (\mathrm{d}y/\mathrm{d}x)\,\Delta x$ (see Section 15.1.4), we see that

$$PT = \sqrt{(PR^2 + RT^2)} = \sqrt{\left[(\Delta x)^2 + \left(\frac{\mathrm{d}y}{\mathrm{d}x}\right)^2 (\Delta x)^2\right]},$$

and so

$$\frac{PT}{\Delta x} = \sqrt{\left[1 + \left(\frac{\mathrm{d}y}{\mathrm{d}x}\right)^2\right]}, \tag{3}$$

which is constant as $\Delta x \to 0$. Finally,

$$TQ = RQ - RT = \Delta y - \frac{\mathrm{d}y}{\mathrm{d}x}\Delta x,$$

and so

$$\frac{TQ}{\Delta x} = \frac{\Delta y}{\Delta x} - \frac{\mathrm{d}y}{\mathrm{d}x} \to 0 \qquad \text{as } \Delta x \to 0. \tag{4}$$

Letting $\Delta x \to 0$ in the inequality (1) we see, using (2)–(4), that

$$\frac{\mathrm{d}s}{\mathrm{d}x} = \sqrt{\left[1 + \left(\frac{\mathrm{d}y}{\mathrm{d}x}\right)^2\right]}. \tag{15.1}$$

We shall assume s is measured so as to increase when x increases; then $\mathrm{d}s/\mathrm{d}x$ is always positive and only the positive value of the square root is to be taken in (15.1).

15.2.4 Formula for Radius of Curvature The slope of the tangent PT (Fig. 15.15) is $\tan\psi$. Hence $\mathrm{d}y/\mathrm{d}x = \tan\psi$, and so, differentiating both sides with respect to x, we obtain

$$\frac{\mathrm{d}^2y}{\mathrm{d}x^2} = \sec^2\psi\,\frac{\mathrm{d}\psi}{\mathrm{d}x} = (1 + \tan^2\psi)\frac{\mathrm{d}\psi}{\mathrm{d}x} = \left[1 + \left(\frac{\mathrm{d}y}{\mathrm{d}x}\right)^2\right]\frac{\mathrm{d}\psi}{\mathrm{d}x},$$

that is

$$\frac{\mathrm{d}\psi}{\mathrm{d}x} = \frac{\mathrm{d}^2y/\mathrm{d}x^2}{1 + (\mathrm{d}y/\mathrm{d}x)^2}.$$

The curvature $\varkappa$ is thus given by

$$\varkappa = \frac{\mathrm{d}\psi}{\mathrm{d}s} = \frac{\mathrm{d}\psi}{\mathrm{d}x}\frac{\mathrm{d}x}{\mathrm{d}s} = \frac{\mathrm{d}^2y/\mathrm{d}x^2}{1 + (\mathrm{d}y/\mathrm{d}x)^2}\,\frac{1}{\mathrm{d}s/\mathrm{d}x},$$

and on using (15.1) this gives

$$\varkappa = \frac{\mathrm{d}^2y/\mathrm{d}x^2}{[1 + (\mathrm{d}y/\mathrm{d}x)^2]^{3/2}}.$$

The radius of curvature ρ is therefore given by

$$\rho = \frac{[1 + (\mathrm{d}y/\mathrm{d}x)^2]^{3/2}}{\mathrm{d}^2y/\mathrm{d}x^2}.$$

Only the positive value of the root in the numerator is to be taken, which means that ρ has the same sign as $\mathrm{d}^2y/\mathrm{d}x^2$ and is therefore positive or negative according as the curve is concave upwards or convex upwards (Section 14.4.8).

Example 19 Find the radius of curvature at any point on the curve $y = \sqrt{x}$.

We have $y = x^{1/2}$, $\mathrm{d}y/\mathrm{d}x = \frac{1}{2}x^{-1/2}$, $\mathrm{d}^2y/\mathrm{d}x^2 = -\frac{1}{4}x^{-3/2}$. Hence

$$1 + \left(\frac{\mathrm{d}y}{\mathrm{d}x}\right)^2 = 1 + \frac{1}{4x},$$

and so

$$\rho = -\frac{(1 + 1/4x)^{3/2}}{\frac{1}{4}x^{-3/2}} = -\tfrac{1}{2}(4x + 1)^{3/2}.$$

For example, putting $x = 0$ we see that at the point $(0, 0)$ the radius of curvature is $-\frac{1}{2}$, and putting $x = 2$ we see that at the point $(2, \sqrt{2})$ the radius of curvature is $-27/2$.

Exercises 15(d)

1. Sketch the curve $y(1 + x^2) = x - x^2$ and find the radius of curvature at the origin.
2. Show that the radius of curvature of the curve $x^4 + y^4 = a^4$ at its points of intersection with the lines $y = \pm x$ is $2^{1/4}\, a/3$.
3. Determine the points of the hyperbola $xy = c^2$ at which the radius of curvature takes its numerically least value.
4. Prove that, for the curve $axy = x^3 - 2a^3$ $(a > 0)$, the ordinate, y, has a minimum value at the point $(-a, 3a)$. Find the radius of curvature at this point.
5. Prove that the curve $y^3 = x(x + 2y)$ has a minimum ordinate when $x = 1$, and show that the radius of curvature at this minimum point is $\frac{1}{2}$.
6. Suppose that the equation of a curve is given in parametric form, and dots denote derivatives with respect to the parameter. Assuming that ρ has the same sign as $\mathrm{d}^2y/\mathrm{d}x^2$, show that

$$\rho = \pm\frac{(\dot{x}^2 + \dot{y}^2)^{3/2}}{\dot{x}\ddot{y} - \ddot{x}\dot{y}},$$

where the plus or minus sign is taken according as $\dot{x}$ is positive or negative.

7. Show that the radius of curvature at the point $t = 0$ on the curve

$$x = a \sin 2t, \quad y = b \cos^3 t$$

is $-4a^2/3b$.

8. A curve is given parametrically by the equations

$$x = a(2 \cos \theta - \cos 2\theta), \quad y = a(2 \sin \theta - \sin 2\theta),$$

where $a > 0$. Show that the radius of curvature at the point $\theta = \pi/2$ is $-4\sqrt{2}\,a/3$.

9. Show that, at the point $(a \cos \theta, b \sin \theta)$ on the ellipse $x^2/a^2 + y^2/b^2 = 1$,

$$a^2b^2\rho^2 = (a^2 \sin^2 \theta + b^2 \cos^2 \theta)^3.$$

10. Show that the radius of curvature at the point $t = \pi/4$ on the curve

$$x = \cos^3 t, y = \sin^3 t$$

is $\frac{3}{2}$.

15.3 Limits as x Tends to Infinity

Definition 4. We say that $f(x) \to l$ as $x \to \infty$ (x tends to infinity) if the values of $f(x)$ can be made to differ from l by as little as we please by choosing x sufficiently large.

Definition 4a. We say that $f(x) \to l$ as $x \to \infty$ if it is possible to find $N(\varepsilon)$, defined for all $\varepsilon > 0$ (however small), such that $|f(x) - l| < \varepsilon$ when $x > N(\varepsilon)$.

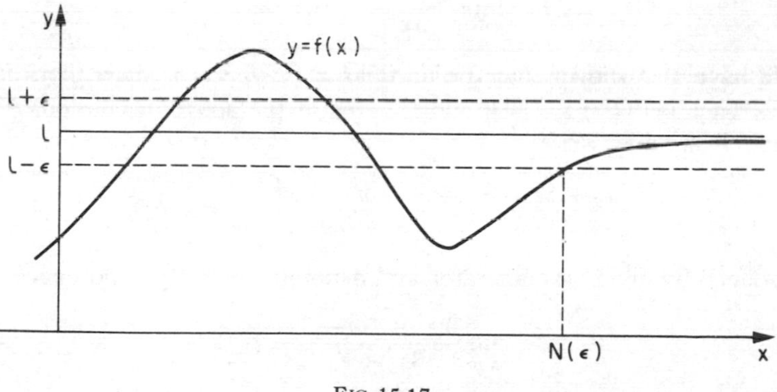

FIG. 15.17

The above definition may be regarded as an amended form of Definition 1 (Section 14.1), requiring that $f(x)$ be close to l *when x is large* rather than *when x is close to a.* On the basis of previous definitions it is now a straightforward matter to construct definitions (the reader should try this) for the cases where

(a) $f(x) \to l$ as $x \to -\infty$,
(b) $f(x) \to \infty$ as $x \to \infty$,
(c) $f(x) \to \infty$ as $x \to -\infty$,
(d) $f(x) \to -\infty$ as $x \to \infty$,
(e) $f(x) \to -\infty$ as $x \to -\infty$.

For example $x^n \to \infty$ as $x \to \infty$ if $n > 0$, $x^n \to 0$ as $x \to \infty$ if $n < 0$, $x^n \to \infty$ as $x \to -\infty$ if n is an even positive integer, $x^n \to -\infty$ as $x \to -\infty$ if n is an odd positive integer, $x^n \to 0$ as $x \to -\infty$ if $n < 0$, $\tan^{-1} x \to \pi/2$ as $x \to \infty$, $\tan^{-1} x \to -\pi/2$ as $x \to -\infty$.

The behaviour of $f(x)$ as $x \to \infty(-\infty)$ is identical with the behaviour of $f(1/t)$ as $t \to 0+$ $(0-)$.

Example 20 Find

$$\lim_{x\to\infty} \frac{2x^2 + 5x - 3}{3x^2 + 4}.$$

Put $x = 1/t$ and the expression becomes

$$\frac{2/t^2 + 5/t - 3}{3/t^2 + 4} = \frac{2 + 5t - 3t^2}{3 + 4t^2} \quad \text{for all } t \neq 0$$
$$\to \tfrac{2}{3} \text{ as } t \to 0.$$

Hence

$$\lim_{x\to\infty} \frac{2x^2 + 5x - 3}{3x^2 + 4} = \tfrac{2}{3}.$$

We have also shown that the limit as $x \to -\infty$ is $\frac{2}{3}$, since there is no difference between $t \to 0+$ and $t \to 0-$ in the above discussion. Alternatively we may write

$$\frac{2x^2 + 5x - 3}{3x^2 + 4} = \frac{2 + 5/x - 3/x^2}{3 + 4/x^2} \quad \text{if } x \neq 0,$$

obtained by dividing numerator and denominator by x^2, and hence

$$\lim_{x\to\infty} \frac{2x^2 + 5x - 3}{3x^2 + 4} = \tfrac{2}{3}$$

since $5/x, 3/x^2, 4/x^2$ all $\to 0$ as $x \to \infty$(also as $x \to -\infty$).

Example 21 Find

$$\lim_{x\to\infty} \frac{x - 1}{x^2 - x + 1}.$$

Put $x = 1/t$ and the expression becomes

$$\frac{1/t - 1}{1/t^2 - 1/t + 1} = \frac{t - t^2}{1 - t + t^2} \quad \text{for all } t \neq 0$$
$$\to 0 \text{ as } t \to 0.$$

Hence

$$\lim_{x \to \infty} \frac{x - 1}{x^2 - x + 1} = 0 \quad \left(= \lim_{x \to -\infty} \frac{x - 1}{x^2 - x + 1} \right).$$

Example 22 Discuss the behaviour of $(x^2 + 1)/(2x - 3)$ as $x \to \pm\infty$.

Put $x = 1/t$ and the expression becomes

$$\frac{1/t^2 + 1}{2/t - 3} = \frac{1 + t^2}{2t - 3t^2} \quad \text{for all } t \neq 0$$
$$\to \infty \text{ as } t \to 0+, \to -\infty \text{ as } t \to 0-.$$

Hence

$$\frac{x^2 + 1}{2x - 3} \to \infty \text{ as } x \to \infty, \to -\infty \text{ as } x \to -\infty.$$

Consider the rational function of x defined by

$$f(x) = \frac{a_0 x^n + a_1 x^{n-1} + \cdots + a_n}{b_0 x^m + b_1 x^{m-1} + \cdots + b_m} \quad (a_0, b_0 \neq 0);$$

then if $m = n$, $f(x) \to a_0/b_0$ as $x \to \pm\infty$,
if $m > n$, $f(x) \to 0$ as $x \to \pm\infty$,
if $m < n$, $|f(x)| \to \infty$ as $x \to \pm\infty$, the behaviour of $f(x)$ being the same as that of $(a_0/b_0)\, x^{n-m}$.

Example 23 Find

$$\lim_{x \to \infty} [\sqrt{(x^2 + 2x)} - x].$$

If $x = 1/t$, then

$$\sqrt{(x^2 + 2x)} - x = \sqrt{(1/t^2 + 2/t)} - 1/t = \frac{\sqrt{(1 + 2t)} - 1}{t}$$
$$= \frac{2t}{t\{\sqrt{(1 + 2t)} + 1\}} \quad \text{(rationalising the numerator)}$$
$$= \frac{2}{\sqrt{(1 + 2t)} + 1} \quad \text{for all } t \neq 0$$
$$\to 1 \text{ as } t \to 0.$$

Hence

$$\lim_{x\to\infty} [\sqrt{(x^2 + 2x)} - x] = 1.$$

15.3.1 Asymptotes An *asymptote* to a curve is a straight line to which the shape of the curve approximates at a great distance from the origin, i.e. the distance from a point P on the curve to the line tends to zero as the distance OP from the origin to P tends to infinity.

We distinguish between three types of asymptote:

(a) *Asymptotes parallel to the x-axis.* The line $y = l$ is an asymptote to the curve $y = f(x)$ if $f(x) \to l$ as $x \to \infty$ or if $f(x) \to l$ as $x \to -\infty$.

For example the line $y = \frac{2}{3}$ is an asymptote to the curve

$$y = \frac{2x^2 + 5x - 3}{3x^2 + 4} \quad \text{(see Example 20)}$$

and the x-axis is an asymptote to the curve

$$y = \frac{x - 1}{x^2 - x + 1} \quad \text{(see Example 21).}$$

(b) *Asymptotes parallel to the y-axis.* The line $x = a$ is an asymptote to the curve $y = f(x)$ if $f(x) \to \pm\infty$ as $x \to a+$ or if $f(x) \to \pm\infty$ as $x \to a-$.

For example the lines $x = \pm\pi/2, \pm 3\pi/2, \ldots$ are all asymptotes to the curve $y = \tan x$, and the y-axis is an asymptote to the curve $y = 1/x$. In particular, the line $x = a$ is an asymptote to the curve $y = f(x)$ if $f(x) = g(x)/h(x)$ and $g(x) \to l \neq 0$ while $h(x) \to 0$ as $x \to a$, e.g. the curve

$$y = \frac{x}{(x - 2)(x - 3)}$$

has asymptotes $x = 2$ and $x = 3$.

(c) *Slant asymptotes.* The line $y = mx + c$ is an asymptote to the curve $y = f(x)$ if $f(x) - mx - c \to 0$ as $x \to \infty$ or if $f(x) - mx - c \to 0$ as $x \to -\infty$.

For example, the line $y = x + 1$ is an asymptote to the curve $y = \sqrt{(x^2 + 2x)}$ since $\sqrt{(x^2 + 2x)} - x - 1 \to 0$ as $x \to \infty$ (see Example 23).

In particular, an expression of the form

$$\frac{Ax^2 + Bx + C}{ax + b}$$

can, by ordinary division, be written in the form

$$mx + c + \frac{R}{ax + b},$$

where R is a constant and so

$$\frac{R}{ax+b} \to 0 \text{ as } x \to \infty;$$

this enables us to find an asymptote $y = mx + c$ for the curve

$$y = \frac{Ax^2 + Bx + C}{ax + b}.$$

For example, if

$$y = \frac{x(1-x)}{1+x},$$

we can write

$$y = -x + 2 - \frac{2}{1+x},$$

showing that the line $y = -x + 2$ is an asymptote.

Exercises 15(e)

Find the limits as $x \to \infty$ of the following:

1. $\dfrac{2x^4 + 3x^2 - 7}{x^4 + 6x + 9}$
2. $\dfrac{2x^3 + 3x^2 - 1}{5x + 6}$
3. $\dfrac{7x - 1}{3x^3 + 4x}$
4. $\sqrt{(4x^2 + 3)} - 2x$
5. $\sqrt{(4x^2 + 3x)} - 2x$
6. $\sqrt{(4x^2 + 3x + 3)} - 2x$.

Find slant asymptotes for the following curves:

7. $y = \dfrac{2x^2 + x - 17}{2x - 5}$
8. $y = \dfrac{3x^2 + 4}{2x + 3}$
9. $y = \dfrac{x^3 - 1}{x^2 + 1}$

15.4 Curve Sketching

As a starting point we assume familiarity with the graphs of the elementary functions. Next suppose we wish to sketch the curve $y = f(x) + k$, given the curve $y = f(x)$. The required curve can easily be obtained by regarding it as the image of the given curve under a mapping that maps the point (a, b) onto the point $(a, b + k)$ for all a, b; thus the given curve is translated k units in the positive y-direction.

Similarly the curve $y = f(x + k)$ can be obtained by regarding it as the image of the given curve under the mapping $(a, b) \mapsto (a - k, b)$, i.e. the given curve is translated k units in the negative x-direction.

Two other mappings are useful. For the curve $y = k\,f(x)$ apply the mapping $(a, b) \mapsto (a, kb)$ and we see that the given curve is *stretched* in the y-direction if $k > 1$, *contracted* if $0 < k < 1$; there is also a reflection in the x-axis if $k < 0$. Likewise, for the curve $y = f(kx)$ apply the mapping $(a, b) \mapsto (a/k, b)$ and we see that the given curve is *contracted* in the

x-direction if $k > 1$, *stretched* if $0 < k < 1$; there is also a reflection in the y-axis if $k < 0$.

Thus the curves

(a) $y = f(x) + k$, (b) $y = f(x + k)$,
(c) $y = kf(x)$, (d) $y = f(kx)$,

are easily derived from the curve $y = f(x)$.

An important example is the curve $y = A \sin(\omega x + \alpha)$. We first write this equation in the form $y = A \sin[\omega(x + \alpha/\omega)]$. From the familiar curve $y = \sin x$ we get the curve $y = \sin \omega x$ as for (d) above; the effect is best described by saying that the usual *wavelength* 2π is changed to a wavelength $2\pi/\omega$. Next, from the curve $y = \sin \omega x$ we get the curve $y = \sin[\omega(x + \alpha/\omega)]$ as for (b) above; the effect is to translate the curve α/ω units in the negative x-direction, and this is usually called a *phase change* of α/ω with this curve. Finally, from the curve $y = \sin(\omega x + \alpha)$ we get the curve $y = A \sin(\omega x + \alpha)$ as for (c) above; the effect is best described as a change in the *amplitude* of the waves. Figure 15.18 shows the final sketch.

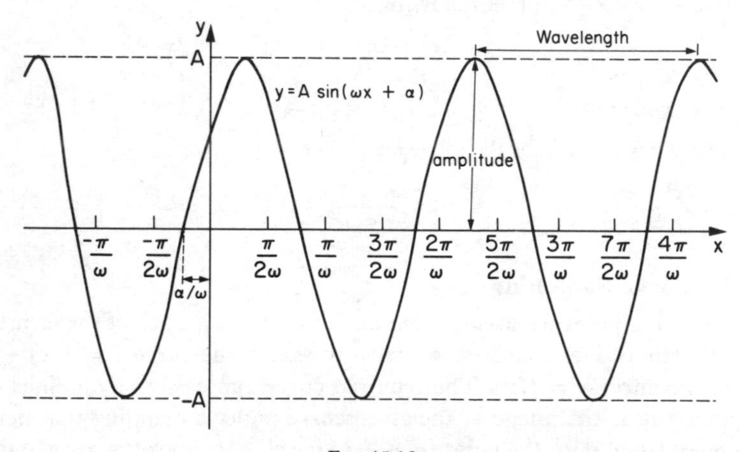

FIG. 15.18

The curves $y = [f(x)]^2$, $y = 1/f(x)$, $y = \sqrt{[f(x)]}$ may be obtained, at least to a useful first approximation, by applying the respective mappings $(a, b) \mapsto (a, b^2)$, $(a, b) \mapsto (a, 1/b)$, $(a, b) \mapsto (a, \sqrt{b})$ to the curve $y = f(x)$. The slope of the curve, found by using the derivative, can then be used to improve the sketch.

Example 24 Sketch the curves $y = \sin^2 x$, $y = \operatorname{cosec} x$, $y = \sqrt{(\sin x)}$.

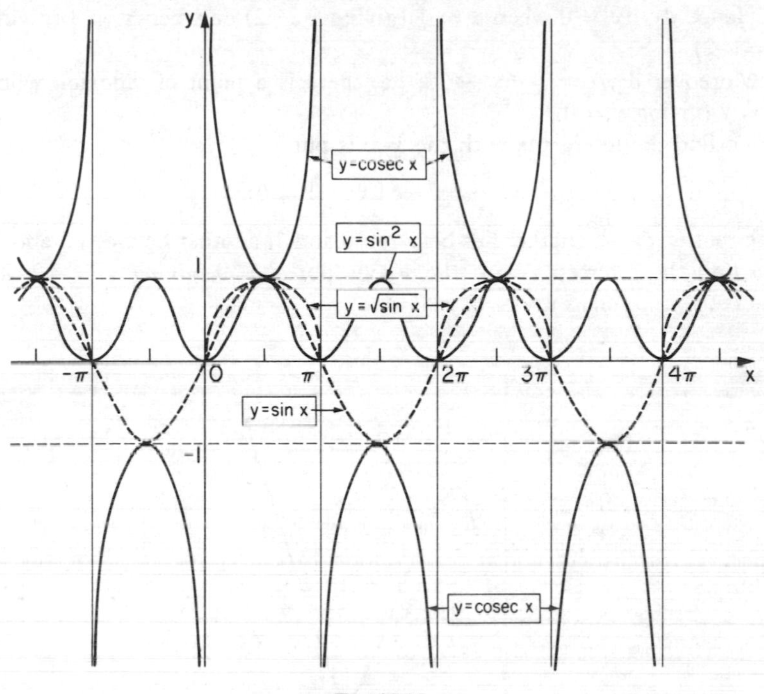

FIG. 15.19

Note that parts of the curve $y = \sin x$ that lie *below* the x-axis have their image *above* the x-axis under the mapping $(a, b) \mapsto (a, b^2)$ and have *no* image under the mapping $(a, b) \mapsto (a, \sqrt{b})$. The flatness of the curve $y = \sin^2 x$ at the points $x = 0, \pm\pi, \pm 2\pi, \ldots$, and the steepness of the curve $y = \sqrt{(\sin x)}$ at these same points, is verified from the derivative in each case. For if $y = \sin^2 x$, then $dy/dx = 2 \sin x \cos x = \sin 2x$ and so $dy/dx = 0$ when $x = 0, \pm\pi/2, \pm 3\pi/2, \pm\pi, \ldots$; likewise if $y = \sqrt{(\sin x)}$, then $dy/dx = \cos x/2\sqrt{(\sin x)}$ and so $dy/dx \to \infty$ as $x \to 0, \pm\pi, \pm 2\pi, \ldots$.

15.4.1 Polynomials The curve $y = P(x)$, where P is a polynomial function, can be sketched directly from information about y and dy/dx easily obtainable from the equation. The following example illustrates the method.

Example 25 Sketch the curve $y = x^3 - 6x^2 + 9x - 2$.

We have

$$\frac{dy}{dx} = 3x^2 - 12x + 9 = 3(x-1)(x-3).$$

Hence $dy/dx = 0$ when $x = 1$ (giving $y = 2$) or when $x = 3$ (giving $y = -2$).

Moreover $d^2y/dx^2 = 6x - 12$, i.e. there is a point of inflexion when $x = 2$ (giving $y = 0$).

To find the intercepts with the x-axis put

$$x^3 - 6x^2 + 9x - 2 = 0;$$

one root is $x = 2$, another lies between 0 and 1, another between 3 and 4. To find the intercepts with the y-axis put $x = 0$, giving $y = -2$. As $x \to \infty, y \to \infty$ and as $x \to -\infty, y \to -\infty$.

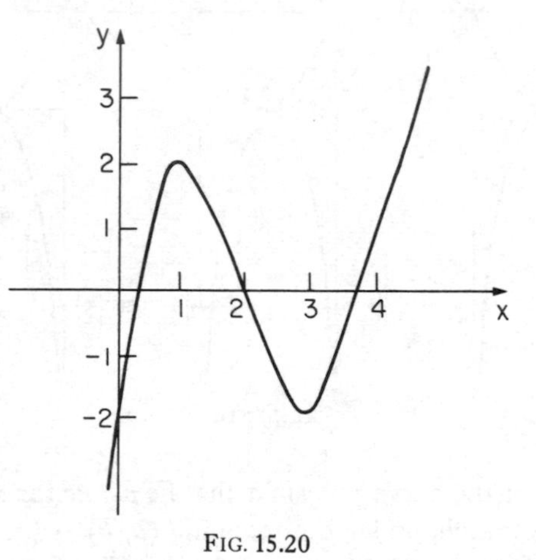

FIG. 15.20

15.4.2 Rational Functions For the curve $y = P(x)/Q(x)$, where P and Q are polynomial functions, it is important first to solve the equation $Q(x) = 0$ to locate asymptotes parallel to the y-axis, as these represent discontinuities where the graph breaks up into separate parts. There may also be another asymptote, either slant or parallel to the x-axis.

Example 26 Sketch the curve

$$y = \frac{x}{(x-2)(x-3)}.$$

Put $(x - 2)(x - 3) = 0$ giving $x = 2, 3$. The lines $x = 2$, $x = 3$ are asymptotes.

Intercept with x-axis: put the numerator equal to zero, i.e. $x = 0$.

Intercept with y-axis: put $x = 0$, giving $y = 0$.

As $x \to \pm\infty, y \to 0$; hence the line $y = 0$ is an asymptote. We have

$$\frac{dy}{dx} = \frac{(x^2 - 5x + 6) - x(2x - 5)}{(x^2 - 5x + 6)^2} = \frac{6 - x^2}{(x^2 - 5x + 6)^2}.$$

Hence $dy/dx = 0$ when $x = \pm\sqrt{6}$.

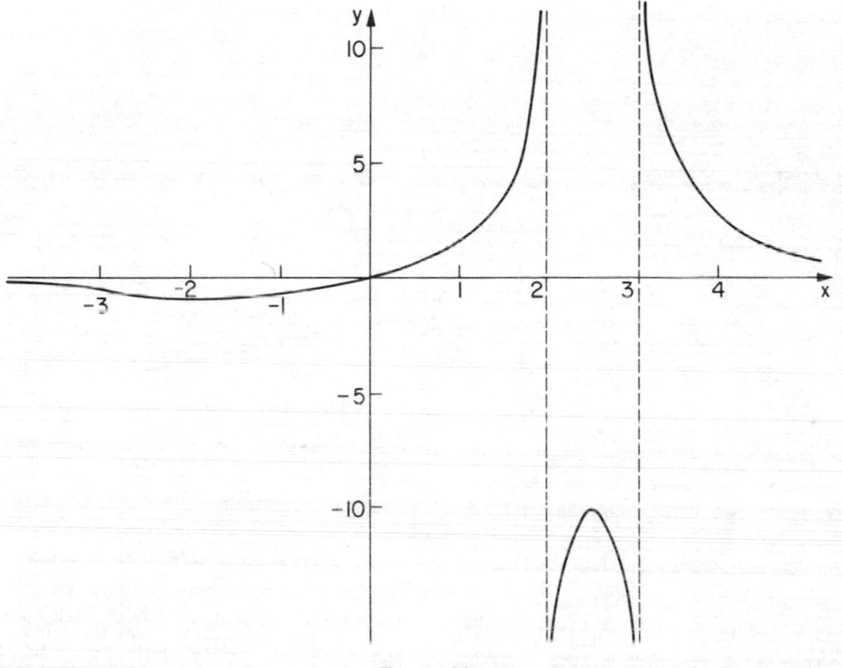

FIG. 15.21

Example 27

Sketch the curve $y = x(1 - x)/(1 + x)$.

Putting the denominator equal to zero, we see that the line $x = -1$ is an asymptote.
By division

$$y = -x + 2 - \frac{2}{1 + x}$$

and so the line $y = -x + 2$ is an asymptote.
Intercept with x-axis: put $x(1 - x) = 0$, giving $x = 0$ or 1.

Intercept with y-axis: put $x = 0$, giving $y = 0$.
We have

$$\frac{dy}{dx} = \frac{(1 + x)(1 - 2x) - (x - x^2)}{(1 + x)^2} = \frac{1 - 2x - x^2}{(1 + x)^2}.$$

Hence $dy/dx = 0$ when $x = -1 \pm \sqrt{2}$.

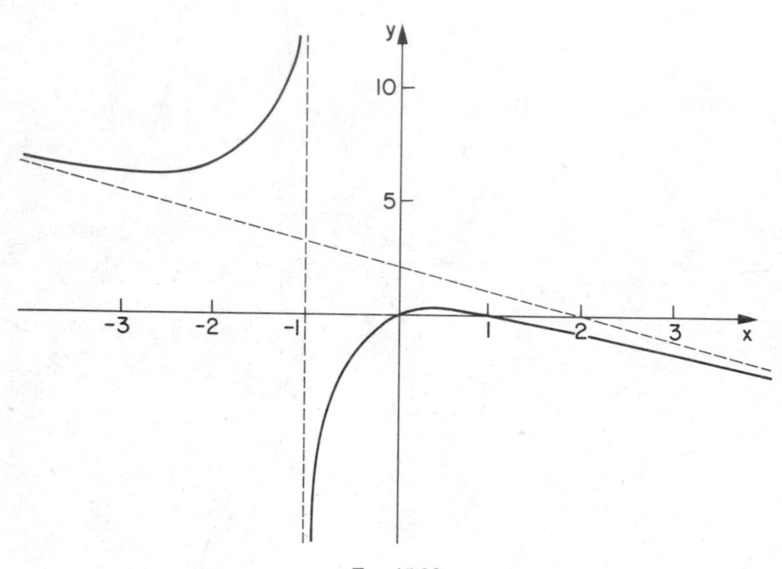

FIG. 15.22

15.4.3 Algebraic Functions Proceeding to other algebraic equations, we note that if the equation is of second degree in x and y, that is of the form

$$ax^2 + 2hxy + by^2 + 2gx + 2fy + c = 0,$$

then the curve is a conic section—this case has been fully dealt with in Chapter 2. Note that such equations represent *two* functions, e.g. the equation $x^2 + y^2 = 1$ gives $y = \pm\sqrt{(1 - x^2)}$, and in general an equation of nth degree in y may represent as many as n functions. We shall not normally consider equations of more than second degree in y.

Note 1. If only even powers of y occur in the equation, then the curve is symmetric about the x-axis. Likewise if only even powers of x occur, the curve is symmetric about the y-axis.

Note 2. Before attempting to sketch the curve, we should determine if there are values of x corresponding to which y is not defined, e.g. $y = \sqrt{(1 - x^2)}$ gives no value of y for $|x| > 1$. This determines the region of the xy-plane in which the curve lies.

Example 28 Sketch the curve $y^2 = x(1 - x)^2$.

Write $y = \pm\sqrt{x}(1 - x)$. Here y is not defined if $x < 0$ and so the curve lies entirely in the half-plane $x \geqslant 0$. By symmetry we may consider only the curve $y = \sqrt{x}(1 - x)$, and then add its reflection in the x-axis.

From $y = \sqrt{x}(1 - x)$ we get

$$\frac{dy}{dx} = \frac{1 - 3x}{2\sqrt{x}}.$$

Hence $dy/dx = 0$ when $x = \frac{1}{3}$. Also $dy/dx \to \infty$ as $x \to 0+$.

Intercepts with x-axis: put $\sqrt{x}(1 - x) = 0$, giving $x = 0$ or 1.

Intercepts with y-axis: put $x = 0$, giving $y = 0$.

As $x \to \infty, y \to -\infty$.

Also

$$\frac{d^2y}{dx^2} = -\frac{(1 + 3x)}{4x^{3/2}},$$

which is negative at all points on the curve.

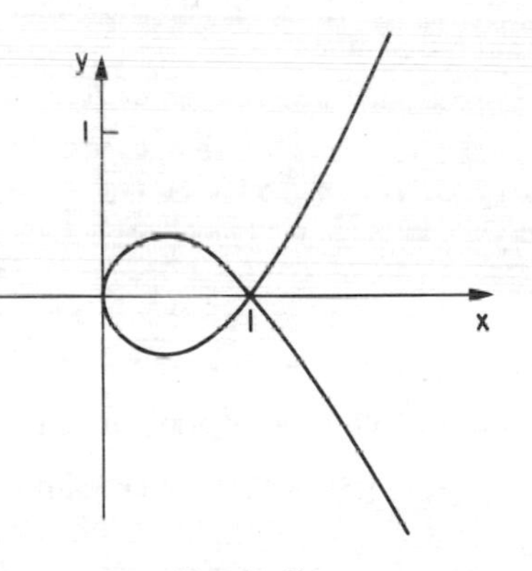

FIG. 15.23

Exercises 15(f)

1. Draw a separate sketch for each of the following:

   (a) a portion of a curve where y and dy/dx are both positive;
   (b) a portion of a curve where y is positive, and dy/dx is negative;
   (c) a portion of a curve where y is negative, and dy/dx changes from positive to negative;

(d) a portion of a curve where x and y are both negative, and d^2y/dx^2 changes from negative to positive.

2. Sketch a continuous curve $y = f(x)$ having the following characteristics:

$$f(-3) = 2, f(0) = 1, f(1) = 0, f(2) = 1,$$
$$f'(x) > 0 \text{ for all } x < -3 \text{ and all } x > 1,$$
$$f'(x) < 0 \text{ for } -3 < x < 1, f'(1) = f'(-3) = 0.$$

3. Sketch the graph of $f(x)$ on the interval (a, b), given that $f'(x) > 0$ on the interval and $f''(x)$ changes sign at $x = \frac{1}{2}(a + b)$ from positive to negative.
4. Sketch the graph of $f(x)$ in the neighbourhood of the point $(a, f(a))$, given that $f(a) = f'(a) = f''(a) = 0$ and $f'''(a) > 0$.
5. Find the turning points on the curve $y = 8x^3 - 11x^2 + 4x + 1$ and distinguish between them.

   Find also the x-co-ordinate of the point of inflexion, and sketch the curve.
6. Find the co-ordinates of the turning points and the x-co-ordinates of the points of inflexion on the curve $y = 6x^4 - 4x^3 - 6x^2 - 1$, and sketch the curve.
7. Show that the curve $y = x^2/16 + 1/x$ has a point of inflexion where it meets the x-axis. Show also that it has one stationary point and sketch the curve.
8. Given $y = x/(x^2 - 9)$, show that dy/dx is always negative and that $d^2y/dx^2 = 0$ when $x = 0$. Sketch the curve $y = x/(x^2 - 9)$.
9. Find the turning points and points of inflexion on the curve $y = x/(1 + x^2)$. Sketch the curve.
10. Sketch the curve $y = (x - 2)/(x + 2)^2$ and show that $(x - 2)/(x + 2)^2$ cannot exceed $1/16$ for any value of x.
11. Sketch the curve

$$y = \frac{4x + 3}{x^2 - 2x + 3}$$

and show that

$$\frac{4x + 3}{x^2 - 2x + 3}$$

always lies between $-\frac{1}{2}$ and 4. Draw also a rough sketch of the curve

$$y = \sqrt{[(4x + 3)/(x^2 - 2x + 3)]}.$$

12. Sketch the curves

$$y = \frac{4x - 3}{x^2 - 1}, \quad y = \frac{4x - 3}{x^2 + 1}$$

and show that $(4x - 3)/(x^2 - 1)$ may have any value, but the value of $(4x - 3)/(x^2 + 1)$ must lie between -4 and 1.

13. Sketch the curve

$$y = \frac{x^2 - 4x + 3}{x^2 - 2x + 2}$$

and show that

$$\frac{x^2 - 4x + 3}{x^2 - 2x + 2}$$

lies in the range $\frac{1}{2}(1 - \sqrt{5})$ to $\frac{1}{2}(1 + \sqrt{5})$.

14. The graph of

$$y = \frac{ax + b}{(x - 1)(x - 4)}$$

has a turning point at $P(2, -1)$. Determine the values of a and b, and show that y is a maximum at P. Find the value of x for which y is a minimum.
   Sketch the curve, showing clearly the asymptotes and turning points.

15. Sketch the curves $y^2 = x(x - 1)^2$, $y^2 = x^2(1 - x)$, showing that each curve has a loop.

15.4.4 Curve Sketching from Polar Equations In general, it is useful to tabulate some values of r and θ when a curve is given by an equation of the form $r = f(\theta)$, but much labour may be saved by applying the following considerations:

I. Symmetry. If the equation of the curve is unaltered when θ is replaced by $(-\theta)$, the curve is symmetric about the line $\theta = 0$. In particular, if r is a function of $\cos\theta$ alone, the curve is symmetric about the initial line.

If the value of r is altered in sign but not in magnitude when θ is replaced by $(-\theta)$ in the equation of the curve, the curve is symmetric about the line $\theta = \frac{1}{2}\pi$. Again, if the equation of the curve is unaltered when $(\pi - \theta)$ is substituted for θ, the curve is symmetric about $\theta = \frac{1}{2}\pi$. In particular, if r is a function of $\sin\theta$ only, the curve is symmetric about $\theta = \frac{1}{2}\pi$.

If only even powers of r occur in its equation, the curve is symmetric about the pole.

II. Form of the Curve at the Pole. In general, if the curve $r = f(\theta)$ passes through the pole, the directions of tangents to the curve at the pole are found by solving the equation $f(\theta) = 0$, since as $r \to 0$ the curve approaches the pole.

III. Limitations on the Value of r and θ. These are readily seen in an equation of the form $r = f(\theta)$. For example, the curve $r = 2 + \sin\theta$ lies entirely within the concentric circles $r = 1$ and $r = 3$, while for the curve $r^2 = a^2 \cos 2\theta$, $r^2 \leqslant a^2$ so that the curve lies wholly within the circle $r = a$; also when $\cos 2\theta < 0$, i.e. when $\frac{1}{4}\pi < \theta < \frac{3}{4}\pi$ and when $\frac{5}{4}\pi < \theta < \frac{7}{4}\pi$ there is no real value of r.

IV. Direction of the Tangent. The relation $\tan\phi = r\,d\theta/dr$ determines the angle between the radius vector and the tangent to the curve at any point.

To see this let $P(r, \theta)$ and $Q(r + \Delta r, \theta + \Delta\theta)$ be two points on the curve $r = f(\theta)$, and let PN be the perpendicular drawn from P onto OQ where O is the pole. Then

$$\begin{aligned} PN &= r \sin \Delta\theta, \\ NQ &= OQ - ON = r + \Delta r - r \cos \Delta\theta \\ &= \Delta r + 2r \sin^2 \tfrac{1}{2}\Delta\theta. \end{aligned}$$

Hence

$$\begin{aligned} \tan(\angle PQO) = \frac{PN}{NQ} &= \frac{r \sin \Delta\theta}{\Delta r + 2r \sin^2 \frac{1}{2}\Delta\theta} \\ &= \frac{r\,\Delta\theta\,(\sin \Delta\theta)/\Delta\theta}{\Delta r + 2r \sin^2 \frac{1}{2}\Delta\theta} \\ &\to r\frac{d\theta}{dr} \text{ as } \Delta\theta \to 0. \end{aligned}$$

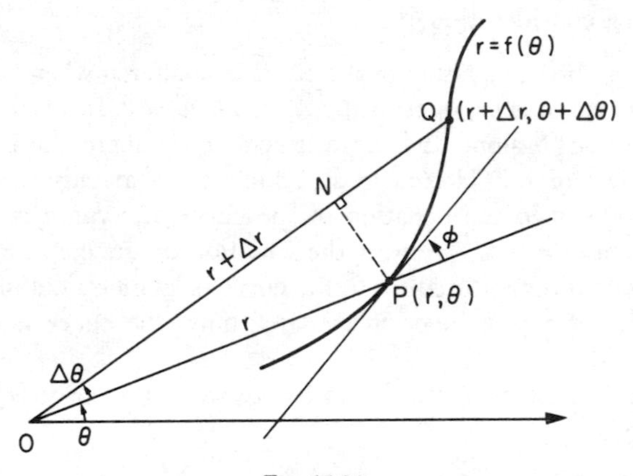

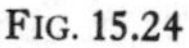
FIG. 15.24

But as $\Delta\theta \to 0$ the line PQ approaches the tangent at P and $\angle PQO \to \phi$, where ϕ denotes the angle between OP and the tangent at P drawn in the direction of increasing θ.

Polar equations of straight line, circle and conic have been dealt with in Section 2.8. We look now at some other standard polar equations.

1. The Limaçon and Cardioid. By drawing the circle $r = a \cos \theta$ and extending the radius vector corresponding to each value of θ by an amount b we construct the curve (known as the limaçon)

$$r = b + a \cos \theta. \tag{1}$$

In this formula a and b can be either positive or negative. Figure 15.25(a) shows the limaçon when $a > b > 0$; Fig. 15.25(b) shows the limaçon when $0 < a < b$. Curves of this type are also obtained from an equation of the form $r = b + a \sin \theta$.

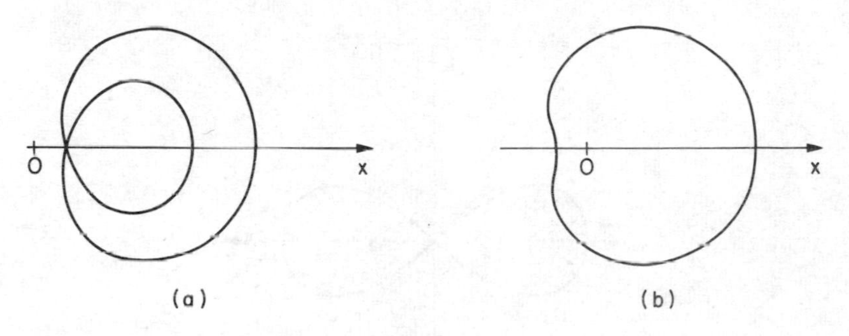

FIG. 15.25

If in (1) we write $b = a$, we obtain the equation $r = a(1 + \cos \theta)$. This curve, known as the *cardioid*, is shown in Fig. 15.26.

The equation of a cardioid may appear in any of the forms

$$r = a(1 - \cos \theta), \quad r = a(1 + \sin \theta), \quad r = a(1 - \sin \theta)$$

and the corresponding graphs are obtained by rotating Fig. 15.26 in its own plane about the pole through π, $+\frac{1}{2}\pi$ and $-\frac{1}{2}\pi$ radians respectively.

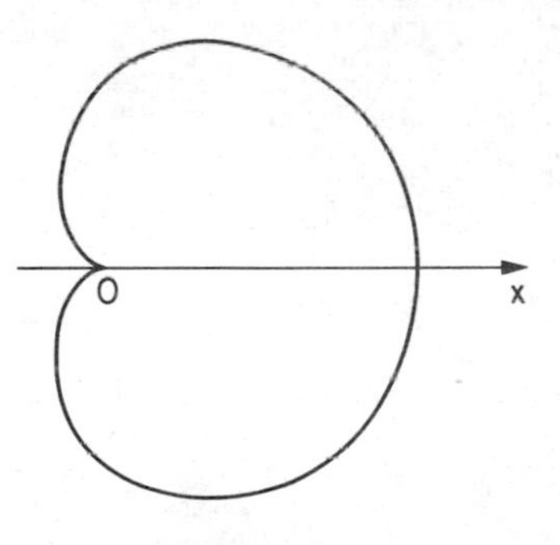

FIG. 15.26

2. *The Lemniscate* $r^2 = a^2 \cos 2\theta$. Since $\cos 2\theta$ may be expressed as a function of $\cos \theta$ only, or as a function of $\sin \theta$ only, this curve is symmetric about the lines $\theta = 0$, $\theta = \frac{1}{2}\pi$ and we need consider only values of θ between 0 and $\frac{1}{2}\pi$. As θ increases from 0 to $\frac{1}{4}\pi$, r decreases from a to 0, and the line $\theta = \frac{1}{4}\pi$ is a tangent to the curve at the pole. Between

$\theta = \frac{1}{4}\pi$ and $\theta = \frac{3}{4}\pi$, where $\cos 2\theta$ (and hence r^2) is negative, the curve does not exist. At the point $(a, 0)$, $\phi = \frac{1}{2}\pi$. The curve is shown in Fig. 15.27. The lemniscate $r^2 = a^2 \sin 2\theta$ is obtained by rotating Fig. 15.27 in its own plane through $\frac{1}{4}\pi$ radians about O.

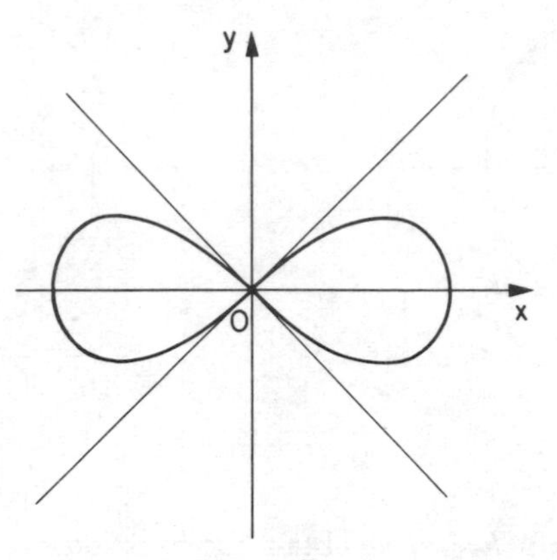

FIG. 15.27

3. *The n-leaved Rose.* The curves $r = a \sin n\theta$, $r = a \cos n\theta$ consist of n leaves if n is odd and $2n$ leaves if n is even. The leaves are equal in size and are spaced at equal intervals round the pole.

Figure 15.28(a) shows the curve $r = a \sin 3\theta$; Fig. 15.28(b) shows the curve $r = a \cos 2\theta$.

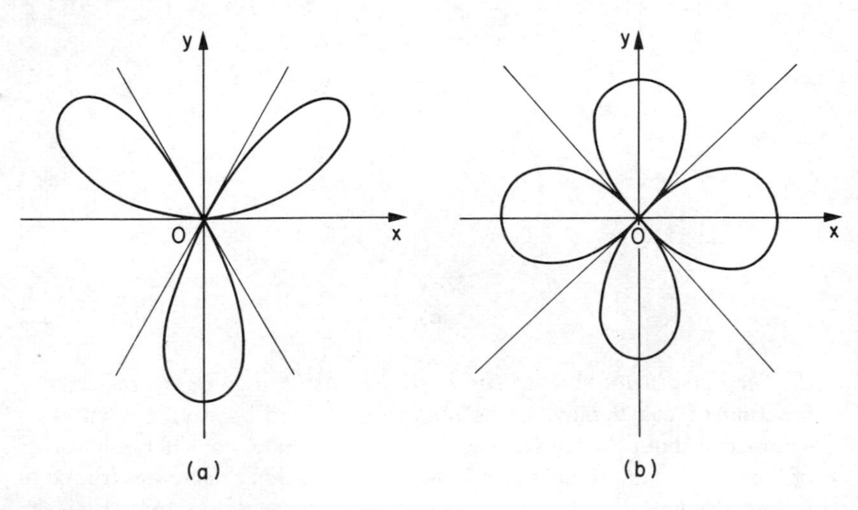

FIG. 15.28

4. *Spirals.* These are curves in which, as θ increases without limit, r either steadily increases or steadily decreases so that the curves wind round and round the pole.

The *Archimedean spiral* $r = a\theta$ $(a > 0)$ is shown in Fig. 15.29.

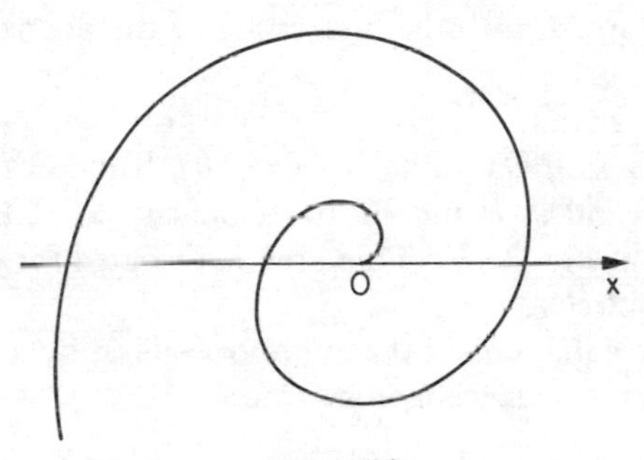

FIG. 15.29

15.4.5 Roulettes A *roulette* is the curve traced out by a point carried by a plane curve which rolls without slipping on a fixed curve in its own plane. We consider here some important examples of curves of this type.

1. *The Cycloid.* The *cycloid* is the curve traced out by a point on the circumference of a circle which rolls without slipping along a fixed straight line. The parametric equations of this curve can be found from Fig. 15.30,

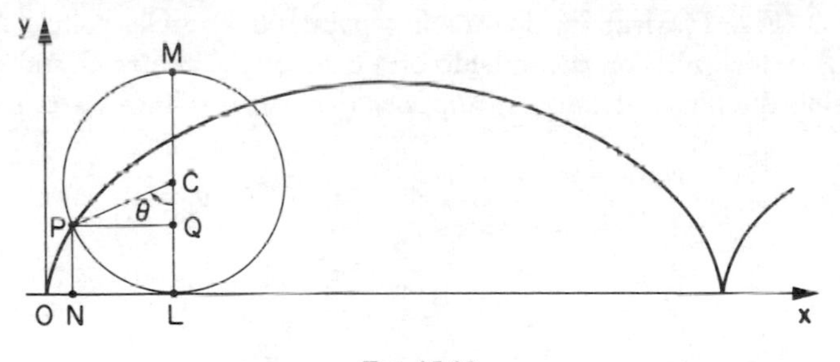

FIG. 15.30

which shows a circle with centre C and radius a moving along a horizontal line OX which we take as x-axis. The tracing point P originally coincided with the origin O, but the circle has rolled through an angle θ so that $\angle PCL = \theta$, where L is the point of the circle now in contact with Ox. Because there is no slipping, $OL = \text{arc } PL = a\theta$, and if P is the point (x, y)

$$x = OL - PQ = a(\theta - \sin\theta), \quad y = CL - CQ = a(1 - \cos\theta).$$

These are the parametric equations of the cycloid. From them we obtain

$$\frac{dy}{dx}=\frac{\sin\theta}{1-\cos\theta}=\cot\tfrac{1}{2}\theta,$$

and so the tangent to the cycloid at P makes with Ox an angle $\frac{1}{2}(\pi-\theta)$. But if P is joined to M, the other extremity of the diameter LC,

$$\angle PML=\tfrac{1}{2}\angle PCL=\tfrac{1}{2}\theta$$

and so PM makes with Ox an angle $\frac{1}{2}(\pi-\theta)$. It follows that PM is the tangent to the cycloid at P and PL is the normal at P. One arch of the cycloid is shown in Fig. 15.30. This arch is repeated for each revolution of the generating circle.

The parametric equations of the curve traced out by a point Q on CP or CP produced at a distance k from C are

$$x=a\theta-k\sin\theta,\quad y=a-k\cos\theta.$$

This curve is known as a *trochoid.*

2. The Epicycloid and the Hypocycloid. When one of two coplanar circles rolls without slipping on the other circle which is fixed, any point on the circumference of the rolling circle traces an *epicycloid* or a *hypocycloid* according as the rolling circle is outside or inside the fixed circle. An epicycloid in which the rolling circle surrounds the fixed circle is sometimes called a *pericycloid.*

Let P (Fig. 15.31(a) be the tracing point on a circle, centre B and radius b, which rolls on the outside of a fixed circle, centre O and radius a, I being the point of contact. Suppose that when rolling starts the line

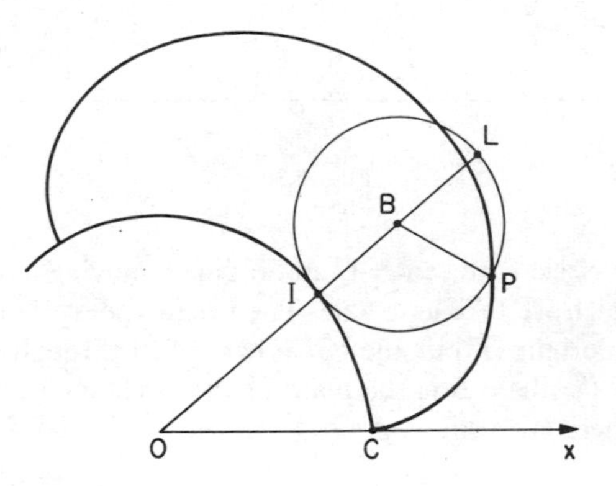

FIG. 15.31(a)

OB is horizontal and the points P and C are in contact. Take OC as x-axis and the vertical through O as y-axis.

Since there is no slipping, arc IC = arc PI and so, if $\angle COI = \theta$ and $\angle IBP = \phi$,

$$a\theta = b\phi \tag{1}$$

and BP makes with Ox an angle

$$\theta + \phi. \tag{2}$$

Then if P is the point (x, y)

$$\begin{aligned} x &= \text{sum of projections of } \overrightarrow{OB} \text{ and } \overrightarrow{BP} \text{ on } Ox \\ &= OB \cos\theta + BP \cos[\pi - (\theta + \phi)], \quad \text{by (2)} \\ &= (a + b)\cos\theta - b\cos\frac{a+b}{b}\theta, \quad \text{by (1)}. \end{aligned}$$

By considering the projections of $\overrightarrow{OB}$ and $\overrightarrow{BP}$ on Oy we obtain

$$y = (a + b)\sin\theta - b\sin\frac{a+b}{b}\theta.$$

Thus the parametric equations of the epicycloid are

$$x = (a + b)\cos\theta - b\cos\frac{a+b}{b}\theta,$$

$$y = (a + b)\sin\theta - b\sin\frac{a+b}{b}\theta.$$

As in the case of the cycloid, we may show that if P is joined to L, the other extremity of the diameter IB, PL and PI are the tangent and normal respectively to the epicycloid at P.

The equations of the corresponding hypocycloid are obtained by changing the sign of b. They are

$$x = (a - b)\cos\theta + b\cos\frac{a-b}{b}\theta,$$

$$y = (a - b)\sin\theta - b\sin\frac{a-b}{b}\theta.$$

As a particular case of the epicycloid we have, if $b = a$,

$$x = 2a\cos\theta - a\cos 2\theta, \quad y = 2a\sin\theta - a\sin 2\theta.$$

This is the equation of a cardioid, for if r is the distance from P to C, the point $(a, 0)$,

$$\begin{aligned} r^2 &= (x - a)^2 + y^2 \\ r^2/a^2 &= (2\cos\theta - 2\cos^2\theta)^2 + (2\sin\theta - 2\sin\theta\cos\theta)^2 \\ &= 4(1 - \cos\theta)^2 \end{aligned}$$

and so

$$r = 2a(1 - \cos\theta). \tag{3}$$

Since $a = b$, $\theta = \phi$ and CP is parallel to OB, i.e. θ is the angle between OC and CP. Thus (3) is the polar equation of a cardioid referred to C as pole and OC as initial line.

The curve is shown in Fig. 15.31(b).

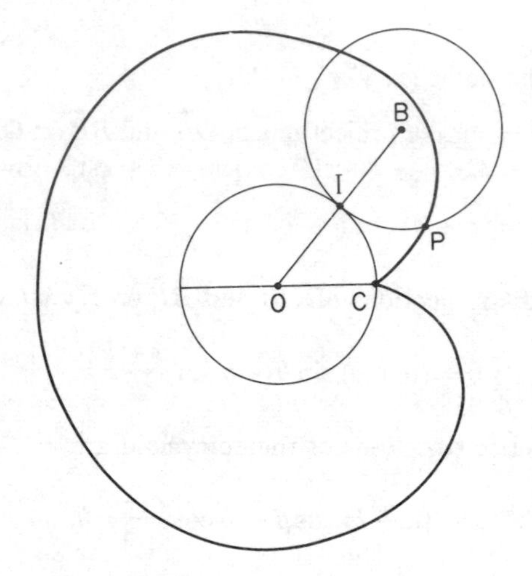

FIG. 15.31(b)

As a particular case of the hypocycloid we have, if $a = 2b$,

$$x = 2b\cos\theta, \quad y = 0,$$

i.e. the curve becomes the diameter through C of the fixed circle (Fig. 15.32(a)).

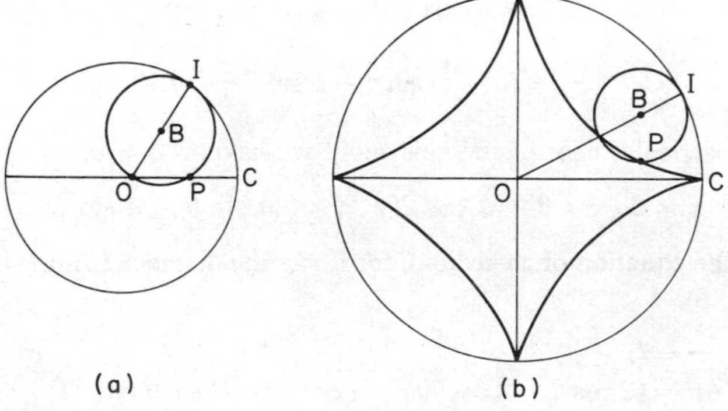

FIG. 15.32

Also, if $a = 4b$ we have

$$x = \tfrac{1}{4}a(3\cos\theta + \cos 3\theta) = a\cos^3\theta, \quad y = a\sin^3\theta,$$

which is equivalent to the cartesian equation

$$x^{2/3} + y^{2/3} = a^{2/3}.$$

This curve is called the *astroid* and is shown in Fig. 15.32(b).

Exercises 15(g)

1. A point moves so that the product of its distances from two fixed points A and B is constant and equal to $3c^2$, where $c = \tfrac{1}{2}AB$. By taking the mid-point of AB as pole, show that the polar equation of the locus can be expressed in the form

$$r^4 - 2c^2r^2\cos 2\theta - 8c^4 = 0.$$

2. In the following curves find ϕ, the angle between the tangent and the radius vector, in terms of θ:

(a) $r = \tfrac{1}{2}a\sec^2\tfrac{1}{2}\theta$, (b) $r^2\cos 2\theta = a^2$, (c) $r^2 = a^2\sin 2\theta$,
(d) $r = a\sin^n(\theta/n)$, (e) $r^n = a^n\sin n\theta$, (f) $r^3 = \cos 3\theta$,
(g) $r = a(1 + \cos\theta)$.

3. Draw a rough sketch of the curve $r^3 = a^3\sin 3\theta$, showing that it consists of three loops. If P is the point (r, θ) on the curve, show that the tangent at P makes an angle 4θ with the initial line.
4. Sketch the curve $r = 1 + 2\cos\theta$, showing that it consists of two loops.
5. For the cardioid $r = a(1 - \cos\theta)$, show that the tangents at the points where $\theta = \alpha, \theta = \alpha + \tfrac{2}{3}\pi, \theta = \alpha + \tfrac{4}{3}\pi$ are parallel.
6. P and P' are points on the curve $r^2 = a^2\cos 2\theta$ whose vectorial angles θ, θ' are each positive and less than $\tfrac{1}{4}\pi$ $(\theta' > \theta)$. If the tangents at P and P' are perpendicular to each other, show that $\theta' = \theta + \tfrac{1}{6}\pi$.
7. Sketch the lemniscate $r^2 = \cos 2\theta$. Given that ϕ is the angle between the tangent and radius vector, show that $\phi = 2\theta + \tfrac{1}{2}\pi$, and hence, or otherwise, determine the greatest width of a loop, measured at right angles to the line $\theta = 0$.
8. A wheel rolls along the x-axis. It may be assumed that the point P on the circumference of the wheel, initially at the origin, moves on the cycloid

$$x = a(\theta - \sin\theta), \quad y = a(1 - \cos\theta),$$

where a is the radius of the wheel and θ is the angle through which P has rotated about the centre of the wheel. Find an expression for the distance of P from the origin, and find the rate at which this distance is increasing when $\theta = \pi$ if at that instant the wheel is revolving at the rate of one revolution per second.

CHAPTER 16

INTEGRALS AND ANTIDERIVATIVES

16.1 The Idea of an Integral

Figure 16.1 illustrates a method of finding an approximate value for the area *ABCD* bounded by the lines $x = 1, y = 0, x = 2, y = \frac{1}{2}x$. The area is divided into n strips by the lines

$$x = 1 + \frac{1}{n}, x = 1 + \frac{2}{n}, \ldots, x = 1 + \frac{n-1}{n}$$

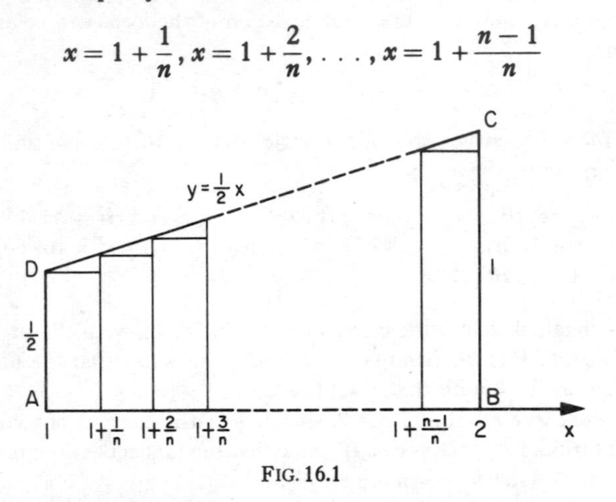

FIG. 16.1

so that each strip has width $1/n$, and each strip is approximately rectangular as shown in the diagram. Hence

$$\begin{aligned} \text{area } ABCD &\simeq \tfrac{1}{2}\frac{1}{n} + \tfrac{1}{2}\left(1 + \frac{1}{n}\right)\frac{1}{n} + \tfrac{1}{2}\left(1 + \frac{2}{n}\right)\frac{1}{n} + \cdots \\ &\quad + \tfrac{1}{2}\left(1 + \frac{n-1}{n}\right)\frac{1}{n} \\ &= \frac{1}{2n}\left[1 + \left(1 + \frac{1}{n}\right) + \left(1 + \frac{2}{n}\right) + \cdots + \left(1 + \frac{n-1}{n}\right)\right] \\ &= \frac{1}{2n}\left(1 + \frac{n-1}{2n}\right)n \quad \text{(using the result for the sum of an A.P.)} \\ &= \frac{3}{4} - \frac{1}{4n}. \end{aligned}$$

Thus if 5 strips are used ($n = 5$) we get an approximate area of 0·70, if ten strips are used ($n = 10$) we get an approximate area of 0·725, and so on. If we take the limit of the approximation formula as $n \to \infty$ we get $\frac{3}{4}$, and this is the exact value of the area as is easily verified.

One way of anticipating that the limit would give the exact area is to consider the error in the approximation. In each strip a small right-angled triangle at the top is neglected, and each such triangle has base $1/n$, height $1/2n$. Thus the error in the area of each strip is $1/4n^2$, and for all n strips this gives a total error of $1/4n$. Clearly the total error tends to zero as $n \to \infty$.

In Fig. 16.2 a similar approach is used for the area $ABCD$ bounded by the lines $x = 0$, $y = 0$, $x = 2$ and the arc CD of the curve $y = 8x^2 + 20$. The base AB has length 2 and is divided into n equal parts each of length $2/n$. The resulting strips of width $2/n$ are approximated by rectangles (the shaded rectangles in Fig. 16.2), and the areas of these rectangles are added together to give an approximate value for the area $ABCD$. Hence

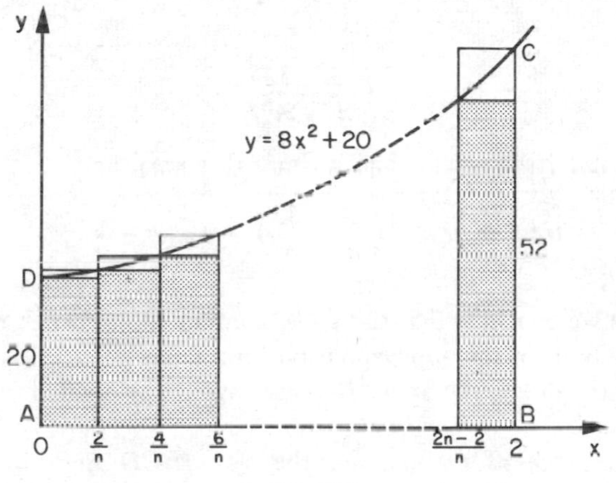

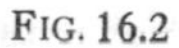
FIG. 16.2

$$\text{area } ABCD \simeq \frac{2}{n}\left\{20 + \left[8\left(\frac{2}{n}\right)^2 + 20\right] + \left[8\left(\frac{4}{n}\right)^2 + 20\right] + \cdots + \left[8\left(\frac{2n-2}{n}\right)^2 + 20\right]\right\}$$

$$= \frac{2}{n}\left\{20n + 8\left[\frac{2^2}{n^2} + \frac{4^2}{n^2} + \cdots + \frac{(2n-2)^2}{n^2}\right]\right\}$$

$$= \frac{2}{n}\left\{20n + \frac{8 \cdot 2^2}{n^2}\left[1^2 + 2^2 + \cdots + (n-1)^2\right]\right\}$$

$$= 40 + \frac{64}{n^3} \cdot \frac{1}{6}(n-1)n(2n-1) \qquad \text{(see Section 4.4)}$$

$$= \frac{184}{3} - \frac{32(3n-1)}{3n^2}.$$

This approximation formula has a limit as $n \to \infty$, viz. $\frac{184}{3}$. There is no *exact* value with which to compare this limit since as yet no rules have been given for the evaluation of areas with curved boundaries (other than circular). But even if we cannot attach precise values to such areas we have some notions about them, and we can readily agree that the areas of the small bits missed out at the tops of the strips are less than those of the corresponding small non-shaded rectangles enclosing them in Fig. 16.2. Thus the total error in the approximation, assuming that a correct value has been nominated and that it agrees with our intuitive notions, would be less than the sum of n rectangles, each of length $2/n$, widths varying from $8(2/n)^2$ (see first strip) to

$$52 - 8\left(\frac{2n-2}{n}\right)^2 - 20,$$

i.e. $64/n - 32/n^2$ (see last strip). Thus the largest of these rectangles has area

$$\frac{2}{n}\left(\frac{64}{n} - \frac{32}{n^2}\right)$$

and so the total error is less than n times this area, i.e.

$$\text{total error} < 2\left(\frac{64}{n} - \frac{32}{n^2}\right) \to 0 \text{ as } n \to \infty.$$

From this we conclude that the value nominated as *correct* must be $\frac{184}{3}$ which is the limit of the approximation formula as $n \to \infty$. By attaching this particular value to the area $ABCD$ we avoid any conflict with our intuitive notions.

The above process for calculating the area $ABCD$ might be described as a summation process since the area is regarded as a sum of a number of component strips, but a limit process is also involved and the term *integration* is used for this combined process. The process of integration is valuable in calculating many quantities, but first we look more formally at its application to area.

16.1.1 Area and Integration Assume that $f(x) \geqslant 0$ on the interval $a \leqslant x \leqslant b$. The area $ABCD$ (see Fig. 16.3) bounded by the lines $x = a$, $y = 0$, $x = b$ and the arc CD of the curve $y = f(x)$ is normally referred to as *the area under the curve* $y = f(x)$ *from* $x = a$ *to* $x = b$. Let this area be divided into n strips by the lines $x = x_2, x = x_3, \ldots, x = x_n$ as shown in the diagram; we shall write $x_1 = a$, $x_{n+1} = b$ and refer to the strip between x_r and x_{r+1} as the rth strip ($r = 1, 2, \ldots, n$). However the area of a region

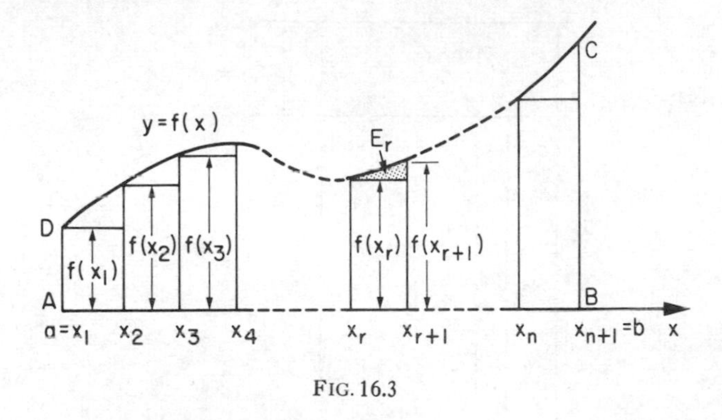

FIG. 16.3

with a curved boundary is defined, it seems reasonable to demand that it have two properties:

Property 1. Area is *additive*, i.e. if a region is the union of several non-overlapping component parts then its area equals the sum of the areas of the component parts.

Property 2. If X and Y are two regions such that $X \subset Y$, then the area of $X \leqslant$ the area of Y.

From Property 1 the area $ABCD$ should equal the sum of the areas of the n strips. That is, writing A for the area of the region $ABCD$ and ΔA_r for the area of the rth strip we have

$$A = \sum_{r=1}^{n} \Delta A_r.$$

Another consequence of Property 1 is that the area of the rth strip is the sum of two areas, one rectangular and the other the shaded area E_r shown in Fig. 16.4. Thus, writing $x_{r+1} - x_r = \Delta x_r$, we have

$$\Delta A_r = f(x_r)\Delta x_r + E_r$$

and so

$$A = \sum_{r=1}^{n} f(x_r)\Delta x_r + \sum_{r=1}^{n} E_r.$$

Suppose now that in the interval $x_r \leqslant x \leqslant x_{r+1}$, $f(x)$ has a least value m_r and a greatest value M_r; then by Property 2 it follows that

$$E_r \leqslant (M_r - m_r)\Delta x_r,$$

and hence

$$\sum_{r=1}^{n} E_r \leqslant \sum_{r=1}^{n} (M_r - m_r)\Delta x_r.$$

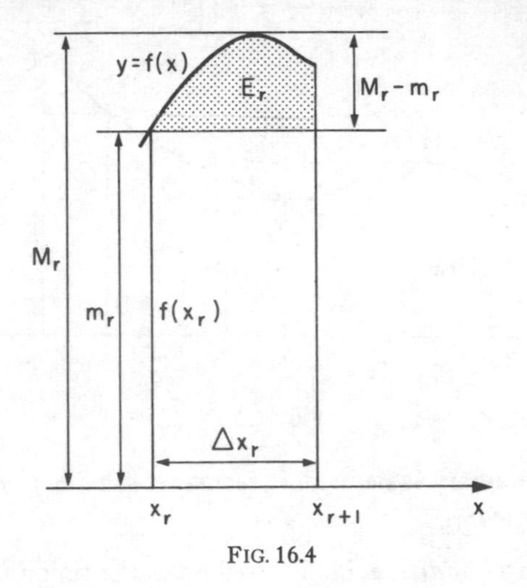

FIG. 16.4

Denoting by d the largest value of $M_r - m_r$ for the n strips, i.e.

$$d = \max_{r=1, 2, \ldots, n} (M_r - m_r),$$

we now have

$$\sum_{r=1}^{n} E_r \leqslant d \sum_{r=1}^{n} \Delta x_r = d(b-a).$$

If $f(x)$ is continuous on the interval $a \leqslant x \leqslant b$ it can be shown that $d \to 0$ as $n \to \infty$ provided that each strip width $\Delta x_r \to 0$. A formal proof is difficult although the reader should have little difficulty in accepting this result as plausible. It then follows that $\Sigma_{r=1}^{n} E_r \to 0$.

Again, if $f(x)$ is continuous on the interval $a \leqslant x \leqslant b$ it can be shown that as each strip width $\Delta x_r \to 0$ the expression $\Sigma_{r=1}^{n} f(x_r) \Delta x_r$ tends to a limit. We therefore assign a value to the area under the curve by the definition

$$A = \lim_{\Delta x_r \to 0} \sum_{r=1}^{n} f(x_r) \Delta x_r,$$

where it is to be understood that *each* $\Delta x_r \to 0$, and in all future discussion of area it will be assumed that the boundary curve is continuous.

16.1.2 Situations where Integration is Used If $f(x) = k$, where k is a constant, then the area under the curve $y = f(x)$ from $x = a$ to $x = b$ is rectangular and has the value $k(b - a)$. The discussion in Section 16.1.1

extends this simple idea by defining a value for the area when f is any continuous function. The situation is typical of many where

(a) corresponding to an interval $a \leqslant x \leqslant b$, we wish to define a value $Q_{a.b}$ (say) of some quantity Q (in the previous discussion Q referred to area),

(b) $Q_{a.b}$ depends on the values of a continuous function q, and

$$Q_{a.b} = k(b - a)$$

if $q(x) = k$ where k is a constant,

(c) for intuitive reasons we require that Q be additive, i.e. if the interval $a \leqslant x \leqslant b$ is subdivided by points $x_1, x_2, \ldots, x_n, x_{n+1}$, where

$$a = x_1 < x_2 < x_3 < \cdots < x_n < x_{n+1} = b,$$

then

$$Q_{a,b} = Q_{x_1,x_2} + Q_{x_2,x_3} + \cdots + Q_{x_r,x_{r+1}} + \cdots + Q_{x_n,x_{n+1}}$$

or writing $\Delta Q_r = Q_{x_r,x_{r+1}}$,

$$Q_{a.b} = \sum_{r=1}^{n} \Delta Q_r,$$

(d) for intuitive reasons we require that if $m_r \leqslant q(x) \leqslant M_r$ for values of x in the interval $x_r \leqslant x \leqslant x_{r+1}$ then $m_r \Delta x_r \leqslant \Delta Q_r \leqslant M_r \Delta x_r$ where $\Delta x_r = x_{r+1} - x_r$.

The particular case where $Q_{a.b}$ is the area under the curve $y = q(x)$ from $x = a$ to $x = b$ is discussed in Section 16.1.1. In all such cases we define

$$Q_{a.b} = \lim_{\Delta x_r \to 0} \sum_{r=1}^{n} q(x_r)\,\Delta x_r.$$

16.1.3 Work Done by a Variable Force If a constant force F acts on a body which moves a distance s in the direction of the force, then we say that the product Fs is the work done by the force on the body. This definition must be extended if we wish to deal with non-constant forces, e.g. the force required to extend a spring. Consider a body that moves in the positive direction along the x-axis and suppose that it is acted upon by a force F where $F = f(x)$. Denote by $W_{a.b}$ the work done by F on the body while the latter moves from position $x = a$ to position $x = b$, and following the discussion in Section 16.1.2 we define

$$W_{a.b} = \lim_{\Delta x_r \to 0} \sum_{r=1}^{n} f(x_r)\,\Delta x_r;$$

in effect the move has been broken up into n small steps of length Δx_r ($r = 1, 2, \ldots, n$) in which F does not vary greatly from the value $f(x_r)$.

If the force F is in the positive x-direction for some values of x and in the opposite direction for other values of x, then $f(x)$ takes both positive and negative values and the above definition of $W_{a,b}$ still applies.

16.1.4 Definition of an Integral The process of integration results in an expression of the form

$$\lim_{\Delta x_r \to 0} \sum_{r=1}^{n} q(x_r)\,\Delta x_r$$

in the notation described in Section 16.1.2. Such an expression is called an *integral*, or more specifically *the integral of $q(x)$ over the interval* $a \leqslant x \leqslant b$, and is denoted in an abbreviated manner by $\int_a^b q(x)\,\mathrm{d}x$. Thus

$$\int_a^b q(x)\,\mathrm{d}x = \lim_{\Delta x_r \to 0} \sum_{r=1}^{n} q(x_r)\,\Delta x_r.$$

Assuming that $q(x)$ (usually called the *integrand*) is continuous on the interval $a \leqslant x \leqslant b$, the required limit exists and so we say that *the integral exists*. Note that $q(x)$ may take both positive and negative values.

Again if $q(x)$ is continuous on the interval $a \leqslant x \leqslant b$ and if x_r^* is a value of x anywhere in the interval $x_r \leqslant x \leqslant x_{r+1}$, then

$$\lim_{\Delta x_r \to 0} \sum_{r=1}^{n} q(x_r^*)\,\Delta x_r \tag{1}$$

exists and has the same value for all choices of x_r^*; for this reason (1) is sometimes given as the definition of $\int_a^b q(x)\,\mathrm{d}x$. The definition above chooses $x_r^* = x_r$ in each case, i.e. the first point in the interval is always taken; another possibility is to choose the last point of each interval, i.e. $x_r^* = x_{r+1}$ so that

$$\int_a^b q(x)\,\mathrm{d}x = \lim_{\Delta x_r \to 0} \sum_{r=1}^{n} q(x_{r+1})\,\Delta x_r.$$

In practice it is normally convenient to make either of these choices in order to evaluate an integral; for example we have already evaluated the integrals $\int_1^2 \frac{1}{2}x\,\mathrm{d}x$, $\int_0^2 (8x^2 + 20)\mathrm{d}x$ using $x_r^* = x_r$ (see Figs. 16.1 and 16.2), and the reader should repeat the calculation using $x_r^* = x_{r+1}$ which means in effect levelling off each strip from the top of its right-hand edge rather than from the top of the left-hand edge. Also, since we may allow each width Δx_r to tend to zero in any way we please, we normally choose them all equal, i.e.

$$\Delta x_r = \frac{1}{n}(b - a) \qquad \text{for } r = 1, 2, \ldots, n.$$

Nevertheless few integrals can be evaluated along these lines because of the fundamental difficulty of summing a series to n terms. We return to the problem of evaluating integrals in Section 16.3.

The above discussion of integrals assumes that $b > a$. It is convenient to complete the definition of an integral by writing

$$\int_a^b f(x)\,dx = -\int_b^a f(x)\,dx$$

whenever $b < a$. If $b = a$ the integral is interpreted as being zero. In the integral $\int_a^b f(x)\,dx$ we refer to a, b as the *lower* and *upper limits of integration* respectively.

Exercises 16(a)

1. Show that $\int_0^1 (1 - x^2)\,dx = \frac{2}{3}$ by dividing the *range of integration* (i.e. the interval $0 \leqslant x \leqslant 1$) into n equal parts and making use of the formula

$$\sum_{r=1}^{n} r^2 = \tfrac{1}{6}n(n+1)(2n+1).$$

2. Show that $\int_0^a \cos x\,dx = \sin a$ by dividing the range of integration into n equal parts and making use of the formula

$$\sum_{r=1}^{n} \cos r\theta = \frac{\cos[\frac{1}{2}(n+1)\theta]\sin(\frac{1}{2}n\theta)}{\sin\frac{1}{2}\theta}.$$

3. Show that $\int_0^a \sin x\,dx = 1 - \cos a$ by dividing the range of integration into n equal parts and making use of the formula

$$\sum_{r=1}^{n} \sin r\theta = \frac{\sin[\frac{1}{2}(n+1)\theta]\sin(\frac{1}{2}n\theta)}{\sin\frac{1}{2}\theta}.$$

16.2 The Idea of an Antiderivative

Suppose that $f(x)$ is the derivative of $F(x)$, i.e. $f(x) = F'(x)$. Equally, $f(x)$ is the derivative of $F(x) + c$ where c is a constant; if we regard c as a parameter or arbitrary constant as it is usually called, then the expression $F(x) + c$ represents a *set* of expressions, corresponding to the possible values of c, each with the same derivative $f(x)$.

It is of fundamental importance to know if there could be further expressions, outside this set, also having derivative $f(x)$.

Theorem 1. Let $F_1(x)$, $F_2(x)$ both be differentiable on the interval $a < x < b$ and suppose that $F_1'(x) = F_2'(x)$ for all values of x in this interval. Then $F_2(x) = F_1(x) + c$, where c is a constant.

Proof. Put $F_3(x) = F_2(x) - F_1(x)$. Then

$$F_3'(x) = F_2'(x) - F_1'(x) = 0$$

for all values of x in the interval $a < x < b$. Let x_1, x_2 be distinct points in the interval such that $a < x_1 < x_2 < b$; by the mean value theorem (Theorem 8, Section 14.4.10), there must be a point x_3 (say) such that

$$F_3'(x_3) = \frac{F_3(x_2) - F_3(x_1)}{x_2 - x_1}.$$

But $F_3'(x_3) = 0$ and so $F_3(x_2) = F_3(x_1)$. This holds for any pair of distinct points x_1, x_2 in the interval so that $F_3(x)$ is constant in the interval. It follows that $F_2(x) = F_1(x) + \text{constant}$.

From Theorem 1 we conclude that *all* expressions having derivative $f(x)$ are included in the set $F(x) + c$. Any expression in this set is called an *antiderivative* of $f(x)$ with respect to x. Following the notation introduced in Section 14.4.1 we may write $f(x) = DF(x)$, and likewise the operator D^{-1} may be used for the inverse relation, i.e. $D^{-1}f(x) = F(x) + c$.

For example

$$D^{-1}(2x) = x^2 + c, \quad D^{-1}(\cos x) = \sin x + c, \quad D^{-1}(\sec^2 x) = \tan x + c.$$

Note that if we take any antiderivative in the set $F(x) + c$ and subtract its value at $x = a$ from its value at $x = b$ we get the same answer $F(b) - F(a)$ whichever antiderivative we take. We write

$$F(b) - F(a) = [D^{-1}f(x)]_a^b.$$

16.2.1 To Find a Curve of Given Slope The derivative $F'(x)$ may be interpreted geometrically as the slope of the curve $y = F(x)$. Hence the set of antiderivatives $F(x) + c$ may be interpreted as a set of curves $y = F(x) + c$ each with the same slope $f(x)$ (Fig. 16.5). Thus the problem

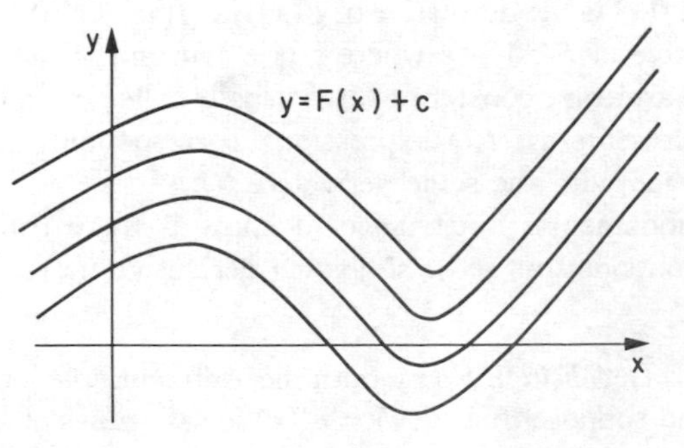

FIG. 16.5

of finding a curve with given slope $f(x)$ may be identified with the problem of finding $D^{-1}f(x)$; obviously the solution is not unique unless some further information about the curve is given.

Example 1 Find the curve with the property that at each point its slope is the square of the x-co-ordinate, and which passes through the origin.

At the point (x, y) on the curve we have $dy/dx = x^2$. Hence

$$y = D^{-1}(x^2) = \tfrac{1}{3}x^3 + c.$$

The set of curves $y = \frac{1}{3}x^3 + c$ all possess the required property about slope; to find the curve that also passes through the origin we put $x = 0$, $y = 0$ in this equation and find that we must have $c = 0$.

Hence $y = \frac{1}{3}x^3$ is the required curve.

16.2.2 A Problem in One-dimensional Motion Using the notation established in Section 14.5.1 let us suppose that the velocity v is known in terms of the time t. Since $v = ds/dt$ it follows that s is an antiderivative with respect to t of v, i.e. $s = D^{-1}v$, where the same operator D^{-1} has been used although it now refers to antidifferentiation with respect to the variable t. If s_a refers to the position of the moving point at time $t = a$ and s_b refers to its position at time $t = b$, then the distance between these two positions is $s_b - s_a$, i.e. $[D^{-1}v]_a^b$.

Example 2 At time t h a train has velocity v km/h, where $v = 8t^2 + 20$. If the train passes through station A at time $t = 0$ and through station B at time $t = 2$, what is the distance between these two stations?

In the usual notation we have $ds/dt = 8t^2 + 20$. Hence

$$s = D^{-1}(8t^2 + 20) = \tfrac{8}{3}t^3 + 20t + c.$$

Now station A is at distance s_0 along the line and station B is at distance s_2 along the line, where, in general, s_k denotes the value of s when $t = k$. Hence the distance between the stations is

$$s_2 - s_0 = [D^{-1}(8t^2 + 20)]_0^2 = [\tfrac{8}{3}t^3 + 20t]_0^2 = \tfrac{184}{3}\text{ km}.$$

16.2.3 Connection between Antiderivatives and Integrals An approximate solution to the problem discussed in Section 16.2.2 may be obtained by using the following approach. Write $v = f(t)$ and subdivide the interval $a \leqslant t \leqslant b$ into n smaller intervals by the points

$$a = t_1 < t_2 < t_3 < \cdots < t_n < t_{n+1} = b.$$

Thus the rth sub-interval is the interval $t_r \leqslant t \leqslant t_{r+1}$ of length $t_{r+1} - t_r = \Delta t_r$ say, and if Δt_r is sufficiently small a reasonable approximation is to assume that v has the constant value $f(t_r)$ throughout that interval. If Δs_r denotes the change in value of s during the rth sub-interval, then Δs_r approximately equals $f(t_r)\,\Delta t_r$, and for the entire interval $a \leqslant t \leqslant b$ the change in value of s is therefore approximately

$$\Sigma_{r=1}^{n} f(t_r)\,\Delta t_r.$$

This approximate result suggests that integration might be used to get an *exact* result. Write $s_{a,b} = s_b - s_a$, where as before s_a, s_b are the values of s when $t = a$, $t = b$ respectively. Obviously $s_{a,b}$ depends on the values of $f(t)$ for values of t in the interval $a \leqslant t \leqslant b$ and in the particular case where $f(t) = k$ (constant) we have $s_{a,b} = k(b - a)$. Furthermore s is additive so that, in the notation used above,

$$s_{a,b} = \sum_{r=1}^{n} \Delta s_r.$$

Finally, if $m_r \leqslant f(t) \leqslant M_r$ for $t_r \leqslant t \leqslant t_{r+1}$, then $m_r\,\Delta t_r \leqslant \Delta s_r \leqslant M_r\,\Delta t_r$. Following earlier discussions (Section 16.1.1) we are led to the conclusion that the exact value of $s_{a.b}$ is given by

$$s_{a,b} = \lim_{\Delta t_r \to 0} \sum_{r=1}^{n} f(t_r)\,\Delta t_r = \int_a^b f(t)\,\mathrm{d}t.$$

The approach used in Section 16.2.2 gave the result

$$s_{a,b} = [D^{-1} f(t)]_a^b,$$

and so we must have

$$\int_a^b f(t)\,\mathrm{d}t = [D^{-1} f(t)]_a^b.$$

In particular, note that Fig. 16.2 shows the velocity–time curve for Example 2 if the axes are relabelled as axes of t and v. The calculation of the area $ABCD$ in Fig. 16.2 is identical with the calculation of the distance between the two stations in Example 2 using the integration method, and the answer obtained agrees with that obtained by the antiderivative method, i.e.

$$\int_0^2 (8t^2 + 20)\,\mathrm{d}t = [D^{-1}(8t^2 + 20)]_0^2 = \tfrac{184}{3}.$$

Exercises 16(b)

1. Find antiderivatives for the following:

(a) $(2x^4 - x)/x^3$, (b) $(x^2 + 3)^2$, (c) $(x^2 - 2x - 1)/\sqrt{x}$.

2. Find the curve through the point (1, 1) with the property that at each point on the curve the slope of the normal equals the square of the x-co-ordinate.
3. The velocity v of a body moving in a straight line is given in terms of the time t by the equation $3v = t^2 - 10t + 36$. Find the distance between the positions at $t = 0$ and $t = 2$.
4. A body moves on a line AB of length 432 m, starting from A with velocity 36 m/s, and its acceleration at time t s after starting is $6(t - 4)$ m/s^2 for any t. Show that the body comes to rest at two points 32 m apart and that, 12 s after leaving A, it passes through B at a speed of 180 m/s.
5. A particle moves according to the law $v = A \cos \omega t$ where A, ω are constants, v denotes velocity and t denotes time. Show that the distance travelled from time $t = 0$ until it first comes to rest is A/ω.
6. From time $t = 1$ to time $t = 5$ water pours into a vessel at a rate inversely proportional to t^2. Given that at time $t = 1$ the rate is r and the vessel is empty, show that at time $t = 5$ the volume of water in the vessel is $\frac{4}{5}r$.
7. From time $t = 1$ water pours from a vessel at a rate inversely proportional to $\sqrt{t}$. Given that at time $t = 1$ this rate is r and the volume of water in the vessel is $12r$, find when the vessel will be empty.

16.3 The Fundamental Theorem of Integral Calculus

This is the name given to the following theorem:

Theorem 2 If $f(x)$ is continuous on the interval $a \leqslant x \leqslant b$, then

$$\int_a^b f(x)\,dx = [D^{-1}f(x)]_a^b,$$

that is if $F(x)$ is an antiderivative of $f(x)$, then

$$\int_a^b f(x)\,dx = F(b) - F(a).$$

We arrived at this result in Section 16.2.3 by investigating one-dimensional motion with velocity $v = f(t)$. Its main significance is that it enables us to evaluate integrals (see Section 16.1.4) without summing a series to n terms; instead we find an antiderivative (if possible) and apply the above theorem. Thus, for example,

$$\int_0^2 (x^2 + 5x - 7)\,dx = [\tfrac{1}{3}x^3 + \tfrac{5}{2}x^2 - 7x]_0^2 = -\tfrac{4}{3},$$

$$\int_1^4 \left(\sqrt{x} + \frac{1}{\sqrt{x}}\right) dx = [\tfrac{2}{3}x^{3/2} + 2x^{1/2}]_1^4 = \tfrac{20}{3}.$$

We look next at another important consequence of the fundamental theorem. By taking different upper limits of integration, keeping the lower limit fixed, we get a set of integrals each of the form

$$\int_a^\xi f(x)\,dx$$

where ξ can have any value for which this integral exists. By the fundamental theorem the values of these integrals are given by

$$\int_a^\xi f(x)\,dx = F(\xi) - F(a).$$

Hence, regarding ξ as a variable we see that the l.h.s. is differentiable with respect to ξ and

$$\frac{d}{d\xi}\left[\int_a^\xi f(x)\,dx\right] = F'(\xi) = f(\xi).$$

At this stage we may, without too much risk of confusion, adopt the traditional habit of using x in the double role of integration variable and variable for the upper limit of integration. Thus we write

$$\frac{d}{dx}\left[\int_a^x f(x)\,dx\right] = f(x).$$

In words, what we have shown is that $\int_a^x f(x)\,dx$ is an antiderivative of $f(x)$; clearly all antiderivatives of $f(x)$ are of the form $\int_a^x f(x)\,dx + c$. This means that antiderivatives may be expressed as integrals, and for this reason it is normal practice to use the term *integral* to mean *antiderivative*, the latter term being dropped from the vocabulary of the subject. More particularly, the integral $\int_a^x f(x)\,dx$, with variable upper limit of integration, is called an *indefinite integral*, a term that may be regarded as the replacement for the term *antiderivative*, whereas the integral $\int_a^b f(x)\,dx$, with fixed limits of integration, is called a *definite integral*. Furthermore the operator symbol D^{-1} is replaced by $\int \ldots dx$, i.e. an integral sign without limits of integration, so that we now write

$$\int f(x)\,dx = F(x) + c.$$

16.3.1 Method for Obtaining Integral Formulae In Section 16.1.2 we have described the situation in which the value $Q_{a,b}$ of a quantity Q is given by the integral $\int_a^b q(x)\,dx$. In practice the required integrand $q(x)$ is usually found by seeking an approximate value of the form $q(x_r)\,\Delta x_r$ for ΔQ_r. For example, in the case of the area under the curve $y = f(x)$ (see Section 16.1.1) the most obvious approximation for the area ΔA_r of the

rth strip is

$$\Delta A_r \simeq f(x_r)\,\Delta x_r, \tag{1}$$

and from this approximation $f(x)$ emerges as the likely integrand.

Now if

$$Q_{a.b} = \int_a^b q(x)\,\mathrm{d}x,$$

then writing

$$Q_{a.x} = \int_a^x q(x)\,\mathrm{d}x$$

we have

$$\frac{\mathrm{d}}{\mathrm{d}x} Q_{a.x} = q(x). \tag{2}$$

Hence if we write $\Delta Q_r = q(x_r)\,\Delta x_r + E_r$, that is

$$\frac{\Delta Q_r}{\Delta x_r} = q(x_r) + \frac{E_r}{\Delta x_r},$$

then letting $\Delta x_r \to 0$ we see that this agrees with (2) if, and only if, $E_r/\Delta x_r \to 0$. This shows how good the approximation to ΔQ_r must be; it is not enough that the error E_r should tend to zero as $\Delta x_r \to 0$ and instead we require that $E_r/\Delta x_r \to 0$.

For example, in the notation used in Section 16.1.1 we saw that the error E_r resulting from the approximation (1) satisfies the inequality

$$E_r \leqslant (M_r - m_r)\,\Delta x_r,$$

and so

$$\frac{E_r}{\Delta x_r} \leqslant M_r - m_r \to 0 \text{ as } \Delta x_r \to 0 \text{ if } f(x) \text{ is continuous at } x = x_r.$$

We therefore summarise Section 16.1.2 in the following form:

If $Q_{a.b} = \Sigma_{r=1}^{n} \Delta Q_r$ and $\Delta Q_r \simeq q(x_r)\,\Delta x_r$, the error E_r in the approximation satisfying the condition that $E_r/\Delta x_r \to 0$ as $\Delta x_r \to 0$, then $Q_{a.b} = \int_a^b q(x)\,\mathrm{d}x$.

16.4 Rules of Integration

The antiderivatives of $f(x)$ are denoted by $\int f(x)\,\mathrm{d}x$ (see Section 16.3) and we are here using the term *integration* for the operation of finding these antiderivatives rather than in the original sense of the term (Section 16.1).

The derivatives of the elementary functions provide some ready-made results which are called *standard integrals*. For example the differentiation results

$$\frac{d}{dx}(\sin x) = \cos x, \quad \frac{d}{dx}(\cos x) = -\sin x,$$

$$\frac{d}{dx}(x^{n+1}) = (n+1)x^n$$

may be interpreted as integration results, giving the standard integrals

$$\int \cos x \, dx = \sin x + c, \quad \int \sin x \, dx = -\cos x + c,$$

$$\int x^n \, dx = \frac{1}{n+1} x^{n+1} + c \qquad (n \neq -1).$$

The derivatives of $\sin^{-1} x$, $\cos^{-1} x$ give two results for the same integral, viz.

$$\int \frac{1}{\sqrt{(1-x^2)}} dx = \sin^{-1} x + c \quad \text{or} \quad -\cos^{-1} x + c.$$

These two results are consistent since by (12.1)

$$-\cos^{-1} x = \sin^{-1} x - \pi/2.$$

Obviously there is no need to include both forms of the result in our list of standard integrals; in fact the form usually given is

$$\int \frac{1}{\sqrt{(k^2 - x^2)}} dx = \sin^{-1}\left(\frac{x}{k}\right) + c \qquad (k \text{ is a constant})$$

which includes many integrals, e.g.

$$\int \frac{dx}{\sqrt{(9-x^2)}} = \sin^{-1}\left(\frac{x}{3}\right) + c, \quad \int \frac{dx}{\sqrt{(7-x^2)}} = \sin^{-1}\left(\frac{x}{\sqrt{7}}\right) + c.$$

Likewise we have the standard integral

$$\int \frac{1}{x^2 + k^2} dx = \frac{1}{k} \tan^{-1}\left(\frac{x}{k}\right) + c \qquad (k \text{ a constant}),$$

which is easily verified by differentiation, and so, in particular,

$$\int \frac{dx}{x^2 + 9} = \tfrac{1}{3} \tan^{-1}\left(\frac{x}{3}\right) + c, \quad \int \frac{dx}{x^2 + 7} = \frac{1}{\sqrt{7}} \tan^{-1}\left(\frac{x}{\sqrt{7}}\right) + c.$$

A list of standard integrals is given in Section 18.1.

We now look at ways in which the *rules* of differentiation can help in finding integrals. Obviously

$$\int kf(x)\,dx = k\int f(x)\,dx \qquad \text{if } k \text{ is a constant,}$$

and

$$\int [f(x) + g(x)]\,dx = \int f(x)\,dx + \int g(x)\,dx.$$

Next, suppose that $\int f(x)\,dx = F(x) + c$. Then $F'(x) = f(x)$ and it follows from the chain rule that, for any other function g,

$$\frac{d}{dx}F \circ g(x) = F' \circ g(x) \times g'(x) = f \circ g(x) \times g'(x).$$

Hence

$$\left.\begin{aligned} &\int f \circ g(x) \times g'(x)\,dx = F \circ g(x) + c, \\ \text{that is} \qquad &\int f(z)\frac{dz}{dx}\,dx = F(z) + c, \end{aligned}\right\} \tag{16.1}$$

using z in place of $g(x)$.

The following examples illustrate the use of (16.1):

$$\int 2x(x^2+1)^5\,dx = \tfrac{1}{6}(x^2+1)^6 + c \quad \left(\text{put } z = g(x) = x^2 + 1, \frac{dz}{dx} = g'(x) = 2x, f(z) = z^5, F(z) = \tfrac{1}{6}z^6\right),$$

$$\int 3x^2(x^3+1)^5\,dx = \tfrac{1}{6}(x^3+1)^6 + c \quad \left(\text{put } z = x^3 + 1, \frac{dz}{dx} = 3x^2, f(z) = z^5, F(z) = \tfrac{1}{6}z^6\right),$$

$$\int 2x\cos(x^2)\,dx = \sin(x^2) + c \quad \left(\text{put } z = x^2, \frac{dz}{dx} = 2x, f(z) = \cos z, F(z) = \sin z\right),$$

$$\int \sin^7 x\cos x\,dx = \tfrac{1}{8}\sin^8 x + c \quad \left(\text{put } z = \sin x, \frac{dz}{dx} = \cos x, f(z) = z^7, F(z) = \tfrac{1}{8}z^8\right).$$

Note the very special form of the integrand when this rule applies. The following integrals, differing in only a single term from those in the above examples, cannot be found from this rule:

$$\int 2x^2(x^2+1)^5\,dx, \quad \int 3x(x^3+1)^5\,dx, \quad \int 2\cos(x^2)\,dx, \quad \int \sin^7 x\,dx.$$

However, if the integrand is of the required form *apart from a constant factor* then the rule may still be used, e.g.

$$\int x(x^2+1)^5\,dx = \tfrac{1}{12}(x^2+1)^6 + c, \quad \int x^2(x^3+1)^5\,dx = \tfrac{1}{18}(x^3+1)^6 + c,$$

$$\int x\cos(x^2)\,dx = \tfrac{1}{2}\sin(x^2) + c.$$

The rule is particularly simple with $z = ax + b$ (a, b constants), viz.

$$\int f(ax+b)\,a\,dx = F(ax+b) + c,$$

that is

$$\int f(ax+b)\,dx = \frac{1}{a}F(ax+b) + c. \qquad (16.2)$$

For example,

$$\int \cos(2x+3)\,dx = \tfrac{1}{2}\sin(2x+3) + c, \quad \int (7x+6)^4\,dx = \tfrac{1}{35}(7x+6)^5 + c.$$

Exercises 16(c)

1. Find the integrals (a) $\int 16x^3\,dx$, (b) $\int 16/x^3\,dx$, (c) $\int \sqrt[3]{(x^{26})}\,dx$, (d) $\int \sqrt[5]{(10/x^7)}\,dx$, (e) $\int 1/\sqrt[4]{(3x^5)}\,dx$, and hence find the integrals (f) $\int 16(2x+1)^3\,dx$, (g) $\int 16/(2x+1)^3\,dx$, (h) $\int \sqrt[3]{(5x-7)^{26}}\,dx$, (i) $\int \sqrt[5]{(10/\sin^7 x)}\cos x\,dx$, (j) $\int \sec^2 x/\sqrt[4]{(3\tan^5 x)}\,dx$.

Find the following integrals:

2. $\int (2x+3)^5\,dx$
3. $\int x(2x^2+3)^5\,dx$
4. $\int \frac{1}{(3x+5)^2}\,dx$
5. $\int \frac{1}{\sqrt{(5+6x)}}\,dx$
6. $\int \operatorname{cosec}^2 5x\,dx$
7. $\int x(x^2+1)^5\,dx$
8. $\int x\sqrt{(9-4x^2)}\,dx$
9. $\int \frac{\sin x}{\cos^2 x}\,dx$
10. $\int \cos x\operatorname{cosec}^7 x\,dx$
11. $\int_0^2 (x+1)(x^2+2x+4)^{-1/2}\,dx$
12. $\int_0^2 x\sqrt{(4-x^2)}\,dx$
13. $\int_0^{\pi/4} \sin^2 2x\cos 2x\,dx$
14. $\int_{\pi/3}^{\pi/2} \sin x\sqrt{(1-2\cos x)}\,dx$
15. $\int \frac{1}{(5+2x)^2+9}\,dx$
16. $\int \frac{1}{\sqrt{[9-(3+4x)^2]}}\,dx$
17. $\int \frac{(\sin^{-1} x)^2}{\sqrt{(1-x^2)}}\,dx$

18. If $\sqrt{(x^2-5)}\,dy/dx = x$ and $y = 3$ when $x = 3$, show that $x^2 - y^2 + 2y = 6$.
19. If $dy/dx = x\sqrt{(9+x^2)}$ and $y = 40$ when $x = 4$, show that

$$(3y+5)^2 = (9+x^2)^3.$$

20. Find the equation of the curve passing through the point $(1, \frac{1}{12})$ such that at all points on the curve $(5-3x)^2\,dy/dx = 1$.

16.4.1 Substitution Rule The rule for changing the variable in integration (from x to z, say, where we know the relationship $x = g(z)$ between these two variables) is found by interchanging x and z in (16.1) and writing the result as

$$\left.\begin{aligned}\int f(x)\,dx &= \int f \circ g(z) \times g'(z)\,dz,\\ \text{that is}\qquad & \\ \int f(x)\,dx &= \int f(x)\frac{dx}{dz}\,dz.\end{aligned}\right\}\qquad(16.3)$$

In words, we must

(a) multiply the integrand by dx/dz to get a new integrand,
(b) express the new integrand entirely in terms of z,

and we may then integrate with respect to the variable z.

Example 3 Find $\int x/\sqrt{(x-1)}\,dx$ by substituting $x = z+1$.

Since $dx/dz = 1$ we get

$$\int \frac{x}{\sqrt{(x-1)}}\,dx = \int \frac{z+1}{z^{1/2}}\,dz = \int (z^{1/2} + z^{-1/2})\,dz = \tfrac{2}{3}z^{3/2} + 2z^{1/2} + c$$
$$= \tfrac{2}{3}(x-1)^{3/2} + 2(x-1)^{1/2} + c.$$

Example 4 Find $\int \sqrt{(x+1)}\,dx$ by substituting $x = z^2 - 1$.

Since $dx/dz = 2z$, we get

$$\int \sqrt{(x+1)}\,dx = \int \sqrt{(x+1)}\,2z\,dz = \int 2z^2\,dz = \tfrac{2}{3}z^3 + c = \tfrac{2}{3}(x+1)^{3/2} + c.$$

Note that in effect we substitute $(dx/dz)\,dz$ for dx; e.g. in Example 3 we put $x = z+1$, $dx = dz$, and in Example 4 we put $x = z^2 - 1$, $dx = 2z\,dz$.

Exercises 16(d)

Find the following integrals by using the given substitutions:

1. $\int x(x-3)^5\,dx$, $\quad z = x-3$
2. $\int x\sqrt{(3+x)}\,dx$, $\quad z = 3+x$

3. $\int x\sqrt{(x+1)}\,dx$, $\quad z^2 = x + 1$ 4. $\int \frac{x}{(2x+1)^3}\,dx$, $\quad z = 2x + 1$

5. $\int \frac{x^2}{1+x^6}\,dx$, $\quad z = x^3$ 6. $\int \frac{x}{9+x^4}\,dx$, $\quad z = x^2$

7. $\int_1^2 \frac{dx}{x^2\sqrt{(5x^2-4)}}$, $\quad x^2 = 1/z$ 8. $\int_{1/2}^{2/3} \frac{dx}{x\sqrt{[(1-x)(2x-1)]}}$, $\quad x = \frac{2}{z+3}$.

16.4.2 Integration by Parts From the product rule of differentiation, viz.

$$\frac{d}{dx}(uv) = u\frac{dv}{dx} + v\frac{du}{dx},$$

we get that

$$uv = \int u\frac{dv}{dx}\,dx + \int v\frac{du}{dx}\,dx,$$

and hence

$$\int u\frac{dv}{dx}\,dx = uv - \int v\frac{du}{dx}\,dx. \tag{16.4}$$

Some integrals can be found by use of (16.4) and this method is called the method of *integration by parts*. The method applies when

(a) the integrand is the product of two expressions $f(x)$ and $g(x)$, say,
(b) the integral of one of these expressions is known, say

$$\int g(x)\,dx = G(x) + c,$$

(c) the integral of the product $G(x)f'(x)$ is also known.

In terms of the notation used in (16.4) we have written $u = f(x)$, $dv/dx = g(x)$, $v = G(x)$, and in this notation (16.4) becomes

$$\int f(x)g(x)\,dx = f(x)G(x) - \int G(x)f'(x)\,dx.$$

Example 5 Find $\int x \cos x\,dx$.

Write $u = x$, $dv/dx = \cos x$. Then $du/dx = 1$, $v = \sin x$. Hence

$$\int x\cos x\,dx = x\sin x - \int \sin x\,dx = x\sin x + \cos x + c.$$

Exercises 16(e)

Use integration by parts to show that

1. $\int x\sin x\,dx = -x\cos x + \sin x + c.$

2. $\int (3x + 5) \sin x \, dx = -(3x + 5) \cos x + 3 \sin x + c.$

3. $\int x \sin 2x \, dx = \frac{1}{4}(\sin 2x - 2x \cos 2x) + c.$

4. $\int (3x + 5) \sin (2x + 3) \, dx = -\frac{1}{2}(3x + 5) \cos (2x + 3) + \frac{3}{4} \sin (2x + 3) + c.$

5. $\int x \tan^{-1} 4x \, dx = \frac{1}{32}[(16x^2 + 1) \tan^{-1} (4x) - 4x] + c.$

6. $\int \frac{x \sin^{-1} x}{\sqrt{(1 - x^2)}} \, dx = -\sqrt{(1 - x^2)} \sin^{-1} x + x + c.$

7. $\int 4x(1 + x^2) \tan^{-1} x \, dx = (1 + x^2)^2 \tan^{-1} x - x - \frac{1}{3}x^3 + c.$

16.5 Improper Integrals

An *improper integral* is so called because it is not an integral in the strict sense of the definition of Section 16.1.4, but rather in the sense of an extended form of that definition. There are two main types of improper integrals:

(a) integrals of the type

$$\int_a^{\infty} f(x) \, dx, \quad \int_{-\infty}^{a} f(x) \, dx, \quad \text{or} \quad \int_{-\infty}^{\infty} f(x) \, dx,$$

where the limits of integration are not both finite, and

(b) integrals of the type

$$\int_a^b f(x) \, dx$$

where $|f(x)| \to \infty$ as $x \to c$ and $a \leqslant c \leqslant b$.

Such integrals are not covered by the definition of Section 16.1.4, and in Sections 16.5.1 and 16.5.2 we show how the definition is extended to allow consideration of integrals of type (a) and type (b) respectively.

16.5.1 Infinite Limits of Integration If

$$\int_a^X f(x) \, dx$$

exists when $X \geqslant a$, and has a limit as $X \to \infty$, then we say that

$$\int_a^{\infty} f(x) \, dx$$

exists and we write

$$\int_a^{\infty} f(x) \, dx = \lim_{X \to \infty} \int_a^X f(x) \, dx.$$

Similarly, if $\int_{-X}^{a} f(x)\,dx$ exists when $-X \leqslant a$, and has a limit as $X \to \infty$, we say that $\int_{-\infty}^{a} f(x)\,dx$ exists and we write

$$\int_{-\infty}^{a} f(x)\,dx = \lim_{X \to \infty} \int_{-X}^{a} f(x)\,dx.$$

The two cases are illustrated diagrammatically in Fig. 16.6, where in each case the improper integral corresponds to the limit of the shaded area as $X \to \infty$.

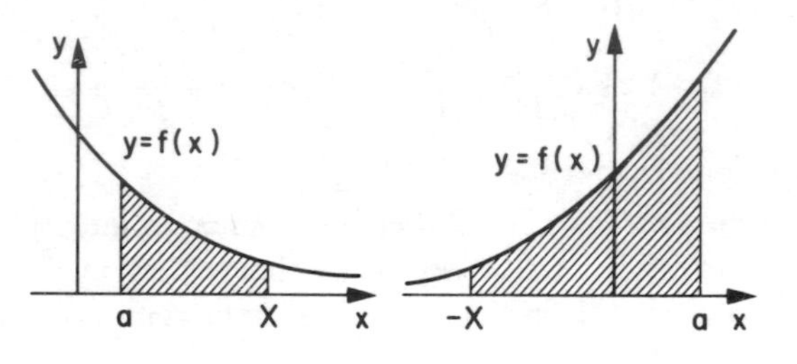

FIG. 16.6

For example,

$$\int_{1}^{\infty} \frac{1}{x^2}\,dx = \lim_{X\to\infty}\int_{1}^{X} \frac{1}{x^2}\,dx = \lim_{X\to\infty}\left[-\frac{1}{x}\right]_1^X = \lim_{X\to\infty}\left(1-\frac{1}{X}\right) = 1,$$

whereas $\displaystyle\int_{1}^{\infty} \frac{1}{\sqrt{x}}\,dx$ does not exist since

$$\int_{1}^{X} \frac{1}{\sqrt{x}}\,dx = \left[2\sqrt{x}\right]_1^X = 2\sqrt{X} - 2 \to \infty \text{ as } X \to \infty.$$

Example 6 Evaluate $\displaystyle\int_{0}^{\infty} \frac{x}{(x^2+1)}\,dx.$

Substituting $z = x^2 + 1$, $dz = 2x\,dx$ we get

$$\int \frac{x}{(x^2+1)^2}\,dx = \int \frac{1}{2z^2}\,dz = -\frac{1}{2z} + c = -\frac{1}{2(x^2+1)} + c.$$

Hence

$$\int_{0}^{X} \frac{x}{(x^2+1)^2}\,dx = \tfrac{1}{2} - \frac{1}{2(X^2+1)} \to \tfrac{1}{2} \text{ as } X \to \infty.$$

Thus

$$\int_{0}^{\infty} \frac{x}{(x^2+1)^2}\,dx = \tfrac{1}{2}.$$

If $\int_{-\infty}^{0} f(x)\,dx$ and $\int_0^\infty f(x)\,dx$ both exist we say that $\int_{-\infty}^{\infty} f(x)\,dx$ exists and we write

$$\int_{-\infty}^{\infty} f(x)\,dx = \int_{\infty}^{0} f(x)\,dx + \int_0^\infty f(x)\,dx.$$

Example 7 Evaluate $\displaystyle\int_{-\infty}^{\infty} \frac{1}{x^2+1}\,dx.$

Since

$$\int_0^X \frac{1}{x^2+1}\,dx = \Big[\tan^{-1} x\Big]_0^X = \tan^{-1} X \to \pi/2 \text{ as } X \to \infty,$$

and

$$\int_{-X}^{0} \frac{1}{x^2+1}\,dx = -\tan^{-1}(-X) = \tan^{-1} X \to \pi/2 \text{ as } X \to \infty,$$

we have

$$\int_{-\infty}^{\infty} \frac{1}{x^2+1}\,dx = \pi.$$

16.5.2 Infinite Integrand Suppose $|f(x)| \to \infty$ as $x \to c$. If (a) $c = a$, or (b) $c = b$, we write

(a) $\displaystyle\int_a^b f(x)\,dx = \lim_{\varepsilon\to 0} \int_{a+\varepsilon}^{b} f(x)\,dx,$ or

(b) $\displaystyle\int_a^b f(x)\,dx = \lim_{\varepsilon\to 0} \int_{a}^{b-\varepsilon} f(x)\,dx,$

and if (c) $a < c < b$ we write

(c) $\displaystyle\int_a^b f(x)\,dx = \lim_{\varepsilon\to 0} \int_{a}^{c-\varepsilon} f(x)\,dx + \lim_{\varepsilon\to 0} \int_{c+\varepsilon}^{b} f(x)\,dx$

provided, of course, that the integrals and limits on the right-hand sides exist.

In Fig. 16.7(iii) $\int_a^b f(x)\,dx$ is represented as the sum of two areas, each of which is to be given its limiting value as $\varepsilon \to 0$; similar diagrams may be drawn to illustrate cases where $f(x) \to -\infty$ as $x \to c-$, or as $x \to c+$.

Example 8 Evaluate $\displaystyle\int_0^1 \frac{1}{\sqrt{x}}\,dx$

Since

$$\int_\varepsilon^1 \frac{1}{\sqrt{x}}\,dx = \Big[2\sqrt{x}\Big]_\varepsilon^1 = 2 - 2\sqrt{\varepsilon} \to 2 \text{ as } \varepsilon \to 0,$$

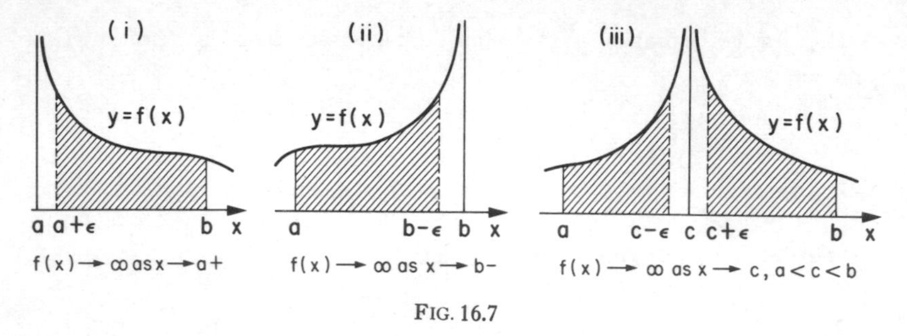

FIG. 16.7

we have

$$\int_0^1 \frac{1}{\sqrt{x}}\, dx = 2.$$

Example 9 Show that $\int_0^1 \frac{1}{x^2}\, dx$ does not exist.

Since

$$\int_\varepsilon^1 \frac{1}{x^2}\, dx = \left[-\frac{1}{x}\right]_\varepsilon^1 = \frac{1}{\varepsilon} - 1 \to \infty \text{ as } \varepsilon \to 0,$$

it follows that the given integral does not exist.

Example 10 Evaluate $\int_{-1}^1 \frac{1}{x^{2/3}} dx$.

Since

$$\int_{-1}^{-\varepsilon} \frac{1}{x^{2/3}}\, dx = \left[3x^{1/3}\right]_{-1}^{-\varepsilon} = 3(1 - \varepsilon^{1/3}) \to 3 \text{ as } \varepsilon \to 0$$

and

$$\int_\varepsilon^1 \frac{1}{x^{2/3}}\, dx = \left[3x^{1/3}\right]_\varepsilon^1 = 3(1 - \varepsilon^{1/3}) \to 3 \text{ as } \varepsilon \to 0,$$

it follows that

$$\int_{-1}^1 \frac{1}{x^{2/3}}\, dx = 6.$$

Example 11 Show that $\int_{-1}^1 \frac{1}{x^{4/3}}\, dx$ does not exist.

We have

$$\int_\varepsilon^1 \frac{1}{x^{4/3}}\, dx = \left[-\frac{3}{x^{1/3}}\right]_\varepsilon^1 = 3\left(\frac{1}{\varepsilon^{1/3}} - 1\right) \to \infty \text{ as } \varepsilon \to 0,$$

and similarly

$$\int_{-1}^{-\varepsilon} 1/x^{4/3}\, dx \text{ does not tend to a limit as } \varepsilon \to 0.$$

Note. When these integrals exist they may be evaluated by finding an indefinite integral $F(x)$ and evaluating $F(b) - F(a)$ in the usual way, without introducing ε and taking the limit as $\varepsilon \to 0$. Thus in Example 8 we get the correct answer by evaluating

$$\left[2\sqrt{x}\right]_0^1 = 2 - 0 = 2,$$

and in Example 10 by evaluating

$$\left[3x^{1/3}\right]_{-1}^{1} = 3 - (-3) = 6.$$

Unfortunately this procedure may give an answer when the integral does not exist, e.g. in Example 11 we may evaluate

$$\left[-\frac{3}{x^{1/3}}\right]_{-1}^{1} = -3 + (-3) = -6$$

although the given integral does not exist. This possibility occurs in case (c) where $|f(x)| \to \infty$ as $x \to c$ with $a < c < b$; in such cases it is essential to verify that the integral exists in accordance with the definition.

Exercises 16(f)

Find, when they exist, the values of the following integrals:

1. $\int_1^{\infty} \frac{dx}{x^4}$
2. $\int_1^{\infty} \frac{dx}{x^{1/3}}$
3. $\int_1^{\infty} \frac{dx}{x^{1/4}}$
4. $\int_0^{\infty} \frac{x}{(x^2+1)^2}\, dx$
5. $\int_0^{1} \frac{1}{\sqrt{(1-x^2)}}\, dx$
6. $\int_0^{1} \frac{x}{\sqrt{(1-x^2)}}\, dx$
7. $\int_0^{\infty} \frac{dx}{16+x^2}$
8. $\int_0^{3} \frac{dx}{\sqrt{(9-x^2)}}$
9. $\int_0^{\infty} \frac{dx}{a^2+b^2x^2}$
10. $\int_2^{4} \frac{dx}{\sqrt{[1-(x-3)^2]}}$
11. $\int_4^{\infty} \frac{dx}{(x+1)^2+25}$
12. $\int_0^{\pi/2} \frac{\sin x}{\sqrt{(\cos x)}}\, dx.$

CHAPTER 17

LOGARITHMIC, EXPONENTIAL, HYPERBOLIC AND INVERSE HYPERBOLIC FUNCTIONS

17.1 The Natural Logarithm

The integration result $\int x^n \, dx = [1/(n+1)]\, x^{n+1} + c$ does not apply when $n = -1$, in which case the right hand side is meaningless. In fact the problem of finding a function with derivative $1/x$ cannot be solved in terms of the elementary functions discussed so far in this book.

Consider $\int_a^x f(x)\, dx$, where a is a constant and a and x are both positive. This expression has derivative $f(x)$ (see Section 16.3) and in this way, putting $f(x) = 1/x$, we may define a function with derivative $1/x$. Other useful properties emerge if we put $a = 1$, and the function so defined is called the ***natural logarithm***, usually written ln.

Definition. $\ln x = \int_1^x (1/x)\, dx$ for all positive values of x. It follows that if $x > 1$, $\ln x$ is the area under the curve $y = 1/x$ from the point 1 to the point x on the x-axis, if $0 < x < 1$, $\ln x$ is minus the area under this curve from the point x to the point 1, and $\ln 1 = 0$ (see Fig. 17.1).

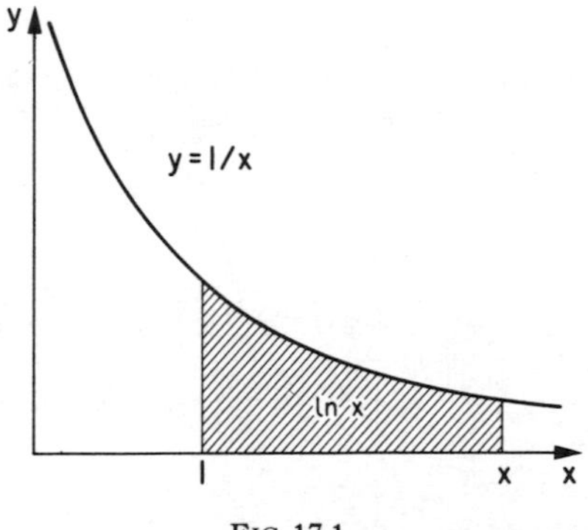

FIG. 17.1

17.1.1 Properties of ln *x* It follows immediately from the above definition that $\ln x$ is differentiable for all positive values of x and that

$$\frac{d}{dx} \ln x = \frac{1}{x}. \tag{17.1}$$

Example 1 Differentiate with respect to x the following: (a) $\ln(x^2 + 5)$, (b) $\ln(3 - 2x)$, (c) $x^2 \ln(1 + 1/x)$.

(a)
$$\frac{\mathrm{d}}{\mathrm{d}x} \ln(x^2 + 5) = \frac{2x}{x^2 + 5},$$

using (17.1) together with the chain rule,

(b)
$$\frac{\mathrm{d}}{\mathrm{d}x} \ln(3 - 3x) = \frac{-2}{3 - 2x},$$

from (17.1) together with the chain rule,

(c)
$$\frac{\mathrm{d}}{\mathrm{d}x}\left[x^2 \ln\left(1 + \frac{1}{x}\right)\right] = x^2 \frac{1}{1 + 1/x}\left(-\frac{1}{x^2}\right) + 2x \ln\left(1 + \frac{1}{x}\right)$$
$$= 2x \ln\left(1 + \frac{1}{x}\right) - \frac{x}{x + 1},$$

on using (17.1), the product rule and the chain rule.

Assume that $x > 0$ and let b denote a positive constant. Then

$$\frac{\mathrm{d}}{\mathrm{d}x} \ln(bx) = \frac{1}{bx} b - \frac{1}{x}$$

and so

$$\frac{\mathrm{d}}{\mathrm{d}x} \ln(bx) = \frac{\mathrm{d}}{\mathrm{d}x} \ln x.$$

Hence by Theorem 1, Section 16.2, $\ln(bx) = \ln x + c$, where c is a constant. Putting $x = 1$ we see that

$$\ln b = \ln 1 + c$$

and since $\ln 1 = 0$, we get

$$c = \ln b.$$

Hence

$$\ln(bx) = \ln x + \ln b,$$

or, in other words, if a and b are any two positive numbers

$$\ln(ab) = \ln a + \ln b. \tag{17.2}$$

Furthermore we have

$$\frac{\mathrm{d}}{\mathrm{d}x} \ln\left(\frac{1}{x}\right) = x\left(-\frac{1}{x^2}\right) = -\frac{1}{x}$$

and so

$$\frac{d}{dx}\ln\left(\frac{1}{x}\right)=\frac{d}{dx}(-\ln x).$$

Hence by Theorem 1, Section 16.2, $\ln(1/x)=-\ln x+c$, where c is a constant. Putting $x=1$ we find that we must have $c=0$, which means that

$$\ln(1/x)=-\ln x.$$

Thus if a and b are any two positive numbers,

$$\ln(a/b)=\ln a+\ln(1/b) \quad \text{by (17.2)}$$

and so from the last result above

$$\ln\left(\frac{a}{b}\right)=\ln a-\ln b. \tag{17.3}$$

Now let n be a rational number. Then

$$\frac{d}{dx}\ln(x^n)=\frac{1}{x^n}nx^{n-1}=\frac{n}{x}$$

and so

$$\frac{d}{dx}\ln(x^n)=\frac{d}{dx}(n\ln x).$$

Hence

$$\ln(x^n)=n\ln x+c$$

and putting $x=1$ we find that $c=0$. Thus

$$\ln(x^n)=n\ln x. \tag{17.4}$$

From (17.2), (17.3) and (17.4) we see that if $a, b, c, \ldots$ are positive numbers and $n_1, n_2, n_3, \ldots$ are positive or negative rational numbers, then

$$\ln(a^{n_1}b^{n_2}c^{n_3}\cdots)=n_1\ln a+n_2\ln b+n_3\ln c+\cdots, \text{ e.g.}$$

$$\ln\left(\frac{40}{\sqrt{3}}\right)=\ln\left(\frac{5.2^3}{\sqrt{3}}\right)=\ln 5+3\ln 2-\tfrac{1}{2}\ln 3.$$

Example 2 Differentiate with respect to x the following:

(a) $\ln\left(\dfrac{2x-1}{2x+1}\right)$, (b) $\ln\sqrt{(x^2+x+1)}$, (c) $\ln\left[\dfrac{(x-1)^7}{\sqrt{(2x+1)}}\right]$.

(a) Put

$$y=\ln\left(\frac{2x-1}{2x+1}\right)=\ln(2x-1)-\ln(2x+1).$$

Then

$$\frac{dy}{dx} = \frac{2}{2x-1} - \frac{2}{2x+1} = \frac{4}{4x^2-1}.$$

(b) Put

$$y = \ln \sqrt{(x^2+x+1)} = \tfrac{1}{2}\ln(x^2+x+1).$$

Then

$$\frac{dy}{dx} = \frac{1}{2}\,\frac{2x+1}{x^2+x+1}.$$

(c) Put

$$y = \ln\left[\frac{(x-1)^7}{\sqrt{(2x+1)}}\right] = 7\ln(x-1) - \tfrac{1}{2}\ln(2x+1).$$

Then

$$\frac{dy}{dx} = \frac{7}{x-1} - \frac{1}{2x+1}.$$

Since ln 10 is the area under the curve $y = 1/x$ from $x = 1$ to $x = 10$ it is easily verified that $\ln 10 \simeq 2{\cdot}3$. Hence $\ln 100 = \ln 10^2 = 2\ln 10 \simeq 4{\cdot}6$ and in general $\ln 10^n \simeq 2{\cdot}3n$ for any rational number n. The graph of $\ln x$ is shown in Fig. 17.2; more accurate values of $\ln x$ can be obtained from mathematical tables. Since

$$\frac{d^2}{dx^2}\ln x = -\frac{1}{x^2} < 0$$

it follows that the graph is everywhere concave downwards. However the

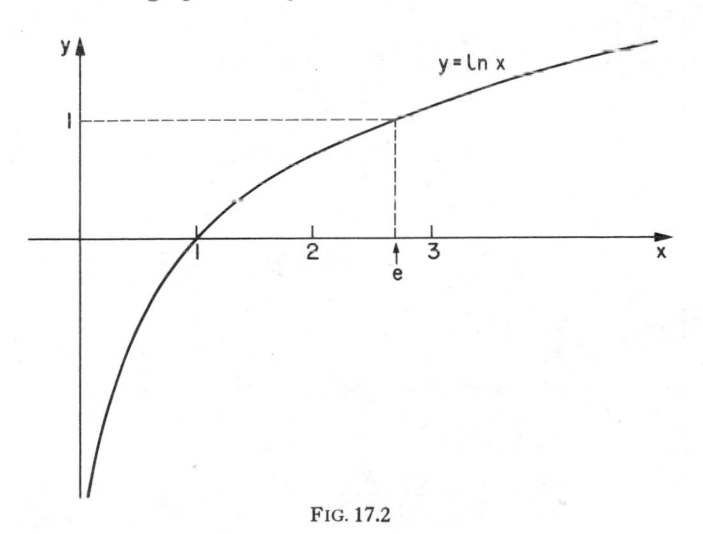

FIG. 17.2

result $\ln 10^n \simeq 2{\cdot}3n$ shows that the values of $\ln x$ increase without limit and $\ln x \to \infty$ as $x \to \infty$; likewise $\ln x \to -\infty$ as $x \to 0+$.

Since $\ln x$ is differentiable for all positive values of x it follows that $\ln x$ is continuous for all positive values of x. By Theorem 5, Section 14.2.1, there must be a value of x, say $x = e$, such that $\ln e = 1$ (see Fig. 17.2). The graph shows that $e \simeq 2{\cdot}718$.

17.1.2 Logarithmic Differentiation If $y = f(x)$, where $f(x)$ involves products, quotients and/or indices, it may be simpler to find $\mathrm{d}y/\mathrm{d}x$ from the equation $\ln y = \ln f(x)$. The method is illustrated by the following examples.

Example 3 Differentiate with respect to x

(a) $\dfrac{x(1+x^2)^3}{\sqrt[3]{(1+x^3)}}$, (b) $\sqrt{\left(\dfrac{a^2+ax+x^2}{a^2-ax+x^2}\right)}$.

(a) Put

$$y = \frac{x(1+x^2)^3}{\sqrt[3]{(1+x^3)}}.$$

Then

$$\ln y = \ln x + 3\ln(1+x^2) - \tfrac{1}{3}\ln(1+x^3).$$

Hence on differentiation with respect to x,

$$\frac{1}{y}\frac{\mathrm{d}y}{\mathrm{d}x} = \frac{1}{x} + \frac{6x}{1+x^2} - \frac{x^2}{1+x^3} = \frac{1+7x^2+6x^5}{x(1+x^2)(1+x^3)}.$$

Hence

$$\frac{\mathrm{d}y}{\mathrm{d}x} = \frac{(1+7x^2+6x^5)(1+x^2)^2}{(1+x^3)^{4/3}}.$$

(b) Put

$$y = \sqrt{\left(\frac{a^2+ax+x^2}{a^2-ax+x^2}\right)}.$$

Then

$$\ln y = \tfrac{1}{2}[\ln(a^2+ax+x^2) - \ln(a^2-ax+x^2)],$$

giving

$$\frac{1}{y}\frac{\mathrm{d}y}{\mathrm{d}x} = \frac{1}{2}\left[\frac{a+2x}{a^2+ax+x^2} - \frac{(-a+2x)}{a^2-ax+x^2}\right]$$
$$= \frac{1}{2}\left[\frac{2a(a^2+x^2) - 4ax^2}{(a^2+ax+x^2)(a^2-ax+x^2)}\right],$$
$$\frac{\mathrm{d}y}{\mathrm{d}x} = \frac{a(a^2-x^2)}{\sqrt{[(a^2+ax+x^2)(a^2-ax+x^2)^3]}}.$$

Example 4 Establish a rule for differentiating a triple product uvw, where u, v, w are functions of x.

Put $y = uvw$. Then $\ln y = \ln u + \ln v + \ln w$, and so

$$\frac{1}{y}\frac{dy}{dx} = \frac{1}{u}\frac{du}{dx} + \frac{1}{v}\frac{dv}{dx} + \frac{1}{w}\frac{dw}{dx}.$$

Hence

$$\frac{dy}{dx} = vw\frac{du}{dx} + uw\frac{dv}{dx} + uv\frac{dw}{dx}.$$

This is an extension of the usual product rule.

17.1.3 Properties of ln $|x|$ Consider $\ln(-x)$. Its domain is the negative x-axis (see Fig. 17.3) and for all x in this domain

$$\frac{d}{dx}\ln(-x) = \frac{1}{-x}.(-1) = \frac{1}{x}.$$

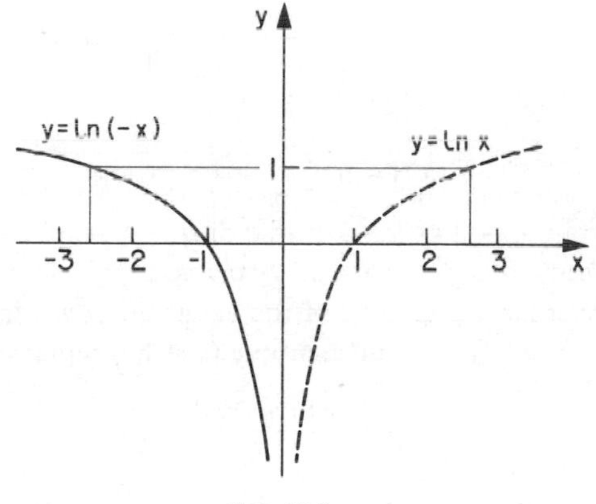

FIG. 17.3

Now

$$\ln|x| = \begin{cases} \ln x & \text{if } x > 0, \\ \ln(-x) & \text{if } x < 0. \end{cases}$$

Hence the domain of $\ln|x|$ is the entire x-axis except for the point $x = 0$ (see Fig. 17.4) and for all x in this domain

$$\frac{d}{dx}\ln|x| = \frac{1}{x}. \tag{17.5}$$

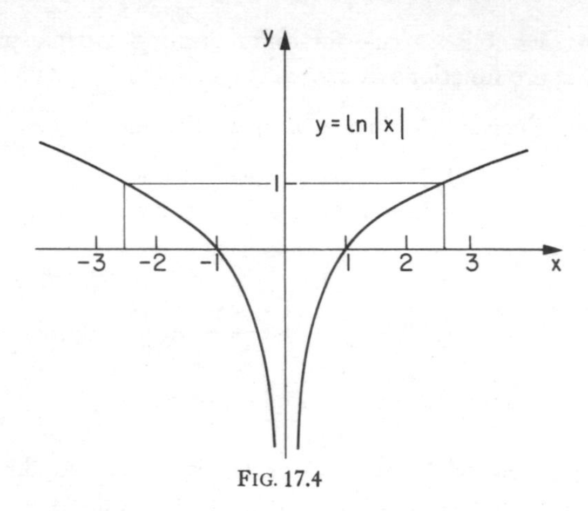

FIG. 17.4

Moreover

$$\ln|ab| = \ln|a||b| = \ln|a| + \ln|b|, \tag{17.6}$$

$$\ln\left|\frac{a}{b}\right| = \ln\frac{|a|}{|b|} = \ln|a| - \ln|b|, \tag{17.7}$$

and

$$\ln|x^n| = \ln(|x|^n) = n\ln|x|. \tag{17.8}$$

Note that the method of logarithmic differentiation (Section 17.1.2) is strictly valid only if logarithms of ***positive*** expressions are used. This restriction is overcome if instead of the equation $\ln y = \ln f(x)$ we use the equation $\ln|y| = \ln|f(x)|$. For example (see Example 3) if

$$y = \frac{x(1+x^2)^3}{\sqrt[3]{(1+x^3)}},$$

then y is differentiable for all x other than $x = -1$ and the derivative obtained in Example 3 is correct for all such values although the method used is valid only if $x > 0$. But results (17.5), (17.6), (17.7), (17.8) show that we have

$$\ln|y| = \ln|x| + 3\ln|1+x^2| - \tfrac{1}{3}\ln|1+x^3|$$

and hence

$$\frac{1}{y}\frac{dy}{dx} = \frac{1}{x} + \frac{6x}{1+x^2} - \frac{x^2}{1+x^3},$$

which establishes the previous result without the restriction that $x > 0$. In practice we usually proceed as in Section 17.1.2.

Exercises 17(a)

Differentiate with respect to x the expressions in Nos. 1–7:

1. $\ln\left(\dfrac{1+\sin x}{1-\sin x}\right)$ 2. $\ln\left(\dfrac{\sqrt{(1+x^2)}-1}{\sqrt{(1+x^2)}+1}\right)$ 3. $\ln(\sec x+\tan x)$
4. $\ln\sqrt{\left(\dfrac{x^2-1}{x^2+1}\right)}$ 5. $\ln[\sec^2(x/a)]$ 6. $\ln(\ln x)$
7. $\sec[\frac{1}{2}\{\ln(a^2+x^2)\}]$.
8. Given $y=x^n\ln x$, show that $x\,dy/dx=ny+x^n$.
9. Differentiate with respect to x

$$\ln(1+\sin 2x)+2\ln\{\sec(\pi/4-x)\},$$

and express the result in its simplest form. Explain why the result is of such simple form.

10. If $y=A\cos(\ln x)+B\sin(\ln x)$, where A, B are constants, show that

$$x^2\frac{d^2y}{dx^2}+x\frac{dy}{dx}+y=0.$$

11. Given $y=\sin[\ln(1+x)]$, show that

$$(1+x)^2\frac{d^2y}{dx^2}+(1+x)\frac{dy}{dx}+y=0.$$

12. Given $y=\ln[(1-x)/(1+x)]$, show that

$$\frac{d^ny}{dx^n}=-(n-1)!\,[(1-x)^{-n}+(-1)^{n-1}(1+x)^{-n}].$$

13. If $x=y\ln(xy)$ show that

$$\frac{dy}{dx}=\frac{xy-y^2}{xy+x^2}.$$

14. Illustrate graphically that if $f(0)=0$ and $f'(x)\geqslant 0$ when $x\geqslant 0$ then $f(x)\geqslant 0$ when $x>0$.

    Deduce that $\ln(1+x)\leqslant x-\frac{1}{2}x^2+\frac{1}{3}x^3$ when $x>0$.
15. By considering the derivative of the function f, where

$$f(x)=\sin x\tan x-2\ln(\sec x),$$

show that $f(x)$ steadily increases as x increases from 0 to $\pi/2$.

    Show also that $2\sin x\tan x-5\ln(\sec x)$ has one minimum value in the interval $0<x<\pi/2$.
16. Find the derivative of the function f, where $f(x)=\ln x-(x-1)/\sqrt{x}$, and hence show that, if $x>1$, $\ln x<(x-1)/\sqrt{x}$.
17. Use logarithmic differentiation to differentiate with respect to x the following:

(a) $\dfrac{x^2+2x+7}{(3x-1)^{1/2}}$, (b) $\dfrac{(3x+1)^5}{(2-x)^{10}}$, (c) $\dfrac{1-x}{(1-x^3)^{1/2}}$, (d) $\sqrt{\left(\dfrac{2+\sin^2 x}{1-\sin x}\right)}$.

18. Given that $y = (n + 1 + x)^{n+1}/(n + x)^n$, n is a fixed positive integer and x is positive, find dy/dx by logarithmic differentiation and show that y increases with x.
19. Find the x-co-ordinates of the stationary points on the curve

$$y = \tfrac{1}{5}x + \tan^{-1} x - \tfrac{1}{2}\ln(1 + x^2)$$

and determine the nature of these stationary points.

Evaluate y at the maximum turning points using mathematical tables and giving your answer to two decimal places.

20. Verify that $\sin x \ln(1 - x)$ has a stationary point at $x = 0$ and determine its nature.

17.1.4 Use of ln *x* in integration From (17.1) we get the standard integral

$$\int \frac{1}{x}\,dx = \ln x + c. \tag{17.9}$$

This may be used to evaluate, for example, $\int_2^3 (1/x)\,dx$, giving the answer ln 3 − ln 2, i.e. ln 3/2, which can be further evaluated as 0·4055 (approximately) using tables of natural logarithms. If, however, we wish to evaluate a definite integral with negative limits of integration, then we require the result

$$\int \frac{1}{x}\,dx = \ln|x| + c \tag{17.10}$$

which follows from (17.5). Thus

$$\int_{-8}^{-5} \frac{1}{x}\,dx = \ln 5 - \ln 8 = \ln\frac{5}{8} \simeq -0{\cdot}4700.$$

Although (17.10) includes (17.9) and is the logical result to take as standard, nevertheless (17.9) is usually quoted as a standard integral for simplicity of writing, answers being interpreted in the light of (17.10) when necessary.

From (17.9) we get, by applying (16.1)

$$\left.\begin{aligned} &\int \frac{g'(x)}{g(x)}\,dx = \ln g(x) + c, \\ \text{that is} \quad &\int \frac{1}{z}\frac{dz}{dx}\,dx = \ln z + c. \end{aligned}\right\} \tag{17.11}$$

In particular

$$\int \frac{dx}{ax + b} = \frac{1}{a}\ln(ax + b) + c.$$

Example 5 Find the following integrals:

(a) $\displaystyle\int \frac{dx}{2x+3}$, (b) $\displaystyle\int \frac{dx}{1-x}$, (c) $\displaystyle\int \frac{2x}{x^2+1}\,dx$, (d) $\displaystyle\int \frac{x^2}{x^3-2}\,dx$,

(e) $\displaystyle\int \frac{\cos x}{1+\sin x}\,dx$.

On using (17.11) the answers are as follows:

(a) $\frac{1}{2}\ln(2x+3)+c$, (b) $-\ln(1-x)+c$, (c) $\ln(x^2+1)+c$,
(d) $\frac{1}{3}\ln(x^3-2)+c$, (e) $\ln(1+\sin x)+c$.

In (b) it is convenient to introduce a different arbitrary constant k by writing $c = \ln k$; the answer is then $-\ln(1-x)+\ln k$, i.e. $\ln[k/(1-x)]$.

17.1.5 Integration of ln x On using integration by parts (see) (16.4) with $u = \ln x$, $dv/dx = 1$, we get

$$\int \ln x\,dx = x\ln x - \int x\,\frac{1}{x}\,dx = x\ln x - x + c. \qquad (17.12)$$

17.1.6 Integration of x^n ln x (n a rational number) On using integration by parts with $u = \ln x$, $dv/dx = x^n$, we get

$$\int x^n \ln x\,dx = \frac{1}{n+1}x^{n+1}\ln x - \int \frac{1}{n+1}x^{n+1}\frac{1}{x}\,dx \qquad (\text{if } n \neq -1)$$
$$= \frac{1}{n+1}x^{n+1}\ln x - \frac{1}{(n+1)^2}x^{n+1} + c.$$

For the case $n = -1$ we may substitute $z = \ln x$, $dz = (1/x)\,dx$; thus

$$\int \frac{1}{x}\ln x\,dx = \int z\,dz = \tfrac{1}{2}z^2 + c = \tfrac{1}{2}(\ln x)^2 + c.$$

Exercises 17(b)

Find the integrals in Nos. 1–5:

1. $\displaystyle\int \frac{x}{1+x^2}\,dx$
2. $\displaystyle\int \frac{x+2}{x^2+4x-5}\,dx$
3. $\displaystyle\int \frac{\cos x}{\sin x}\,dx$
4. $\displaystyle\int \frac{\sin x}{2+3\cos x}\,dx$
5. $\displaystyle\int \frac{\sec^2 x}{1+2\tan x}\,dx$

Verify the results in Nos. 6–10:

6. $\displaystyle\int_{4/3}^{20} \frac{dx}{3x+4} = \ln 2$
7. $\displaystyle\int_0^4 \frac{x\,dx}{x^2+9} = \ln(5/3)$
8. $\displaystyle\int_0^1 \frac{2x+5}{x^2+5x+3} = \ln 3$
9. $\displaystyle\int_0^{\pi/3} \frac{\sin x}{\cos x}\,dx = \ln 2$

10. $\int_{\pi/4}^{\pi/2} \frac{\sin x - \cos x}{\sin x + \cos x} dx = \frac{1}{2} \ln 2$

Use integration by parts to find the integrals in Nos 11–12:

11. $\int \frac{\ln x}{x^3} dx$

12. $\int \frac{\ln x}{\sqrt{x}} dx$

Use the substitution $z = \ln x$ to find the integrals in Nos. 13–14:

13. $\int \frac{(\ln x)^n}{x} dx$

14. $\int \frac{\ln (x^n)}{x} dx$

17.1.7 Connection between Natural Logarithm and Algebraic Logarithm From the properties of $\ln x$ the reader will immediately suspect that the natural logarithm has some connection with the logarithms of elementary algebra. In algebra the idea of logarithm has its origins in the notion of an *index*, which we summarise as follows:

(a) if n is a positive integer, then $b^n = b.b.b. \cdots$ (n factors),
(b) if n is a negative integer, say $n = -m$ where $m > 0$, then $b^n = 1/b^m$,
(c) $b^0 = 1$,
(d) if n is rational, say $n = p/q$, where p and q are integers, then $b^n = \sqrt[q]{(b^p)}$.

Note. This attaches no meaning to b^n if the index n is not rational, e.g. $b^{\sqrt{2}}$ is undefined.

From this idea of index a logarithm is defined as follows: if $x = b^n$, then n is called the logarithm of x to base b and we write $n = \log_b x$. It follows that *only rational numbers n are logarithms in this sense.*

In Fig. 17.5 several points on the graph of $\log_b x$ are shown and the remaining points have been filled in as though they formed a continuous curve. *But there must be no points on this curve with irrational values of n*; for example the line $n = \sqrt{2}$ must not meet the curve which must therefore have a gap (discontinuity) as indicated in the figure, and corresponding to this gap there is a value of x for which $\log_b x$ is not defined. For every irrational value of n there must be a similar gap in the curve, and it would be desirable to extend the definition of logarithm so that these gaps would not exist.

In Section 17.1.1 it was shown that there is a number e with the property that $\ln e = 1$. If we consider logarithms to base e, then $n = \log_e x$ means that $x = e^n$ and so from (17.4) $\ln x = n \ln e = n$, i.e.

$$\log_e x = \ln x = n.$$

Summarising, some (but not all) positive numbers x may be expressed in

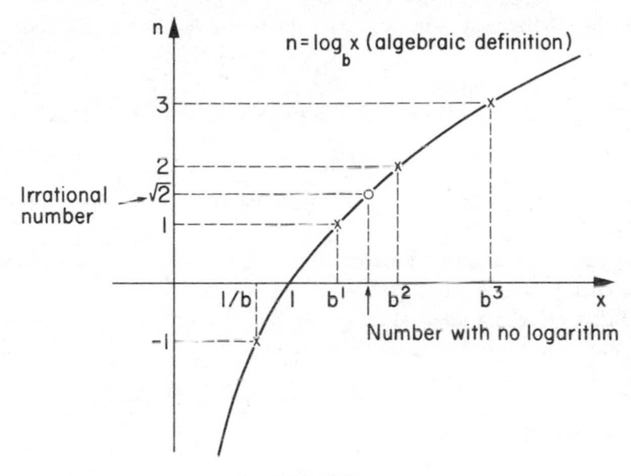

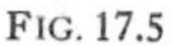
FIG. 17.5

the form e^n and for these numbers $\ln x$ and $\log_e x$ are equal; but $\ln x$ is defined *for all positive numbers* x and so may be regarded as the natural extension of $\log_e x$. We therefore abandon the algebraic approach to logarithms and redefine $\log_e x$ by saying

$$\log_e x = \ln x \quad \text{for all positive values of } x.$$

17.1.8 Differentiation of $\log_b x$ From the algebraic definition of logarithm it follows that

$$\log_b x = \frac{\log_e x}{\log_e b}.$$

Since $\log_e x$ is now defined for all positive values of x, we may use this equation to extend the definition of $\log_b x$ so that it, too, is defined for all positive values of x. Extended in this way $\log_b x$ is differentiable and

$$\frac{\mathrm{d}}{\mathrm{d}x}(\log_b x) = \frac{1}{x \ln b}.$$

Exercises 17(c)

1. Given

$$\ln\left(\frac{x+1}{x-1}\right) = n,$$

where n is a rational number, show that

$$x = (e^n + 1)/(e^n - 1).$$

2. Show that the equation $\ln(x^2 + 2x + 1) = 4$ has the single root $x = e^2 - 1$.

3. Show that the equation $\ln (x + 1) - \ln x = 2$ has the single root
$$x = 1/(e^2 - 1).$$
4. Show that the roots of the equation $\ln (x + 2) + \ln x = 2$ are
$$x = -1 \pm \sqrt{(e^2 + 1)}.$$
5. Verify that $x^2 \ln x$ has a minimum value $-1/2e$ which occurs when $x = 1/\sqrt{e}$.
6. Verify that $x^4 \ln x$ has a minimum value $-1/4e$ which occurs when $x = e^{-1/4}$.
7. Show that $\int_1^e x \ln x \, dx = \frac{1}{4}(e^2 + 1)$.
8. Show that $\int_1^{e^n} x \ln x \, dx = \frac{1}{4}[(2n - 1)e^{2n} + 1]$, where n is a rational number.
9. Find the stationary points on the curve
$$y = 6 \ln (x/7) + (x - 7)(x - 1).$$
Deduce that the equation
$$6 \ln (x/7) + (x - 7)(x - 1) = 0$$
has only one real root and state its value.

17.2 The Exponential Function

From the graph of $\ln x$ (Fig. 17.2) we see that the natural logarithm has an inverse function the graph of which is shown in Fig. 17.6. This inverse

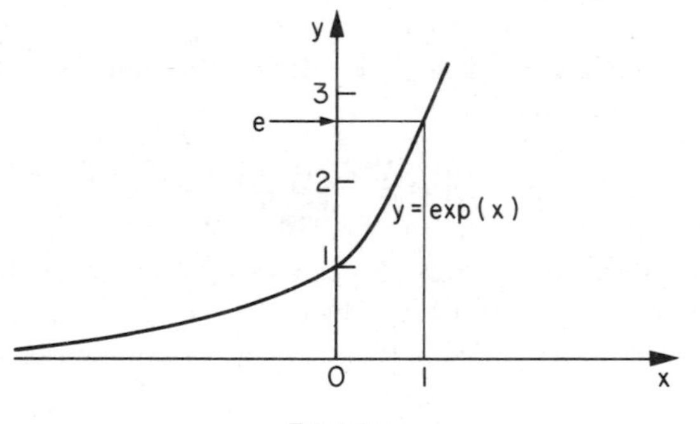

FIG. 17.6

function is called the *exponential* function. Thus
$$y = \exp (x) \Leftrightarrow x = \ln y,$$
where $\exp (x)$ denotes the image of x under the exponential function.

If x is rational then
$$x = \ln y \Leftrightarrow y = e^x \qquad \text{(see Section 17.1.7).}$$
Hence e^x and $\exp (x)$ are equivalent *for all rational values of* x; e^x has not been defined for an irrational index x whereas $\exp (x)$ is defined for

all x. This means that $\exp(x)$ is the natural extension of e^x in the same way as $\ln x$ is the natural extension of the algebraic logarithm to base e. We therefore redefine e^x by saying

$$e^x = \exp(x) \quad \text{for all } x,$$

thereby removing the restriction that the index must be rational. This also means that in all further discussions about the exponential function we may replace $\exp(x)$ by e^x.

17.2.1 Properties of e^x The main properties of e^x follow easily from those of $\ln x$. Its graph is shown in Fig. 17.6 and we note that $e^x \to \infty$ as $x \to \infty$, $e^x \to 0$ as $x \to -\infty$.

If $y = e^x$, then $x = \ln y$ and so

$$\frac{dx}{dy} = \frac{1}{y}, \quad \text{that is} \quad \frac{dy}{dx} = y = e^x.$$

Hence

$$\frac{d}{dx}(e^x) = e^x. \tag{17.13}$$

Example 6 Differentiate with respect to x the following:

$$\text{(a) } e^{(x^2+x+1)}, \quad \text{(b) } e^{1/x}, \quad \text{(c) } e^{\sin x}.$$

On using (17.13) together with the chain rule we get the following answers:

$$\text{(a) } (2x+1)e^{(x^2+x+1)}, \quad \text{(b) } -(1/x^2)\,e^{1/x}, \quad \text{(c) } \cos x\, e^{\sin x}.$$

Write $y_1 = e^a$, $y_2 = e^b$ so that $a = \ln y_1$, $b = \ln y_2$. Now

$$a + b = \ln y_1 + \ln y_2 = \ln y_1 y_2,$$
$$a - b = \ln y_1 - \ln y_2 = \ln (y_1/y_2).$$

Hence

$$y_1 y_2 = e^{a+b}, \quad y_1/y_2 = e^{a-b},$$

that is,

$$e^a . e^b = e^{a+b}, \tag{17.14}$$
$$e^a/e^b = e^{a-b}. \tag{17.15}$$

Also, if $y = e^x$ then $x = \ln y$ and so

$$nx = n \ln y = \ln y^n = \ln (e^x)^n,$$

that is

$$(e^x)^n = e^{nx}. \tag{17.16}$$

Properties (17.14), (17.15) and (17.16) show that the usual index laws apply when the notion of index is extended as in Section 17.2.

17.2.2 Differentiation of b^x If b is any positive number we have $b = e^{\ln b}$. As yet, b^x is defined only for rational values of x, and

$$b^x = (e^{\ln b})^x = e^{x \ln b} \qquad \text{by (17.16).}$$

Since $e^{x \ln b}$ is defined for all x, this result suggests an obvious way of extending the definition of b^x; we simply redefine b^x by saying

$$b^x = e^{x \ln b} \qquad \text{for all } x.$$

Extended in this way b^x is differentiable and

$$\frac{\mathrm{d}}{\mathrm{d}x}(b^x) = \ln b \, e^{x \ln b} = b^x \ln b;$$

this result need not be remembered since in practice we normally use the method of logarithmic differentiation (Section 17.1.2).

Example 7 Differentiate with respect to x the following: (a) 5^x, (b) 3^{x^2+x+1}, (c) x^x, (d) $(x^2 + 1)^{\sin x}$.

(a) Put $y = 5^x$. Then

$$\ln y = x \ln 5, \quad \frac{1}{y}\frac{\mathrm{d}y}{\mathrm{d}x} = \ln 5, \quad \text{and so } \frac{\mathrm{d}y}{\mathrm{d}x} = 5^x \ln 5.$$

(b) Put $y = 3^{x^2+x+1}$. Then

$$\ln y = (x^2 + x + 1) \ln 3, \quad \frac{1}{y}\frac{\mathrm{d}y}{\mathrm{d}x} = (2x + 1) \ln 3,$$

and so

$$\frac{\mathrm{d}y}{\mathrm{d}x} = 3^{x^2+x+1}(2x + 1) \ln 3.$$

(c) Put $y = x^x$. Then

$$\ln y = x \ln x, \quad \frac{1}{y}\frac{\mathrm{d}y}{\mathrm{d}x} = x\frac{1}{x} + \ln x,$$

and so

$$\frac{\mathrm{d}y}{\mathrm{d}x} = x^x(1 + \ln x).$$

(d) Put $y = (x^2 + 1)^{\sin x}$. Then

$$\ln y = \sin x \ln (x^2 + 1),$$

$$\frac{1}{y}\frac{\mathrm{d}y}{\mathrm{d}x} = \sin x \frac{2x}{x^2 + 1} + \cos x \ln (x^2 + 1),$$

and so

$$\frac{dy}{dx} = (x^2 + 1)^{\sin x} \left[\frac{2x \sin x}{x^2 + 1} + \cos x \ln (x^2 + 1) \right].$$

17.2.3 Use of e^x in Integration From (17.13) we get the standard integral

$$\int e^x \, dx = e^x + c. \tag{17.17}$$

On applying (16.1), this gives

$$\left. \begin{aligned} &\int e^{g(x)} \, g'(x) \, dx = e^{g(x)} + c, \\ \text{that is} \quad &\int e^z \frac{dz}{dx} \, dx = e^z + c. \end{aligned} \right\} \tag{17.18}$$

In particular

$$\int e^{ax+b} \, dx = \frac{1}{a} e^{ax+b} + c.$$

Example 8 Find the following integrals: (a) $\int e^{2x+3} \, dx$, (b) $\int e^{2-x} \, dx$, (c) $\int x e^{x^2} \, dx$, (d) $\int \sin x \, e^{\cos x} \, dx$.

By using (17.18) we get the answers (a) $\frac{1}{2} e^{2x+3} + c$, (b) $-e^{2-x} + c$, (c) $\frac{1}{2} e^{x^2} + c$, (d) $-e^{\cos x} + c$.

17.2.4 Integration of $x^n e^x$ (n a positive integer) By using integration by parts we get

$$\int x^n e^x \, dx = x^n e^x - \int n x^{n-1} e^x \, dx.$$

Again by use of integration by parts, the integral on the r.h.s. may be replaced by

$$n \, x^{n-1} e^x - \int n \, (n - 1) x^{n-2} e^x \, dx.$$

By continuing in this way the integration may be completed after using integration by parts n times. For example,

$$\begin{aligned} \int x^2 e^x \, dx &= x^2 e^x - \int 2x \, e^x \, dx \\ &= x^2 e^x - 2x \, e^x + \int 2e^x \, dx \\ &= x^2 e^x - 2x e^x + 2e^x + c. \end{aligned}$$

Exercises 17(d)

Differentiate with respect to x the expressions in Nos. 1–15:

1. e^{x^2}
2. xe^x
3. $e^{2x}\sin 3x$
4. $(x^2 - x)/e^x$
5. $1/(e^x + 1)$
6. $e^{-2x}\sqrt{(1 + 4x)}$
7. $(e^x - 1)/(e^x + 1)$
8. $x^2\, 10^x$
9. $x^{1/x}$
10. $e^{-x}(3x + 5)/(7x - 1)$
11. $(e^{\cos x} - 1)/(e^{\cos x} + 1)$
12. $(1 + e^{2x})/(1 - e^{2x})$
13. $(\ln x)^x$
14. $\ln\sqrt{(e^x)} - e^{\ln\sqrt{x}}$
15. $\ln\{e^x[(x - 1)/(x + 1)]^{1/2}\}$
16. Evaluate
$$\frac{d}{dx}(x^2 e^{2x} \ln 2x)$$
when $x = \frac{1}{2}$.
17. Evaluate
$$\frac{d^3}{dx^3}(e^{2x}\tan^{-1} x)$$
when $x = 0$.
18. Given $y = x\, e^{-x^2}$, show that
$$\frac{d^2y}{dx^2} + 2x\frac{dy}{dx} + 4y = 0.$$
19. Given $y = e^x/(1 + x^2)$, show that y has a stationary value when $x = 1$, but that $dy/dx > 0$ for all $x \neq 1$.
20. Find the turning points between $x = 0$ and $x = 2\pi$ on the curve
$$y = e^{-x}\sin x.$$
21. Verify that

(a) $\int_0^4 e^{x/2}\, dx = 2(e^2 - 1)$, (b) $\int_0^{\pi/2} e^{\sin x}\cos x\, dx = e - 1$,

(c) $\int_0^1 \frac{e^x}{e^x + 1}\, dx = \ln\left(\frac{e + 1}{2}\right)$, (d) $\int_0^1 \frac{1 - e^{-x}}{x + e^{-x}}\, dx = \ln(e + 1) - 1$

22. Show that
$$\int x^3 e^x\, dx = e^x(x^3 - 3x^2 + 6x - 6) + c.$$
23. Determine the slopes of the tangents to the curve $y = e^{2x} - 12e^x + 4x^2$ at the points of inflexion on the curve.
24. Show that the maximum value of $e^{tx}/(e^x + 1)$ is $t^t(1 - t)^{1-t}$, where t is a constant such that $0 < t < 1$.
25. Show that the expression $e^{ax}/(1 + x^2)$, where a is a constant, has a maximum or a minimum value if $|a| < 1$, but that there are no turning points if $|a| \geqslant 1$.

17.3 Hyperbolic Functions

We define the *hyperbolic sine* of x (written $\sinh x$) and the *hyperbolic cosine* of x (written $\cosh x$) by the relations

$$\sinh x = \tfrac{1}{2}(e^x - e^{-x}), \qquad \cosh x = \tfrac{1}{2}(e^x + e^{-x}). \tag{17.19}$$

The graphs of these functions are easily derived from the graph of e^x and are shown in Fig. 17.7.

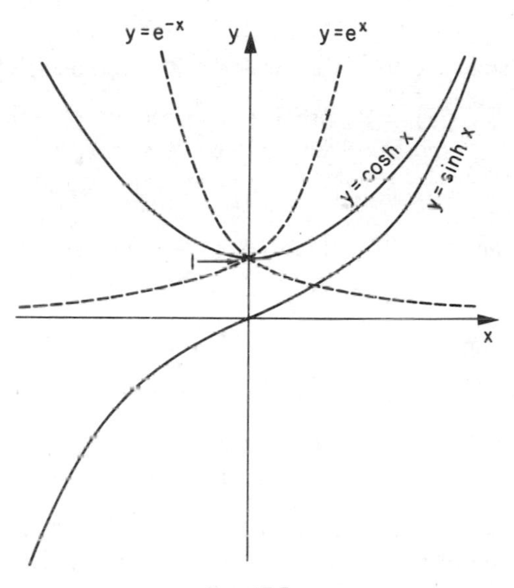

FIG. 17.7

From (17.19) we see that $\sinh(-x) = -\sinh x$, $\cosh(-x) = \cosh x$. Also,

$$\cosh x + \sinh x = e^x, \quad \cosh x - \sinh x = e^{-x}, \tag{17.20}$$

from which we readily obtain

$$\cosh^2 x - \sinh^2 x = 1. \tag{17.21}$$

The identity (17.21) gives a clue to the reason for calling these functions *hyperbolic*; for if $x = \cosh t$, $y = \sinh t$, then as t varies the point (x, y) describes a hyperbola, since we may eliminate t by use of (17.21) to get $x^2 - y^2 = 1$. Compare this with the situation where $x = \cos t, y = \sin t$; as t varies the point (x, y) describes the circle $x^2 + y^2 = 1$, and elimination of t in this case uses the familiar identity $\cos^2 t + \sin^2 t = 1$ for the circular functions sine and cosine.

Squaring and adding corresponding sides in (17.20) we have

$$\cosh^2 x + \sinh^2 x = \tfrac{1}{2}(e^{2x} + e^{-2x}) = \cosh 2x \qquad \text{by definition,}$$

that is

$$\cosh 2x = 2\cosh^2 x - 1 = 2\sinh^2 x + 1 \qquad \text{by (17.21)}$$

Again from (17.20), by squaring and subtracting corresponding sides we get

$$\sinh 2x = 2\sinh x \cosh x.$$

By writing $\sinh(x + y) = \frac{1}{2}(e^x . e^y - e^{-x} . e^{-y})$ and using (17.20) we have

$$\sinh(x + y) = \tfrac{1}{2}[(\cosh x + \sinh x)(\cosh y + \sinh y) - (\cosh x - \sinh x)(\cosh y - \sinh y)].$$

Hence

$$\sinh(x + y) = \sinh x \cosh y + \cosh x \sinh y.$$

Similarly,

$$\cosh(x + y) = \cosh x \cosh y + \sinh x \sinh y.$$

Substituting $-y$ for y in these results we obtain

$$\sinh(x - y) = \sinh x \cosh y - \cosh x \sinh y,$$
$$\cosh(x - y) = \cosh x \cosh y - \sinh x \sinh y.$$

It should now be clear that to every identity involving sines and cosines there corresponds an identity involving hyperbolic sines and cosines.

The analogy is carried still further by the introduction of hyperbolic tangents, cosecants, secants and cotangents defined by the relations

$$\tanh x = \sinh x/\cosh x, \quad \operatorname{cosech} x = 1/\sinh x,$$
$$\operatorname{sech} x = 1/\cosh x, \quad \coth x = 1/\tanh x.$$

On dividing throughout (17.21) by $\cosh^2 x$ and $\sinh^2 x$ in turn, we have

$$\operatorname{sech}^2 x = 1 - \tanh^2 x \quad \text{and} \quad \operatorname{cosech}^2 x = \coth^2 x - 1;$$

from the formulae for $\sinh(x + y)$ and $\cosh(x + y)$ we deduce that

$$\tanh(x \pm y) = \frac{\tanh x \pm \tanh y}{1 \pm \tanh x \tanh y}.$$

The following rule (known as Osborn's rule) enables us to write down an identity involving hyperbolic functions from one involving circular functions: replace each circular function by the corresponding hyperbolic function but change the sign in front of a product or an implied product of *two* sines; for example, in front of $\sin^2 x$, $\tan^2 x$, $\cot^2 x$, $\sin x \sin y$.

The identities $\cos 3\theta = 4\cos^3\theta - 3\cos\theta$, $\sin 3\theta = 3\sin\theta - 4\sin^3\theta$

give $\cosh 3x = 4\cosh^3 x - 3\cosh x$, $\sinh 3x = 3\sinh x + 4\sinh^3 x$; and from the formulae

$$\begin{aligned}
\sin\theta\cos\phi &= \tfrac{1}{2}[\sin(\theta+\phi) + \sin(\theta-\phi)],\\
\cos\theta\cos\phi &= \tfrac{1}{2}[\cos(\theta+\phi) + \cos(\theta-\phi)],\\
\cos\theta\sin\phi &= \tfrac{1}{2}[\sin(\theta+\phi) - \sin(\theta-\phi)],\\
\sin\theta\sin\phi &= \tfrac{1}{2}[\cos(\theta-\phi) - \cos(\theta+\phi)],
\end{aligned}$$

we obtain the relations

$$\begin{aligned}
\sinh x\cosh y &= \tfrac{1}{2}[\sinh(x+y) + \sinh(x-y)],\\
\cosh x\cosh y &= \tfrac{1}{2}[\cosh(x+y) + \cosh(x-y)],\\
\cosh x\sinh y &= \tfrac{1}{2}[\sinh(x+y) - \sinh(x-y)],\\
\sinh x\sinh y &= \tfrac{1}{2}[\cosh(x+y) - \cosh(x-y)].
\end{aligned}$$

All the above hyperbolic identities may be proved from the definitions of the hyperbolic functions. Derivatives of hyperbolic functions should *not* be found by Osborn's rule and are dealt with separately in Section 17.3.1.

If

$$y = \tanh x = \frac{e^x - e^{-x}}{e^x + e^{-x}},$$

it is clear that $y = 0$ when $x = 0$ and that $|y| < 1$. By writing

$$y = \frac{1 - e^{-2x}}{1 + e^{-2x}},$$

we see that $y \to 1$ as $x \to \infty$ and by writing

$$y = \frac{e^{2x} - 1}{e^{2x} + 1}$$

we see that $y \to -1$ as $x \to -\infty$. The graph of $\tanh x$ is shown in Fig. 17.8.

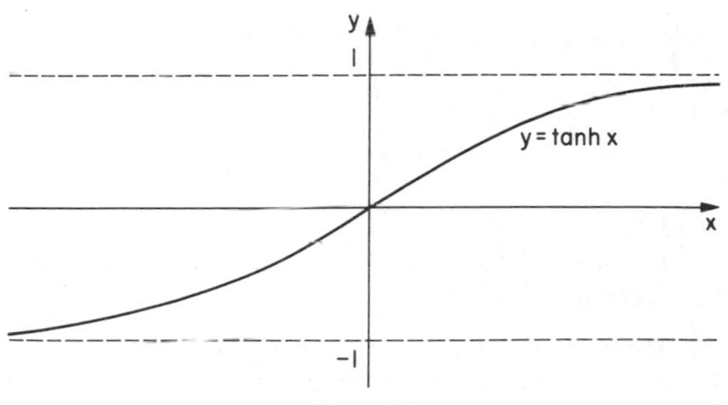

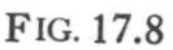
FIG. 17.8

Example 9 Solve for real values of x the equations

(a) $2\cosh 2x - \sinh 2x = 2$, (b) $\cosh(\ln x) = \sinh(\ln \frac{1}{4}x) + \frac{19}{8}$.

(a) If

$$\begin{aligned} 2\cosh 2x - \sinh 2x &= 2, \\ e^{2x} + e^{-2x} - \tfrac{1}{2}(e^{2x} - e^{-2x}) &= 2 \\ e^{2x} - 4 + 3e^{-2x} &= 0 \\ (e^{x} - e^{-x})(e^{x} - 3e^{-x}) &= 0. \end{aligned}$$

Hence

$$e^{x} - e^{-x} = 0, \quad e^{2x} = 1 \quad \text{and so} \quad x = 0;$$

or

$$e^{x} - 3e^{-x} = 0, \quad e^{2x} = 3 \quad \text{and so} \quad x = \tfrac{1}{2}\ln 3.$$

(b)

$$\begin{aligned} \cosh(\ln x) &= \tfrac{1}{2}(e^{\ln x} + e^{-\ln x}) = \tfrac{1}{2}(x + 1/x) \\ \sinh(\ln \tfrac{1}{4}x) &= \tfrac{1}{2}(e^{\ln(1/4)x} - e^{-\ln(1/4)x}) = \tfrac{1}{2}(x/4 - 4/x). \end{aligned}$$

Hence if

$$\begin{aligned} &\cosh(\ln x) = \sinh(\ln \tfrac{1}{4}x) + \tfrac{19}{8}, \\ &\tfrac{1}{2}(x + 1/x) = \tfrac{1}{2}(x/4 - 4/x) + \tfrac{19}{8}, \\ &3x^2 - 19x + 20 = 0, \quad x = \tfrac{4}{3} \text{ or } 5. \end{aligned}$$

17.3.1 Derivatives of Hyperbolic Functions From (17.19) we get

$$\frac{d}{dx}(\sinh x) = \cosh x, \quad \frac{d}{dx}(\cosh x) = \sinh x,$$

and from these results we may readily deduce that

$$\frac{d}{dx}(\tanh x) = \operatorname{sech}^2 x, \quad \frac{d}{dx}(\coth x) = -\operatorname{cosech}^2 x,$$

$$\frac{d}{dx}(\operatorname{sech} x) = -\operatorname{sech} x \tanh x, \quad \frac{d}{dx}(\operatorname{cosech} x) = -\operatorname{cosech} x \coth x.$$

Exercises 17(e)

1. From the definitions of $\sinh x$, $\cosh x$ in terms of the exponential function, prove the identities

   (a) $\sinh x \cosh y = \frac{1}{2}[\sinh(x + y) + \sinh(x - y)]$,
   (b) $\cosh 3x = 4\cosh^3 x - 3\cosh x$,
   (c) $\sinh 3x = 3\sinh x + 4\sinh^3 x$.

2. Solve the equations

   (a) $5 \cosh x - 4 \sinh x = 3$, (b) $8 \sinh x - 4 \cosh x = 1$.

3. Show that $\cosh 2x = 2 \cosh^2 x - 1$. Hence, or otherwise, solve the equation

$$\cosh 4x - 4 \cosh 2x + 3 = 0.$$

4. Define $\tanh x$ in terms of e^x and e^{-x}. Deduce that

$$\tanh (2x) = 2 \tanh x/(1 + \tanh^2 x).$$

5. Express $\sinh^4 x$ in the form $A \cosh 4x + B \cosh 2x + C$.
6. Differentiate with respect to x the following:

   (a) $\sinh (1/x)$, (b) $\tanh (\ln x)$, (c) $\ln (\tanh x)$, (d) $\sinh (\tanh x)$.

7. If $y = \tan^{-1} (\sinh x)$, show that

$$\frac{d^2y}{dx^2} + \tan y\left(\frac{dy}{dx}\right)^2 = 0.$$

8. Find the maximum value of the expression $\sinh (2x - e^x)$.

17.4 Inverse Hyperbolic Functions

From the curves $y = \sinh x$, $y = \cosh x$ (see Fig. 17.7) we can easily sketch the curves $x = \sinh y$, $x = \cosh y$ (see Fig. 17.9). We define the inverse hyperbolic functions $\sinh^{-1}$, $\cosh^{-1}$ as follows:

$$y = \sinh^{-1} x \text{ if } x = \sinh y, \quad y = \cosh^{-1} x \text{ if } y \geqslant 0 \text{ and } x = \cosh y.$$

It follows that the graph of $\cosh^{-1} x$ is the part of the curve $x = \cosh y$ that has been drawn solid in Fig. 17.9.

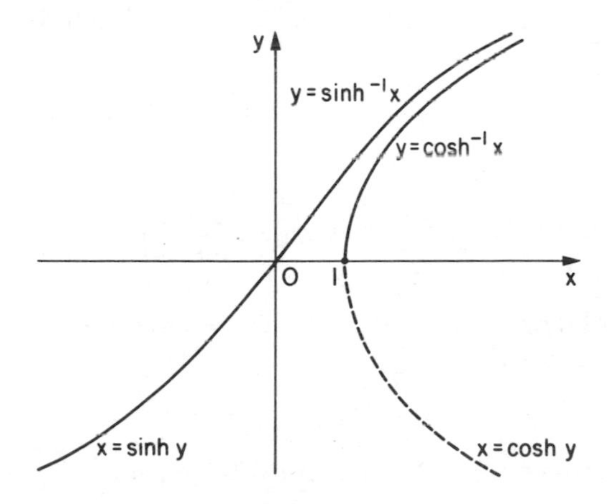

FIG. 17.9

Other inverse hyperbolic functions are defined in a similar way; thus

$$\begin{aligned}
&y = \tanh^{-1} x \quad \text{if} \quad x = \tanh y \quad \text{(see Fig. 17.10)},\\
&y = \coth^{-1} x \quad \text{if} \quad x = \coth y,\\
&y = \operatorname{sech}^{-1} x \quad \text{if} \quad y \geqslant 0 \text{ and } x = \operatorname{sech} y,\\
&y = \operatorname{cosech}^{-1} x \quad \text{if} \quad x = \operatorname{cosech} y.
\end{aligned}$$

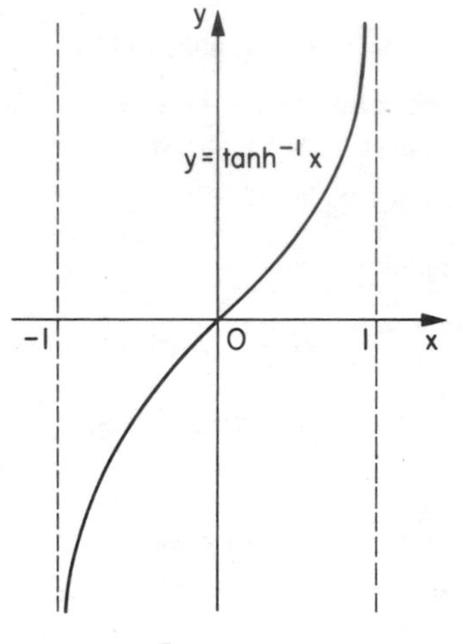

FIG. 17.10

Example 10 Prove that

$$\tanh^{-1} x + \tanh^{-1} y = \tanh^{-1}\frac{x+y}{1+xy}.$$

Let $\tanh^{-1} x = A$ and $\tanh^{-1} y = B$ so that $x = \tanh A$ and $y = \tanh B$. Then

$$\begin{aligned}
\tanh^{-1} x + \tanh^{-1} y = A + B &= \tanh^{-1}[\tanh(A+B)]\\
&= \tanh^{-1}\left(\frac{\tanh A + \tanh B}{1 + \tanh A \tanh B}\right),\\
&= \tanh^{-1}\left(\frac{x+y}{1+xy}\right).
\end{aligned}$$

17.4.1 Relation between Inverse Hyperbolic Functions and Natural Logarithms If $y = \sinh^{-1} x$, then

$$x = \sinh y = \tfrac{1}{2}(e^{y} - e^{-y}).$$

Multiply throughout this equation by $2e^{y}$ to get

$$e^{2y} - 2xe^{y} - 1 = 0.$$

This is a quadratic equation for e^y, giving

$$e^y = x \pm \sqrt{(x^2 + 1)}$$

and since e^y is necessarily positive we must reject the negative square root.

Hence

$$y = \ln [x + \sqrt{(x^2 + 1)}],$$

that is

$$\sinh^{-1} x = \ln [x + \sqrt{(x^2 + 1)}]. \tag{17.22}$$

Similarly if $y = \cosh^{-1} x$, then $y \geqslant 0$ and

$$x = \cosh y = \tfrac{1}{2}(e^y + e^{-y}).$$

Multiply this equation throughout by $2e^y$ to get

$$e^{2y} - 2xe^y + 1 = 0,$$

from which

$$e^y = x \pm \sqrt{(x^2 - 1)}.$$

Since $y \geqslant 0$ we must have $e^y \geqslant 1$; also $\cosh y \geqslant 1$ and so $x \geqslant 1$. It follows that the negative square root must be rejected and so

$$\cosh^{-1} x = \ln [x + \sqrt{(x^2 - 1)}]. \tag{17.23}$$

Finally, if $y = \tanh^{-1} x$, then

$$x = \tanh y = \frac{e^{2y} - 1}{e^{2y} + 1}$$

and hence

$$e^{2y} = \frac{1 + x}{1 - x},$$

that is

$$e^y = \sqrt{\left(\frac{1 + x}{1 - x}\right)}.$$

It follows that

$$\tanh^{-1} x = \tfrac{1}{2} \ln \left(\frac{1 + x}{1 - x}\right). \tag{17.24}$$

17.4.2 Derivatives of Inverse Hyperbolic Functions These may be found by using the inverse rule of differentiation, $dx/dy = 1/(dy/dx)$, or alternatively by using (17.22)–(17.24). The results are

$$\frac{d}{dx}(\sinh^{-1} x) = \frac{1}{\sqrt{(1+x^2)}}, \quad \frac{d}{dx}(\cosh^{-1} x) = \frac{1}{\sqrt{(x^2-1)}},$$
$$\frac{d}{dx}(\tanh^{-1} x) = \frac{1}{1-x^2}.$$

Exercises 17(f)

1. Show that

$$\frac{d}{dx}[\tanh^{-1}\sqrt{(2-x)}] = -\frac{1}{2(x-1)\sqrt{(2-x)}}.$$

2. Show that

$$\frac{d}{dx}[\sinh^{-1}(\tan x)] = \sec x.$$

3. Show that

$$\frac{d}{dx}[\sinh^{-1}(\sin x)] = (\sec^2 x + \tan^2 x)^{-1/2}.$$

4. Differentiate

$$\tanh^{-1}[2\sqrt{x}/(1+x)] + \tan^{-1}[2\sqrt{x}/(1-x)]$$

with respect to x, simplifying your answer.

5. Given $y = \sinh^{-1} x/\sqrt{(1+x^2)}$, show that

$$(1+x^2)\frac{dy}{dx} + xy = 1.$$

6. Given $y = \sin(m \sinh^{-1} x)$, show that

$$(1+x^2)\frac{d^2y}{dx^2} + x\frac{dy}{dx} + m^2y = 0.$$

7. Determine dy/dx in terms of x and y from the relationship

$$y \sinh^{-1} x = x \sinh^{-1} y.$$

8. Show that

$$\tanh^{-1} x = \tfrac{1}{2}\ln[(1+x)/(1-x)], \quad |x| < 1.$$

Given $\tanh u = \cos\theta$ ($\theta \neq n\pi$), show that

(a) $\cosh u = |\operatorname{cosec}\theta|$, (b) $u = \ln|\cot(\theta/2)|$.

9. Show that

$$\coth^{-1} x = \tfrac{1}{2}\ln[(x+1)/(x-1)], \quad |x| > 1,$$

and that

$$\coth^{-1}(1 + 2\cot^2\theta) = -\ln|\cos\theta|, \quad (\theta \neq (2n + 1)\pi/2, n \text{ an integer}).$$

Find the values of θ between 0 and 2π that satisfy the equation

$$2\coth^{-1}(1 + 2\cot^2\theta) = \ln 2.$$

10. Express $\tanh^{-1} x$ in terms of the logarithmic function, and hence, or otherwise, show that

$$\tanh^{-1}\alpha + \tanh^{-1}\beta = \tanh^{-1}\left(\frac{\alpha + \beta}{1 + \alpha\beta}\right).$$

CHAPTER 18

METHODS OF INTEGRATION

18.1 Standard Integrals

The following list includes all the standard integrals encountered in this book:

$$\int x^n \, dx = \frac{1}{n+1} x^{n+1} + c \quad (n \neq -1), \tag{18.1}$$

$$\int \frac{1}{x} \, dx = \ln|x| + c, \tag{18.2}$$

$$\int \sin x \, dx = -\cos x + c, \tag{18.3}$$

$$\int \cos x \, dx = \sin x + c, \tag{18.4}$$

$$\int \sec^2 x \, dx = \tan x + c, \tag{18.5}$$

$$\int \operatorname{cosec}^2 x \, dx = -\cot x + c, \tag{18.6}$$

$$\int \sec x \tan x \, dx = \sec x + c, \tag{18.7}$$

$$\int \operatorname{cosec} x \cot x \, dx = -\operatorname{cosec} x + c, \tag{18.8}$$

$$\int e^x \, dx = e^x + c, \tag{18.9}$$

$$\int \sinh x \, dx = \cosh x + c, \tag{18.10}$$

$$\int \cosh x \, dx = \sinh x + c, \tag{18.11}$$

$$\int \operatorname{sech}^2 x \, dx = \tanh x + c, \tag{18.12}$$

$$\int \operatorname{cosech}^2 x \, dx = -\coth x + c, \tag{18.13}$$

$$\int \operatorname{sech} x \tanh x \, dx = -\operatorname{sech} x + c, \tag{18.14}$$

$$\int \operatorname{cosech} x \coth x \, dx = -\operatorname{cosech} x + c, \tag{18.15}$$

$$\int \frac{dx}{\sqrt{(k^2 - x^2)}} = \sin^{-1}\left(\frac{x}{k}\right) + c \quad (|x| < k), \tag{18.16}$$

$$\int \frac{dx}{x^2 + k^2} = \frac{1}{k}\tan^{-1}\left(\frac{x}{k}\right) + c, \tag{18.17}$$

$$\int \frac{dx}{\sqrt{(x^2 + k^2)}} = \sinh^{-1}\left(\frac{x}{k}\right) + c \quad (k > 0), \quad \text{or} \tag{18.18}$$
$$\ln[x + \sqrt{(x^2 + k^2)}] + c,$$

$$\int \frac{dx}{\sqrt{(x^2 - k^2)}} = \cosh^{-1}\left(\frac{x}{k}\right) + c, \quad \text{or} \tag{18.19}$$
$$\ln[x + \sqrt{(x^2 - k^2)}] + c, \quad (x > k > 0)$$

$$\int \frac{dx}{k^2 - x^2} = \frac{1}{k}\tanh^{-1}\left(\frac{x}{k}\right) + c = \frac{1}{2k}\ln\left(\frac{k + x}{k - x}\right) + c, \quad (|x| < k)$$
$$\int \frac{dx}{x^2 - k^2} = -\frac{1}{k}\coth^{-1}\left(\frac{x}{k}\right) + c = \frac{1}{2k}\ln\left(\frac{x - k}{x + k}\right) + c, \quad (|x| > k > 0). \tag{18.20}$$

18.1.1 Use of Standard Integrals As already pointed out in Section 16.4 the use of standard integrals is greatly extended by virtue of (16.1). Thus the results (18.1)–(18.3) may be generalised as follows:

$$\int [g(x)]^n g'(x)\, dx = \frac{1}{n + 1}[g(x)]^{n+1} + c \qquad (n \neq -1), \tag{18.1a}$$

$$\int \frac{g'(x)}{g(x)}\, dx = \ln|g(x)| + c, \tag{18.2a}$$

$$\int g'(x)\sin[g(x)]\, dx = -\cos[g(x)] + c, \tag{18.3a}$$

where g is any function. All the standard integrals may be similarly generalised.

In particular, putting $g(x) = 1 + x^2$, $g'(x) = 2x$ we see that

$$\int \frac{x\, dx}{(1 + x^2)^2} = -\frac{1}{2(1 + x^2)} + c \qquad \text{by using (18.1a) with } n = -2,$$

$$\int \frac{x\, dx}{1 + x^2} = \tfrac{1}{2}\ln(1 + x^2) + c \qquad \text{by using (18.2a),}$$

$$\int \frac{x\, dx}{\sqrt{(1 + x^2)}} = \sqrt{(1 + x^2)} + c \qquad \text{by using (18.1a) with } n = -\tfrac{1}{2}.$$

In practice it may be useful to make the substitution

$$z = g(x), \; dz = g'(x)\, dx.$$

For example if $I = \int e^x/(1 + e^{2x})\, dx$, we may suspect that I can be found by using (16.1) since the numerator is the derivative of e^x and the denominator is a function of e^x; on putting $u = e^x$, $du = e^x\,dx$, we find

$$I = \int \frac{du}{1+u^2} = \tan^{-1} u + c,$$

and so

$$I = \tan^{-1}(e^x) + c.$$

Standard integrals may be generalised less widely by using (16.2) and these results are especially useful. Thus

$$\int (ax+b)^n\,dx = \frac{1}{a(n+1)}(ax+b)^{n+1} + c \qquad (n \neq -1), \quad (18.1b)$$

$$\int \frac{1}{ax+b}\,dx = \frac{1}{a}\ln|ax+b| + c, \qquad (18.2b)$$

$$\int \sin(ax+b)\,dx = -\frac{1}{a}\cos(ax+b) + c, \qquad (18.3b)$$

and likewise for other standard integrals.

Exercises 18(a)

Use the table of standard integrals to integrate with respect to x the expressions in Nos. 1–18:

1. e^{3x+2}
2. e^{-2x}
3. $\sinh(2x+3)$
4. $\cosh(3-5x)$
5. $1/\sqrt{[(3x-1)^2+2]}$
6. $1/\sqrt{[(2-x)^2-3]}$
7. xe^{x^2}
8. $x^2e^{x^3}$
9. $\cos x\, e^{\sin x}$
10. $(2x+5)\,e^{(x^2+5x+4)}$
11. $x\sinh(x^2+4)$
12. $\cos x/\sqrt{(\sin^2 x+4)}$
13. $e^x/\sqrt{(e^{2x}+1)}$
14. $e^x/\sqrt{(e^{2x}-1)}$
15. $e^{2x}/\sqrt{(e^{4x}+1)}$
16. $\sin(\ln x)/x$
17. $e^{\sqrt{x}}/\sqrt{x}$
18. $(\ln x)^3/x$
19. Show that

(a) $\displaystyle\int_0^4 \frac{x}{(x^2+4)^2}\,dx = \tfrac{1}{10}$, (b) $\displaystyle\int_0^{\pi/6} \frac{\cos x}{\sqrt{(1+\sin x)}}\,dx = \sqrt{6}-2$,

(c) $\displaystyle\int_0^2 \frac{x^2}{(1+x^3)^{3/2}}\,dx = \tfrac{4}{9}$. (d) $\displaystyle\int_0^{1/\sqrt{3}} \frac{dx}{\sqrt{(2-3x^2)}} = \pi/4\sqrt{3}$.

18.1.2 The Integral $\int 1/(Ax^2+Bx+C)\,dx$, where A, B, C are Constants By putting a minus sign in front of the integral if necessary, we can arrange that $A > 0$. By 'completing the square', i.e. writing $Ax^2 + Bx + C = (\sqrt{A}x + B/2\sqrt{A})^2 + C - B^2/4A$, we see that the integral may be found by using (18.17) if $B^2 < 4AC$, or by using (18.20) if $B^2 > 4AC$, or by using (18.1) if $B^2 = 4AC$.

Example 1 Find

$$\int \frac{dx}{5x^2 - 2x + 3}.$$

We have

$$\int \frac{dx}{5x^2 - 2x + 3} = \int \frac{dx}{[\sqrt{5}x - (1/\sqrt{5})]^2 + 14/5}$$

$$= \frac{1}{\sqrt{14}} \tan^{-1}\left(\frac{5x - 1}{\sqrt{14}}\right) + c \qquad \text{by (18.17).}$$

Example 2 Find

$$\int \frac{dx}{5x^2 - 2x - 3}.$$

We have

$$\int \frac{dx}{5x^2 - 2x - 3} = \int \frac{dx}{[\sqrt{5}x - (1/\sqrt{5})]^2 - 16/5}$$

$$= \tfrac{1}{8} \ln \left\{\frac{[\sqrt{5}x - (1/\sqrt{5})] - 4/\sqrt{5}}{[\sqrt{5}x - (1/\sqrt{5})] + 4/\sqrt{5}}\right\} + c \qquad \text{by (18.20)}$$

$$= \tfrac{1}{8} \ln \left(\frac{5x - 5}{5x + 3}\right) + c.$$

Example 3 Find

$$\int \frac{dx}{4x^2 + 12x + 9}.$$

We have

$$\int \frac{dx}{4x^2 + 12x + 9} = \int \frac{dx}{(2x + 3)^2} = -\frac{1}{2}\frac{1}{(2x + 3)} + c \qquad \text{by (18.1).}$$

If $B^2 > 4AC$ then $Ax^2 + Bx + C$ factorises; if the factors have rational coefficients it is a simple matter to express the integrand in partial fractions and thereby find the integral.

Example 4 Do Example 2 by the method of partial fractions.

We have

$$\int \frac{dx}{5x^2 - 2x - 3} = \frac{dx}{(x - 1)(5x + 3)}$$

$$= \int \tfrac{1}{8}\left(\frac{1}{x - 1} - \frac{5}{5x + 3}\right) dx$$

$$= \tfrac{1}{8}[\ln(x - 1) - \ln(5x + 3)] + c = \tfrac{1}{8} \ln\left(\frac{x - 1}{5x + 3}\right) + c.$$

Note that the previous answer may be written as

$$\tfrac{1}{8}\ln\frac{5(x-1)}{5x+3}+c,$$

that is

$$\tfrac{1}{8}\ln\frac{x-1}{5x+3}+\tfrac{1}{8}\ln 5+c;$$

since $\frac{1}{8}\ln 5$ is constant and c is an *arbitrary* constant, the two answers are equivalent.

Exercises 18(b)

Integrate with respect to x the following expressions:

1. $1/(x+1)(x-2)$
2. $1/(x^2-5x+6)$
3. $1/(x^2+5x+6)$
4. $1/(x^2-2x-3)$
5. $1/(x^2+2x+3)$
6. $1/(x^2-x-6)$
7. $1/(x^2+x)$
8. $1/(x^2+8x+16)$
9. $1/(x^2+6x+25)$
10. $1/(x^2+6x-4)$
11. $1/(x^2+2x+26)$
12. $1/(4x^2+16x+25)$
13. $1/(3x^2+6x-2)$
14. $1/(1+6x-2x^2)$

18.1.3 The Integral $\int 1/\sqrt{(Ax^2+Bx+C)}\,dx$ where A, B, C are Constants Again the technique of 'completing the square' is used. If $B^2=4AC$ then Ax^2+Bx+C is a perfect square, say

$$Ax^2+Bx+C=(\alpha x+\beta)^2;$$

the integral then reduces to the trivial case

$$\int\frac{dx}{\alpha x+\beta}=\frac{1}{\alpha}\ln|\alpha x+\beta|+c.$$

Otherwise, if $A>0$ the integral is found by using (18.18) if $B^2<4AC$, or by using (18.19) if $B^2>4AC$; if $A<0$ the integral is unreal if $B^2<4AC$ and is found by using (18.16) if $B^2>4AC$.

Example 5 Find

$$\int\frac{dx}{\sqrt{(x^2+x+1)}}.$$

We have

$$\int\frac{dx}{\sqrt{(x^2+x+1)}}=\int\frac{dx}{\sqrt{[(x+\frac{1}{2})^2+\frac{3}{4}]}}=\sinh^{-1}\left(\frac{2x+1}{\sqrt{3}}\right)+c \quad \text{by (18.18).}$$

Example 6 Find

$$\int\frac{dx}{\sqrt{(2x^2+2x-1)}}.$$

We have

$$\int \frac{dx}{\sqrt{(2x^2+2x-1)}} = \int \frac{dx}{\sqrt{[(\sqrt{2}x+1/\sqrt{2})^2-3/2]}}$$
$$= \frac{1}{\sqrt{2}}\cosh^{-1}\left(\frac{2x+1}{\sqrt{3}}\right)+c \quad \text{by (18.19).}$$

Example 7 Find

$$\int \frac{dx}{\sqrt{(3+2x-2x^2)}}.$$

We have

$$\int \frac{dx}{\sqrt{(3+2x-2x^2)}} = \int \frac{dx}{\sqrt{[\frac{7}{2}-(\sqrt{2}x-1/\sqrt{2})^2}}$$
$$= \frac{1}{\sqrt{2}}\sin^{-1}\left(\frac{2x-1}{\sqrt{7}}\right)+c \quad \text{by (18.16)}$$

Exercises 18(c)

Integrate with respect to x the following expressions:

1. $1/\sqrt{(3-2x-x^2)}$
2. $1/\sqrt{(3+2x-x^2)}$
3. $1/\sqrt{(8-2x-x^2)}$
4. $1/\sqrt{(7-6x-x^2)}$
5. $1/\sqrt{(x^2+2x+26)}$
6. $1/\sqrt{(x^2-4x-21)}$
7. $1/\sqrt{[x(4-x)]}$.

18.2 Integrals of Elementary Functions

Integrals of the elementary functions x^n, $\sin x$, $\cos x$, e^x, $\sinh x$, $\cosh x$ are included in the list of standard integrals in Section 18.1.

We can find $\int \tan x \, dx$, $\int \tanh x \, dx$ by using (18.2) together with (16.1). Thus

$$\int \tan x \, dx = \int \frac{\sin x}{\cos x} dx = -\ln|\cos x| + c = \ln|\sec x| + c,$$
$$\int \tanh x \, dx = \int \frac{\sinh x}{\cosh x} dx = \ln|\cosh x| + c.$$

Since

$$\sec x = \frac{\sec x(\sec x + \tan x)}{\sec x + \tan x} = \frac{\sec^2 x + \sec x \tan x}{\sec x + \tan x}$$

it follows from (18.2a) that

$$\int \sec x \, dx = \ln|\sec x + \tan x| + c.$$

By a similar device we may show that

$$\int \operatorname{cosec} x \, dx = \ln\left|\frac{1}{\operatorname{cosec} x + \cot x}\right| + c = \ln\left|\tan\frac{x}{2}\right| + c.$$

On writing

$$\operatorname{sech} x = \frac{1}{\cosh^2 \tfrac{1}{2}x + \sinh^2 \tfrac{1}{2}x} = \frac{\operatorname{sech}^2 \tfrac{1}{2}x}{1 + \tanh^2 \tfrac{1}{2}x},$$

it follows from (18.17) and (16.1) that

$$\int \operatorname{sech} x \, dx = 2 \tan^{-1} (\tanh \tfrac{1}{2}x) + c.$$

Similarly

$$\operatorname{cosech} x = \frac{1}{2 \sinh \tfrac{1}{2}x \cosh \tfrac{1}{2}x} = \frac{\tfrac{1}{2} \operatorname{sech}^2 \tfrac{1}{2}x}{\tanh \tfrac{1}{2}x},$$

and so

$$\int \operatorname{cosech} x \, dx = \ln|\tanh \tfrac{1}{2}x| + c.$$

The logarithmic, inverse circular and inverse hyperbolic functions are dealt with by using integration by parts. In formula (16.4) for integration by parts, u is taken as the function to be integrated whereas we set $dv/dx = 1$. Thus

$$\begin{aligned}
\int \ln x \, dx &= x \ln x - \int x \frac{1}{x} dx &&= x \ln x - x + c,\\
\int \sin^{-1} x \, dx &= x \sin^{-1} x - \int \frac{x}{\sqrt{(1-x^2)}} dx &&= x \sin^{-1} x + \sqrt{(1-x^2)} + c,\\
\int \cos^{-1} x \, dx &= x \cos^{-1} x + \int \frac{x}{\sqrt{(1-x^2)}} dx &&= x \cos^{-1} x - \sqrt{(1-x^2)} + c,\\
\int \tan^{-1} x \, dx &= x \tan^{-1} x - \int \frac{x}{1+x^2} dx &&= x \tan^{-1} x - \tfrac{1}{2} \ln (1+x^2) + c,\\
\int \sinh^{-1} x \, dx &= x \sinh^{-1} x - \int \frac{x}{\sqrt{(1+x^2)}} dx &&= x \sinh^{-1} x - \sqrt{(1+x^2)} + c,\\
\int \cosh^{-1} x \, dx &= x \cosh^{-1} x - \int \frac{x}{\sqrt{(x^2-1)}} dx &&= x \cosh^{-1} x - \sqrt{(x^2-1)} + c,\\
\int \tanh^{-1} x \, dx &= x \tanh^{-1} x - \int \frac{x}{1-x^2} dx &&= x \tanh^{-1} x + \tfrac{1}{2} \ln|1-x^2| + c.
\end{aligned}$$

18.3 Integration of Rational Functions

Here we are concerned with an integrand of the form $P(x)/Q(x)$, where P and Q are polynomials.

If $Q(x)$ has only a single term, then we may divide out and integrate directly, e.g.

$$\int \frac{x^2 + x - 1}{x^3}\,dx = \int \left(\frac{1}{x} + \frac{1}{x^2} - \frac{1}{x^3}\right) dx = \ln x - \frac{1}{x} + \frac{1}{2x^2} + c.$$

Other cases which we have already dealt with are the following:

(a) $\displaystyle\int \frac{1}{ax+b}\,dx = \frac{1}{a}\ln|ax+b| + c$ (see (18.2b)),

(b) $\displaystyle\int \frac{1}{(ax+b)^r}\,dx = -\frac{1}{a(r-1)}\frac{1}{(ax+b)^{r-1}} + c$ $(r = 2, 3, \ldots)$ (see (18.1b)),

(c′) $\displaystyle\int \frac{1}{Ax^2 + Bx + C}\,dx$ (see Section 18.1.2),

(c″) $\displaystyle\int \frac{2Ax + B}{Ax^2 + Bx + C}\,dx = \ln|Ax^2 + Bx + C| + c$ (see (18.2a)).

By using a combination of (c′) and (c″) we can find any integral of the form

(c) $\displaystyle\int \frac{\alpha x + \beta}{Ax^2 + Bx + C}\,dx$ (α, β constants),

for we can easily determine constants λ, μ such that

$$\alpha x + \beta \equiv \lambda(2Ax + B) + \mu$$

and then

$$\int \frac{\alpha x + \beta}{Ax^2 + Bx + C}\,dx = \lambda \int \frac{2Ax + B}{Ax^2 + Bx + C}\,dx + \mu \int \frac{1}{Ax^2 + Bx + C}\,dx.$$

Example 8 Find

$$\int \frac{2x + 3}{3x^2 + 4x + 7}\,dx.$$

We require to write $2x + 3 \equiv \lambda\,(6x + 4) + \mu$, and clearly the solution is $\lambda = \frac{1}{3}$, $\mu = \frac{5}{3}$.

Hence

$$\int \frac{2x + 3}{3x^2 + 4x + 7}\,dx = \frac{1}{3}\int \frac{6x + 4}{3x^2 + 4x + 7}\,dx + \frac{5}{3}\int \frac{1}{3x^2 + 4x + 7}\,dx$$
$$= \tfrac{1}{3}\ln(3x^2 + 4x + 7) + \tfrac{5}{3}I,$$

where

$$I = \int \frac{1}{(\sqrt{3}x + 2/\sqrt{3})^2 + 17/3}\,dx = \frac{1}{\sqrt{3}}\sqrt{\left(\frac{3}{17}\right)}\tan^{-1}\left(\frac{3x + 2}{\sqrt{17}}\right) + c.$$

Thus

$$\int \frac{2x+3}{3x^2+4x+7}\,dx = \tfrac{1}{3}\ln(3x^2+4x+7) + \frac{5}{3\sqrt{17}}\tan^{-1}\left(\frac{3x+2}{\sqrt{17}}\right) + c.$$

For all other cases we shall use the method of partial fractions. If $P(x)/Q(x)$ is an improper fraction we first divide out to express it as the sum of a polynomial and a *proper* fraction, and then express the proper fraction in partial fractions corresponding to the factors of $Q(x)$. (See Section 4.3.1.) Corresponding to non-repeated linear, repeated linear, and non-repeated quadratic factors of $Q(x)$ we obtain integrals of type (a), (b) and (c) above. We omit discussion of the unusual case of a repeated quadratic factor.

Example 9 Find

$$\int \frac{2x^2-2x+3}{(2x-1)(x^2+1)}\,dx.$$

The integrand is proper. By writing

$$\frac{2x^2-2x+3}{(2x-1)(x^2+1)} \equiv \frac{2}{2x-1} - \frac{1}{x^2+1},$$

we get

$$\int \frac{2x^2-2x+3}{(2x-1)(x^2+1)}\,dx = \ln|2x-1| - \tan^{-1}x + c.$$

Example 10 Find

$$\int \frac{47+x-5x^2}{3(x+2)(x-3)^2}\,dx.$$

The integrand is proper. By writing

$$\frac{47+x-5x^2}{3(x+2)(x-3)^2} \equiv \frac{\frac{1}{3}}{x+2} - \frac{2}{x-3} + \frac{\frac{1}{3}}{(x-3)^2},$$

we get

$$\int \frac{47+x-5x^2}{3(x+2)(x-3)^2}\,dx = \tfrac{1}{3}\ln|x+2| - 2\ln|x-3| - \frac{1}{3(x-3)} + c.$$

Example 11 Find

$$\int \frac{6x-2x^2-8}{(x-1)(x^4-1)}\,dx.$$

The integrand is proper. By writing

$$\frac{6x - 2x^2 - 8}{(x-1)(x^4-1)} \equiv \frac{2}{x-1} - \frac{1}{(x-1)^2} - \frac{2}{x+1} - \frac{3}{x^2+1}$$

we get

$$\int \frac{6x - 2x^2 - 8}{(x-1)(x^4-1)}\,dx = 2\ln|x-1| + \frac{1}{x-1} - 2\ln|x+1| - 3\tan^{-1}x + c.$$

In Examples 9–11 the fractions are proper. The vital step that reduces integration of an improper fraction to integration of a proper fraction is most easily forgotten in simple examples like the following:

Example 12 Find $\int x/(x+1)\,dx$.

The integrand is improper and we write

$$\frac{x}{x+1} = 1 - \frac{1}{x+1}.$$

Hence

$$\int \frac{x}{x+1}\,dx = x - \ln|x+1| + c.$$

Example 13 Find

$$\int \frac{x^3}{x-1}\,dx.$$

The integrand is improper, and on dividing x^3 by $x - 1$ we get $x^2 + x + 1$ with remainder 1. Hence

$$\frac{x^3}{x-1} = x^2 + x + 1 + \frac{1}{x-1},$$

and so

$$\int \frac{x^3}{x-1}\,dx = \tfrac{1}{3}x^3 + \tfrac{1}{2}x^2 + x + \ln|x-1| + c.$$

Exercises 18(d)

Integrate with respect to x the expressions in Nos. 1–15:

1. $x/(2x^2 - x - 3)$
2. $(x-3)/(3x^2 + 2x - 5)$
3. $(2 - 3x)/(3x^2 - 4x + 1)$
4. $(x+3)/(x^3 - 6x^2 + 8x)$
5. $(4x^2 - x + 12)/x(x^2 + 4)$
6. $(3x^2 + 3x + 18)/(3-x)(3+x)^2$
7. $(x^2 + 1)/(x+2)^2$
8. $(x^2 + 3x - 2)/(x^4 - 2x^3 + x^2)$

9. $(4x-2)/(12-6x-x^2)$
10. $(x+1)/(1-x)(x^2+1)$
11. $(4x-3)/(2x-1)(4x^2+1)$
12. $(2x+1)/(x^2+2x+26)$
13. $(4x+1)/(4x^2+16x+25)$
14. $(5x+1)/(3x^2+6x-2)$
15. $(6x-1)/(1+6x-2x^2)$.

Verify the results in Nos. 16–22:

16. $\int_1^2 \frac{x}{(2x-1)^3}\,dx = \frac{5}{18}$
17. $\int_4^5 \frac{3x+7}{(x-3)(x+1)}\,dx = \ln\frac{40}{3}$
18. $\int_4^6 \frac{x-5}{(x-2)(x-3)}\,dx = \ln\frac{8}{9}$
19. $\int_0^1 \frac{x}{(x+1)(x^2+1)}\,dx = (\pi - 2\ln 2)/8$
20. $\int_0^1 \frac{dx}{x^2+2x\cos\alpha+1} = \frac{\alpha}{2\sin\alpha} \quad (0<\alpha<\pi/2)$
21. $\int_0^\infty \frac{x-1}{x^3+1}\,dx = 0$
22. $\int_1^\infty \frac{x^2}{(1+x^2)^2}\,dx = (\pi+2)/8$
23. Find the integral $\int 1/(1+e^x)\,dx$ by using the substitution $z = e^x$, or otherwise.

18.4 Integration of Circular Functions

Integrals of sin x, cos x, tan x, sec x, cosec x, cot x have been dealt with in Sections 18.1 and 18.2. We now deal with more complicated expressions involving these functions.

18.4.1 Odd Powers of sin x, cos x For $n = 1, 2, 3, \ldots$ we have

$$\int \sin^{2n+1} x\,dx = \int (1-\cos^2 x)^n \sin x\,dx$$
$$= -\int (1-u^2)^n\,du, \text{ where } u = \cos x,$$

$$\int \cos^{2n+1} x\,dx = \int (1-\sin^2 x)^n \cos x\,dx$$
$$= \int (1-u^2)^n\,du, \text{ where } u = \sin x.$$

Example 14 Find $\int \sin^5 x\,dx$, $\int \cos^3 x\,dx$.

$$\int \sin^5 x\,dx = \int (1-\cos^2 x)^2 \sin x\,dx = -\int (1-u^2)^2\,du$$
(where $u = \cos x$)

$$= \int (-1+2u^2-u^4)\,du$$
$$= -u + \tfrac{2}{3}u^3 - \tfrac{1}{5}u^5 + c$$
$$= -\cos x + \tfrac{2}{3}\cos^3 x - \tfrac{1}{5}\cos^5 x + c.$$

$$\int \cos^3 x \, dx = \int (1 - \sin^2 x) \cos x \, dx = \int (1 - u^2) \, du \quad (\text{where } u = \sin x)$$
$$= u - \tfrac{1}{3}u^3 + c$$
$$= \sin x - \tfrac{1}{3} \sin^3 x + c.$$

The above method can be used to find integrals of the form

$$\int \sin^m x \cos^n x \, dx$$

provided that *at least one* of the numbers m, n is an odd integer, e.g.

$$\int \sin^5 x \cos^3 x \, dx = \int u^5 (1 - u^2) \, du,$$

$$\int \sqrt{\sin x} \cos^3 x \, dx = \int u^{1/2} (1 - u^2) \, du$$

where in each case $u = \sin x$.

18.4.2 Even Powers of $\sin x$, $\cos x$ We rely on two basic identities, viz.,

$$\sin^2 x = \tfrac{1}{2}(1 - \cos 2x), \quad \cos^2 x = \tfrac{1}{2}(1 + \cos 2x).$$

It follows that

$$\int \sin^2 x \, dx = \tfrac{1}{2}(x - \tfrac{1}{2} \sin 2x) + c, \quad \int \cos^2 x \, dx = \tfrac{1}{2}(x + \tfrac{1}{2} \sin 2x) + c,$$

and for $n = 2, 3, \ldots$ the integrals $\int \sin^{2n} x \, dx$, $\int \cos^{2n} x \, dx$ can be found by using the same basic identities, as illustrated in the following examples.

Example 15 Find $\int \sin^4 x \, dx$.

We have

$$\sin^4 x = (\sin^2 x)^2 = \tfrac{1}{4}(1 - \cos 2x)^2 = \tfrac{1}{4}(1 - 2 \cos 2x + \cos^2 2x)$$
$$= \tfrac{1}{4}[1 - 2 \cos 2x + \tfrac{1}{2}(1 + \cos 4x)]$$
$$= \tfrac{1}{4}(\tfrac{3}{2} - 2 \cos 2x + \tfrac{1}{2} \cos 4x).$$

Hence

$$\int \sin^4 x \, dx = \tfrac{1}{4}(\tfrac{3}{2}x - \sin 2x + \tfrac{1}{8} \sin 4x) + c$$
$$= \tfrac{1}{32}(12x - 8 \sin 2x + \sin 4x) + c.$$

Example 16 Find $\int \cos^6 x \, dx$.

We have

$$\cos^6 x = (\cos^2 x)^3 = \tfrac{1}{8}(1 + \cos 2x)^3$$
$$= \tfrac{1}{8}(1 + 3 \cos 2x + 3 \cos^2 2x + \cos^3 2x)$$
$$= \tfrac{1}{8}[1 + 3 \cos 2x + \tfrac{3}{2}(1 + \cos 4x) + (1 - \sin^2 2x) \cos 2x]$$
$$= \tfrac{1}{8}[\tfrac{5}{2} + 3 \cos 2x + \tfrac{3}{2} \cos 4x + (1 - \sin^2 2x) \cos 2x]$$

Hence

$$\begin{aligned}\int \cos^6 x\,\mathrm{d}x &= \tfrac{1}{8}[\tfrac{5}{2}x + \tfrac{3}{2}\sin 2x + \tfrac{3}{8}\sin 4x + \tfrac{1}{2}(\sin 2x - \tfrac{1}{3}\sin^3 2x)] + c \\ &= \tfrac{1}{192}(60x + 48\sin 2x - 4\sin^3 2x + 9\sin 4x) + c.\end{aligned}$$

18.4.3 Powers of tan x, cot x We rely on the basic identities

$$\tan^2 x = \sec^2 x - 1, \quad \cot^2 x = \operatorname{cosec}^2 x - 1.$$

It follows that

$$\int \tan^2 x\,\mathrm{d}x = \tan x - x + c, \quad \int \cot^2 x\,\mathrm{d}x = -\cot x - x + c,$$

and in general

$$\begin{aligned}\int \tan^n x\,\mathrm{d}x &= \int \tan^{n-2} x(\sec^2 x - 1)\,\mathrm{d}x \\ &= \frac{1}{n-1}\tan^{n-1} x - \int \tan^{n-2} x\,\mathrm{d}x \qquad (n \geqslant 2), \\ \int \cot^n x\,\mathrm{d}x &= \int \cot^{n-2} x(\operatorname{cosec}^2 x - 1)\,\mathrm{d}x \\ &= -\frac{1}{n-1}\cot^{n-1} x - \int \cot^{n-2} x\,\mathrm{d}x \qquad (n \geqslant 2),\end{aligned}$$

that is we can always reduce to the case $n = 1$ or $n = 0$ (see Section 18.7).

Example 17 Find $\int \tan^3 x\,\mathrm{d}x$.

$$\begin{aligned}\int \tan^3 x\,\mathrm{d}x = \int \tan x(\sec^2 x - 1)\,\mathrm{d}x &= \tfrac{1}{2}\tan^2 x - \int \tan x\,\mathrm{d}x \\ &= \tfrac{1}{2}\tan^2 x - \ln|\sec x| + c.\end{aligned}$$

Example 18 Find $\int \tan^4 x\,\mathrm{d}x$.

$$\begin{aligned}\int \tan^4 x\,\mathrm{d}x = \int \tan^2 x(\sec^2 x - 1)\,\mathrm{d}x &= \tfrac{1}{3}\tan^3 x - \int \tan^2 x\,\mathrm{d}x \\ &= \tfrac{1}{3}\tan^3 x - \tan x + x + c.\end{aligned}$$

Example 19 Find $\int \tan^5 x\,\mathrm{d}x$.

$$\begin{aligned}\int \tan^5 x\,\mathrm{d}x = \int \tan^3 x(\sec^2 x - 1)\,\mathrm{d}x &= \tfrac{1}{4}\tan^4 x - \int \tan^3 x\,\mathrm{d}x \\ &= \tfrac{1}{4}\tan^4 x - \tfrac{1}{2}\tan^2 x + \ln|\sec x| + c\end{aligned}$$

(see Example 17).

18.4.4 Powers of sec x, cosec x Odd powers may be dealt with by using the method of Section 18.4.5. Even powers may be dealt with more easily by using the identities

$$\sec^2 x = 1 + \tan^2 x, \quad \operatorname{cosec}^2 x = 1 + \cot^2 x$$

together with the standard results

$$\int \sec^2 x \, dx = \tan x + c, \quad \int \operatorname{cosec}^2 x \, dx = -\cot x + c.$$

Thus for $n = 2, 3, \ldots$ we write

$$\int \sec^{2n} x \, dx = \int \sec^{2(n-1)} x \sec^2 x \, dx$$

$$= \int (1 + \tan^2 x)^{n-1} \sec^2 x \, dx = \int (1 + u^2)^{n-1} \, du,$$

where $u = \tan x$, and

$$\int \operatorname{cosec}^{2n} x \, dx = \int \operatorname{cosec}^{2(n-1)} x \operatorname{cosec}^2 x \, dx$$

$$= \int (1 + \cot^2 x)^{n-1} \operatorname{cosec}^2 x \, dx = -\int (1 + u^2)^{n-1} \, du,$$

where $u = \cot x$.

Example 20 Find $\int \sec^6 x \, dx$.

$$\begin{aligned}\int \sec^6 x \, dx = \int \sec^4 x \sec^2 x \, dx &= \int (1 + \tan^2 x)^2 \sec^2 x \, dx \\ &= \int (1 + u^2)^2 \, du \qquad \text{(where } u = \tan x\text{)} \\ &= \int (1 + 2u^2 + u^4) \, du \\ &= u + \tfrac{2}{3}u^3 + \tfrac{1}{5}u^5 + c \\ &= \tan x + \tfrac{2}{3}\tan^3 x + \tfrac{1}{5}\tan^5 x + c.\end{aligned}$$

18.4.5 Rational Functions of sin x, cos x The substitution $t = \tan \frac{1}{2}x$ transforms these integrals into integrals of rational functions of t which may be found by the methods of Section 18.3. Note that

$$\sin x = \frac{2t}{1 + t^2}, \quad \cos x = \frac{1 - t^2}{1 + t^2}, \quad dx = \frac{2\,dt}{1 + t^2}.$$

Example 21 Find $\int \sec x \, dx$ using the substitution $t = \tan \frac{1}{2}x$.

We have

$$\int \sec x\, dx = \int \frac{1+t^2}{1-t^2}\frac{2\,dt}{1+t^2} = \int \frac{2}{1-t^2}\,dt$$
$$= \int \left(\frac{1}{1-t} + \frac{1}{1+t}\right) dt$$
$$= \ln \left|\frac{1+t}{1-t}\right| + c = \ln \left|\frac{1+\tan \frac{1}{2}x}{1-\tan \frac{1}{2}x}\right| + c.$$

Note that

$$\frac{1+\tan \frac{1}{2}x}{1-\tan \frac{1}{2}x} = \sec x + \tan x.$$

Example 22 Evaluate

$$\int_0^{\pi/2} \frac{dx}{2+\cos x}.$$

$$\int_0^{\pi/2} \frac{dx}{2+\cos x} = \int_0^1 \frac{1}{2+(1-t^2)/(1+t^2)}\frac{2\,dt}{1+t^2}$$

$$= \int_0^1 \frac{2\,dt}{t^2+3} = \frac{2}{\sqrt{3}}\left[\tan^{-1}\left(\frac{t}{\sqrt{3}}\right)\right]_0^1 = \frac{\pi}{3\sqrt{3}}.$$

If only even powers of $\sin x$, $\cos x$, occur then the substitution $t = \tan x$ is simpler. We then have

$$\sin^2 x = \frac{t^2}{1+t^2}, \quad \cos^2 x = \frac{1}{1+t^2}, \quad dx = \frac{dt}{1+t^2} \qquad \text{(see Fig. 18.1).}$$

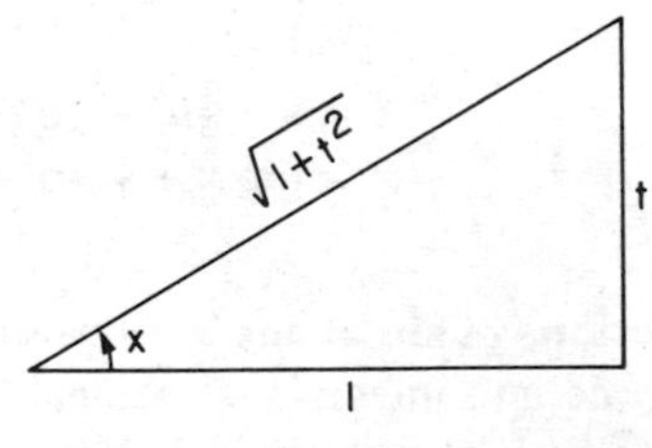

FIG. 18.1

Example 23 Evaluate

$$\int_0^{\pi/4} \frac{dx}{9\cos^2 x - \sin^2 x}.$$

We have

$$\int_0^{\pi/4} \frac{dx}{9\cos^2 x - \sin^2 x} = \int_0^1 \frac{dt/(1+t^2)}{9/(1+t^2) - t^2/(1+t^2)} \quad \text{where } t = \tan x$$

$$= \int_0^1 \frac{dt}{9-t^2} = \tfrac{1}{6}\left[\ln\left(\frac{3+t}{3-t}\right)\right]_0^1 = \tfrac{1}{6}\ln 2.$$

18.4.6 Products of Type sin ax cos bx, sin ax sin bx, cos ax cos bx These may be integrated by using the identities

$$\sin ax \cos bx = \tfrac{1}{2}[\sin(a+b)x + \sin(a-b)x],$$
$$\sin ax \sin bx = -\tfrac{1}{2}[\cos(a+b)x - \cos(a-b)x],$$
$$\cos ax \cos bx = \tfrac{1}{2}[\cos(a+b)x + \cos(a-b)x].$$

For example,

$$\int \sin 6x \cos 2x \, dx = \tfrac{1}{2}\int (\sin 8x + \sin 4x)\, dx = -\tfrac{1}{16}\cos 8x - \tfrac{1}{8}\cos 4x + c.$$

18.4.7 Products of the Type x^n sin ax, x^n cos ax (n a positive integer), e^{ax} sin bx, e^{ax} cos bx The first two integrals may be found by using the method of integration by parts n times. Thus to find $\int x^2 \sin x \, dx$ we use integration by parts twice as follows:

$$\int x^2 \sin x \, dx = -x^2 \cos x + 2\int x \cos x \, dx$$
$$= -x^2 \cos x + 2[x \sin x - \int \sin x \, dx]$$
$$= -x^2 \cos x + 2x \sin x + 2\cos x + c.$$

An *equation* for the other integrals may be found by using integration by parts twice. Let

$$I = \int e^{ax} \sin bx \, dx = \frac{1}{a} e^{ax} \sin bx - \frac{b}{a}\int e^{ax} \cos bx \, dx$$
$$= \frac{1}{a} e^{ax} \sin bx - \frac{b}{a}\left[\frac{1}{a} e^{ax} \cos bx + \frac{b}{a}\int e^{ax} \sin bx \, dx\right]$$
$$= \frac{1}{a} e^{ax} \sin bx - \frac{b}{a^2} e^{ax} \cos bx - \frac{b^2}{a^2} I.$$

Hence

$$\left(1 + \frac{b^2}{a^2}\right) I = \frac{1}{a^2} e^{ax}(a \sin bx - b \cos bx) + c,$$

and so

$$I = \frac{e^{ax}(a \sin bx - b \cos bx)}{a^2 + b^2} + c.$$

Similarly

$$\int e^{ax} \cos bx \, dx = \frac{e^{ax}(a \cos bx + b \sin bx)}{a^2 + b^2} + c.$$

18.5 Integration of Hyperbolic Functions

We may use methods similar to those employed for the corresponding circular functions, but sometimes it is advantageous to express hyperbolic functions in exponential form.

Example 24 Evaluate

$$\int_0^{\ln 2} \frac{dx}{5 \cosh x - 3 \sinh x}.$$

We have

$$5 \cosh x - 3 \sinh x = \tfrac{5}{2}(e^x + e^{-x}) - \tfrac{3}{2}(e^x - e^{-x}) = e^x + 4e^{-x}$$

and so

$$\begin{aligned}\int_0^{\ln 2} \frac{dx}{5 \cosh x - 3 \sinh x} &= \int_0^{\ln 2} \frac{e^x}{e^{2x} + 4} dx \\ &= \int_1^2 \frac{du}{u^2 + 4} \quad \text{(where } u = e^x\text{)} \\ &= [\tfrac{1}{2} \tan^{-1} (\tfrac{1}{2} u)]_1^2 \\ &= \tfrac{1}{2}(\tan^{-1} 1 - \tan^{-1} \tfrac{1}{2}) = \tfrac{1}{2} \tan^{-1} \tfrac{1}{3}.\end{aligned}$$

Exercises 18(e)

Find the following integrals:

1. $\int_0^{\pi/2} \sin^3 x \, dx$
2. $\int_0^{\pi/2} \sin^4 x \, dx$
3. $\int \sin^4 2x \cos 2x \, dx$
4. $\int \frac{\cos x - \sin x}{\cos x + \sin x} dx$
5. $\int_0^{\pi/2} \sin^3 x \cos^3 x \, dx$
6. $\int \sin^4 x \cos^5 x \, dx$
7. $\int_0^{\pi/6} \frac{\cos x}{\cos 2x} dx$
8. $\int_0^{\pi/4} \sin 2x \cos 3x \, dx$
9. $\int \operatorname{cosec} 2x \, dx$
10. $\int \sec 3x \, dx$
11. $\int \tan^2 \tfrac{1}{2}x \, dx$
12. $\int \sin^3 x \sec^4 x \, dx$
13. $\int \sec^4 x \, dx$
14. $\int \cos^2 x \operatorname{cosec}^4 x \, dx$
15. $\int \sec^4 x \operatorname{cosec}^2 x \, dx$
16. $\int \frac{1}{1 + \cos x} dx$
17. $\int \frac{1}{1 + \sin x} dx$
18. $\int \sin 4x \sin 6x \, dx$

19. $\int \cos 3x \cos 5x \, dx$

20. $\int \frac{1}{12 + 13 \cos x} dx$

21. $\int \frac{1}{11 + 61 \sin x} dx$

22. $\int \frac{1}{16 \cos^2 x + 9 \sin^2 x} dx$

23. $\int \frac{2 \sin x + 9 \cos x}{\sin x + 2 \cos x} dx$

24. $\int \frac{1 + \cos x - 3 \sin x}{2 + 2 \cos x - \sin x} dx$

25. $\int_0^{\pi} x \cos \frac{1}{2}x \, dx$

26. $\int_0^{\pi/2} x \sin x \, dx$

27. $\int x^2 \cos 3x \, dx$

28. $\int e^{2x} \sin 3x \, dx$

29. $\int_0^{\pi} x \cos^2 x \, dx$

30. $\int_0^{\infty} e^{-x} \cos x \, dx$

31. $\int_0^{\infty} e^{-x} \sin x \, dx$

32. $\int_0^{1/2} e^{2x} \sin \pi x \, dx$

33. $\int x \sec^2 x \, dx$

34. $\int \sinh^2 x \, dx$

35. $\int \tanh^3 x \, dx$

36. $\int_0^{\ln 2} \frac{dx}{\sinh x + 5 \cosh x}$

37. $\int x \cosh x \, dx$

38. $\int \frac{dx}{\cosh x}$

39. $\int_0^{\tan^{-1}(4/3)} \frac{dx}{3 \cos x + 4 \sin x}$

40. $\int_0^{\pi/4} \frac{3 \sin^2 x + \cos^2 x}{3 \cos^2 x + \sin^2 x} dx.$

18.6 Integration of Irrational Functions

Many irrational functions can be integrated by the methods already given. For example,

$$\int \frac{x^{1/2} + x^{2/3} + x^{3/4}}{x^{1/4}} dx = (x^{1/4} + x^{5/12} + x^{1/2}) \, dx = \tfrac{4}{5}x^{5/4} + \tfrac{12}{17}x^{17/12} + \tfrac{2}{3}x^{3/2} + c,$$

$$\int (x^3 + x^2 + 1)^{1/3} (3x^2 + 2x) \, dx = \tfrac{3}{4}(x^3 + x^2 + 1)^{4/3} + c,$$

$$\int \frac{x}{\sqrt{(x^2 + 1)}} dx = \sqrt{(x^2 + 1)} + c,$$

$$\int \frac{1}{\sqrt{(1 - x^2)}} dx = \sin^{-1} x + c.$$

However, these results do not suggest any general method and unfortunately no very comprehensive rules can be given.

18.6.1 Integrand Containing $\sqrt{(ax + b)}$ An algebraic expression containing only a single irrational expression of the form $\sqrt{(ax + b)}$ can be integrated by means of the substitution $z^2 = ax + b$.

Example 25 Find $\int (4x + 3)\sqrt{(2x + 1)} \, dx$.

Put $z^2 = 2x + 1$, $2z \, dz = 2dx$. Then $x = \frac{1}{2}(z^2 - 1)$ and the integral becomes

$$\int (2z^2 + 1)z^2 \, dz = \int (2z^4 + z^2) \, dz$$
$$= \tfrac{2}{5}z^5 + \tfrac{1}{3}z^3 + c$$
$$= \tfrac{2}{5}(2x + 1)^{5/2} + \tfrac{1}{3}(2x + 1)^{3/2} + c.$$

Example 26 Find $\int (x - 2)/[x\sqrt{(x + 1)}] \, dx$.

Put $z^2 = x + 1$, $2z \, dz = dx$. The integral becomes

$$2\int \frac{z^2 - 3}{z^2 - 1} \, dz = 2\int \left(1 - \frac{2}{z^2 - 1}\right) dz$$
$$= 2\left(z - \ln \left|\frac{z - 1}{z + 1}\right|\right) + c$$
$$= 2\left(\sqrt{(x + 1)} - \ln \left|\frac{\sqrt{(x + 1)} - 1}{\sqrt{(x + 1)} + 1}\right|\right) + c.$$

18.6.2 Integrand Containing $\sqrt{(Ax^2 + Bx + C)}$ A special case of this has already been considered in Section 18.1.3. In general, by completing the square we may write $\sqrt{(Ax^2 + Bx + C)}$ in one of the forms

(a) $\sqrt{[(ax + b)^2 + k^2]}$, (b) $\sqrt{[(ax + b)^2 - k^2]}$, (c) $\sqrt{[k^2 - (ax + b)^2]}$

Consider first the expressions

(a′) $\sqrt{(x^2 + k^2)}$, (b′) $\sqrt{(x^2 - k^2)}$, (c′) $\sqrt{(k^2 - x^2)}$;

when one of these occurs in the integrand we can remove the square root by means of the respective substitutions

(a″) $x = k \tan \theta$, (b″) $x = k \sec \theta$, (c″) $x = k \sin \theta$,

and by assuming that $-\pi/2 \leqslant \theta \leqslant \pi/2$ in (a″) and (c″), $0 \leqslant \theta \leqslant \pi$ in (b″), we obtain

(a‴) $\theta = \tan^{-1} (x/k)$, (b‴) $\theta = \sec^{-1} (x/k)$, (c‴) $\theta = \sin^{-1} (x/k)$.

Example 27 Find $\int \sqrt{(5 - x^2)} \, dx$.

Put $x = \sqrt{5} \sin \theta$, $dx = \sqrt{5} \cos \theta \, d\theta$, and the integral becomes

$$5\int \sqrt{(1 - \sin^2 \theta)} \cos \theta \, d\theta = 5\int \cos^2 \theta \, d\theta = \tfrac{5}{2}(\theta + \tfrac{1}{2} \sin 2\theta) + c.$$

Now

$$\tfrac{1}{2} \sin 2\theta = \sin \theta \cos \theta = \frac{x}{\sqrt{5}} \frac{\sqrt{(5 - x^2)}}{\sqrt{5}} = \tfrac{1}{5}x\sqrt{(5 - x^2)} \qquad \text{(see Fig. 18.2).}$$

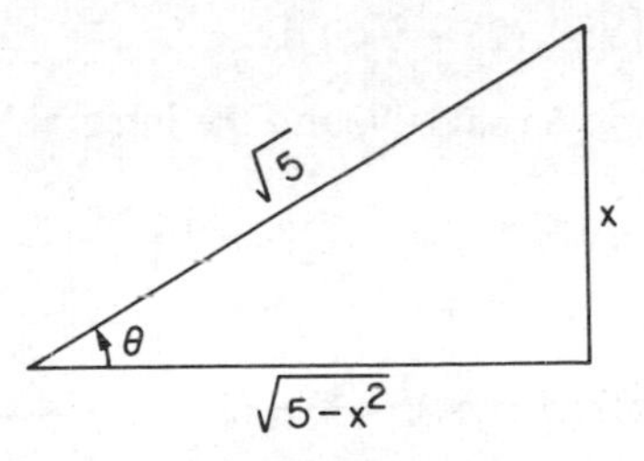

FIG. 18.2

Hence

$$\int \sqrt{(5-x^2)}\,dx = \tfrac{5}{2}\sin^{-1}\left(\frac{x}{\sqrt{5}}\right) + \tfrac{1}{2}x\sqrt{(5-x^2)} + c.$$

Example 28 Evaluate

$$\int_1^3 \frac{dx}{(x^2+3)^{3/2}}.$$

Put $x = \sqrt{3}\tan\theta$, $dx = \sqrt{3}\sec^2\theta\, d\theta$, and the integral becomes

$$\int_{\pi/6}^{\pi/3} \frac{\sqrt{3}\sec^2\theta\, d\theta}{3^{3/2}(\tan^2\theta+1)^{3/2}} = \tfrac{1}{3}\int_{\pi/6}^{\pi/3}\cos\theta\, d\theta = \tfrac{1}{3}\Big[\sin\theta\Big]_{\pi/6}^{\pi/3} = \tfrac{1}{6}(\sqrt{3}-1).$$

Example 29 Find $\int 1/(x^2-5)^{5/2}\, dx$.

Put $x = \sqrt{5}\sec\theta$, $dx = \sqrt{5}\sec\theta\tan\theta\, d\theta$, and the integral becomes

$$\begin{aligned}
\frac{1}{25}\int \frac{1}{\tan^5\theta}\sec\theta\tan\theta\, d\theta &= \frac{1}{25}\int\frac{\cos^3\theta}{\sin^4\theta}\, d\theta \\
&= \frac{1}{25}\int\frac{1-z^2}{z^4}\, dz \\
&\qquad (\text{where } z = \sin\theta,\ dz = \cos\theta\, d\theta) \\
&= \frac{1}{25}\left(-\frac{1}{3z^3}+\frac{1}{z}\right) + c \\
&= \frac{1}{25}\left(-\frac{x^3}{3(x^2-5)^{3/2}} + \frac{x}{(x^2-5)^{1/2}}\right) + c \\
&= \frac{2x^3-15x}{75(x^2-5)^{3/2}} + c.
\end{aligned}$$

Alternative substitutions that may be tried are $x = k\sinh z$ for (a′), $x = k\cosh z$ for (b′), $x = k\tanh z$ for (c′).

Consider next the expressions (a), (b), (c) and the corresponding substitutions

(a) $ax + b = k\tan\theta$, (b) $ax + b = k\sec\theta$, (c) $ax + b = k\sin\theta$.

Example 30 Find $\int x^2/\sqrt{(25-9x^2)}\,dx$.

Put $3x = 5\sin\theta$, $3\,dx = 5\cos\theta\,d\theta$, and the integral becomes

$$\tfrac{25}{27}\int \sin^2\theta\,d\theta = \tfrac{25}{54}(\theta - \tfrac{1}{2}\sin 2\theta) + c = \tfrac{25}{54}\sin^{-1}\left(\frac{3x}{5}\right) - \tfrac{1}{18}x\sqrt{(25-9x^2)} + c.$$

Example 31 Find $\int 1/\sqrt{(x^2+x-6)}\,dx$.

Since $x^2+x-6 = (x+\tfrac{1}{2})^2 - \tfrac{25}{4}$, we put $x+\tfrac{1}{2} = \tfrac{5}{2}\sec\theta$, $dx = \tfrac{5}{2}\sec\theta\tan\theta\,d\theta$. The integral becomes

$$\begin{aligned}\int \sec\theta\,d\theta &= \ln|\sec\theta + \tan\theta| + c\\ &= \ln\left|\frac{2x+1}{5} + \frac{2\sqrt{(x^2+x-6)}}{5}\right| + c\\ &= \ln|2x+1+2\sqrt{(x^2+x-6)}| + c,\end{aligned}$$

on omitting the constant ln 5 from the previous answer.

Example 32 Find $\int\sqrt{(2+2x-3x^2)}\,dx$.

Since $2+2x-3x^2 = \tfrac{7}{3} - (\sqrt{3}x - 1/\sqrt{3})^2$, we put

$$\sqrt{3}x - \frac{1}{\sqrt{3}} = \sqrt{\tfrac{7}{3}}\sin\theta, \quad \sqrt{3}\,dx = \sqrt{\tfrac{7}{3}}\cos\theta\,d\theta.$$

The integral becomes

$$\begin{aligned}\frac{7}{3\sqrt{3}}\int\cos^2\theta\,d\theta &= \frac{7\sqrt{3}}{18}(\theta + \tfrac{1}{2}\sin 2\theta) + c\\ &= \frac{7\sqrt{3}}{18}\sin^{-1}\left(\frac{3x-1}{\sqrt{7}}\right) + \tfrac{1}{6}(3x-1)\sqrt{(2+2x-3x^2)} + c.\end{aligned}$$

Example 33 Find $\int\sqrt{[(x+1)/(x+2)]}\,dx$.

On rationalising the numerator we get

$$\begin{aligned}\int\sqrt{\frac{x+1}{x+2}}\,dx &= \int\frac{x+1}{\sqrt{(x^2+3x+2)}}\,dx\\ &= \int\frac{x+1}{\sqrt{(x+\tfrac{3}{2})^2 - \tfrac{1}{4}}}\,dx\\ &= \tfrac{1}{2}\int(\sec^2\theta - \sec\theta)\,d\theta \qquad (\text{where } x+\tfrac{3}{2} = \tfrac{1}{2}\sec\theta)\\ &= \tfrac{1}{2}\tan\theta - \tfrac{1}{2}\ln|\sec\theta+\tan\theta| + c\\ &= \sqrt{(x^2+3x+2)} - \tfrac{1}{2}\ln|2x+3+2\sqrt{(x^2+3x+2)}| + c.\end{aligned}$$

Exercises 18(f)

Find the following integrals:

1. $\int \frac{(x+1)^2}{\sqrt{x}}\,dx$ 2. $\int x\sqrt{(x+1)}\,dx$ 3. $\int_0^1 \frac{x}{\sqrt{(1+x)}}\,dx$

4. $\int_0^1 \frac{1}{(3-2x)^{3/2}}\,dx$ 5. $\int x(2x^2+5)^{1/3}\,dx$ 6. $\int x^2\sqrt{(x^3+1)}\,dx$

7. $\int_0^2 \frac{x^2}{(1+x^3)^{3/2}}\,dx$ 8. $\int \sqrt{(9-2x)}\,dx$ 9. $\int \frac{1}{\sqrt{(9-4x^2)}}\,dx$

10. $\int \frac{x+3}{\sqrt{(x^2+2x+10)}}\,dx$ 11. $\int \frac{x+3}{\sqrt{(x^2-7x+12)}}\,dx$ 12. $\int \frac{x+1}{(x^2+1)^{3/2}}\,dx$

13. $\int_1^2 \frac{2x-1}{\sqrt{(4x^2+4x+2)}}\,dx$ 14. $\int_3^8 \frac{2x+3}{\sqrt{(16+6x-x^2)}}\,dx$ 15. $\int_0^5 \frac{x}{\sqrt{(x+4)}-2}\,dx$

16. $\int_1^{5/3} \frac{2x+1}{\sqrt{(x^2-1)}}\,dx$ 17. $\int \sqrt{\left(\frac{x}{1-x}\right)}\,dx$ 18. $\int_0^1 \sqrt{\left(\frac{x}{2-x}\right)}\,dx$

19. $\int_2^4 \frac{x}{\sqrt{(6x-8-x^2)}}\,dx$ 20. $\int \sqrt{\left(\frac{x+1}{x-1}\right)}\,dx.$

18.7 Reduction Formulae

In Section 18.4.3 the identity $\tan^2 x = \sec^2 x - 1$ was used to show that

$$\int \tan^n x\,dx = \frac{1}{n-1}\tan^{n-1} x - \int \tan^{n-2} x\,dx \qquad (n \geqslant 2).$$

On writing $I_n = \int \tan^n x\,dx$ the above formula becomes

$$I_n = \frac{1}{n-1}\tan^{n-1} x - I_{n-2}. \tag{1}$$

Note that we are dealing with a *set* of integrals. The formula (1) expresses one such integral in terms of another *with a lower value of n*; such a formula is called a *reduction formula.* Successive applications of formula (1) reduce the problem of finding I_n to that of finding I_1 if n is odd, or I_0 if n is even. Thus an alternative approach to Example 19 (Section 18.4.3) would be as follows:

$$\begin{aligned}\int \tan^5 x\,dx = I_5 &= \tfrac{1}{4}\tan^4 x - I_3 && \text{by (1) with } n = 5\\ &= \tfrac{1}{4}\tan^4 x - \tfrac{1}{2}\tan^2 x + I_1 && \text{by (1) with } n = 3\\ &= \tfrac{1}{4}\tan^4 x - \tfrac{1}{2}\tan^2 x + \ln|\sec x| + c.\end{aligned}$$

The reduction formula (1) was obtained simply by using an identity. Integration by parts is particularly useful for obtaining reduction formulae, e.g. to obtain a reduction formula for $\int x^n e^x\,dx$, write $I_n = \int x^n e^x\,dx$

and use integration by parts to obtain

$$I_n = x^n e^x - \int n x^{n-1} e^x \, dx = x^n e^x - n I_{n-1}.$$

Most reduction formulae involve the use of integration by parts followed by the use of an identity.

18.7.1 Reduction Formulae for $\int \sin^n x \, dx$ and $\int \cos^n x \, dx$ Write

$$S_n = \int \sin^n x \, dx = \int \sin^{n-1} x \frac{d}{dx}(-\cos x) \, dx.$$

Using integration by parts we now get

$$S_n = -\sin^{n-1} x \cos x + (n-1) \int \sin^{n-2} x \cos^2 x \, dx.$$

Replacing $\cos^2 x$ by $1 - \sin^2 x$ in the integrand, we have

$$S_n = -\sin^{n-1} x \cos x + (n-1)(S_{n-2} - S_n).$$

Finally, on solving for S_n we get

$$S_n = -\frac{\sin^{n-1} x \cos x}{n} + \frac{n-1}{n} S_{n-2}.$$

Similarly, if $C_n = \int \cos^n x \, dx$,

$$C_n = \frac{\cos^{n-1} x \sin x}{n} + \frac{n-1}{n} C_{n-2}.$$

Thus an alternative approach to Example 15 (Section 18.4.2) would be as follows:

$$\begin{aligned}\int \sin^4 x \, dx = S_4 &= -\tfrac{1}{4} \sin^3 x \cos x + \tfrac{3}{4} S_2 \\ &= -\tfrac{1}{4} \sin^3 x \cos x + \tfrac{3}{4}(-\tfrac{1}{2} \sin x \cos x + \tfrac{1}{2} S_0) \\ &= -\tfrac{1}{4} \sin^3 x \cos x - \tfrac{3}{8} \sin x \cos x + \tfrac{3}{8} x + c.\end{aligned}$$

18.7.2 Reduction Formula for $\int \sin^m x \cos^n x \, dx$ Here m and n are both non-negative integers. On writing

$$I_{m,n} = \int \sin^m x \cos^n x \, dx$$

we have

$$I_{m,n} = \int \sin^m x \cos^{n-1} x \frac{d}{dx}(\sin x)\, dx$$

$$= \frac{1}{m+1}\int \frac{d}{dx}(\sin^{m+1} x) \cos^{n-1} x\, dx$$

$$= \frac{1}{m+1}\sin^{m+1} x \cos^{n-1} x + \frac{n-1}{m+1}\int \sin^{m+2} x \cos^{n-2} x\, dx$$

$$= \frac{1}{m+1}\sin^{m+1} x \cos^{n-1} x + \frac{n-1}{m+1}\int \sin^{m} x\,(1 - \cos^2 x) \cos^{n-2} x\, dx$$

$$= \frac{1}{m+1}\sin^{m+1} x \cos^{n-1} x + \frac{n-1}{m+1}(I_{m,n-2} - I_{m,n})$$

Solving for $I_{m,n}$ we now get

$$I_{m,n} = \frac{\sin^{m+1} x \cos^{n-1} x}{m+n} + \frac{n-1}{m+n} I_{m,n-2}.$$

By successive applications of this formula the problem of finding $I_{m,n}$ is reduced to that of finding $I_{m,1}$ if n is odd, or $I_{m,0}$ if n is even. Now

$$I_{m,1} = \int \sin^m x \cos x\, dx = \frac{1}{m+1}\sin^{m+1} x + c,$$

while

$$I_{m,0} = \int \sin^m x\, dx,$$

and this can be found by using the appropriate reduction formula from Section 18.7.1.

18.7.3 Wallis's Formulae If the integrals considered in Sections 18.7.1 and 18.7.2 are taken as definite integrals with limits 0 and $\pi/2$, we have

$$S_n = \int_0^{\pi/2} \sin^n x\, dx = \frac{n-1}{n} S_{n-2} \qquad \text{if } n \geqslant 2, \tag{1}$$

$$C_n = \int_0^{\pi/2} \cos^n x\, dx = \frac{n-1}{n} C_{n-2} \qquad \text{if } n \geqslant 2,$$

and

$$I_{m,n} = \int_0^{\pi/2} \sin^m x \cos^n x\, dx = \frac{n-1}{m+n} I_{m,n-2} \qquad \text{if } n \geqslant 2. \tag{2}$$

By repeated application of formula (1) we get

$$S_n = \frac{n-1}{n} S_{n-2} = \frac{n-1}{n}\frac{n-3}{n-2} S_{n-4} = \cdots.$$

If n is even, S_n depends ultimately on S_0, i.e. on $\int_0^{\pi/2} 1 \, dx$, which equals $\pi/2$. If n is odd, S_n depends ultimately on S_1, i.e. on $\int_0^{\pi/2} \sin x \, dx$, which equals 1. In the same way, C_n may be evaluated. The results may be written in the form

$$\left.\begin{aligned} S_n = C_n &= \frac{(n-1)(n-3)\cdots 5.3.1}{n(n-2)\cdots 6.4.2}\frac{\pi}{2} && \text{when } n \text{ is even} \\ &= \frac{(n-1)(n-3)\cdots 6.4.2}{n(n-2)\cdots 7.5.3.1} && \text{when } n \text{ is odd.} \end{aligned}\right\} \quad (18.21)$$

By repeated application of formula (2) we obtain

$$I_{m,n} = \frac{(m-1)(m-3)\cdots(n-1)(n-3)\cdots}{(m+n)(m+n-2)(m+n-4)\cdots} p. \qquad (18.22)$$

where each product is continued until 2 or 1 is reached, $p = \pi/2$ when m and n are both even and $p = 1$ in all other cases.

Formulae (18.21) and (18.22) are known as Wallis's formulae.

Example 34

$$\int_0^{\pi/2} \sin^8 x \, dx = \frac{7.5.3.1}{8.6.4.2}\frac{\pi}{2} = \frac{35}{256}\pi.$$

Example 35

$$\int_0^{\pi/2} \sin^3 x \cos^5 x \, dx = \frac{2.4.2}{8.6.4.2} = \frac{1}{24}.$$

Example 36

$$\int_0^{\pi/2} \sin^6 x \cos^4 x \, dx = \frac{5.3.1.3.1}{10.8.6.4.2}\frac{\pi}{2} = \frac{3\pi}{512}.$$

Example 37

$$I = \int_0^{\pi} \sin^9 x \cos^4 x \, dx = \int_0^{\pi/2} \sin^9 x \cos^4 x \, dx + \int_{\pi/2}^{\pi} \sin^9 x \cos^4 x \, dx.$$

In the second integral put $x = \pi - y$ so that $\sin x = \sin y$, $\cos x = -\cos y$

and $dx = -dy$. Then

$$\int_{\pi/2}^{\pi} \sin^9 x \cos^4 x \, dx = \int_0^{\pi/2} \sin^9 y \cos^4 y \, dy.$$

Hence

$$I = 2\int_0^{\pi/2} \sin^9 x \cos^4 x \, dx = 2\,\frac{8.6.4.2.3.1}{13.11.9.7.5.3} = \frac{256}{15015}.$$

Exercises 18(g)

Find the integrals in Nos. 1–9:

1. $\int_0^{\pi/2} \cos^7 x \, dx$
2. $\int_0^{\pi/2} \sin^6 x \, dx$
3. $\int_0^{\pi/2} \sin^2 x \cos^3 x \, dx$
4. $\int_0^{\pi/2} \sin^4 x \cos^2 x \, dx$
5. $\int_0^{\pi} (1 - \cos x)^4 \, dx$
6. $\int_0^{\pi/2} \sin^4 2x \, dx$
7. $\int \tan^5 x \, dx$
8. $\int_0^{\pi} (\sin x + \cos x)^3 \, dx$
9. $\int_0^1 x^2 (1 - x^2)^{5/2} \, dx$

10. Prove that if

$$I_n = \int_0^{\pi/4} \tan^n x \, dx,$$

then for $n \geqslant 2$, $(n-1)(I_n + I_{n-2}) = 1$, and hence evaluate I_5.

11. Given

$$I_n(t) = \int_0^t x^n e^{-x} \, dx,$$

use integration by parts to show that

$$I_n(t) = -t^n e^{-t} + nI_{n-1}(t).$$

Hence evaluate

$$\int_1^2 x^3 e^{-x} \, dx.$$

12. Given

$$I_n = \int_0^{\infty} x^{n-1} e^{-x} \, dx,$$

where n is positive, prove that $I_{n+1} = nI_n$. Hence evaluate I_n when n is a positive integer.

13. Given

$$I_n = \int_0^{\pi/2} x^n \sin x \, dx \text{ and } n > 1,$$

prove that

$$I_n + n(n-1)I_{n-2} = n(\pi/2)^{n-1}.$$

Evaluate

$$\int_0^{\pi/2} x^5 \sin x \, dx.$$

14. Given

$$I_n = \int_0^{\pi} e^{-x} \sin^n x \, dx,$$

show that for $n > 1$, $(n^2+1)I_n = n(n-1)I_{n-2}$. Evaluate

$$\int_{-\pi/2}^{\pi/2} e^{-x} \cos^3 x \, dx.$$

15. Given

$$I_n = \int x^n (a^2 - x^2)^{1/2} \, dx,$$

show that

$$(n+2)I_n = -x^{n-1}(a^2-x^2)^{3/2} + a^2(n-1)I_{n-2}.$$

(*Hint*: Write the integrand as $\frac{1}{2}x^{n-1}[2x(a^2-x^2)^{1/2}]$ before using integration by parts.)

16. By means of the substitution $1 + x = 2\cos^2\theta$ show that

$$\int_{-1}^{1} (1+x)^m (1-x)^n \, dx = \frac{2^{m+n+1}\, m!\, n!}{(m+n+1)!}.$$

17. Given

$$I_n = \int_0^{\pi/2} x \cos^n x \, dx, \text{ where } n > 1,$$

show that

$$n^2 I_n = n(n-1)I_{n-2} - 1.$$

Evaluate I_4 and I_5.

18. Given

$$u_n(t) = \int_0^t \frac{\sin(2n-1)x}{\sin x} \, dx, \quad v_n(t) = \int_0^t \frac{\sin^2 nx}{\sin^2 x} \, dx$$

show that

$$u_{n+1}(t) - u_n(t) = \frac{1}{n}\sin 2nt, \quad v_{n+1}(t) - v_n(t) = u_{n+1}(t).$$

Deduce that if n is a positive integer,

$$\int_0^{\pi/2} \frac{\sin(2n-1)x}{\sin x} \, dx = \frac{\pi}{2}, \quad \int_0^{\pi/2} \frac{\sin^2 nx}{\sin^2 x} \, dx = n\frac{\pi}{2}.$$

18.8 Numerical Integration

In applications of mathematics to science and engineering it is frequently required to evaluate the definite integral $\int_a^b f(x)\,dx$, where $f(x)$ is such that the indefinite integral $\int f(x)\,dx$ cannot be found by any of the known methods.

Let the interval $a \leqslant x \leqslant b$ be divided into n sub-intervals by the points $x_1(=a), x_2, x_3, \ldots, x_{n+1}(=b)$ (see Fig. 18.3). For simplicity we shall space these points evenly so that each sub-interval has the same length $(b-a)/n = h$, say. Write

$$y_1 = f(x_1),\ y_2 = f(x_2),\ \ldots,\ y_{n+1} = f(x_{n+1})$$

so that y_r is the ordinate at $x = x_r$ and is positive when the curve is above the x-axis, negative when the curve is below the x-axis.

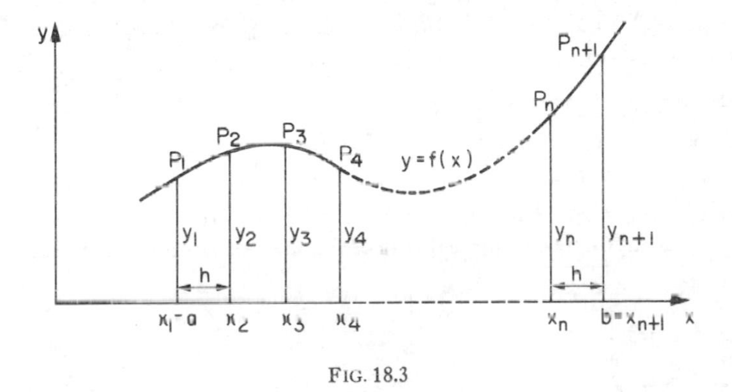

FIG. 18.3

According to the definition of an integral in Section 16.1.4 a good approximation to $\int_a^b f(x)\,dx$ is given by the sum $\sum_{r=1}^{n} hy_r$ when h is small. Let us assume that $f(x) \geqslant 0$ on the interval $a \leqslant x \leqslant b$ so that $y_r \geqslant 0$ for $r = 1, 2, \ldots, n+1$; we can then interpret $\int_a^b f(x)\,dx$ as the area under the curve $y = f(x)$ from $x = a$ to $x = b$. This interpretation enables us to suggest better approximations, but is not essential to the validity of these approximations, and the results obtained in this section apply when $f(x)$ takes both positive and negative values.

The lines $x = x_r$ $(r = 1, 2, \ldots, n+1)$ divide the area under the curve into n 'strips' (see Fig. 18.3), a typical strip being shown on an enlarged scale in Fig. 18.4. Approximations to the total area may be based on some suitable approximation to the area of this typical strip, and this in turn may be based on some suitable replacement for the curved boundary P_rP_{r+1}.

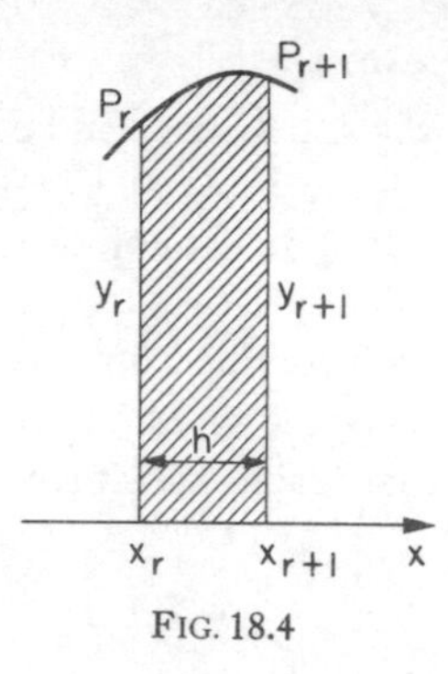

FIG. 18.4

18.8.1 The Trapezoidal Rule Let the arc P_rP_{r+1} be replaced by the straight line P_rP_{r+1} (see Fig. 18.5). In effect the strip is replaced by a trapezium of area $\frac{1}{2}(y_r + y_{r+1})h$, and adding up for all the strips we find that

$$\int_a^b f(x)\,dx \simeq \tfrac{1}{2}(y_1 + y_2)h + \tfrac{1}{2}(y_2 + y_3)h + \cdots + \tfrac{1}{2}(y_n + y_{n+1})h$$
$$= h[\tfrac{1}{2}(y_1 + y_{n+1}) + y_2 + y_3 + \cdots + y_n].$$

This is the *trapezoidal rule* with n strips, or with $(n + 1)$ ordinates, and obviously the larger the value of n the greater the accuracy obtained.

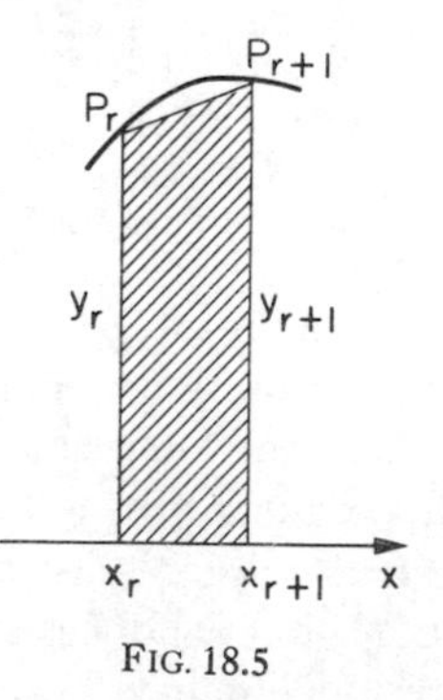

FIG. 18.5

Example 38 Evaluate $\int_0^1 1/(1 + x^2)\,dx$ by using the trapezoidal rule with 10 strips.

The width of each strip must be $\frac{1}{10}$, i.e. $h = 0{\cdot}1$. In the notation used above, $x_1, x_2, \ldots, x_{11}$ are the points $0, 0{\cdot}1, 0{\cdot}2, \ldots, 1$ on the x-axis, and $y_r = 1/(1 + x_r^2)$ so that the values of y_r are given by Table 18.1.

TABLE 18.1

r	1	2	3	4	5	6
y_r	1·0000	0·9901	0·9615	0·9174	0·8621	0·8000
r	7	8	9	10	11	
y_r	0·7353	0·6711	0·6098	0·5525	0·5000	

By the trapezoidal rule

$$\int_0^1 \frac{dx}{1+x^2} \simeq 0{\cdot}1\,(0{\cdot}7500 + 7{\cdot}0998) = 0{\cdot}7850 \text{ (to 4 decimal places).}$$

Note that

$$\int_0^1 \frac{dx}{1+x^2} = \Big[\tan^{-1} x\Big]_0^1 = \frac{\pi}{4} = 0{\cdot}7854 \text{ (to 4 decimal places).}$$

18.8.2 Simpson's Rule Instead of the single strip shown in Fig. 18.4, consider a *pair* of adjacent strips as in Fig. 18.6(a). It is possible to determine constants A, B, C such that the parabola $y = Ax^2 + Bx + C$ passes through the three points P_{r-1}, P_r, P_{r+1}, and this parabolic arc may be used to replace the arc $P_{r-1}P_{r+1}$ of the original curve.

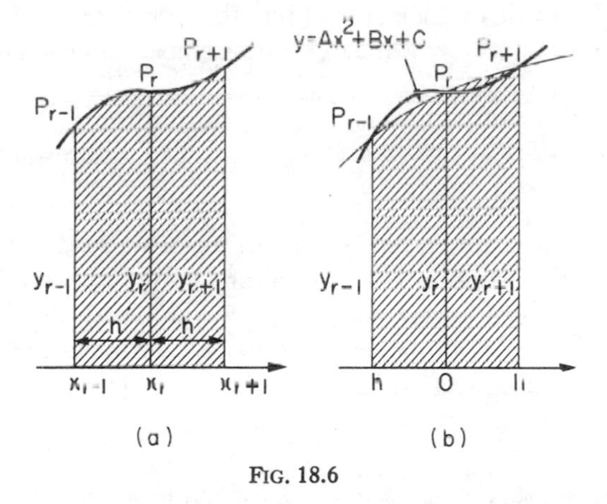

FIG. 18.6

In other words the two strips are replaced by the shaded area in Fig. 18.6(b) where, since the area is independent of the position of the y-axis, it has been assumed for simplicity that the y-axis passes through P_r. We next calculate the shaded area in Fig. 18.6(b) in terms of the ordinates y_{r-1}, y_r, y_{r+1} and the strip width h; since these quantities are all independent of the position of the y-axis, it follows that the result remains valid when the y-axis does not pass through P_r.

In Fig. 18.6(b) P_{r-1}, P_r, P_{r+1} are the points $(-h, y_{r-1})$, $(0, y_r)$, (h, y_{r+1}) respectively, and for the curve $y = Ax^2 + Bx + C$ to pass through these points we require

$$y_{r-1} = Ah^2 - Bh + C, \quad y_r = C, \quad y_{r+1} = Ah^2 + Bh + C$$

giving

$$A = (y_{r+1} - 2y_r + y_{r-1})/2h^2, \quad B = (y_{r+1} - y_{r-1})/2h, \quad C = y_r.$$

Now the shaded area in Fig. 18.6(b) is

$$\int_{-h}^{h} (Ax^2 + Bx + C)\,dx = 2Ah^3/3 + 2Ch = \tfrac{1}{3}h(y_{r-1} + 4y_r + y_{r+1}).$$

Let us now suppose that the area under the curve $y = f(x)$ from $x = a$ to $x = b$ (see Fig. 18.3) has been divided into n strips where n is *even*. Taking the strips in pairs the above approximation gives

$$\begin{aligned}\int_a^b f(x)\,dx &\simeq \tfrac{1}{3}h(y_1 + 4y_2 + y_3) + \tfrac{1}{3}h(y_3 + 4y_4 + y_5) \\ &\quad + \cdots + \tfrac{1}{3}h(y_{n-1} + 4y_n + y_{n+1}) \\ &= \tfrac{1}{3}h[y_1 + y_{n+1} + 4(y_2 + y_4 + \cdots + y_n) \\ &\quad + 2(y_3 + y_5 + \cdots + y_{n-1})].\end{aligned}$$

This is *Simpson's rule* with n strips (note that the number of strips must be even), or with $(n + 1)$ ordinates (the number of ordinates must be odd).

Example 39 Evaluate $\int_0^1 1/(1 + x^2)\,dx$ by using Simpson's rule with ten strips.

We require to evaluate eleven ordinates, as has already been done in Example 38. The rest of the calculation may be set out as in Table 18.2; in the column headed 'factor' we put the factor 1 with the first and last ordinates, the factor 4 with all intervening *even* ordinates, and the factor 2 with all intervening *odd* ordinates.

Hence

$$\int_0^1 \frac{dx}{1 + x^2} \simeq \tfrac{1}{3}(0\cdot 1) \times 23\cdot 5618 = 0\cdot 7854 \quad \text{(to 4 decimal places).}$$

It can be shown that there is a constant K, depending mainly on the function being integrated, such that the error resulting from the use of Simpson's rule is less than Kh^4. The error resulting from the use of the trapezoidal rule is similarly related to h^2. Thus to achieve a given degree of accuracy, fewer strips are required with Simpson's rule than with the trapezoidal rule.

TABLE 18.2

x	$1/(1+x^2)$	Factor	
0·0	1·0000	1	1·0000
0·1	0·9901	4	3·9604
0·2	0·9615	2	1·9230
0·3	0·9174	4	3·6696
0·4	0·8621	2	1·7242
0·5	0·8000	4	3·2000
0·6	0·7353	2	1·4706
0·7	0·6711	4	2·6844
0·8	0·6098	2	1·2196
0·9	0·5525	4	2·2100
1·0	0·5000	1	0·5000
		Total	23·5618

Exercises 18(h)

1. Evaluate $\int_1^2 x^2\,dx$ by using Simpson's rule with two strips. Show that the answer obtained in this way is exactly correct, and suggest a reason for this.
2. Evaluate $\int_0^{1\cdot 2} x^3\,dx$ by using Simpson's rule with six strips.
3. Tables give $\ln 2 = 0\cdot 6931$. From the fact that $\ln 2 = \int_1^2 (1/x)\,dx$, use the trapezoidal rule with four strips to get the approximation 0·6970, and Simpson's rule with four strips to get the approximation 0·6933, for $\ln 2$.
4. Use Simpson's rule with three ordinates to find the approximate value of $\int_1^2 1/(1+x)\,dx$ and use tables of natural logarithms to check your result.
5. Evaluate $\int_0^1 1/(x^2+x+1)\,dx$ by using Simpson's rule with four strips, giving your answer to four decimal places. Check your answer by evaluating the integral by an exact method.
6. Simpson's rule with two strips can be expressed in the form

$$\int_{-h}^{h} f(x)\,dx \simeq \frac{h}{3}[f(-h)+4f(0)+f(h)].$$

Show that if f is a cubic polynomial the value of the integral obtained by this approximation is exactly correct.

7. The pressure p g/cm² and the volume v cm³ of a certain gas obey the law $pv^{1\cdot 4} = 100$. Using Simpson's rule and the table of values below calculate approximately the work done (i.e. $\int p\,dv$) when the gas expands from 20 to 26 cm³.

v	20	21	22	23	24	25	26
p	1·509	1·409	1·320	1·240	1·168	1·104	1·045

Check your result by direct integration.

8. Evaluate $\int_0^{0\cdot 8} x/(1+x^2)\,dx$ to three decimal places by using Simpson's rule with four strips and then by integration and use of appropriate tables.
9. Obtain an approximate value for $\int_0^{\pi} x \sin x\,dx$ by using Simpson's rule with six strips. Use the method of integration by parts to obtain the exact value of this integral.

CHAPTER 19

APPLICATIONS OF INTEGRATION

19.1 Area

A connection between area and integration was established in Chapter 16, where, in the case of a function f such that $f(x) \geqslant 0$ on the interval $a \leqslant x \leqslant b$, the area $A_{a,b}$ under the curve $y = f(x)$ from $x = a$ to $x = b$ was expressed as an integral. Let us recall that

(a) if the area is divided into strips parallel to the y-axis, the strip $MNQP$ in Fig. 19.1 being a typical one, then $A_{a,b}$ is simply the sum of the areas of these strips,

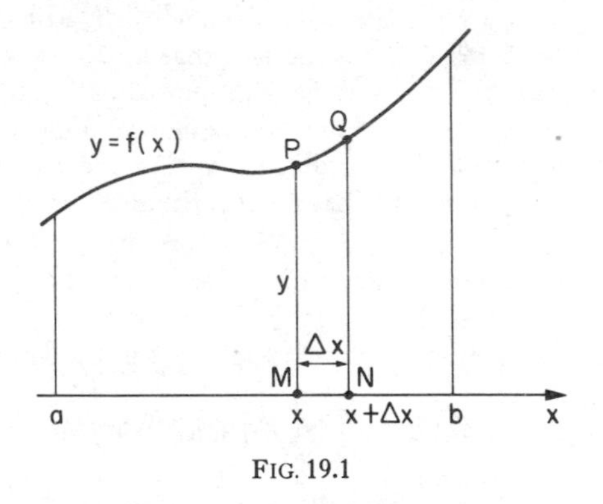

FIG. 19.1

(b) if P, Q are the points (x, y), $(x + \Delta x, y + \Delta y)$ so that $MP = y$, $MN = \Delta x$, then the area $A_{x,x+\Delta x}$ of the typical strip is approximately $y\,\Delta x$; the error E (say) in this approximation is such that $E/\Delta x \to 0$ as $\Delta x \to 0$, and so we are led (see Section 16.3.1) to the result

$$A_{a,b} = \int_a^b y\,\mathrm{d}x.$$

Note that since P is on the curve we may have $y = f(x)$ and we may regard $\int_a^b y\,\mathrm{d}x$ as merely an alternative way of writing $\int_a^b f(x)\,\mathrm{d}x$.

On comparing with the original discussion of area under a curve in Chapter 16, it will be seen that in the above summary the 'typical strip' is not referred to as the 'rth strip' and correspondingly x_r, Δx_r are replaced by simply x, Δx. This kind of licence is quite harmless once the basic principles of Chapter 16 have been fully understood. Throughout this chapter integral formulae are derived in this informal way and verification that approximation errors satisfy the criterion of Section 16.3.1 is usually left as an exercise for the reader.

19.1.1 Area between Curve and Axis We shall refer to the area enclosed by the curve $y = f(x)$, the x-axis and the lines $x = a$, $x = b$ as *the area between the curve* $y = f(x)$ *and the x-axis from* $x = a$ *to* $x = b$; in the present context the phrase 'from $x = a$ to $x = b$' can sometimes be taken for granted to avoid undue repetition.

If $f(x) \leqslant 0$ on the interval $a \leqslant x \leqslant b$, then the strip $MNQP$ (see Fig. 19.2) has the approximate area $(-y)\,\Delta x$, and the area between the curve and the x-axis is

$$-\int_a^b y\,dx, \quad \text{that is} \quad -\int_a^b f(x)\,dx.$$

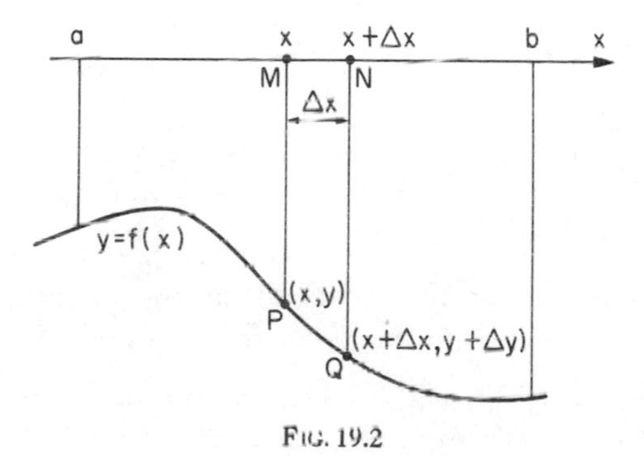

FIG. 19.2

If $f(x)$ takes both positive and negative values on the interval $a \leqslant x \leqslant b$, then $\int_a^b f(x)\,dx$ represents the *signed* area between the curve and the x-axis in the sense that parts above the x-axis are counted positively and parts below the x-axis are counted negatively. The 'true' area is $\int_a^b |f(x)|\,dx$; to evaluate this we must find all points $x = c$, $x = d, \ldots$ (say) in the interval $a < x < b$ where the curve *crosses* the x-axis, evaluate separately $\int_a^c f(x)\,dx$, $\int_c^d f(x)\,dx, \ldots$, changing the sign of all negative answers before adding up.

Example 1 Find (a) the true area, (b) the signed area, between the curve $y = \cos x$ and the x-axis from $x = 0$ to $x = \pi$.

(a) Between 0 and π the curve crosses the x-axis only at $x = \pi/2$. We require the integrals

$$\int_0^{\pi/2} \cos x \, dx, \quad \int_{\pi/2}^{\pi} \cos x \, dx,$$

i.e. 1, −1. Thus the true area is $1 + 1 = 2$.

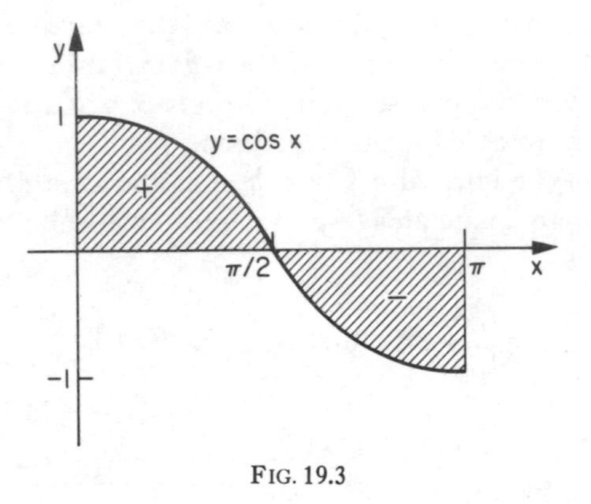

FIG. 19.3

(b) We simply evaluate $\int_0^{\pi} \cos x \, dx$ to get zero. This means that the negative area exactly cancels the positive area (see Fig. 19.3).

Example 2 Find (a) the true area, (b) the signed area, between the curve $y = 6x^2 - 6x - 12$ and the x-axis from $x = -3$ to $x = 3$.

(a) Writing $y = 6(x + 1)(x - 2)$, we see that the curve crosses the x-axis at $x = -1$ and $x = 2$. We require the integrals

$$\int_{-3}^{-1} y \, dx, \quad \int_{-1}^{2} y \, dx, \quad \int_{2}^{3} y \, dx,$$

where $y = 6x^2 - 6x - 12$. Since

$$\int y \, dx = 2x^3 - 3x^2 - 12x + c,$$

the above integrals have the values 52, −27, 11 and so the true area is $52 + 27 + 11$, i.e. 90.

(b) We simply evaluate $\int_{-3}^{3} (6x^2 - 6x - 12) \, dx$ to get 36.

If the equation of the curve is given parametrically, with parameter t, then we write $\int y \, dx = \int y \, (dx/dt) \, dt$ and integrate with respect to t between appropriate limits.

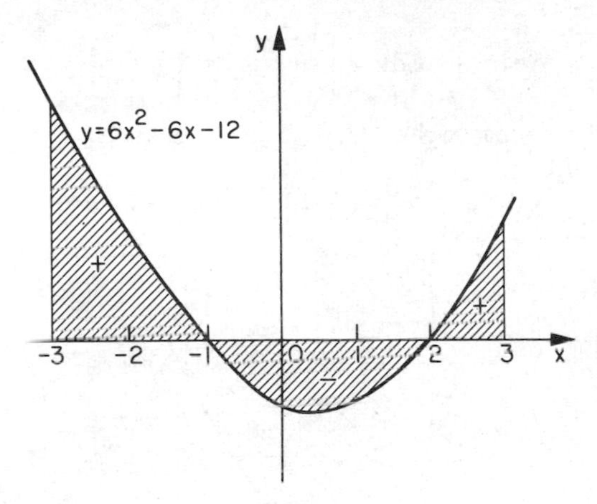

FIG. 19.4

Example 3 Find the area enclosed by the ellipse

$$x = a \cos t, y = b \sin t.$$

By symmetry, we may find the area in the first quadrant and multiply by four. The area in the first quadrant is the area under the ellipse from $x = 0$ to $x = a$, and so the total area enclosed is

$$4 \int_0^a y \, dx = 4 \int_{\pi/2}^0 b \sin t \, (-a \sin t) \, dt = 4ab \int_0^{\pi/2} \sin^2 t \, dt = \pi ab.$$

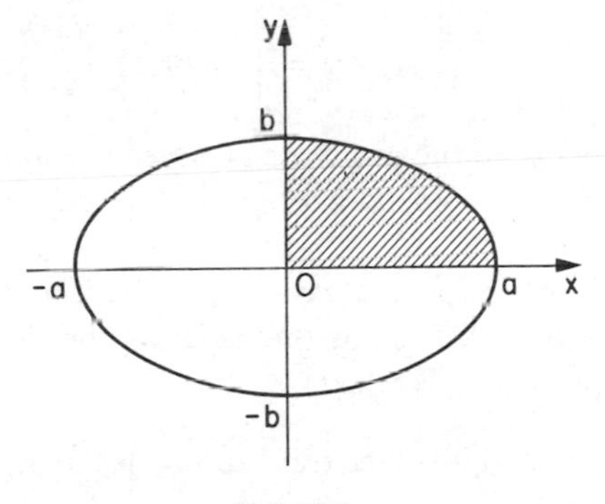

FIG. 19.5

The integral $\int_c^d g(y) \, dy$ is similarly related to the area enclosed by the curve $x = g(y)$, the y-axis, and the lines $y = c$, $y = d$.

Example 4 Find the area between the curve $y = x^2$ and the y-axis from $y = 1$ to $y = 4$.

$$\text{Area} = \int_1^4 x\,dy = \int_1^4 y^{1/2}\,dy = \left[\tfrac{2}{3}y^{3/2}\right]_1^4 = \tfrac{14}{3}.$$

Alternatively we may write

$$\int_1^4 x\,dy = \int_1^2 x\frac{dy}{dx}\,dx = \int_1^2 2x^2\,dx = \left[\tfrac{2}{3}x^3\right]_1^2 = \tfrac{14}{3}.$$

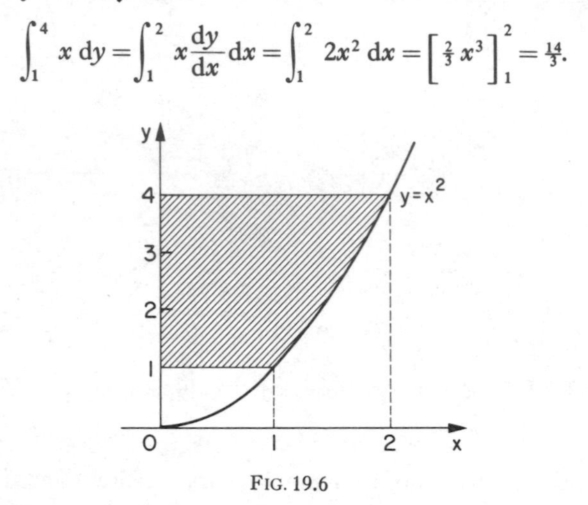

FIG. 19.6

19.1.2 Area between Velocity–Time Curve and t-axis In one-dimensional motion (see Section 15.1.1) the relation between the velocity v and the distance s is that $v = ds/dt$, or $s = \int v\,dt$. In particular, if s_1, s_2 are the values of s at times t_1, t_2 then

$$s_2 - s_1 = \int_{t_1}^{t_2} v\,dt,$$

i.e. the distance between the positions at times t_1 and t_2 equals the signed area between the velocity–time curve $v = f(t)$, say, and the t-axis from $t = t_1$ to $t = t_2$.

If the direction of motion changes during the interval $t_1 < t < t_2$, then the total distance travelled is not the same as the distance between the initial and final positions. We have

$$\text{total distance travelled} = \int_{t_1}^{t_2} |v|\,dt,$$

i.e. the total distance travelled equals the 'true' area between the velocity–time curve and the t-axis from $t = t_1$ to $t = t_2$.

Example 5 Given $v = 6t^2 - 18t + 12$, find (a) the distance between the positions at $t = 0$ and $t = 4$, (b) the total distance travelled between $t = 0$ and $t = 4$.

(a) Distance between positions is

$$\int_0^4 (6t^2 - 18t + 12)\, dt = \Big[2t^3 - 9t^2 + 12t\Big]_0^4 = 32.$$

(b) Since $v = 6(t-1)(t-2)$, we see that v changes sign when $t = 1$ and again when $t = 2$. Hence we must evaluate separately the integrals

$$\int_0^1 v\, dt, \quad \int_1^2 v\, dt, \quad \int_2^4 v\, dt$$

to get the answers 5, -1, 28.

The total distance travelled is thus $5 + 1 + 28 = 34$. (This consists of 5 units forwards, followed by 1 unit backwards, followed by 28 units forwards).

19.1.3 Average Value of $f(x)$ on the Interval $a \leqslant x \leqslant b$ Let $y_1, y_2, \ldots, y_n$ be the values of $f(x)$ corresponding to the values $x = x_1$, $x = x_2$, $\ldots$, $x = x_n$ where

$$a = x_1 < x_2 < x_3 < \cdots < x_{n+1} = b$$

is a subdivision of the interval $a \leqslant x \leqslant b$ into n equal sub-intervals each of length $(b - a)/n = \Delta x$, say.

The set of values $y_1, y_2, \ldots, y_n$ has an *average*, or *mean*, of $(1/n) \sum_{r=1}^{n} y_r$, that is

$$\frac{1}{n}\sum_{r=1}^{n} f(x_r) = \frac{\Delta x}{(b-a)} \sum_{r=1}^{n} f(x_r)$$

$$= \frac{1}{(b-a)} \sum_{r=1}^{n} f(x_r)\, \Delta x \to \frac{1}{(b-a)} \int_a^b f(x)\, dx \text{ as } \Delta x \to 0.$$

This limit is called the *average value*, or *mean value*, of $f(x)$ on the interval $a \leqslant x \leqslant b$. From Example 1, part (b), we see that the average value of $\cos x$ on the interval $0 \leqslant x \leqslant \pi$ is zero.

Example 6 Assuming the velocity v at time t is given by $v = t^2 - 3t + 2$, find the average velocity and the average speed during the time interval $0 \leqslant t \leqslant 3$.

Average velocity

$$= \tfrac{1}{3}\int_0^3 (t^2 - 3t + 2)\, dt = \tfrac{1}{3}\Big[\tfrac{1}{3}t^3 - \tfrac{3}{2}t^2 + 2t\Big]_0^3 = \tfrac{1}{2}.$$

Average speed $= \frac{1}{3}\int_0^3 |v|\, dt$.

Now $v = (t-2)(t-1)$, and so v changes sign when $t = 1$ and when

$t = 2$. We easily see that

$$\int_0^1 v \, dt = \tfrac{5}{6}, \quad \int_1^2 v \, dt = -\tfrac{1}{6}, \quad \int_2^3 v \, dt = \tfrac{5}{6},$$

and so

$$\int_0^3 |v| \, dt = \tfrac{5}{6} + \tfrac{1}{6} + \tfrac{5}{6} = \tfrac{11}{6}.$$

Hence average speed $= \frac{11}{18}$.

Another kind of average, used frequently in statistics, is the *root mean square*. For the numbers $y_1, y_2, \ldots y_n$ this is defined as

$$\left[\frac{1}{n}\sum_{r=1}^{n} y_r^2\right]^{1/2}$$

and so, in the case of a continuous function f, we get the formula

$$\left[\frac{1}{b-a}\int_a^b \{f(x)\}^2 \, dx\right]^{1/2}$$

for the root mean square value of $f(x)$ on the interval $a \leqslant x \leqslant b$.

19.1.4 Area between Two Curves Let $A_{a,b}$ denote the area between the curves $y = f_1(x)$ and $y = f_2(x)$ from $x = a$ to $x = b$. Taking strips parallel to the y-axis we see that the area $A_{x.x+\Delta x}$ of the typical strip shown in Fig. 19.7 is approximately $|y_1 - y_2| \Delta x$, where $y_1 = f_1(x)$, $y_2 = f_2(x)$ and it is assumed that f_1, f_2 are both continuous on the interval $a \leqslant x \leqslant b$. The modulus sign is required since if the curves cross each other

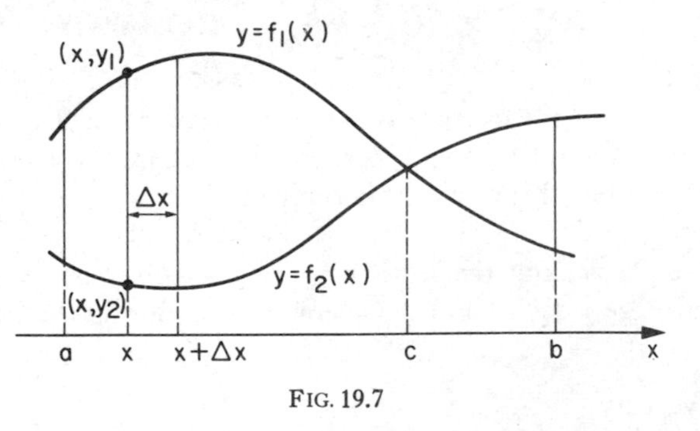

FIG. 19.7

at a point $x = c$ where $a < c < b$ (see Fig. 19.7), then $(y_1 - y_2)$ is positive on one side of this point and negative on the other. Thus

$$A_{a,b} = \int_a^b |y_1 - y_2| \, dx = \int_a^b |f_1(x) - f_2(x)| \, dx.$$

As with previous examples involving an integrand with modulus signs, we must find all points where the integrand changes sign, i.e. all points where the curves cross each other.

Example 7 Find the area enclosed by the curves $y = 2x$, $y = 2x^3 - 3x^2$.

The curves intersect when

$$2x = 2x^3 - 3x^2, \quad \text{or} \quad x(2x + 1)(x - 2) = 0,$$

that is when $x = 0, -\frac{1}{2}, 2$.

Now

$$\int (2x - 2x^3 + 3x^2)\,dx = x^2 - \tfrac{1}{2}x^4 + x^3 + c,$$

and we must evaluate this separately with limits of integration $-\frac{1}{2}$ and 0, and with limits of integration 0 and 2. We get $-\frac{3}{32}$ and 4 respectively; hence

$$\int_{-1/2}^{2} |\,2x - 2x^3 + 3x^2\,|\,dx = \tfrac{3}{32} + 4 = \tfrac{131}{32}.$$

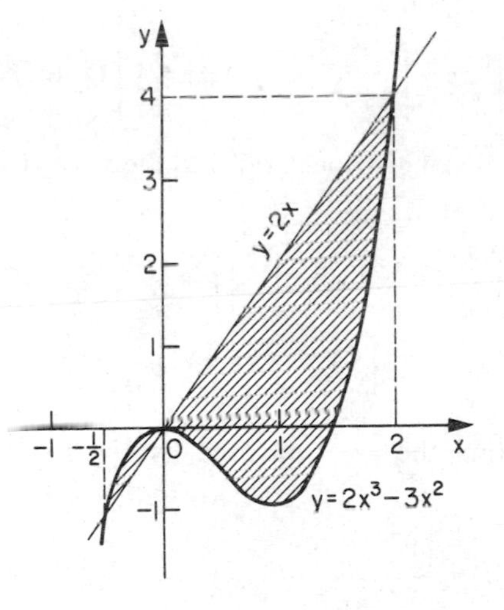

FIG. 19.8

Example 8 Find the area enclosed by the curve $y^2 - 2y = x + 2$ and the line $x = 1$.

First method. Taking strips parallel to the y-axis as was done in Fig.

19.7, we must first solve the given equation for y to get $y = 1 \pm \sqrt{(x+3)}$. The area is then

$$\int_{-3}^{1} |y_1 - y_2|\, dx,$$

where $y_1 = 1 + \sqrt{(x+3)}$ and $y_2 = 1 - \sqrt{(x+3)}$.

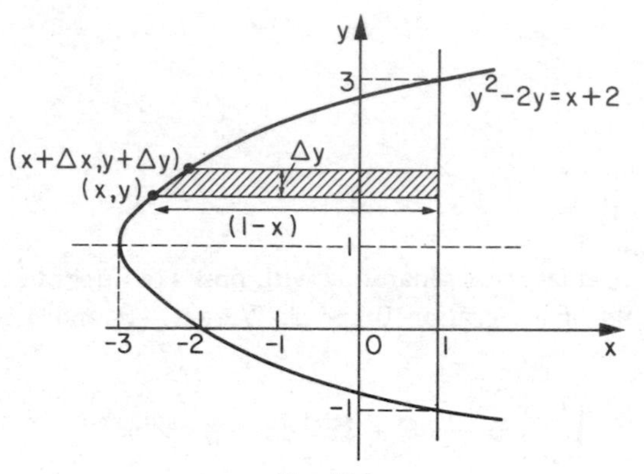

FIG. 19.9

Now

$$\int_{-3}^{1} |y_1 - y_2|\, dx = \int_{-3}^{1} 2\sqrt{(x+3)}\, dx = \tfrac{4}{3}\Big[(x+3)^{3/2}\Big]_{-3}^{1} = \tfrac{32}{3}.$$

Second method. Taking strips parallel to the x-axis (see Fig. 19.9), we see that the area is

$$\int_{-1}^{3} (1-x)\, dy = \int_{-1}^{3} [1 - (y^2 - 2y - 2]\, dy$$
$$= \Big[3y + y^2 - \tfrac{1}{3}y^3\Big]_{-1}^{3} = \tfrac{32}{3}.$$

Example 9 Find the area enclosed by the loop of the curve $y^2 = x(1-x)^2$.

We may write

$$y_1 = \sqrt{x}(1-x), \quad y_2 = -\sqrt{x}(1-x)$$

and consider $\int_0^1 |y_1 - y_2|\, dx$. Alternatively, the area enclosed by the loop is, by symmetry, exactly twice the area under the curve $y = \sqrt{x}(1-x)$ from $x = 0$ to $x = 1$. Either way we are led to the answer

$$2\int_0^1 \sqrt{x}(1-x)\, dx = 2\Big[\tfrac{2}{3}x^{3/2} - \tfrac{2}{5}x^{5/2}\Big]_0^1 = \tfrac{8}{15}.$$

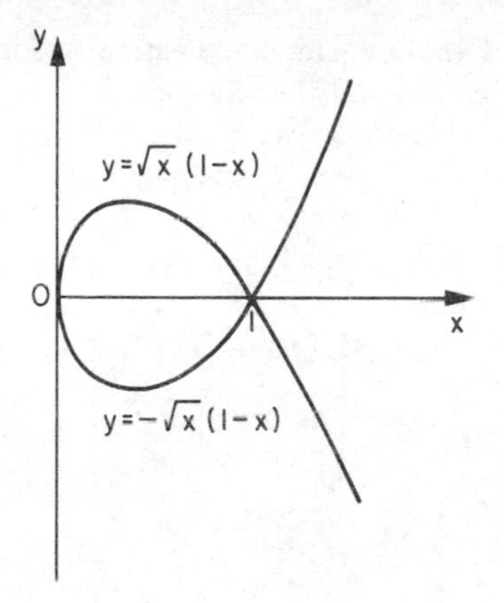

FIG. 19.10

19.1.5 Areas in Polar Co-ordinates Let $A_{\alpha,\beta}$ be the area enclosed by the curve $r = f(\theta)$ and the lines $\theta = \alpha$, $\theta = \beta$. This area may be divided into a number of narrow sectors by lines $\theta =$ constant, a typical sector OPQ being shown in Fig. 19.11. With centre O and radius OP draw an arc of a circle to cut OQ in Q'. The area of the circular sector OPQ' is

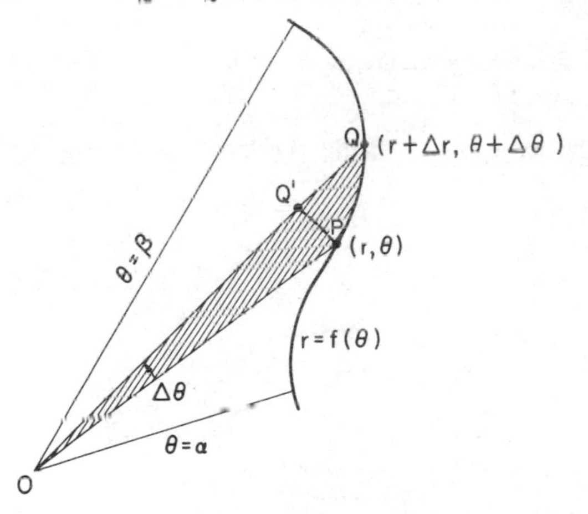

FIG. 19.11

$\frac{1}{2}OP^2\,\Delta\theta$, i.e. $\frac{1}{2}r^2\,\Delta\theta$. Assuming that $f(\theta)$ is continuous on the interval $\alpha \leqslant \theta \leqslant \beta$, the area $A_{\theta,\theta+\Delta\theta}$ of the sector OPQ is approximately $\frac{1}{2}r^2\,\Delta\theta$ and we get

$$A_{\alpha,\beta} = \int_\alpha^\beta \tfrac{1}{2}r^2\,\mathrm{d}\theta = \int_\alpha^\beta \tfrac{1}{2}[f(\theta)]^2\,\mathrm{d}\theta.$$

Example 10 Find the area in the positive quadrant bounded by the cardioid $r = a(1 + \cos\theta)$ and the axes.

$$\begin{aligned}\text{Area} &= \int_0^{\pi/2} \tfrac{1}{2}a^2(1 + \cos\theta)^2 \, d\theta \\ &= \tfrac{1}{4}a^2 \int_0^{\pi/2} (3 + 4\cos\theta + \cos 2\theta) \, d\theta \\ &= \tfrac{1}{4}a^2 \Big[3\theta + 4\sin\theta + \tfrac{1}{2}\sin 2\theta \Big]_0^{\pi/2} \\ &= \tfrac{1}{8}a^2(3\pi + 8).\end{aligned}$$

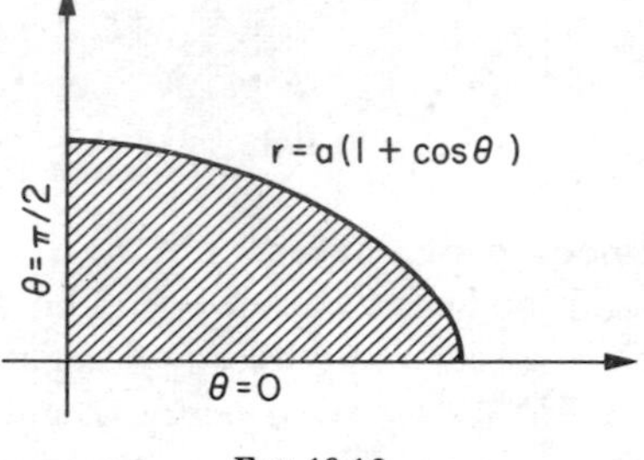

FIG. 19.12

The above formula is frequently used to find the area within a closed curve. There are three cases:

Case 1. If the pole O lies inside the curve as in Fig. 19.13, the area enclosed is

$$\int_0^{2\pi} \tfrac{1}{2}r^2 \, d\theta.$$

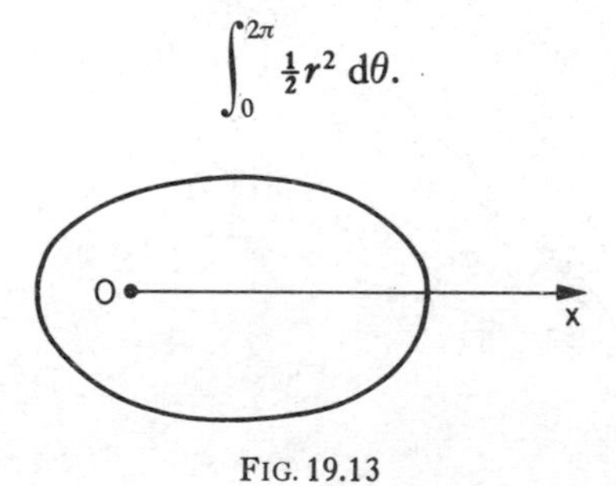

FIG. 19.13

Case 2. If the pole O lies on the curve as in Fig. 19.14, and if the tangents at the pole are the lines $\theta = \alpha$, $\theta = \beta$, then the area enclosed is

$$\int_\alpha^\beta \tfrac{1}{2}r^2 \, d\theta.$$

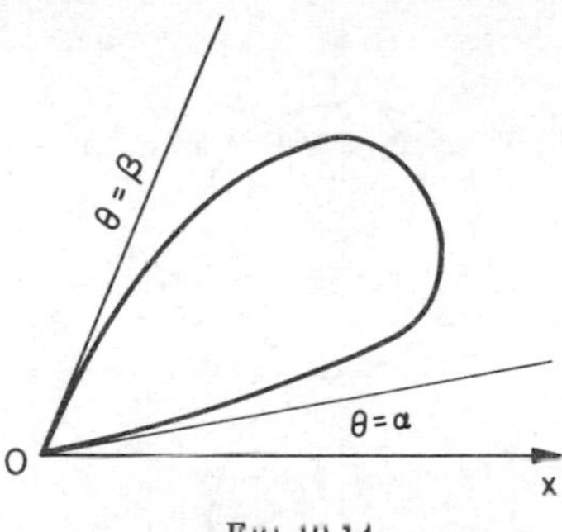

FIG. 19.14

Case 3. If the pole O lies outside the curve as in Fig. 19.15, let the tangents from the pole be the lines $\theta = \alpha$, $\theta = \beta$ meeting the curve at A, B respectively. The points A, B divide the curve into two parts, one with equation $r = f_1(\theta)$ (say), and the other with equation $r = f_2(\theta)$ (say) where $f_1(\theta) \geqslant f_2(\theta)$ on the interval $\alpha \leqslant \theta \leqslant \beta$ (see Fig. 19.15). The area enclosed is

$$\int_\alpha^\beta \tfrac{1}{2}[f_1(\theta)]^2 \, d\theta - \int_\alpha^\beta \tfrac{1}{2}[f_2(\theta)]^2 \, d\theta.$$

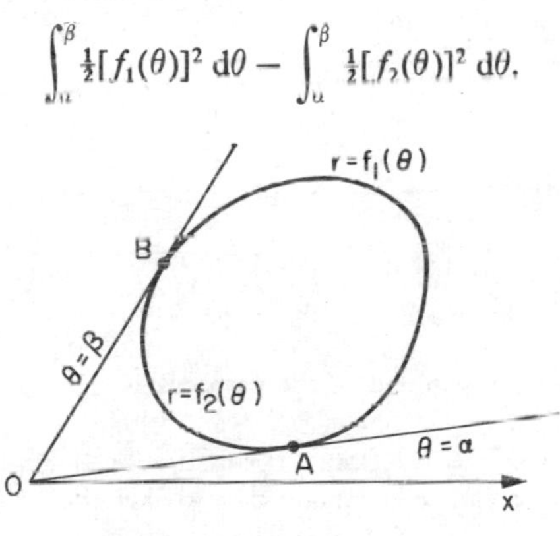

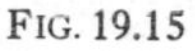

FIG. 19.15

Example 11 Find the area enclosed by one loop of the curve

$$r = a \sin^2 \theta.$$

The curve has two loops as shown in Fig. 19.16. The tangents at the pole are $\theta = 0$, $\theta = \pi$. Hence the area of the upper loop is

$$\begin{aligned} \int_0^\pi \tfrac{1}{2} r^2 \, d\theta &= \int_0^\pi \tfrac{1}{2} a^2 \sin^4 \theta \, d\theta \\ &= \int_0^{\pi/2} a^2 \sin^4 \theta \, d\theta \qquad \text{by symmetry,} \\ &= 3\pi a^2/16 \qquad \text{by (18.21)} \end{aligned}$$

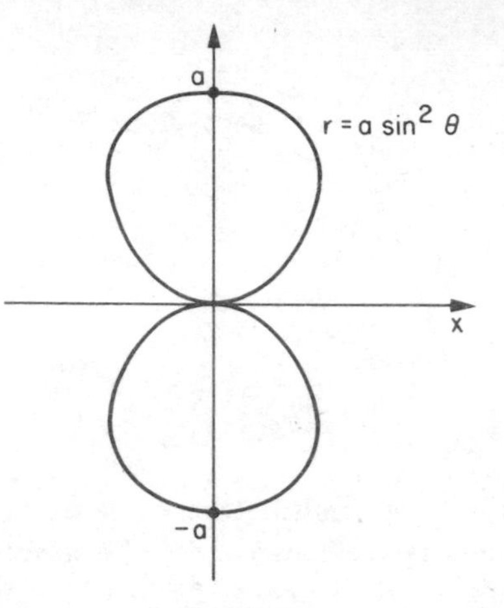

FIG. 19.16

Exercises 19(a)

1. Evaluate the following integrals by considering the areas which they represent, i.e. *without integration*:

   (a) $\int_0^2 \sqrt{(4-x^2)}\,dx$, (b) $\int_1^3 \sqrt{[(x-1)(5-x)]}\,dx$,

   (c) $\int_0^{2\pi} \sin x\,dx$, (d) $\int_0^{\pi/2} \cos^3 2x\,dx$.

2. Show in one diagram that the integral $\int_0^2 [\sqrt{(4-x^2)}-(2-x)]\,dx$ represents the difference between two geometrically known areas and hence evaluate the integral.
3. Find the area of the segment of the parabola $y = 2 + x - x^2$ cut off by the x-axis.
4. Sketch the curve $y^2 = x^3$ from $x = 0$ to $x = 4$. Find the area enclosed between the curve and the ordinate $x = 4$. Find the ordinate which bisects this area.
5. The velocity v m/s of a body moving in a straight line and the time t are connected by the equation $v = 3t^2 - 15t + 18$. Find the total distance travelled by the body from $t = 0$ to $t = 3$.
6. Use Simpson's rule with six strips to calculate the area between the curve $y = \sin(x^2)$ and the x-axis from $x = 0$ to $x = 0{\cdot}6$, giving your answer correct to three decimal places.
7. Find the area of the loop of the curve $y^2 = x(4-x)^2$.
8. Find the area bounded by the curve $y = (x+1)(2-x)$ and the line $y = x + 1$.
9. Find the area bounded by the curve $y = x(4-x)$ and the line $y = x + 2$.
10. Find the area enclosed by the curves

$$y = 3 + 2x - x^2,\quad y = x^2 - 4x + 3.$$

11. Find the area enclosed by the two parabolas

$$y^2 = 4(x + 1), \quad y^2 = 8(2 - x).$$

12. For the family of parabolas $y = ax(b - x)$, where $a > 0$, $b > 0$, find (a) the area bounded by a parabola and the x-axis, (b) the parabola which passes through $(4, 1)$ and gives the smallest area in (a).
13. Find the mean value of (a) $1/(1 + x^2)$ from $x = 0$ to $x = 1$, (b) $1/\sqrt{(1 - x^2)}$ from $x = 0$ to $x = \frac{1}{2}$, (c) $7 \cos x - 3 \sin 2x$ from $x = -\pi/3$ to $x = \pi/3$, (d) $x^2 \cos x$ from $x = 0$ to $x = \pi/2$.
14. Sketch and find the area bounded by the parabola $y = -x^2 + 4x - 3$ and its tangents at the points $(0, -3)$ and $(4, -3)$.
15. Show in a diagram the set of points (x, y) satisfying the inequalities

$$\frac{9}{x^2} \leqslant y \leqslant 10 - x^2.$$

Find the area of this set of points.
16. Find the area between the y-axis and the portion of the curve $x = t - t^3$, $y = 1 - t^4$ corresponding to the interval $0 \leqslant t \leqslant 1$.
17. Show that the loop of the curve $x = a \cos^2 t \sin t$, $y = a \cos t \sin^2 t$ corresponding to the interval $0 \leqslant t \leqslant \frac{1}{2}\pi$ has an area $\pi a^2/32$.
18. Sketch the curve $y^2(a + x) = x^2(3a - x)$, and show that the area of the loop is $9a^2/\sqrt{3}$.
19. Find the area between the curve $a(a - x)y = x^3$, the axis of x and the line $2x = a$.
20. Prove that the parabola $y = a(x - a)^2$ touches the curve $27y = 4x^3$, and find the equation of the tangent at the point of contact. Show that the area bounded by this tangent and the parabola $y = 4a(x - 4a)^2$ is $18a^4$.
21. Sketch the curve given by the equation $y^2(1 - x)(x - 2) = x^2$, and prove that the area between the curve and its asymptotes is 3π.
22. Show that the area bounded by the curve

$$r^2(a^2 \sin^2 \theta + b^2 \cos^2 \theta) = (a^2 - b^2)b^2 \cos^2 \theta,$$

where $a > b > 0$, is $\pi b(a - b)$.
23. The straight line $r \cos \theta = 2$ intersects the initial line OX at A and the curve $r = 3 + 2 \cos \theta$ at B and C. Show that the triangle OAB has area $2\sqrt{3}$. Find also the area of the smaller of the two portions into which the line BC divides the area enclosed by the curve $r = 3 + 2 \cos \theta$.
24. Find the area of a loop of the curve $r = \cos^2 \theta$.
25. Show that the curve $r = 1 + 2 \sin \theta$ consists of an outer and an inner loop, and that the area of the inner loop is $\frac{1}{2}(2\pi - 3\sqrt{3})$.

19.2 Volume

Let $A(x)$ be the area of the cross-section of a solid by a plane perpendicular to Ox at a distance x from O. Denote by $V_{a,b}$ the volume of the solid enclosed between planes perpendicular to Ox at $x = a$ and $x = b$, and let this volume be divided up into a number of thin slices by planes perpendicular to Ox. Assuming that $A(x)$ is continuous on the interval $a \leqslant x \leqslant b$, the volume $V_{x,x+\Delta x}$ of the typical slice shown in Fig. 19.17 is

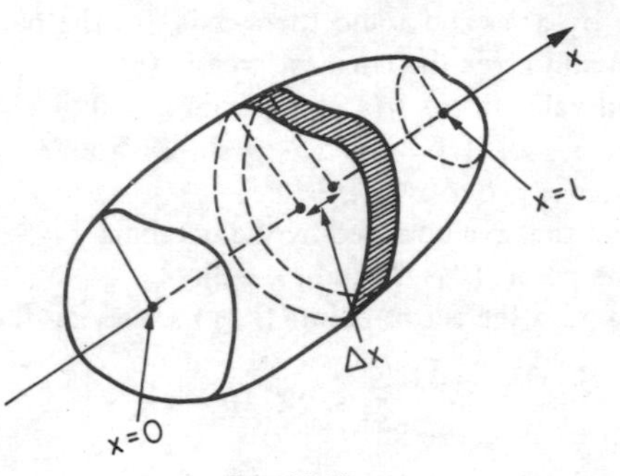

FIG. 19.17

approximately $A(x)\,\Delta x$, and we get

$$V_{a,b} = \int_a^b A(x)\,\mathrm{d}x.$$

Example 12 Find the volume of a right pyramid of height h standing on a square base of side b.

The cross-section at distance x from the vertex (see Fig. 19.18) is a

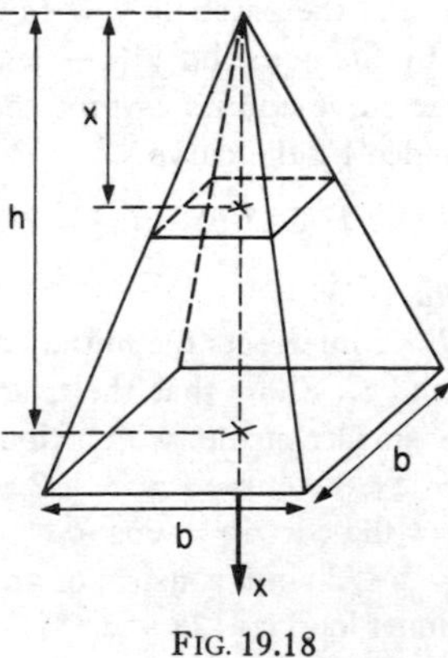

FIG. 19.18

square of side bx/h. The area of this square is b^2x^2/h^2 and so the volume of the pyramid is

$$\int_0^h \frac{b^2}{h^2}\,x^2\,\mathrm{d}x = \tfrac{1}{3}b^2\,h.$$

Example 13 Find the volume of a pyramid given that the base has area B and the height is h.

Consider two corresponding dimensions D, d in the base and in the cross-section respectively (see Fig. 19.19). From similar triangles we see that

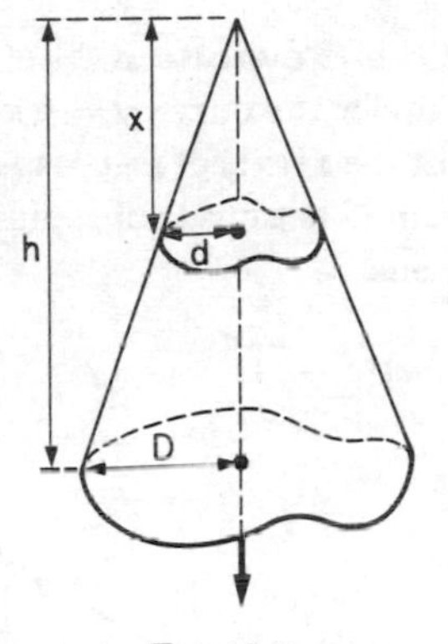

FIG. 19.19

$d = (x/h)D$. It follows that the area $A(x)$ of the cross-section is given by $A(x) = (x/h)^2 B$. Hence the volume of the pyramid is

$$\int_0^h (x/h)^2 B \, dx = \frac{B}{h^2}\left[\tfrac{1}{3}x^3\right]_0^h = \tfrac{1}{3}Bh.$$

Example 14 A solid has two plane faces meeting at right angles to form a straight edge. The length of the straight edge is π and each cross-section perpendicular to this edge is a right-angled isoceles triangle, the right angle having its vertex on the edge and the equal sides having length $\sin x$, where x is the distance of the section from one end of the edge. Find the volume of the solid.

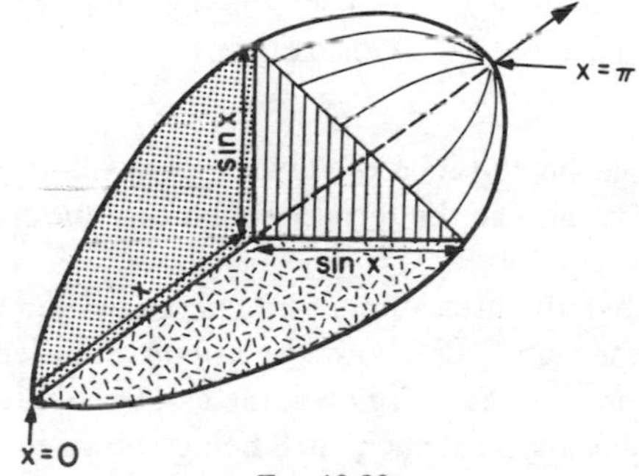

FIG. 19.20

The cross-sectional area $A(x)$ at distance x from one end is given by $A(x) = \frac{1}{2}\sin^2 x$ (see Fig. 19.20). Hence

$$\text{volume} = \int_0^{\pi} \tfrac{1}{2}\sin^2 x \, dx = \tfrac{1}{4}\int_0^{\pi} (1 - \cos 2x) \, dx = \pi/4.$$

19.2.1 Volume of a Solid of Revolution Let $V_{a,b}$ denote the volume swept out when the area under the curve $y = f(x)$ from $x = a$ to $x = b$ is rotated through 360° about the x-axis. The cross-section perpendicular to the x-axis at distance x from O is a circle of radius y, where $y = f(x)$. The area of this circle is πy^2, and so

$$V_{a,b} = \int_a^b \pi y^2 \, dx.$$

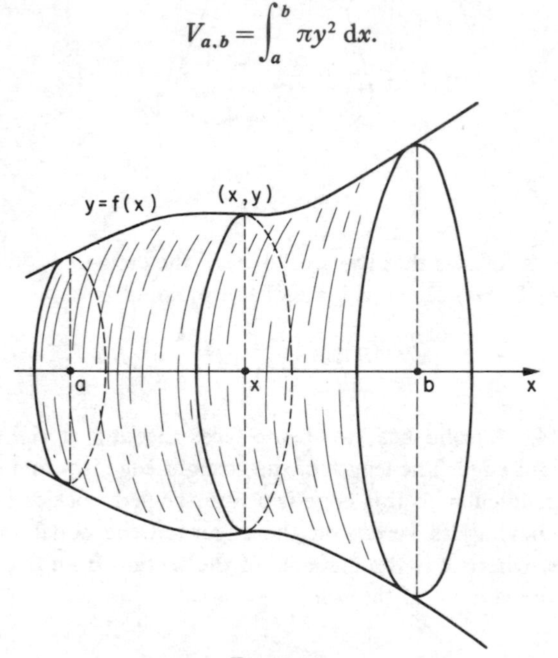

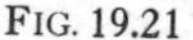
FIG. 19.21

This result is based on the idea of dividing the volume into thin slices perpendicular to Ox, and so the term 'circular disc method' is sometimes used.

Suppose next that the area under the curve $y = f(x)$ from $x = a$ to $x = b$ is rotated through 360° about the y-axis, and denote by $\tilde{V}_{a,b}$ the volume swept out in this case. This volume can be divided into a number of 'cylindrical shells', a typical such shell being shown in Fig. 19.22—this shell is swept out by the rotation of the typical strip of area between x

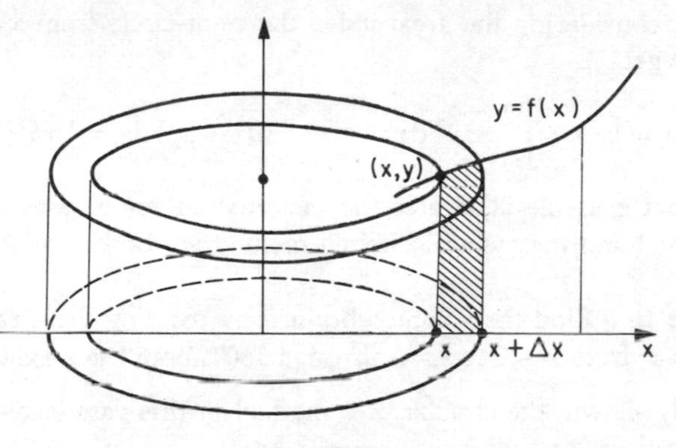

FIG. 19.22

and $x + \Delta x$. The volume of the shell is approximately the product of its inner circumference, inner height, and thickness, i.e. $2\pi xy\ \Delta x$. Hence

$$\tilde{V}_{a,b} = \int_a^b 2\pi xy\ dx.$$

In contrast to the circular disc method used to find $V_{a,b}$, the above method of finding $\tilde{V}_{a,b}$ is called the 'cylindrical shell method'.

Similarly, if the area between the curve $x = g(y)$ and the y-axis from $y = c$ to $y = d$ is rotated through 360° about the x-axis the volume swept out is $\int_c^d 2\pi yx\ dy$; if it is rotated through 360° about the y-axis the volume swept out is $\int_c^d \pi x^2\ dy$.

Example 15 Find the volume of a sphere of radius a, and also of a spherical cap of height h.

The sphere is swept out by the rotation of the area under the semi-circle $y = \sqrt{(a^2 - x^2)}$ about the x-axis. Hence

$$\text{volume of sphere} = 2\int_0^a \pi y^2\ dx = 2\pi \int_0^a (a^2 - x^2)\ dx = \tfrac{4}{3}\pi a^3.$$

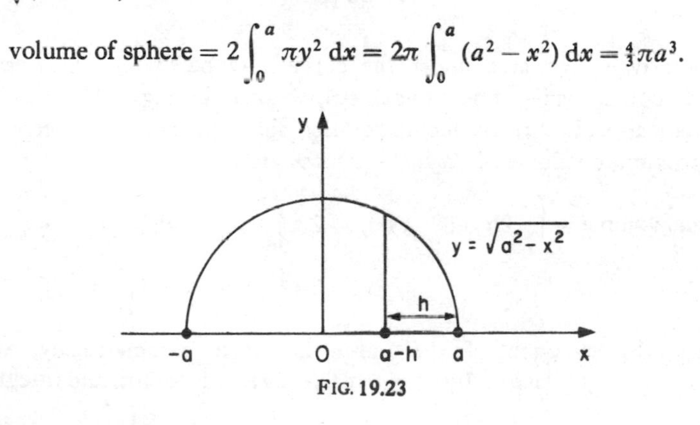

FIG. 19.23

Likewise, considering the area under the semi-circle from $x = a - h$ to $x = a$, we get

$$\text{volume of cap} = \int_{a-h}^{a} \pi y^2 \, dx = \pi \int_{a-h}^{a} (a^2 - x^2) \, dx = \tfrac{1}{3}\pi h^2(3a - h).$$

The next example illustrates the fact that in some cases the circular disc method and the cylindrical shell method are both available.

Example 16 Find the volume obtained by rotating the area under the curve $y = x^2$ from $x = 0$ to $x = 1$ through 360° about the x-axis.

As already shown, the circular disc method in this case leads to the formula $\int_0^1 \pi y^2 \, dx$, and we have

$$\int_0^1 \pi y^2 \, dx = \int_0^1 \pi x^4 \, dx = \pi/5.$$

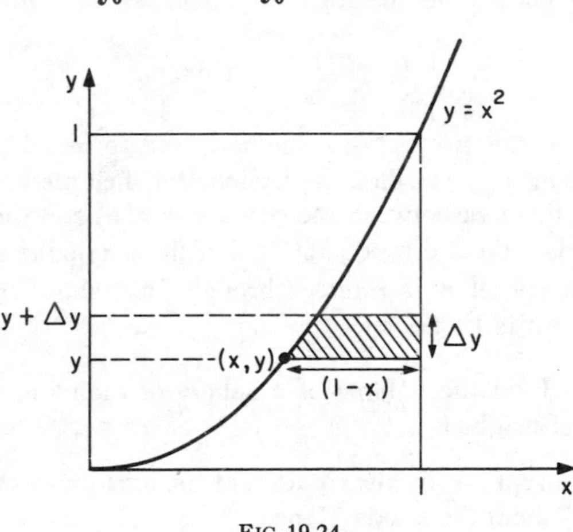

FIG. 19.24

Alternatively, the area under the curve may be divided into strips parallel to the x-axis. The typical strip shown in Fig. 19.24 has approximate area $(1 - x)\,\Delta y$, and on rotation about the x-axis it sweeps out an approximate volume of $2\pi y(1 - x)\,\Delta y$. Hence

$$\text{total volume} = \int_0^1 2\pi y(1 - x) \, dy = 2\pi \int_0^1 y(1 - y^{1/2}) \, dy$$

$$= 2\pi \left[\tfrac{1}{2}y^2 - \tfrac{2}{5}y^{5/2}\right]_0^1 = \pi/5.$$

When the equation of the curve is given parametrically, with parameter t, we replace dx by $(dx/dt)\,dt$, or dy by $(dy/dt)\,dt$, and integrate with respect to t between appropriate limits. This point is illustrated in Example 17, which also shows how Simpson's rule may be used when exact integration is not possible.

Example 17 The curve C has parametric equations $x = (1/\pi)\tan^{-1} t$, $y = e^t$. The area bounded by C and the lines $x = 0, y = 0, x = \frac{1}{4}$ is rotated through 2π radians about the x-axis. Find approximately the volume swept out.

The volume V is given by

$$V = \int_0^{1/4} \pi y^2 \, dx = \int_0^1 \pi y^2 \frac{dx}{dt} dt = \int_0^1 \frac{e^{2t}}{1+t^2} dt.$$

We now evaluate this integral using Simpson's rule with four strips (Table 19.1). From the table $V \simeq \frac{1}{12}(26{\cdot}722) = 2{\cdot}23$ (to two decimal places).

TABLE 19.1

t	$1 + t^2$	e^{2t}	$e^{2t}/(1 + t^2)$	Factor	
0	1·0000	1·000	1·000	1	1·000
0·25	1·0625	1·649	1·552	4	6·208
0·5	1·2500	2·718	2·174	2	4·348
0·75	1·5625	4·482	2·868	4	11·472
1·0	2·0000	7·389	3·694	1	3·694
				Total	26·722

Example 18 Sketch the curve given by the equation

$$y^2 = a^2x/(2a - x).$$

Prove that the area enclosed by the above curve and the line $x = a$ is $(\pi - 2)a^2$, and that the volume traced out by rotating this area about the axis $y = 0$ through two right angles is

$$\pi(\ln 4 - 1)a^3.$$

This curve, which is symmetrical about Ox, touches Oy at the origin, and has $x = 2a$ as vertical asymptote. There are no real values of y when $x < 0$ or when $x > 2a$. Logarithmic differentiation gives

$$\frac{dy}{dx} = \frac{ay}{x(2a - x)} = \pm \frac{a^2}{\sqrt{[x(2a - x)^3]}},$$

from which we see that there is no point at which $dy/dx = 0$. The form of the curve when $a > 0$ is shown in Fig. 19.25.

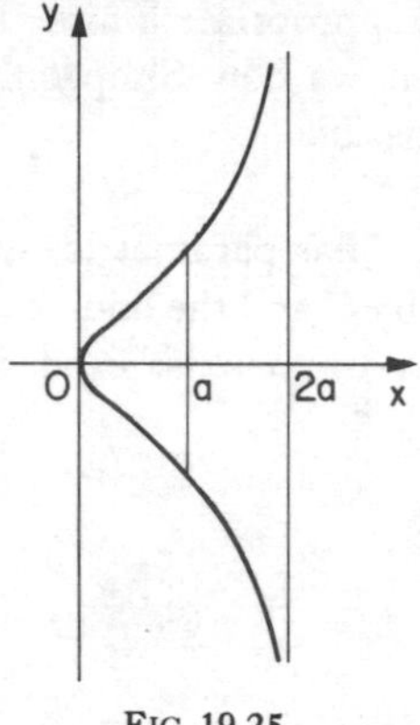

FIG. 19.25

The area enclosed by the curve and the line $x = a$ is given by

$$I_1 = 2\int_0^a \frac{a\sqrt{x}}{\sqrt{(2a-x)}}\,dx.$$

Put $x = 2a\sin^2\theta$, $dx = 4a\sin\theta\cos\theta\,d\theta$. Then

$$I_1 = 8a^2\int_0^{\pi/4} \sin^2\theta\,d\theta = 4a^2\left[\theta - \tfrac{1}{2}\sin 2\theta\right]_0^{\pi/4} = (\pi - 2)\,a^2.$$

The volume traced out by rotating this area about Ox is given by

$$\begin{aligned} I_2 &= \pi\int_0^a \frac{a^2x}{2a-x}\,dx \\ &= \pi a^2\int_a^{2a}\left(\frac{2a}{t} - 1\right) dt, \quad t = 2a - x \\ &= \pi a^2\left[2a\ln t - t\right]_a^{2a} = \pi a^3(\ln 4 - 1). \end{aligned}$$

Example 19 Find the area enclosed between the ellipse $3x^2 + 4y^2 = 3$ and the parabola $8y^2 = 9x$, and the volume generated when this area revolves through four right angles about the y-axis.

From $8y^2 = 9x$ substitute for y in the equation $3x^2 + 4y^2 = 3$ to get

$$3x^2 + \tfrac{9}{2}x = 3,$$

which gives $x = -2$ or $\frac{1}{2}$. Since the equation $8y^2 = 9x$ implies that $x \geqslant 0$, we conclude that $(\frac{1}{2}, \frac{3}{4})$ and $(\frac{1}{2}, -\frac{3}{4})$ are the points of intersection of the two curves. Dividing the area between the two curves into strips parallel to the x-axis, we see that the typical strip shown in Fig. 19.26 has approximate area $(x_2 - x_1)\,\Delta y$, where $3x_2^2 + 4y^2 = 3$ and $9x_1 = 8y^2$. Hence

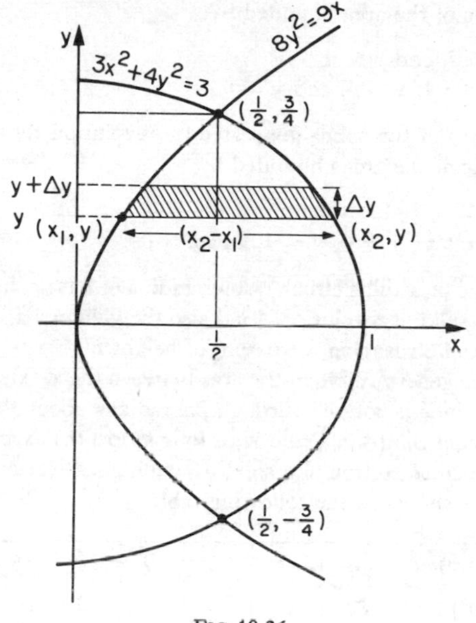

FIG 19.26

the total area is

$$2\int_0^{3/4}(x_2-x_1)\,dy=2\int_0^{3/4}[\sqrt{(1-\tfrac{4}{3}y^2)}-\tfrac{8}{9}y^2]\,dy.$$

Now

$$\int_0^{3/4}\sqrt{(1-\tfrac{4}{3}y^2)}\,dy=\frac{\sqrt{3}}{2}\int_0^{\pi/3}\cos^2\theta\,d\theta\quad\left(y=\frac{\sqrt{3}}{2}\sin\theta\right),$$
$$=\frac{\sqrt{3}}{4}\Big[\theta+\tfrac{1}{2}\sin 2\theta\Big]_0^{\pi/3}=\frac{\sqrt{3}}{4}\left(\frac{\pi}{3}+\frac{\sqrt{3}}{4}\right),$$

and $\int_0^{3/4}\frac{8}{9}y^2\,dy=\frac{1}{8}$.

Thus the total area is $(\sqrt{3}/24)(4\pi+\sqrt{3})$. If this area rotates about Oy, the volume generated is

$$V=2\pi\int_0^{3/4}(x_2^2-x_1^2)\,dy$$
$$=2\pi\int_0^{3/4}(1-\tfrac{4}{3}y^2-\tfrac{64}{81}y^4)\,dy=2\pi\Big[y-\tfrac{4}{9}y^3-\tfrac{64}{405}y^5\Big]_0^{3/4}=\frac{21\pi}{20}.$$

Exercises 19(b)

1. Find the volumes of the solids generated by revolution through 2π radians

about the x-axis of the area bounded by

(a) $y = x^4$, $x = 2$, and $y = 0$,
(b) $y^4 = 16x$, $x = 1$, $x = 4$, and $y = 0$.

2. Find the volumes of the solids generated by revolution through 2π radians about the y-axis of the areas bounded by

(a) $y^2 = x^3$, $y = 8$, and $x = 0$,
(b) $y^2 = (1 - x^2)/x^2$, $y = 0$, $y = 1$, and $x = 0$.

3. Find the volume of a right circular cone, radius of base r, height h, by considering it as a solid of revolution. Find also the volume left when the top of this cone is sliced off leaving a frustum of height h'.
4. Find the volume generated when the area between the x-axis and one arch of the curve $y = \sin x$ is rotated through 2π radians about the x-axis (a) approximately, using Simpson's rule with four strips, (b) exactly.
5. The area of the cross-section of a solid, 18 cm high, at various heights above its plane base is shown in the following table:

Height (cm)	0	3	6	9	12	15	18
Area (cm^2)	20	24	30	38	23	10	0

Use Simpson's rule to find the approximate volume of the solid.

6. The volume of a barrel is estimated by measuring its circumference at various heights above the bottom. Given that the height of the barrel is 80 cm, estimate the capacity of the barrel from the following table (neglecting the thickness of the wood) by means of Simpson's rule with four strips:

Height from bottom (cm)	0	20	40	60	80
Circumference (cm)	150	155	160	150	140

(It may be assumed that cross-sections of the barrel are circular).

7. Find the volume of the solid generated by revolution through 2π radians about the line $x = 1$ of the area bounded by $y = x^3$, $y = 0$, and $x = 1$.
8. The curve $y^2 = x(4 - x)^2$ has a loop of area $\frac{256}{15}$ units. Show that the volume of the solid formed when this area is revolved through 180° about the x-axis is $64\pi/3$.
9. Sketch the curve $y^2 = x^2(4 - x^2)$. Find the area enclosed by the curve and the volume of the solid formed when this area is rotated through 180° about the x-axis.
10. Show that the curve $y = 4x - x^2$ is symmetric about the line $x = 2$. Sketch the area in the first quadrant bounded by this curve and the x-axis. Find the volume of the solid formed when this area is revolved through 360° about (a) the x-axis, (b) the y-axis.
11. A solid has a circular base of radius 4. Each plane section perpendicular to a fixed diameter is an equilateral triangle. Find the volume of the solid.

12. Use integration to derive the formula πab for the area enclosed by the ellipse

$$\frac{x^2}{a^2}+\frac{y^2}{b^2}=1.$$

The base of a certain solid is a triangle ABC, where $BC = 2b$ and h is the length of the altitude from A to BC. Cross-sections of the solid perpendicular to this altitude are semi-ellipses, the major axis lying in the base of the solid, the minor axis being half the major axis. Find the volume of the solid.

13. A solid sitting on a horizontal shelf has its base and top at heights h_1 and h_2, respectively, above the floor. Find the volume V of the solid given that at height h above the floor ($h_1 \leqslant h \leqslant h_2$) the area of horizontal cross-section is $ah^2 + bh + c$, where a, b, c are constants. Verify the formula

$$V=\tfrac{1}{6}H(B + 4M + T)$$

where B, M, T are the areas of horizontal cross-section at the bottom, middle, and top respectively, while H is the height of the solid.

Show that a sphere is a particular example of such a solid and that the above formula can be used to calculate the volume of a sphere.

14. A solid has a circular base, radius a, and each cross-section perpendicular to a certain diameter of the base is an isosceles right-angled triangle with its hypotenuse in the plane base. Find the volume of the solid.

15. A four-sided solid is formed by sawing a corner off a rectangular block. The base of the solid is a right-angled triangle with sides 3 cm and 4 cm meeting at the right angle, and the altitude of the solid is 5 cm. Find its volume.

16. The base of a certain solid is the circle $x^2 + y^2 = a^2$. Each plane section of the solid perpendicular to the x-axis is a square with one edge of the square in the base of the solid. Find the volume of the solid.

17. Find the volume of the right cone of height h whose base is an ellipse of major axis $2a$ and minor axis $2b$.

18. The closed boundary C of a certain region consists of part of a parabola and a straight line perpendicular to the axis of the parabola. Find the area of the enclosed region, given that the length of the straight boundary is $2l$ and its distance from the vertex of the parabola is l.

A solid has its base in the form of an ellipse with major and minor axes $2a$ and $2b$ respectively. Find the volume of the solid if each section perpendicular to the major axis of the base is similar to C with straight portion in the base.

19. The interior of a certain vessel is the space swept by the area bounded by the curve $y=\sqrt{x-1}$ and the lines $x = 0, y = 0, y = 2$, when this area is rotated through 2π radians about the y-axis. Given that the y-axis is vertical with the x-axis in the base of the vessel and the vessel is filled with water to a depth h ($0 < h < 2$), find a formula in terms of h for the volume of water. Find the relation between the rate of change of this volume and the rate of change of h when water is being poured into the vessel. At what variable rate must water be poured in to maintain a constant rate of increase in water level?

20. Sketch the curve $ay^2 = x^2(a - x)$, where $a > 0$. Find the area of the loop of the curve and the volume generated when the loop rotates through π radians about the axis of x.
21. Sketch the curves

$$ay^2 = (x - a)(x - 2a)^2, \quad ay^2 = (x - a)(x - 2a)(x - 3a)$$

for a positive value of the constant a.

Find the area of the loop of the first curve, and the volume of the solid formed by the revolution of the loop of the latter curve round the axis of x.
22. Prove that, if $a > 0$, the curves $y^2 = 4ax$ and $27ay^2 = 4(x - 2a)^3$ intersect where $x = 8a$, and that the area bounded by the curves is $352\sqrt{2a^2}/15$. Prove that if this area revolves through two right angles about the axis of x, then the volume generated is $80\pi a^3$.
23. Sketch the curve $y^2 = x^2(a - x)/(a + x)$, $a > 0$, and show that the tangents to the curve at the origin are perpendicular. Show also that the area enclosed by the loop of the curve is $(4 - \pi)a^2/2$. Show that the volume generated when the area enclosed by the loop is rotated through π about the x-axis is $\pi a^3(6 \ln 2 - 4)/3$.
24. Prove that the volume obtained by revolving the cycloid

$$x = a(\theta - \sin\theta), \quad y = a(1 - \cos\theta)$$

about the axis of x, between the points where $x = 0$ and $x = 2\pi$ is $5\pi^2a^3$.
25. The curve $y = ae^{-x/a}$ passes through the point (b, c). The area bounded by this curve, the axes, and the line $x = b$ is rotated through four right angles about the line $x = b$. Show that the volume swept out is $2\pi a^2(c + b - a)$.
26. Sketch roughly the curve whose equation is $y^2 = x^3/(a - x)$ where $a > 0$, and find the area included between the curve and its asymptote. Find also the volume generated on revolving the curve about its asymptote.
27. Sketch in the same diagram the curves $xy^2 = a^2(a - x)$ and $(a - x)y^2 = a^2x$. Prove that they enclose an area $(\pi - 2)a^2$, and find the volume of the solid obtained by rotating this area through two right angles about the line $y = 0$.
28. Find the volume generated when the area contained between the ellipse $x^2 + 2y^2 = 2$ and the two branches of the hyperbola $2x^2 - 2y^2 = 1$ revolves through two right angles about the y-axis.
29. Given that the area bounded by the curve $r = f(\theta)$ and the radii vectors $\theta = \theta_1$ and $\theta = \theta_2$ is rotated through four right angles about the initial line, prove that the volume of the resulting solid of revolution is given by

$$\tfrac{2}{3}\pi \int_{\theta_1}^{\theta_2} r^3 \sin\theta \, d\theta.$$

19.3 Arc Length

Let $s_{a,b}$ denote the length of the curve $y = f(x)$ from the point A where $x = a$ to the point B where $x = b$. The arc AB of the curve may be divided into small pieces by lines parallel to the y-axis, and we denote by $s_{x,x+\Delta x}$ the length of the typical piece PQ shown in Fig. 19.27. We shall take the

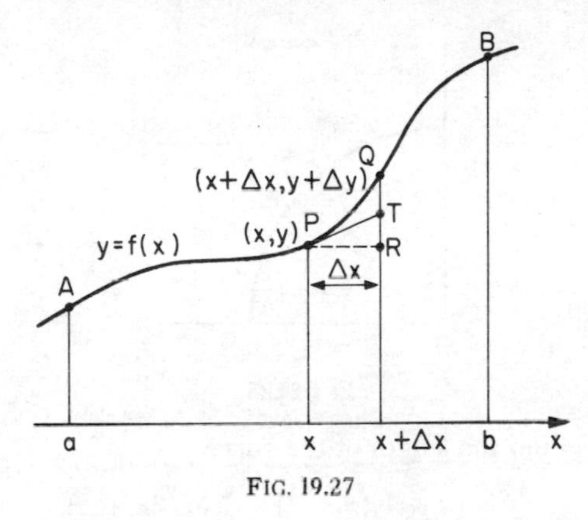

FIG. 19.27

length of the piece of tangent PT as an approximation to $s_{x,x+\Delta x}$. This approximation satisfies the criterion of Section 16.3.1, for the difference between $s_{x,x+\Delta x}$ and PT is less than TQ ($PT + TQ > s_{x,x+\Delta x}$) and

$$TQ = RQ - RT = \Delta y - f'(x)\,\Delta x,$$

so that

$$\frac{TQ}{\Delta x} = \frac{\Delta y}{\Delta x} - f'(x) \to 0 \text{ as } \Delta x \to 0.$$

(The chord PQ is also a sufficiently close approximation to $s_{x,x+\Delta x}$.)
From the right-angled triangle PRT we have

$$PT^2 = (\Delta x)^2 + [f'(x)\,\Delta x]^2, \quad \text{that is} \quad PT = \sqrt{\{1 + [f'(x)]^2\}}\,\Delta x.$$

Since $s_{x,x+\Delta x} \simeq \sqrt{\{1 + [f'(x)]^2\}}\,\Delta x$, we get

$$s_{a,b} = \int_a^b \sqrt{\{1 + [f'(x)]^2\}}\,\mathrm{d}x.$$

Example 20 Find the circumference of a circle of radius a.
The curve $y = \sqrt{(a^2 - x^2)}$ from $x = 0$ to $x = a$ represents a quarter of the total circumference. From the equation of the curve we get

$$1 + \left(\frac{\mathrm{d}y}{\mathrm{d}x}\right)^2 = \frac{a^2}{a^2 - x^2},$$

and hence

$$\begin{aligned}\text{total circumference} &= 4\int_0^a \sqrt{\left[1 + \left(\frac{\mathrm{d}y}{\mathrm{d}x}\right)^2\right]}\,\mathrm{d}x \\ &= 4\int_0^a \frac{a}{\sqrt{(a^2 - x^2)}}\,\mathrm{d}x = 4a\left[\sin^{-1}\left(\frac{x}{a}\right)\right]_0^a \\ &= 2a\pi.\end{aligned}$$

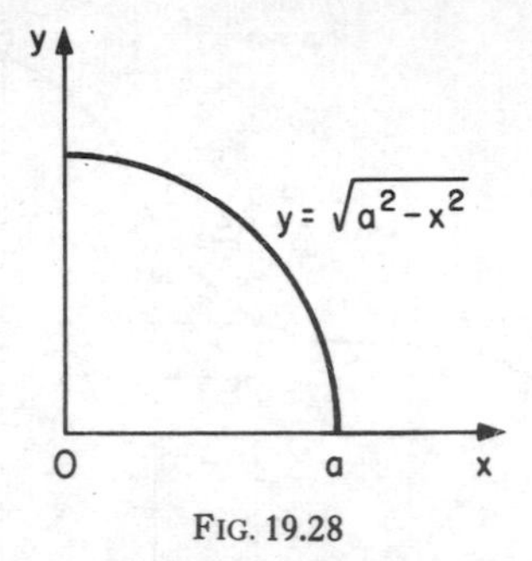

FIG. 19.28

Example 21 Find the length of the curve

$$y = (x + 1)(x + 2) - \tfrac{1}{8} \ln (2x + 3)$$

between the points where $x = 1$ and $x = 2$.

From

$$y = (x + 1)(x + 2) - \tfrac{1}{8} \ln (2x + 3),$$

we get

$$\frac{dy}{dx} = 2x + 3 - \frac{1}{4(2x + 3)}, \quad \text{and}$$

$$1 + \left(\frac{dy}{dx}\right)^2 = \left[(2x + 3) + \frac{1}{4(2x + 3)}\right]^2.$$

Hence the required length is

$$\int_1^2 \left[(2x + 3) + \frac{1}{4(2x + 3)}\right] dx = \left[x^2 + 3x + \tfrac{1}{8} \ln (2x + 3)\right]_1^2$$
$$= 6 + \tfrac{1}{8} \ln (\tfrac{7}{5}).$$

Likewise the length of the curve $x = g(y)$ from the point where $y = c$ to the point where $y = d$ is

$$\int_c^d \sqrt{\left[1 + \left(\frac{dx}{dy}\right)^2\right]}\, dy.$$

For instance in Example 20 the equation of the curve may be written as $x = \sqrt{(a^2 - y^2)}$, giving

$$1 + \left(\frac{dx}{dy}\right)^2 = \frac{a^2}{a^2 - y^2},$$

and so the total circumference is

$$4 \int_0^a \frac{a}{\sqrt{(a^2 - y^2)}}\, dy = 2a\pi.$$

If x and y are given in terms of a parameter t then, writing $\dot{x} = \mathrm{d}x/\mathrm{d}t$, $\dot{y} = \mathrm{d}y/\mathrm{d}t$, we have $\mathrm{d}y/\mathrm{d}x = \dot{y}/\dot{x}$ and so

$$\int_a^b \sqrt{\left[1 + \left(\frac{\mathrm{d}y}{\mathrm{d}x}\right)^2\right]}\mathrm{d}x = \int_{t_1}^{t_2} \sqrt{\left(1 + \frac{\dot{y}^2}{\dot{x}^2}\right)}\, \dot{x}\, \mathrm{d}t$$

if t increases steadily from t_1 to t_2 as x increases from a to b ($\dot{x} > 0$), or

$$-\int_{t_1}^{t_2} \sqrt{\left(1 + \frac{\dot{y}^2}{\dot{x}^2}\right)}\, \dot{x}\, \mathrm{d}t$$

if t decreases steadily from t_1 to t_2 as x increases from a to b ($\dot{x} < 0$). Thus

$$\text{arc length} = \left| \int_{t_1}^{t_2} \sqrt{(\dot{x}^2 + \dot{y}^2)}\, \mathrm{d}t \right| .$$

For instance in Example 20 the equation of the curve may be written in the parametric form $x = a \cos \theta$, $y = a \sin \theta$ giving

$$\sqrt{(\dot{x}^2 + \dot{y}^2)} = \sqrt{(a^2 \sin^2 \theta + a^2 \cos^2 \theta)} = a,$$

and so the total circumference is

$$4\left|\int_{\pi/2}^{0} a\, \mathrm{d}\theta\right| = 4|-a\pi/2| = 2a\pi.$$

Example 22 Find the length of one arch of the cycloid

$$x = a(\theta - \sin \theta), \quad y = a(1 - \cos \theta).$$

The curve is shown in Fig. 15.30. We have

$$\dot{x} = a(1 - \cos \theta), \quad \dot{y} = a \sin \theta$$

and so

$$\dot{x}^2 + \dot{y}^2 = 2a^2(1 - \cos \theta) = 4a^2 \sin^2 (\theta/2).$$

Hence the length of one arch is

$$\int_0^{2\pi} 2a \sin (\theta/2)\, \mathrm{d}\theta = \Big[-4a \cos (\theta/2)\Big]_0^{2\pi} = 8a.$$

19.3.1 Arc Length in Polar Co-ordinates Let $s_{\alpha,\beta}$ denote the length of the curve $r = f(\theta)$ from the point where $\theta = \alpha$ to the point where $\theta = \beta$, and let $s_{\theta,\theta+\Delta\theta}$ denote the length of the typical piece PQ shown in Fig. 19.29. The length of the chord PQ can be found by appling the cosine rule to the triangle OPQ, as follows:

$$PQ^2 = (r + \Delta r)^2 + r^2 - 2r(r + \Delta r) \cos \Delta\theta$$
$$= 2r^2 + 2r\,\Delta r + (\Delta r)^2 - 2r(r + \Delta r)\left[1 - 2\sin^2\left(\frac{\Delta\theta}{2}\right)\right]$$
$$= (\Delta r)^2 + 4r(r + \Delta r)\sin^2\left(\frac{\Delta\theta}{2}\right)$$
$$= (\Delta\theta)^2\left[\left(\frac{\Delta r}{\Delta\theta}\right)^2 + r(r + \Delta r)\left(\frac{\sin(\Delta\theta/2)}{(\Delta\theta/2)}\right)^2\right].$$

As $\Delta\theta \to 0$ we have

$$\frac{\Delta r}{\Delta\theta} \to \frac{dr}{d\theta}, \quad \frac{\sin(\Delta\theta/2)}{(\Delta\theta/2)} \to 1$$

and so

$$s_{\theta,\theta+\Delta\theta} \simeq PQ \simeq \sqrt{\left[r^2 + \left(\frac{dr}{d\theta}\right)^2\right]}\,\Delta\theta.$$

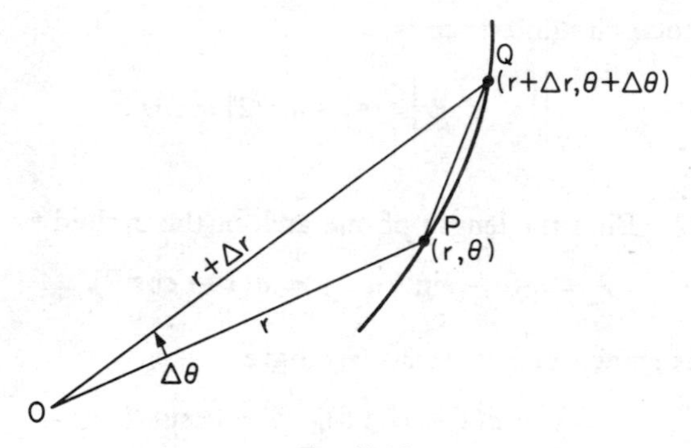

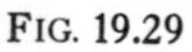
FIG. 19.29

Hence

$$s_{\alpha,\beta} = \int_\alpha^\beta \sqrt{\left[r^2 + \left(\frac{dr}{d\theta}\right)^2\right]}\,d\theta.$$

For instance, in Example 20 the polar equation of the circle is $r = a$, giving $dr/d\theta = 0$ and $r^2 + (dr/d\theta)^2 = a^2$; hence the circumference of the circle is

$$\int_0^{2\pi} a\,d\theta = 2a\pi.$$

Example 23 Find the total length of the cardioid $r = a(1 + \cos\theta)$.

The curve is shown in Fig. 15.26. In this case $dr/d\theta = -a\sin\theta$, and so

$$r^2 + \left(\frac{dr}{d\theta}\right)^2 = a^2(1+\cos\theta)^2 + a^2\sin^2\theta$$
$$= 2a^2(1+\cos\theta) = 4a^2\cos^2(\theta/2).$$

The curve is symmetric about the initial line and so its total length is

$$2\int_0^{\pi} 2a\cos(\theta/2)\,d\theta = 8a.$$

Exercises 19(c)

1. For the curve $y = \cosh x$, show that $1 + (dy/dx)^2 = y^2$ and hence find the length of this curve from the point where $x = 0$ to the point where $x = 1$.
2. For the curve $x^4 - 6xy + 3 = 0$, show that
$$1 + \left(\frac{dy}{dx}\right)^2 = \frac{(x^4+1)^2}{4x^4},$$
and hence find the length of this curve from the point where $x = 1$ to the point where $x = 2$.
3. Find the length of the curve $y = \ln[x + \sqrt{(x^2-1)}]$ between the points whose x-co-ordinates are 1 and 2.
4. Find the length of the curve $8x^2y = x^6 + 2$ between the points whose x-co-ordinates are 1 and 2.
5. Show that the length of the portion of the curve $y^2 = 5x^3$ from the point $(0, 0)$ to the point in the positive quadrant $(x > 0, y > 0)$ where it meets the circle $x^2 + y^2 = 6$ is $\frac{67}{27}$.
6. Show that the length of the curve $9y^2 = x(3-x)^2$ from its minimum turning point to the point $(3, 0)$ is $2\sqrt{3} - \frac{4}{3}$.
7. For the curve $y = \frac{1}{3}x^3 + 1/4x$, show that
$$1 + \left(\frac{dy}{dx}\right)^2 = \left(x^2 + \frac{1}{4x^2}\right)^2.$$
Find the arc length of the portion of the curve where $1 \leqslant x \leqslant 2$.
8. Given $y = \frac{1}{5}x^5 + 1/12x^3$, show that
$$1 + \left(\frac{dy}{dx}\right)^2 = \left(x^4 + \frac{1}{4x^4}\right)^2.$$
Show that the length of the curve $y = \frac{1}{5}x^5 + 1/12x^3$ from the point where $x = 1$ to the point where $x = 2$ is $3011/480$.
9. Find the length of the curve $y = (x-1)^{2/3}$ between the points where $x = 1$ and $x = 2$. (*Hint*: Express as an integral with respect to y.)
10. Find the length of the curve $x = at^2$, $y = at^3$ between the points $(0, 0)$ and (a, a).

11. Find the length of the curve $x = \ln t, y = \frac{1}{2}(t + 1/t)$ between the points (0, 1) and $(\ln 2, \frac{5}{4})$.
12. For the curve $x = 2a \cos t + a \cos 2t$, $y = 2a \sin t - a \sin 2t$, show that $\dot{x}^2 + \dot{y}^2 = 16a^2 \sin^2 (3t/2)$, where dots denote derivatives with respect to the parameter t. Find the length of the curve from the point where $t = 0$ to the point where $t = 2\pi/3$.
13. The parametric equations of a curve are $x = 3t^2, y = 3t^3 - t$; show that the cartesian equation is $3y^2 = x(x - 1)^2$ and sketch the curve. Find the length of the portion of the curve which corresponds to the interval $0 \leqslant t \leqslant 1$.
14. A curve is defined in terms of a parameter θ by the equations

$$x = a \cos^3 \theta, \quad y = a \sin^3 \theta,$$

where a is a constant. Prove that $dy/dx = -\tan \theta$, and show that the length of the curve which lies in the first quadrant of the xy-plane is $3a/2$.
15. Sketch the curve $16y^2 = x^2(2 - x^2)$ and find the area of one loop. Show that the total length of the curve is 2π.
16. Trace roughly the curve $8a^2y^2 = x^2(a^2 - 2x^2)$, and show that its whole length of arc is πa.

    Show that the area enclosed by the curve is two-thirds of that of the circumscribing rectangle whose sides are parallel to the axes of co-ordinates.
17. Sketch the curve $y = -\ln (1 - x^2)$, and find its length from the origin to the point where $x = x_1$ $(0 < x_1 < 1)$.

    Find also the area bounded by the curve, the x-axis, and the ordinate $x = x_1$. Show that, as $x_1 \to 1$, this area tends to the limit $2 - 2 \ln 2$.
18. Find the length of the curve $r = a \cos^3 \frac{1}{3}\theta$ from $\theta = 0$ to $\theta = 3\pi$.
19. Find the length of the equiangular spiral $r = ae^{\theta \cot \alpha}$ between the radii vectors whose lengths are r_1 and r_2 $(r_1 < r_2)$.

19.4 Surface Area

Denote by $S_{a,b}$ the surface area swept over by the curve $y = f(x)$ from $x = a$ to $x = b$ when it revolves through 360° about the x-axis. Slicing up this surface area by planes perpendicular to the x-axis, we divide it into a number of bands typified by that shown in Fig. 19.30. Let $\Delta s = s_{x,x+\Delta x}$ be the length of the arc PQ of the curve, so that

$$\Delta s \simeq \sqrt{\left[1 + \left(\frac{dy}{dx}\right)^2\right]} \Delta x.$$

The typical band of surface area shown in Fig. 19.30 is the area swept over by the arc PQ; if this area is denoted by $S_{x,x+\Delta x}$ we have

$$S_{x,x+\Delta x} \simeq 2\pi y \,\Delta s \simeq 2\pi y \sqrt{\left[1 + \left(\frac{dy}{dx}\right)^2\right]} \Delta x.$$

It follows that

$$S_{a,b} = \int_a^b 2\pi y \sqrt{\left[1 + \left(\frac{dy}{dx}\right)^2\right]} dx.$$

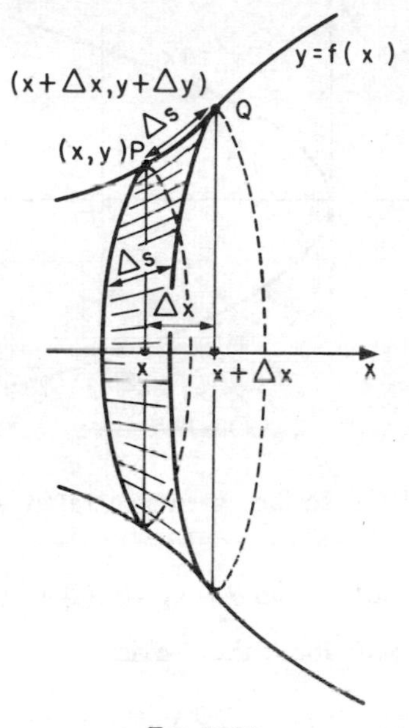

FIG. 19.30

(It should be noted that the approximation $2\pi y\,\Delta x$ for $S_{x,x+\Delta x}$ does *not* satisfy the criterion of Section 16.3.1.)

Example 24 Find the surface area swept over by the portion of the parabola $9y^2 = 4x$ lying between the lines $x = 0$ and $x = 1$, when it revolves through two right angles about the x-axis.

By symmetry we may consider only the upper half of the parabola and suppose that it revolves through *four* right angles about the x-axis. We then have $y = \frac{2}{3}x^{1/2}$, $dy/dx = 1/3x^{1/2}$, and so

$$1 + \left(\frac{dy}{dx}\right)^2 = 1 + \frac{1}{9x},$$

$$y\sqrt{\left[1 + \left(\frac{dy}{dx}\right)^2\right]} = \tfrac{2}{3}x^{1/2}\,\frac{\sqrt{(9x+1)}}{3x^{1/2}} = \tfrac{2}{9}\sqrt{(9x+1)}.$$

Hence the required surface area is

$$\frac{4\pi}{9}\int_0^1 \sqrt{(9x+1)}\,dx = \frac{8\pi}{243}\Big[(9x+1)^{3/2}\Big]_0^1 = \frac{8\pi}{243}(10^{3/2} - 1).$$

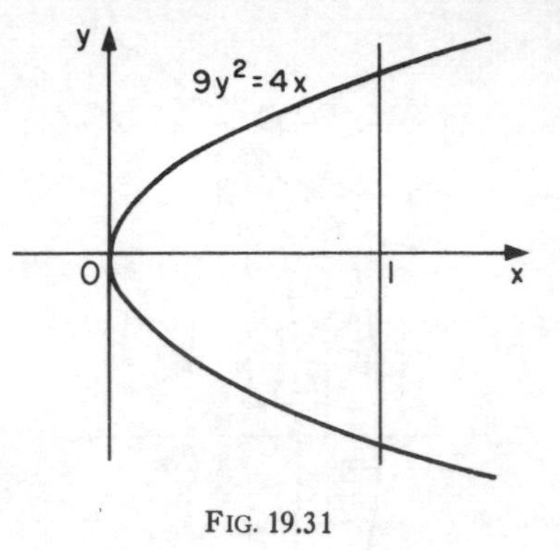

FIG. 19.31

Example 25 Find the surface area generated when one arch of the cycloid

$$x = a(\theta - \sin\theta), \quad y = a(1 - \cos\theta)$$

is revolved through 360° about the x-axis.

$$\begin{aligned}
\text{Surface area} &= \int_0^{2a\pi} 2\pi y \sqrt{\left[1 + \left(\frac{dy}{dx}\right)^2\right]} \, dx \\
&= \int_0^{2\pi} 2\pi y \sqrt{(\dot{x}^2 + \dot{y}^2)} \, d\theta \text{ where } \dot{x} = \frac{dx}{d\theta}, \dot{y} = \frac{dy}{d\theta} \\
&= \int_0^{2\pi} 2\pi a(1 - \cos\theta) \, 2a \sin(\theta/2) \, d\theta \quad \text{from Example 22} \\
&= \int_0^{2\pi} 8\pi a^2 \sin^3(\theta/2) \, d\theta \\
&= 16\pi a^2 \int_0^{\pi} \sin^3\left(\frac{\theta}{2}\right) d\theta \quad \text{by symmetry,} \\
&= 32\pi a^2 \int_0^{\pi/2} \sin^3\gamma \, d\gamma \quad \text{where } \gamma = \theta/2 \\
&= \tfrac{64}{3}\pi a^2 \quad \text{by (18.21).}
\end{aligned}$$

Example 26 Find approximately the length of the curve $y = x^4$ between the points (0, 0) and (1, 1), and the surface area generated when this arc is revolved through 360° about the x-axis.

We have $dy/dx = 4x^3$ and so

$$1 + \left(\frac{dy}{dx}\right)^2 = 1 + 16x^6.$$

Denoting the required length by l and the required surface area by S, we have

$$l = \int_0^1 \sqrt{(1 + 16x^6)}\, dx, \quad S = \int_0^1 2\pi x^4 \sqrt{(1 + 16x^6)}\, dx.$$

We are unable to evaluate these integrals by the usual method of finding indefinite integrals, so we use Simpson's rule. Table 19.2 sets out the calculation from Simpson's rule with four strips.

TABLE 19.2

x	x^4	$\sqrt{(1+16x^6)}$	Factor		$x^4\sqrt{(1+16x^6)}$	Factor	
0	0	1·000	1	1·000	0	1	0
0·25	0·0039	1·002	4	4·008	0·004	4	0·016
0·50	0·0625	1·118	2	2·236	0·070	2	0·140
0·75	0·3164	1·962	4	7·848	0·621	4	2·484
1·00	1·0000	4·123	1	4·123	4·123	1	4·123
			Total	19·215		Total	6·763

From Table 19.2, $l \simeq \frac{1}{12}(19{\cdot}215) \simeq 1{\cdot}60$, $S \simeq (\pi/6)(6{\cdot}763) \simeq 3{\cdot}54$.

Example 27 The region bounded by a quadrant of a circle of radius a and the tangents at its extremities revolves through 360° about one of these tangents. Show that the volume of the solid thus generated is $(\frac{5}{3} - \pi/2)\pi a^3$, and that the area of its curved surface is $\pi(\pi - 2)a^2$.

Let AC (Fig. 19.32) be the tangent about which rotation takes place. The small strip shown in Fig. 19.32 generates an approximate volume of $\pi(a - x)^2\, \Delta y$ with approximate curved surface area

$$2\pi(a - x)\sqrt{\left[1 + \left(\frac{dx}{dy}\right)^2\right]}\, \Delta y.$$

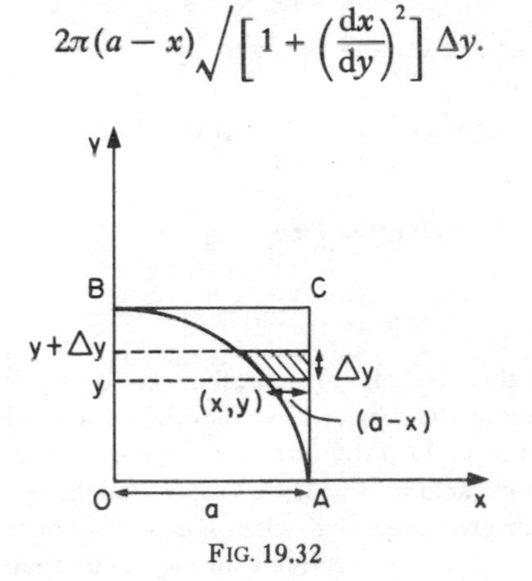

FIG. 19.32

Hence the total volume V and the total curved surface area S are given by

$$V = \int_0^a \pi(a-x)^2\,dy, \quad S = \int_0^a 2\pi(a-x)\sqrt{\left[1+\left(\frac{dx}{dy}\right)^2\right]}\,dy.$$

Writing the equation of the circle in parametric form, $x = a\cos\theta$, $y = a\sin\theta$, we have

$$V = \pi a^3 \int_0^{\pi/2} (1-\cos\theta)^2 \cos\theta\,d\theta$$

$$= \pi a^3 \int_0^{\pi/2} (\cos\theta - 2\cos^2\theta + \cos^3\theta)\,d\theta$$

$$= (\tfrac{5}{3} - \pi/2)\pi a^3.$$

Likewise,

$$S = 2\pi a^2 \int_0^{\pi/2} (1-\cos\theta)\,d\theta = \pi(\pi-2)a^2.$$

Example 28 The area in the first quadrant bounded by the axes and the curve $x = a\cos^3\theta$, $y = a\sin^3\theta$ is rotated through four right angles about Ox. Show that the surface area generated is $\frac{6}{5}\pi a^2$.

The curve is shown in Fig. 15.32(b). Writing $\dot{x} = dx/d\theta$, $\dot{y} = dy/d\theta$, we have

$$\dot{x} = -3a\cos^2\theta\sin\theta, \quad \dot{y} = 3a\sin^2\theta\cos\theta$$

and so

$$\dot{x}^2 + \dot{y}^2 = 9a^2\cos^2\theta\sin^2\theta(\cos^2\theta + \sin^2\theta) = 9a^2\cos^2\theta\sin^2\theta.$$

The required surface area S is given by

$$S = 2\pi \int_0^{\pi/2} y\sqrt{(\dot{x}^2 + \dot{y}^2)}\,d\theta = 2\pi \int_0^{\pi/2} a\sin^3\theta\,(3a\cos\theta\sin\theta)\,d\theta$$

$$= 6\pi a^2 \int_0^{\pi/2} \sin^4\theta\cos\theta\,d\theta = \tfrac{6}{5}\pi a^2.$$

Exercises 19(d)

1. The portion of the parabola $y^2 = 4ax$ between $x = 0$ and $x = 3a$ is rotated through 180° about the x-axis. Find the surface area generated.
2. Use integration to find (a) the total surface area of a sphere of radius a, (b) the curved surface area of a spherical cap of height h.
3. Find the surface area generated when the portion of the curve $y = \cosh x$ between $x = 0$ and $x = 1$ is rotated through 360° about the x-axis.

4. Find the surface area generated when the portion of the curve $x^4 - 6xy + 3 = 0$ between $x = 1$ and $x = 2$ is rotated through 360° about the x-axis.
5. The portion of the curve $x = a \cos^3 t, y = a \sin^3 t$ which corresponds to the interval $0 \leqslant t \leqslant \pi/2$ is revolved through 360° about the x-axis. Find the surface area generated.
6. The parabolic segment S is bounded by the parabola $y^2 = 2lx$ and the chord $x = l$. A rectangle R is inscribed in S with one side on the chord. Show that the maximum area of R is $1/\sqrt{3}$ times the area of S. If S is rotated through π radians about the axis of the parabola show that the curved surface area swept out is $(2\pi l^2/3)(3\sqrt{3} - 1)$.
7. The area bounded by the two parabolas $y^2 = 4ax$ and $x^2 = 4ay$ is rotated through 360° about the x-axis. Find the superficial area and the volume of the solid so formed.
8. Sketch the curve whose equation is $y = e^{-x} \sin x$, and show that the areas included between the axis of x and semi-undulations of the curve form a decreasing geometric progression.

   The curve $y = \sin x$ is rotated about the x-axis; find the superficial area generated by the portion of the curve lying between the lines $x = 0$ and $x = \pi$.
9. The curve C is represented by the equation

$$9ay^2 = (a - x)(x + 2a)^2 \quad (a > 0).$$

   Find the co-ordinates of the points on C at which the tangent is parallel to the x-axis, and also of the point at which C has two distinct tangents. Give a sketch of the curve.

   Find the area of the surface of revolution obtained by rotating the closed portion of this curve through two right angles about the x-axis.
10. The curve $y = c \cosh (x/c)$ cuts the axis Oy in the point C and the straight line $y = 2c$ in the points A and B. Prove that the volume of the solid formed by rotating the area ABC about the line AB through four right angles is

$$\pi[9 \ln (2 + \sqrt{3}) - 6\sqrt{3}]c^3.$$

   Prove also that the area of the surface of the solid is

$$\pi[4\sqrt{3} - 2 \ln (2 + \sqrt{3})]c^2.$$

11. A and B are the points on the curve $y = c \cosh (x/c)$ at which x has the values a and b respectively ($a < b$). The region bounded by the arc AB, the ordinates at A and B, and the x-axis is rotated through 360° about the x-axis. Given that V is the volume of the solid generated and S is its curved surface area, show that

$$V = \tfrac{1}{2}cS = \tfrac{1}{2}\pi c^2[b - a + \tfrac{1}{2}c \sinh (2b/c) - \tfrac{1}{2}c \sinh (2a/c)].$$

CHAPTER 20

FUNCTIONS OF TWO VARIABLES

20.1 Continuous Functions of Two Variables

A mapping from n-dimensional space R^n (see Section 11.3) into R is called a *function of n real variables*. For simplicity we shall deal with functions of two variables, but the methods used and the results obtained may be extended to functions of more than two variables. As examples of simple functions of two variables, consider the following:

(a) Let the image of the point (x, y) in R^2 be the distance r of the point from the origin, so that $r = \sqrt{(x^2 + y^2)}$. Then R^2 is mapped onto the set of non-negative real numbers.

(b) Let the image of the point (x, y) in R^2 be the area A of a rectangle with sides x and y, so that $A = xy$. In this case the subset

$$\{(x, y) : x \geqslant 0, y \geqslant 0\}$$

of R^2 is mapped onto the set of non-negative real numbers.

(c) Let the image of the point (r, h) in R^2 be the volume V of a cone with base radius r and height h, so that $V = \frac{1}{3}\pi r^2 h$. The domain and range are the same as in (b).

(d) Under certain conditions the pressure p, volume v and temperature T of a gas are related by the equation $pv = RT$, where R is a constant. In this connection we may wish to consider any of the mappings (i) $(v, T) \to p$, where $p = RT/v$, (ii) $(p, T) \to v$, where $v = RT/p$, (iii) $(p, v) \to T$, where $T = pv/R$.

The image of the point (x, y) under the mapping f is denoted by $f(x, y)$; note that whereas (x, y) is in R^2, $f(x, y)$ is a real number. The set of points (x, y, z) with $z = f(x, y)$ lie on a surface (Section 8.6) and this surface is a useful representation of the function f. The domain of f is represented by part, or all, of the xy-plane, and the image of the point P' is represented by the perpendicular height $P'P$ from P' to the surface (see Fig. 20.1).

As an example consider

$$f : (x, y) \mapsto \sqrt{(a^2 - x^2 - y^2)}.$$

Here

$$f(x, y) = \sqrt{(a^2 - x^2 - y^2)}$$

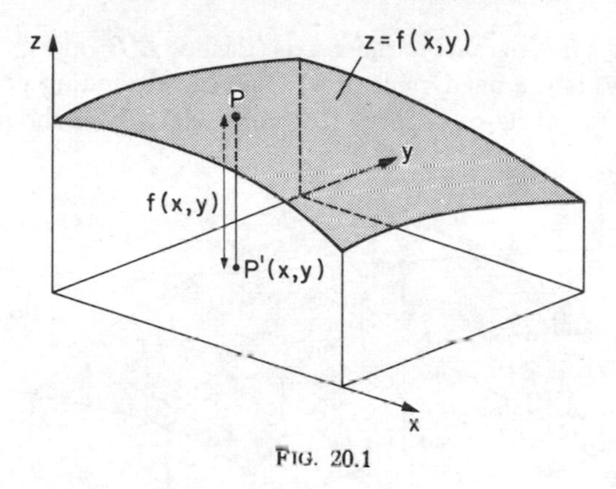

FIG. 20.1

and we shall assume as usual that only the positive square root is intended. The domain of f is the circular region

$$\{(x, y) : x^2 + y^2 \leqslant a^2\}$$

of the xy-plane and f is represented by the hemispherical surface $z=\sqrt{(a^2-x^2-y^2)}$, i.e. the upper half of the sphere $x^2+y^2+z^2=a^2$ (see Fig. 20.2).

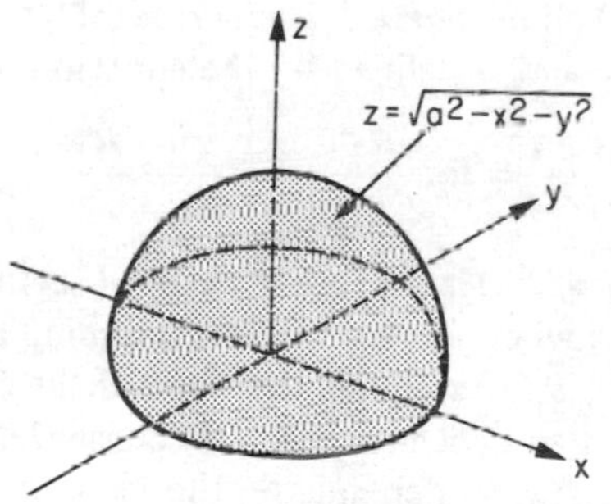

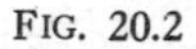
FIG. 20.2

We say $f(x, y)$ is continuous at (a, b) if $f(a, b)$ is defined and if $\lim f(x, y) = f(a, b)$ as $x \to a$ and $y \to b$; more precisely, for every positive number ε, however small, we must be able to find a number δ such that

$$|f(x, y) - f(a, b)| < \varepsilon \quad \text{when} \quad |x - a| < \delta \text{ and } |y - b| < \delta.$$

We shall assume that the functions under consideration in this chapter are continuous at all points in their domains.

20.2 Partial Derivatives

If y is held fixed while x changes, then the point $P'(x, y)$ in the domain of

f moves along a line parallel to the x-axis (the line $A'B'$ in Fig. 20.3 where $OA' = k$ (say) is the fixed value of y). The corresponding points on the surface $z = f(x, y)$ lie on a curve (the curve AB where the plane $y = k$

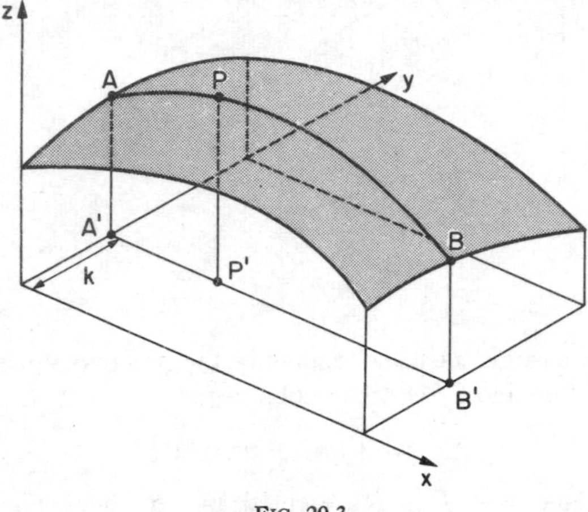

FIG 20.3

meets the surface). The *partial derivative* of f with respect to x is denoted by $\partial f/\partial x$ and is defined by the equation

$$\frac{\partial f}{\partial x} = \lim_{\Delta x \to 0} \frac{f(x + \Delta x, y) - f(x, y)}{\Delta x},$$

that is $\partial f/\partial x$ (or $\partial z/\partial x$) is the rate of change (Section 14.3) of $f(x, y)$ (or z) with respect to x when y is held fixed. Assuming the fixed value of y to be k (see Fig. 20.3), $\partial f/\partial x$ equals the slope of the curve AB; this varies along the curve and so depends on x. Of course the particular curve AB on the surface $z = f(x, y)$ depends on the value k chosen for y, and so $\partial f/\partial x$ depends on y also. Thus $\partial f/\partial x$ is a function of the two variables x and y; its value at the point P' (x, y) measures the slope of the surface $z = f(x, y)$ observed from P looking in the positive x-direction.

Similarly we define

$$\frac{\partial f}{\partial y} = \lim_{\Delta y \to 0} \frac{f(x, y + \Delta y) - f(x, y)}{\Delta y}$$

so that $\partial f/\partial y$ (or $\partial z/\partial y$) is the rate of change of $f(x, y)$ (or z) with respect to y when x is held fixed. The value of $\partial f/\partial y$ at the point P' (x, y) measures the slope of the surface $z = f(x, y)$ observed from P looking in the positive y-direction.

The notation $f_x = \partial f/\partial x$, $f_y = \partial f/\partial y$ is commonly used. To find

$\partial f/\partial x$, we regard $f(x, y)$ as a function of the single variable x, treating y as a constant, and we differentiate with respect to x in the usual way. Similarly we can find $\partial f/\partial y$ by ordinary differentiation with respect to y, treating x as a constant.

Example 1 If

$$z = x^2 + y^2 + 5xy + xy^2 + 2x^2y,$$

then

$$\frac{\partial z}{\partial x} = 2x + 5y + y^2 + 4xy, \quad \frac{\partial z}{\partial y} = 2y + 5x + 2xy + 2x^2.$$

Example 2 If $z = y/x + x/y$, then

$$\frac{\partial z}{\partial x} = -y/x^2 + 1/y, \quad \frac{\partial z}{\partial y} = 1/x - x/y^2.$$

Example 3 If $z = e^{x^2+y^2}$, then

$$\frac{\partial z}{\partial x} = 2xe^{x^2+y^2}, \quad \frac{\partial z}{\partial y} = 2ye^{x^2+y^2}.$$

In this context the chain rule (Section 14.4.3) may be stated as follows: if $z = g(u)$ where $u = h(x, y)$, then z is a function of x and y, and

$$\frac{\partial z}{\partial x} = g'(u)\frac{\partial h}{\partial x}, \quad \frac{\partial z}{\partial y} = g'(u)\frac{\partial h}{\partial y},$$

that is

$$\frac{\partial z}{\partial x} = \frac{dz}{du}\frac{\partial u}{\partial x}, \quad \frac{\partial z}{\partial y} = \frac{dz}{du}\frac{\partial u}{\partial y}.$$

Example 4 If $z = \tan^{-1}(y/x)$, then writing $u = y/x$ we have $z = \tan^{-1} u$ and so

$$\frac{\partial z}{\partial x} = \frac{1}{1+u^2}(-y/x^2) \quad = \frac{-y}{x^2+y^2},$$

$$\frac{\partial z}{\partial y} = \frac{1}{1+u^2}\left(\frac{1}{x}\right) \quad = \frac{x}{x^2+y^2}.$$

Example 5 If $z = 1/(2x + y)$, then writing $u = 2x + y$ we have $z = 1/u$ and so

$$\frac{\partial z}{\partial x} = -\frac{1}{u^2}2 = -\frac{2}{(2x+y)^2}, \quad \frac{\partial z}{\partial y} = -\frac{1}{u^2}1 = -\frac{1}{(2x+y)^2}.$$

Example 6 If $z = y + x \ln (x/y)$, then, using the product rule to differentiate $x \ln (x/y)$,

$$\frac{\partial z}{\partial x} = x \frac{\partial}{\partial x}[\ln (x/y)] + \ln (x/y),$$

and without formally substituting $u = x/y$ in order to differentiate $\ln (x/y)$, we see that

$$\frac{\partial z}{\partial x} = x \frac{1}{(x/y)} \frac{1}{y} + \ln (x/y) = 1 + \ln (x/y).$$

Likewise

$$\frac{\partial z}{\partial y} = 1 + \frac{x}{(x/y)} (-x/y^2) = 1 - x/y.$$

Example 7 Show that if $z = x^n f(y/x)$, where f denotes an arbitrary function, then

$$x \frac{\partial z}{\partial x} + y \frac{\partial z}{\partial y} = nz.$$

Writing $u = y/x$, we have $z = x^n f(u)$ and so

$$\frac{\partial z}{\partial x} = x^n f'(u) (-y/x^2) + nx^{n-1} f(u),$$

$$x \frac{\partial z}{\partial x} = -x^{n-1} y f'(u) + nx^n f(u). \tag{1}$$

Likewise

$$\frac{\partial z}{\partial y} = x^n f'(u) (1/x),$$

$$y \frac{\partial z}{\partial y} = x^{n-1} y f'(u). \tag{2}$$

The result follows immediately from (1) and (2). This result is known as Euler's first theorem for a homogeneous function of degree n.

The idea of partial differentiation extends easily to functions of more than two variables. For example, if z is a function of the three variables x, y, t then z has three partial derivatives

$$\frac{\partial z}{\partial x}, \quad \frac{\partial z}{\partial y}, \quad \frac{\partial z}{\partial t};$$

to obtain $\partial z/\partial x$ we differentiate with respect to x treating both y and t as constants, and similarly for $\partial z/\partial y$, $\partial z/\partial t$.

20.2.1 Partial Derivatives of Higher Order If $z = f(x, y)$, then $\partial z/\partial x$ and $\partial z/\partial y$ are themselves functions of x and y, and (for all the functions we shall be concerned with in this chapter) have partial derivatives with respect to x and y. We define the second-order partial derivatives of z as follows:

$$\frac{\partial^2 z}{\partial x^2} = \frac{\partial}{\partial x}\left(\frac{\partial z}{\partial x}\right), \quad (1) \qquad \frac{\partial^2 z}{\partial y^2} = \frac{\partial}{\partial y}\left(\frac{\partial z}{\partial y}\right), \quad (2)$$

$$\frac{\partial^2 z}{\partial x\, \partial y} = \frac{\partial}{\partial x}\left(\frac{\partial z}{\partial y}\right), \quad (3) \qquad \frac{\partial^2 z}{\partial y\, \partial x} = \frac{\partial}{\partial y}\left(\frac{\partial z}{\partial x}\right), \quad (4)$$

assuming that the limits implied in these definitions exist. The second-order derivatives (1)–(4) are also denoted by $f_{xx}, f_{yy}, f_{xy}, f_{yx}$ respectively.

Subject to certain conditions concerning the continuity of z and its partial derivatives it may be shown that

$$\frac{\partial^2 z}{\partial x\, \partial y} = \frac{\partial^2 z}{\partial y\, \partial x}. \quad (20.1)$$

This result, known as the commutative property of partial derivatives, may be assumed to be true for all functions encountered at this stage.

Partial derivatives of any order are defined in a similar manner and the result (20.1) may be generalised; for example, to find $\partial^5 z/\partial x^3\, \partial y^2$ we differentiate with respect to x three times and with respect to y twice, and the result is independent of the order in which these differentiations are carried out.

Example 8 Show that if $z = x^3 - 3xy^2$, then

$$\frac{\partial^2 z}{\partial x\, \partial y} = \frac{\partial^2 z}{\partial y\, \partial x} \quad \text{and} \quad \frac{\partial^2 z}{\partial x^2} + \frac{\partial^2 z}{\partial y^2} = 0.$$

We have $\partial z/\partial x = 3x^2 - 3y^2$ and so

$$\frac{\partial^2 z}{\partial x^2} = 6x, \quad \frac{\partial^2 z}{\partial y\, \partial x} = -6y.$$

Also $\partial z/\partial y = -6xy$, and so

$$\frac{\partial^2 z}{\partial y^2} = -6x, \quad \frac{\partial^2 z}{\partial x\, \partial y} = -6y.$$

The results follow.

Example 9 If $z = (x^2 + y^2) \tan^{-1}(y/x)$, then

$$\frac{\partial z}{\partial x} = 2x \tan^{-1}(y/x) - y, \quad \frac{\partial z}{\partial y} = 2y \tan^{-1}(y/x) + x.$$

(See Example 4.) Hence

$$\frac{\partial^2 z}{\partial x^2} = 2\tan^{-1}\left(\frac{y}{x}\right) - \frac{2xy}{x^2+y^2}, \quad \frac{\partial^2 z}{\partial y^2} = 2\tan^{-1}\left(\frac{y}{x}\right) + \frac{2xy}{x^2+y^2}.$$

Also

$$\frac{\partial}{\partial y}\left(\frac{\partial z}{\partial x}\right) = \frac{2x^2}{x^2+y^2} - 1, \quad \frac{\partial}{\partial x}\left(\frac{\partial z}{\partial y}\right) = -\frac{2y^2}{x^2+y^2} + 1$$

so that

$$\frac{\partial^2 z}{\partial y\,\partial x} = \frac{\partial^2 z}{\partial x\,\partial y} = \frac{x^2-y^2}{x^2+y^2}.$$

Exercises 20(a)

Find f_x, f_y for the expressions $f(x, y)$ in Nos. 1–6:

1. $x^2/y - y^2/x$
2. $(x-y)/(x+y)$
3. xe^{2x+3y}
4. $y\sqrt{(x^2-y^2)}$
5. $\tan^{-1}(x/y)$
6. $e^{x-y}\ln(y-x)$.

Find f_{xx}, f_{yy} for the expressions $f(x, y)$ in Nos. 7–9 and verify in each case that $f_{xy} = f_{yx}$:

7. $x^2 \sin y + y^2 \cos x$
8. $(y/x) \ln x$
9. $\sin^{-1}(y/x)$.
10. Given $z = (x+y)/\sqrt{(x^2+y^2)}$, find $x\,\partial z/\partial x + y\,\partial z/\partial y$.
11. Given $f(x, y) = \ln(x^2+y^2)$, prove that
$$\frac{\partial^2 f}{\partial x^2} + \frac{\partial^2 f}{\partial y^2} = 0.$$
12. Given $z = x\ln(x^2+y^2) - 2y\tan^{-1}(y/x)$, show that
$$\frac{\partial^2 z}{\partial x^2} + \frac{\partial^2 z}{\partial y^2} = 0.$$
13. Given $z = f(x^2+y^2)$, prove that $x\,\partial z/\partial y = y\,\partial z/\partial x$.
14. Given $z = f(y/x)$, prove that $x\,\partial z/\partial x + y\,\partial z/\partial y = 0$.
15. Given $z = (y/x) f(\theta)$, where $\theta = xy$ and f is an arbitrary function, prove that $2z + xz_x - yz_y = 0$.
16. Given $z = x^2 f(y/x)$, where f is an arbitrary function, show that
$$x\frac{\partial z}{\partial x} + y\frac{\partial z}{\partial y} = 2z.$$
17. Given $z^2 = xy\,F(x^2-y^2)$, prove that
$$2xy\left(x\frac{\partial z}{\partial y} + y\frac{\partial z}{\partial x}\right) = z(x^2+y^2).$$
18. Given $z = y^3 f(x/y)$, prove that
$$x^2\frac{\partial^2 z}{\partial x^2} + 2xy\frac{\partial^2 z}{\partial x\,\partial y} + y^2\frac{\partial^2 z}{\partial y^2} = 6z.$$

19. Given $u = f(y/x) + 2xy$, where f denotes an arbitrary function, prove that

$$x^2 \frac{\partial^2 u}{\partial x^2} + 2xy \frac{\partial^2 u}{\partial x\, \partial y} + y^2 \frac{\partial^2 u}{\partial y^2} = 4xy.$$

20. If $z = f(x + y) + g(xy)$, prove that

$$(y - x)\left[x \frac{\partial^2 z}{\partial x^2} - (x + y) \frac{\partial^2 z}{\partial x\, \partial y} + y \frac{\partial^2 z}{\partial y^2}\right] = (x + y)\left(\frac{\partial z}{\partial y} - \frac{\partial z}{\partial x}\right).$$

21. Given $u = x^n[f(y + x) + g(y - x)]$, where f and g are arbitrary functions, prove that

$$\frac{\partial^2 u}{\partial x^2} - \frac{\partial^2 u}{\partial y^2} - \frac{2n}{x}\frac{\partial u}{\partial x} + \frac{n(n + 1)}{x^2} u = 0.$$

22. Given $u = f(z)$ and $v = \phi(z)$ where $z = px^2 + 2qxy + ry^2$ and p, q, r are constants, show that

$$\frac{\partial u}{\partial x}\frac{\partial v}{\partial y} = \frac{\partial u}{\partial y}\frac{\partial v}{\partial x}.$$

23. Given $V = e^{(r-x)/l}$, where $r = x^2 + y^2$ and l is constant, prove

(a) $$\left(\frac{\partial V}{\partial x}\right)^2 + \left(\frac{\partial V}{\partial y}\right)^2 + \frac{2V}{l}\frac{\partial V}{\partial x} = 0,$$

(b) $$\frac{\partial^2 V}{\partial x^2} + \frac{\partial^2 V}{\partial y^2} + \frac{2}{l}\frac{\partial V}{\partial x} = \frac{V}{lr}.$$

24. Given $z = f(x, y) + g(u)$, where $u = xy$, and f and g are arbitrary functions, show that

$$w = x\frac{\partial z}{\partial x} - y\frac{\partial z}{\partial y}$$

is independent of the choice of g. Find w when $f(x, y) = xye^{x-y}$.

25. Given $\phi = f(\rho)$, where f is an arbitrary function and $\rho = (x^2 + y^2)^{n/2}$, prove that

(a) $$x\frac{\partial \phi}{\partial x} + y\frac{\partial \psi}{\partial y} = n\rho f'(\rho);$$

(b) $$(x^2 + y^2)\left(\frac{\partial^2 \phi}{\partial x^2} + \frac{\partial^2 \phi}{\partial y^2}\right) = n^2\rho\frac{\mathrm{d}}{\mathrm{d}\rho}\left(\rho\frac{\mathrm{d}\phi}{\mathrm{d}\rho}\right).$$

26. Given $z = (x + y)\phi(y/x)$, where ϕ is an arbitrary function, prove that

$$x\frac{\partial z}{\partial x} + y\frac{\partial z}{\partial y} = z$$

and that

$$x^2\frac{\partial^2 z}{\partial x^2} + 2xy\frac{\partial^2 z}{\partial x\, \partial y} + y^2\frac{\partial^2 z}{\partial y^2} = 0.$$

27. Given $V = f(x^2 + y^2)$, where f is any function, show that

$$y\frac{\partial V}{\partial x} - x\frac{\partial V}{\partial y} = 0$$

and

$$y^2\frac{\partial^2 V}{\partial x^2} - 2xy\frac{\partial^2 V}{\partial x\,\partial y} + x^2\frac{\partial^2 V}{\partial y^2} = x\frac{\partial V}{\partial x} + y\frac{\partial V}{\partial y}.$$

28. Given $u = (x^2 - y^2)f(t)$, where $t = xy$, prove that

$$\frac{\partial^2 u}{\partial x\,\partial y} = (x^2 - y^2)[tf''(t) + 3f'(t)].$$

Deduce that $\partial^2 u/\partial x\,\partial y = 0$ if $f(t) = A + B/t^2$, where A and B are constants.

29. Given $U = f(x^2 + y^2 + z^2)$, prove that

$$\frac{\partial^2 U}{\partial x^2} + \frac{\partial^2 U}{\partial y^2} + \frac{\partial^2 U}{\partial z^2}$$
$$= 4(x^2 + y^2 + z^2)f''(x^2 + y^2 + z^2) + 6f'(x^2 + y^2 + z^2).$$

30. Given

$$u = \frac{1}{r}f(ct - r) + \frac{1}{r}F(ct + r),$$

where f and F denote arbitrary functions, show that

$$\frac{\partial^2 u}{\partial t^2} = c^2\left(\frac{\partial^2 u}{\partial r^2} + \frac{2}{r}\frac{\partial u}{\partial r}\right).$$

31. Given $z = f(x + y)g(x - y)$, where f and g are arbitrary functions, prove that

$$z\frac{\partial^2 z}{\partial x^2} - z\frac{\partial^2 z}{\partial y^2} = \left(\frac{\partial z}{\partial x}\right)^2 - \left(\frac{\partial z}{\partial y}\right)^2.$$

32. Prove that the partial differential equation

$$x^2\frac{\partial^2 z}{\partial x^2} + 2xy\frac{\partial^2 z}{\partial x\,\partial y} + y^2\frac{\partial^2 z}{\partial y^2} + x\frac{\partial z}{\partial x} + y\frac{\partial z}{\partial y} = z$$

is satisfied by $z = x\phi(y/x) + y^{-1}\psi(y/x)$, where ϕ and ψ denote arbitrary functions.

33. If $u = x^n F(x/y)$, where F denotes an arbitrary function, show that

$$x\frac{\partial u}{\partial x} + y\frac{\partial u}{\partial y} = nu,$$

and hence that

$$x^2\frac{\partial^2 u}{\partial x^2} + 2xy\frac{\partial^2 u}{\partial x\,\partial y} + y^2\frac{\partial^2 u}{\partial y^2} = n(n - 1)u.$$

20.3 Differentials

If y is a function of the single variable x, say $y = f(x)$, then for small increments Δx, Δy we have the approximation (see Section 15.1.4)

$$\Delta y \simeq f'(x)\,\Delta x. \tag{1}$$

It is frequently convenient to denote the small increment in x by dx instead of Δx, at the same time denoting the above approximate value of Δy by dy, i.e.

$$dy = f'(x)\,dx. \tag{2}$$

In this notation the approximation (1) becomes the equation (2), in which we refer to dx and dy as *differentials*.

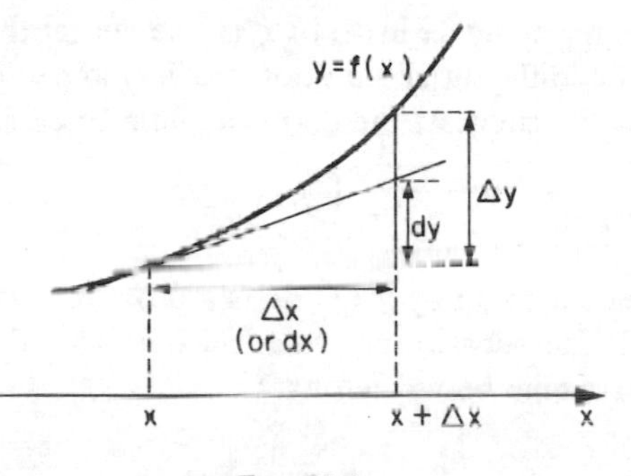

FIG. 20.4

For example, if $y = x^2$, then $dy = 2x\,dx$, or we may write $d(x^2) = 2x\,dx$. Likewise $d(\sin x) = \cos x\,dx$, $d(\ln x) = (1/x)\,dx$, and so on.

Note that in the case of the dependent variable y there is a distinction between the increment Δy and the differential dy; the two are not equal but approximately equal. However, when it is intended to use the approximation (1), or when it is intended subsequently to let $\Delta x \to 0$, the distinction between Δy and dy is unimportant and the phrase 'small increment dy' appears in some books.

From equation (2) we get

$$\frac{dy}{dx} = f'(x), \tag{3}$$

where the l.h.s. is the ratio of the differentials dy, dx. The symbol dy/dx, introduced in Section 14.4 to denote a single entity (the derivative of y with respect to x), may now in effect be regarded as denoting the ratio of two separate entities dy and dx.

The interpretation of the derivative as the ratio of two differentials has certain advantages when it comes to writing out the rules of differentiation. The product rule (Section 14.4.3) may be written as

$$\mathrm{d}(uv) = \left(u\frac{\mathrm{d}v}{\mathrm{d}x} + v\frac{\mathrm{d}u}{\mathrm{d}x}\right)\mathrm{d}x,$$

or simply

$$\mathrm{d}(uv) = u\,\mathrm{d}v + v\,\mathrm{d}u,$$

and likewise the quotient rule may be written as

$$\mathrm{d}\left(\frac{u}{v}\right) = \frac{v\,\mathrm{d}u - u\,\mathrm{d}v}{v^2}.$$

If $x = g(z)$, then replacing $\mathrm{d}x$ in (2) by $g'(z)\,\mathrm{d}z$ we get the chain rule. The advantages of the 'differential' notation are also apparent in integration. Equation (2) readily suggests the corresponding integral form

$$y = \int f(x)\,\mathrm{d}x,$$

and the replacement of $\mathrm{d}x$ by $g'(z)\,\mathrm{d}z$ is a procedure already familiar in connection with the substitution rule (Section 16.4.1). The rule of integration by parts may be written as

$$\int u\,\mathrm{d}v = uv - \int v\,\mathrm{d}u.$$

20.3.1 Differential of a Function of Two Variables Let P, Q be the points on the surface $z = f(x, y)$ corresponding to the points

$$P'(x, y), \quad Q'(x + \Delta x, y + \Delta y)$$

in the domain of f (see Fig. 20.5). Write

$$\Delta z = f(x + \Delta x, y + \Delta y) - f(x, y)$$

so that Δz is the difference in height (i.e. difference in z-co-ordinate) between the points P and Q.

Consider the point $R'(x + \Delta x, y)$ in the domain of f and the corresponding point R on the surface. Write

$$\Delta_1 z = f(x + \Delta x, y) - f(x, y),$$
$$\Delta_2 z = f(x + \Delta x, y + \Delta y) - f(x + \Delta x, y)$$

so that $\Delta_1 z$, $\Delta_2 z$ are the differences in height between the points P and R, and the points R and Q, respectively. Then $\Delta z = \Delta_1 z + \Delta_2 z$.

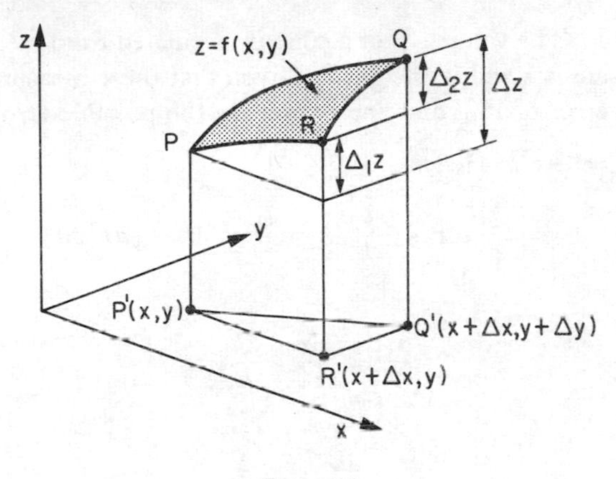

FIG. 20.5

Since P' and R' differ only in x-co-ordinate, we may calculate $\Delta_1 z$ by regarding $f(x, y)$ as a function of x only; thus $\Delta_1 z \simeq (\partial f/\partial x)\,\Delta x$ (Section 15.1.4), and similarly $\Delta_2 z \simeq (\partial f/\partial y)\,\Delta y$. We are concerned here with the value of $\partial f/\partial x$ *at* P' $(x, \ y)$ and the value of $\partial f/\partial y$ at R' $(x + \Delta x, y)$; nevertheless, for a function f with continuous partial derivatives, we may conveniently take the values of both $\partial f/\partial x$ and $\partial f/\partial y$ at P' (x, y). Adding, we get

$$\Delta z \simeq \frac{\partial f}{\partial x}\Delta x + \frac{\partial f}{\partial y}\Delta y.$$

In practice we frequently denote the small increments Δx, Δy by $\mathrm{d}x$, $\mathrm{d}y$ respectively, and then define $\mathrm{d}z$ by the equation

$$\mathrm{d}z = \frac{\partial f}{\partial x}\mathrm{d}x + \frac{\partial f}{\partial y}\mathrm{d}y,$$

that is

$$\mathrm{d}z = \frac{\partial z}{\partial x}\mathrm{d}x + \frac{\partial z}{\partial y}\mathrm{d}y, \tag{20.1}$$

in which $\mathrm{d}x$, $\mathrm{d}y$, $\mathrm{d}z$ are called *differentials*. If x varies while y remains constant, then $\mathrm{d}y = 0$ and so $\mathrm{d}z = (\partial z/\partial x)\,\mathrm{d}x$. Thus, $(\partial z/\partial x)\,\mathrm{d}x$ is the differential of z corresponding to a variation in x alone, and similarly $(\partial z/\partial y)\,\mathrm{d}y$ is the differential of z corresponding to a variation in y alone; these terms are called the *partial differentials* of z and their sum is called the *total differential* of z.

Example 10 The volume V of a cone is calculated from measurements of its base radius r and its height h. Given that these measurements are liable to an error of 1%, find approximately the possible error in V.

We have $V = \frac{1}{3}\pi r^2 h$. Hence

$$dV = \frac{\partial V}{\partial r}\,dr + \frac{\partial V}{\partial h}\,dh = \tfrac{2}{3}\pi rh\,dr + \tfrac{1}{3}\pi r^2 dh$$

and so

$$\frac{dV}{V} = \frac{2dr}{r} + \frac{dh}{h}.$$

Putting

$$\frac{dr}{r} = \frac{dh}{h} = \pm\frac{1}{100},$$

we see that dV/V lies between $-\frac{3}{100}$ and $\frac{3}{100}$. Since $\Delta V \simeq dV$, there is a possible error of approximately 3% in V.

Example 11 The diameter of a circle from which a segment has been cut is determined from the length of the chord a and the maximum height b of the segment. Given that the measurements of a and b are slightly inaccurate, each to an extent of p %, find the approximate percentage error in the calculated value of the diameter. Prove that this error is also p %, provided (a) that a and b are measured both in excess or both in defect of their actual values, or (b) that if a and b are measured one in excess and the other in defect of their actual values, the segment is a semicircle.

Let C (Fig. 20.6) be the mid-point of the arc AB of a circle and let CD, the diameter through C, meet AB at O. Then OC is the maximum height of the segment ABC and

$$OC.OD = OA.OB.$$

If $AB = a$, $OC = b$ and $CD = x$, we have

$$b(x-b) = \tfrac{1}{4}a^2,$$
$$x = b + \tfrac{1}{4}a^2/b. \tag{1}$$

If Δx is the error in x caused by small errors Δa and Δb in a and b respectively, we have

$$\Delta x \simeq \frac{\partial x}{\partial a}\Delta a + \frac{\partial x}{\partial b}\Delta b,$$

that is

$$\Delta x \simeq \tfrac{1}{2}(a/b)\,\Delta a + (1 - \tfrac{1}{4}a^2/b^2)\,\Delta b.$$

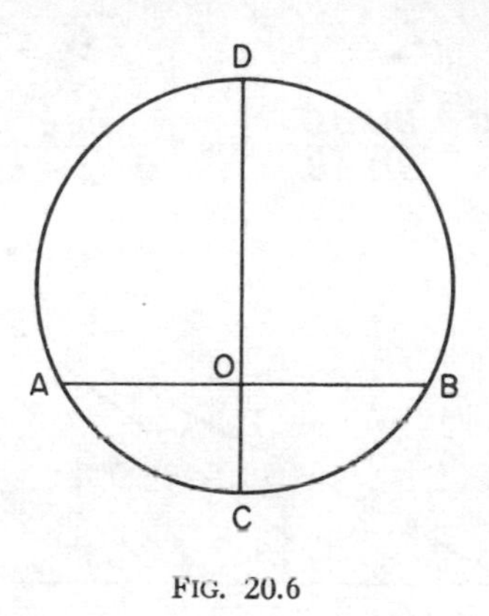

FIG. 20.6

If the percentage error in a and b is p per cent, then $100(\Delta a/a) = \pm p$, $100(\Delta b/b) = \pm p$, and ε, the percentage error in x, is given by

$$\varepsilon = 100(\Delta x/x) = \pm\tfrac{1}{4}p[2a^2 \pm (4b^2 - a^2)]/bx \text{ approximately.}$$

If a and b are measured both in excess or both in defect of their actual values, Δa and Δb have the same signs and so

$$\varepsilon \simeq \pm\tfrac{1}{4}p[2a^2 + (4b^2 - a^2)]/bx = \pm p[b + \tfrac{1}{4}a^2/b]/x$$

that is

$$\varepsilon \simeq p, \quad \text{by (1).}$$

If a and b are measured one in excess and the other in defect of their actual values, Δa and Δb have opposite signs and

$$\varepsilon \simeq \pm\tfrac{1}{4}p[3a^2 - 4b^2]/bx.$$

If, in addition, the segment ABC is a semicircle, $x = a = 2b$ and so

$$\varepsilon \simeq \pm p.$$

Example 12 The points A and B, at a distance a apart on a horizontal plane, are in line with the base C of a vertical tower and on the same side of C. The elevations of the top of the tower from A and B are observed to be α and β ($\alpha < \beta$). Show that the distance BC is

$$a \sin \alpha \cos \beta \operatorname{cosec} (\beta - \alpha).$$

Show that if the observations of the angles of elevation are uncertain by 4 minutes, then the maximum possible percentage error in the calculated value of BC is approximately

$$\pi \sin (\alpha + \beta)/[27 \sin \alpha \cos \beta \tan (\beta - \alpha)].$$

In Fig. 20.7,

$$\frac{BC}{AB} = \frac{BC}{BD}\frac{BD}{AB} = \cos\beta\,\frac{\sin\alpha}{\sin(\beta-\alpha)}.$$

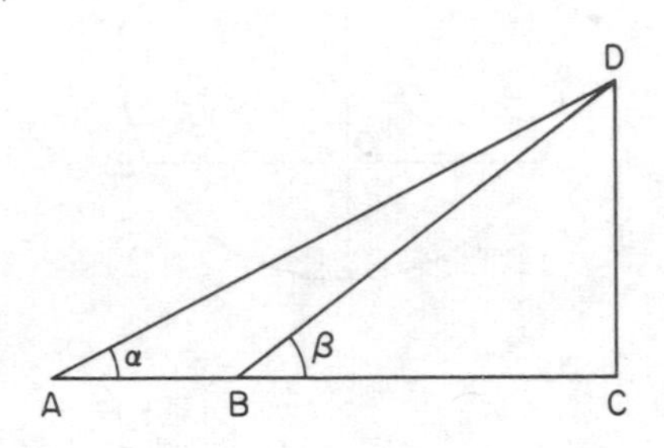

FIG. 20.7

Hence

$$BC = a\cos\beta\sin\alpha\,\operatorname{cosec}(\beta-\alpha),$$

and denoting BC by x we have

$$\ln x = \ln a + \ln\cos\beta + \ln\sin\alpha + \ln\operatorname{cosec}(\beta-\alpha). \qquad (1)$$

If Δx is the error in x caused by small errors $\Delta\alpha$ and $\Delta\beta$ in α and β respectively, the proportional error in x is $(\Delta x/x)$ where, by (1),

$$\Delta x/x \simeq -\tan\beta\,\Delta\beta + \cot\alpha\,\Delta\alpha - (\Delta\beta - \Delta\alpha)\cot(\beta-\alpha),$$

that is

$$\Delta x/x \simeq [\cot\alpha + \cot(\beta-\alpha)]\,\Delta\alpha - [\tan\beta + \cot(\beta-\alpha)]\,\Delta\beta.$$

Now $\Delta\alpha = \pm\pi/2700$ radians and $\Delta\beta = \pm\pi/2700$ radians. Hence if ε is the percentage error in x,

$$\varepsilon = 100(\Delta x/x) \simeq \pm\frac{\pi}{27}\{[\cot\alpha + \cot(\beta-\alpha)] \mp [\tan\beta + \cot(\beta-\alpha)]\}.$$

Since α and β are acute and $\beta > \alpha$, the maximum numerical value of ε is obtained by taking the positive sign inside the curly brackets.

$$\begin{aligned}\varepsilon_{\max} &= \frac{\pi}{27}\left\{\frac{\cos\alpha}{\sin\alpha} + \frac{\sin\beta}{\cos\beta} + \frac{2}{\tan(\beta-\alpha)}\right\},\\ &= \pi[\sin(\beta-\alpha) + 2\sin\alpha\cos\beta]/27\sin\alpha\cos\beta\tan(\beta-\alpha),\\ &= \pi\sin(\alpha+\beta)/[27\sin\alpha\cos\beta\tan(\beta-\alpha)] \text{ approximately.}\end{aligned}$$

The extension of result (20.1) to functions of more than two variables is straightforward; e.g., if z is a function of the three variables x, y, t, then

$$dz = \frac{\partial z}{\partial x}dx + \frac{\partial z}{\partial y}dy + \frac{\partial z}{\partial t}dt.$$

Exercises 20(b)

1. The height and base radius of a cone are 14 cm and 4 cm respectively. Find the maximum error in the volume $V = \frac{1}{3}\pi hr^2$ if this is calculated from values of h and r both subject to errors of ± 0.03 cm.
2. The formula $p = A\rho^{1.4}$ is used to find A. If the percentage errors in p and ρ are 1% and -2%, respectively, what is the approximate percentage error in A?
3. A container is in the form of a right circular cylinder of radius r and height h surmounted by a hemisphere of radius r. If r and h are measured as $6 \pm 0{\cdot}02$ cm and $4 \pm 0{\cdot}01$ cm respectively, show that the maximum error in V, calculated from the formula $V = \pi r^2 h + \frac{2}{3}\pi r^3$, is less than 1%.
4. The volume, V, bounded by a hemisphere of radius r and a right circular cone of height h $(h < r)$, whose base coincides with that of the hemisphere, is given by

$$V = \frac{2}{3}\pi r^3 - \frac{1}{3}\pi r^2 h.$$

Suppose that, for a solid of this type, r and h are measured as $9{\cdot}0$ and $6{\cdot}0$ units, respectively. If each measurement may be in error by $\pm 0{\cdot}01$ units calculate V and find, approximately, the maximum percentage error in the calculation.

5. A physical quantity R is given by the relation $R = a\, b^{4/3}\, c^{-1/3}$, and a, b, c are measured with percentage errors $\pm 0{\cdot}5\%$, $\pm 1\%$, and $\pm 1{\cdot}5\%$. Find approximately the maximum percentage error in R.
6. The formula $z = x^2\, y^{2/3}\, s^{4/5}$ is used to calculate z from observed values of x, y and s. If the measurements of x, y and s are accurate to 1%, 1% and 2%, respectively, find, approximately, the maximum percentage error in the calculated value of z.
7. In a triangle ABC, the angle A is accurately known, but the measurement of the side b is in error to the extent Δb, and that of the side c to an extent Δc. Find the error in calculating the value of a from b, c and A.

   What is the best shape for triangle ABC in order to minimise the effect of the error Δb?
8. The area of a triangle ABC is calculated from measurements of the sides b, c with a possible error of $\frac{1}{2}\%$ in each, and of the angle A, correct to the nearest half degree. Find an approximate expression for the proportional error in the area in terms of the errors Δb, Δc, ΔA in the measured values.

   Given that the measured value of A is 60°, determine approximately the maximum proportional error in the area.
9. The side BC of a triangle ABC is to be determined from measurements of the sides AB and AC and of the angle BAC. The measured values of the

sides are liable to a small proportional error θ and the angle BAC to a small absolute error ΔA. Show that the calculated value of BC is liable to a proportional error $\theta + [(bc/a^2) \sin A] \Delta A$.

The measured values of b, c and A are 4, 5 and 120° respectively and are liable to errors of $\frac{1}{2}$%, $\frac{1}{2}$% and 1° respectively. Show that the calculated value of a is liable to an error of approximately 1%.

10. Two triangles have equal bases, each of length a, and their base angles are B, C and $B + \Delta B$, $C + \Delta C$ respectively, where ΔB and ΔC are small. Prove that their areas differ approximately by

$$\tfrac{1}{2}a^2 (\sin^2 C \, \Delta B + \sin^2 B \, \Delta C) \operatorname{cosec}^2 A.$$

The angles of a triangle whose sides are proportional to 3:4:5 are 36° 52′, 53° 8′ and 90°. Given that a radian is 57° 18′, show that the area of a triangle, whose base is 60 m and whose base angles are 54° and 89° 8′, is approximately 2427 m².

11. Find the diameter D of the circumcircle of the triangle with sides a, a, $2b$. Calculate the approximate change in D due to small changes Δa and Δb in the values of a and b respectively. Deduce that, if $a = \sqrt{3}b$, there is no change in the value of D when a slightly increases and b slightly decreases in the same ratio.

12. Show that the volume of a segment of a sphere is $\frac{1}{6}\pi h (h^2 + 3R^2)$, where h is the height of the segment and R is the radius of its base.

If the measurement of h is too large by a small amount a, and that of R is too small by an equal amount, show that the calculated volume is too large by an amount $\frac{1}{2}\pi a (h - R)^2$ approximately.

If the segment is a hemisphere, show that the error in the calculated volume is $\frac{2}{3}\pi a^3$ exactly.

13. If $f(x, y) = xe^{xy}$, and the values of x and y are slightly changed from 1 and 0 to $1 + \Delta x$ and Δy respectively so that Δf, the change in f, is very nearly $3 \Delta x$, show that Δy must be very nearly $2 \Delta x$.

14. If $z = \sin \theta \sin \phi / \sin \psi$ and z is calculated for the values $\theta = 30°$, $\psi = 60°$, $\phi = 45°$, find approximately the change in the value of z if each of the angles θ and ψ is increased by the same small angle $\alpha°$ and ϕ is decreased by $\frac{1}{2}\alpha°$.

20.4 Extended Chain Rule

Suppose that $z = f(x, y)$ and that x, y are both functions of two other variables u, v. Then z may be expressed in terms of u, v and $\partial z/\partial u$, $\partial z/\partial v$ may be found by normal methods of partial differentiation.

For example, if $z = 2x^2 + 3y^2$ and $x = u + v$, $y = uv$, then

$$z = 2(u + v)^2 + 3u^2v^2,$$

giving

$$\frac{\partial z}{\partial u} = 4(u + v) + 6uv^2, \quad \frac{\partial z}{\partial v} = 4(u + v) + 6u^2v.$$

Alternatively $\partial z/\partial u$, $\partial z/\partial v$ may be found by using a 'chain rule'. The following equations are all examples of the expression (20.1) for a total differential given in Section 20.3.1:

$$dz = \frac{\partial z}{\partial x}dx + \frac{\partial z}{\partial y}dy, \tag{1}$$

$$dx = \frac{\partial x}{\partial u}du + \frac{\partial x}{\partial v}dv, \tag{2}$$

$$dy = \frac{\partial y}{\partial u}du + \frac{\partial y}{\partial v}dv, \tag{3}$$

$$dz = \frac{\partial z}{\partial u}du + \frac{\partial z}{\partial v}dv. \tag{4}$$

Substituting from (2) and (3) into (1) we get

$$dz = \frac{\partial z}{\partial x}\left(\frac{\partial x}{\partial u}du + \frac{\partial x}{\partial v}dv\right) + \frac{\partial z}{\partial y}\left(\frac{\partial y}{\partial u}du + \frac{\partial y}{\partial v}dv\right),$$

that is

$$dz = \left(\frac{\partial z}{\partial x}\frac{\partial x}{\partial u} + \frac{\partial z}{\partial y}\frac{\partial y}{\partial u}\right)du + \left(\frac{\partial z}{\partial x}\frac{\partial x}{\partial v} + \frac{\partial z}{\partial y}\frac{\partial y}{\partial v}\right)dv.$$

Comparison of this equation with (4) gives

$$\left.\begin{aligned}\frac{\partial z}{\partial u} &= \frac{\partial z}{\partial x}\frac{\partial x}{\partial u} + \frac{\partial z}{\partial y}\frac{\partial y}{\partial u},\\ \frac{\partial z}{\partial v} &= \frac{\partial z}{\partial x}\frac{\partial x}{\partial v} + \frac{\partial z}{\partial y}\frac{\partial y}{\partial v}.\end{aligned}\right. \tag{20.2}$$

Equations (20.2) may be regarded as an extension of the chain rule of Section 14.4.3. In the example above we have

$$\frac{\partial x}{\partial u} = \frac{\partial x}{\partial v} = 1, \quad \frac{\partial y}{\partial u} = v, \quad \frac{\partial y}{\partial v} = u$$

and so from (20.2) we get

$$\frac{\partial z}{\partial u} = 4x1 + 6yv = 4(u+v) + 6uv^2,$$

$$\frac{\partial z}{\partial v} = 4x1 + 6yu = 4(u+v) + 6u^2v.$$

Example 13 Given $z = f(x, y)$ and $x = \frac{1}{2}(u^2 - v^2)$, $y = uv$, show that

(a) $$u\frac{\partial z}{\partial v} - v\frac{\partial z}{\partial u} = 2\left(x\frac{\partial z}{\partial y} - y\frac{\partial z}{\partial x}\right);$$

(b) $$\frac{\partial^2 z}{\partial u^2}+\frac{\partial^2 z}{\partial v^2}=(u^2+v^2)\left(\frac{\partial^2 z}{\partial x^2}+\frac{\partial^2 z}{\partial y^2}\right).$$

If

$$x=\tfrac{1}{2}(u^2-v^2) \quad \text{and} \quad y=uv \tag{1}$$

$$\frac{\partial x}{\partial u}=u, \quad \frac{\partial x}{\partial v}=-v, \quad \frac{\partial y}{\partial u}=v, \quad \frac{\partial y}{\partial v}=u.$$

By (20.2)

$$\frac{\partial z}{\partial u}=u\frac{\partial z}{\partial x}+v\frac{\partial z}{\partial y} \quad \text{and} \quad \frac{\partial z}{\partial v}=-v\frac{\partial z}{\partial x}+u\frac{\partial z}{\partial y},$$

giving

$$u\frac{\partial z}{\partial v}-v\frac{\partial z}{\partial u}=(u^2-v^2)\frac{\partial z}{\partial y}-2uv\frac{\partial z}{\partial x}=2\left(x\frac{\partial z}{\partial y}-y\frac{\partial z}{\partial x}\right), \quad \text{by (1);}$$

$$\frac{\partial^2 z}{\partial u^2}=\frac{\partial}{\partial u}\left(u\frac{\partial z}{\partial x}+v\frac{\partial z}{\partial y}\right)$$

$$=\frac{\partial z}{\partial x}+u\frac{\partial}{\partial u}\left(\frac{\partial z}{\partial x}\right)+v\frac{\partial}{\partial u}\left(\frac{\partial z}{\partial y}\right). \tag{2}$$

Now if V is any function of x and y, by (20.2)

$$\frac{\partial V}{\partial u}=\frac{\partial V}{\partial x}\frac{\partial x}{\partial u}+\frac{\partial V}{\partial y}\frac{\partial y}{\partial u}=u\frac{\partial V}{\partial x}+v\frac{\partial V}{\partial y}.$$

Put $V=\partial z/\partial x$; then

$$\frac{\partial}{\partial u}\left(\frac{\partial z}{\partial x}\right)=u\frac{\partial^2 z}{\partial x^2}+v\frac{\partial^2 z}{\partial x\,\partial y}.$$

Put $V=\partial z/\partial y$; then

$$\frac{\partial}{\partial u}\left(\frac{\partial z}{\partial y}\right)=u\frac{\partial^2 z}{\partial x\,\partial y}+v\frac{\partial^2 z}{\partial y^2}.$$

Substituting these values in (2), we have

$$\frac{\partial^2 z}{\partial u^2}=\frac{\partial z}{\partial x}+u^2\frac{\partial^2 z}{\partial x^2}+2uv\frac{\partial^2 z}{\partial x\,\partial y}+v^2\frac{\partial^2 z}{\partial y^2}.$$

Similarly,

$$\frac{\partial^2 z}{\partial v^2}=-\frac{\partial z}{\partial x}+v^2\frac{\partial^2 z}{\partial x^2}-2uv\frac{\partial^2 z}{\partial x\,\partial y}+u^2\frac{\partial^2 z}{\partial y^2}.$$

Hence

$$\frac{\partial^2 z}{\partial u^2} + \frac{\partial^2 z}{\partial v^2} = (u^2 + v^2)\left(\frac{\partial^2 z}{\partial x^2} + \frac{\partial^2 z}{\partial y^2}\right).$$

We have assumed above that $\partial z/\partial x$, $\partial z/\partial y$ are the partial derivatives of z when z is regarded as a function of x and y, whereas $\partial z/\partial u$, $\partial z/\partial v$ are the partial derivatives of z when z is regarded as a function of u and v. However, it is possible to regard z as a function of any two of the four variables; if, for example, z is expressed in terms of x and u we may differentiate with respect to x treating u as constant, and the result is denoted by $(\partial z/\partial x)_{u \text{ const.}}$. Other partial derivatives, e.g.

$$\left(\frac{\partial z}{\partial x}\right)_{v \text{ const.}}, \quad \left(\frac{\partial z}{\partial y}\right)_{u \text{ const.}}, \quad \left(\frac{\partial z}{\partial u}\right)_{x \text{ const.}}$$

are similarly defined.

Example 14 Given $z = 2x^2 + 3y^2$ and $x = u + v$, $y = uv$ find $(\partial z/\partial x)_{u \text{ const.}}$, $(\partial z/\partial x)_{v \text{ const.}}$, $(\partial z/\partial y)_{u \text{ const.}}$, $(\partial z/\partial y)_{v \text{ const.}}$.

We have

$$z = 2x^2 + 3y^2 = 2x^2 + 3u^2v^2 = 2x^2 + 3u^2(x - u)^2.$$

Hence

$$\left(\frac{\partial z}{\partial x}\right)_{u \text{ const.}} = 4x + 6u^2(x - u).$$

Similarly,

$$z = 2x^2 + 3v^2(x - v)^2$$

and so

$$\left(\frac{\partial z}{\partial x}\right)_{v \text{ const.}} = 4x + 6v^2(x - v).$$

Also,

$$z = 2(u + v)^2 + 3y^2 = 2\left(u + \frac{y}{u}\right)^2 + 3y^2,$$

and so

$$\left(\frac{\partial z}{\partial y}\right)_{u \text{ const.}} = \frac{4}{u}\left(u + \frac{y}{u}\right) + 6y.$$

Similarly

$$z = 2\left(\frac{y}{v} + v\right)^2 + 3y^2,$$

giving

$$\left(\frac{\partial z}{\partial y}\right)_{v\ \text{const.}} = \frac{4}{v}\left(\frac{y}{v}+v\right) + 6y.$$

The application of eqs. (20.2) is less straightforward when, instead of knowing x, y in terms of u, v, we know u, v in terms of x, y. In this case it may not be easy, or even possible, to express x, y in terms of u, v in order to find $\partial x/\partial u$, $\partial x/\partial v$, $\partial y/\partial u$, $\partial y/\partial v$. [Note carefully that in general $\partial x/\partial u$ is *not* the reciprocal of $\partial u/\partial x$; this pitfall is avoided if we describe these derivatives more fully as $(\partial x/\partial u)_{v\ \text{const.}}$ and $(\partial u/\partial x)_{y\ \text{const.}}$, showing that a different variable is held constant in each case, and whereas the 'inverse rule' of Section 14.4.3 implies that

$$\left(\frac{\partial x}{\partial u}\right)_{v\ \text{const.}} = 1\Big/\left(\frac{\partial u}{\partial x}\right)_{v\ \text{const.}},$$

it says nothing about the relation between $(\partial x/\partial u)_{v\ \text{const.}}$ and $(\partial u/\partial x)_{y\ \text{const.}}$.] Writing

$$du = \frac{\partial u}{\partial x}\,dx + \frac{\partial u}{\partial y}\,dy, \quad dv = \frac{\partial v}{\partial x}\,dx + \frac{\partial v}{\partial y}\,dy,$$

we may solve these equations for the differentials dx, dy and hence find $\partial x/\partial u$, $\partial x/\partial v$, $\partial y/\partial u$, $\partial y/\partial v$ as illustrated in Example 15.

Example 15 Given $u = x + y$ and $v = xy$, find $\partial x/\partial u$, $\partial x/\partial v$, $\partial y/\partial u$, $\partial y/\partial v$.

By (20.2)

$$du = dx + dy, \quad dv = y\,dx + x\,dy.$$

Hence

$$(x-y)\,dx = x\,du - dv, \quad (x-y)\,dy = dv - y\,du,$$

so that

$$dx = \left(\frac{x}{x-y}\right)du + \left(\frac{-1}{x-y}\right)dv,$$

$$dy = \left(\frac{-y}{x-y}\right)du + \left(\frac{1}{x-y}\right)dv.$$

Comparing these equations with (20.1) we conclude that

$$\frac{\partial x}{\partial u} = \frac{x}{x-y}, \quad \frac{\partial x}{\partial v} = \frac{-1}{x-y}, \quad \frac{\partial y}{\partial u} = \frac{-y}{x-y}, \quad \frac{\partial y}{\partial v} = \frac{1}{x-y}.$$

Example 16 Given $V = f(x, y)$ and $x = e^u \cos v$, $y = e^u \sin v$, prove that

$$\frac{\partial^2 V}{\partial u^2} + \frac{\partial^2 V}{\partial v^2} = e^{2u}\left(\frac{\partial^2 V}{\partial x^2} + \frac{\partial^2 V}{\partial y^2}\right).$$

From (20.2),

$$\frac{\partial V}{\partial u} = \frac{\partial V}{\partial x}\frac{\partial x}{\partial u} + \frac{\partial V}{\partial y}\frac{\partial y}{\partial u} = e^u \cos v \frac{\partial V}{\partial x} + e^u \sin v \frac{\partial V}{\partial y}.$$

Hence

$$\frac{\partial V}{\partial u} = x\frac{\partial V}{\partial x} + y\frac{\partial V}{\partial y}. \tag{1}$$

Similarly

$$\frac{\partial V}{\partial v} = -y\frac{\partial V}{\partial x} + x\frac{\partial V}{\partial y}.$$

The symbol $\partial/\partial u$ may be regarded as an operator which obtains from V its derivative $\partial V/\partial u$. If we write (1) in the form

$$\frac{\partial}{\partial u}(V) = \left(x\frac{\partial}{\partial x} + y\frac{\partial}{\partial y}\right)V,$$

we see that the operator $\partial/\partial u$ is equivalent to the operator

$$x\frac{\partial}{\partial x} + y\frac{\partial}{\partial y}$$

and, in the same way, the operator $\partial/\partial v$ is equivalent to

$$-y\frac{\partial}{\partial x} + x\frac{\partial}{\partial y}.$$

Again,

$$\begin{aligned}
\frac{\partial^2 V}{\partial u^2} &= \frac{\partial}{\partial u}\left(\frac{\partial V}{\partial u}\right) \\
&= \left(x\frac{\partial}{\partial x} + y\frac{\partial}{\partial y}\right)\left(x\frac{\partial V}{\partial x} + y\frac{\partial V}{\partial y}\right) \\
&= x\left(x\frac{\partial^2 V}{\partial x^2} + \frac{\partial V}{\partial x} + y\frac{\partial^2 V}{\partial x\,\partial y}\right) + y\left(x\frac{\partial^2 V}{\partial x\,\partial y} + y\frac{\partial^2 V}{\partial y^2} + \frac{\partial V}{\partial y}\right) \\
&= x^2\frac{\partial^2 V}{\partial x^2} + 2xy\frac{\partial^2 V}{\partial x\,\partial y} + y^2\frac{\partial^2 V}{\partial y^2} + x\frac{\partial V}{\partial x} + y\frac{\partial V}{\partial y}.
\end{aligned}$$

Similarly,

$$\frac{\partial^2 V}{\partial v^2} = y^2 \frac{\partial^2 V}{\partial x^2} - 2xy \frac{\partial^2 V}{\partial x \, \partial y} + x^2 \frac{\partial^2 V}{\partial y^2} - x \frac{\partial V}{\partial x} - y \frac{\partial V}{\partial y},$$

and so

$$\frac{\partial^2 V}{\partial u^2} + \frac{\partial^2 V}{\partial v^2} = (x^2 + y^2)\left(\frac{\partial^2 V}{\partial x^2} + \frac{\partial^2 V}{\partial y^2}\right) = e^{2u}\left(\frac{\partial^2 V}{\partial x^2} + \frac{\partial^2 V}{\partial y^2}\right).$$

20.4.1 Total Derivative A simple extension of the chain rule applies when $z = f(x, y)$ and x and y are both functions of the single variable u. Putting derivatives with respect to v equal to zero in (20.2) we see that eqs. (20.2) reduce to the single equation

$$\frac{dz}{du} = \frac{\partial z}{\partial x}\frac{dx}{du} + \frac{\partial z}{\partial y}\frac{dy}{du}. \tag{20.3}$$

Note that in this case z is a function of the single variable u, and dz/du (sometimes called the *total derivative* of z) is not a *partial* derivative.

Example 17 Find dz/du given that $z = x^2 + y^2$ and $x = 2u^2$, $y = 4u^3$.

Eliminating x and y, we have $z = 4u^4 + 16u^6$ and hence

$$\frac{dz}{du} = 16u^3 + 96u^5.$$

Alternatively, by (20.3)

$$\frac{dz}{du} = 2x \,.\, 4u + 2y \,.\, 12u^2 = 16u^3 + 96u^5.$$

Example 18 Given that the radius of a right circular cone is increasing at the rate of 1 cm/min and the height is increasing at the rate of 2 cm/min, find the rate at which the volume is increasing when the radius is 12 cm and the height 36 cm.

Since $V = \frac{1}{3}\pi r^2 h$, we have by (20.3)

$$\frac{dV}{dt} = (\tfrac{2}{3}\pi r h)\frac{dr}{dt} + (\tfrac{1}{3}\pi r^2)\frac{dh}{dt}.$$

Now $dr/dt = 1$, $dh/dt = 2$ and so when $r = 12$, $h = 36$ we have

$$\frac{dV}{dt} = \tfrac{1}{3}\pi \,.\, 12(72 + 24) = 384\pi.$$

Thus the volume is increasing at the rate of 384π cm^3/min.

In particular, if $z = f(x, y)$ and y is a function of x then from (20.3) we get

$$\frac{dz}{dx} = \frac{\partial z}{\partial x} + \frac{\partial z}{\partial y}\frac{dy}{dx}. \tag{20.4}$$

20.4.2 Implicit Functions We have previously found dy/dx using an implicit equation of the form $f(x, y) = 0$ (see Section 14.4.4). We can now give a general formula; differentiating both sides of the equation $f(x, y) = 0$ with respect to x, by (20.4) we get

$$f_x + f_y \frac{dy}{dx} = 0,$$

that is

$$\frac{dy}{dx} = -\frac{f_x}{f_y}.$$

We may extend this idea to the equation $F(x, y, z) = 0$ which expresses z implicitly as a function of the two variables x and y. Differentiating the equation $F(x, y, z) = 0$ partially, first with respect to x and then with respect to y, by (20.4) we get

$$F_x + F_z \frac{\partial z}{\partial x} = 0, \quad F_y + F_z \frac{\partial z}{\partial y} = 0.$$

Hence

$$\frac{\partial z}{\partial x} = -\frac{F_x}{F_z}, \quad \frac{\partial z}{\partial y} = -\frac{F_y}{F_z}.$$

Example 19 Find dy/dx at any point of the curve $x^3 + y^3 - 3xy = a^3$.

Here

$$f(x, y) = x^3 + y^3 - 3xy - a^3,$$

giving

$$f_x = 3(x^2 - y), \quad f_y = 3(y^2 - x),$$

and so

$$\frac{dy}{dx} = -\frac{x^2 - y}{y^2 - x}.$$

Example 20 Find $\partial z/\partial x$, $\partial z/\partial y$ at any point on the surface

$$x^3 + y^3 + z^3 - 3xyz = a^3.$$

Here

$$F(x, y, z) = x^3 + y^3 + z^3 - 3xyz - a^3,$$

giving

$$F_x = 3(x^2 - yz), \quad F_y = 3(y^2 - xz), \quad F_z = 3(z^2 - xy),$$

and so

$$\frac{\partial z}{\partial x} = -\frac{x^2 - yz}{z^2 - xy}, \quad \frac{\partial z}{\partial y} = -\frac{y^2 - xz}{z^2 - xy}.$$

Exercises 20(c)

1. Given that z is a function of x and y, where x and y are defined in terms of the independent variables u and v by the equations $x = uv$, $y = u/v$, show that

$$2xz_x = uz_u + vz_v, \quad 2yz_y = uz_u - vz_v.$$

2. Given that z is a function of u and v, where u and v are defined in terms of the independent variables x and y by the equations

$$u = x \cosh y, \quad v = x \sinh y,$$

show that

$$\left(\frac{\partial z}{\partial x}\right)^2 - \frac{1}{x^2}\left(\frac{\partial z}{\partial y}\right)^2 = \left(\frac{\partial z}{\partial u}\right)^2 - \left(\frac{\partial z}{\partial v}\right)^2.$$

3. The equations $ux = v^2 + y$ and $vy = x^2 + 2u^2$ define u and v as functions of the independent variables x and y. Show that

$$\frac{\partial u}{\partial x} = \frac{uy - 4xv}{8uv - xy} \quad \text{and} \quad \frac{\partial v}{\partial x} = \frac{4u^2 - 2x^2}{8uv - xy}.$$

4. Given that u, v are functions of X and Y, and X, Y are functions of x and y, prove that

$$\frac{\partial u}{\partial x}\frac{\partial v}{\partial y} - \frac{\partial u}{\partial y}\frac{\partial v}{\partial x} = \left(\frac{\partial u}{\partial X}\frac{\partial v}{\partial Y} - \frac{\partial u}{\partial Y}\frac{\partial v}{\partial X}\right)\left(\frac{\partial X}{\partial x}\frac{\partial Y}{\partial y} - \frac{\partial X}{\partial y}\frac{\partial Y}{\partial x}\right).$$

5. Given that $z = f(x, y)$ and $u = x + y$, $v = y(x + y)$, prove that

$$u\frac{\partial z}{\partial u} = \left(u + \frac{v}{u}\right)\frac{\partial z}{\partial x} - \frac{v}{u}\frac{\partial z}{\partial y}, \quad u\frac{\partial z}{\partial v} = \frac{\partial z}{\partial y} - \frac{\partial z}{\partial x}.$$

6. Given that z is a function of x and y and $x = e^u \sin v$, $y = e^u \cos v$, show that

$$\left(\frac{\partial z}{\partial u}\right)^2 + \left(\frac{\partial z}{\partial v}\right)^2 = e^{2u}\left[\left(\frac{\partial z}{\partial x}\right)^2 + \left(\frac{\partial z}{\partial y}\right)^2\right].$$

7. Given that $x(1/y + 1/z) =$ constant, prove that

$$z^2\left(\frac{\partial y}{\partial x}\right)_z = y^2\left(\frac{\partial z}{\partial x}\right)_y,$$

the suffix indicating the quantity that is kept constant.

8. Given that x, y, u, v are variables connected by the equations

$$x^2 = au^{1/2} + bv^{1/2}, \quad y^2 = au^{1/2} - bv^{1/2},$$

where a, b are constants, show that

$$\left(\frac{\partial u}{\partial x}\right)_y \left(\frac{\partial x}{\partial u}\right)_v = \tfrac{1}{2} = \left(\frac{\partial v}{\partial y}\right)_x \left(\frac{\partial y}{\partial v}\right)_u,$$

where the suffix indicates the variable which remains constant in each partial differentiation.

9. Given that the variables x, y, z are connected by the equations

$$f(x, y, z) = 0, \quad x^2 + y^2 + z^2 = \text{constant},$$

prove that

$$\mathrm{d}y/\mathrm{d}x = -(zf_x - xf_z)/(zf_y - yf_z).$$

10. The pairs of variables x, y and u, v are connected by the relations

$$x = \frac{au + bv}{u^2 + v^2}, \quad y = \frac{bu - av}{u^2 + v^2}.$$

Prove that

$$v\frac{\partial x}{\partial u} - u\frac{\partial x}{\partial v} = -y,$$

and, by expressing u, v in terms of x, y, or otherwise, show that

$$\frac{\partial^2 u}{\partial x^2} + \frac{\partial^2 u}{\partial y^2} = 0.$$

11. (a) Given that $f(x, y) \equiv F(u, v)$, where $u = x^2 - y^2$ and $v = 2xy$, show that

$$\left(\frac{\partial f}{\partial x}\right)^2 + \left(\frac{\partial f}{\partial y}\right)^2 = 4(u^2 + v^2)^{1/2}\left[\left(\frac{\partial F}{\partial u}\right)^2 + \left(\frac{\partial F}{\partial v}\right)^2\right],$$

and

$$\frac{\partial^2 f}{\partial x^2} + \frac{\partial^2 f}{\partial y^2} = 4(u^2 + v^2)^{1/2}\left[\frac{\partial^2 F}{\partial u^2} + \frac{\partial^2 F}{\partial v^2}\right].$$

(b) Given that x, y and z satisfy the relations $f(x, y, z) = $ constant, and $xyz = $ constant, show that

$$\frac{\mathrm{d}y}{\mathrm{d}x} = -\frac{y}{x}\left(x\frac{\partial f}{\partial x} - z\frac{\partial f}{\partial z}\right)\Big/\left(y\frac{\partial f}{\partial y} - z\frac{\partial f}{\partial z}\right).$$

12. Given that $\xi = x + y, \eta = \sqrt{(xy)}$, and z is a function of x and y, show that

$$x\frac{\partial z}{\partial x} + y\frac{\partial z}{\partial y} = \xi\frac{\partial z}{\partial \xi} + \eta\frac{\partial z}{\partial \eta}.$$

13. Given that $z(x, y) = f(u, v)$ where $u = x\phi(y/x), v = y\phi(x/y)$, prove that

(a)
$$x\frac{\partial z}{\partial x} + y\frac{\partial z}{\partial y} = u\frac{\partial f}{\partial u} + v\frac{\partial f}{\partial v},$$

(b) $$x^2\frac{\partial^2 z}{\partial x^2}+2xy\frac{\partial^2 z}{\partial x\,\partial y}+y^2\frac{\partial^2 z}{\partial y^2}=u^2\frac{\partial^2 f}{\partial u^2}+2uv\frac{\partial^2 f}{\partial u\,\partial v}+v^2\frac{\partial^2 f}{\partial v^2}.$$

14. Given that (r, θ) are the polar co-ordinates of a point in a plane, show that $V = r\cos\theta$ and $V = (\cos\theta)/r$ both satisfy the equation

$$\frac{\partial^2 V}{\partial x^2}+\frac{\partial^2 V}{\partial y^2}=0.$$

Given that

$$V = V_0 = br\cos\theta + c\,(\cos\theta)/r, \qquad r \geqslant a,$$
$$V = V_1 = dr\cos\theta, \qquad r \leqslant a,$$

where a and b are known constants, find the values of c and d, being given that when $r = a$

(a) $V_0 = V_1$; (b) $4\,\partial V_0/\partial r = 3\,\partial V_1/\partial r$ for all values of θ.

15. Express $xy/(x^2+y^2)^2$ in polar co-ordinates, and show that in the cartesian form the expression satisfies the equation

$$\frac{\partial^2 V}{\partial x^2}+\frac{\partial^2 V}{\partial y^2}=0,$$

and in the polar form it satisfies the equation

$$\frac{\partial^2 V}{\partial r^2}+\frac{1}{r}\frac{\partial V}{\partial r}+\frac{1}{r^2}\frac{\partial^2 V}{\partial \theta^2}=0.$$

16. Given that $x = r\cos\theta$, $y = r\sin\theta$ and $V = f(x, y)$, show that

$$\frac{\partial V}{\partial x}=\cos\theta\frac{\partial V}{\partial r}-\frac{\sin\theta}{r}\frac{\partial V}{\partial \theta},\quad \frac{\partial V}{\partial y}=\sin\theta\frac{\partial V}{\partial r}+\frac{\cos\theta}{r}\frac{\partial V}{\partial \theta},$$

and that

$$\frac{\partial^2 V}{\partial x^2}+\frac{\partial^2 V}{\partial y^2}=\frac{\partial^2 V}{\partial r^2}+\frac{1}{r}\frac{\partial V}{\partial r}+\frac{1}{r^2}\frac{\partial^2 V}{\partial \theta^2}.$$

20.5 Directional Derivatives

In Section 20.2 we discussed the slope of the surface $z = f(x, y)$ observed from the point $P(x, y, z)$ on the surface, looking in the positive x-direction (slope $= \partial z/\partial x$) or in the positive y-direction (slope $= \partial z/\partial y$). It is now possible to determine the slope of the surface in *any* direction. Consider the direction making an angle α with the positive x-direction; then if x'-, y'-axes are obtained by rotating the x-, y-axes through α so that the direction under consideration is the positive x'-direction, the required slope is $\partial z/\partial x'$.

Now from (2.4) (Section 2.6),

$$x = x'\cos\alpha - y'\sin\alpha, \quad y = x'\sin\alpha + y'\cos\alpha,$$

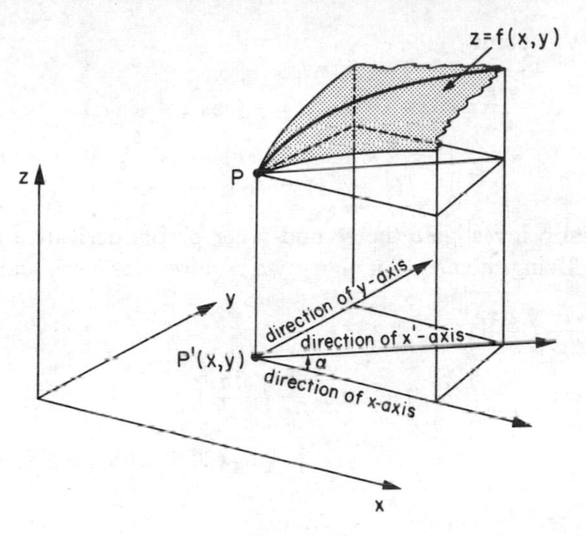

FIG. 20.8

and so

$$\frac{\partial z}{\partial x'} = \frac{\partial z}{\partial x}\frac{\partial x}{\partial x'} + \frac{\partial z}{\partial y}\frac{\partial y}{\partial x'} \qquad \text{(from (20.2))}$$

$$= \cos\alpha \frac{\partial z}{\partial x} + \sin\alpha \frac{\partial z}{\partial y}. \qquad (20.5)$$

The expression

$$\cos\alpha \frac{\partial z}{\partial x} + \sin\alpha \frac{\partial z}{\partial y}$$

is called the *directional derivative* of z in the direction making an angle α with the positive x-direction.

Example 21 Find the slope at the point $(a/\sqrt{3}, a/\sqrt{3}, a/\sqrt{3})$ on the hemisphere $z = \sqrt{(a^2 - x^2 - y^2)}$ in the direction making an angle (a) $\pi/4$, (b) $\pi/3$, with the positive x-axis.

We have

$$\frac{\partial z}{\partial x} = \frac{-x}{\sqrt{(a^2 - x^2 - y^2)}}, \quad \frac{\partial z}{\partial y} = \frac{-y}{\sqrt{(a^2 - x^2 - y^2)}}$$

and so at the point $(a/\sqrt{3}, a/\sqrt{3}, a/\sqrt{3})$ we get

$$\frac{\partial z}{\partial x} = \frac{\partial z}{\partial y} = -1.$$

Hence at this point

$$\frac{\partial z}{\partial x}\cos\alpha + \frac{\partial z}{\partial y}\sin\alpha = -(\cos\alpha + \sin\alpha)$$
$$= \begin{cases} -\sqrt{2} \text{ when } \alpha = \pi/4 \\ -\tfrac{1}{2}(1+\sqrt{3}) \text{ when } \alpha = \pi/3. \end{cases}$$

We may also investigate the second-order partial derivative in a given direction. Using the notation above we require $\partial^2 z/\partial x'^2$, and

$$\frac{\partial^2 z}{\partial x'^2} = \frac{\partial}{\partial x'}\left(\frac{\partial z}{\partial x'}\right)$$
$$= \cos\alpha\,\frac{\partial}{\partial x}\left(\frac{\partial z}{\partial x'}\right) + \sin\alpha\,\frac{\partial}{\partial y}\left(\frac{\partial z}{\partial x'}\right)$$
$$\left(\text{using (20.5) with } z \text{ replaced by } \frac{\partial z}{\partial x'}\right)$$
$$= \cos\alpha\,\frac{\partial}{\partial x}\left[\cos\alpha\,\frac{\partial z}{\partial x} + \sin\alpha\,\frac{\partial z}{\partial y}\right]$$
$$+ \sin\alpha\,\frac{\partial}{\partial y}\left[\cos\alpha\,\frac{\partial z}{\partial x} + \sin\alpha\,\frac{\partial z}{\partial y}\right]$$
$$= \cos^2\alpha\,\frac{\partial^2 z}{\partial x^2} + 2\sin\alpha\cos\alpha\,\frac{\partial^2 z}{\partial x\,\partial y} + \sin^2\alpha\,\frac{\partial^2 z}{\partial y^2}. \tag{20.6}$$

20.5.1 Maxima and Minima A point P on the surface $z = f(x, y)$ is a *stationary point* if the directional derivative of z is zero *for all directions* at P. This means that P is a stationary point on any curve on the surface passing through P. The condition for a stationary point is that

$$\cos\alpha\,\frac{\partial z}{\partial x} + \sin\alpha\,\frac{\partial z}{\partial y} = 0$$

for any α in the interval $0 \leqslant \alpha < 2\pi$, that is

$$\frac{\partial z}{\partial x} = \frac{\partial z}{\partial y} = 0.$$

We say P is a *maximum turning point* (minimum turning point) on the surface if it is a max. T.P. (min. T.P.) on *every* curve on the surface through P. It follows that P is a maximum turning point (minimum turning point) on the surface if it is a stationary point and if, in addition, the directional second derivative is negative (positive) for all directions at P. A stationary point at which the directional second derivative is positive for some directions and negative for others is called a *saddle point* (see Fig. 20.10).

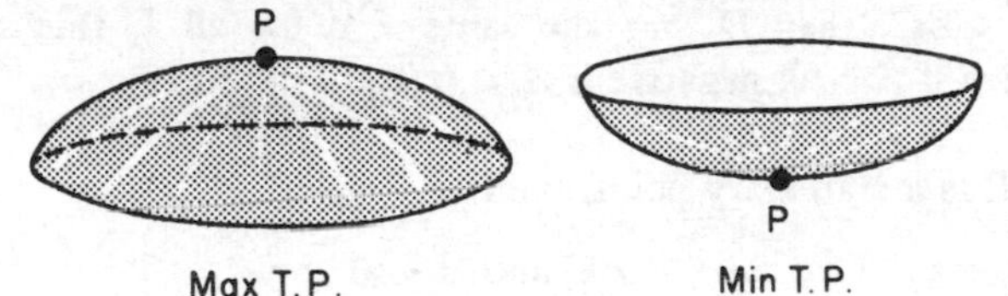

FIG. 20.9

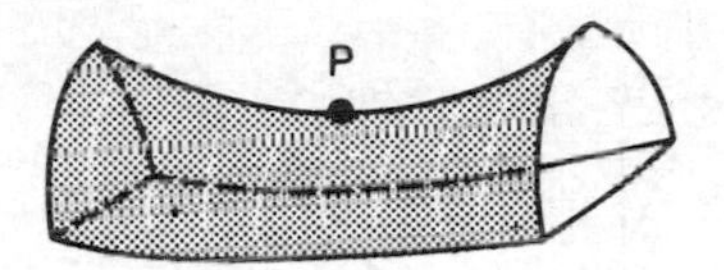

FIG. 20.10

From (20.6) the value D of the directional second derivative at P is given by

$$D = A \cos^2 \alpha + 2B \sin \alpha \cos \alpha + C \sin^2 \alpha$$

where A, B, C are the values of $\partial^2 z/\partial x^2$, $\partial^2 z/\partial x\, \partial y$, $\partial^2 z/\partial y^2$ respectively at P. We may investigate the sign of D by considering D', where

$$D' = D/\sin^2 \alpha = A \cot^2 \alpha + 2B \cot \alpha + C,$$

and since $\sin^2 \alpha$ is positive, D' has the same sign as D. Writing $\lambda = \cot \alpha$, we see that D' is a quadratic polynomial in λ, namely $D' = A\lambda^2 + 2B\lambda + C$, and it follows that

(1) if $B^2 > AC$, then D' may be positive or negative, depending on λ (see Fig. 20.11);

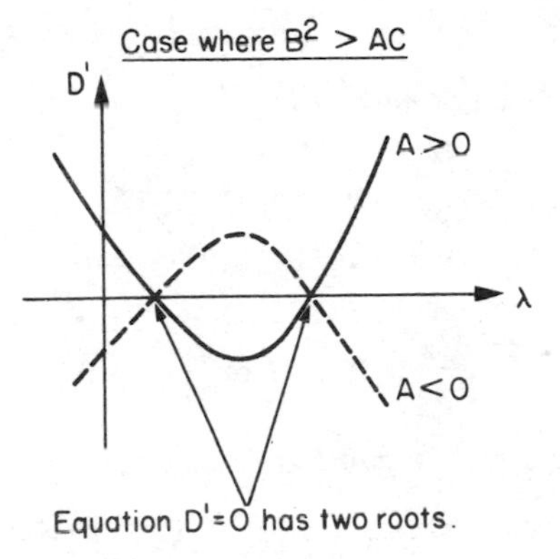

FIG. 20.11

(2) if $B^2 < AC$, then D' has the same sign for all λ, this sign being positive if $A > 0$, negative if $A < 0$ (see Fig. 20.12).

Thus, if P is a stationary point, then

(a) P is a max. T.P. if $B^2 < AC$ and $A < 0$,
(b) P is a min. T.P. if $B^2 < AC$ and $A > 0$,
(c) P is a saddle-point if $B^2 > AC$.

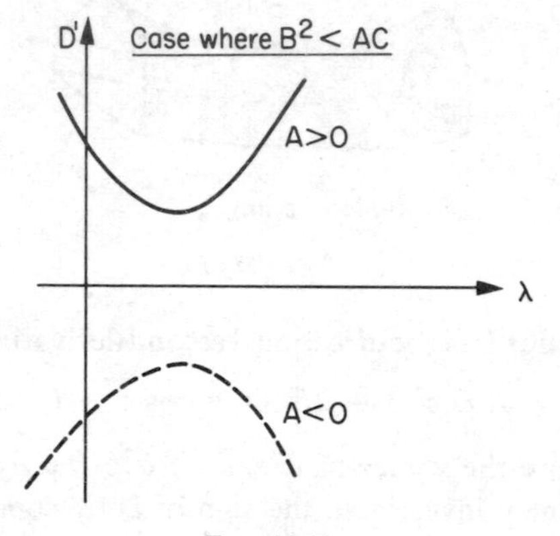

FIG. 20.12

Since $AC > B^2 \Rightarrow AC > 0 \Rightarrow A$ and C have the same sign, in (a) we may replace $A < 0$ by $C < 0$, and in (b) we may replace $A > 0$ by $C > 0$. Summarising, the conditions for a ***turning point*** are

$$\frac{\partial z}{\partial x} = \frac{\partial z}{\partial y} = 0, \quad \left(\frac{\partial^2 z}{\partial x\,\partial y}\right)^2 < \frac{\partial^2 z}{\partial x^2}\frac{\partial^2 z}{\partial y^2},$$

the turning-point being a ***max. T.P.*** if

$$\frac{\partial^2 z}{\partial x^2} < 0 \qquad \left(\text{or } \frac{\partial^2 z}{\partial y^2} < 0\right)$$

and a ***min. T.P.*** if

$$\frac{\partial^2 z}{\partial x^2} > 0 \qquad \left(\text{or } \frac{\partial^2 z}{\partial y^2} > 0\right);$$

the conditions for a ***saddle point*** are

$$\frac{\partial z}{\partial x} = \frac{\partial z}{\partial y} = 0, \quad \left(\frac{\partial^2 z}{\partial x\,\partial y}\right)^2 > \frac{\partial^2 z}{\partial x^2}\frac{\partial^2 z}{\partial y^2}.$$

Special difficulties arise if

$$\left(\frac{\partial^2 z}{\partial x\,\partial y}\right)^2 = \frac{\partial^2 z}{\partial x^2}\frac{\partial^2 z}{\partial y^2}$$

and we shall not discuss this exceptional case.

Example 22 Find the nature of the stationary points on the surface $z = x^3 - xy + y^3$.

Since

$$\frac{\partial z}{\partial x} = 3x^2 - y, \quad \frac{\partial z}{\partial y} = -x + 3y^2,$$

for a stationary point we require

$$3x^2 - y = 0, \quad 3y^2 - x = 0.$$

Eliminating y, we obtain $27x^4 = x$ so that $x = 0$ or $\frac{1}{3}$. When $x = 0$ we find that $y = 0$, and when $x = \frac{1}{3}$ we find that $y = \frac{1}{3}$, so the stationary points are $(0, 0)$ and $(\frac{1}{3}, \frac{1}{3})$. Now

$$\frac{\partial^2 z}{\partial x^2} = 6x, \quad \frac{\partial^2 z}{\partial y^2} = 6y, \quad \frac{\partial^2 z}{\partial x\,\partial y} = -1,$$

giving $(\partial^2 z/\partial x\,\partial y)^2 = 1$ at all points on the surface, and

$$\frac{\partial^2 z}{\partial x^2}\frac{\partial^2 z}{\partial y^2} = 36xy = \begin{cases} 0 \text{ at } (0, 0) \\ 4 \text{ at } (\frac{1}{3}, \frac{1}{3}). \end{cases}$$

It follows that $(0, 0)$ is a saddle-point and $(\frac{1}{3}, \frac{1}{3})$ is a turning point. Further, $\partial^2 z/\partial x^2 > 0$ when $x = \frac{1}{3}$ and so $(\frac{1}{3}, \frac{1}{3})$ is a min. T.P. The minimum value of z is $(\frac{1}{3})^3 - (\frac{1}{3})^2 + (\frac{1}{3})^3$, i.e. $-\frac{1}{27}$.

Example 23 A closed rectangular box is made from sheet metal and has a fixed volume V. Find the dimensions of the box that requires the minimum amount of metal for its construction.

Let the sides of the box be x, y, z. It is a necessary condition that $xyz = V$ and so $z = V/xy$, i.e. we may regard x and y as the only independent variables.

The area of sheet metal required, A say, is given by

$$A = 2xy + 2yz + 2zx = 2xy + 2V/x + 2V/y.$$

Hence

$$\frac{\partial A}{\partial x} = 2y - 2V/x^2, \quad \frac{\partial A}{\partial y} = 2x - 2V/y^2,$$

so for stationary points we require

$$y - V/x^2 = 0, \quad x - V/y^2 = 0,$$

that is $x = y = V^{1/3}$.

Further,

$$\frac{\partial^2 A}{\partial x^2} = \frac{4V}{x^3}, \quad \frac{\partial^2 A}{\partial y^2} = \frac{4V}{y^3}, \quad \frac{\partial^2 A}{\partial x\, \partial y} = 2,$$

giving

$$\left(\frac{\partial^2 A}{\partial x\, \partial y}\right)^2 = 4, \quad \frac{\partial^2 A}{\partial x^2}\frac{\partial^2 A}{\partial y^2} = \frac{16V^2}{x^3 y^3} = 16 \quad \text{when } x = y = V^{1/3}.$$

Thus we have a turning point when $x = y = V^{1/3}$, and since $\partial^2 A/\partial x^2 > 0$ at this point it is a min T.P. The required dimensions are

$$x = y = z = V^{1/3}.$$

Exercises 20(d)

1. Find the (infinitely many) stationary points of the surface

$$z = \sin(\pi x/2) \sin(\pi y/2).$$

Show that these points are saddle points when both x and y have even integer values.

Find the position and nature of the stationary points on the following surfaces:

2. $z = x^2 + y^2 - (x - 1)(y - 2)$
3. $z = x^3 + y^3 - y(3x^2 - 1)$
4. $z = x^2y^2 - x^2 - y^2 + 1$
5. $z = x^3 - y^2x + x + y + 1$
6. $z = x^2 + y(y - 1)(x + 1)$
7. $z = xy^2 - 2y^2 - 2x^2 + 4x + 1$
8. $z = x^3 + 2x^2 + 2xy - y^2 - 11x + 2y + 7$
9. $z = 4xy - 5x^2 - 8y^2 + 4x + 8y - 4$.

20.6 Integration of Functions of Two Variables

Given $f(x, y)$ suppose that we wish to find $F(x, y)$ such that

$$\partial F/\partial x = f(x, y).$$

Clearly $F(x, y) + g(y)$, where g is an arbitrary function, is also a solution, since

$$\frac{\partial}{\partial x}[F(x, y) + g(y)] = f(x, y).$$

Thus we write

$$\int f(x, y)\,dx = F(x, y) + g(y).$$

For example,

$$\int (x^2y + xy^2)\,dx = \tfrac{1}{3}x^3y + \tfrac{1}{2}x^2y^2 + g(y),$$

$$\int \cos xy\,dx = \frac{1}{y}\sin xy + g(y),$$

$$\int \frac{1}{x^2 + y^2}\,dx = \frac{1}{y}\tan^{-1}\left(\frac{x}{y}\right) + g(y),$$

and it will be seen that we are integrating with respect to x in the usual way, treating y as a constant. Limits of integration may also be used; thus

$$\int_a^b f(x, y)\,dx = F(b, y) - F(a, y),$$

and, in particular

$$\int_1^2 \frac{1}{x^2 + y^2}\,dx = \frac{1}{y}\left[\tan^{-1}\left(\frac{2}{y}\right) - \tan^{-1}\left(\frac{1}{y}\right)\right],$$

showing that the value of the definite integral depends on y.

Similarly if $\partial G/\partial y = f(x, y)$ we write

$$\int f(x, y)\,dy = G(x, y) + g(x),$$

where g is again an arbitrary function, and

$$\int_c^d f(x, y)\,dy = G(x, d) - G(x, c).$$

For example,

$$\int (x^2y + xy^2)\,dy = \tfrac{1}{2}x^2y^2 + \tfrac{1}{3}xy^3 + g(x),$$

$$\int \cos xy\,dy = \frac{1}{x}\sin xy + g(x),$$

$$\int_1^2 \frac{1}{x^2 + y^2}\,dy = \left[\frac{1}{x}\tan^{-1}\left(\frac{y}{x}\right)\right]_{y=1}^{y=2} = \frac{1}{x}\left[\tan^{-1}\left(\frac{2}{x}\right) - \tan^{-1}\left(\frac{1}{x}\right)\right].$$

Again we see that no new principles of integration are involved and all the techniques of Chapter 18 are available in this new situation.

20.6.1 Repeated Integrals Since $\int f(x, y)\,dx$ is a function of the two variables x and y, it may be integrated with respect to y to give the repeated integral

$$\int \left[\int f(x, y)\,dx \right] dy,$$

usually written

$$\int dy \int f(x, y)\,dx.$$

The same process of repeated integration is possible with limits of integration a and b for x, where a and b may depend on y, and limits of integration c and d for y, where c and d are constants; for $\int_a^b f(x, y)\,dx$ is a function of y (this is true even if a and b depend on y) and we may integrate with respect to y to get

$$\int_c^d dy \int_a^b f(x, y)\,dx.$$

Example 24 Evaluate I, where

$$I = \int_0^1 dy \int_{y^2}^{y} (x^2 y + x y^2)\,dx.$$

We have

$$I = \int_0^1 \left[\tfrac{1}{3}x^3 y + \tfrac{1}{2}x^2 y^2 \right]_{x=y^2}^{x=y} dy$$

$$= \int_0^1 (\tfrac{1}{3}y^4 + \tfrac{1}{2}y^4 - \tfrac{1}{3}y^7 - \tfrac{1}{2}y^6)\,dy$$

$$= \left[\tfrac{1}{6}y^5 - \tfrac{1}{24}y^8 - \tfrac{1}{14}y^7 \right]_0^1$$

$$= \tfrac{3}{56}.$$

Example 25 Evaluate I, where

$$I = \int_1^e dy \int_0^{1/2y} \cos \pi x y\,dx.$$

We have

$$I = \int_1^e \left[\frac{1}{\pi y} \sin \pi xy\right]_{x=0}^{x=1/2y} dy$$

$$= \int_1^e \frac{1}{\pi y}\left(\sin\frac{\pi}{2} - \sin 0\right) dy$$

$$= \frac{1}{\pi}\Big[\ln y\Big]_1^e = \frac{1}{\pi}.$$

Similarly we have repeated integrals $\int_a^b dx \int_c^d f(x, y)\, dy$ in which integration is carried out first with respect to y and finally with respect to x. In this case the limits c and d may depend on x.

Example 26 Evaluate I, where

$$I = \int_1^3 dx \int_0^x \frac{1}{x^2 + y^2}\, dy.$$

We have

$$I = \int_1^3 \left[\frac{1}{x} \tan^{-1}\left(\frac{y}{x}\right)\right]_{y=0}^{y=x} dx$$

$$= \int_1^3 \frac{1}{x}(\tan^{-1} 1 - \tan^{-1} 0)\, dx$$

$$= \frac{\pi}{4}\Big[\ln x\Big]_1^3 = \frac{\pi}{4} \ln 3.$$

20.6.2 Separable Variables Suppose that $f(x, y)$ is the product of a function of x and a function of y, say $f(x, y) = f_1(x) f_2(y)$. We then say that the variables are ***separable***. For example, in the expressions

$$x^2y^3, \quad (x^2 + 2x + 1) \sin y, \quad e^{3x} \ln y$$

the variables are separable, whereas in the expressions

$$x^2y + xy^2, \quad \cos xy, \quad 1/(x^2 + y^2)$$

the variables are *not* separable.

Now if $f(x, y) = f_1(x) f_2(y)$ it follows that

$$\int f(x, y)\, dx = f_2(y) \int f_1(x)\, dx,$$

since in this integration $f_2(y)$ is treated as a constant. Hence in this case we write

$$\int dy \int f(x, y)\, dx = \int f_2(y)\, dy \int f_1(x)\, dx,$$

indicating that $f_2(y)$ does not enter into the process of integration with respect to x.

Introducing limits of integration, we have

$$\int_c^d dy \int_a^b f(x, y)\, dx = \int_c^d f_2(y)\, dy \int_a^b f_1(x)\, dx.$$

Example 27 Evaluate I where

$$I = \int_0^1 dy \int_{y-1}^{y+1} x^2 y^3\, dx.$$

We write

$$\begin{aligned} I &= \int_0^1 y^3\, dy \int_{y-1}^{y+1} x^2\, dx \\ &= \int_0^1 y^3\, dy \left[\tfrac{1}{3}x^3\right]_{y-1}^{y+1} \\ &= \int_0^1 y^3\,(2y^2 + \tfrac{2}{3})\, dy \\ &= \left[\tfrac{1}{3}y^6 + \tfrac{1}{6}y^4\right]_0^1 = \tfrac{1}{2}. \end{aligned}$$

If a and b do not depend on y, then $\int_b^a f_1(x)\, dx$ is a constant and does not enter into the process of integration with respect to y. Thus

$$\int_c^d dy \int_a^b f_1(x) f_2(y)\, dx = \left[\int_c^d f_2(y)\, dy\right] \times \left[\int_a^b f_1(x)\, dx\right] \qquad (20.7)$$

if a, b, c, d are constants, i.e. the integral may be evaluated as a product of two independent integrals.

Example 28 Evaluate I where

$$I = \int_0^1 dy \int_0^{\pi/2} e^{3y} \sin 2x\, dx.$$

We have

$$I = \int_0^1 e^{3y}\,dy \int_0^{\pi/2} \sin 2x\,dx$$

$$= \left[\tfrac{1}{3}e^{3y}\right]_0^1 \times \left[-\tfrac{1}{2}\cos 2x\right]_0^{\pi/2}$$

$$= \tfrac{1}{3}(e^3 - 1).$$

Exercises 20(e)

Verify the following results:

1. $\int_1^5 dy \int_3^6 dx = 12$

2. $\int_1^2 dx \int_0^x (x+y)\,dy = \frac{7}{2}$

3. $\int_1^2 \frac{1}{y^2}\,dy \int_0^{y^{3/2}} x\,dx = \frac{3}{4}$

4. $\int_1^2 y\,dy \int_0^{y^2} e^x\,dx = \frac{1}{2}(e^2 - e - 3)$

5. $\int_0^1 y\,dy \int_{y^2}^{y} x^2\,dx = \frac{1}{40}$

6. $\int_0^1 dy \int_y^{\sqrt{y}} (x^2 + y^2)\,dx = \frac{3}{35}$

7. $\int_1^2 dy \int_1^y \left(\frac{1}{y} + \frac{1}{x}\right) dx = \ln 2$

8. $\int_0^{\pi/2} \sin^2\theta\,d\theta \int_0^{\cos\theta} r^2\,dr = \frac{2}{45}$

9. $\int_0^{\pi/2} \sin^2\theta \cos^2\theta\,d\theta \int_0^{\sin\theta\cos\theta} r\,dr = \frac{3\pi}{512}$

10. $\int_0^{2\pi} \cos^2\theta\,d\theta \int_0^{1+\cos\theta} r^3\,dr = \frac{49\pi}{32}.$

20.7 Double Integrals

Let D be a region of the xy-plane and let f be a function of the two variables x and y such that the domain of f includes D. Suppose that a network of curves divides D into N small regions, or *interstices*, $D_1, D_2, \ldots, D_N$ from which the points $P_1, P_2, \ldots, P_N$ respectively have been selected. Denote by $\Delta A_1, \Delta A_2, \ldots, \Delta A_N$ the areas of the interstices $D_1, D_2, \ldots, D_N$ respectively, and by $(x_1, y_1), (x_2, y_2), \ldots, (x_N, y_N)$ the coordinates of the points $P_1, P_2, \ldots, P_N$ respectively. Consider the sum

$$\sum_{r=1}^{N} f(x_r, y_r)\,\Delta A_r; \tag{S}$$

such a sum can be associated with *any* subdivision of D into a number of small regions, and if, as $N \to \infty$ in such a way that each $\Delta A_r \to 0$, the sum tends to a limit (this limit being independent of the mode of subdivision and of the selection of the points P_r) then the limit is called the *double integral* of f over D, written $\iint_D f(x, y)\,dA$.

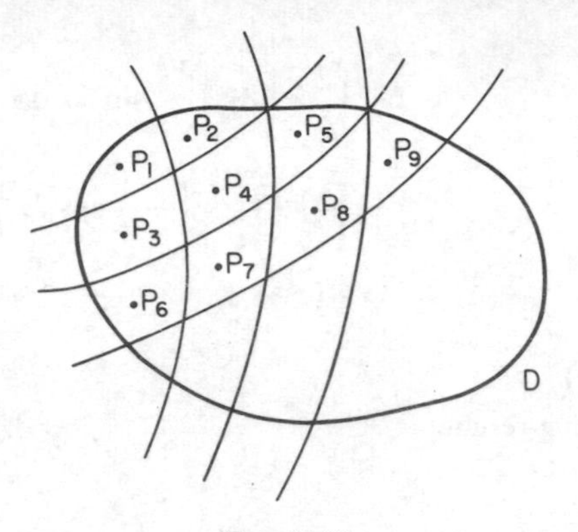

FIG. 20.13

The region D is called the *area of integration*, or *field of integration*. It is convenient to assume that $f(x, y) \geqslant 0$ for all points (x, y) in D; we can then readily identify the double integral with the volume of a solid with base D, bounded on top by the surface $z = f(x, y)$, and with sides formed by drawing perpendiculars from the boundary of D (Fig. 20.14).

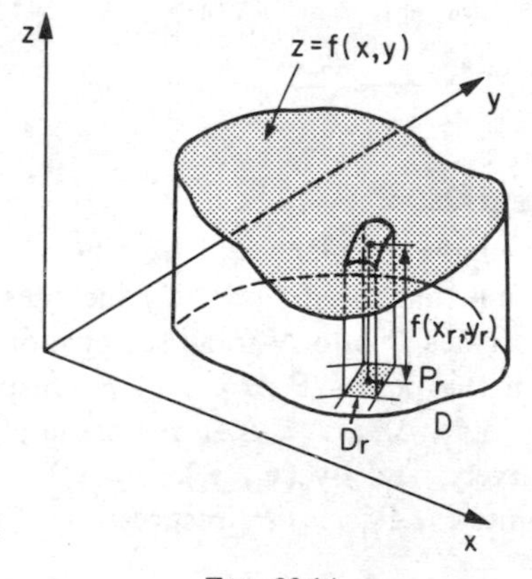

FIG. 20.14

The volume directly above the interstice D_r is approximately $f(x_r, y_r)\Delta A_r$ and so the sum (S) is the approximate total volume of the solid. It is reasonable to expect that, for most surfaces $z = f(x, y)$ that we are likely to encounter, the sum (S) has a limit $\iint_D f(x, y)\, dA$ which may be taken as the exact volume of the solid.

If $f(x, y) = 1$ for all (x, y) in D the solid has a uniform height of 1 unit and its volume equals the area of its base, i.e.

$$\text{area of } D = \iint_D 1 \, dA.$$

If $f(x, y)$ takes both positive and negative values, $\iint_D f(x, y) \, dA$ can similarly be interpreted in terms of volume, with volume below the xy-plane being counted negatively.

20.7.1 Double Integral in Cartesian Co-ordinates Let D be divided by the network of lines $x = x_1$, $x = x_2$, ..., $x = x_n$ and $y = y_1$, $y = y_2$, ..., $y = y_m$ into small rectangular interstices as shown in Fig. 20.15. The area ΔA_{ij} of the interstice bounded by the lines $x = x_i$,

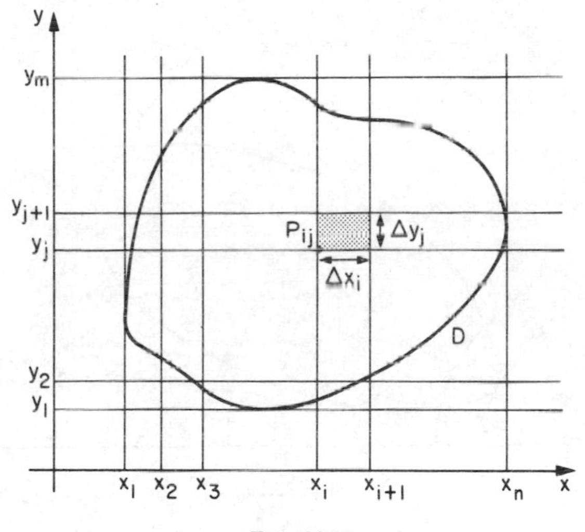

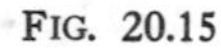
FIG. 20.15

$x = x_{i+1}$, $y = y_j$, $y = y_{j+1}$ is given by

$$\Delta A_{ij} = \Delta x_i \, \Delta y_j,$$

where $\Delta x_i = x_{i+1} - x_i$, $\Delta y_j = y_{j+1} - y_j$. Let the point P_{ij} with co-ordinates (x_i, y_j) be selected from this interstice for the purpose of forming the sum (S), which now becomes

$$\Sigma f(x_i, y_j) \, \Delta x_i \, \Delta y_j, \tag{S$'$}$$

summation being over all interstices lying within D. It follows that

$\iint_D f(x,y)\,dA$ is the limit of (S′) as m and n both $\to\infty$ in such a way that each Δx_i and each $\Delta y_j \to 0$. Accordingly this integral is usually written as

$$\iint_D f(x, y)\,dx\,dy.$$

20.7.2 Double Integral in Polar Co-ordinates Let D be divided by the network of circles $r = r_1, r = r_2, \ldots, r = r_n$ and lines $\theta = \theta_1, \theta = \theta_2, \ldots, \theta = \theta_m$. The area ΔA_{ij} of the interstice bounded by the circles $r = r_i$, $r = r_{i+1}$ and the lines $\theta = \theta_j$, $\theta = \theta_{j+1}$ is given approximately by $\Delta A_{ij} \simeq r_i\,\Delta\theta_j\,\Delta r_i$ (see Fig. 20.16) where $\Delta r_i = r_{i+1} - r_i$, $\Delta\theta_j = \theta_{j+1} - \theta_j$.

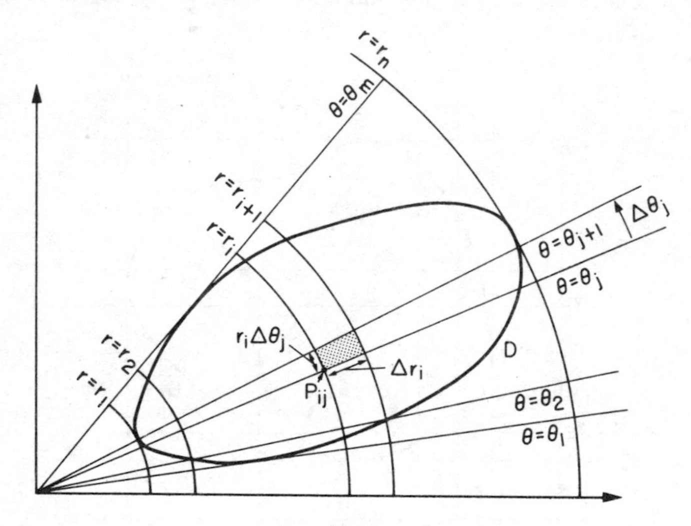

FIG. 20.16

Let the point P_{ij} with polar co-ordinates (r_i, θ_j) be selected from this interstice for the purpose of forming the sum (S), which now becomes

$$f(r_i \cos\theta_j, r_i \sin\theta_j)\,r_i\,\Delta\theta_j\,\Delta r_i, \qquad \text{(S}''\text{)}$$

summation being over all interstices within D. Thus $\iint_D f(x, y)\,dA$ is the limit of (S″) as m and n both $\to\infty$ in such a way that each Δr_i and each $\Delta\theta_j \to 0$.

Let D' be the region of the $r\theta$-plane corresponding to the region D of the xy-plane, i.e. D' is the image of D under the mapping $(x, y) \mapsto (r, \theta)$ where $x = r\cos\theta$, $y = r\sin\theta$ (see Fig. 20.17).

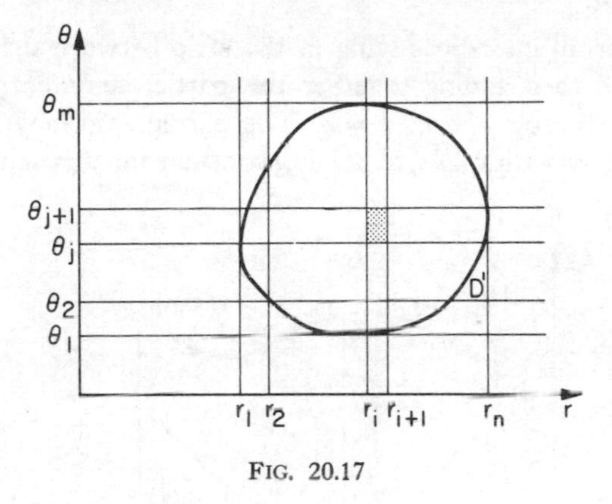

FIG. 20.17

Comparing (S″) with (S′) we see that

$$\iint_D f(x, y)\, dx\, dy = \iint_{D'} g(r, \theta)\, dr\, d\theta,$$

where $g(r, \theta) = r\, f(r \cos \theta, r \sin \theta)$.

As a simple example, consider $\iint_D x^2y^3\, dx\, dy$, where D is the region bounded by the circles $x^2 + y^2 = 1$, $x^2 + y^2 = 4$ and the lines $y = x$, $x = 0$; now $x^2y^3 = r^5 \cos^2 \theta \sin^3 \theta$ and so the given integral is equal to the double integral of $r^6 \cos^2 \theta \sin^3 \theta$ over the rectangle bounded by the lines $r = 1$, $r = 2$, $\theta = \pi/4$, $\theta = \pi/2$ (see Fig. 20.18).

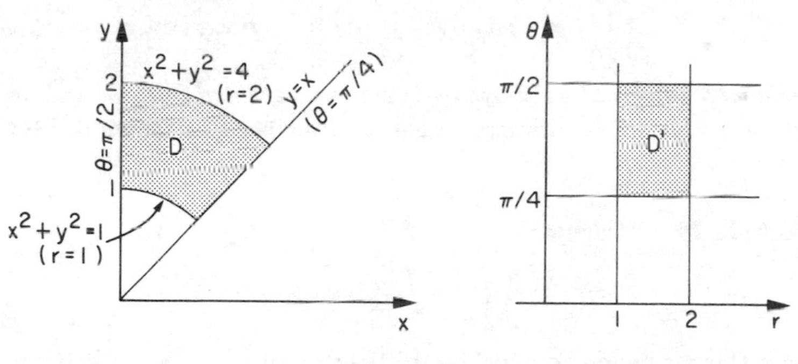

FIG. 20.18

20.7.3 Evaluation of Double Integrals (1) Suppose that D is the region between the curves $y = \phi_1(x)$ and $y = \phi_2(x)$ from $x = a$ to $x = b$ (Fig. 20.19). Imagine the sum (S) to be evaluated by first obtaining a partial sum over all interstices lying in the strip between the lines $x = x_i$, $x = x_{i+1}$ and then adding together the partial sums corresponding to all the strips from $x = a$ to $x = b$. The partial sum may be written as $\Delta x_i \Sigma f(x_i, y_j) \Delta y_j$ with x_i, Δx_i remaining constant and y_j ranging from $\phi_1(x_i)$

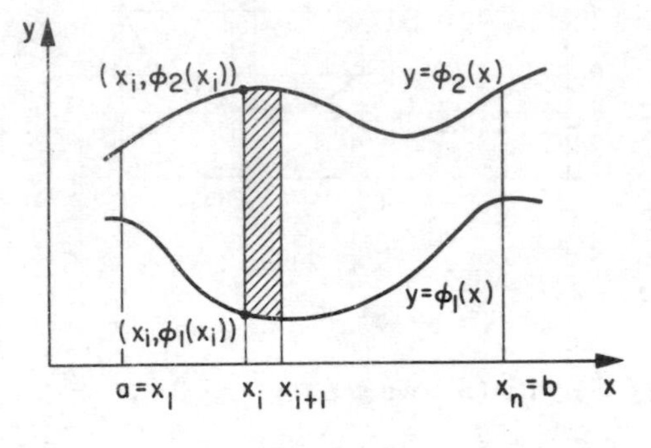

FIG. 20.19

at the bottom of the strip to $\phi_2(x_i)$ at the top; as each $\Delta y_j \to 0$ the partial sum becomes

$$\Delta x_i \int_{\phi_1(x_i)}^{\phi_2(x_i)} f(x_i, y)\,\mathrm{d}y = F(x_i)\,\Delta x_i \qquad \text{(say).}$$

The process of adding all the partial sums and letting each $\Delta x_i \to 0$ now gives $\int_a^b F(x)\,\mathrm{d}x$; thus

$$\iint_D f(x, y)\,\mathrm{d}x\,\mathrm{d}y = \int_a^b \mathrm{d}x \int_{\phi_1(x)}^{\phi_2(x)} f(x, y)\,\mathrm{d}y.$$

The *double integral* has been expressed as a *repeated integral* and hence may be evaluated by ordinary integration methods as shown in Section 20.6.1.

Example 29 Evaluate

$$\iint_D \frac{1}{x^2 + y^2}\,\mathrm{d}x\,\mathrm{d}y,$$

where D is the region bounded by the lines $y = 0$, $y = x$, $x = 1$ and $x = 3$.

D is the region between the lines $y = 0$ and $y = x$ from $x = 1$ to $x = 3$. Hence the given integral is equal to

$$\int_1^3 \mathrm{d}x \int_0^x \frac{1}{x^2 + y^2}\,\mathrm{d}y = \frac{\pi}{4} \ln 3 \qquad \text{(by Example 26).}$$

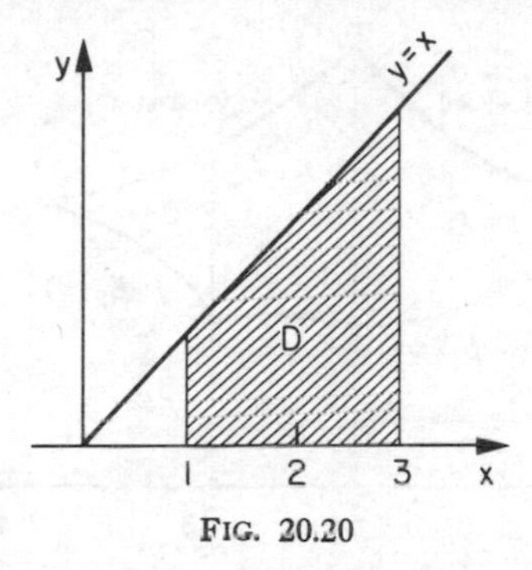

FIG. 20.20

Example 30 Evaluate $\iint_D xy\,dx\,dy$, where D is the first quadrant of the circle $x^2 + y^2 = a^2$.

D is the region between the x-axis and the curve $y = \sqrt{(a^2 - x^2)}$ from $x = 0$ to $x = a$. Hence the given integral is equal to

$$\int_0^a dx \int_0^{\sqrt{(a^2 - x^2)}} xy\,dy = \int_0^a x\,dx \left[\tfrac{1}{2}y^2\right]_0^{\sqrt{(a^2 - x^2)}}$$

$$= \int_0^a \tfrac{1}{2}x(a^2 - x^2)\,dx = \tfrac{1}{8}a^4.$$

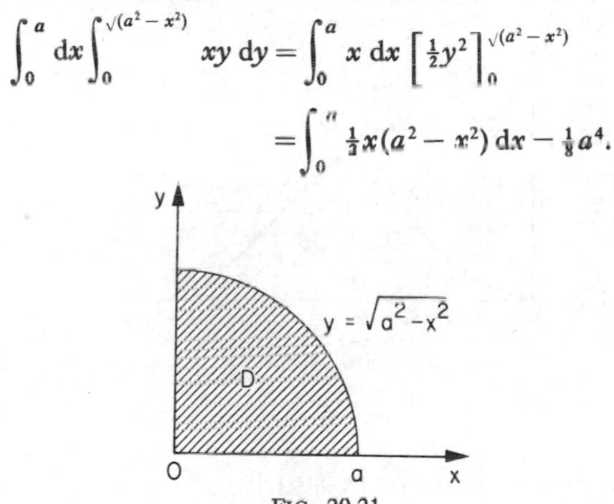

FIG. 20.21

However many curves $y = \phi_1(x)$, $y = \phi_2(x)$, . . . are included in the boundary of D, it is usually a simple matter to divide D into a number of regions of the type discussed above. Figure 20.22 shows a region with four boundary curves $y = \phi_1(x)$, $y = \phi_2(x)$, $y = \phi_3(x)$, $y = \phi_4(x)$, and in this case

$$\iint_D f(x, y)\,dx\,dy = \int_{a_1}^{a_2} dx \int_{\phi_1(x)}^{\phi_4(x)} f(x, y)\,dy$$

$$+ \int_{a_2}^{a_3} dx \int_{\phi_1(x)}^{\phi_3(x)} f(x, y)\,dy + \int_{a_3}^{a_4} dx \int_{\phi_2(x)}^{\phi_3(x)} f(x, y)\,dy.$$

A diagram should always be used.

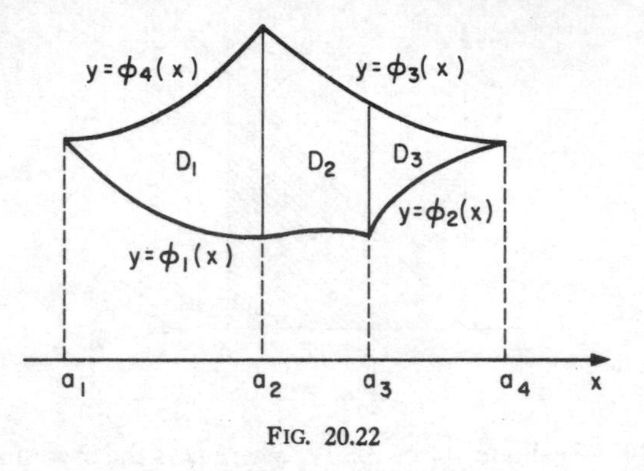

FIG. 20.22

Example 31 Evaluate $\iint_D (x + y)\,dx\,dy$, where D is the region bounded by the lines $y = 0$, $x + y = 2$ and the curve $y = x^2$.

The line $x = 1$ divides D into two regions D_1, D_2 as shown in Fig. 20.23.

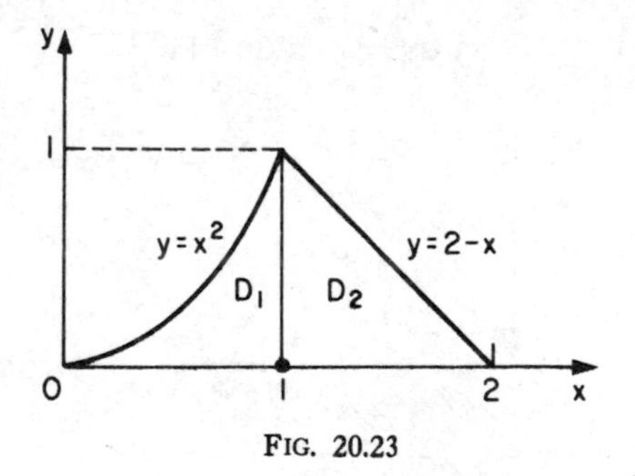

FIG. 20.23

Now

$$\iint_{D_1} (x + y)\,dx\,dy = \int_0^1 dx \int_0^{x^2} (x + y)\,dy$$

$$= \int_0^1 dx \left[xy + \tfrac{1}{2}y^2\right]_{y=0}^{y=x^2}$$

$$= \int_0^1 (x^3 + \tfrac{1}{2}x^4)\,dx = \tfrac{7}{20}.$$

Also

$$\iint_{D_2} (x + y)\,dx\,dy = \int_1^2 dx \int_0^{2-x} (x + y)\,dy$$

$$= \int_1^2 [x(2 - x) + \tfrac{1}{2}(2 - x)^2]\,dx = \tfrac{5}{6}.$$

Hence

$$\iint_D (x+y)\,dx\,dy = \tfrac{7}{20} + \tfrac{5}{6} = \tfrac{71}{60}.$$

20.7.4 Evaluation of Double Integrals (2) Similarly if D is the region between the curves $x = \psi_1(y)$ and $x = \psi_2(y)$ from $y = c$ to $y = d$, then

$$\iint_D f(x, y)\,dx\,dy = \int_c^d dy \int_{\psi_1(y)}^{\psi_2(y)} f(x, y)\,dx,$$

or if D can be divided into several such regions then the double integral over D may be evaluated as the sum of the corresponding repeated integrals.

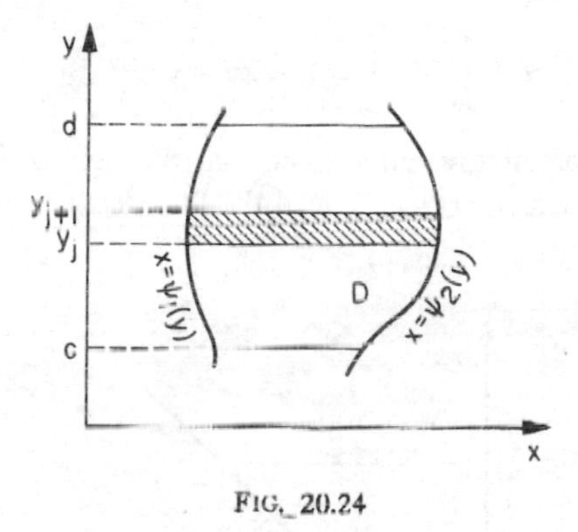

FIG. 20.24

Example 32 Do Example 31 by this method.

In this case D is the region between the curves $x = y^{1/2}$ and $x = 2 - y$ from $y = 0$ to $y = 1$. Hence

$$\iint_D (x+y)\,dx\,dy = \int_0^1 dy \int_{y^{1/2}}^{2-y} (x+y)\,dx = \int_0^1 dy \Big[\tfrac{1}{2}x^2 + xy\Big]_{x=y^{1/2}}^{x=2-y}$$

$$= \int_0^1 [\tfrac{1}{2}(2-y)^2 + y(2-y) - \tfrac{1}{2}y - y^{3/2}]\,dy = \tfrac{71}{60}.$$

20.7.5 Changing the Order of Integration The repeated integral

$$\int_a^b dx \int_{\phi_1(x)}^{\phi_2(x)} f(x, y)\,dy \tag{1}$$

is equivalent to a double integral (see Section 20.7.3) and hence it may be possible (by Section 20.7.4) to express it as a repeated integral, or the sum of several repeated integrals, of the type

$$\int_c^d \mathrm{d}y \int_{\psi_1(y)}^{\psi_2(y)} f(x, y)\,\mathrm{d}x. \tag{2}$$

Likewise a repeated integral of type (2) may possibly be converted into a repeated integral, or the sum of several repeated integrals, of type (1).

This procedure, called *changing the order of integration*, is important since it can sometimes be used to simplify the evaluation of an integral and in some cases the evaluation is impossible otherwise. Drawing a sketch of the field of integration is a vital step in the procedure.

Example 33 Change the order of integration in the integral I, and hence evaluate I, where

$$I = \int_0^1 \mathrm{d}x \int_0^x (1 + 2y - y^2)^{1/2}\,\mathrm{d}y.$$

From the limits of integration we see that the field of integration is the region between the x-axis ($y = 0$) and the line $y = x$ from $x = 0$ to $x = 1$.

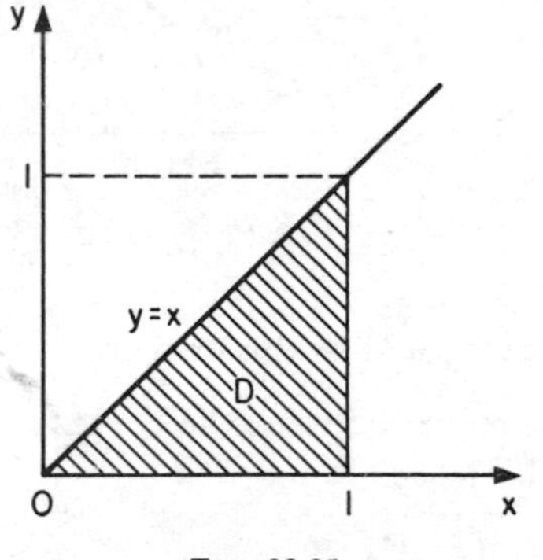

FIG. 20.25

This may alternatively be described as the region between the lines $x = y$ and $x = 1$ from $y = 0$ to $y = 1$. Hence

$$I = \int_0^1 (1 + 2y - y^2)^{1/2}\,\mathrm{d}y \int_y^1 \mathrm{d}x$$

$$= \int_0^1 (1 - y)(1 + 2y - y^2)^{1/2}\,\mathrm{d}y = \left[\tfrac{1}{3}(1 + 2y - y^2)^{3/2}\right]_0^1 = \tfrac{1}{3}(2^{3/2} - 1).$$

Example 34 Evaluate

$$\int_0^1 dy \int_{\sqrt{y}}^1 e^{x^3}\, dx.$$

Since $\int e^{x^3}\, dx$ is unknown we investigate the effect of changing the order of integration. The field of integration is shown in Fig. 20.26 and

$$\int_0^1 dy \int_{\sqrt{y}}^1 e^{x^3}\, dx = \int_0^1 e^{x^3}\, dx \int_0^{x^2} dy$$

$$= \int_0^1 x^2 e^{x^3}\, dx = \left[\tfrac{1}{3}e^{x^3}\right]_0^1 = \tfrac{1}{3}(e-1).$$

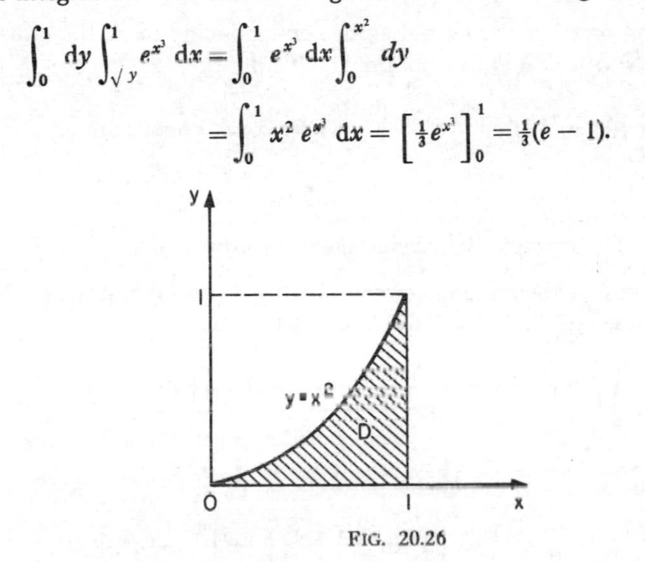

FIG. 20.26

20.7.6 Evaluation of Double Integrals (3) When the field of integration D is the region between two curves $r = \phi_1(\theta)$, $r = \phi_2(\theta)$ and between the lines $\theta = \alpha$, $\theta = \beta$, it is usually simpler to evaluate a double integral by the use of polar co-ordinates.

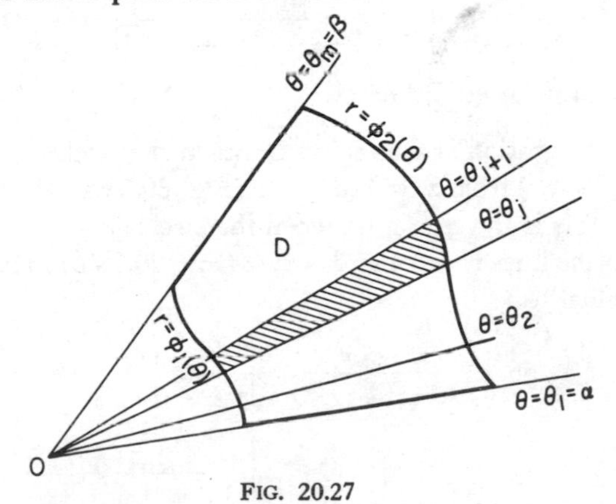

FIG. 20.27

Imagine the sum (S″) to be evaluated by first obtaining a partial sum over all the interstices lying in the radial strip between the lines $\theta = \theta_j$ and $\theta = \theta_{j+1}$. Letting each $\Delta r_i \to 0$ this partial sum becomes

$$\Delta\theta_j \int_{\phi_1(\theta_j)}^{\phi_2(\theta_j)} r f(r \cos\theta_j, r \sin\theta_j)\, dr = F(\theta_j)\, \Delta\theta_j \qquad \text{(say).}$$

Finally, adding together the partial sums corresponding to all the radial strips from $\theta = \alpha$ to $\theta = \beta$, we obtain $\int_\alpha^\beta F(\theta)\, d\theta$, that is,

$$\iint_D f(x, y)\, dx\, dy = \int_\alpha^\beta d\theta \int_{\phi_1(\theta)}^{\phi_2(\theta)} r f(r \cos\theta, r \sin\theta)\, dr.$$

Example 35 Do Example 30 by using polar co-ordinates.

In this problem D is the region between the origin ($r = 0$) and the circle $r = a$, and between the lines $\theta = 0$ and $\theta = \pi/2$. Hence

$$\iint_D xy\, dx\, dy = \int_0^{\pi/2} d\theta \int_0^a r^3 \sin\theta \cos\theta\, dr$$

$$= \int_0^{\pi/2} \sin\theta \cos\theta\, d\theta \int_0^a r^3\, dr$$

$$= \left[\tfrac{1}{2} \sin^2\theta\right]_0^{\pi/2} \times \left[\tfrac{1}{4} r^4\right]_0^a = \tfrac{1}{8} a^4.$$

Example 36 Evaluate

$$\int_0^a dx \int_{\sqrt{(ax-x^2)}}^{\sqrt{(a^2-x^2)}} \frac{dy}{\sqrt{(a^2 - x^2 - y^2)}}$$

by changing to polar co-ordinates.

The field of integration is the region between the circles $y = \sqrt{(ax - x^2)}$ and $y = \sqrt{(a^2 - x^2)}$ from $x = 0$ to $x = a$ (Fig. 20.28(a)). In terms of polar co-ordinates this is the region between the circles $r = a \cos\theta$ and $r = a$, and between the lines $\theta = 0$ and $\theta = \pi/2$ (Fig. 20.28(b)). Hence the given integral is equal to

$$\int_0^{\pi/2} d\theta \int_{a\cos\theta}^a \frac{r\, dr}{\sqrt{(a^2 - r^2)}} = \int_0^{\pi/2} d\theta \left[-\sqrt{(a^2 - r^2)}\right]_{a\cos\theta}^a$$

$$= \int_0^{\pi/2} a \sin\theta\, d\theta = a.$$

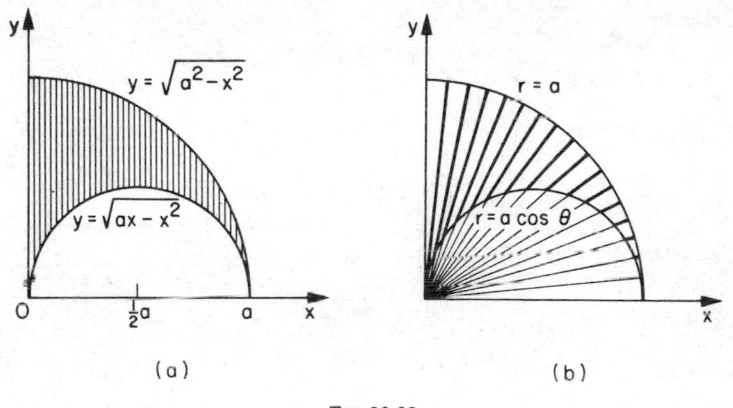

FIG. 20.28

Exercises 20(f)

1. Evaluate $\iint_D x^{1/2}\,dy\,dx$, where D is the region bounded by the parabola $x = y^2$ and the line $x = 2$.
2. Evaluate $\iint_D x^{5/2}\,dy\,dx$, where D is the region bounded by the parabola $y^2 = 4x$ and the line $x = 4$.
3. Evaluate $\iint_D x\,dy\,dx$, where D is the set of points

$$\{(x, y) : x + y \geqslant a,\ x^2 + y^2 \leqslant a^2\}.$$

4. Evaluate $\iint_D x^2y\,dy\,dx$, where D is the region in the first quadrant bounded by the lines $x = 1$, $y = 1$ and the circle $x^2 + y^2 = 1$.

In Nos. 5–8 evaluate by changing to polar co-ordinates:

5. Evaluate $\iint_D x^2y\,dy\,dx$, where D is the region in the first quadrant between the circles $x^2 + y^2 = 1$ and $x^2 + y^2 = 4$.
6. Evaluate

$$\iint_D \frac{x}{\sqrt{(x^2 + y^2)}}\,dy\,dx,$$

where D is the region bounded by the lines $y = x$, $y = 0$ and $x = 1$.

7. Evaluate $\iint_D x\,dy\,dx$, where D is the smaller segment of the circle $x^2 + y^2 = 2y$ cut off by the line $y = x$.
8. Evaluate

$$\iint_D \frac{x^3y^3}{(x^2 + y^2)^{3/2}}\,dy\,dx,$$

where D is the region bounded by the lines $y = 0$, $y = x$ and $x = 1$.

9. Evaluate $\iint_D y^2\,dx\,dy$, where D is the region lying inside the ellipse $x^2 + 4y^2 = 4$ and in the first quadrant.

In Nos. 10–13 evaluate by changing the order of integration:

10. Evaluate

$$\int_0^1 dx \int_{x^3}^1 \frac{x^2 y}{4 + y^3}\, dy.$$

11. Evaluate

$$\int_0^1 dy \int_{-\sqrt{(1-y)}}^{\sqrt{(1-y)}} (1 - x^2)^{-3/2}\, dx.$$

12. Evaluate

$$\int_0^1 dy \int_{y^2}^1 y e^{x^2}\, dx.$$

13. Evaluate

$$\int_0^1 y^3\, dy \int_{y^2}^1 \frac{x\sqrt{(1 - x^2)}}{\sqrt{(1 - x^2 + y^4)}}\, dx.$$

14. Show by considering the fields of integration that

$$\int_0^1 dx \int_x^{4\sqrt{x}} \frac{1}{1 + y^2}\, dy + \int_1^4 dx \int_x^4 \frac{1}{1 + y^2}\, dy = \int_0^4 dy \int_{y^2/16}^{y} \frac{1}{1 + y^2}\, dx,$$

and evaluate the integral on the right.

15. Sketch the field of integration S where

$$\iint_S xye^{-(x^2+y^2)^2}\, dx\, dy = \int_0^1 dx \int_{\sqrt{(1-x^2)}}^{\infty} xye^{-(x^2+y^2)^2}\, dy + \int_1^{\infty} dx \int_0^{\infty} xye^{-(x^2+y^2)^2}\, dy.$$

Evaluate this integral by transforming to polar co-ordinates.

16. Sketch the fields of integration of the two integrals

$$I_1 = \int_0^1 dy \int_{\sqrt{(1-y^2)}}^{\infty} x^2 e^{-(x^2+y^2)}\, dx$$

and

$$I_2 = \int_1^{\infty} dy \int_0^{\infty} x^2 e^{-(x^2+y^2)}\, dx.$$

By changing to polar co-ordinates write $\mathcal{J} = I_1 + I_2$ as one integral, and evaluate $\mathcal{J}$.

20.7.7 Change of Variable in Double Integrals The change from cartesian to polar co-ordinates was discussed in Section 20.7.2, and although this is the main change of variables with which readers are likely to be concerned at this stage, it is worth pointing out that the result of Section 20.7.2 is a particular case of the following more general result:

If D' is the image of D under the mapping $(x, y) \mapsto (u, v)$, $x = g(u, v)$, $y = h(u, v)$ where g and h are given functions, then

$$\iint_D f(x, y)\, dx\, dy = \iint_{D'} f(g(u, v), h(u, v)) \left| \frac{\partial(x, y)}{\partial(u, v)} \right| du\, dv \tag{20.7}$$

where

$$\frac{\partial(x, y)}{\partial(u, v)} = \begin{vmatrix} \dfrac{\partial x}{\partial u} & \dfrac{\partial y}{\partial u} \\ \\ \dfrac{\partial x}{\partial v} & \dfrac{\partial y}{\partial v} \end{vmatrix}, \tag{20.8}$$

and it is assumed that $\partial(x, y)/\partial(u, v) \neq 0$ on D'.

To give some justification for this result let us imagine the region D of the xy-plane divided by the network of curves $u = u_1, u = u_2, \ldots, u = u_n$ and $v = v_1, v = v_2, \ldots, v = v_m$. On curves $u =$ constant we have

$$\Delta x \simeq \frac{\partial x}{\partial v} \Delta v, \quad \Delta y \simeq \frac{\partial y}{\partial v} \Delta v$$

and on curves $v =$ constant we have

$$\Delta x \simeq \frac{\partial x}{\partial u} \Delta u, \quad \Delta y \simeq \frac{\partial y}{\partial u} \Delta u,$$

so that the interstice bounded by the curves $u = u_i$, $u = u_{i+1}$, $v = v_j$, $v = v_{j+1}$ (see Fig. 20.29) is approximately a parallelogram whose adjacent sides are the vectors $\mathbf{a}$, $\mathbf{b}$ given by

$$\mathbf{a} = \frac{\partial x}{\partial u} \Delta u_i\, \mathbf{i} + \frac{\partial y}{\partial u} \Delta u_i\, \mathbf{j},$$

$$\mathbf{b} = \frac{\partial x}{\partial v} \Delta v_j\, \mathbf{i} + \frac{\partial y}{\partial v} \Delta v_j\, \mathbf{j},$$

the partial derivatives being evaluated at the point P_{ij}. Hence the area ΔA_{ij} of this interstice is given approximately by

$$\Delta A_{ij} \simeq |\mathbf{a} \wedge \mathbf{b}| \qquad \text{(see Section 7.7.2)},$$

where

$$|\mathbf{a} \wedge \mathbf{b}| = \begin{vmatrix} \dfrac{\partial x}{\partial u} & \dfrac{\partial y}{\partial u} \\ \dfrac{\partial x}{\partial v} & \dfrac{\partial y}{\partial v} \end{vmatrix} \Delta u_i\, \Delta v_j.$$

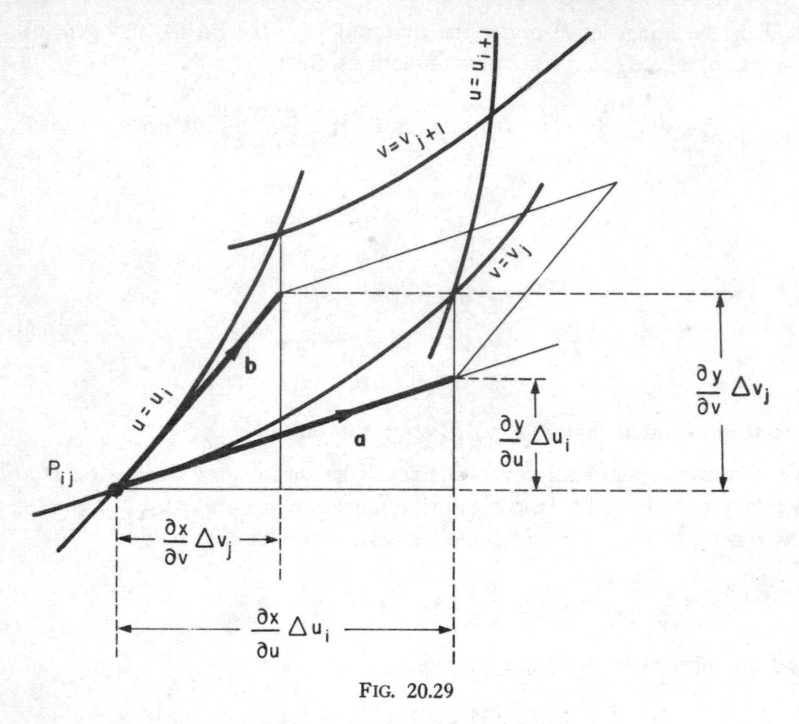

FIG. 20.29

By forming the sum (S) for this network and comparing with (S′) we arrive at the result (20.7).

The determinant denoted by $\partial(x, y)/\partial(u, v)$ (see (20.8)) is called the *Jacobian* of x, y with respect to u, v.

In the case of changing from cartesian to polar co-ordinates we have $x = r\cos\theta, y = r\sin\theta$, giving

$$\frac{\partial(x, y)}{\partial(r, \theta)} = \begin{vmatrix} \dfrac{\partial x}{\partial r} & \dfrac{\partial y}{\partial r} \\ \dfrac{\partial x}{\partial \theta} & \dfrac{\partial y}{\partial \theta} \end{vmatrix} = \begin{vmatrix} \cos\theta & \sin\theta \\ -r\sin\theta & r\cos\theta \end{vmatrix} = r.$$

Hence from (20.7)

$$\iint_D f(x, y)\,dx\,dy = \iint_{D'} f(r\cos\theta, r\sin\theta)\,r\,dr\,d\theta,$$

which agrees with the result in Section 20.7.2.

Example 37 Evaluate $\iint_D (x + y)\,dx\,dy$, where D is the parallelogram with vertices at the points (0, 0), (3, 1), (4, 4) and (1, 3).

The sides of D have equations $3y - x = 0$, $3x - y = 8$, $3y - x = 8$ and $3x - y = 0$. Putting

$$u = 3x - y, \quad v = 3y - x,$$

that is

$$x = \tfrac{1}{8}(3u + v), \quad y = \tfrac{1}{8}(u + 3v),$$

we have a mapping such that the image of D is the square D' with vertices (0, 0), (8, 0), (8, 8) and (0, 8) in the uv-plane.

Now

$$\frac{\partial(x, y)}{\partial(u, v)} = \begin{vmatrix} \frac{3}{8} & \frac{1}{8} \\ \frac{1}{8} & \frac{3}{8} \end{vmatrix} = \frac{1}{8},$$

and so, by (20.7),

$$\iint_D (x + y)\, dx\, dy = \iint_{D'} \tfrac{1}{2}(u + v)\, \tfrac{1}{8}\, du\, dv$$

$$= \tfrac{1}{16} \int_0^8 du \int_0^8 (u + v)\, dv = 32.$$

CHAPTER 21

DIFFERENTIAL EQUATIONS OF THE FIRST ORDER

21.1 Definitions

Let x be an independent variable and let y be a function of x. Then an equation involving x, y and the derivatives dy/dx, d^2y/dx^2, ... is called an *ordinary differential equation.*

For example, the equations

$$x\frac{dy}{dx} + y^2 = \sin x, \tag{1}$$

$$x^3\left(\frac{d^2y}{dx^2}\right)^2 + y^2\left(\frac{dy}{dx}\right)^3 = 1, \tag{2}$$

$$x\frac{d^3y}{dx^3} + \left(\frac{dy}{dx}\right)^2 = x^2 + y^2, \tag{3}$$

are ordinary differential equations.

The term *ordinary* is used to distinguish these equations from *partial differential equations* which involve partial derivatives. Since partial differential equations are not discussed here we may omit the term ordinary without risk of confusion. Frequently we shall refer to a differential equation as simply a D.E.

The *order* of a D.E. is the order of the highest derivative occurring in it. Equations (1), (2), (3) are of first, second, and third order respectively.

The commonest type of D.E. is of the form

$$F\left(\frac{dy}{dx}, \frac{d^2y}{dx^2}, \ldots\right) = 0,$$

where F is a polynomial with functions of x and y as coefficients (e.g. eqs. (1)–(3) above); the highest power of the highest order derivative is then called the *degree* of the D.E.

Thus equations (1) and (3) are both of first degree, and equation (2) is of the second degree.

When we wish to emphasise that an equation is free from derivatives we shall do so by calling it a *non-differential equation.*

A non-differential equation is a solution of a D.E. if the D.E. can be derived from it by the process of differentiation.

For example, the equation $x^5 + y^3 = 1$ is a solution of the first order D.E. $5x^4 + 3y^2\, dy/dx = 0$ obtained from it by a single differentiation, and is also a solution of the second order D.E.

$$20x^3 + 3y^2 \frac{d^2y}{dx^2} + 6y\left(\frac{dy}{dx}\right)^2 = 0$$

obtained from it by differentiating twice.

It should be emphasised that if an equation is a solution of a D.E. then it satisfies the D.E. *for all x in some interval I of positive length.* Thus

(a) $y = x^2$ satisfies the D.E. $dy/dx = 2x$ for all x.
(b) $y = x^{3/2}$ satisfies the D.E. $dy/dx = \frac{3}{2}x^{1/2}$ for all x in the interval $\{x: x \geqslant 0\}$ and outside this interval neither the equation nor the D.E. is meaningful.
(c) $y = \ln x$ satisfies the D.E. $dy/dx = 1/x$ for all x in the interval $\{x: x > 0\}$, while for $x < 0$ the D.E. is meaningful but $\ln x$ is not.

Consider now the equation $y = 3x$, which gives $dy/dx = 3$; although we can say that $dy/dx = 3x^2$ when $x = \pm 1$, nevertheless $dy/dx \neq 3x^2$ for all other values of x, i.e. $y = 3x$ is *not* a solution of the D.E. $dy/dx = 3x^2$ since it does not satisfy this D.E. on an interval *of positive length.* Solving, or *integrating*, a D.E. is the process of deriving its solution.

For example, an equation of the form $dy/dx = f(x)$ may be solved by the process of ordinary integration, and an equation of the form $d^2y/dx^2 = f(x)$ may be solved by performing ordinary integration twice. As we have seen, ordinary integration is not always possible in terms of elementary functions, and in general we must expect that many D.E.'s cannot be solved in terms of elementary functions.

21.2 Formation of Differential Equations

The importance of D.E.'s lies in the fact that they arise naturally in a wide variety of practical situations. Some evidence for this has already been provided by the applications of ordinary integration, studied in Chapter 19. We look now at some situations where D.E.'s of a less elementary type arise.

21.2.1 Motion in a Straight Line We have already seen (Section 15.1.1) that the acceleration a, the velocity v, and the distance s measured from a fixed point on the line, are related as follows:

$$v = \frac{ds}{dt}, \tag{21.1}$$

$$a = \frac{dv}{dt} = \frac{d^2s}{dt^2}. \tag{21.2}$$

Then if the acceleration is known in terms of t, say $a = f(t)$, we have a first-order D.E. for v, viz. $dv/dt = f(t)$, and a second order D.E. for s, viz. $d^2s/dt^2 = f(t)$, both of which may be solved by ordinary integration.

In some situations we may wish to regard a and v as functions of s. Note that

$$a = \frac{dv}{dt} = \frac{dv}{ds}\frac{ds}{dt} \quad \text{by the chain rule}$$

$$= v\frac{dv}{ds} \quad \text{by (21.1),}$$

giving

$$a = v\frac{dv}{ds}. \tag{21.3}$$

(A) *The acceleration is known in terms of s.* Let $a = f(s)$; then from (21.3) we have a first-order D.E. for v, viz. $v\,dv/ds = f(s)$, and from (21.2) we have a second order D.E. for s, viz. $d^2s/dt^2 = f(s)$.

Example 1 For a mass under the action of a spring as shown in Fig. 21.1, provided the displacement s remains small, it is a fairly accurate assumption (Hooke's law) that the deceleration due to the spring is proportional to s, say $a = -ks$.

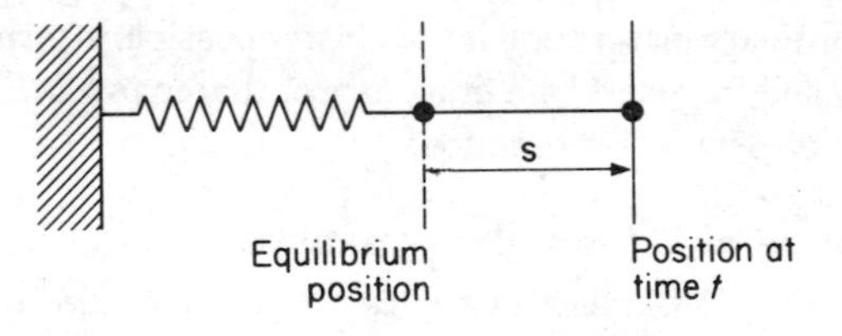

FIG. 21.1

(Here k is a positive constant associated with the stiffness of the particular spring being used, and the minus sign is necessary because the action of the spring is in opposition to motion away from the equilibrium position.)

Thus from (21.3)

$$v\frac{dv}{ds} = -ks, \tag{21.4}$$

and from (21.2)

$$\frac{d^2s}{dt^2} = -ks. \tag{21.5}$$

Example 2 For a rocket shot vertically upwards from a point on the earth's surface, we assume that the deceleration due to the earth's gravity is inversely proportional to the square of the distance from the earth's centre. (Newton's law of gravitation.) Thus, neglecting other effects (e.g. air resistance), we have $a = -k/s^2$ where k is some positive constant, and

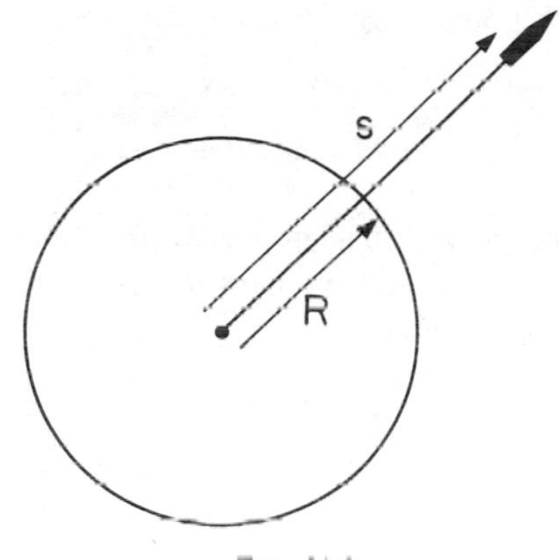

FIG. 21.2

s is measured from the earth's centre. We can relate k to the well-known deceleration g due to gravity *at the earth's surface*; thus if R denotes the radius of the earth, then $a = -g$ when $s = R$, so that

$$-g = -\frac{k}{R^2},$$

i.e. $k = gR^2$, and so $a = -gR^2/s^2$. From (21.3) we now get

$$v\frac{dv}{ds} = -\frac{gR^2}{s^2}. \tag{21.6}$$

(B) *The acceleration is known in terms of v.* Let $a = f(v)$; then from (21.2) we get $dv/dt = f(v)$, which is a first-order D.E. that we may try to solve in order to get v in terms of t, or alternatively from (21.3) we get $v\,dv/ds = f(v)$, which is a first-order D.E. that we may try to solve for v in terms of s.

Example 3 For a parachute falling towards the earth, gravity is not the only major force and air resistance must also be taken into account. It is easily verified that the faster a parachute is dragged through the air, the greater the resistance offered by the air, and in many cases a reasonable assumption is that the deceleration due to air resistance is proportional to the velocity. Thus, assuming the acceleration due to gravity has the constant value g, we have

$$a = g - kv$$

where we are now measuring s downwards, i.e. in the direction of motion. From (21.2) we now get

$$\frac{dv}{dt} = g - kv \qquad (21.7)$$

and from (20.3) we get

$$v\frac{dv}{ds} = g - kv. \qquad (21.8)$$

For a parachute of different design a more realistic assumption might well be that the deceleration due to air resistance is proportional to v^2, in which case the equations would be

$$\frac{dv}{dt} = g - kv^2, \qquad (21.9)$$

$$v\frac{dv}{ds} = g - kv^2. \qquad (21.10)$$

21.2.2 Other Physical Examples The D.E.'s occurring in Section 21.2.1, part (B), are of a type that arise in a number of physical situations in which a quantity, Q say, is changing subject to a law relating its rate of change at any instant to its value at that instant, i.e. $dQ/dt = f(Q)$.

Radioactive decay. Suppose a substance disintegrates at a rate that is always proportional to the amount present. Then if m denotes the mass present at time t, we have

$$\frac{dm}{dt} = -km, \qquad (21.11)$$

where k is a positive constant, and the negative sign indicates that m is decreasing.

Newton's law of cooling. Suppose that a substance, surrounded by air at a lower temperature, is cooling down at a rate that is always proportional to the temperature difference. Let T denote the temperature of the substance at time t, and let T_a denote the temperature of the surrounding air —we shall assume that T_a is constant. Then

$$\frac{dT}{dt} = -k(T - T_a). \qquad (21.12)$$

21.2.3 Geometrical Examples For the curve $y = f(x)$ the slope of the tangent at any point equals the value of dy/dx at that point.

Sometimes we wish to determine all curves possessing some given property; if the given property concerns the slope of the tangent, this amounts to solving a first-order D.E.

Example 4 Suppose the tangent at any point (x, y) on a curve is perpendicular to the line joining (x, y) to the origin.

The slope of the line is y/x, the slope of the tangent is dy/dx, and so the two are perpendicular if

$$\frac{dy}{dx} = -\frac{x}{y}. \qquad (21.13)$$

This is the D.E. for finding curves with this particular property. The D.E. for the 'orthogonal trajectories' of a given family of curves is discussed in Section 21.6.

21.2.4 Non-physical Examples In recent years mathematical methods have been adopted in various fields previously considered non-mathematical. Consider, for example, a simple version of the national economy in which the following assumptions are made:

(a) The annual national savings S are proportional to the annual national output Y, say $S = \alpha Y$.
(b) the rate of growth of the national output is proportional to the national savings, say $dY/dt = \beta S$. We then have

$$\frac{dY}{dt} = \alpha\beta Y, \qquad (21.14)$$

which is a first-order D.E. for Y.

21.3 The Complete Solution of a Differential Equation

Experience of ordinary integration teaches that a D.E. does not have a unique solution. We shall use the term *complete solution* to denote the set of *all* solutions of a D.E. Thus for an elementary D.E. of the type $dy/dx = f(x)$, the complete solution may be represented by a formula $y = F(x) + c$ in which the constant of integration c is a parameter generating the set of possible solutions. Likewise, for a D.E. of the type $d^2y/dx^2 = f(x)$, the complete solution may be represented by a formula containing two constants of integration, and so on. For various reasons one might expect something similar to happen with D.E.'s in general, i.e.

Conjecture: One expects that if an nth-order D.E. can be solved in terms of elementary functions, the complete solution can be represented by a single equation containing n arbitrary constants.

One reason for this conjecture is illustrated by Examples 5 and 6 which show how, from an equation containing n arbitrary constants, we can derive an nth-order D.E. free from arbitrary constants.

Example 5 Eliminate the arbitrary constant c from the equation $x^2 + y^2 = c^2$.

Now

$$x^2 + y^2 = c^2 \Leftrightarrow \frac{d}{dx}(x^2 + y^2) = 0$$

$$\Leftrightarrow 2x + 2y\frac{dy}{dx} = 0$$

$$\Leftrightarrow \frac{dy}{dx} = -\frac{x}{y}$$

which is a first-order D.E.

The equation $x^2 + y^2 = c^2$ represents a family of concentric circles, centres at the origin; c is constant for each circle but varies from circle to circle. The D.E. $dy/dx = -x/y$ represents a property common to all members of the family, viz. the tangent at any point is perpendicular to the radius drawn to that point (see Section 21.2.3, Example 4). Note that in the derivation of the D.E. the steps are reversible, i.e. starting from the D.E. $dy/dx = -x/y$ we can derive the equation $x^2 + y^2 = c^2$ as the complete solution; thus the only curves with the property stated in Example 4 are the circles $x^2 + y^2 = c^2$.

Example 6 Eliminate the arbitrary constants a, b from the equation $y = ax + b/x$.

Now

$$y = ax + \frac{b}{x} \Leftrightarrow yx - ax^2 = b$$

$$\Leftrightarrow \frac{d}{dx}(yx - ax^2) = 0$$

$$\Leftrightarrow y + x\frac{dy}{dx} - 2ax = 0$$

$$\Leftrightarrow \frac{y}{x} + \frac{dy}{dx} = 2a$$

$$\Leftrightarrow \frac{d}{dx}\left(\frac{y}{x} + \frac{dy}{dx}\right) = 0$$

$$\Leftrightarrow \frac{x\,(dy/dx) - y}{x^2} + \frac{d^2y}{dx^2} = 0$$

$$\Leftrightarrow x^2\frac{d^2y}{dx^2} + x\frac{dy}{dx} - y = 0,$$

which is a second-order D.E. As indicated, the steps are reversible so that the given equation is the complete solution of the D.E.

It will be seen that the conjecture discussed above is true for the vast majority of D.E.'s that we deal with. There are important exceptions (see Section 21.5) where a solution containing n arbitrary constants may be found *not* to include all solutions.

In any case where a solution of an nth-order D.E. can be found in the form of an equation with n arbitrary constants, the set of solutions represented by this equation is called the *general solution* (G.S.) of the D.E. Normally the G.S. is the *complete* solution, but in exceptional cases this may not be so and other solutions called *singular solutions* (i.e. solutions not included in the G.S.) may exist.

Notes. (1) In referring to an equation with n arbitrary constants, we always assume that n is the *smallest* number of arbitrary constants with which the equation can be written.

For example the equation $y = ax + b + c$ with 3 arbitrary constants and the equation $y = ax + b$ with 2 arbitrary constants, both represent the *same* set of equations. Likewise the equation $y = ae^{x+b}$, with two arbitrary constants, can be written in the form $y = ae^b\, e^x$ showing that it represents the same set of equations as $y = ce^x$ which contains only *one* arbitrary constant.

(2) Expressing the G.S. as an equation with n arbitrary constants *is possible in different ways.*

For example $y = ax^2 + bx$, $y = c(2x^2 + 3x) + d(5x^2 - 7x)$ both represent the same set of equations; for corresponding to any given values of a and b we get the same equation by choosing c and d so that $2c + 5d = a$, $3c - 7d = b$, and in the same way we can choose a, b to correspond to any given values of c, d. Likewise $y = a \cos x + b \sin x$, $y = c \sin (x + d)$ both represent the same set of equations.

21.3.1 Particular Integrals In contrast to the term G.S. we use the term *particular solution*, or *particular integral* (P.I.), to denote an individual solution, i.e. a solution not containing arbitrary constants. Thus a P.I. is either an individual member of the set of solutions comprising the G.S., in which case it can be obtained from the G.S. by giving appropriate values to all the arbitrary constants, or else it is a singular solution.

Exercises 21(a)

1. Obtain a differential equation of the second order by eliminating the constants A and B from the equation $\phi = A/r + B$.

2. Given that a and b are arbitrary constants, find the second-order differential equation whose solution is $y = a \sec x + b \tan x$.

In Nos. 3–8 express the given statement in the form of a D.E.:

3. An object is heated by a flame at a rate proportional to the difference in temperature between the flame and the object.
4. Flame spreads over a certain substance in such a way that the area of substance alight is increasing at any instant at a rate proportional to the area already alight at that instant.
5. A body moves in a straight line so as to decelerate at a rate proportional to the sum of the distance from the origin and the cube of the velocity.
6. A vessel contains a volume V_0 of liquid. The liquid is poured out so that the volume remaining in the vessel is always decreasing at a rate proportional to the volume already poured out.
7. On a certain curve the slope of the tangent at any point equals the cube of the distance of the point from the origin.
8. On a certain curve the slope of the normal at any point equals the square of the distance of the point from the x-axis.

21.4 Differential Equations of the First Order and First Degree: Methods for Solving

For this type of D.E. the method of attack will be to reduce the problem of solving the D.E. to a matter of ordinary integration. In this way the complete solution is found merely by including the constant of integration, and P.I.'s, if required, may be singled out thereafter. This seems the ideal approach, but unfortunately has to be abandoned with higher-order D.E.'s as we shall see in Chapter 23.

Here we are concerned with equations of the form

$$\frac{dy}{dx} = f(x, y). \tag{21.15}$$

Alternatively, this equation may be described as having either of the forms

$$\begin{aligned} M(x, y) + N(x, y)\frac{dy}{dx} &= 0, \\ M(x, y)\,dx + N(x, y)\,dy &= 0. \end{aligned} \tag{21.16}$$

Thus the equation $dy/dx = 2xy/(x^2 + y^2)$ may be expressed in the alternative forms

$$2xy - (x^2 + y^2)\frac{dy}{dx} = 0, \quad 2xy\,dx - (x^2 + y^2)\,dy = 0.$$

At this point it is useful to consider the D.E. that results from differentiating once a non-differential equation. From the equation $F(x, y) = 0$ we obtain, on differentiating with respect to x by the use of (20.4) (Section 20.4.1),

$$\frac{\partial F}{\partial x} + \frac{\partial F}{\partial y}\frac{dy}{dx} = 0. \tag{21.17}$$

(e.g. in Example 5, we have $F(x, y) = x^2 + y^2 - c^2$, etc.) Conversely, the complete solution of the D.E. (21.17) is $F(x, y) =$ constant. Comparing (21.16) with (21.17) we see that (21.16) can be solved completely if there is some function F such that

$$\frac{\partial F}{\partial x} = M(x, y), \quad \frac{\partial F}{\partial y} = N(x, y)$$

and in that case the complete solution is $F(x, y) =$ constant; we then say that (21.16) is *exact*. If (21.16) is exact we have

$$\frac{\partial M}{\partial y} = \frac{\partial^2 F}{\partial x\, \partial y} = \frac{\partial N}{\partial x},$$

so that

$$\frac{\partial M}{\partial y} = \frac{\partial N}{\partial x}. \tag{21.18}$$

The converse is also true (but harder to prove) and so (21.18) is the condition for (21.16) to be exact.

Example 7 Solve the equation $(3x^2 + y + 1)\,dx + (3y^2 + x + 1)\,dy = 0$.

This equation has the form (21.16) with $M = 3x^2 + y + 1, N = 3y^2 + x + 1$, and since $\partial M/\partial y = \partial N/\partial x = 1$ the equation is exact. This means that the equation is equivalent to $dF(x,y)/dx = 0$, where F is a function to be determined from

$$\frac{\partial F}{\partial x} = 3x^2 + y + 1, \tag{1}$$

$$\frac{\partial F}{\partial y} = 3y^2 + x + 1. \tag{2}$$

Integration of (1) with respect to x gives

$$F = x^3 + xy + x + \phi(y), \tag{3}$$

where $\phi(y)$ is an arbitrary function of y.

Differentiating (3) partially with respect to y, we obtain

$$\frac{\partial F}{\partial y} = x + \phi'(y),$$

which, by comparison with (2), requires that

$$\phi'(y) = 3y^2 + 1, \quad \text{that is} \quad \phi(y) = y^3 + y + c. \tag{4}$$

It is sufficient to put $c = 0$, thereby getting the simplest function F satisfying our requirements, i.e.

$$F(x, y) = x^3 + xy + x + y^3 + y$$

from (3) and (4). Hence the solution is

$$x^3 + y^3 + xy + x + y = C.$$

Exercises 21(b)

For brevity y' is written for $(\mathrm{d}y/\mathrm{d}x)$. Show that the following equations are exact and integrate them:

1. $xy' + y = e^x$.
2. $(1/x)y' - y/x^2 = 2x$.
3. $(ax + hy + g)\,\mathrm{d}x + (hx + by + f)\,\mathrm{d}y = 0$.
4. $(1 - \cos 2x)\,\mathrm{d}y + 2y \sin 2x\,\mathrm{d}x = 0$.
5. $(3x^2 + 2y + 1)\,\mathrm{d}x + (2x + 6y^2 + 2)\,\mathrm{d}y = 0$.

21.4.1 Integrating Factors If eq. (21.16) becomes exact when multiplied throughout by a suitable factor, this factor is known as an *integrating factor* (I.F.).

For example the equation $3y + 2x\,(\mathrm{d}y/\mathrm{d}x) = 0$ is not exact since, in the usual notation, $\partial M/\partial y = 3$, $\partial N/\partial x = 2 \neq \partial M/\partial y$. Multiplying by x^2y we get the equation $3x^2y^2 + 2x^3y\,(\mathrm{d}y/\mathrm{d}x) = 0$, which is exact, and may be written as $\mathrm{d}(x^3y^2)/\mathrm{d}x = 0$. Thus x^2y is an I.F. for this equation.

Note that $x^{1/2}$ is also an I.F. for this equation; multiplying by $x^{1/2}$ gives

$$3x^{1/2}y + 2x^{3/2}\frac{\mathrm{d}y}{\mathrm{d}x} = 0, \quad \text{that is} \quad \frac{\mathrm{d}}{\mathrm{d}x}(2x^{3/2}y) = 0.$$

The solution obtained using this I.F., viz. $2x^{3/2}y = \text{constant}$, is easily seen to be the same as the solution obtained using the first I.F., viz. $x^3y^2 = \text{constant}$.

Example 8 Show that the equation

$$(3xy - 2ay^2)\,\mathrm{d}x + (x^2 - 2axy)\,\mathrm{d}y = 0$$

has an I.F. which is a function of x alone. Solve the equation.

Suppose that the I.F. is $f(x)$; then the equation

$$(3xy - 2ay^2)f(x)\,dx + (x^2 - 2axy)f(x)\,dy = 0 \tag{1}$$

is exact, and so

$$\frac{\partial}{\partial y}[(3xy - 2ay^2)f(x)] = \frac{\partial}{\partial x}[(x^2 - 2axy)f(x)],$$

giving

$$(3x - 4ay)f(x) = (x^2 - 2axy)f'(x) + 2(x - ay)f(x),$$

that is,

$$\frac{f'(x)}{f(x)} = \frac{1}{x},$$

or

$$\frac{d}{dx}\ln f(x) = \frac{1}{x},$$

giving

$$\ln f(x) = \ln x + c.$$

Putting $c = 0$ we get the simplest I.F. of this form, viz. $f(x) = x$. Substituting $f(x) = x$ in (1) leads to the exact equation

$$(3x^2y - 2axy^2)\,dx + (x^3 - 2ax^2y)\,dy = 0,$$

that is

$$\frac{d}{dx}F(x, y) = 0,$$

where $F(x, y) = x^3y - ax^2y^2$ as may be found by intelligent guesswork, or else by the formal procedure used in Example 7.

Hence the solution of the given equation is $x^3y - ax^2y^2 = C$.

Example 9 Show that the equation $dx + [1 + (x + y)\tan y]\,dy = 0$ has an I.F. of the form $(x + y)^n$, where n is a constant. Solve the equation.

If $(x + y)^n$ is an I.F., then

$$\frac{\partial}{\partial y}(x + y)^n = \frac{\partial}{\partial x}[\{1 + (x + y)\tan y\}(x + y)^n],$$

that is

$$\begin{aligned} n(x + y)^{n-1} &= n(x + y)^{n-1}[1 + (x + y)\tan y] + (x + y)^n \tan y \\ &= (x + y)^{n-1}[n + (n + 1)(x + y)\tan y], \end{aligned}$$

that is

$$n = n + (n + 1)(x + y) \tan y,$$

which is satisfied by choosing $n = -1$.

Hence $(x + y)^{-1}$ is an I.F. and so the equation

$$\frac{dx}{x+y} + \left(\frac{1}{x+y} + \tan y\right) dy = 0$$

is exact. As with previous examples, we can show that this equation is

$$\frac{d}{dx}[\ln \{(x + y) \sec y\}] = 0,$$

and so the solution of the given equation is

$$\ln [(x + y) \sec y] = \text{constant},$$

that is

$$x + y = C \cos y.$$

21.4.2 The Linear First-order Equation Suppose that in eq. (21.16) N is a function of x only while $M(x, y) = a(x)y + b(x)$. Then (21.16) involves y and dy/dx to the first degree only and is called a first-order *linear* equation. Dividing throughout by N, we see that a first-order linear equation has the form

$$p(x)y + q(x) + \frac{dy}{dx} = 0. \tag{21.19}$$

Once the equation has been put into this form it is always possible to find an I.F. that is a function of x only. For, using condition (21.18), $f(x)$ is an I.F. if

$$\frac{\partial}{\partial y}[f(x)p(x)y + f(x)q(x)] = \frac{\partial}{\partial x}f(x),$$

that is if

$$f(x)p(x) = f'(x),$$

or

$$\frac{f'(x)}{f(x)} = p(x),$$

$$\ln f(x) = \int p(x)\, dx,$$

$$f(x) = e^{\int p(x)\, dx}.$$

Thus $e^{\int p(x)\,dx}$ is an I.F. and in practice we can omit the constant of integration to get the simplest I.F.

The following results are useful for expressing the I.F. in its simplest form:

$$e^{\ln x} = x, \quad e^{n \ln x} = (e^{\ln x})^n = x^n, \quad e^{-n \ln x} = \frac{1}{e^{n \ln x}} = \frac{1}{x^n}.$$

Once the I.F. has been found, instead of expressing the equation in the form $dF(x, y)/dx = 0$, it is simpler when $q(x) \neq 0$ to express it in the form $dF(x, y)/dx =$ function of x, as shown in the following examples.

Example 10 Solve the equation $y \cot x - 2 \cos x + dy/dx = 0$.

Here

$$p(x) = \cot x, \quad \int p(x)\,dx = \ln \sin x, \quad e^{\int p(x)\,dx} = \sin x.$$

Multiplying the given equation by $\sin x$, we get

$$y \cos x - \sin 2x + (\sin x)\frac{dy}{dx} = 0,$$

that is

$$\frac{d}{dx}(y \sin x) - \sin 2x = 0.$$

We may now integrate to get the solution,

$$y \sin x + \tfrac{1}{2} \cos 2x = C.$$

Example 11 Solve the equation

$$y + x(x+1)\frac{dy}{dx} = x(x+1)^2 e^{-x^2}.$$

To express the given equation in the form (21.19) we must divide by $x(x + 1)$, giving

$$\frac{1}{x(x+1)}y + \frac{dy}{dx} = (x+1)e^{-x^2}. \tag{1}$$

Hence

$$p(x) = \frac{1}{x(x+1)} = \frac{1}{x} - \frac{1}{x+1},$$

so that

$$\int p(x)\,dx = \ln\left(\frac{x}{x+1}\right), \quad e^{\int p(x)\,dx} = \frac{x}{x+1}.$$

Multiplying (1) by $x/(x+1)$, we obtain

$$\frac{1}{(x+1)^2}y+\frac{x}{x+1}\frac{dy}{dx}=xe^{-x^2},$$

that is

$$\frac{d}{dx}\left(\frac{x}{x+1}y\right)=xe^{-x^2},$$

or

$$\frac{x}{x+1}y=C-\tfrac{1}{2}e^{-x^2},$$

$$xy=(x+1)(C-\tfrac{1}{2}e^{-x^2}).$$

Note. Multiplying by the I.F. enables us to express the terms involving y and dy/dx in the form

$$d(e^{\int p(x)\,dx}y)/dx.$$

Exercises 21(c)

Solve the differential equations given in Nos. 1–15:

1. $y'\sin x-2y\cos x=e^x\sin^3 x$.
2. $y'\cos x-4y\sin x=6\cos^2 x\sin x$.
3. $x^2y'+xy=\ln x$.
4. $y'\tan x+2y=x\operatorname{cosec} x$.
5. $x(1+x)y'-y=3x^4$.
6. $(1-x^2)y'+xy=(1-x^2)^{3/2}e^{\cos x}\sin x$.
7. $y'+xy-x^3=0$, where $y=0$ when $x=0$.
8. $(x-2)(x-3)y'+2y=(x-1)(x-2)$.
9. $x(1-x^2)y'+(3x^2+1)y=(1+x)^3$.

For each of the following equations, find the solution that remains finite as x tends to zero:

10. $(x-1)y'-2y=(x-1)^4\cos^2 x$.
11. $x^2y'+x(3+2x)y=e^{-2x}+e^{-3x}$.
12. $x\,y'+(1+x)y=x\sin x$.
13. $(1+x^2)y'-xy=(1+x^2)x^2$.
14. $(1+3x)y'+(3-9x)y=3$.
15. $x(x+1)y'-y=3x^4$, with $y=1$ when $x=1$.
16. $xy'-y=x^3\cos x$, with $y=0$ when $x=\pi$.

21.4.3 Variables Separable Suppose that in eq. (21.16) M is a function of x only and N is a function of y only. We then say the variables are *separable.* In this case $\partial M/\partial y = \partial N/\partial x = 0$ and the equation is exact. It is particularly easy to find $F(x, y)$ such that $\partial F/\partial x = M$, $\partial F/\partial y = N$; we simply put $F(x, y) = \int M \, dx + \int N \, dy$.

Thus the equation can be written as

$$\frac{d}{dx}\left(\int M \, dx + \int N \, dy\right) = 0$$

and so the solution is

$$\int M \, dx + \int N \, dy = \text{constant.}$$

In practice this means that the solution may be obtained by the following procedure:

(a) Arrange the equation so that the terms in x are together with dx and the terms in y are together with dy.
(b) Integrate the terms in x with respect to x and integrate the terms in y with respect to y.

Example 12 Solve the equation

$$(1 - x^2)^{1/2} \frac{dy}{dx} + 1 + y^2 = 0.$$

The equation may be written in the form

$$\frac{dy}{1 + y^2} + \frac{dx}{(1 - x^2)^{1/2}} = 0$$

whence, on integration,

$$\tan^{-1} y + \sin^{-1} x = C.$$

Example 13 Solve the equation

$$xy(1 + x^2) \frac{dy}{dx} - (1 + y^2) = 0.$$

The equation can be written in the form

$$\frac{y \, dy}{1 + y^2} - \frac{dx}{x(1 + x^2)} = 0,$$

that is

$$\frac{y \, dy}{1 + y^2} - \left(\frac{1}{x} - \frac{x}{1 + x^2}\right) dx = 0.$$

Integrating, we get

$$\tfrac{1}{2}\ln(1+y^2) - \ln x + \tfrac{1}{2}\ln(1+x^2) = \text{constant} = \ln C \quad \text{(say)},$$

so that

$$\tfrac{1}{2}\ln(1+x^2)(1+y^2) = \ln Cx,$$

$$\sqrt{[(1+x^2)(1+y^2)]} = Cx.$$

Note. When logarithmic functions occur in the solution of a D.E. the constant of integration is frequently written in the form $\ln C$ in order to simplify the form of the G.S.

Example 14 Solve eq. (21.7), Example 3, given that the parachute opens at time $t = 0$ when the velocity of fall is v_0.

The equation can be written in the form

$$\frac{dv}{kv - g} + dt = 0.$$

Since k and g are constants, this means that the variables v, t are separable and we may now integrate to obtain

$$\frac{1}{k}\ln(kv - g) + t = C,$$

giving

$$kv - g = e^{k(C-t)},$$

that is

$$v = g/k + \frac{1}{k}e^{kC}\,e^{-kt}, \quad \text{or} \quad v = g/k + A\,e^{-kt} \qquad \text{(G.S.)},$$

where $A = (1/k)\,e^{kC}$ and we may regard A as an arbitrary constant since C is arbitrary.

Now $v = v_0$ when $t = 0$, giving $v_0 = g/k + A$, that is

$$A = v_0 - g/k.$$

Hence the required P.I. is $v = g/k + (v_0 - g/k)e^{-kt}$, found by giving A the appropriate value in the G.S.

As $t \to \infty$, $e^{-kt} \to 0$ and so $v \to g/k$ (called the terminal velocity). Assuming $v_0 > g/k$, we see that v decreases steadily towards this terminal velocity.

Example 15 Solve eqs. (21.11), (21.12) and (21.14).

Equation (21.11) can be written in the form

$$\frac{dm}{m} + k\,dt = 0,$$

giving

$$\ln m + kt = \text{constant}, \quad \text{or} \quad m = Ce^{-kt}.$$

Equation (21.12) can be written in the form

$$\frac{dT}{T - T_a} + k\,dt = 0,$$

giving

$$\ln (T - T_a) + kt = \text{constant},$$

that is

$$T = T_a + Ce^{-kt}.$$

Equation (21.14) can be written in the form

$$\frac{dY}{Y} - \alpha\beta\,dt = 0,$$

giving

$$\ln Y - \alpha\beta t = \text{constant},$$

that is

$$Y = Ce^{\alpha\beta t}.$$

Exercises 21(d)

Solve the differential equations in Nos. 1–8:

1. $y' = x^2/y$.
2. $y' = e^x \tan y$.
3. $y' = y/(x^2 - 4)$.
4. $(x^2 + 1)y' = y^2 + 4$.
5. $xyy' = 1 + x^2 + y^2 + x^2y^2$.
6. $yy' = e^{x+2y} \sin x$.
7. $2x^3 y' = y^2 + 3xy^2$, given that $y = 1$ when $x = 1$.
8. $y' = x\sqrt{(1-y^2)}\cos x$, given that $y = 0$ when $x = 0$.
9. Use the method of separation of variables to obtain the particular solution of the differential equation

$$e^x \frac{dy}{dx} + xy^2 = 0$$

that satisfies the condition that $y \to \frac{1}{2}$ as $x \to \infty$.

10. A point moves along the x-axis with velocity v at time t given by

$$v = x(\cot t - 1).$$

Write this as a differential equation connecting x and t. Determine x as a function of t, if $x = 1$ when $t = \pi/2$, and find the time interval between successive passages through the origin.

11. A body having a temperature of 50°C is immersed in water having a temperature of 25°C, and after 2 min the temperature of the body has dropped to 34°C. Assuming that Newton's law of cooling applies to the temperature of the body in relation to that of the surrounding water, find the temperature of the body after 3 min.

12. A firm has a monthly sales volume s when the unit sale price is p. Assuming that s decreases with respect to p at a rate directly proportional to s and inversely proportional to $(p + A)$, where A is a constant, give an expression for $\mathrm{d}s/\mathrm{d}p$. Hence express s in terms of p.

13. In an investigation of the spread of infection in a culture it is found that the time rate at which the infected area increases is directly proportional to the size of the infected area. Initially one quarter of the whole area is infected and after 12 h the infection has spread to seven tenths of the whole area.

    Set up a differential equation relating x, the infected proportion of the whole area, and the time t. Solve this differential equation subject to the given conditions and show that the whole area is infected in less than 17 h.

14. A culture of bacteria grows so that at time t days the population is p and the rate of change of p at any instant is proportional to the value of p itself at that instant. Express this statement in the form of a differential equation and solve for p given that when $t = 0$, $p = 10^2$ and when $t = 2$, $p = 10^5$.

15. At any time t the amount of active ferment in a culture of yeast is increasing at a rate which is directly proportional to the amount of active ferment already in the culture. Express this law in the form of a differential equation and solve this equation. Given that the amount doubles between the times $t = 0$ and $t = 1$, at what time will the amount have reached four times its original value?

16. It is assumed that, at any point of time, the rate of growth of population in a certain city is k times the population of the city at that time, k being a constant. Set down a differential equation that can be solved to obtain an estimate of the population at any time. Solve the differential equation, and show that if the population will multiply by a factor m in the next n years, then it will multiply by a factor m^2 in the next $2n$ years.

17. In a certain chemical reaction the concentration of a substance is at any instant reducing at a rate proportional to the cube of its value at that instant. Express this statement in the form of a differential equation. Hence determine the concentration at any time t, given that when $t = 0$ the concentration has the value a while its rate of reduction is then b.

18. The rate at which water will flow through a 1 cm hole at a depth h cm below the surface is approximately $(\pi/120)\sqrt{h}$ cm^3/s. Find the approximate time

required to empty an upright cylindrical vessel, height 9 cm and radius 1 cm, through a 1 cm hole in the bottom.

19. The point P with co-ordinates (X, Y) lies on the curve $y = f(x)$ and $f'(x) > 0$ for all x. Given that N is the foot of the ordinate at P and T is the point where the tangent to the curve at P meets the x-axis, show that the length of TN equals $Y/f'(X)$.

   Given that TN has constant length k for all positions (x, y) of P on the curve, express dy/dx in terms of y and solve this differential equation to determine the equation of the curve, subject to the additional condition that the curve passes through the point $(0, 5)$.

20. Show that $\int \sec\theta \, d\theta = \ln|\sec\theta + \tan\theta| + c$.

   In a certain navigational problem the distance x that a ship has sailed in a given direction is related to the angle θ through which it has turned by the differential equation

$$(d - x)\frac{d\theta}{dx} = \tfrac{1}{2}\cos\theta,$$

   where d is a constant.

   Solve this differential equation and find a value of the constant of integration such that $\theta = 0$ when $x = 0$. Verify that $x = 2(\sqrt{2} - 1)d$ when $\theta = \pi/4$.

21.4.4 Homogeneous Coefficients Suppose that in eq. (21.16) M and N are both homogeneous polynomials of the same degree in x and y. In this case $M(x, y)/N(x, y)$ is a function of y/x and so (21.16) can be written in the form

$$\frac{dy}{dx} = f(y/x).$$

Putting $y/x = v$, so that $y = vx$ and $dy/dx = v + x\, dv/dx$, we get

$$v + x\frac{dv}{dx} = f(v),$$

that is

$$\frac{dv}{f(v) - v} - \frac{dx}{x} = 0.$$

Thus when the variables v, x are used instead of y, x we see that the variables are separable and the equation can be solved by the method of Section 21.4.3, the original variables being restored by substituting y/x for v in the solution.

Example 16 Solve the equation

$$\frac{dy}{dx} = \frac{(x^2 + 2y^2)}{xy}$$

and find the particular solution for which $y = 0$ when $x = 1$.

If $y = vx$, the equation becomes

$$v + x\frac{dv}{dx} = \frac{1 + 2v^2}{v},$$

giving

$$x\frac{dv}{dx} - \frac{1 + v^2}{v} = 0,$$

that is

$$\frac{v\,dv}{1 + v^2} - \frac{dx}{x} = 0.$$

Integrating, we obtain

$$\tfrac{1}{2}\ln(1 + v^2) - \ln x = \ln C,$$

that is

$$\sqrt{(1 + v^2)} = Cx,$$
$$\sqrt{(x^2 + y^2)} = Cx^2.$$

If $y = 0$ when $x = 1$ we must have $C = 1$. Hence the required P.I. is $\sqrt{(x^2 + y^2)} = x^2$.

21.4.5 First-degree Coefficients Suppose that in eq. (21.16) M and N are both of first degree in x and y, i.e. eq. (21.16) can be written in the form

$$\frac{dy}{dx} = \frac{ax + by + c}{a'x + b'y + c'}.$$

If $a/a' \neq b/b'$ we may solve by writing

$$Y = ax + by + c, \quad X = a'x + b'y + c'.$$

Then

$$\frac{dy}{dx} = \frac{Y}{X} \quad \text{and} \quad \frac{dY}{dX} = \frac{dY/dx}{dX/dx} = \frac{a + b\,dy/dx}{a' + b'\,dy/dx} = \frac{a + bY/X}{a' + b'Y/X},$$

that is

$$\frac{dY}{dX} = \frac{aX + bY}{a'X + b'Y}.$$

This D.E. in the variables Y, X can be solved by the method of Section 21.4.4.

If $a/a' = b/b'$, the substitution $z = ax + by$ produces a D.E. with separable variables.

Example 17 Solve the equations

(a) $\dfrac{dy}{dx} = \dfrac{4x - 2y + 4}{2x + y - 2}$, (b) $\dfrac{dy}{dx} = \dfrac{2x + 3y + 2}{4x + 6y - 3}$.

(a) Let $Y = 4x - 2y + 4$ and $X = 2x + y - 2$; then the given equation becomes

$$\frac{dY}{dX} = \frac{4X - 2Y}{2X + Y}$$

and the substitution $Y = vX$ gives

$$v + X\frac{dv}{dX} = \frac{4 - 2v}{2 + v},$$

$$X\frac{dv}{dx} = \frac{4 - 4v - v^2}{2 + v},$$

that is

$$\frac{(2 + v)}{v^2 + 4v - 4}\,dv + \frac{dX}{X} = 0.$$

Integrating, we have

$$\tfrac{1}{2}\ln(v^2 + 4v - 4) + \ln X = \ln C,$$

$$X\sqrt{(v^2 + 4v - 4)} = C, \quad \text{or} \quad Y^2 + 4XY - 4X^2 = C^2.$$

Restoring the original variables we obtain

$$4(2x - y + 2)^2 + 8(2x + y - 2)(2x - y + 2) - 4(2x + y - 2)^2 = C^2$$

and simplifying,

$$4x^2 - 4xy - y^2 + 8x + 4y = C', \quad \text{where } C' = \tfrac{1}{8}C^2 + 4.$$

(b) $$\frac{dy}{dx} = \frac{2x + 3y + 2}{4x + 6y - 3}.$$

Here, since the coefficients of x, y in the numerator and denominator are in the same ratio, we let $z = 2x + 3y$; then the given equation becomes

$$\tfrac{1}{3}\left(\frac{dz}{dx} - 2\right) = \frac{z + 2}{2z - 3},$$

which leads to

$$\frac{dz}{dx} = \frac{7z}{2z - 3},$$

that is

$$(2 - 3/z)\, dz = 7\, dx.$$

Integrating, we have

$$2z - 3 \ln z = 7x + C,$$

that is

$$\ln (2x + 3y) = 2y - x + C' \qquad (C' = -C/3).$$

Exercises 21(e)

Solve the differential equations:

1. $x(y - 3x)y' = 2y^2 - 9xy + 8x^2$.
2. $(x^2 + y^2)y' = 2xy$.
3. $xy^2 y' = x^3 + y^3$.
4. $x(y + 4x)y' + y(x + 4y) = 0$.
5. $(2x + y)y' = 6y - 4x$.
6. $(x + 2y - 3)\, y' = 2x - y + 1$.
7. $(x + y - 2)\, y' = x + y + 2$.
8. $(2x - y)^2 y' + (x - 2y)^2 = 0$.
9. $(3x + y + 3)\, y' + 2(x + 3) = 0$.
10. $(2x - 4y - 8)y' = 3x - 5y - 9$.
11. $y' = y/x + x \sin (y/x)$.
12. $4x^2y' = 4xy - x^2 + y^2$, given that when $x = 1$, $y = 2$.
13. $xyy' = x^2 + y^2$, given that $y = 1$ when $x = 1$.
14. $xyy' = x^2 + 2y^2$, given that $y = 0$ when $x = 1$.
15. $4x^2y' = 4xy - x^2 + y^2$, given that $y = 2$ when $x = 1$.

21.4.6 Bernoulli Equations Suppose that in eq. (21.16) N is a function of x only while $M(x, y) = a(x)y + b(x)y^n$, where n is a constant. Such an equation is called a Bernoulli equation and is reducible to linear form. Dividing throughout by N we see that it has the form

$$p(x)y + q(x)y^n + \frac{dy}{dx} = 0. \tag{21.20}$$

If $n = 1$ this equation may be solved by separation of variables. If $n \neq 1$, divide throughout by y^n to get

$$p(x)\, y^{1-n} + q(x) + y^{-n} \frac{dy}{dx} = 0.$$

Now put $v = y^{1-n}$ so that

$$\frac{dv}{dx} = (1 - n)y^{-n} \frac{dy}{dx}$$

and the equation may now be written as

$$p(x)v + q(x) + \frac{1}{1-n}\frac{dv}{dx} = 0.$$

This is a first-order linear equation in the variables v, x and may be solved by the method of Section 21.4.2.

Example 18 Solve the equation

$$x\frac{dy}{dx} + y = y^2x^2 \ln x.$$

This is the Bernoulli equation corresponding to $n = 2$. To put it in the form (21.20) we must divide throughout by x, and following the above procedure we must also divide throughout by y^2. We thus obtain

$$\frac{1}{y^2}\frac{dy}{dx} + \frac{1}{xy} = x \ln x.$$

Put $v = y^{-1}$ so that $dv/dx = -y^{-2}\, dy/dx$, and the equation becomes

$$\frac{dv}{dx} - \frac{v}{x} = -x \ln x.$$

The I.F. for this equation is

$$e^{\int -(1/x)\,dx} = e^{-\ln x} = 1/x;$$

thus

$$\frac{1}{x}\frac{dv}{dx} - \frac{v}{x^2} = -\ln x,$$

that is

$$\frac{d}{dx}\left(\frac{v}{x}\right) = -\ln x.$$

Integrating, we have

$$\frac{v}{x} = C - (x \ln x - x), \quad \text{that is} \quad \frac{1}{xy} = C + x(1 - \ln x).$$

Exercises 21(f)

Solve the differential equations given in Nos. 1–8:

1. $y' + y \cot x = y^2 \sin^2 x$.
2. $y' = y \tan x + y^3 \tan^3 x$.
3. $y' = x/y + y$.
4. $2y' \sin x - y \cos x = y^3 \sin x \cos x$. Also find the particular solution for which

$y = -1$ when $x = \pi/2$.

5. $y' + y = xy^3$.
6. $y' = 2y \tan x + y^2 \tan^2 x$.
7. $2y' - y(2x + 1)/(x^2 + x + 1) = xy^3/(1 - x)$.
8. $(x^2 - 1)y' + xy = e^x y^{-2} \sqrt{(x^2 - 1)}$.
9. The tangent at a point P on a curve meets the y-axis at T and O is the origin. PN is the ordinate at P. Show that if, for all positions of P on the curve, $OT = PN^3$, then the curve satisfies a Bernoulli differential equation. Hence find the curve, given that it passes through the point $(1, \frac{1}{2}\sqrt{2})$.

21.4.7 Change of Variable D.E.'s of the first order and first degree that are not of the foregoing types may sometimes be solved by a suitable change of variable.

Thus, for example, in an equation of the form

$$\frac{dy}{dx} = f(ax + by + c)$$

the substitution $u = ax + by + c$ is indicated.

Suppose that in eq. (21.16) M is a function of y only while

$$N(x, y) = a(y)x + b(y).$$

Then we can regard y as the independent and x as the dependent variable, using the fact that $dy/dx = 1/(dx/dy)$, and in this way the equation becomes linear.

A few substitutions that are frequently useful are listed below, but in most cases an appropriate substitution is suggested by the expressions occurring in the equations under consideration. Thus, for example, the Bernoulli equation solved in Example 18 may also be solved by the substitution $u = xy$. This substitution is suggested by the presence of the expression x^2y^2 and by the expression $y + x(dy/dx)$ which is equivalent to $d(xy)/dx$.

The substitution $u = x^2 + y^2$ is suggested by $x\,dx + y\,dy$ and the substitution $u = y/x$ by $x\,dy - y\,dx$.

If both $(x\,dx + y\,dy)$ and $(x\,dy - y\,dx)$ occur, simplification may be obtained by a change to polar co-ordinates, for we have

$$x^2 + y^2 = r^2 \quad \text{and} \quad y/x = \tan\theta,$$

so that

$$x\,dx + y\,dy = r\,dr \quad \text{and} \quad x\,dy - y\,dx = r^2\,d\theta.$$

Example 19 Solve the equation $dy/dx = (x + y)/(x - y)$.

This equation may be solved by the method of Section 21.4.4. Alternatively it may be written in the form

$$x\,dx + y\,dy = x\,dy - y\,dx,$$

which, expressed in terms of polar co-ordinates, reduces to $dr/r = d\theta$.

The general solution, $r = Ce^{\theta}$, is the equation of a family of equiangular spirals.

Example 20 Given that $dy/dx = (x + y)^2$ and $y = \frac{1}{2}$ when $x = \frac{1}{2}$, calculate to three significant figures the value of y when $x = 0{\cdot}7$.

The substitution $x + y = u$, $dy/dx = du/dx - 1$ leads to the equation

$$\frac{du}{dx} = u^2 + 1.$$

The general solution is

$$\tan^{-1} u = x + C,$$

that is

$$\tan^{-1}(x + y) = x + C,$$

But when $x = \frac{1}{2}$, $y = \frac{1}{2}$ and so $C = \frac{1}{4}\pi - \frac{1}{2}$.

Hence

$$\tan^{-1}(x + y) = x + \tfrac{1}{4}\pi - \tfrac{1}{2},$$

that is

$$y = \tan\left(\tfrac{1}{4}\pi + x - \tfrac{1}{2}\right) - x.$$

When $x = 0{\cdot}7$,

$$y = \tan\left(\tfrac{1}{4}\pi + 0{\cdot}2\right) - 0{\cdot}7 = 0{\cdot}808 \qquad \text{from tables.}$$

Example 21 Transform the equation

$$(2xyy' + x^2 - y^2)(x^2 + y^2)^{1/2} = xyy' - y^2,$$

where $y' = dy/dx$, into one involving θ, r and $dr/d\theta$. Hence, or otherwise, solve the equation.

The given equation may be written in the form

$$(x^2 + y^2)^{1/2}\{2xy\,dy + (x^2 - y^2)\,dx\} = xy\,dy - y^2\,dx,$$

that is

$$(x^2 + y^2)^{1/2}[x(y\,dy + x\,dx) + y(x\,dy - y\,dx)] = y(x\,dy - y\,dx),$$

and on changing to polar co-ordinates we obtain

$$r\,[r\cos\theta(r\,dr) + r\sin\theta(r^2\,d\theta)] = r\sin\theta(r^2\,d\theta),$$

which reduces to

$$dr/d\theta + r\tan\theta = \tan\theta.$$

Separating the variables we have

$$\frac{dr}{r-1} + \tan\theta\,d\theta = 0,$$

which leads to

$$(r-1)\sec\theta = C \quad \text{or} \quad r = 1 + C\cos\theta.$$

The solution is generally left in polar form.

Exercises 21(g)

Using the methods discussed so far in this chapter, solve the differential equations given in Nos. 1–24:

1. (a) $(4y + 3x)y' + y - 3x = 0$;
   (b) $xy' - 2y = x + 1$,
   given that when $x = \frac{1}{2}$, $y = 1$.
2. (a) $(5y + x)y' = 5x + y$;
   (b) $y' \sin x - y \cos x = \sin^3 x$.
3. (a) $\dfrac{1}{t}\dfrac{dx}{dt} = 1 - \dfrac{x}{1-t^2}$;
   (b) $4(y - x)y' = 3(3x + 4y)$.
4. (a) $y' = y \tan x - 2 \sin x$;
   (b) $(x^2 - 3y^2)x\,dx = (y^2 - 3x^2)y\,dy$.
5. (a) $y' + y \tan x = \sin 2x$;
   (b) $x(x + y)y' = x^2 + y^2$;
   (c) $(\cos x - x \cos y)\,dy - (\sin y + y \sin x)\,dx = 0$.
6. (a) $(x^2 + 1)y' + 4xy = 5(x^2 - 1)$;
   (b) $x(x^2 + 3y^2)y' = y(3x^2 + y^2)$.
7. (a) $y' \sin x - y \cos x = \sin x - (1 + x) \cos x$;
   (b) $xyy' = 2y^2 - 3xy + 2x^2$.
8. (a) $y' + y \tan x = \cos 2x$;
   (b) $(4x - 3y)y' = 3x + 4y$.
9. (a) $4x^2y' = 4xy - x^2 + y^2$;
   (b) $(1 + x^2)y' = 2x(1 + y + x^2)$;
   (c) $y' = (x + 4y)^2$.
10. (a) $xy' = 2y + x^{n+1}y - x^n y^{3/2}$;
    (b) $(x + 3y + 8)y' = 3x + y$.
11. (a) $y' + (2x + 1)y = 2x^2 + x + 1$;
    (b) $(2x - 2y - 4)y' = 2x + 7y + 5$.

12. (a) $(4x + 2y + 1)y' = 2x + y - 3$;
    (b) $y' + y \cot x = y^2 \cos^2 x$.
13. (a) $x(1 - x^2)y' + (2x^2 - 1)y = x^3y^3$;
    (b) $xy' - y = (x^2 + y^2)^{1/2}$.
14. (a) $(3x + 2y - 1)y' = x + 2y - 3$;
    (b) $4y' \sin^2 x - y \sin 2x = y^3 \cos x$.
15. (a) $(x^2 + y^2)y' = x^2 + xy$;
    (b) $(x - y)y' = 2x + y - 3$;
    (c) $2(1 + x)y' - (1 + 2x)y = x^2(1 + x)^{1/2}$.
16. (a) $(3x + 2y - 4)y' = 3y - 2x + 7$;
    (b) $y' + y \tan x = y^3 \sec^6 x$.
17. (a) $x(1 - x^2)y' + (1 - 2x - x^2)\, y = 1$;
    (b) $y^2y' = x^2 + xy - y^2$.
18. (a) $y' = 2x - y$;
    (b) $(x + 2y - 5)y' = 2x - y$.
19. (a) $y' \cos x + y \sin x = x \sin 2x + x^2$;
    (b) $(x - 8y + 7)y' = x - y$.
20. (a) $(3y - 2x)y' = 2x + 3y$;
    (b) $y' + y = xy^3$.
21. (a) $(x^2 + xy)y' = x^2 + xy - y^2$;
    (b) $y' + y \cot x = (y \sin x)^{1/2}$.

Show also that the only solution of the latter equation that remains finite as $x \to 0$ is $y \sin x = \sin^4 \frac{1}{2}x$.

22. (a) $(x^2 - x)y' + y = (x^2 - x) \ln x$;
    (b) $xy' - y = (x^2 + y^2)^{1/2}$;
    (c) $xy' + 3y = x^2y^2$.
23. (a) $x(y^3 + x^3)y' = y^4 + 2x^3y - x^4$;
    (b) $y' \sin x + 2y \cos x = \cos x$.
24. (a) $xyy' = y^2 + x^2e^{y/x}$;
    (b) $x(1 + x)y' + (2 + x)y + 3x + 2x^2 = 0$.
25. (a) Solve the equation $(x^2 + y^2)y' + 2x(x + y) = 0$.
    (b) Find the integral curves of the equation $(y' - y)e^x + 1 = 0$. Show that, in general, every curve of the system has *either* a real point of inflexion, on the line $y = 0$, *or* a real point for which y is a minimum, on the curve $y = e^{-x}$.
26. (a) Solve the differential equation $x^2 y' + 1 + (1 - 2x)y = 0$.
    (b) Prove that the differential equation $M\, dx + N\, dy = 0$, where M and N are functions of x and y, is exact if $\partial M/\partial y = \partial N/\partial x$.

    Show that a constant a can be found so that $(x + y)^a$ is an integrating factor of

$$(4x^2 + 2xy + 6y)\, dx + (2x^2 + 9y + 3x)\, dy = 0,$$

    and hence integrate the equation.
27. Show how to solve the differential equation $y' + Py = Qy^n$, where P and Q are functions of x only.

Find the solution of the equation $xy^2 y' - 2y^3 = 2x^3$, which is such that $y = 1$ when $x = 1$.

28. (a) It is known that, when multiplied by a certain power of x, the equation

$$(5x^2 + 12xy - 3y^2)\,dx + (3x^2 - 2xy)\,dy = 0$$

becomes exact. Find this integrating factor and solve the equation.

(b) Obtain the solution of the equation $(x - 1)y' + xy = (x - 1)e^x$ for which y and x vanish together.

29. (a) Solve the equation $xyy' = (x + y)^2$.

(b) Show that the equation $x^2 y' = 1 - 2x^2y^2$ may be reduced to a linear differential equation of the first order by the substitution $y = 1/x + 1/z$. Hence, or otherwise, solve the equation.

30. (a) Solve the equation $x(x + 1)y' + y = 2x$.

(b) By means of the substitution $y^2 = u - x$, reduce the equation

$$y^3 y' + x + y^2 = 0$$

to homogeneous form and hence, or otherwise, solve it.

31. (a) Reduce the differential equation $(x + 1)yy' - y^2 = x$ to a linear form by writing $z = y^2$, and solve it, given that $y = 1$ when $x = 0$.

(b) Solve the equation $(2x + y)y' = x + 2y$, by writing $y = vx$, given that $y = 0$ when $x = 1$.

21.5 Clairaut's Equation

It is not intended to discuss in this book equations of the first order that are not also first degree. However, it is of interest to look at the case where the equation has the form

$$y = px + f(p), \tag{21.21}$$

where $p = dy/dx$.

Differentiating (21.21) with respect to x we obtain

$$p = x\frac{dp}{dx} + p + f'(p)\frac{dp}{dx},$$

that is

$$\frac{dp}{dx}[x + f'(p)] = 0. \tag{21.22}$$

Not all solutions of (21.22) are also solutions of (21.21); hence if we solve (21.22) we must check the solutions using (21.21). In other words, the equation (21.22) is a second-order D.E. and the G.S. would contain two arbitrary constants; we could then regard (21.21) as a condition that reduces the number of arbitrary constants to one, as is appropriate for a first-order D.E.

In practice it is simpler to note that (21.21) and (21.22) are both D.E.'s, since they involve p, and elimination of p between them gives the required solution.

From (21.22) we see that either

$$\frac{dp}{dx} = 0, \quad \text{that is} \quad p = C, \tag{21.23}$$

or

$$x + f'(p) = 0. \tag{21.24}$$

Hence elimination of p between (21.21) and (21.22) gives the result that either

$$y = Cx + f(C) \tag{21.25}$$

obtained by substituting from (21.23) into (21.21), or

$$\phi(x, y) = 0 \tag{21.26}$$

where (21.26) denotes the result of eliminating p between (21.24) and (21.21).

Note that (21.25) contains an arbitrary constant and so is the G.S., whereas (21.26) contains no arbitrary constants and is a *singular* solution (see Section 21.3). The complete solution is given by (21.25) together with (21.26).

The G.S. (21.25) may be interpreted as a family of straight lines. The singular solution may likewise be interpreted as a curve with equation (21.26), which may be expressed in parametric form as eqs. (21.21) and (21.24) with p as parameter. It is easily verified that the family of straight lines (G.S.) is the family of all tangents to the curve (singular solution) as indicated in Fig. 21.3. (We say the curve is the *envelope* of the family of

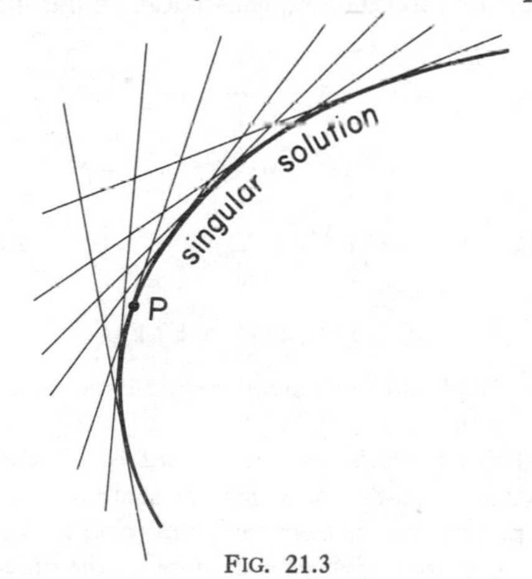

FIG. 21.3

lines.) At a point P on the curve the co-ordinates (x, y) and the slope dy/dx are the same (and hence satisfy the same first-order D.E.), whether P is regarded as a point on the curve or a point on the tangent line through P.

Example 22 Find the G.S. and the singular solution of the equation $y = px - \ln p$ when p denotes dy/dx.

Differentiating the given equation with respect to x, we have

$$p = x\frac{dp}{dx} + p - \frac{1}{p}\frac{dp}{dx},$$

giving

$$\frac{dp}{dx}(x - 1/p) = 0,$$

so that

$$\frac{dp}{dx} = 0 \quad \text{or} \quad p = 1/x,$$

that is

$$p = C \quad \text{or} \quad p = 1/x.$$

Substituting in the given equation we get $y = Cx - \ln C$ (G.S.) or $y = 1 + \ln x$ (singular solution).

Exercises 21(h)

Obtain the general solution and the singular solution of the differential equations in Nos. 1–7:

1. $y = px + a/p \quad (p = dy/dx)$.
2. $y = px + 2\sqrt{(ap)}$.
3. $y = px + p^3$.
4. $y = px + a\sqrt{(1 + p^2)}$.
5. $y = px + p - p^2$.
6. $e^{y+p} = (1 + p)e^{px}$.
7. $2y = 2px - \ln \sec^2 p$.
8. By means of the substitution $x^2 = X$, $y^2 = Y$ (or otherwise), reduce the equation

$$x^2 + y^2 - xy(p + 1/p) = c^2$$

to Clairaut's form and find the general solution and singular solution.

9. Show that the equation $(px - y)(px - 2y) + x^3 = 0$, where p denotes dy/dx, may be reduced to Clairaut's form by means of the substitution $y = vx$. Hence find its general solution and singular solution.
10. The feet of the perpendiculars from the point $(c, 0)$ to the tangents to a certain curve lie on the circle $x^2 + y^2 = a^2$. Obtain the differential equation of

the curve in the form

$$y^2 - 2xyp + (x^2 - a^2)p^2 + c^2 - a^2 = 0,$$

where $p = dy/dx$, and show that $y = mx \pm \sqrt{[a^2(1 + m^2) - c^2]}$ is a solution.

21.6 Orthogonal Trajectories

Two families of curves, such that each member of one family cuts every member of the other family at right angles, are called *orthogonal trajectories* of one another.

From the equation $f(x, y, c) = 0$ representing a one-parameter family of curves we can form a differential equation of the first order

$$F(x, y, y') = 0 \tag{1}$$

which is the differential equation of the family.

Now through any point $P(x, y)$ there passes a curve C of the given family and a curve T that is a member of the family of orthogonal trajectories. Let m and μ be the gradients of the tangents at P to the curves C and T respectively (Fig. 21.4).

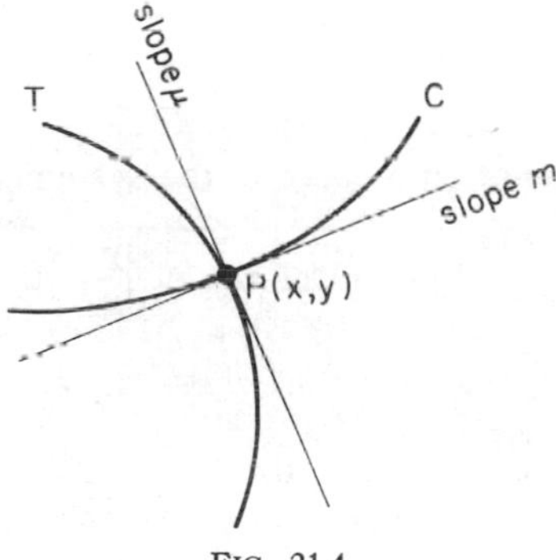

FIG. 21.4

Since C is a curve of the given family, $F(x, y, m) = 0$, by (1). But C and T cut orthogonally and so $m = -1/\mu$. Hence

$$F(x, y, -1/\mu) = 0. \tag{2}$$

Now, by definition of μ, $\mu = y'$ for the curve T and at P, x and y are the same for both curves. Hence if we write y' for μ in (2) we shall obtain the differential equation of the family of orthogonal trajectories. This equation is

$$F(x, y, -1/y') = 0.$$

It is obtained by writing $-1/y'$ for y' in the differential equation of the given family.

Example 23 The normal at a point P of a curve meets the x-axis at G, and N, the foot of the ordinate of P, lies between G and the origin O. Given that $OG = OP$, find the differential equation of the system of curves for which this condition holds, and integrate it. Find also the equation of the orthogonal trajectories of the system and show that they are parabolas.

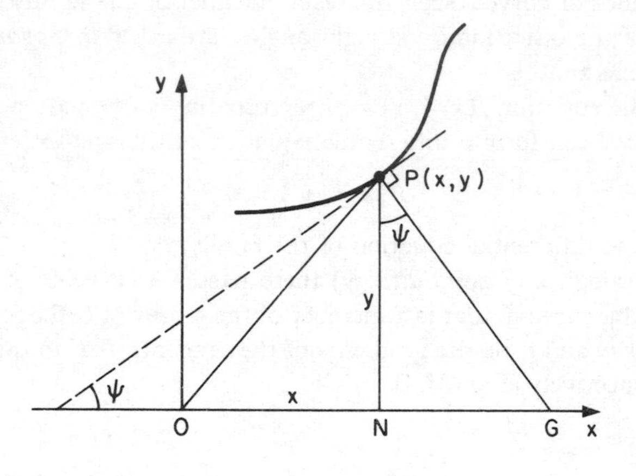

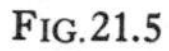

FIG. 21.5

Let $P(x, y)$ be any point on a curve of the system. Then

$$NG = y \tan \psi = y \frac{dy}{dx},$$

and so

$$OG = ON + NG = x + y \frac{dy}{dx}.$$

Also we have $OP = \sqrt{(x^2 + y^2)}$.

If

$$OG = OP,$$

then

$$y\, dy/dx + x = \sqrt{(x^2 + y^2)}. \tag{1}$$

This is the differential equation of the system. To integrate it let

$$x^2 + y^2 = r^2,$$

so that

$$x + y\, dy/dx = r\, dr/dx.$$

The equation then becomes

$$dr/dx = 1,$$
$$r = x + C,$$

that is

$$x^2 + y^2 = (x + C)^2$$
$$y^2 = C(C + 2x). \tag{2}$$

This is the equation of a family of parabolas with a common focus at the origin and common axis Ox.

It follows that the differential equation of the orthogonal trajectories of the system is

$$x - y\, dx/dy = \sqrt{(x^2 + y^2)}$$
$$dy/dx = y/[x - \sqrt{(x^2 + y^2)}] = -[x + \sqrt{(x^2 + y^2)}]/y.$$

Thus

$$y\, dy/dx + x = -\sqrt{(x^2 + y^2)}$$

and, comparing this equation with (1), we obtain the solution

$$y^2 = A(A - 2x). \tag{3}$$

This equation represents a family of parabolas with a common focus at the origin and common axis Ox; also, if we write $A = -C$ we see that (3) represents the same system as (2). On account of this property the given system is said to be self-orthogonal.

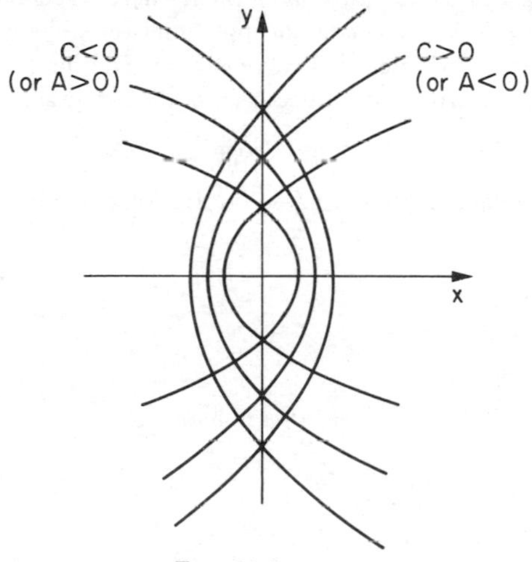

FIG. 21.6

Curves (2) and (3) intersect in real points only when A and C have the same signs, and so each member of the system intersects orthogonally an infinite number (but not all) of the members of the system (see Fig. 21.6).

Exercises 21(i)

1. Find the orthogonal trajectories of the family of curves defined by the equation $3x^2 + y^2 + \lambda x = 0$, where λ is a variable parameter.
2. Find the orthogonal trajectories of the system of ellipses

$$x^2 - xy + y^2 = c.$$

3. $P(x, y)$ is a point on a curve C. The tangent at P meets the axis OX at T and has gradient p. The normal at P meets the axis OY at G. Prove that the gradient of TG is $(x + py)/(y - px)$.

   Show that if, for all positions of P for which T and G are determinate, TG is perpendicular to OP, C is a circle of the system of circles touching OX at O.

   Write down the differential equation of the system of orthogonal trajectories of this system of circles.
4. The tangent and normal to a curve at the point P intersect the x-axis at A and B, and the foot of the perpendicular from P to the x-axis is M. Find and solve the differential equation of the family of curves for which $2MP^2 = OM.AB$, where O is the origin.

   Find also the differential equation of the orthogonal trajectories to these curves.
5. A plane curve has the property that the tangents from any point on the y-axis to the curve are of constant length, a. Find the differential equation of the family to which the curve belongs and integrate it.

   Show that the orthogonal trajectories of the curves are circles.
6. The tangent at a point $P(x, y)$ on a plane curve meets the x-axis at T. Given that $PT = ay^2$, where a is a constant, find the differential equation of the family of curves to which this curve belongs and integrate the equation.

   Show that the orthogonal trajectories of the family are the curves

$$ay = \cosh(ax + b).$$

Find the equations of the two curves, one from each family, that pass through the point $(0, 1/a)$.

CHAPTER 22

INFINITE SERIES

22.1 Convergence and Divergence

The series

$$u_1 + u_2 + u_3 + \cdots + u_n + \cdots$$

in which u_r is defined for *all* positive integers r, so that the series has no last term, is called an *infinite* series. The equation

$$S_n = u_1 + u_2 + \cdots + u_n$$

defines a *partial sum* S_n (the sum of the first n terms of the series) for *any* positive integer n. If S_n tends to a limit S as $n \to \infty$, then the series is said to be *convergent* and S is called its *sum to infinity* or, more briefly, its *sum*; we write

$$S = u_1 + u_2 + \cdots + u_n + \cdots \quad \text{or} \quad S = \sum_{r=1}^{\infty} u_r.$$

For example, if $u_r = 1/r(r+1)$ we may write

$$u_r = \frac{1}{r} - \frac{1}{r+1},$$

so that

$$S_n = \left(1 - \frac{1}{2}\right) + \left(\frac{1}{2} - \frac{1}{3}\right) + \left(\frac{1}{3} - \frac{1}{4}\right) + \cdots + \left(\frac{1}{n} - \frac{1}{n+1}\right)$$

$$= 1 - \frac{1}{n+1} \to 1 \text{ as } n \to \infty.$$

Hence the series is convergent with sum 1, and we write

$$\sum_{r=1}^{\infty} \frac{1}{r(r+1)} = \frac{1}{1.2} + \frac{1}{2.3} + \cdots + \frac{1}{n(n+1)} + \cdots = 1.$$

Example 1 Find the sum to infinity of the series

$$\frac{1}{1.3.5} + \frac{1}{3.5.7} + \frac{1}{5.7.9} + \cdots.$$

We have seen (Example 23, Chapter 4) that for this series

$$S_n = \frac{1}{4}\left[\frac{1}{3} - \frac{1}{(2n+1)(2n+3)}\right].$$

Hence $S_n \to \frac{1}{12}$ as $n \to \infty$, and so

$$\sum_{r=1}^{\infty} \frac{1}{(2r-1)(2r+1)(2r+3)} = \frac{1}{12}.$$

Example 2 Find the sum to infinity of the series whose rth term is $1/r(r+1)(r+3)$.

We have seen (Example 24, Chapter 4) that for this series

$$S_n = \frac{1}{2}\left[\frac{1}{6} - \frac{1}{(n+2)(n+3)}\right] + \frac{2}{3}\left[\frac{1}{6} - \frac{1}{(n+1)(n+2)(n+3)}\right].$$

Hence, as $n \to \infty$, $S_n \to \frac{1}{12} + \frac{1}{9} = \frac{7}{36}$ and so

$$\sum_{r=1}^{\infty} 1/r(r+1)(r+3) = \tfrac{7}{36}.$$

If the partial sums S_n do not tend to a limit as $n \to \infty$ the series is said to be *divergent*. The following examples illustrate ways in which this may happen.

(a) The series $1 + 2 + 3 + \ldots$ is divergent since

$$S_n = 1 + 2 + 3 + \ldots + n = \tfrac{1}{2}n(n+1)$$

and so $S_n \to \infty$ as $n \to \infty$. We sometimes say that the series diverges to $+\infty$ in a case like this.

(b) The series $1 - 1 + 1 - 1 + \cdots$ is divergent since $S_n = 0$ if n is even and $S_n = 1$ if n is odd so that S_n does not tend to a limit as $n \to \infty$. This series is said to *oscillate finitely*.

(c) The series $1 - 1 + 3 - 3 + 5 - 5 + \cdots$ is divergent since $S_n = 0$ if n is even and $S_n = n$ if n is odd so that S_n does not tend to a limit as $n \to \infty$. This series is said to *oscillate infinitely*.

Example 3 Show that the series $1^2 + 2^2 + 3^2 + \cdots$ is divergent.

We have

$$S_n = 1^2 + 2^2 + 3^2 + \cdots + n^2 = \tfrac{1}{6}n(n+1)(2n+1)$$

and clearly $S_n \to \infty$ as $n \to \infty$.

Example 4 Show that the series $2.3 + 3.4 + 4.5 + \cdots$ is divergent.

We have seen (Example 20, Chapter 4) that for this series

$$S_n = \tfrac{1}{3}n(n^2 + 6n + 11).$$

Hence $S_n \to \infty$ and the series is divergent.

22.1.1 The Geometrical Progression

The geometrical progression

$$a + ax + ax^2 + \cdots + ax^{n-1} + \cdots$$

where $a \neq 0$, provides an important example of the behaviour of an infinite series.

When $x \neq 1$, the partial sum S_n is given by

$$S_n = \frac{a(1-x^n)}{1-x}.$$

If $|x| < 1$, $x^n \to 0$ as $n \to \infty$; hence S_n tends to the finite limit $a/(1-x)$ and the series converges.

If $x > 1$, $x^n \to \infty$ as $n \to \infty$; hence $S_n \to +\infty$ if $a > 0$, $S_n \to -\infty$ if $a < 0$, and so the series diverges.

If $x < -1$, x^n oscillates infinitely; hence S_n oscillates infinitely and the series cannot converge.

If $x = 1$, the series is $a + a + a + \cdots$ and $S_n = na$; hence $S_n \to +\infty$ if $a > 0$, $S_n \to -\infty$ if $a < 0$; and so the series diverges.

If $x = -1$, the series is $a - a + a - a + \cdots$; hence $S_n = 0$ or a according as n is even or odd, and the series oscillates finitely.

It is important to note that the geometrical progression converges only when the common ratio x (i.e. ratio of each term to the preceding term) is numerically less than unity.

22.1.2 General Theorems on Convergence

The theorems on limits given in Section 14.1.1 lead to some general theorems on the convergence of series:

Theorem 1 If Σu_r is convergent with sum S, then Σku_r, where k is any constant, is convergent with sum kS. Similarly, if Σu_r is divergent, so also is Σku_r unless $k = 0$.

Theorem 2 If $\Sigma_{r=1}^{\infty} u_r$ is convergent with sum S then $\Sigma_{r=m+1}^{\infty} u_r$ is convergent with sum $S - S_m$, m being any given positive integer and $S_m = \Sigma_{r=1}^{m} u_r$. Similarly, if $\Sigma_{r=1}^{\infty} u_r$ diverges, so also does $\Sigma_{r=m+1}^{\infty} u_r$.

The significance of Theorem 2 is that in discussing the convergence of a series, any finite number of terms at the beginning of the series may be ignored. Thus from Example 1 it follows that the series

$$\frac{1}{5.7.9}+\frac{1}{7.9.11}+\frac{1}{9.11.13}+\cdots$$

in which the first two terms of the original series have been omitted, and the series

$$5.7.9+3.5.7+1.3.5+\frac{1}{1.3.5}+\frac{1}{3.5.7}+\cdots$$

in which three terms have been added to the original series, are both convergent with sums

$$\tfrac{1}{12}-\left(\frac{1}{1.3.5}+\frac{1}{3.5.7}\right)=\tfrac{1}{140}$$

and

$$\tfrac{1}{12}+5.7.9+3.5.7+1.3.5=435\tfrac{1}{12}$$

respectively.

Theorem 3 If Σu_r and Σv_r are two convergent series with sums S and T respectively, then the series $\Sigma(pu_r \pm qv_r)$ is also convergent with sum $pS \pm qT$, p and q being any constants.

From the definition of the partial sum S_n we easily see that

$$S_n - S_{n-1} = u_n,$$

and so, for a convergent series, $\lim_{n\to\infty} u_n = 0$ since

$$\lim_{n\to\infty} S_n = \lim_{n\to\infty} S_{n-1} = S.$$

We state this result as

Theorem 4 If Σu_r is convergent, then $u_r \to 0$ as $r \to \infty$.

Theorem 4 may sometimes be used to establish the divergence of a series; for if u_r does not tend to zero as $r \to \infty$ it follows from Theorem 4 that the series is divergent. Thus in Example 3 we have $u_r = r^2 \to \infty$ as $r \to \infty$, giving an alternative proof that the series is divergent; likewise in Example 4 we have $u_r = (r+1)(r+2) \to \infty$ as $r \to \infty$. However, it is important to realise that Theorem 4 *cannot be used to establish the convergence of a series* since there are many divergent series also having the property that $u_r \to 0$ as $r \to \infty$. For example, if $u_r = 1/\sqrt{r}$ then

$$S_n = \frac{1}{\sqrt{1}} + \frac{1}{\sqrt{2}} + \frac{1}{\sqrt{3}} + \cdots + \frac{1}{\sqrt{n}} > n . \frac{1}{\sqrt{n}} = \sqrt{n}$$

so that $S_n \to \infty$ as $n \to \infty$ and the series is divergent although $u_r \to 0$ as $r \to \infty$.

22.1.3 The Harmonic Series

The series

$$\Sigma(1/r) = 1 + \tfrac{1}{2} + \tfrac{1}{3} + \tfrac{1}{4} + \tfrac{1}{5} + \cdots$$

is known as the *harmonic* series. Now

$$\tfrac{1}{3} + \tfrac{1}{4} > \tfrac{1}{4} + \tfrac{1}{4} = \tfrac{1}{2}, \tfrac{1}{5} + \tfrac{1}{6} + \tfrac{1}{7} + \tfrac{1}{8} > \tfrac{4}{8} = \tfrac{1}{2},$$

and so on.

Hence, if S_n is the sum of n terms of the harmonic series $S_1 = 1$, $S_2 = 1 + \frac{1}{2}$, $S_4 > 1 + \frac{1}{2} + \frac{1}{2}$, $S_8 > 1 + \frac{1}{2} + \frac{1}{2} + \frac{1}{2}$, and if $n = 2^p$, $S_n > 1 + \frac{1}{2}p$ ($p = 2, 3, \ldots$). When $n \to \infty, p \to \infty$, and so $S_n \to \infty$. Hence $\Sigma(1/r)$ diverges to $+\infty$, although $u_r = 1/r \to 0$ as $r \to \infty$.

22.1.4 Series of Positive Terms

The investigation of the convergence of a series is simplified if the terms are all positive, for we can then apply the following theorem which we do not prove but which we illustrate graphically:

Theorem 5 If S_n is the sum of the first n terms of a series of positive terms and if for all values of n, S_n remains less than a fixed number k, independent of n, then the series converges, and its sum to infinity is less than or equal to k.

On a straight line Ox (Fig. 22.1) following the usual sign conventions

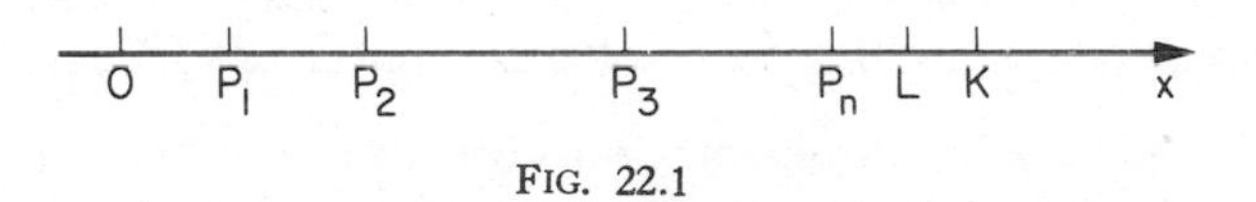

FIG. 22.1

we mark the points $P_1, P_2, P_3, \ldots$ such that

$$OP_1 = S_1, OP_2 = S_2, OP_3 = S_3, \ldots.$$

Let $OK = k$. Then since the series contains only positive terms, S_n increases with n, and so $P_1, P_2, P_3, \ldots$ advance steadily from left to right along Ox. For all values of n, $S_n < k$ and so every point P_n lies to the left of K. The theorem states that there exists a point L either to the left of K

or coincident with K beyond which P_n never passes and such that $P_nL \to 0$ as $n \to \infty$. Thus if $OL = l$, $\lim_{n \to \infty} S_n = l$, where $l \leqslant k$, and so the series converges.

If no such number k can be found, S_n increases without limit and the series diverges to $+\infty$. A series of positive terms cannot oscillate.

A series of negative terms either converges, or diverges to $-\infty$.

A series whose terms are all positive only after the mth term (say) is said to be *ultimately positive*. The convergence of such a series is unaffected by the removal of the first m terms (Theorem 2) and we can apply to the series

$$u_{m+1} + u_{m+2} + u_{m+3} + \cdots$$

tests for a series of positive terms.

22.2 Tests for Convergence and Divergence

Unfortunately there are comparatively few series for which a general formula for the partial sum S_n in terms of n can be obtained. We now consider tests that can be used to establish the convergence or divergence of a series without recourse to a formula for S_n. The first such test, based on Theorem 4, was given in Section 22.1.2, and may be summarised as follows:

Test 1 If u_n does not tend to zero as $n \to \infty$, then $\Sigma_{n=1}^{\infty} u_n$ is divergent.

Note that when partial sums are not being used we frequently write the series as $\Sigma_{n=1}^{\infty} u_n$ instead of $\Sigma_{r=1}^{\infty} u_r$ as before.

22.2.1 Comparison Tests

Test 2 A series of positive terms converges if its terms are less than (or equal to) the corresponding terms of some convergent series.

Let Σu_r and Σv_r be two series of positive terms. Let Σv_r be convergent with sum V, and let $u_r \leqslant v_r$ for all values of r.

Then

$$u_1 + u_2 + \cdots + u_r \leqslant v_1 + v_2 + \cdots + v_r < V$$

for all values of r; and so by Theorem 5, Σu_r is convergent.

For example, the terms of the series

$$\frac{1}{1!} + \frac{1}{2!} + \frac{1}{3!} + \frac{1}{4!} + \cdots$$

are less than or equal to the corresponding terms of the convergent series

$$1 + \frac{1}{2} + \frac{1}{2^2} + \frac{1}{2^3} + \cdots$$

Hence the given series converges.

Corollary to Test 2 If $0 < u_r \leqslant kv_r$, where k is a positive constant and Σv_r converges, then Σu_r is also convergent, for Σkv_r converges (see Theorem 1).

Test 3 A series of positive terms diverges if its terms are greater than (or equal to) the corresponding terms of a divergent series of positive terms.

Let Σu_r and Σv_r be two series of positive terms; let Σv_r be divergent and let $u_r \geqslant v_r$ for all values of r.

Then

$$u_1 + u_2 + \cdots + u_r \geqslant v_1 + v_2 + \cdots + v_r$$

for all values of r. But

$$v_1 + v_2 + \cdots + v_r \to \infty \quad \text{when } r \to \infty$$

and so

$$u_1 + u_2 + \cdots + u_r \to \infty \text{ when } r \to \infty.$$

Hence the series Σu_r diverges.

For example, if $r > 0$, $r(r + 1) < (r + 1)^2$, and so

$$\frac{1}{\sqrt{[r(r+1)]}} > \frac{1}{r+1}.$$

But $\Sigma\, 1/(r + 1)$ is a divergent series (see Section 22.1.3); hence

$$\Sigma \frac{1}{\sqrt{[r(r+1)]}}$$

is divergent.

Corollary to Test 3 If $u_r \geqslant kv_r > 0$, where k is a positive constant and Σv_r diverges, then Σu_r is also divergent, for Σkv_r is divergent (see Theorem 1).

Tests 2 and 3 can be combined in the following form, given here without proof:

Test 4 If Σu_n and Σv_n are series of positive terms and if $\lim_{n\to\infty} (u_n/v_n)=L$ where $L\neq 0$, then Σu_n and Σv_n are either both convergent of both divergent.

Example 5 Test for convergence the series Σu_r, where

(a) $u_r = (3r + 1)/(3r^2 - 2)$, (b) $u_r = (3r - 1)/(3r^2 + 2)$.

(a)

$$\frac{3r+1}{3r^2-2} > \frac{3r}{3r^2} = \frac{1}{r},$$

and $\Sigma(1/r)$ is a divergent series. Hence, by Test 3, Σu_r is divergent.

(b)

$$\frac{3r-1}{3r^2+2} = \frac{(3-1/r)}{r(3+2/r^2)} \simeq \frac{1}{r}$$

when r is large. Hence we compare the given series Σu_r with Σv_r where $v_r = 1/r$.

We have

$$\frac{u_r}{v_r} = \frac{r(3r-1)}{3r^2+2} = \frac{3-1/r}{3+2/r^2} \to 1 \text{ as } r \to \infty.$$

But Σv_r is divergent. Hence, by Test 4, Σu_r is divergent.

Two series are regarded as standard and are frequently used in applications of the comparison tests:

(1) The geometrical series $\Sigma_{n=1}^{\infty} x^{n-1}$, which (as has been shown in Section 22.1.1) converges only when $|x| < 1$.
(2) The series $\Sigma_{r=1}^{\infty} (1/r^p)$, which we now show to converge if $p > 1$ and diverge if $p \leqslant 1$.

(a) Let $p > 1$.

$$S_n = \frac{1}{1^p} + \frac{1}{2^p} + \frac{1}{3^p} + \cdots + \frac{1}{n^p}.$$

Now

$$\frac{1}{2^p} + \frac{1}{3^p} < \frac{2}{2^p} = 2^{1-p},$$

$$\frac{1}{4^p} + \frac{1}{5^p} + \frac{1}{6^p} + \frac{1}{7^p} < \frac{4}{4^p} = 4^{1-p},$$

$$\frac{1}{8^p} + \frac{1}{9^p} + \cdots + \frac{1}{15^p} < \frac{8}{8^p} = 8^{1-p},$$

and so on. Hence

$$S_n < 1 + \frac{1}{2^{p-1}} + \frac{1}{4^{p-1}} + \frac{1}{8^{p-1}} + \cdots$$

The right-hand side of this inequality is a geometrical series whose common ratio $(\frac{1}{2})^{p-1}$ is less than unity when $p > 1$; it follows that

$$S_n < \frac{1}{1-(\frac{1}{2})^{p-1}} \text{ for all values of } n$$

and so by Theorem 5 the series converges.

(b) Let $p = 1$. The series is then $\Sigma(1/r)$, which has been shown in Section 22.1.3 to be divergent.

(c) Let $p < 1$.

When $p < 1$, $1/r^p > 1/r$ if $r > 1$.

Hence each term of $\Sigma(1/r^p)$ after the first exceeds the corresponding term of the divergent series $\Sigma(1/r)$ and so, by Test 3, $\Sigma(1/r^p)$ is divergent when $p < 1$.

Thus the series $\Sigma(1/r^p)$ converges when $p > 1$ and diverges when $p \leqslant 1$.

22.2.2 The Ratio Test for a Series of Positive Terms

Test 5 If Σu_n is a series of positive terms such that $\lim_{n\to\infty}(u_{n+1}/u_n) = p$, then Σu_n converges if $p < 1$ and diverges if $p > 1$. (This test gives no conclusive result if $p = 1$.)

(a) Let $p < 1$ and suppose that q is a number such that $p < q < 1$. Then since u_{n+1}/u_n can be made to differ from p by as small a quantity as we please by making n sufficiently large, we can find a number N such that when $n > N$, $u_{n+1}/u_n < q < 1$.

Then $u_{N+1} < qu_N$, $u_{N+2} < qu_{N+1} < q^2u_N$, and so on. It follows that

$$u_N + u_{N+1} + u_{N+2} + \cdots u_{N+K} < u_N(1 + q + q^2 + \cdots + q^K) < \frac{u_N}{1-q}$$

for all values of K however large, since $0 < q < 1$. Hence

$$u_1 + u_2 + \cdots + u_{N-1} + u_N + \cdots + u_{N+K}$$
$$< u_1 + u_2 + \cdots + u_{N-1} + \frac{u_N}{1-q}$$

for all values of K.

But the right-hand side is a positive number independent of K, and so Σu_r is convergent.

(b) Let $p > 1$. Then since $\lim_{n\to\infty} (u_{n+1}/u_n) = p$, $u_{n+1}/u_n > 1$ ultimately, and so u_n does not tend to zero. Hence Σu_r is divergent (Test 1).

Example 6 Test for convergence the series

(a) $\Sigma a^r/r!$ when $a > 0$, (b) $\Sigma\, 1.3.5 \ldots (2r - 1)/3.6.9 \ldots (3r)$.

(a)

$$\frac{u_{n+1}}{u_n} = \frac{a}{n+1} \to 0 \text{ as } n \to \infty.$$

Hence $\Sigma\, a^r/r!$ converges.

(b)

$$\frac{u_{n+1}}{u_n} = \frac{2n+1}{3(n+1)} = \frac{2 + 1/n}{3(1 + 1/n)} \to \tfrac{2}{3} \text{ as } n \to \infty.$$

Hence the series converges.

Example 7 Test for convergence the series $\Sigma(1/r^p)$.

$$\frac{u_{n+1}}{u_n} = \left(\frac{n}{n+1}\right)^p \to 1 \text{ as } n \to \infty.$$

Now we have shown in Section 22.2.1 that $\Sigma(1/r^p)$ is convergent when $p > 1$ and divergent when $p \leqslant 1$. It is therefore seen that if $\lim_{n\to\infty}(u_{n+1}/u_n) = 1$ for a given series we cannot, from this fact, draw any conclusion as to the convergence of series Σu_r.

22.2.3 A Test for Alternating Series An *alternating* series is one in which the terms are alternately positive and negative. Such a series may be written as

$$u_1 - u_2 + u_3 - u_4 + \cdots \qquad (1)$$

where $u_n > 0$ for all n.

Test 6 If $u_n > u_{n+1}$ for all n and if $u_n \to 0$ as $n \to \infty$, then the series (a) is convergent.

Proof. We consider the partial sum S_n firstly when n is even, say $n = 2m$, and then when n is odd, say $n = 2m + 1$. We have

$$S_{2m} = (u_1 - u_2) + (u_3 - u_4) + \cdots + (u_{2m-1} - u_{2m}),$$

and since the expression in each bracket is positive it follows that S_{2m} increases when m increases. Also

$$S_{2m} = u_1 - (u_2 - u_3) - \cdots - (u_{2m-2} - u_{2m-1}) - u_{2m}$$

and so $S_{2m} \leqslant u_1$. Hence (see Section 22.1.4) S_{2m} tends to a limit as

$m \to \infty$. Now

$$S_{2m+1} = S_{2m} + u_{2m+1}$$

and so

$$\lim_{m\to\infty} S_{2m+1} = \lim_{m\to\infty} S_{2m}, \quad \text{since} \quad \lim_{m\to\infty} u_{2m+1} = 0.$$

Thus S_{2m} and S_{2m+1} both tend to limits as $m \to \infty$ and these limits are equal. Hence S_n tends to a limit as $n \to \infty$ and the given series is convergent.

Example 8 (a) The alternating series $1 - \frac{1}{2} + \frac{1}{3} - \frac{1}{4} + \cdots$ is convergent, since $u_n > u_{n+1}$ and $u_n \to 0$ as $n \to \infty$.

(b) The alternating series $1{\cdot}1 - 1{\cdot}01 + 1{\cdot}001 - 1{\cdot}0001 + \cdots$ does not converge because $\lim_{n\to\infty} u_n \neq 0$ (Test 1). In this case

$$S_{2n} = \frac{1}{11}\left[1 - (0{\cdot}1)^{2n}\right] \to \frac{1}{11} \text{ as } n \to \infty,$$

$$S_{2n+1} = 1 + \frac{1}{11}\left[1 + (0{\cdot}1)^{2n+1}\right] \to \frac{12}{11} \text{ as } n \to \infty.$$

Hence the series oscillates finitely.

22.3 Absolute Convergence

If Σu_r contains positive and negative terms, then Σu_r is said to be *absolutely convergent* if $\Sigma |u_r|$ is convergent.

Theorem 6 A series that is absolutely convergent is also convergent.

To prove this result for a series Σu_r we construct two series of positive terms Σv_r and Σw_r by taking

$$v_r = u_r \text{ when } u_r \geqslant 0, \quad v_r = 0 \text{ when } u_r < 0;$$

$$w_r = -u_r \text{ when } u_r < 0, \quad w_r = 0 \text{ when } u_r \geqslant 0.$$

Then

$$|u_r| = v_r + w_r \tag{1}$$

and

$$u_r = v_r - w_r. \tag{2}$$

Since $v_r \geqslant 0$ and $w_r \geqslant 0$ we have from (1)

$$v_r \leqslant |u_r| \quad \text{and} \quad w_r \leqslant |u_r|.$$

But, by hypothesis, $\Sigma|u_r|$ is convergent; hence by Test 2 Σv_r and Σw_r are both convergent. It follows from (2) that Σu_r is convergent (Theorem 3).

From this important theorem it follows that if a series of positive terms is convergent and a new series is obtained by changing the signs of any of the terms of the given series, then the new series is convergent. For example, since the series

$$1 + \frac{1}{2^2} + \frac{1}{3^2} + \frac{1}{4^2} + \cdots$$

is convergent, so also is the series

$$1 - \frac{1}{2^2} - \frac{1}{3^2} + \frac{1}{4^2} - \frac{1}{5^2} - \frac{1}{6^2} + \cdots$$

or any other series obtained from it by changing only the signs.

If Σu_r is a convergent series of positive and negative terms such that $\Sigma|u_r|$ is divergent, we say that Σu_r is *conditionally convergent*. For example, the series $1 - \frac{1}{2} + \frac{1}{3} - \frac{1}{4} + \cdots$ is conditionally convergent since it is convergent (see Example 8) whereas the corresponding series of positive terms, viz. $1 + \frac{1}{2} + \frac{1}{3} + \cdots$ is divergent (see Section 22.1.3).

22.3.1 Tests for Absolute Convergence Any test applicable to series of positive terms can be used as a test for absolute convergence of the series Σu_r since $\Sigma|u_r|$ is a series of positive terms. For example, applying the ratio test we see that if $|u_{n+1}/u_n| \to p$ as $n \to \infty$, Σu_n is absolutely convergent if $p < 1$ and is not convergent if $p > 1$. For if $p < 1$, $\Sigma|u_r|$ is convergent while if $p > 1$, the terms of Σu_r ultimately increase numerically and so Σu_r cannot converge (Test 1). The ratio test is inconclusive when $p = 1$.

Example 9 Examine for absolute convergence the series

$$\text{(a)}\quad \Sigma(-1)^{n+1}\, n(n+1)/3^n \qquad \text{(b)}\quad \Sigma(-1)^{n+1}/\sqrt{(n^2+1)}.$$

(a) Here

$$|u_n| = n(n+1)/3^n, \quad |u_{n+1}| = (n+1)(n+2)/3^{n+1}$$

and

$$\left|\frac{u_{n+1}}{u_n}\right| = \tfrac{1}{3}(1 + 2/n) \to \tfrac{1}{3} \text{ when } n \to \infty.$$

Hence by the ratio test, Σu_n is absolutely convergent.

(b)

$$\left|\frac{u_{n+1}}{u_n}\right| = \frac{\sqrt{(n^2+1)}}{\sqrt{(n^2+2n+2)}} = \frac{\sqrt{(1+1/n^2)}}{\sqrt{(1+2/n+2/n^2)}} \to 1 \text{ when } n \to \infty.$$

Hence the ratio test fails.

Now Σu_n is an alternating series in which each term is numerically less than the preceding one and $u_n \to 0$ as $n \to \infty$. Hence Σu_n is convergent (see Section 22.2.3). Also $|u_n| \simeq 1/n$ when n is large and so we compare $\Sigma|u_n|$ with Σv_n where $v_n = 1/n$. Then the comparison test in limit form (Test 4) gives

$$\frac{|u_n|}{v_n} = \frac{n}{\sqrt{(n^2+1)}} \to 1 \text{ as } n \to \infty.$$

Hence $\Sigma|u_n|$ is divergent since Σv_n is divergent, and so Σu_n is conditionally convergent.

22.4 Power Series

A series of the form $a_0 + a_1x + a_2x^2 + a_3x^3 + \cdots$, where the coefficients $a_0, a_1, a_2, \ldots$ are independent of x, is called a *power series* in x. Such a series cannot diverge for all values of x since, when $x = 0$, it converges with sum a_0.

Associated with any power series there is a number R such that the series converges absolutely if $|x| < R$ but does not converge if $|x| > R$. R is called the *radius of convergence* of the series and the interval containing all the values of x for which the series converges is called the *interval of convergence* of the series. Except for its endpoints this interval

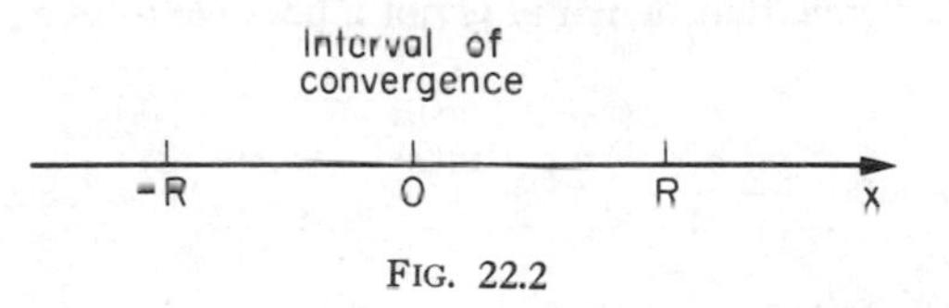

FIG. 22.2

may often be found by the ratio test. Other methods must be used to determine the behaviour of the series at the end-points of the interval.

Example 10 Find the values of x for which the exponential series

$$1 + \frac{x}{1!} + \frac{x^2}{2!} + \frac{x^3}{3!} + \cdots$$

is convergent.

The test ratio is $|u_{n+1}/u_n| = |x|/n$. Thus $\lim_{n\to\infty} |u_{n+1}/u_n| = 0$ for any finite value of x, and so the exponential series is convergent for any finite value of x. The interval of convergence is $(-\infty, \infty)$.

Example 11 Find the radius and interval of convergence of the logarithmic series

$$x - \frac{x^2}{2} + \frac{x^3}{3} - \frac{x^4}{4} + \cdots$$

In this case

$$\left|\frac{u_{n+1}}{u_n}\right| = \left(\frac{n}{n+1}\right)|x| = \left(1 - \frac{1}{n+1}\right)|x|,$$

and so

$$\lim_{n\to\infty} \left|\frac{u_{n+1}}{u_n}\right| = |x|.$$

Hence the logarithmic series converges when $|x| < 1$ and does not converge when $|x| > 1$. The radius of convergence is 1 and values of x in the range $-1 < x < 1$ belong to the interval of convergence. The end-points of this interval, i.e. $x = \pm 1$, must be tested individually.

When $x = 1$ the series is $1 - \frac{1}{2} + \frac{1}{3} - \frac{1}{4} + \cdots$ which converges (see Example 8).

When $x = -1$ the series is $-(1 + \frac{1}{2} + \frac{1}{3} + \frac{1}{4} + \cdots)$ which diverges (see Section 22.1.3).

Hence the interval of convergence of the series is $-1 < x \leqslant 1$.

Example 12 Prove that, when m is not a positive integer, the binomial series

$$1 + mx + \frac{m(m-1)}{2!}x^2 + \frac{m(m-1)(m-2)}{3!}x^3 + \cdots$$

converges absolutely if $-1 < x < 1$.

The test ratio of this series for $n > 1$ has the form

$$\frac{u_{n+1}}{u_n} = \frac{m(m-1)(m-2)\cdots(m-n+1)}{n!}x^n$$

$$\div \frac{m(m-1)(m-2)\cdots(m-n+2)}{(n-1)!}x^{n-1}$$

that is

$$\frac{u_{n+1}}{u_n} = \frac{m-n+1}{n}x.$$

Hence

$$\lim_{n\to\infty}\left|\frac{u_{n+1}}{u_n}\right| = |x|,$$

and so by the ratio test the binomial series converges absolutely when $|x| < 1$.

22.4.1 Properties of Power Series

Property 1 From Theorem 3 it follows that two power series may be added or subtracted term by term for all values of x for which both series are convergent. Hence if $\Sigma a_n x^n = f(x)$ when $|x| < R_1$ and $\Sigma b_n x^n = g(x)$ when $|x| < R_2$, $R_1 < R_2$, then

$$\Sigma(a_n + b_n)x^n = f(x) + g(x) \text{ when } |x| < R_1.$$

Property 2 It can also be shown that the above series may be multiplied together like polynomials for any value of x for which both series are absolutely convergent (and thus for every value of x within the smaller of their intervals of convergence), i.e.

$$f(x)\,g(x) = a_0 b_0 + (a_0 b_1 + a_1 b_0)x + (a_0 b_2 + a_1 b_1 + a_2 b_0)x^2 + \cdots.$$

We state without proof further properties of power series:

Property 3 If two power series have the same sum over some interval $|x| < R$, the coefficients of corresponding powers of x in these series are identical.

Property 4 The sum of a power series is a continuous function in the interval of convergence of the series. It follows that, if the radius of convergence is not zero

$$\lim_{x\to 0}(a_0 + a_1 x + a_2 x^2 + \cdots) = a_0.$$

Property 5 If $\Sigma_{n=0}^{\infty} a_n x^n = f(x)$ with radius of convergence $R > 0$, then $f(x)$ is differentiable on the open interval $-R < x < R$ and

$$f'(x) = \sum_{n=1}^{\infty} n\, a_n x^{n-1} \qquad (-R < x < R).$$

Similarly

$$\sum_{n=0}^{\infty}\frac{1}{n+1}\, a_n x^{n+1} = \int_0^x f(t)\,\mathrm{d}t \qquad (-R < x < R).$$

This property is usually described by saying that the series may be differentiated or integrated *term by term*.

The power series $1 + x + x^2 + \cdots$ is a G.P.; its interval of convergence is the interval $-1 < x < 1$ and within that interval it has the sum $1/(1-x)$. Hence we write

$$1 + x + x^2 + \cdots = \frac{1}{1-x} \qquad (-1 < x < 1). \tag{1}$$

Similarly

$$1 + 2x + 4x^2 + \cdots = \frac{1}{1-2x} \qquad (-\tfrac{1}{2} < x < \tfrac{1}{2}). \tag{2}$$

By Property 3 the power series in (1) is the *only* power series in x having the sum $1/(1-x)$; likewise the power series in (2) is the only power series in x having the sum $1/(1-2x)$.

By Property 1 we may add (1) and (2) to find that the power series $2 + 3x + 5x^2 + \cdots$ has sum $1/(1-x) + 1/(1-2x)$ within the interval $-\frac{1}{2} < x < \frac{1}{2}$.

By Property 2 we may multiply (1) and (2) to find that the power series $1 + 3x + 7x^2 + \cdots$ has sum $1/(1-x)(1-2x)$ within the interval $-\frac{1}{2} < x < \frac{1}{2}$.

By Property 5 we may differentiate (1) to find that the power series $1 + 2x + 3x^2 + \cdots$ has sum $1/(1-x)^2$ within the interval $-1 < x < 1$; also we may integrate (1) between the limits 0 and x to find that the power series $x + \frac{1}{2}x^2 + \frac{1}{3}x^3 + \cdots$ has sum $-\ln(1-x)$ within the interval $-1 < x < 1$.

Exercises 22(a)

1. Discuss the convergence of the series whose nth term is given by

(a) $\dfrac{2^n}{(n+1)(n+3)}$, (b) $\dfrac{1}{n(n+2)3^n}$,

(c) $\dfrac{1}{2n(2n+1)}$, (d) $\dfrac{2n+1}{2.4.\cdots(2n+2)}$,

(e) $\dfrac{1+3n^2}{1+n^2}$, (f) $\dfrac{1-n+n^2}{n!}$,

(g) $\dfrac{n}{\sqrt{(4n^3+1)}}$, (h) $\sqrt{\left(\dfrac{n}{4n^4+7}\right)}$,

(i) $\dfrac{n^p}{n!}$ (p constant), (j) $\dfrac{1}{\sqrt{(n+2)}-\sqrt{n}}$,

(k) $\dfrac{\sqrt{(n^2+n+1)}+\sqrt{(n^2-n+1)}}{n}$, (l) $\dfrac{\sqrt{(n+1)}-\sqrt{n}}{n^2}$.

2. Determine for what values of x the following series are (a) absolutely convergent, (b) conditionally convergent:

(i) $\sum \frac{r}{(r+1)(r+2)} x^r$, (ii) $\sum \frac{r!}{(2r)!} x^r$,

(iii) $\sum \frac{x^r}{(r^2+1)^{1/3}}$, (iv) $\sum \frac{r^{10}x^r}{r!}$,

(v) $\sum \frac{(-1)^r x^{2r+1}}{2r+1}$.

3. Show that $\Sigma_{r=1}^{\infty} (\cos rx/r^2)$ is convergent.
4. Prove that the following alternating series are convergent:

(a) $1 - \frac{1}{2} + \frac{1}{4} - \frac{1}{8} + \cdots$,

(b) $\frac{1}{3^2} - \frac{2}{3^3} + \frac{3}{3^4} - \frac{4}{3^5} + \cdots$,

(c) $\frac{3}{2} - \frac{4}{3}\left(\frac{1}{2}\right) + \frac{5}{4}\left(\frac{1}{2^2}\right) - \frac{6}{5}\left(\frac{1}{2^3}\right) + \cdots$,

(d) $\frac{1}{3} - \frac{1.2}{3.5} + \frac{1.2.3}{3.5.7} - \frac{1.2.3.4}{3.5.7.9} + \cdots$,

(e) $\frac{1}{3} - \frac{1.3}{3.6} + \frac{1.3.5}{3.6.9} - \frac{1.3.5.7}{3.6.9.12} + \cdots$.

5. Show that the following series are conditionally convergent:

(a) $\frac{2}{1.3} - \frac{3}{2.4} + \frac{4}{3.5} - \frac{5}{4.6} + \cdots$,

(b) $1 - \frac{1}{\sqrt{2}} + \frac{1}{\sqrt{3}} - \frac{1}{\sqrt{4}} + \cdots$,

(c) $\frac{1}{2} - \frac{2}{5} + \frac{3}{10} - \frac{4}{17} + \cdots$.

6. Show that the series

$$1 + \frac{1}{\sqrt{2}} + \frac{1}{\sqrt{3}} + \cdots + \frac{1}{\sqrt{n}}$$

is divergent, and prove that if $\Sigma_{n=1}^{\infty} u_n$ is absolutely convergent so also is $\Sigma_{n=1}^{\infty} u_n^2$.

7. Determine the interval of convergence of the series $\Sigma_{r=1}^{\infty} [x^r/r(r+1)]$.

22.5 Expansions in Series

There are relatively few cases where we are able to find the sum of a given power series. However, given $f(x)$ we are normally able (see Section 22.5.1) to find a power series for which $f(x)$ is the sum, say

$$f(x) = a_0 + a_1x + a_2x^2 + \cdots \quad (x \text{ in } I), \tag{1}$$

where I is the interval of convergence of the series.

We refer to the series as the *expansion* of $f(x)$ as a power series in x, and by Property 3 of Section 22.4.1 we see that this expansion is unique.

One of the main advantages in having an expansion for $f(x)$ is that it enables us to approximate $f(x)$ by a polynomial; for, by the definition of the sum of an infinite series it follows from (1) that, when n is large,

$$f(x) \simeq a_0 + a_1x + a_2x^2 + \cdots + a_nx^n.$$

This approximation is of limited value unless we are able to estimate the error involved, and for this reason the expansion (1) is sometimes given in the form

$$f(x) = a_0 + a_1x + a_2x^2 + \cdots + a_nx^n + R_n(x), \tag{1a}$$

together with a statement of a formula for $R_n(x)$, the sum of the terms following the term in x^n. We describe (1a) as the expansion of $f(x)$ as a power series in x *with remainder*. For example, we may write

$$\frac{1}{1-x} = 1 + x + x^2 + \cdots + x^n + R_n(x)$$

where $R_n(x) = x^{n+1} + x^{n+2} + \cdots = x^{n+1}/(1-x)$.

22.5.1 Maclaurin's Series By Property 5 of Section 22.4.1, if

$$f(x) = a_0 + a_1x + a_2x^2 + \cdots + a_nx^n + \cdots \qquad (-R < x < R)$$

then

$$f'(x) = a_1 + 2a_2x + 3a_3x^2 + \cdots + na_nx^{n-1} + \cdots, \qquad (-R < x < R)$$
$$f''(x) = 2a_2 + 2.3a_3x + \cdots + (n-1)na_nx^{n-2} + \cdots, \qquad (-R < x < R)$$
$$f'''(x) = 2.3a_3 + 2.3.4a_4x + \cdots + (n-2)(n-1)na_nx^{n-3} + \cdots, \qquad (-R < x < R)$$

and so on. Substituting $x = 0$ in these results in turn, we get

$$a_0 = f(0),\ a_1 = f'(0),\ a_2 = \frac{f''(0)}{2!},\quad a_3 = \frac{f'''(0)}{3!}, \cdots, a_n = \frac{f^{(n)}(0)}{n!}.$$

Hence the only possible expansion of $f(x)$ as a power series in x is the expansion

$$f(x) = f(0) + xf'(0) + \frac{x^2}{2!}f''(0) + \cdots + \frac{x^n}{n!}f^{(n)}(0) + \cdots \quad (x \text{ in } I) \tag{22.1}$$

where I is the interval of convergence. It is beyond the scope of this book to discuss the conditions on the function f under which $f(x)$ can be expanded as a power series in x. We shall assume that $f(x)$ can be so expanded in some interval I if the derivatives of f of all orders exist at $x = 0$, and the expansion is then given by (22.1). The series in (22.1) is called the Maclaurin series for $f(x)$.

We may also expand $f(x)$ as a series with remainder as follows:

$$f(x) = f(0) + xf'(0) + \frac{x^2}{2!}f''(0) + \cdots + \frac{x^n}{n!}f^{(n)}(0) + R_n(x), \tag{22.2}$$

where

$$R_n(x) = \frac{x^{n+1}}{(n+1)!}f^{(n+1)}(\alpha)$$

and α lies between 0 and x. We shall not discuss the derivation of this formula for $R_n(x)$, but it should be noted that it is incomplete in the sense that α is unknown; usually the information that α lies between 0 and x is adequate for the purpose of estimating bounds for the value of $R_n(x)$.

22.5.2 Series for sin x and cos x Putting $f(x) = \sin x$ we obtain

$$f'(x) = \cos x,\ f''(x) = -\sin x, \ldots, f^{(n)}(x) = \sin (x + \tfrac{1}{2}n\pi),$$

and so

$$f(0) = 0,\ f'(0) = 1,\ f''(0) = 0, \ldots, f^{(n)}(0) = \sin \tfrac{1}{2}n\pi.$$

Hence, by (22.1),

$$\sin x = x - \frac{x^3}{3!} + \frac{x^5}{5!} - \frac{x^7}{7!} + \cdots \tag{22.3}$$

and similarly

$$\cos x = 1 - \frac{x^2}{2!} + \frac{x^4}{4!} - \frac{x^6}{6!} + \cdots \tag{22.4}$$

It is easily verified by the ratio test that both series are convergent for all values of x, and so we assume that the expansions are valid for all values of x.

Example 13 Evaluate the sine of 0·1 radians approximately, using the first two terms of the series for sin x, and estimate the error involved.

We have $\sin (0{\cdot}1) \simeq 0{\cdot}1 - (0{\cdot}1)^3/3! \simeq 0{\cdot}1 - 0{\cdot}00017 = 0{\cdot}09983$. In the notation of (22.2) the error is $R_3(0{\cdot}1)$ where

$$R_3(0{\cdot}1) = \frac{(0{\cdot}1)^4}{4!} \sin \alpha, \quad 0 < \alpha < 0{\cdot}1.$$

Whatever the value of α we must have $|\sin \alpha| < 1$, and so

$$R_3(0{\cdot}1) < \frac{0{\cdot}0001}{24} < 0{\cdot}000005.$$

Example 14 Use series expansions to find

(a) $\lim_{x \to 0} \dfrac{\sin x}{x}$, (b) $\lim_{x \to 0} \dfrac{1 - \cos x}{x^2}$.

(a) $$\frac{\sin x}{x} = \frac{x - \frac{1}{6}x^3 + \cdots}{x} = 1 - \tfrac{1}{6}x^2 + \cdots$$

$$\to 1 \text{ as } x \to 0 \qquad \text{by Property 4, Section 22.4.1.}$$

(b) $$\frac{1 - \cos x}{x^2} = \frac{1 - (1 - \frac{1}{2}x^2 + \frac{1}{24}x^4 - \cdots)}{x^2}$$

$$= \tfrac{1}{2} - \tfrac{1}{24}x^2 + \cdots \to \tfrac{1}{2} \text{ as } x \to 0.$$

22.5.3 Series for e^x Putting $f(x) = e^x$ we obtain $f^{(n)}(x) = e^x$ for all n, giving $f(0) = 1, f^{(n)}(0) = 1$ for $n = 1, 2, 3, \ldots$. Hence, by (22.1),

$$e^x = 1 + x + \frac{x^2}{2!} + \frac{x^3}{3!} + \cdots + \frac{x^n}{n!} + \cdots \tag{22.5}$$

the expansion being valid for all x (see Example 10).

22.5.4 Binomial Expansion Putting $f(x) = (1 + x)^k$, where k is any rational number, we obtain

$$f'(x) = k(1 + x)^{k-1}, \quad f''(x) = k(k - 1)(1 + x)^{k-2}, \ldots$$
$$f^{(n)}(x) = k(k - 1) \cdots (k - n + 1)(1 + x)^{k-n},$$

and so

$$f^{(n)}(0) = k(k - 1) \cdots (k - n + 1).$$

Hence the Maclaurin series for $(1 + x)^k$ is

$$1 + kx + \frac{k(k-1)}{2!}x^2 + \frac{k(k-1)(k-2)}{3!}x^3 + \cdots + \frac{k(k-1)\cdots(k-n+1)}{n!}x^n + \cdots.$$

If k is a positive integer the series is finite, e.g. if $k = 2$ there are only three terms since all the remaining terms contain $k - 2$, which is zero; furthermore this finite series is simply the binomial expansion we met in Section 4.1.

If k is not a positive integer, the series is infinite and converges when $|x| < 1$ (see Example 12); hence in this case

$$(1 + x)^k = 1 + kx + \frac{k(k-1)}{2!} + \cdots + \frac{k(k-1)\cdots(k-n+1)}{n!}x^n + \cdots \quad (-1 < x < 1) \tag{22.6}$$

For example

$$\frac{1}{1-x} = [1+(-x)]^{-1} = 1 + (-1)(-x) + \frac{(-1)(-2)}{2!}(-x)^2 + \cdots$$

$$(|-x| < 1)$$

$$= 1 + x + x^2 + x^3 + \cdots \qquad (|x| < 1)$$

and similarly

$$\frac{1}{1-2x} = [1+(-2x)]^{-1} = 1 + 2x + 4x^2 + 8x^3 + \cdots \qquad (|x| < \tfrac{1}{2}).$$

(These two results were already known by virtue of the fact that the series are G.P.'s.)

Also, when $|x| < 1$,

$$(1+x)^{1/2} = 1 + \tfrac{1}{2}x - \frac{1}{2.4}x^2 + \frac{1.3}{2.4.6}x^3 - \cdots,$$

$$(1-x)^{-1/2} = 1 + \tfrac{1}{2}x + \frac{1.3}{2.4}x^2 + \frac{1.3.5}{2.4.6}x^3 + \cdots.$$

When $|b| < |a|$,

$$(a-b)^{-2} = [a(1-b/a)]^{-2} = \frac{1}{a^2}[1 + 2b/a + 3b^2/a^2 + 4b^3/a^3 + \cdots]$$

When $|a| < |b|$,

$$(a+b)^{-3} = [b(1+a/b)]^{-3} = \frac{1}{b^3}[1 - 3a/b + \frac{3.4}{2}a^2/b^2 - \frac{4.5}{2}a^3/b^3 + \cdots].$$

Example 15 Given $f(x) = 25x/(1-x)^2(1-6x)$, express $f(x)$ in partial fractions and hence expand $f(x)$ as a power series in x, stating the range of values of x for which the expansion is valid.

By the methods of Section 4.3.1 we find

$$f(x) = \frac{6}{1-6x} - \frac{1}{1-x} - \frac{5}{(1-x)^2}.$$

When $|x| < \frac{1}{6}$,

$$\frac{1}{1-6x} = 1 + 6x + (6x)^2 + \cdots + (6x)^n + \cdots.$$

When $|x| < 1$,

$$\frac{1}{1-x} = 1 + x + x^2 + \cdots + x^n + \cdots.$$

When $|x| < 1$,

$$\frac{1}{(1-x)^2} = 1 + 2x + 3x^2 + \cdots + (n+1)x^n + \cdots.$$

Hence, when $|x| < \frac{1}{6}$,

$$f(x) = 25x + 200x^2 + \cdots + [6^{n+1} - 5(n+1) - 1]\,x^n + \cdots.$$

Example 16 Evaluate $\lim_{x\to\infty}[x\sqrt{(x^2+a^2)} - \sqrt{(x^4+a^4)}]$.

In order to find the limit as $x \to \infty$ we expand the given expression as a power series in $(1/x)$. Now

$$\begin{aligned} x\sqrt{(x^2+a^2)} &= x^2(1 + a^2/x^2)^{1/2} \\ &= x^2\left(1 + \frac{1}{2}\frac{a^2}{x^2} - \frac{1}{8}\frac{a^4}{x^4} + \cdots\right) \quad \text{when } x^2 > a^2. \end{aligned}$$

Also

$$\begin{aligned} \sqrt{(x^4+a^4)} &= x^2\left(1 + \frac{a^4}{x^4}\right)^{1/2} \\ &= x^2\left(1 + \frac{1}{2}\frac{a^4}{x^4} + \cdots\right) \quad \text{when } x^2 > a^2. \end{aligned}$$

Hence

$$x\sqrt{(x^2+a^2)} - \sqrt{(x^4+a^4)} = \tfrac{1}{2}a^2 - \frac{5}{8}\frac{a^4}{x^2} + \cdots \to \tfrac{1}{2}a^2 \text{ as } x \to \infty.$$

22.5.5 Series for ln (1 + x) Note that if we put $f(x) = \ln x$, then we find $f'(x) = 1/x$, $f''(x) = -1/x^2$, etc. and it is clear that the derivatives of $\ln x$ do not exist at $x = 0$. Hence $\ln x$ cannot be expanded as a power series in x. However, if we put $f(x) = \ln(1+x)$, then

$$\begin{aligned} f'(x) = 1/(1+x),\ f''(x) &= -1/(1+x)^2, \ldots, \\ f^{(n)}(x) &= (-1)^{n-1}(n-1)!/(1+x)^n, \end{aligned}$$

and the Maclaurin series for $\ln(1+x)$ is

$$x - \frac{x^2}{2} + \frac{x^3}{3} - \frac{x^4}{4} + \cdots + (-1)^{n-1}\frac{x^n}{n} + \cdots.$$

We have seen (Example 11) that this series is convergent for $-1 < x \leqslant 1$. Thus

$$\ln(1+x) = x - \frac{x^2}{2} + \frac{x^3}{3} - \frac{x^4}{4} + \cdots \qquad (-1 < x \leqslant 1). \qquad (22.7)$$

If we replace x by $(-x)$ in (22.7) we obtain

$$\ln(1-x) = -\left(x + \frac{x^2}{2} + \frac{x^3}{3} + \frac{x^4}{4} + \cdots\right) \qquad (-1 \leqslant x < 1).$$

Many other expansions can be obtained from (22.7) by using the properties of the logarithmic function, e.g.

$$\ln(1+x)^3 = 3\ln(1+x) - 3\left(x - \frac{x^2}{2} + \cdots\right) \qquad (-1 < x \leqslant 1)$$

and

$$\ln\left(\frac{1+x}{1-x}\right) = \ln(1+x) - \ln(1-x) = 2\left(x + \frac{x^3}{3} + \frac{x^5}{5} + \cdots\right) \ (-1 < x < 1).$$

Example 17 Evaluate $\ln \frac{3}{2}$ by using four terms in the expansion of $\ln(1+x)$, and estimate the error involved.

From (22.2) we have

$$\ln(1+x) = x - \frac{x^2}{2} + \frac{x^3}{3} - \frac{x^4}{4} + R_4(x),$$

where

$$R_4(x) = \frac{x^5}{5!}\frac{4!}{(1+\alpha)^5}, \text{ and } \alpha \text{ lies between 0 and } x.$$

Hence

$$\ln \tfrac{3}{2} = \tfrac{1}{2} - \tfrac{1}{8} + \tfrac{1}{24} - \tfrac{1}{64} + R_4(\tfrac{1}{2}),$$

where

$$R_4(\tfrac{1}{2}) = \frac{1}{160}\frac{1}{(1+\alpha)^5}, \qquad 0 < \alpha < \tfrac{1}{2}.$$

Whatever the value of α in this interval it is clear that $R_4(\frac{1}{2}) < 1/160$. Thus $\ln(3/2) \simeq 77/192$, with an error not exceeding $1/160$.

22.5.6 Series for sinh x and cosh x It is easily verified that, for all values of x,

$$\sinh x = x + \frac{x^3}{3!} + \frac{x^5}{5!} + \cdots, \tag{22.8}$$

$$\cosh x = 1 + \frac{x^2}{2!} + \frac{x^4}{4!} + \cdots. \tag{22.9}$$

Note that these results may be obtained from the definitions of sinh x and cosh x (Section 17.3), by using (22.5).

22.5.7 Use of Implicit Differentiation in Finding Expansions In order to find the Maclaurin series for $f(x)$ it is not essential to find explicit expressions for $f'(x), f''(x)$, etc. The following example illustrates the fact that much labour may be avoided by the use of implicit differentiation.

Example 18 Find the Maclaurin series for $\sin^{-1} x/\sqrt{(1-x^2)}$, as far as the term in x^5.

Write $y = \sin^{-1} x/(1-x^2)^{1/2}$, so that $(1-x^2)^{1/2}y = \sin^{-1} x$. By differentiating with respect to x we can show that $(1-x^2)y' - xy = 1$. By differentiating repeatedly with respect to x, and writing $y_r = \mathrm{d}^r y/\mathrm{d}x^r$, we obtain

$$\begin{aligned}
(1-x^2)y_2 - 3xy_1 - y &= 0,\\
(1-x^2)y_3 - 5xy_2 - 4y_1 &= 0,\\
(1-x^2)y_4 - 7xy_3 - 9y_2 &= 0,\\
(1-x^2)y_5 - 9xy_4 - 16y_3 &= 0,
\end{aligned}$$

and so on. We now see that when $x = 0$ we have $y = 0, y_1 = 1, y_2 = 0, y_3 = 4, y_4 = 0, y_5 = 64$ and hence that the Maclaurin series for y is

$$x + \frac{x^3}{3!}4 + \frac{x^5}{5!}64 + \cdots,$$

that is

$$x + \tfrac{2}{3}x^3 + \tfrac{8}{15}x^5 + \cdots.$$

22.5.8 Other Methods of Obtaining Expansions The expansions (22.3)–(22.9) are of basic importance and may be used to derive further expansions, as illustrated in the following examples.

Example 19 Find a series expansion for $e^x \sin x$, as far as the term in x^5.

We may formally multiply the series in (22.3) and (22.5) to obtain

$$e^x \sin x = x + x^2 + \tfrac{1}{3}x^3 - \tfrac{1}{30}x^5 + \cdots.$$

This is valid for all values of x since (22.3) and (22.5) are both valid for all x.

Example 20 Find a series expansion for $\tan x$, as far as the term in x^5.

We have

$$\tan x = \frac{\sin x}{\cos x} = \sin x(\cos x)^{-1}.$$

Now

$$(\cos x)^{-1} = (1 - \tfrac{1}{2}x^2 + \tfrac{1}{24}x^4 - \cdots)^{-1} = [1 - (\tfrac{1}{2}x^2 - \tfrac{1}{24}x^4 + \cdots)]^{-1}$$
$$= 1 + (\tfrac{1}{2}x^2 - \tfrac{1}{24}x^4 + \cdots) + (\tfrac{1}{2}x^2 - \tfrac{1}{24}x^4 + \cdots)^2 + \cdots \quad \text{by (22.6)}$$
$$= 1 + (\tfrac{1}{2}x^2 - \tfrac{1}{24}x^4 + \cdots) + \tfrac{1}{4}x^4 + \cdots$$
$$= 1 + \tfrac{1}{2}x^2 + \tfrac{5}{24}x^4 + \cdots.$$

This expansion is valid if $|\tfrac{1}{2}x^2 - \tfrac{1}{24}x^4 + \cdots| < 1$, i.e. if $|1 - \cos x| < 1$, which is true for values of x such that $\cos x > 0$.

We now have

$$\tan x = \sin x\,(1 + \tfrac{1}{2}x^2 + \tfrac{5}{24}x^4 + \cdots)$$
$$= (x - \tfrac{1}{6}x^3 + \tfrac{1}{120}x^5 - \cdots)(1 + \tfrac{1}{2}x^2 + \tfrac{5}{24}x^4 + \cdots)$$
$$= x + \tfrac{1}{3}x^3 + \tfrac{2}{15}x^5 + \cdots.$$

Example 21 Find a series expansion for $\sqrt{(3 + \sec^2 x)}$, as far as the term in x^4.

We have, when $\cos x > 0$,

$$\sec x = (\cos x)^{-1} = 1 + \tfrac{1}{2}x^2 + \tfrac{5}{24}x^4 + \cdots \qquad \text{(see Example 20).}$$

Hence

$$3 + \sec^2 x = 3 + (1 + \tfrac{1}{2}x^2 + \tfrac{5}{24}x^4 + \cdots)^2$$
$$= 3 + 1 + 2(\tfrac{1}{2}x^2 + \tfrac{5}{24}x^4 + \cdots) + (\tfrac{1}{2}x^2 + \tfrac{5}{24}x^4 + \cdots)^2$$
$$= 4 + x^2 + \tfrac{2}{3}x^4 \cdots.$$

Finally

$$(3 + \sec^2 x)^{1/2} = [4(1 + \frac{x^2}{4} + \frac{x^4}{6} + \cdots)]^{1/2}$$
$$= 2[1 + \tfrac{1}{2}(\tfrac{1}{4}x^2 + \tfrac{1}{6}x^4 + \cdots) + \frac{\tfrac{1}{2}(-\tfrac{1}{2})}{2!}(\tfrac{1}{4}x^2 + \cdots)^2 + \cdots]$$
$$= 2[1 + \tfrac{1}{8}x^2 + \tfrac{1}{12}x^4 + \cdots - \tfrac{1}{8}(\tfrac{1}{16}x^4 + \cdots) + \cdots]$$
$$= 2 + \tfrac{1}{4}x^2 - \tfrac{29}{192}x^4 + \cdots.$$

For convergence we require, in addition to $\cos x > 0$, that

$$\frac{x^2}{4} + \frac{x^4}{6} + \cdots < 1, \quad \text{that is} \quad \tfrac{1}{4}(\sec^2 x - 1) < 1,$$

and so the expansion is valid when $\cos x > 1/\sqrt{5}$.

Example 22 Find the first four terms of the series for $\ln(1 + \sin x)$ in powers of x.

By (22.7),

$$\ln(1 + \sin x) = \sin x - \tfrac{1}{2}\sin^2 x + \tfrac{1}{3}\sin^3 x - \tfrac{1}{4}\sin^4 x + \cdots,$$

when $-1 < \sin x \leqslant 1$. Also,

$$\sin x = x - \frac{x^3}{3!} + \frac{x^5}{5!} - \cdots = x\left(1 - \frac{x^2}{6} + \frac{x^4}{120} - \cdots\right)$$

for all values of x. Hence, if x is not an odd multiple of $\pi/2$,

$$\begin{aligned}\ln(1 + \sin x) &= x\left(1 - \frac{x^2}{6} + \frac{x^4}{120} - \cdots\right) - \tfrac{1}{2}x^2\left(1 - \frac{x^2}{6} + \cdots\right)^2 \\ &\quad + \tfrac{1}{3}x^3\left(1 - \frac{x^2}{6} + \cdots\right)^3 - \tfrac{1}{4}x^4\left(1 - \frac{x^2}{6} + \cdots\right)^4 + \cdots \\ &= \left(x - \frac{x^3}{6} + \cdots\right) - \left(\tfrac{1}{2}x^2 - \tfrac{1}{6}x^4 + \cdots\right) + \left(\tfrac{1}{3}x^3 - \cdots\right) \\ &\quad - \left(\tfrac{1}{4}x^4 + \cdots\right) + \cdots \\ &= x - \frac{x^2}{2} + \frac{x^3}{6} - \frac{x^4}{12} + \cdots.\end{aligned}$$

Example 23 By integrating the appropriate binomial expansions show that, when x is small,

$$\tan^{-1} x = x - \frac{x^3}{3} + \frac{x^5}{5} - \frac{x^7}{7} + \cdots,$$

$$\sin^{-1} x = x + \frac{1}{2}\frac{x^3}{3} + \frac{1.3}{2.4}\frac{x^5}{5} + \frac{1.3.5}{2.4.6}\frac{x^7}{7} + \cdots.$$

Evaluate

$$\lim_{x \to 0} \frac{2\sin^{-1} x + \tan^{-1} x - 3x(1 + x^4)^{1/5}}{x^5}.$$

We have

$$\begin{aligned}\frac{d}{dt}(\tan^{-1} t) &= \frac{1}{1 + t^2} \\ &= 1 - t^2 + t^4 - t^6 + \cdots \qquad \text{when } |t| < 1\end{aligned}$$

by the binomial theorem.

On integrating both sides of this equation between the limits 0 and x, where x is small, we have by Property 5, Section 22.4.1,

$$\tan^{-1} x = x - \frac{x^3}{3} + \frac{x^5}{5} - \frac{x^7}{7} + \cdots. \tag{1}$$

This series, known as Gregory's series for $\tan^{-1} x$, can be shown, by using the ratio test and the test for the convergence of an alternating series, to converge when $-1 \leqslant x \leqslant 1$.

Similarly

$$\frac{d}{dt}(\sin^{-1} t) = \frac{1}{\sqrt{(1-t^2)}} = 1 + \tfrac{1}{2}t^2 + \frac{1.3}{2.4}t^4 + \frac{1.3.5}{2.4.6}t^6 + \cdots,$$

when $|t| < 1$.

On integrating between the limits 0 and x, where x is small, we have

$$\sin^{-1} x = x + \frac{1}{2}\frac{x^3}{3} + \frac{1.3}{2.4}\frac{x^5}{5} + \frac{1.3.5}{2.4.6}\frac{x^7}{7} + \cdots \quad (2)$$

Substitution from (1) and (2) in

$$2 \sin^{-1} x + \tan^{-1} x - 3x(1 + x^4)^{1/5}$$

gives

$$2(x + \tfrac{1}{6}x^3 + \tfrac{3}{40}x^5 + \cdots) + (x - \tfrac{1}{3}x^3 + \tfrac{1}{5}x^5 - \cdots) - 3x(1 + \tfrac{1}{5}x^4 + \cdots)$$
$$= -\tfrac{1}{4}x^5 + \text{terms involving higher powers of } x.$$

Hence by Property 4, Section 22.4.1,

$$\lim_{x \to 0} \frac{2 \sin^{-1} x + \tan^{-1} x - 3x(1 + x^4)^{1/5}}{x^5} = -\tfrac{1}{4}.$$

22.5.9 Taylor's Series It is sometimes advantageous to expand $f(x)$ as a power series in $(x - a)$, where a is a constant, say

$$f(x) = b_0 + b_1(x - a) + b_2(x - a)^2 + \cdots + b_n(x - a)^n + \cdots. \quad (1)$$

For, if the radius of convergence of this power series is R, then the series is convergent in the interval $|x - a| < R$, i.e. $a - R < x < a + R$; thus

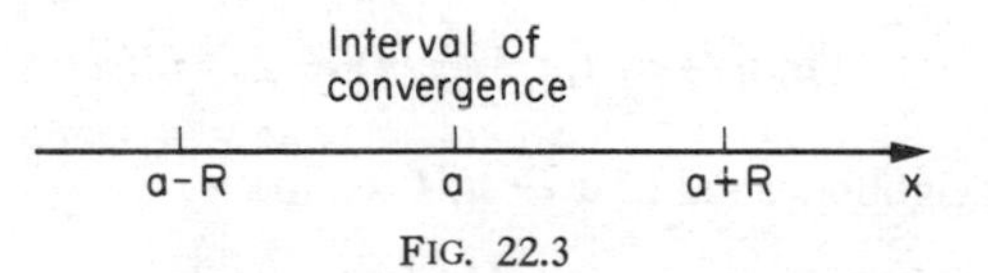

FIG. 22.3

when x is close to a this series may be convergent whereas the expansion of $f(x)$ in powers of x may, for values of x close to a, be divergent or very slowly convergent (needing a lot of terms for a good approximation to $f(x)$). By successive differentiation of (1), as in Section 22.5.1, and putting $x = a$ in each of the derived series, we find

$$b_0 = f(a),\ b_1 = f'(a),\ b_2 = \frac{f''(a)}{2!}, \ldots, b_n = \frac{f^{(n)}(a)}{n!}, \ldots.$$

Hence, if $f(x)$ has an expansion of the type (1), it has the form

$$f(x) = f(a) + (x-a)f'(a) + \frac{(x-a)^2}{2!}f''(a) + \cdots + \frac{(x-a)^n}{n!}f^{(n)}(a) + \cdots \quad (x \text{ in } I) \tag{22.10}$$

where I is the interval of convergence. This series is called the Taylor series for $f(x)$, or the expansion of $f(x)$ *in the neighbourhood of* $x = a$.

It can also be shown that

$$f(x) = f(a) + (x-a)f'(a) + \cdots + \frac{(x-a)^n}{n!}f^{(n)}(a) + R_n(x), \tag{22.11}$$

where

$$R_n(x) = \frac{(x-a)^{n+1}}{(n+1)!}f^{(n+1)}(\alpha),$$

α lying between a and x.

Maclaurin's series may be obtained as a special case of Taylor's series by putting $a = 0$, and is sometimes called the expansion of $f(x)$ in the neighbourhood of $x = 0$.

Example 24 Use the first three terms in the expansion of tan x in the neighbourhood of $x = \pi/4$ to find the approximate value of tan 46°.

Put $f(x) = \tan x$. Then

$$f'(x) = \sec^2 x = 1 + \tan^2 x,$$

$$f''(x) = 2\tan x \sec^2 x = 2(\tan x + \tan^3 x),$$

and hence

$$f(\pi/4) = \tan \pi/4 = 1, \quad f'(\pi/4) = 2, \quad f''(\pi/4) = 4.$$

Thus in the neighbourhood of $x = \pi/4$ we have

$$\tan x = 1 + (x - \pi/4)2 + \frac{(x-\pi/4)^2}{2!}4 + \cdots. \tag{1}$$

Now $\pi/4$ rad $= 45°$ and so $46° = \pi/4 + \pi/180$ rad
$= \pi/4 + 0{\cdot}0175$ rad.

Hence, on putting $x = \pi/4 + 0{\cdot}0175$ in (1),

$$\tan 46° \simeq 1 + 2(0{\cdot}0175) + 2(0{\cdot}0175)^2 \simeq 1{\cdot}0356.$$

Exercises 22(b)

1. Write down the first four terms of series whose sums to infinity are respectively

$$\text{(a)}\quad e - 1/e, \qquad \text{(b)}\quad e^2 + 1/e^2, \qquad \text{(c)}\quad e^3.$$

Show that

$$\tfrac{1}{2}(1 + e)^2/e = 2 + \frac{1}{2!} + \frac{1}{4!} + \frac{1}{6!} + \cdots.$$

2. Express as power series in x giving the first four terms and the general term

$$\text{(a)}\quad e^{2x}, \qquad \text{(b)}\quad (1 + x)e^x, \qquad \text{(c)}\quad (1 - x^2)/e^x.$$

3. Expand e^{x+x^2} as a power series in x as far as the term in x^4.

4. Verify the following expansions:

(a) $\ln(1 + 2x) = 2x - 2x^2 + \frac{8}{3}x^3 - 4x^4 + \cdots \qquad (-\frac{1}{2} < x \leqslant \frac{1}{2})$

(b) $\ln(1 - \frac{1}{4}x) = -\left(\dfrac{x}{4} + \dfrac{x^2}{32} + \dfrac{x^3}{192} + \dfrac{x^4}{1024} + \cdots\right) \qquad (-4 \leqslant x < 4)$

(c) $\ln(3 + x) = \ln 3 + \dfrac{x}{3} - \dfrac{x^2}{18} + \dfrac{x^3}{81} - \dfrac{x^4}{324} + \cdots) \qquad (-3 < x \leqslant 3)$

(d) $\ln(2 - 3x) = \ln 2 - (\frac{3}{2}x + \frac{9}{8}x^2 + \frac{9}{8}x^3 + \frac{81}{64}x^4 + \cdots) \qquad (-\frac{2}{3} \leqslant x < \frac{2}{3})$

(e) $\ln(4 + x)^2 = 2\ln 4 + 2\left(\dfrac{x}{4} - \dfrac{x^2}{32} + \dfrac{x^3}{192} - \dfrac{x^4}{1024} + \cdots\right) \qquad (-4 < x \leqslant 4)$

(f) $\ln\sqrt{(1 - x - 2x^2)} = -\frac{1}{2}(x + \frac{5}{2}x^2 + \frac{7}{3}x^3 + \frac{17}{4}x^4 + \cdots) \qquad (-\frac{1}{2} \leqslant x < \frac{1}{2})$

(g) $\ln(1 + x + x^2) = x + \frac{1}{2}x^2 - \frac{2}{3}x^3 + \frac{1}{4}x^4 + \cdots \qquad (-1 \leqslant x < 1)$

(h) $\ln\left(\dfrac{1 + 2x}{1 - x}\right) = 3x - \frac{3}{2}x^2 + 3x^3 - \frac{15}{4}x^4 + \cdots. \qquad (-\frac{1}{2} < x \leqslant \frac{1}{2})$

5. Prove that, if m and n have the same sign,

$$\ln\frac{m}{n} = 2\left(\frac{m-n}{m+n} + \frac{1}{3}\left(\frac{m-n}{m+n}\right)^3 + \frac{1}{5}\left(\frac{m-n}{m+n}\right)^5 + \cdots\right)$$

and state why m and n must have the same sign for the series to converge.

6. Show that, if θ is not a multiple of π,

$$\ln \operatorname{cosec}\theta = \tfrac{1}{2}\cos^2\theta + \tfrac{1}{4}\cos^4\theta + \tfrac{1}{6}\cos^6\theta + \cdots.$$

7. Using the binomial expansion, verify that

(a) $\dfrac{1 - x}{(1 - 2x^2)(1 - 2x)} = 1 + x + 4x^2 + 6x^3 + 16x^4 + 28x^5 + \cdots \qquad (|x| < \frac{1}{2})$

(b) $\dfrac{11x - 2}{(x - 2)^2(x^2 + 1)} = -\frac{1}{2} + \frac{9}{4}x + \frac{23}{8}x^2 - \frac{7}{16}x^3 - \frac{53}{32}x^4 + \frac{77}{64}x^5 + \cdots \quad (|x| < 1)$

(c) $\dfrac{1}{(x^2 + 1)(x - 2)} = -\frac{1}{2} - \frac{1}{4}x + \frac{3}{8}x^2 + \frac{3}{16}x^3 - \frac{13}{32}x^4 + \cdots \qquad (|x| < 1)$

8. Evaluate the following limits:

(a) $\lim_{x \to 0} \dfrac{(2-x)\ln(1-x)+(2+x)\ln(1+x)}{x^4}$,

(b) $\lim_{x \to 1+} \dfrac{\ln x}{x-1}$,

(c) $\lim_{x \to 0} \dfrac{\ln(1+\frac{1}{2}x)-(1+x)^{1/2}+1}{x^3}$,

(d) $\lim_{x \to 0} \dfrac{\ln(1-\frac{1}{3}x)-(1-x)^{1/3}+1}{x^2}$,

(e) $\lim_{x \to 0} \dfrac{e^{-x}-1+x}{x-\ln(1+x)}$,

(f) $\lim_{x \to 0} \dfrac{1-\cos 2x}{1-\cos 4x}$,

(g) $\lim_{x \to 0} \dfrac{\sin x}{\sinh x}$,

(h) $\lim_{x \to 0} \dfrac{\cot 4x}{\cot x}$,

(i) $\lim_{x \to 0} (\operatorname{cosec} x - \cot x)$,

(j) $\lim_{t \to \pi} \dfrac{1+\cos t}{\tan^2 t}$.

9. Expand $\cos^2 x \sin^2 x$ as far as the term involving x^6 and evaluate

$$\lim_{x \to 0} (\cos^2 x \sin^2 x - x^2(1-x^2)^{4/3})/x^6.$$

10. Evaluate

$$\lim_{x \to 0} \frac{3x + 2\sin\frac{1}{2}x - 16\sin\frac{1}{4}x}{\sin^5 x}.$$

11. Write down the power series for e^x and for $\sin x$ and use them to show that

$$e^{\sin x} = 1 + x + \tfrac{1}{2}x^2 - \tfrac{1}{8}x^4 + \cdots.$$

12. From the series for $\cos x$ deduce that

$$\sec x = 1 + \tfrac{1}{2}x^2 + \tfrac{5}{24}x^4 + \tfrac{61}{720}x^6 + \cdots.$$

13. Express $1 + \cos x$ in terms of $\cos\frac{1}{2}x$ and show that, if x is small,

$$\ln(1+\cos x) \simeq \ln 2 - \tfrac{1}{4}x^2 - \tfrac{1}{96}x^4.$$

14. Find the power series expansion of $[(1-\cos x)/x^2]^{1/3}$ as far as and including the term in x^4.

15. Use the series

$$\tan x = x + \tfrac{1}{3}x^3 + \tfrac{2}{15}x^5 + \cdots$$

to prove that

$$x\cot x = 1 - \tfrac{1}{3}x^2 - \tfrac{1}{45}x^4 + \cdots$$

and

$$\ln\sec x = \tfrac{1}{2}x^2 + \tfrac{1}{12}x^4 + \tfrac{1}{45}x^6 + \cdots.$$

16. Prove that

$$x/(e^x - 1) = 1 - \tfrac{1}{2}x + \tfrac{1}{12}x^2 - \tfrac{1}{720}x^4 + \cdots$$

and that

$$x/(e^x + 1) = \tfrac{1}{2}x - \tfrac{1}{4}x^2 + \tfrac{1}{48}x^4 - \cdots.$$

17. Prove by Maclaurin's theorem that

$$\sin(x + a) = \sin a + x \cos a - \frac{x^2}{2!} \sin a - \frac{x^3}{3!} \cos a + \cdots$$

and that

$$\tan(\tfrac{1}{4}\pi + x) = 1 + 2x + 2x^2 + \tfrac{8}{3}x^3 + \tfrac{10}{3}x^4 + \cdots.$$

18. By integrating the binomial series for $1/\sqrt{(1 + x^2)}$ prove that

$$\sinh^{-1} x = x - \frac{1}{2}\frac{x^3}{3} + \frac{1.3}{2.4}\frac{x^5}{5} - \frac{1.3.5}{2.4.6}\frac{x^7}{7} + \cdots.$$

19. Given $y = e^x \cos x$, show by induction that

$$\frac{d^n y}{dx^n} = 2^{n/2} e^x \cos(x + \tfrac{1}{4}n\pi)$$

and prove that

$$\frac{d^4 y}{dx^4} + 4y = 0.$$

Deduce that

$$e^x \cos x = 1 + x - \frac{2x^3}{3!} - \frac{2^2 x^4}{4!} - \frac{2^2 x^5}{5!} + \frac{2^3 x^7}{7!} + \cdots,$$

20. Use Maclaurin's expansion to prove that

$$\ln(1 + e^x) = \ln 2 + \tfrac{1}{2}x + \tfrac{1}{8}x^2 - \tfrac{1}{192}x^4 + \cdots.$$

21. By Maclaurin's theorem, or otherwise, find the expansion of

$$y = \ln[1 - \ln(1 - x)]$$

as far as the term in x^3. By the substitution $x = t/(1 + t)$ deduce the expansion of $\ln[1 + \ln(1 + t)]$ in powers of t as far as the term in t^3.

22. Given that $f(x)$ possesses a Maclaurin series and $f(x) = f(-x)$, prove that the series contains only even powers of x.

Expand $x \operatorname{cosec} x$ as far as the fourth power of x inclusive.

23. Given

$$y = \tfrac{2}{3}\sqrt{3} \tan^{-1}[x\sqrt{3}/(2 + x)],$$

prove that

$$\frac{dy}{dx} = 1/(1 + x + x^2).$$

Prove by assuming that y can be expanded in a series of ascending powers of x, and using the equation

$$(1-x^3)\frac{dy}{dx}=1-x,$$

or otherwise, that if $-1 < x < 1$, then

$$y=x-\frac{x^2}{2}+\frac{x^4}{4}-\frac{x^5}{5}+\frac{x^7}{7}-\cdots.$$

24. (a) Obtain the first two terms of the expansion of $\ln(1+e^{-x^2})$ as a power series in x.
(b) Find the coefficient of x^8 in the power series expansion of $x^3 \sin^3 x$.

22.6 Series of Complex Terms

Consider the infinite series $\Sigma_{r=1}^{\infty}(u_r+iv_r)$, where u_r and v_r are real, and $i^2=-1$.

Write

$$S_n=\sum_{r=1}^{n}(u_r+iv_r)=\sum_{r=1}^{n}u_r+i\sum_{r=1}^{n}v_r$$
$$=U_n+iV_n, \quad \text{say.}$$

If there is a complex number $S=U+iV$ such that $|S-S_n|\to 0$ as $n\to\infty$, then we say that the series is convergent and we write $S=\Sigma_{r=1}^{\infty}(u_r+iv_r)$.

Now

$$|S-S_n|=|(U-U_n)+i(V-V_n)|=\sqrt{[(U-U_n)^2+(V-V_n)^2]},$$

and so $|S-S_n|\to 0$ if, and only if, $U-U_n\to 0$ and $V-V_n\to 0$. This means that the series $\Sigma(u_r+iv_r)$ is convergent with sum $U+iV$ if, and only if, the real series Σu_r, Σv_r are both convergent with sums U, V respectively.

For example, the series

$$(\tfrac{1}{2}+\tfrac{1}{3}i)+(\tfrac{1}{4}+\tfrac{1}{9}i)+\cdots+\left(\frac{1}{2^n}+\frac{1}{3^n}i\right)+\cdots$$

is convergent with sum $1+\frac{1}{2}i$, since

$$\sum_{n=1}^{\infty}\frac{1}{2^n}=1, \quad \sum_{n=1}^{\infty}\frac{1}{3^n}=\tfrac{1}{2},$$

both these series being G.P.'s.

The series $\Sigma(u_r+iv_r)$ is said to be ***absolutely convergent*** if the real series $\Sigma|u_r+iv_r|$ is convergent. Thus the series

$$\sum_{n=1}^{\infty}\frac{(1+\sqrt{3}i)^n}{n!}$$

is absolutely convergent since the series

$$\sum_{n=1}^{\infty} \frac{|1+\sqrt{3}i|^n}{n!} = \sum_{n=1}^{\infty} \frac{2^n}{n!}$$

is convergent, the sum of the latter series being $e^2 - 1$ from (22.5).

A series that is convergent, but not absolutely convergent, is said to be *conditionally convergent*. On the other hand, a series that is absolutely convergent *must* also be convergent; for

$$|u_r| \leqslant |u_r + iv_r|, \quad |v_r| \leqslant |u_r + iv_r|$$

and if $\Sigma|u_r + iv_r|$ is convergent so also are both series Σu_r, Σv_r by Test 2 and Theorem 6.

If the terms of a series involve the variable z, where $z = x + iy$, then the sum of the series is a function of z, $f(z)$ say, the domain of f being the set of points z at which the series is convergent.

22.6.1 Power Series in z A series of the form

$$a_0 + a_1 z + a_2 z^2 + \cdots + a_n z^n + \cdots$$

where the coefficients a_n are (complex) constants, is called a power series in z. It can be shown that any such series has a radius of convergence $R \geqslant 0$ such that the series is absolutely convergent when $|z| < R$ and is divergent when $|z| > R$ (R may be infinite). The region $|z| < R$ on the Argand diagram is a circle, called the *circle of convergence* of the series. No general information can be given about the convergence of the series at points z lying on the circumference of this circle.

We shall be mainly concerned with power series in which the coefficients a_n are purely real. In this case, for points z lying on the real axis, the complex series $\Sigma a_n z^n$ can be identified with the real series $\Sigma a_n x^n$; if the real series $\Sigma a_n x^n$ has radius of convergence R, so also has the complex series $\Sigma a_n z^n$.

For example, since the exponential series in (22.5) has infinite radius of convergence, so also has the complex series

$$1 + z + \frac{z^2}{2!} + \cdots + \frac{z^n}{n!} + \cdots,$$

and since the logarithmic series in (22.7) has radius of convergence 1, so also has the complex series

$$z - \frac{z^2}{2} + \frac{z^3}{3} - \frac{z^4}{4} + \cdots.$$

(It can be shown that the last mentioned series is convergent at all points on the circumference of its circle of convergence except for the single point $z = -1$).

We may, without further definitions, replace x by z in Properties 1–3 of Section 22.4.1 and the properties still hold.

22.6.2 Expansions of Algebraic Functions of z

Consider the geometrical progression

$$1 + z + z^2 + \cdots + z^n + \cdots.$$

Since complex numbers form a field (Section 9.2.1) the result

$$S_n(z) = 1 + z + z^2 + \cdots + z^{n-1} = \frac{1 - z^n}{1 - z}$$

is true, exactly as for a real G.P. Putting $S(z) = 1/(1 - z)$, we find

$$|S(z) - S_n(z)| = |z|^n/|1 - z| \to 0 \text{ as } n \to \infty \qquad \text{if } |z| < 1.$$

Thus the formula for the sum to infinity is the same as for the corresponding real series. In general, expansions of algebraic functions carry over from the real to the complex field in this way; further examples are

$$\frac{1}{(1 - z)^2} = 1 + 2z + 3z^2 + \cdots \qquad (|z| < 1), \text{ (cf. Section 22.4.1)}$$

$$\frac{25z}{(1 - z)^2(1 - 6z)} = 25z + 200z^2 + \cdots \qquad (|z| < 1/6), \text{ (see Example 15)}$$

22.7 Transcendental Functions of a Complex Variable

No meaning has so far been attached to e^z, sin z, cos z, sinh z, cosh z, ln z, where z is complex. In this section we give definitions that satisfy two important criteria:

(a) when z is purely real, say $z = x$, e^x, sin x, ... can be identified with the known functions e^x, sin x, ... of the real variable x, and
(b) the basic properties of e^x, sin x, ..., where x is a real variable, also hold for e^z, sin z, ...

22.7.1 Exponential Function of z

We have seen (Section 22.6.1) that the series

$$1 + z + \frac{z^2}{2!} + \cdots + \frac{z^n}{n!} + \cdots$$

is absolutely convergent for all values of z. Hence the sum of this series is a function of z with the entire z-plane as domain. We define e^z by writing

$$e^z = 1 + z + \frac{z^2}{2!} + \cdots + \frac{z^n}{n!} + \cdots. \tag{22.12}$$

If z is purely real, say $z = x$, then (22.12) is in agreement with (22.5). If z is purely imaginary, say $z = iy$, then (22.12) becomes

$$\begin{aligned} e^{iy} &= 1 + iy + \frac{i^2y^2}{2!} + \cdots + \frac{i^ny^n}{n!} + \cdots \\ &= \left(1 - \frac{y^2}{2!} + \frac{y^4}{4!} - \cdots\right) + i\left(y - \frac{y^3}{3!} + \frac{y^5}{5!} - \cdots\right) \quad \text{(see Section 9.2.2)} \\ &= \cos y + i \sin y \qquad \text{by (22.3) and (22.4)} \end{aligned} \tag{22.13}$$

The result (22.13) has already been used in Sections 9.4 and 9.4.1, where important properties of e^{iy}, repeated here for completeness, were shown to follow from Demoivre's theorem; thus, if n is a positive integer,

$$(e^{iy})^n = (\cos y + i \sin y)^n = \cos(ny) + i \sin(ny) = e^{iny},$$

$$(e^{iy})^{-n} = (\cos y + i \sin y)^{-n} = \cos(-ny) + i \sin(-ny) = e^{-iny},$$

and similarly if p and q are integers

$$(e^{iy})^{p/q} \text{ has the } q \text{ values } e^{i(py + 2k\pi)/q} \qquad (k = 0, 1, \ldots, q-1),$$

where we assume that the fraction p/q is in its lowest terms, and q is positive. Putting $y = \theta$ and $y = -\theta$ in turn in (22.13) we can deduce that

$$\cos\theta = \tfrac{1}{2}(e^{i\theta} + e^{-i\theta}), \tag{22.14}$$

$$\sin\theta = \frac{1}{2i}(e^{i\theta} - e^{-i\theta}). \tag{22.15}$$

22.7.2 Index Laws for e^z From the definition (22.12) it follows by multiplication of series that

$$\begin{aligned} e^{z_1}e^{z_2} &= \left(1 + z_1 + \frac{z_1^2}{2!} + \cdots\right)\left(1 + z_2 + \frac{z_2^2}{2!} + \cdots\right) \\ &= 1 + z_1 + z_2 + \frac{z_1^2 + 2z_1z_2 + z_2^2}{2!} + \cdots \\ &= 1 + (z_1 + z_2) + \frac{(z_1 + z_2)^2}{2!} + \cdots = e^{(z_1+z_2)}. \end{aligned}$$

Also $e^ze^{-z} = 1$, so that $e^{-z} = 1/e^z$ and hence

$$e^{z_1}/e^{z_2} = e^{z_1}e^{-z_2} = e^{z_1 - z_2}.$$

Writing $z = x + iy$, we now have

$$\begin{aligned} e^z = e^{x+iy} &= e^x e^{iy} \\ &= e^x(\cos y + i \sin y) \quad \text{by (22.13).} \end{aligned} \tag{22.16}$$

Hence

$$\operatorname{Re}(e^z) = e^x \cos y, \quad \operatorname{Im}(e^z) = e^x \sin y.$$

In fact, (22.16) expresses e^z in polar form and shows that

$$|e^z| = e^x, \quad \arg(e^z) = y$$

(not necessarily the principal value). Using the polar form (22.16) we see that if n is an integer

$$\begin{aligned} (e^z)^n &= (e^x)^n(\cos ny + i \sin ny) \\ &= e^{nx} e^{iny} = e^{n(x+iy)} = e^{nz}, \end{aligned}$$

and similarly, if p and q are integers,

$$(e^z)^{p/q} \text{ has the } q \text{ values } e^{(pz+2k\pi i)/q} \quad (k = 0, 1, 2, \ldots, q-1).$$

Definition. We say the function f is periodic with period p if, for all z in the domain of f, $f(z + p) = f(z)$.

Thus e^z is periodic with period $2\pi i$ since

$$e^{z+2\pi i} = e^z e^{2\pi i} = e^z.$$

22.7.3 Circular Functions of z In eqs. (22.14), (22.15), θ is a real variable. We define $\cos z$, $\sin z$ by the analogous equations

$$\cos z = \tfrac{1}{2}(e^{iz} + e^{-iz}), \tag{22.17}$$

$$\sin z = \frac{1}{2i}(e^{iz} - e^{-iz}), \tag{22.18}$$

and we write

$$\tan z = \sin z/\cos z, \quad \operatorname{cosec} z = 1/\sin z,$$

$$\sec z = 1/\cos z, \quad \cot z = 1/\tan z.$$

From (22.17) and (22.18),

$$(\cos z + i \sin z)(\cos z - i \sin z) = e^{iz}e^{-iz}$$

and hence

$$\cos^2 z + \sin^2 z = 1.$$

Other identities, e.g.

$$\sin(z_1 + z_2) = \sin z_1 \cos z_2 + \cos z_1 \sin z_2,$$

$$\cos(z_1 + z_2) = \cos z_1 \cos z_2 - \sin z_1 \sin z_2,$$

are easily verified. It also follows that $\sin z$ and $\cos z$ are periodic functions with period 2π, while $\tan z$ is periodic with period π.

If z is purely imaginary, say $z = iy$, then from (22.17) and (22.18) we find that

$$\cos(iy) = \tfrac{1}{2}(e^{i^2y} + e^{-i^2y}) = \tfrac{1}{2}(e^{-y} + e^{y}) = \cosh y, \tag{22.19}$$

$$\sin(iy) = \frac{1}{2i}(e^{-y} - e^{y}) = \frac{i}{2}(e^{y} - e^{-y}) = i \sinh y. \tag{22.20}$$

Writing $z = x + iy$, we now have

$$\begin{aligned} \sin z = \sin(x + iy) &= \sin x \cos(iy) + \cos x \sin(iy) \\ &= \sin x \cosh y + i \cos x \sinh y \quad \text{by (22.19) and (22.20).} \end{aligned}$$

Hence

$$\mathrm{Re}(\sin z) = \sin x \cosh y, \quad \mathrm{Im}(\sin z) = \cos x \sinh y,$$

and similarly

$$\mathrm{Re}(\cos z) = \cos x \cosh y, \quad \mathrm{Im}(\cos z) = -\sin x \sinh y.$$

Example 25 Solve the equation $\cos z = 2$.

We have

$$\mathrm{Re}(\cos z) = 2, \quad \mathrm{Im}(\cos z) = 0,$$

that is

$$\cos x \cosh y = 2, \quad \sin x \sinh y = 0.$$

The last equation shows that either $\sin x = 0$, i.e. $x = n\pi$, or $\sinh y = 0$, i.e. $y = 0$. But $y = 0$ would require that $\cos x = 2$ from the previous equation, and so we must have $x = n\pi$ together with $\cosh y = 2/\cos n\pi$. Since $\cosh y$ is always positive, we must restrict n to even integers so that $\cosh y = 2$, i.e. $y = \pm\cosh^{-1} 2 \simeq \pm 1{\cdot}317$.

Hence

$$z \simeq 2n\pi \pm 1{\cdot}317i \qquad (n = 0, \pm 1, \pm 2, \ldots).$$

22.7.4 Hyperbolic Functions of z We define $\sinh z$, $\cosh z$ by the equations

$$\sinh z = \tfrac{1}{2}(e^z - e^{-z}), \tag{22.21}$$

$$\cosh z = \tfrac{1}{2}(e^z + e^{-z}), \tag{22.22}$$

and we write $\tanh z = \sinh z/\cosh z$, $\operatorname{cosech} z = 1/\sinh z$, $\operatorname{sech} z = 1/\cosh z$, $\coth z = 1/\tanh z$.

From (22.21) and (22.22)

$$(\cosh z + \sinh z)(\cosh z - \sinh z) = e^z e^{-z},$$

that is

$$\cosh^2 z - \sinh^2 z = 1.$$

Other familiar identities are easily verified. The functions are periodic, sinh and cosh having period $2\pi i$ while tanh has period πi.

If z is purely imaginary, say $z = iy$, eqs. (22.21) and (22.22) become

$$\sinh (iy) = \tfrac{1}{2}(e^{iy} - e^{-iy}) = i \sin y, \tag{22.23}$$

$$\cosh (iy) = \tfrac{1}{2}(e^{iy} + e^{-iy}) = \cos y. \tag{22.24}$$

Hence

$$\begin{aligned} \sinh z = \sinh (x + iy) &= \sinh x \cosh (iy) + \cosh x \sinh (iy) \\ &= \sinh x \cos y + i \cosh x \sin y, \end{aligned}$$

and similarly

$$\cosh z = \cosh x \cos y + i \sinh x \sin y.$$

22.7.5 Connection between Circular and Hyperbolic Functions Equations (22.19), (22.20), (22.23), (22.24) need not be restricted to the real variable y. In general

$$\sin (iz) = i \sinh z, \quad \cos (iz) = \cosh z, \tag{22.25}$$

and

$$\sinh (iz) = i \sin z, \quad \cosh (iz) = \cos z, \tag{22.26}$$

as is easily verified from the appropriate definitions. Equations (22.25) justify Osborn's rule (given in Section 17.3) for deducing hyperbolic identities from the corresponding trigonometric identities.

22.7.6 Logarithmic Function of z If w is any complex number such that $z = e^w$, then w is called a *natural logarithm* of z and we write $w = \operatorname{Ln} z$.

Writing $w = u + iv$, we have

$$z = e^{u+iv} = e^u(\cos v + i \sin v).$$

Hence

$$|z| = e^u, \quad \arg z = v \pm 2n\pi \qquad (n = 0, 1, 2, \ldots),$$

giving

$$u = \ln |z|, \quad v = \arg z \pm 2n\pi,$$

that is

$$\operatorname{Ln} z = u + iv = \ln |z| + i(\arg z \pm 2n\pi).$$

The *principal value* of Ln z, denoted by ln z, is defined by the relation

$$\ln z = \ln |z| + i \arg z. \tag{22.27}$$

For example

$$\ln (1 + i) = \ln \sqrt{2} + i\pi/4 = \tfrac{1}{2} \ln 2 + i\,\pi/4,$$
$$\ln (-1 - i) = \tfrac{1}{2} \ln 2 - i\,3\pi/4.$$

Example 26 Solve the equation $\tan z = 3i$.

We have

$$\frac{e^{iz} - e^{-iz}}{i(e^{iz} + e^{-iz})} = 3i,$$

and, on multiplying numerator and denominator on l.h.s. by e^{iz},

$$\frac{e^{2iz} - 1}{i(e^{2iz} + 1)} = 3i.$$

Solving for e^{2iz} we find that $e^{2iz} = -\frac{1}{2}$.

Hence

$$\begin{aligned} 2iz = \operatorname{Ln}(-\tfrac{1}{2}) &= \ln(\tfrac{1}{2}) + i(\pi \pm 2n\pi) \qquad & n = 0, 1, 2, \ldots \\ &= -\ln 2 + i(2n + 1)\pi \qquad & n = 0, \pm 1, \pm 2, \ldots, \end{aligned}$$

giving

$$z = \tfrac{1}{2}[(2n + 1)\pi + i \ln 2].$$

22.7.7 Expansions of Transcendental Functions of z The expansion (22.12) of e^z is a matter of definition, and from (22.21), (22.22) it then follows that

$$\sinh z = z + \frac{z^3}{3!} + \frac{z^5}{5!} + \cdots,$$

$$\cosh z = 1 + \frac{z^2}{2!} + \frac{z^4}{4!} + \cdots.$$

By continuing in this way it is easily shown that all the expansions (22.3)–(22.9) are valid when x is replaced by the complex variable z.

When $f(z)$ can be expanded as a power series in z with real coefficients, then, as for polynomials with real coefficients (see Section 10.2),

$$f(x + iy) = X + iY \Rightarrow f(x - iy) = X - iY.$$

For example,

$$\sin(x + iy) = \sin x \cosh y + i \cos x \sinh y \qquad \text{(see Section 22.7.3)}$$

and we may therefore assert that

$$\sin(x - iy) = \sin x \cosh y - i \cos x \sinh y.$$

Example 27 Given $x + iy = c \tanh(u + iv)$, where x, y, c, u and v are all real, determine x and y in terms of u, v and c.

Prove that this relationship implies both

$$x^2 + y^2 + c^2 - 2cx \coth 2u = 0$$

and

$$x^2 + y^2 - c^2 + 2cy \cot 2v = 0.$$

If $x + iy = c \tanh(u + iv)$, then $x - iy = c \tanh(u - iv)$,

so that

$$2x/c = \tanh(u + iv) + \tanh(u - iv)$$

$$= \frac{\sinh(u + iv)\cosh(u - iv) + \cosh(u + iv)\sinh(u - iv)}{\cosh(u + iv)\cosh(u - iv)}$$

$$= \frac{2 \sinh 2u}{\cosh 2u + \cosh 2iv},$$

$$x = \frac{c \sinh 2u}{\cosh 2u + \cos 2v}.$$

Similarly

$$2iy = \frac{2c \sinh 2iv}{\cosh 2u + \cos 2v}, \quad y = \frac{c \sin 2v}{\cosh 2u + \cos 2v}.$$

Now since

$$c \tanh(u + iv) = x + iy \quad \text{and} \quad c \tanh(u - iv) = x - iy,$$

$$c \tanh[(u + iv) + (u - iv)] = \frac{c[\tanh(u + iv) + \tanh(u - iv)]}{1 + \tanh(u + iv)\tanh(u - iv)},$$

that is

$$c \tanh 2u = \frac{2x}{1 + (x^2 + y^2)/c^2},$$

$$\tanh 2u = \frac{2cx}{x^2 + y^2 + c^2},$$

and so

$$x^2 + y^2 + c^2 - 2cx \coth 2u = 0.$$

Similarly,

$$c \tanh [(u + iv) - (u - iv)] = \frac{2iy}{1 - (x^2 + y^2)/c^2}.$$

Hence

$$\tanh 2iv = i \tan 2v = \frac{-2ciy}{x^2 + y^2 - c^2}$$

and so

$$x^2 + y^2 - c^2 + 2cy \cot 2v = 0.$$

22.7.8 Use of Complex Variables in Summation of Series By using the identities

$$\cos \theta + i \sin \theta = e^{i\theta}, \quad \cos n\theta + i \sin n\theta = e^{in\theta} \quad \text{(see (22.13))}$$

we can sometimes express a trigonometrical series as the real part, or imaginary part, of a complex series with known sum. The method is best illustrated by examples.

Example 28 Find the sum of the series

$$\sin \theta + \sin 2\theta + \cdots + \sin n\theta.$$

Write

$$S_n = \sin \theta + \sin 2\theta + \cdots + \sin n\theta$$

and introduce the corresponding cosine series by putting

$$C_n = \cos \theta + \cos 2\theta + \cdots + \cos n\theta.$$

Then

$$\begin{aligned} C_n + iS_n &= (\cos \theta + i \sin \theta) + (\cos 2\theta + i \sin 2\theta) + \cdots \\ &\qquad + (\cos n\theta + i \sin n\theta) \\ &= e^{i\theta} + e^{2i\theta} + \cdots + e^{ni\theta}. \end{aligned}$$

This last series is a G.P. with common ratio $e^{i\theta}$. Hence

$$C_n + iS_n = \frac{e^{i\theta}(1 - e^{in\theta})}{1 - e^{i\theta}}.$$

To find S_n we must now find the imaginary part of the expression on the r.h.s. This expression may be written in the form

$$\frac{e^{i(n+1)\theta/2}(e^{-in\theta/2} - e^{in\theta/2})}{e^{-i\theta/2} - e^{i\theta/2}},$$

that is

$$[\cos\{(n+1)\theta/2\} + i\sin\{(n+1)\theta/2\}]\frac{\sin(n\theta/2)}{\sin(\theta/2)}$$

Hence

$$S_n = \frac{\sin\{(n+1)\theta/2\}\sin(n\theta/2)}{\sin(\theta/2)}$$

Note that we have also obtained the result

$$C_n = \frac{\cos\{(n+1)\theta/2\}\sin(n\theta/2)}{\sin(\theta/2)}$$

Example 29 Find the sum to infinity of the series

$$\cos\theta + \frac{\cos 2\theta}{2!} + \frac{\cos 3\theta}{3!} + \cdots.$$

Write

$$C = \cos\theta + \frac{\cos 2\theta}{2!} + \frac{\cos 3\theta}{3!} + \cdots,$$

$$S = \sin\theta + \frac{\sin 2\theta}{2!} + \frac{\sin 3\theta}{3!} + \cdots.$$

Then

$$\begin{aligned} C + iS &= e^{i\theta} + \frac{e^{2i\theta}}{2!} + \frac{e^{3i\theta}}{3!} + \cdots \\ &= e^{(e^{i\theta})} - 1 \qquad \text{by (22.5)} \\ &= e^{(\cos\theta + i\sin\theta)} - 1 \\ &= e^{\cos\theta} e^{i\sin\theta} - 1 \\ &= e^{\cos\theta}[\cos(\sin\theta) + i\sin(\sin\theta)] - 1 \end{aligned}$$

Hence

$$C = e^{\cos\theta}\cos(\sin\theta) - 1.$$

Exercises 22(c)

1. Express in the form $a + ib$:

   (a) $\ln(1 - i\sqrt{3})$, (b) $\cos(\pi/4 + it)$, (c) $e^{2-i\pi}$,
   (d) $\sinh(t - i\pi/3)$, (e) $\sin(-\pi/6 + i\ln 2)$, (f) $\cosh(-t + i\pi)$.

2. Show that $z = \pi/4 + i$ satisfies the equation

$$\cos z = \tfrac{1}{4}\sqrt{2}[e(1 - i) + e^{-1}(1 + i)],$$

   and solve the equation $\cos z = \frac{5}{4}$.

3. Prove that, if $\tan z = \cos\alpha + i\sin\alpha$ where α is real and acute, then

$$z = (n + \tfrac{1}{4})\pi + \tfrac{1}{2}i\ln\tan(\tfrac{1}{4}\pi + \tfrac{1}{2}\alpha),$$

   where n is any integer.

4. Determine the general values of the complex number z for which (a) e^z and (b) $\cos z$ have real values.

5. Show that all the points in the Argand diagram that represent the values of $\mathrm{Ln}(1 + i)$ lie on a straight line parallel to the imaginary axis. What is the distance between consecutive points?

6. Solve the equations (a) $\sin z = 3$, (b) $\tan z = 1 + i$.

7. Given $z = x + iy = \tanh(u + \frac{1}{4}i\pi)$, where u is real, find x and y in terms of u and show that for all values of u, the point z lies on the circle $x^2 + y^2 = 1$.

8. Given $\sin(u + iv) = x + iy$, where u, v, x, y are all real, find x and y in terms of u and v.

   Show that, if v is constant and u varies, the point whose co-ordinates are (x, y) describes an ellipse, while, if u is constant and v varies, it describes a hyperbola.

9. If u, v, x, y are real numbers such that $u + iv = e^{x+iy}$, prove that $u^2 + v^2 = e^{2x}$ and $v/u = \tan y$.

   Draw a sketch of the path of the point (u, v) when the point (x, y) describes the rectangle formed by the axes of reference and the lines $x = 1$, $y = \frac{1}{2}\pi$.

10. Given $x + iy = c\cosh(u + iv)$, where u, v, x, y and c are all real, prove that

$$x^2\sinh^2 u + y^2\cosh^2 u = c^2\sinh^2 u\cosh^2 u.$$

11. Given $x + iy = \tan(u + iv)$, where x, y, u, v are all real, show that

$$\tan 2u = \frac{2x}{1 - x^2 - y^2} \quad \text{and} \quad \tanh 2v = \frac{2y}{1 + x^2 + y^2}.$$

12. Express the modulus and argument of e^z, where z is a complex number, in terms of the modulus and argument of z.

   Find the polar equation of the curve in the Argand diagram described by the point z when it varies so that ze^z is real.

   Sketch that part of the curve where $\arg(ze^z) = 0$ and $-\pi < \arg z < \pi$, indicating the asymptotes.

13. Given

$$\cos x + i\sin x = \sin(u + iv),$$

where x, u and v are real and $i = \sqrt{-1}$, and $0 < x < \frac{1}{2}\pi$ whilst v is positive, prove that

$$u = \sin^{-1}\sqrt{(1 - \sin x)} \quad \text{and} \quad v = \ln[\sqrt{(\sin x)} + \sqrt{(1 + \sin x)}].$$

14. Given

$$\tan(x + iy) = \sin(p + iq),$$

where x, y, p, q are real, prove that

$$\tan p \sinh 2y = \tanh q \sin 2x.$$

15. Given that θ is real, find the real and imaginary parts of

$$\frac{\cosh(\theta + i\pi/3)}{(1 + \sqrt{3i})^{1/2}}.$$

16. Show that

$$x \sin\theta + \frac{x^3 \sin 3\theta}{3!} + \frac{x^5 \sin 5\theta}{5!} + \cdots = \cosh(x\cos\theta)\sin(x\sin\theta).$$

17. Show that, provided $|r| < 1$,

$$r\sin\theta + r^2\sin 2\theta + r^3\sin 3\theta + \cdots = \frac{r\sin\theta}{1 - 2r\cos\theta + r^2}.$$

18. Show that, if x is not an integral multiple of π,

$$\tfrac{1}{2} + \cos 2x + \cos 4x + \cdots + \cos 2nx = \frac{\sin(2n+1)x}{2\sin x},$$

and

$$\sin x + \sin 3x + \cdots + \sin(2n-1)x = \frac{\sin^2 nx}{\sin x}.$$

19. Show that

$$\sum_{n=1}^{\infty} \frac{r^{2n}\sin(2n\theta)}{(2n)!} = \sin(r\sin\theta)\sinh(r\cos\theta).$$

20. Show that

$$\sum_{r=1}^{n} {}^nC_r \sin 2r\theta = 2^n \cos^n\theta \sin n\theta.$$

21. Show that

$$\sum_{r=1}^{n} r\sin r\theta = \frac{(n+1)\sin n\theta - n\sin(n+1)\theta}{2(1 - \cos\theta)}.$$

22. Show that

$$\sum_{r=1}^{\infty} x^r \cos rx = \frac{x(\cos x - x)}{1 - 2x\cos x + x^2}:$$

CHAPTER 23

HIGHER-ORDER LINEAR DIFFERENTIAL EQUATIONS

23.1 Functions and Linear Dependence

It is sometimes helpful to regard the idea of function as an extension of the idea of vector. The 3-dimensional vector (y_1, y_2, y_3) is completely specified by the three real numbers y_1, y_2, y_3; hence it may be represented as in Fig. 23.1 by three ordinates equally spaced (say) on some interval I of positive length on the x-axis. Likewise we may represent the 5-dimensional vector $(y_1, y_2, y_3, y_4, y_5)$ by five ordinates as in Fig. 23.2.

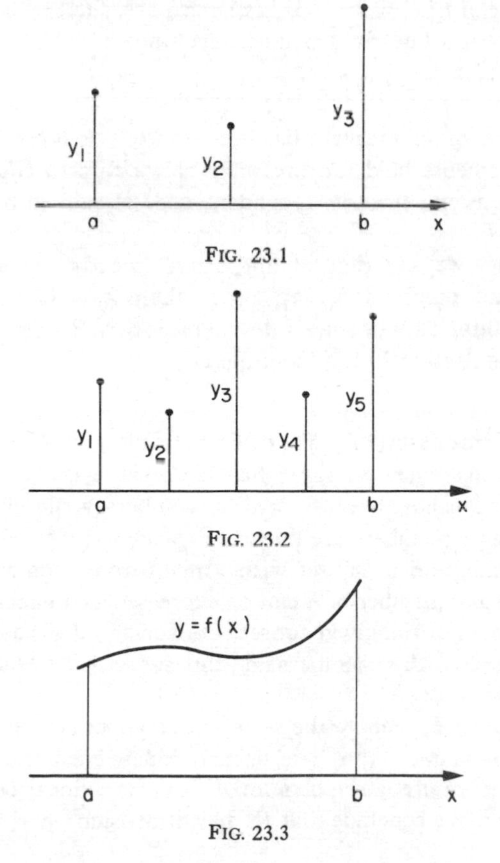

Fig. 23.1

Fig. 23.2

Fig. 23.3

Continuing in this way we are led to think of a function f with domain I as being a vector of infinitely many dimensions; for the values of $f(x)$ are represented by all the ordinates that can be drawn from points in I to the curve $y = f(x)$, suggesting the idea of a vector with infinitely many components.

We say that the function f is a *linear combination* of the functions f_1, $f_2, \ldots, f_m$ on I if

$$f(x) = k_1 f_1(x) + k_2 f_2(x) + \cdots + k_m f_m(x)$$

for all x in I, where $k_1, k_2, \ldots, k_m$ are constants, not all zero.

A set of functions $\{f_1, f_2, \ldots, f_m\}$ is said to be a *linearly dependent* set on I if constants $k_1, k_2, \ldots, k_m$ can be found, not all zero, such that

$$k_1 f_1(x) + k_2 f_2(x) + \cdots + k_m f_m(x) = 0$$

for all x in I.

For example, if $f_1(x) = 2x + 3, f_2(x) = x^2 + 4, f_3(x) = 2x^2 + 10x + 23$ then $\{f_1, f_2, f_3\}$ is a linearly dependent set since

$$5f_1(x) + 2f_2(x) - f_3(x) \equiv 0.$$

Here mention of an interval I has been omitted, as is common practice, since the statements hold for any interval we care to choose. A set of functions that is not linearly dependent on I is said to be *linearly independent* on I.

For example, we say that x^2 and x are linearly independent since we cannot find numbers k_1, k_2, other than $k_1 = k_2 = 0$, such that $k_1x^2 + k_2x = 0$ for all x in some interval of positive length. (This will be verified later, in Section 23.1.2, Example 1.)

23.1.1 The Dimension of a Set of Functions Let $F = \{f_1, f_2, \ldots\}$ be a set of functions. Then we say F has *dimension* s if F contains at least one linearly independent subset with s members while all subsets of F with more than s members are linearly dependent (cf. Section 11.3).

A linearly independent subset with s members is then a *basis* for F in the sense that any member of F can be expressed as a linear combination of the s members of the given subset. Conversely, if F has a linearly independent subset with s members and this subset is a basis for F, then F has dimension s.

For example let P_m denote the set of all polynomials with degree $\leqslant m$. The subset $\{1, x, x^2, \ldots, x^m\}$ is linearly independent and has $(m + 1)$ members; furthermore, any polynomial in P_m is a linear combination of $1, x, x^2, \ldots, x^m$. We conclude that P^m has dimension $(m + 1)$, and we say that $\{1, x, \ldots, x^m\}$ is a basis for polynomials in x of degree $\leqslant m$.

With n-dimensional vectors we can find sets of any dimension up to and including n. We should therefore expect to be able to find sets of functions of infinite dimension, and this is indeed the case; for example

$$\{1, x, x, \ldots, x^n, \ldots\},$$

$$\{\sin x, \sin 2x, \ldots, \sin nx, \ldots\}$$

are both linearly independent sets with infinitely many members.

If F denotes the set of all functions for which derivatives of all orders exist at $x = 0$, then Maclaurin's theorem shows that $1, x, x, \ldots, x^n, \ldots$ is a basis for F.

23.1.2 Condition for Linear Independence We saw that two vectors $(y_1, y_2, \ldots, y_n)$, $(\eta_1, \eta_2, \ldots, \eta_n)$ are linearly independent if at least one of the ${}^nC_2$ 2×2 determinants that can be formed from the component block

$$\begin{matrix} y_1 & y_2 & \ldots & y_n \\ \eta_1 & \eta_2 & \ldots & \eta_n \end{matrix}$$

is not zero (Section 11.3, Theorem 3). If we think of each value of $f(x)$ as a 'component' of f, the corresponding condition in the case of two functions f, g is as follows: f and g are linearly independent on I if there is at least one pair of numbers x_1, x_2 in I such that

$$\begin{vmatrix} f(x_1) & f(x_2) \\ g(x_1) & g(x_2) \end{vmatrix} \neq 0.$$

Since there are infinitely many choices of x_1, x_2 in an interval I of positive length, this is not a very convenient condition to apply. A more convenient condition may be obtained in the following manner:

If f, g are linearly dependent on I, say

$$k_1 f(x) + k_2 g(x) = 0 \qquad \text{for all } x \text{ in } I,$$

then

$$k_1 f'(x) + k_2 g'(x) = 0 \qquad \text{for all } x \text{ in } I,$$

and so, unless $k_1 = k_2 = 0$, we must have

$$\begin{vmatrix} f(x) & g(x) \\ f'(x) & g'(x) \end{vmatrix} = 0 \qquad \text{for all } x \text{ in } I \qquad \text{(Theorem 8, Section 11.4.2).}$$

Thus, writing

$$\Delta(f, g) = \begin{vmatrix} f(x) & g(x) \\ f'(x) & g'(x) \end{vmatrix} = \begin{vmatrix} f(x) & f'(x) \\ g(x) & g'(x) \end{vmatrix},$$

we see that if $\Delta(f, g) \not\equiv 0$ on I (i.e. $\Delta(f, g)$ is non-zero for some value of x in I) then f and g must be linearly independent on I.

The condition that *three* functions f, g, h be linearly independent on I is that

$$\begin{vmatrix} f(x) & f'(x) & f''(x) \\ g(x) & g'(x) & g''(x) \\ h(x) & h'(x) & h''(x) \end{vmatrix} \not\equiv 0 \text{ on } I,$$

and so on. A determinant formed in this way from a set of functions is called the *Wronskian* of that set.

Example 1 Show that x, x^2 are linearly independent.

Here the Wronskian is the determinant

$$\begin{vmatrix} x & 1 \\ x^2 & 2x \end{vmatrix}$$

which equals $2x^2 - x^2$, i.e. x^2, which is zero only at the single point $x = 0$.

Example 2 Show that $\sin x$, $\cos x$ are linearly independent.

$$\begin{vmatrix} \sin x & \cos x \\ \cos x & -\sin x \end{vmatrix} = -\sin^2 x - \cos^2 x = -1.$$

Thus the Wronskian is non-zero for all x.

23.2 Linear Differential Operators

The symbol d/dx is frequently regarded as an *operator* in the sense that any expression prefixed by this symbol undergoes the operation of differentiation with respect to x. A more convenient symbol for this operator is D; thus, for example

$$Dx^2 = 2x, \quad D \sin x = \cos x, \quad D(x + 1/x) = 1 - 1/x^2,$$
$$Df(x) = f'(x), \quad Dy = \frac{dy}{dx}.$$

Here it is understood that $D \equiv d/dx$; on another occasion we may wish to put $D \equiv d/dt$, and so on, but this need not cause any confusion.

It follows that $D[Df(x)] = Df'(x) = f''(x)$; if we now agree to abbreviate $D[Df(x)]$ to the simpler form $D^2 f(x)$, we see that $D^2 \equiv d^2/dx^2$. Continuing in this way we have $D^r \equiv d^r/dx^r$ for any positive integer r. The use of an index r in this way is particularly appropriate, since clearly

$$D^m[D^n f(x)] = D^n[D^m f(x)] = D^{m+n} f(x).$$

Two rules of differentiation, expressed in this notation, are

$$D[f(x) + g(x)] = Df(x) + Dg(x), \quad D[k f(x)] = kDf(x) \quad (k \text{ a constant})$$

and combining these two rules we see that

$$D[\alpha f(x) + \beta g(x)] = \alpha Df(x) + \beta Dg(x)$$

for any constants α, β. It is for this reason that the operator D is called *linear*.

Consider now the operator L defined by the equation

$$Lf(x) = a(x)f'(x) + b(x)f(x).$$

Here it is convenient to write $L \equiv a(x)D + b(x)$; thus, for example,

$$(x^2D + x) \sin x = x^2 \cos x + x \sin x,$$
$$(\sin x\, D + \cos x)x^2 = 2x \sin x + x^2 \cos x.$$

Likewise we may introduce a second-order differential operator L, writing

$$L \equiv a(x)D^2 + b(x)D + c(x);$$

for example,

$$\left(xD^2 + x^3D - \frac{1}{x}\right) \ln x = x(-1/x^2) + x^3\frac{1}{x} - \frac{1}{x}\ln x$$
$$= x^2 - (1/x)(1 + \ln x).$$

More generally, the operator L, where

$$L \equiv a_0(x)D^n + a_1(x)D^{n-1} + \cdots + a_{n-1}(x)D + a_n(x), \qquad (23.1)$$

is called an ***nth-order linear differential operator***; it is easily verified that

$$L[\alpha f(x) + \beta g(x)] = \alpha Lf(x) + \beta Lg(x)$$

for any constants α, β, justifying the use of the term linear. If $a_0, a_1, \ldots, a_n$ are constants (i.e. independent of x) we say that L is a *constant-coefficient* linear differential operator; in this case (23.1) shows that L is a polynomial in D, say $L \equiv \phi(D)$.

Example 3 Find $(D^3 - D^2 + 2)e^{3x}$, and show that for any constant-coefficient operator $\phi(D)$ we have $\phi(D)e^{kx} = \phi(k)e^{kx}$, where k is a constant.

$$(D^3 - D^2 + 2)e^{3x} = (27 - 9 + 2)e^{3x} = 20\, e^{3x}.$$

In general, $D^r e^{kx} = k^r e^{kx}$, and so

$$\begin{aligned}(a_0D^n + a_1D^{n-1} + \cdots + a_{n-1}D + a_n)e^{kx} \\ = (a_0k^n + a_1k^{n-1} + \cdots + a_{n-1}k + a_n)e^{kx} = \phi(k)e^{kx}.\end{aligned}$$

Example 4 Show that for a constant-coefficient operator $\phi(D)$ we have

$$\phi(D)[e^{kx}f(x)] = e^{kx}\,\phi(D+k)f(x),$$

where k is a constant, and use this result to find

$$(D-1)(e^{3x}\sin x) \quad \text{and} \quad (D^2-4)[e^{2x}(x^2+1)].$$

By the product rule for differentiation

$$D[e^{kx}f(x)] = e^{kx}f'(x) + ke^{kx}f(x) = e^{kx}(D+k)f(x).$$

More generally, Leibniz's theorem (Section 14.4.7) gives

$$D^r[e^{kx}f(x)] = e^{kx}(D+k)^r f(x).$$

It follows that if $\phi(D)$ is a polynomial in D, then

$$\phi(D)[e^{kx}f(x)] = e^{kx}\,\phi(D+k)f(x).$$

Hence

$$(D-1)(e^{3x}\sin x) = e^{3x}(D+2)\sin x = e^{3x}(\cos x + 2\sin x).$$

Also

$$\begin{aligned}(D^2-4)[e^{2x}(x^2+1)] &= e^{2x}[(D+2)^2 - 4](x^2+1) \\ &= e^{2x}(D^2+4D)(x^2+1) \\ &= e^{2x}(8x+2).\end{aligned}$$

23.2.1 Multiplication and Factorisation of Operators Let L_1, L_2 denote two linear differential operators. Then $L_2[L_1 f(x)]$ denotes the result of applying L_2 to $L_1 f(x)$. Thus if

$$L_1 \equiv a_1(x)D + b_1(x), \quad L_2 \equiv a_2(x)D + b_2(x),$$

then

$$\begin{aligned}L_2[L_1 f(x)] &= [a_2(x)D + b_2(x)][a_1(x)f'(x) + b_1(x)f(x)] \\ &= a_2(x)a_1(x)f''(x) + [a_2(x)a_1'(x) + a_1(x)b_2(x) \\ &\quad + a_2(x)b_1(x)]f'(x) + [a_2(x)b_1'(x) + b_2(x)b_1(x)]f(x)\end{aligned}$$

$$= \{a_2(x)a_1(x)D^2 + [a_2(x)a_1'(x) + a_1(x)b_2(x) + a_2(x)b_1(x)]D + [a_2(x)b_1'(x) + b_2(x)b_1(x)]\} f(x) \quad (23.2)$$

For convenience we shall define the operator L_1L_2 by the equation $L_1L_2f(x)=L_1[L_2f(x)]$. We see from (23.2) that when L_1 and L_2 are first-order operators, then L_1L_2 is a second-order operator.

Example 5 Apply the operators

(a) $(xD + 1)[(1/x) D+ 2]$, (b) $[(1/x) D + 2](xD + 1)$,

to e^{3x}.

(a)

$$\left(\frac{1}{x}D + 2\right)e^{3x} = (3/x + 2)e^{3x},$$

and so

$$\begin{aligned}(xD + 1)\left(\frac{1}{x}D + 2\right)e^{3x} &= (xD + 1)[(3/x + 2)e^{3x}]\\ &= x(9/x + 6 - 3/x^2)e^{3x} + (3/x + 2)e^{3x}\\ &= (6x + 11)e^{3x}.\end{aligned}$$

(b)

$$\begin{aligned}\left(\frac{1}{x}D + 2\right)(xD + 1)e^{3x} &= \left(\frac{1}{x}D + 2\right)[(3x + 1)e^{3x}]\\ &= (1/x)(9x + 6)e^{3x} + (6x + 2)e^{3x}\\ &= (6x + 11 + 6/x)e^{3x}.\end{aligned}$$

Example 5 illustrates that, in general, $L_2L_1 \neq L_1L_2$. In the particular case of constant-coefficient operators (23.2) reduces to the simpler result that

$$\begin{aligned}(a_2D + b_2)(a_1D + b_1) &\equiv a_2a_1D^2 + (a_2b_1 + a_1b_2)D + b_2b_1\\ &\equiv (a_1D + b_1)(a_2D + b_2)\end{aligned}$$

(by interchange of suffices 1 and 2).

It follows that constant-coefficient operators may be multiplied together in any order using the familiar algebra of real numbers and treating D as a real variable. Conversely, we may use the algebra of real numbers to factorise a higher-order constant-coefficient operator into first-order factors, arranging the factors in any order.

For example,

$$(D^2 - 2D - 3)(\cos x + \sin x) = -6 \cos x - 2 \sin x,$$

and the reader should verify that

$$(D-3)(D+1)(\cos x+\sin x)=(D+1)(D-3)(\cos x+\sin x)$$
$$=-6\cos x-2\sin x.$$

It should of course be observed that some second-order operators, e.g. D^2+D+1, which correspond to non-factorising quadratics, cannot be factorised into two first-order operators with constant coefficients.

23.3 Linear Differential Equations

The D.E.

$$a_0(x)\frac{d^n y}{dx^n}+a_1(x)\frac{d^{n-1}y}{dx^{n-1}}+\cdots+a_{n-1}(x)\frac{dy}{dx}+a_n(x)y=b(x)$$

is said to be *linear*. Dividing through by $a_0(x)$ we see that there is no loss of generality if we suppose that $a_0(x)\equiv 1$. If $b(x)\equiv 0$ the equation is *homogeneous*, otherwise it is called *non-homogeneous*; this is a distinction of major importance in what follows.

The methods given in this chapter apply equally to equations of any order. For simplicity these methods are discussed in relation to second-order equations, where the main applications occur, but the extension to higher-order equations is straightforward. Thus we consider the equation

$$\left.\begin{aligned}\frac{d^2y}{dx^2}+p(x)\frac{dy}{dx}+q(x)y=h(x),\\ \text{that is}\qquad Ly=h(x),\end{aligned}\right\}\qquad(23.3)$$

where $L\equiv D^2+p(x)D+q(x)$, a linear differential operator. This equation is homogeneous if $h(x)\equiv 0$, non-homogeneous if $h(x)\not\equiv 0$.

For example

$$\frac{d^2y}{dx^2}-x\frac{dy}{dx}+(\tan x)y=0$$

is a homogeneous linear D.E., and

$$\frac{d^2y}{dx^2}+\frac{1}{x}\frac{dy}{dx}+xy=e^x$$

is a non-homogeneous linear D.E.

23.3.1 Complete Solution by Integration

The case of the first-order equation has already been dealt with in Section 21.4.2. In following this, it would be natural to try to reduce the problem of solving (23.3) to a problem of *two* integrations.

Example 6 Solve the equation

$$x^3 \frac{d^2y}{dx^2} + 4x^2 \frac{dy}{dx} + 2xy = x^5.$$

It so happens that the equation can be written in the form

$$\frac{d}{dx}\left(x^3 \frac{dy}{dx} + x^2y\right) = x^5.$$

Hence, by integration,

$$x^3 \frac{dy}{dx} + x^2y = \frac{1}{6}x^6 + c.$$

This is a first-order linear equation and can be solved by the method of Section 21.4.2.

We have

$$x \frac{dy}{dx} + y = \frac{1}{6}x^4 + \frac{c}{x^2},$$

that is

$$\frac{d}{dx}(xy) = \frac{1}{6}x^4 + \frac{c}{x^2}.$$

Hence, by integration,

$$xy = \frac{1}{30}x^5 - \frac{c}{x} + d.$$

In terms of operators, the equation in Example 6 can be written as $(x^3D^2 + 4x^2D + 2x)y = x^5$; the key to solving it was to express it as $D(x^3D + x^2)y = x^5$. More generally, eq. (23.3) can be solved whenever L is expressed as a product of two first-order factors, say $L = L_1L_2$. For eq. (23.3) is then $L_1L_2\,y = h(x)$, and writing

$$Y = L_2y \tag{1}$$

this gives

$$L_1Y = h(x). \tag{2}$$

Equation (2) is first-order linear and so Y can be determined by the method of Section 21.4.2. Substituting for Y in (1), we then have another first-order linear equation to solve for y.

Example 7 Solve the equation $(xD - 1)(xD + 2)y = x^3$.

Put

$$Y = (xD + 2)y. \qquad (1)$$

Then

$$(xD - 1)Y = x^3,$$

that is

$$x\frac{\mathrm{d}Y}{\mathrm{d}x} - Y = x^3. \qquad (2)$$

Solving (2) by the method of Section 21.4.2, we get $Y = \frac{1}{2}x^3 + cx$. Hence (1) becomes

$$(xD + 2)y = \tfrac{1}{2}x^3 + cx,$$

that is

$$x\frac{\mathrm{d}y}{\mathrm{d}x} + 2y = \tfrac{1}{2}x^3 + cx.$$

Again using the method of Section 21.4.2, we get $y = \frac{1}{10}x^3 + \frac{1}{3}cx + d/x^2$, that is

$$y = \tfrac{1}{10}x^3 + c'x + d/x^2 \qquad (c' = \tfrac{1}{3}c).$$

The main difficulties with solving (23.3) by the foregoing method are

(1) L may not factorise,
(2) it may be too difficult to spot the factors of L, even when they exist.

However, if L is a constant-coefficient operator then (2) does not arise, since factorisation is particularly simple in this case. We conclude this section by showing how the method applies to the homogeneous equation

$$(D^2 + pD + q)y = 0, \qquad (23.4)$$

where p, q are constants. Suppose the equation

$$m^2 + pm + q = 0$$

(called the auxiliary equation for (23.4)) has real roots m_1, m_2, then

$$m^2 + pm + q = (m - m_1)(m - m_2)$$

and the operator $D^2 + pD + q$ has corresponding factors $(D - m_1)(D - m_2)$; hence (23.4) can be written in the form $(D - m_1)(D - m_2)y = 0$.

Write $Y = (D - m_2)y$ so that $(D - m_1)Y = 0$, i.e. $Y = ce^{m_1 x}$(see Section 21.4.3). Thus $(D - m_2)y = ce^{m_1x}$, that is

$$\frac{\mathrm{d}}{\mathrm{d}x}(e^{\;m_2x}y) = ce^{(m_1-m_2)x}$$

(e^{-m^2x} is an I.F., see Section 21.4.2).

Case 1. $m_1 \neq m_2$. The solution is

$$e^{-m_2x}y = Ae^{(m_1-m_2)x} + B \qquad (A = c/(m_1 - m_2))$$

that is

$$y = Ae^{m_1x} + Be^{m_2x}.$$

Case 2. $m_1 = m_2$. Then

$$\frac{\mathrm{d}}{\mathrm{d}x}(e^{-m_1x}y) = c$$

giving

$$e^{-m_1x}y = cx + d,$$

that is

$$y = (cx + d)e^{m_1x}.$$

These solutions are easily remembered and in practice we solve equations of the form (23.4) by finding the roots of the auxiliary equation, or the factors of the operator (which amounts to the same thing), and writing down the appropriate solution from memory.

Example 8 Solve the equation $(D^2 - 3D - 4)y = 0$.

The equation is $(D + 1)(D - 4)y = 0$. Hence the solution is

$$y = Ae^{-x} + Be^{4x}.$$

Example 9 Solve the equation $(D^2 - 4)y = 0$.

The equation is $(D - 2)(D + 2)y = 0$. Hence the solution is

$$y = Ae^{2x} + Be^{-2x}.$$

There is an alternative form of solution, for

$$e^{2x} = \cosh 2x + \sinh 2x \quad \text{and} \quad e^{-2x} = \cosh 2x - \sinh 2x;$$

hence

$$y = (A + B) \cosh 2x + (A - B) \sinh 2x, \quad \text{or} \quad y = a \cosh 2x + b \sinh 2x$$

where a, b are arbitrary constants.

Example 10 Solve the equation $(D^2 + 2D - 5)y = 0$.

The auxiliary equation is $m^2 + 2m - 5 = 0$, with roots $m = -1 \pm \sqrt{6}$.
Hence the solution of the given equation is

$$y = Ae^{(-1+\sqrt{6})x} + Be^{-(1+\sqrt{6})x}.$$

Example 11 Solve the equation $(D^2 + 6D + 9)y = 0$.

The equation is $(D + 3)^2 y = 0$. Thus the auxiliary equation has equal roots (Case 2 above) and the solution of the given equation is

$$y = (Ax + B)e^{-3x}.$$

23.3.2 The Initial-value Problem The problem of solving the second-order linear equation (23.3) together with the conditions that

$$y = a \text{ when } x = 0, \quad \mathrm{d}y/\mathrm{d}x = b \text{ when } x = 0, \tag{23.5}$$

where a and b are known constants, is called a second-order initial-value problem.

For example, eq. (21.5) governs the motion in the simple vibrating system discussed in Example 1, Section 21.2.1. To determine the motion given the initial displacement (value of s when $t = 0$) and the initial velocity (value of $\mathrm{d}s/\mathrm{d}t$ when $t = 0$) would be an initial-value problem. In this, as in most applications, it is obvious from the nature of the problem that it must have a solution and that the solution is unique. Although a strictly mathematical proof is possible we shall simply assume

Lemma 1 The initial-value problem consisting of eq. (23.3) together with the conditions (23.5) has a unique solution.

Example 12 Solve the initial-value problem $(D^2 - 3D - 4)y = 0$, $y = 1$ when $x = 0$, $\mathrm{d}y/\mathrm{d}x = -11$ when $x = 0$.

From Example 8 we see that $y = Ae^{-x} + Be^{4x}$. Hence

$$\mathrm{d}y/\mathrm{d}x = -Ae^{-x} + 4Be^{4x}.$$

The initial conditions now give

$$1 = A + B \quad \text{and} \quad -11 = -A + 4B.$$

Solving for A and B we get the unique solution $A = 3$, $B = -2$; hence

$$y = 3e^{-x} - 2e^{4x}.$$

As a general rule we would expect that two conditions should determine a unique solution of eq. (23.3), but if two conditions are not of the form (23.5), then it is possible that a solution may not exist, or may not be unique, as the next two examples illustrate.

Example 13 Show that the solution of the problem $(D^2 + 1)y = 0$, $y = 0$ when $x = 0$, $y = 0$ when $x = \pi$, is not unique.

By trial we see that $y = A \sin x$ is a solution for *any* constant A.

Example 14 Solve the problem $(D^2 - 1)y = 0$, $y = 1$ when $x = 0$, $d^2y/dx^2 = 3$ when $x = 0$.

We have $(D - 1)(D + 1)y = 0$ and so $y = Ae^x + Be^{-x}$. Hence

$$\frac{dy}{dx} = Ae^x - Be^{-x}, \quad \frac{d^2y}{dx^2} = Ae^x + Be^{-x}.$$

The required conditions are that

$$1 = A + B \quad \text{and} \quad 3 = A + B.$$

Clearly no values of A and B satisfy these conditions. Thus the problem does not have a solution.

23.3.3 The Homogeneous Equation Suppose eq. (23.3) has the form

$$Ly = 0, \tag{23.6}$$

where $L \equiv D^2 + p(x)D + q(x)$. Let S denote the set of all functions that satisfy (23.6), i.e. f is in the set S if $L f(x) = 0$.

Lemma 2 If f, g are both in S, then if α, β are any constants, the function $\alpha f + \beta g$ is also in S.

By the linearity property of the operator L (see Section 23.2) we have

$$\begin{aligned} L[\alpha f(x) + \beta g(x)] &= \alpha\, Lf(x) + \beta Lg(x) \\ &= \alpha . 0 + \beta . 0 \qquad \text{since } f, g \text{ both satisfy (23.6)} \\ &= 0. \end{aligned}$$

In Section 23.1 we introduced the idea of the 'dimension' of a set of functions. We now make an important application of this concept.

Theorem 1 The set S has dimension 2.

Proof. By Lemma 1 the initial-value problem

$$Ly = 0, \quad y = 1 \text{ when } x = 0, \quad dy/dx = 0 \text{ when } x = 0,$$

has a unique solution, say $y = f(x)$. Thus

$$Lf(x) = 0, \quad f(0) = 1, \quad f'(0) = 0.$$

Likewise the initial-value problem

$$Ly = 0, \quad y = 0 \text{ when } x = 0, \quad dy/dx = 1 \text{ when } x = 0,$$

has a unique solution, say $y = g(x)$. Thus

$$Lg(x) = 0, \quad g(0) = 0, \quad g'(0) = 1.$$

Now let ϕ be any member of S, i.e. $L\,\phi(x) = 0$, and suppose that $\phi(0) = a$, $\phi'(0) = b$. Then $y = \phi(x)$ is the unique solution of the initial-value problem

$$Ly = 0, \quad y = a \text{ when } x = 0, \quad dy/dx = b \text{ when } x = 0.$$

But $y = af(x) + bg(x)$ is a solution of this problem, for by Lemma 2 $L[af(x) + bg(x)] = 0$ and the initial conditions are easily checked.

Hence $\phi(x) = af(x) + bg(x)$. We have thus proved that *all* solutions of eq. (23.6) may be expressed in the form $af(x) + bg(x)$, with suitable constants a, b, i.e. f, g is a ***basis*** for S. It remains to verify that f, g are linearly independent; suppose that k_1, k_2 are constants and that

$$k_1 f(x) + k_2 g(x) = 0.$$

By differentiation,

$$k_1 f'(x) + k_2 g'(x) = 0.$$

Putting $x = 0$ in these equations we get $k_1 = 0$, $k_2 = 0$. Hence f, g are linearly independent and the dimension of S is 2.

It now follows that any two solutions of (23.6), provided they are linearly independent, form a basis for the complete solution. This enables us to determine the complete solution more easily, as is shown in the following example.

Example 14 Solve the equation $(x^2D^2 - 4xD + 6)y = 0$.

From the nature of the operator one might guess that a solution in the form of a power of x is possible. Try putting $y = x^m$ and the equation becomes

$$[m(m-1) - 4m + 6]x^m = 0,$$

that is

$$(m^2 - 5m + 6)x^m = 0.$$

Hence $y = x^m$ is a solution if $m^2 - 5m + 6 = 0$, i.e. if $m = 2$ or 3.

Since $y = x^2$ and $y = x^3$ are both solutions, and are linearly independent, they are a basis for the complete solution, which can now be represented by the equation $y = Ax^2 + Bx^3$.

23.3.4 The Non-homogeneous Equation Suppose that $y = \psi(x)$ is a solution of eq. (23.3), i.e. $L\psi(x) = h(x)$. Suppose also that $y = \phi(x)$ is another solution, i.e. $L\,\phi(x) = h(x)$. Then

$$L[\phi(x) - \psi(x)] = L\,\phi(x) - L\,\psi(x) = h(x) - h(x) = 0.$$

This means that $y = \phi(x) - \psi(x)$ is a solution of the homogeneous equation (23.6). Thus $\phi - \psi$ is in the set S of Section 23.3.3. Hence if we know two linearly independent solutions of (23.6), say $y = f(x)$ and $y = g(x)$, then

$$\phi(x) - \psi(x) = af(x) + bg(x)$$

for some constants a, b; i.e.

$$\phi(x) = \psi(x) + af(x) + bg(x).$$

This means that all solutions of (23.3) are known once we know one such solution ($\psi(x)$) and the complete solution ($af(x) + bg(x)$) of the homogeneous equation (23.6). We now state this useful result as

Theorem 2 If $y = \psi(x)$ is a particular solution of eq. (23.3) and $y = af(x) + bg(x)$ (a, b arbitrary constants) is the complete solution of the homogeneous equation (23.6), then

$$y = \psi(x) + af(x) + bg(x)$$

is the complete solution of (23.3).

We say that $\psi(x)$ is a P.I. of (23.3) (see Section 21.3.1). We shall call $af(x) + bg(x)$ the Complementary Function (C.F.) of (23.3).

Example 15 Solve the equation $(x^2D^2 - 4xD + 6)y = 36x^6$.

It would be reasonable (cf. Example 14) to suggest a trial solution of the form $y = kx^6$. This gives

$$30kx^6 - 24kx^6 + 6kx^6 = 36x^6, \quad \text{that is} \quad 12kx^6 = 36x^6,$$

and so the equation is satisfied if $k = 3$.

Thus $y = 3x^6$ is a P.I. From Example 14 we know that the complete solution of the corresponding homogeneous equation is $y = Ax^2 + Bx^3$. Hence, by Theorem 2, the complete solution of the given equation is $y = 3x^6 + Ax^2 + Bx^3$.

23.4 Linear Equations with Constant Coefficients

Consider the equation

$$(D^2 + pD + q)y = h(x), \tag{23.7}$$

where p, q are constants.

Homogeneous case. If $h(x) \equiv 0$, we find the roots m_1, m_2 of the auxiliary equation $m^2 + pm + q = 0$. The solution can then be written down as follows (see Section 23.3.1):

(a) if $m_1 \neq m_2$ and m_1, m_2 both real, the solution is

$$y = Ae^{m_1 x} + Be^{m_2 x}.$$

(b) if $m_1 = m_2$ the solution is $y = (Ax + B)e^{m_1 x}$.

(c) if m_1, m_2 are complex, say $m_1 = \alpha + i\beta$, $m_2 = \alpha - i\beta$, then the method used to obtain (a) and (b) does not apply, but if we write tentatively

$$y = e^{(\alpha + i\beta)x}, \quad \text{that is} \quad y = e^{\alpha x}(\cos \beta x + i \sin \beta x),$$

we are led to the idea of trial solutions of the form

$$y = e^{\alpha x} \cos \beta x, \quad y = e^{\alpha x} \sin \beta x;$$

these can be shown to satisfy the equation, and since they are linearly independent we can represent the complete solution by the equation

$$y = e^{\alpha x}(A \cos \beta x + B \sin \beta x).$$

Example 16 Solve the equation $(D^2 + 9)y = 0$.

The auxiliary equation is $m^2 + 9 = 0$; its roots are $m = \pm 3i$. Hence the solution is $y = A \cos 3x + B \sin 3x$.

Example 17 Solve the equation $(D^2 + 4D + 13)y = 0$.

The auxiliary equation is $m^2 + 4m + 13 = 0$; its roots are $m = -2 \pm 3i$. Hence the solution is $y = e^{-2x}(A \cos 3x + B \sin 3x)$.

Example 18 Solve the equation $(D^4 + 8D^2 + 16)y = 0$.

The auxiliary equation is $(m^2 + 4)^2 = 0$; its roots are

$$m = +2i, +2i, -2i, -2i.$$

Hence, by an extension of the above rules, the solution is

$$y = (A + Bx) \cos 2x + (C + Dx) \sin 2x.$$

Non-homogeneous case. The C.F. is obtained by solving the homogeneous equation obtained when $h(x)$ is replaced by zero; a P.I. is usually obtained most easily by trial. The form of trial solution to be used depends on the form of $h(x)$ and in the following sections we consider the most important cases. We shall write $\phi(D) = D^2 + pD + q$ (p, q constants).

23.4.1 h a Polynomial It is easily verified that the result of applying $\phi(D)$ to a polynomial of degree n is again a polynomial of degree n unless $q = 0$, in which case it is a polynomial of degree $n - 1$ (assume p is not also zero, since (23.7) is then trivial). Hence, if h is a polynomial of degree n a suitable trial solution would be a polynomial of degree n if $q \neq 0$, or degree $(n + 1)$ if $q = 0$.

Example 19 Solve the equation $(5D^2 - 4D + 16)y = 32x^2$.

The auxiliary equation is $5m^2 - 4m + 16 = 0$, and its roots are $m = \frac{2}{5} \pm \frac{2}{5}i\sqrt{19}$. Hence the C.F. is

$$e^{2x/5} (A \cos \tfrac{2}{5}\sqrt{19}x + B \sin \tfrac{2}{5}\sqrt{19}x.$$

To find a P.I. put $y = ax^2 + bx + c$. Then

$$Dy = 2ax + b, \quad D^2y = 2a,$$

and so

$$(5D^2 - 4D + 16)y = 16ax^2 + (16b - 8a)x + (16c - 4b + 10a).$$

We require

$$16a = 32, \quad 16b - 8a = 0, \quad 16c - 4b + 10a = 0,$$

giving

$$a = 2, \quad b = 1, \quad c = -1.$$

Thus $y = 2x^2 + x - 1$ is a P.I. and the complete solution is

$$y = 2x^2 + x - 1 + e^{2x/5} (A \cos \tfrac{2}{5}\sqrt{19}x + B \sin \tfrac{2}{5}\sqrt{19}x).$$

Example 20 Solve the equation $(D^2 + 2D)y = x^3$.

The auxiliary equation is $m^2 + 2m = 0$; its roots are $m = 0, -2$. Hence the C.F. is $A + Be^{-2x}$.

To find a P.I. put $y = ax^4 + bx^3 + cx^2 + dx + e$. (In this example we have $q = 0$.)

Then

$$Dy = 4ax^3 + 3bx^2 + 2cx + d, \quad D^2y = 12ax^2 + 6bx + 2c,$$

and so

$$(D^2 + 2D)y = 8ax^3 + (6b + 12a)x^2 + (4c + 6b)x + (2d + 2c).$$

We require

$$8a = 1, \quad 6b + 12a = 4c + 6b = 2d + 2c = 0,$$

giving

$$a = \tfrac{1}{8}, \quad b = -\tfrac{1}{4}, \quad c = \tfrac{3}{8}, \quad d = -\tfrac{3}{8}. \quad \text{Put } e = 0.$$

Thus

$$y = \tfrac{1}{8}(x^4 - 2x^3 + 3x^2 - 3x)$$

is a P.I. and the complete solution is

$$y = \tfrac{1}{8}(x^4 - 2x^3 + 3x^2 - 3x) + A + Be^{-2x}.$$

23.4.2 $h(x) = ke^{\lambda x}$, k and λ Constants, $\phi(\lambda) \neq 0$ By differentiation we see that

$$(D^2 + pD + q)e^{\lambda x} = (\lambda^2 + p\lambda + q)e^{\lambda x}, \quad \text{that is} \quad \phi(D)e^{\lambda x} = \phi(\lambda)e^{\lambda x}.$$

Hence, to find a P.I. for the equation $\phi(D)y = ke^{\lambda x}$ we put $y = ae^{\lambda x}$ and the equation becomes $a\,\phi(\lambda)e^{\lambda x} = ke^{\lambda x}$, which is satisfied if $a = k/\phi(\lambda)$.

Example 21 Solve the equation $(D^2 - 8D + 16)y = e^{3x}$.

The auxiliary equation is $(m - 4)^2 = 0$; its roots are $m = 4, 4$. Hence the C.F. is $(A + Bx)e^{4x}$.

To find a P.I. put $y = ae^{3x}$. Then $Dy = 3ae^{3x}$, $D^2y = 9ae^{3x}$, so that $(D^2 - 8D + 16)y = ae^{3x}$. Thus we require $a = 1$.

It follows that $y = e^{3x}$ is a P.I. and the complete solution is

$$y = e^{3x} + (A + Bx)e^{4x}.$$

Alternatively, we may find a P.I. directly from the formula $[1/\phi(3)]e^{3x}$,

that is

$$\frac{1}{3^2 - 8.3 + 16} e^{3x} = e^{3x}.$$

23.4.3 $h(x) = ke^{\lambda x}$, k and λ Constants, $\phi(\lambda) = 0$ In this case the method of Section 23.4.2 fails. We have shown (Example 4, Section 23.2) that

$$\phi(D)[e^{\lambda x} f(x)] = e^{\lambda x} \phi(D + \lambda) f(x).$$

Hence, if we write $y = e^{\lambda x}\tilde{y}$, the equation $\phi(D)y = ke^{\lambda x}$ becomes $e^{\lambda x}\phi(D + \lambda)\tilde{y} = ke^{\lambda x}$, and it is now sufficient to solve the equation $\phi(D + \lambda)\tilde{y} = k$.

Example 22 Solve the equation $(D^2 + D - 2)y = e^{-2x}$.

The equation is $(D - 1)(D + 2)y = e^{-2x}$. The C.F. is $Ae^x + Be^{-2x}$. Since $\phi(-2) = (-2 - 1)(-2 + 2) = 0$, the method of Section 23.4.2 fails.

If we write $y = e^{-2x}\tilde{y}$, the equation becomes

$$e^{-2x}(D - 3)D\tilde{y} = e^{-2x} \qquad \text{(replacing } D \text{ by } D - 2\text{)},$$

that is

$$(D^2 - 3D)\tilde{y} = 1.$$

A P.I. may now be found by the method of Section 23.4.1 (trial solution $\tilde{y} = ax + b$) giving $\tilde{y} = -\frac{1}{3}x$, i.e. $y = -\frac{1}{3}xe^{-2x}$. The complete solution is

$$\begin{aligned} y &= -\tfrac{1}{3}xe^{-2x} + Ae^x + Be^{-2x} \\ &= (B - \tfrac{1}{3}x)e^{-2x} + Ae^x. \end{aligned}$$

23.4.4 $h(x) = e^{\lambda x} f(x)$ More generally, the method of Section 23.4.3 may be applied to equations of this type. On substituting $y = \tilde{y}\, e^{\lambda x}$, the equation $\phi(D)y = e^{\lambda x} f(x)$ becomes

$$e^{\lambda x} \phi(D + \lambda)\tilde{y} = e^{\lambda x} f(x),$$

and it is now sufficient to solve the equation

$$\phi(D + \lambda)\tilde{y} = f(x).$$

Example 23 Solve the equation $(D^2 + 6D + 8)y = x^3 e^{-2x}$.

The C.F. is $Ae^{-2x} + Be^{-4x}$.

To find a P.I. put $y = e^{-2x}\tilde{y}$, and the equation becomes

$$e^{-2x}[(D-2)^2 + 6(D-2) + 8]\tilde{y} = x^3 e^{-2x},$$

that is

$$(D^2 + 2D)\tilde{y} = x^3,$$

for which (Example 20) a P.I. is $\tilde{y} = \frac{1}{8}(x^4 - 2x^3 + 3x^2 - 3x)$.

Hence a P.I. of the given equation is

$$y = \tfrac{1}{8}e^{-2x}(x^4 - 2x^3 + 3x^2 - 3x)$$

and the complete solution is

$$y = \tfrac{1}{8}e^{-2x}(x^4 - 2x^3 + 3x^2 - 3x) + Ae^{-2x} + Be^{-4x}.$$

23.4.5 Use of Complex-valued Functions Let f_1, f_2 be two real-valued functions. Writing $F(x) = f_1(x) + i f_2(x)$ we see that the values of $F(x)$ are complex numbers. As usual we write Re, Im to denote real part, imaginary part respectively, that is

$$\text{Re}\,[F(x)] = f_1(x), \quad \text{Im}\,[F(x)] = f_2(x).$$

We now define a process of differentiation for complex-valued functions by writing

$$DF(x) = f_1'(x) + i f_2'(x), \quad D^2 F(x) = f_1''(x) + i f_2''(x), \text{ etc.}$$

Consider the complex-valued function F, where $F(x) = \cos\beta x + i\sin\beta x$. Then

$$DF(x) = -\beta\sin\beta x + i\beta\cos\beta x = i\beta F(x),$$

but $F(x) = e^{i\beta x}$ and so this result may be written as $De^{i\beta x} = i\beta e^{i\beta x}$. More generally, if we put $F(x) = e^{(\alpha+i\beta)x}$ we can similarly show that

$$De^{(\alpha+i\beta)x} = (\alpha + i\beta)e^{(\alpha+i\beta)x},$$

that is

$$De^{\lambda x} = \lambda e^{\lambda x} \qquad \text{if } \lambda \text{ is real or complex.}$$

From the definition of differentiation, it easily follows that

$$\phi(D)\,F(x) = \phi(D)f_1(x) + i\,\phi(D)f_2(x).$$

Hence if $H(x) = h_1(x) + i\,h_2(x)$, and $\phi(D)\,F(x) = H(x)$, then

$$\phi(D)f_1(x) = h_1(x), \quad \phi(D)f_2(x) = h_2(x);$$

in other words, if $y = F(x)$ satisfies the equation $\phi(D)y = H(x)$, then

$y = \text{Re}\,[F(x)]$ satisfies the equation $\phi(D)y = \text{Re}\,[H(x)]$; and $y = \text{Im}\,[F(x)]$ satisfies the equation $\phi(D)y = \text{Im}\,[H(x)]$.

In particular $\phi(D)e^{\lambda x} = \phi(\lambda)e^{\lambda x}$ even if λ is complex, so the methods of Sections 23.4.2–4 apply as illustrated in the next section.

23.4.6 $h(x)$ Involving Trigonometric Functions The following examples illustrate how the ideas of Section 23.4.5 may be used in various situations involving trigonometric functions.

Example 24 Solve the equation $(D^2 + 2D + 2)y = 5 \cos x$.

The C.F. is $e^{-x} (A \cos x + B \sin x)$.

Writing the equation as $(D^2 + 2D + 2)y = \text{Re}\,(5e^{ix})$, we first find a complex-valued P.I. for the equation $(D^2 + 2D + 2)y = 5e^{ix}$; the real part of this complex-valued P.I. will then be a P.I. for the given equation.

Put $y = a\, e^{ix}$. Then $Dy = ia\, e^{ix}$, $D^2y = i^2ae^{ix} = -ae^{ix}$, and so

$$(D^2 + 2D + 2)y = (1 + 2i)a\, e^{ix}.$$

Thus we require $(1 + 2i)a = 5$, giving

$$a = \frac{5}{1 + 2i} = 1 - 2i.$$

Hence $y = (1 - 2i)e^{ix}$ is a P.I.

Writing

$$y = (1 - 2i)(\cos x + i \sin x),$$

we see that the real part is $\cos x + 2 \sin x$; hence $y = \cos x + 2 \sin x$ is the required P.I. The complete solution is given by

$$y = \cos x + 2 \sin x + e^{-x} (A \cos x + B \sin x).$$

Example 25 Solve the equation $(D^2 + 2D + 5)y = 4e^{-x} \sin 2x$.

The auxiliary equation is $m^2 + 2m + 5 = 0$ with roots $m = -1 \pm 2i$. Hence the C.F. is $e^{-x} (A \cos 2x + B \sin 2x)$.

Since

$$4e^{-x} \sin 2x = \text{Im}\,(4e^{-x}\, e^{2ix}) = \text{Im}\,(4e^{(-1+2i)x}),$$

we first seek a P.I. of the equation $(D^2 + 2D + 5)y = 4e^{(-1+2i)x}$. But $-1 + 2i$ is a root of the auxiliary equation, so we put $y = e^{(-1+2i)x}\tilde{y}$.

Writing the equation as

$$(D + 1 - 2i)(D + 1 + 2i)y = 4e^{(-1+2i)x}$$

it now becomes

$$D(D + 4i)\tilde{y} = 4 \qquad \text{(replacing } D \text{ by } D - 1 + 2i\text{)}.$$

Put $\tilde{y} = ax + b$. Then $D\tilde{y} = a$, $D^2\tilde{y} = 0$, so that $(D^2 + 4iD)\tilde{y} = 4ia$. Thus we require $4ia = 4$, i.e. $a = -i$, giving $\tilde{y} = -ix$, that is

$$y = -ixe^{(-1+2i)x} = -ixe^{-x}(\cos 2x + i \sin 2x).$$

Then $\text{Im}(y) = -xe^{-x}\cos 2x$, so that $y = -xe^{-x}\cos 2x$ is a P.I. of the given equation. The complete solution is now given by

$$\begin{aligned} y &= -xe^{-x}\cos 2x + e^{-x}(A\cos 2x + B\sin 2x) \\ &= e^{-x}[(A - x)\cos 2x + B\sin 2x]. \end{aligned}$$

Example 26 Solve the equation $(D^2 + 2D + 2)y = 50x\cos 2x$.

The C.F. is $e^{-x}(A\cos x + B\sin x)$. Also

$$50x\cos 2x = \text{Re}(50xe^{2ix}),$$

so we consider the equation

$$(D^2 + 2D + 2)y = 50xe^{2ix}.$$

Put $y = e^{2ix}\tilde{y}$, and the equation becomes

$$[(D + 2i)^2 + 2(D + 2i) + 2]\tilde{y} = 50x,$$

that is

$$[D^2 + (2 + 4i)D - (2 - 4i)]\tilde{y} = 50x.$$

Try $\tilde{y} = ax + b$ (see Section 23.4.1). Then

$$\begin{aligned} [D^2 + (2 + 4i)D - (2 - 4i)]\tilde{y} &= (2 + 4i)a - (2 - 4i)(ax + b) \\ &= -(2 - 4i)ax + [(2 + 4i)a - (2 - 4i)b]. \end{aligned}$$

Thus we require

$$-(2 - 4i)a = 50, \quad (2 + 4i)a - (2 - 4i)b = 0,$$

giving

$$a = -5(1 + 2i), \quad b = 11 + 2i.$$

This gives

$$\tilde{y} = -5(1 + 2i)x + 11 + 2i = (11 - 5x) + i(2 - 10x),$$

that is

$$y = e^{2ix}\tilde{y} = [(11 - 5x) + i(2 - 10x)](\cos 2x + i\sin 2x),$$

that is

$$\text{Re}(y) = (11 - 5x)\cos 2x - (2 - 10x)\sin 2x.$$

Hence the solution of the given equation is

$$y = (11 - 5x)\cos 2x - (2 - 10x)\sin 2x + e^{-x}(A\cos x + B\sin x).$$

23.5 Theory of Small Oscillations

In Section 21.2.1, Example 1, the equation of motion of a mass attached to a spring was given in the form (21.5), i.e.

$$\frac{d^2s}{dt^2} + ks = 0.$$

This equation applies whenever the distance (s) of the mass from its equilibrium position is small. The constant k, associated with the stiffness of the spring, is positive and so we may conveniently write $k = \omega^2$. Then

$$(D^2 + \omega^2)s = 0$$

where $D = d/dt$. The auxiliary equation is $m^2 + \omega^2 = 0$ with roots $m = \pm i\omega$. Hence the solution is

$$s = A\cos \omega t + B\sin \omega t.$$

Another form for this solution is $s = R\cos(\omega t - \alpha)$, where R, α are arbitrary constants. This shows that the mass oscillates between two points at distance R on each side of the equilibrium position (Fig. 23.4). R is called the *amplitude* of the oscillation, α the *phase-angle*. A complete oscillation is performed each time t increases by $2\pi/\omega$, which is therefore called the *period* of the oscillations. The reciprocal of the period, viz. $\omega/2\pi$, is called the *frequency* of the oscillations; this represents the number of oscillations taking place in unit time.

The motion described above is called simple harmonic motion (S.H.M.). An analogous situation arises in an electrical circuit of inductance L and capacitance C; the current i in such a circuit satisfies the equation

$$(D^2 + \omega^2)i = 0,$$

where $\omega^2 = 1/LC$, and so it follows that

$$i = R\cos(\omega t - \alpha)$$

where R, α are constants.

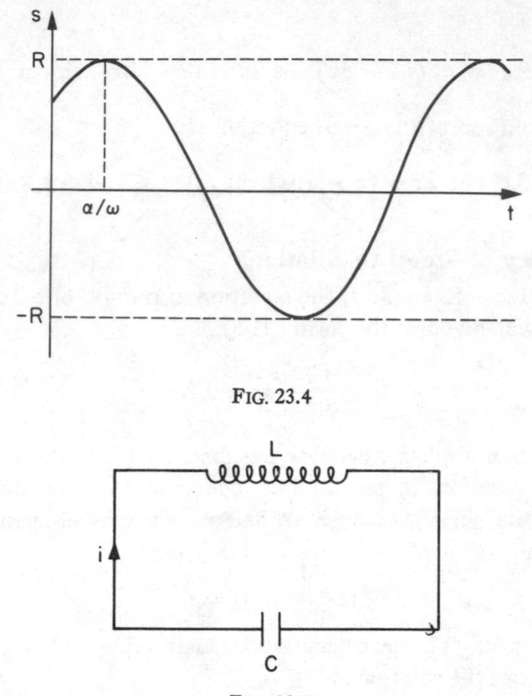

FIG. 23.4

FIG. 23.5

23.5.1 Damped Oscillations We saw that S.H.M. results when a mass attached to a spring is acted upon by no forces other than that of the spring. In theory this means that once the oscillations have been started they continue indefinitely. In practical applications (e.g. vehicle springs) there is normally a damping effect (shock absorbers), the resistance offered to the motion varying so as to be always proportional to the velocity. Let us suppose there is a deceleration of $2\rho\, \mathrm{d}s/\mathrm{d}t$ due to this resistance, where ρ is a constant (the constant of proportionality has been taken as 2ρ for later convenience); the equation of motion is then

$$\frac{\mathrm{d}^2 s}{\mathrm{d}t^2} + 2\rho \frac{\mathrm{d}s}{\mathrm{d}t} + \omega^2 s = 0,$$

that is

$$(D^2 + 2\rho D + \omega^2)s = 0.$$

The auxiliary equation is $m^2 + 2\rho m + \omega^2 = 0$ with roots

$$m = -\rho \pm \sqrt{(\rho^2 - \omega^2)}.$$

Case 1: slight damping. Suppose $\rho < \omega$. Then the auxiliary equation has complex roots $m = -\rho \pm i\sqrt{(\omega^2 - \rho^2)}$. The solution is

$$s = e^{-\rho t}\{A \cos\sqrt{(\omega^2 - \rho^2)}t + B \sin\sqrt{(\omega^2 - \rho^2)}t\},$$

or, in another form,

$$s = e^{-\rho t} R \cos\{\sqrt{(\omega^2 - \rho^2)}t - \alpha\}.$$

Again the mass oscillates about its equilibrium position, but the amplitude of the oscillations is steadily decreasing due to the factor $e^{-\rho t}$ (see Fig. 23.6)

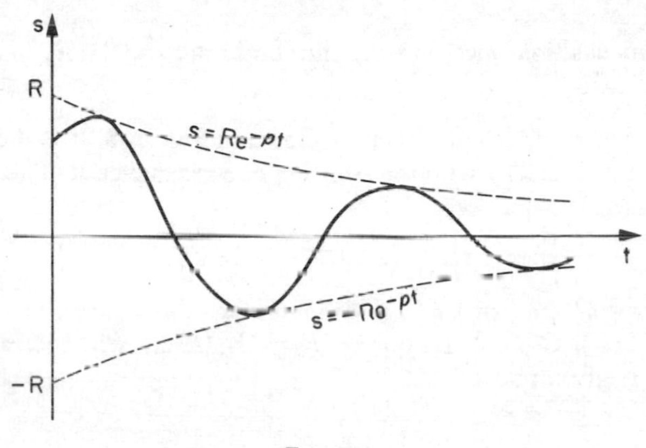

FIG. 23.6

Case 2: heavy damping. Suppose $\rho > \omega$. Then the auxiliary equation has real roots $m = -\rho \pm \sqrt{(\rho^2 - \omega^2)}$, and the solution is

$$s = Ae^{\{-\rho+\sqrt{(\rho^2-\omega^2)}\}t} + Be^{\{-\rho-\sqrt{(\rho^2-\omega^2)}\}t},$$

that is,

$$s = Ae^{-k_1 t} + Be^{-k_2 t}, \qquad k_1 \text{ and } k_2 \text{ both positive.}$$

The motion is not oscillatory. As $t \to \infty$ we see that $s \to 0$, i.e. the mass approaches its equilibrium position. If A and B have opposite signs, the mass may pass through its equilibrium position once as illustrated by Fig. 23.7.

Case 3: critical damping. Suppose $\rho = \omega$. Then the auxiliary equation has equal roots $m = -\rho, -\rho$, and the solution is

$$s = (A + Bt)e^{-\rho t}.$$

If A and B have opposite signs, the mass may pass through its

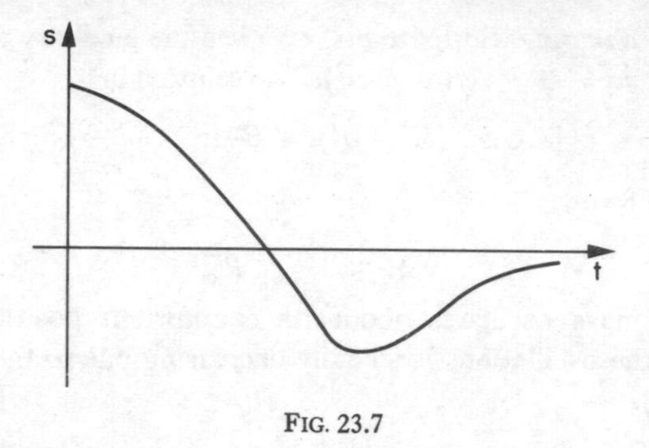

FIG. 23.7

equilibrium position once, but the motion is not oscillatory and is very similar to Case 2.

The analogous electrical circuit is shown in Fig. 23.8. It is the same as that shown in Fig. 23.5 with the addition of a resistance R. The current i now satisfies the equation

$$(D^2 + 2\rho D + \omega^2)i = 0,$$

where $\rho = R/2L$, $\omega^2 = 1/LC$. Thus there are three cases, according as $R < 2\sqrt{(L/C)}$, $R > 2\sqrt{(L/C)}$, or $R = 2\sqrt{(L/C)}$ with corresponding solutions as given above.

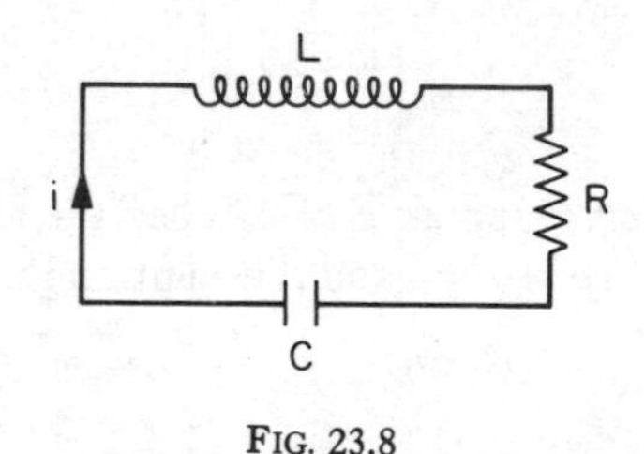

FIG. 23.8

23.5.2 Forced Oscillations If the mass in Section 23.5.1, in addition to the forces already described, is driven by another force which depends only on t, then the equation of motion has the form

$$(D^2 + 2\rho D + \omega^2)s = h(t).$$

The C.F. is the solution of the equation discussed in Section 23.5.1 and so it represents the motion that would occur in the absence of this additional driving force, i.e. it represents the *free oscillations* of the system.

The P.I., which is added to the C.F. to give the complete solution, is thus the part of the solution that is attributable to the additional driving force, and so it represents the *forced oscillations.*

Since in all cases (see Section 23.5.1), the free oscillations die out as $t \to \infty$, the forced oscillations (P.I.) represent the dominant part of the solution when t is large. For this reason the P.I. is sometimes called the *steady state* and the C.F. the *transient.*

Example 27 Solve the equation $(D^2 + 4D + 29)s = \cos 5t$ and show that this represents a motion that is approximately simple harmonic with period $2\pi/5$ when t is large.

The C.F. is $e^{-2t}(A \cos 5t + B \sin 5t)$; this represents the free oscillations and because of the factor e^{-2t} it dies out as $t \to \infty$.

A P.I. is given by the real part of $[1/4(1 + 5i)]\, e^{5it}$, namely

$$\tfrac{1}{104}(\cos 5t + 5 \sin 5t).$$

This represents the ultimate steady state, and is simple harmonic with period $2\pi/5$.

Example 28 Solve the equation $(D^2 + \omega^2)s = k \cos pt$, given that $s = ds/dt = 0$ when $t = 0$, and discuss the motion represented by this equation when $p = \omega$.

The C.F. is $A \cos \omega t + B \sin \omega t$. This represents the free oscillations which in this case are not damped.

(a) If $p \neq \omega$, then we get a P.I. of the equation $(D^2 + \omega^2)s = ke^{ipt}$ by using a trial solution $s = a\, e^{ipt}$. This gives $a = k/(\omega^2 - p^2)$, so that

$$s = \frac{k}{\omega^2 - p^2}\, e^{ipt}$$

is a P.I., and taking the real part we get a P.I., viz

$$s = \frac{k}{\omega^2 - p^2} \cos pt,$$

of the given equation.

Hence the solution is

$$s = \frac{k}{\omega^2 - p^2} \cos pt + A \cos \omega t + B \sin \omega t,$$

and from $s = ds/dt = 0$ when $t = 0$ we get $B = 0, A = -k/(\omega^2 - p^2)$. Thus

$$s = \frac{k}{\omega^2 - p^2} (\cos pt - \cos \omega t).$$

(b) If $p = \omega$, then we write $s = e^{i\omega t} \tilde{s}$ and the equation $(D^2 + \omega^2)s = ke^{i\omega t}$ becomes $D(D + 2i\omega)\tilde{s} = k$. A.P.I. for this equation is

$$\tilde{s} = -(k/2\omega)\, it, \text{ i.e. } s = -(k/2\omega)\, it\, e^{i\omega t}.$$

Taking the real part we get the P.I. $s = (k/2\omega)\, t \sin \omega t$ of the given equation. Hence the solution is

$$s = \frac{k}{2\omega} t \sin \omega t + A \cos \omega t + B \sin \omega t,$$

and from $s = ds/dt = 0$ when $t = 0$ we get $A = B = 0$. Thus

$$s = \frac{k}{2\omega} t \sin \omega t.$$

This represents an oscillation with ever-increasing amplitude (see Fig. 23.9). For a mass attached to a spring such oscillations are impossible since the amplitude tends to infinity. However, we must remember that

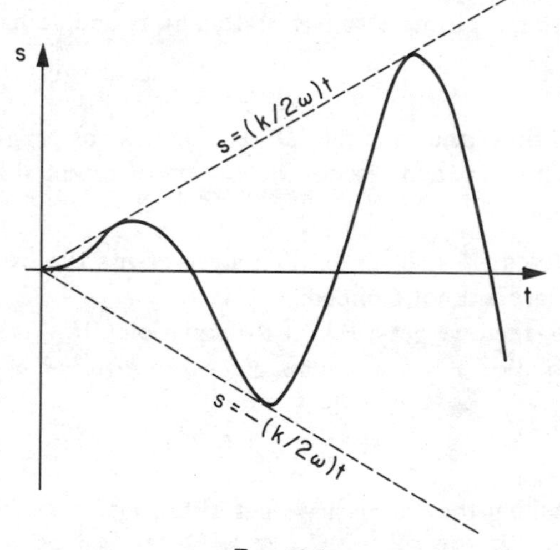

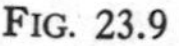
FIG. 23.9

the equation of motion remains valid only so long as s is small, and hence the solution we have found could not be valid except for a limited time. In practice some kind of damping is always present in any case, and this affects the amplitude of the forced oscillations. Nevertheless it is certainly true that when the period of the forced oscillations is close to that of the free oscillations (p close to ω) abnormally large oscillations do occur, a phenomenon known as *resonance*.

The electrical circuit shown in Fig. 23.10 is the same as that shown in

Fig. 23.8 with the addition of an applied voltage $E(t)$. The current i satisfies the equation

$$(D^2 + 2\rho D + \omega^2)i = \frac{1}{L}E'(t),$$

where $\rho = R/2L$, $\omega^2 = 1/LC$, $E'(t) = \mathrm{d}E/\mathrm{d}t$.

It follows that the ideas discussed in this section apply equally to the current in such a circuit.

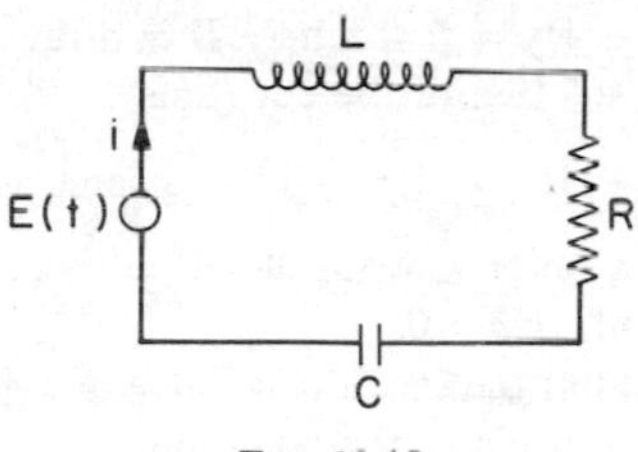

FIG. 23.10

Exercises 23(a)

For brevity, y'', y' are written for $\mathrm{d}^2y/\mathrm{d}x^2$, $\mathrm{d}y/\mathrm{d}x$, and $\ddot{x}$, $\dot{x}$ for $\mathrm{d}^2x/\mathrm{d}t^2$, $\mathrm{d}x/\mathrm{d}t$ respectively. Unless otherwise stated, $D \equiv \mathrm{d}/\mathrm{d}x$. Solve the differential equations in Nos. 1–36:

1. $\ddot{x} + 9x = e^{-t}$, given that $x = 0$ and $\dot{x} = 1$ when $t = 0$.
2. $y'' + 4y = x - \sin 3x$.
3. $y'' + y = e^{-x} \sin x$.
4. $\ddot{x} - 4\dot{x} + 3x = t + e^{2t}$.
5. $y'' + y' = 1 + x^2$.
6. $y'' + 4y' + 4y = 18xe^{-2x}$.
7. $y'' - 2y' + y = e^x \cos x$.
8. $y'' + 4y' + 3y = \sin x$.
9. $y'' - 5y' + 4y = xe^x$.
10. $y'' + 3y' + 2y = 2xe^{-x}$.
11. $y'' + 4y' + 13y = e^{-2x} \cos x$.
12. $y'' - 3y' + 2y = e^x(1 + x)$.
13. $y'' - 4y' - 5y = \cos x$.
14. $4\ddot{y} - 4\dot{y} + 5y = 17 \cos t$, given that $y = 2$, $\dot{y} = -7\frac{1}{2}$ when $t = 0$.
15. $y'' - 5y' + 4y = xe^x$.
16. $y'' - 2y' + 2y = \sin 2x$.
17. $a^2y'' - 2aby' + 4b^2y = 4b^3x^2$, where a and b are non-zero constants.
18. $y'' + 4y' + 4y = \cosh 2x$.
19. $y'' - 2y' + y = (x + 1)^2e^{2x}$.
20. $y'' + 2y' + 3y = e^{-x} + \cos 2x$.
21. $y'' + 2y' + 2y = x + \sin x$.
22. $y'' - 9y = \cosh 3x + x^2$.
23. $\ddot{s} + 6\dot{s} + 25s = 24 \cos 4t$.
24. $y'' - 4y' + 5y = xe^{2x} \sin x$.
25. $y'' - 2y' + 2y = e^x \sin x$, given that $y = 0$ when $x = 0$, and $y' = 0$ when $x = 2\pi$.
26. $y'' + 4y' + 5y = (1 + e^{-2x}) \cos x$.
27. $2y'' + 3y' - 2y = x^2 + \sin 2x$.
28. $(D^3 + 4D)y = \sin x - x$.
29. $(D^3 + 2D^2 + 4D + 8)y = \cos x + 2 \sin x + 5 + 6x$.
30. $(D^5 - 3D^3 - 4D)y = 2x^3 + 3 + \sin 2x$.

31. (a) $(D^2 + 4)(D - 1)^2y = 16e^{2x} + 6 \cos x$;
(b) $(D^3 + D^2 + D + 1)y = e^{-x} + x^4$.
32. $(D^4 + 3D^2 - 4)y = 170e^x \sin^2 x$.
33. $(D^3 + D^2 + D + 1)x = \cos t$, where $D \equiv d/dt$, with the initial conditions $x = Dx = D^2x = 0$ at $t = 0$.
34. $(D^3 - 3D^2 + 7D - 5)y = 5x^2 - 14x + 6$, where $y = Dy = D^2y = 0$ when $x = 0$.
35. (a) $(D^4 - 1)y = \sinh x + \sin 2x$;
(b) $D^2(D + 1)y = x + \sin x$.
36. (a) $(D^2 - 2D + 5)y = \cos 2t$, where $D \equiv d/dt$;
(b) $(D^3 - D^2 + 4D - 4)y = 5e^t$, where $D \equiv d/dt$.
37. (a) Find the general solution of the equation

$$(D^4 + 3D^2 - 4)y = x^2 + 6 \sin x + 24e^{2x}.$$

(b) The abscissa of a point moving along the x-axis satisfies the equation $\dot{x} + 2x = 4ae^{-4t}$, where $a > 0$.
When $t = 0$, $x = a$. Find the maximum value of x for $t > 0$.
38. (a) Find the general solution of the equation

$$(D^5 - D)y = 1 + 6e^{2x} + \sin 2x.$$

(b) The abscissa of a point moving along the x-axis satisfies the equation $\ddot{x} + 9n^2x = -6an^2 \cos nt$, where a, n are positive constants.
It is given that $x = -2a/3$, $\dot{x} = 0$ when $t = 0$. Find the maximum speed of the point.
39. Find the particular integral of

$$\ddot{x} + 2h\dot{x} + (h^2 + p^2)x = ke^{-ht} \cos pt,$$

where x is the distance, t the time, and h, p, k are constants. Show that it represents a vibration of variable amplitude, and that this amplitude is a maximum when $ht = 1$.
40. The differential equation of a vibrating system is

$$\ddot{x} + 2\dot{x} + 5x = 4e^{-t} \cos 2t,$$

where x represents distance and t time.
Solve the equation completely and find the particular solution satisfying the initial conditions $x = \dot{x} = 0$ when $t = 0$. Show that the particular solution represents an oscillation of varying amplitude, the maximum value of the amplitude being $1/e$.
41. If $V = (Ar + B/r)f(\theta)$, where A and B are constants, satisfies the equation

$$r^2 \frac{\partial^2 V}{\partial r^2} + r \frac{\partial V}{\partial r} + \frac{\partial^2 V}{\partial \theta^2} = 0,$$

find the form of $f(\theta)$.

23.6 Change of Variable

There are two types of change of variable that may be used in connection with differential equations:

(a) change of dependent variable,
(b) change of independent variable.

In (a) we replace the dependent variable, y say, by another dependent variable, v say, that is related to y in some way. The object is to change the given D.E., which involves y and its derivatives dy/dx, $d^2y/dx^2, \ldots$, into a D.E. involving v and its derivatives dv/dx, $d^2v/dx^2, \ldots$.

In (b) we replace the independent variable, x say, by another independent variable, u say, that is related to x in some way. The object is to change the given D.E., which involves y and its derivatives with respect to x (dy/dx, d^2y/dx^2, ...), into a D.E. involving y and its derivatives with respect to u (dy/du, d^2y/du^2, ...).

23.6.1 Change of Dependent Variable The most usual change of dependent variable is of the form

$$y = f(x)v. \tag{1}$$

From this the product rule of differentiation gives

$$\frac{dy}{dx} = f(x)\frac{dv}{dx} + f'(x)v, \tag{2}$$

and

$$\frac{d^2y}{dx^2} = f(x)\frac{d^2v}{dx^2} + 2f'(x)\frac{dv}{dx} + f''(x)v. \tag{3}$$

For a second-order D.E. we can now make the required transformation by substituting from (1)–(3) in the given D.E.

Example 29 Solve the equation

$$x^2\frac{d^2y}{dx^2} + 2x(x+2)\frac{dy}{dx} + 2(x+1)^2y = e^{-x}\cos x$$

by using the change of variable $y = x^{-2}v$.

Here

$$\frac{dy}{dx} = x^{-2}\frac{dv}{dx} - 2x^{-3}v,$$

$$\frac{d^2y}{dx^2} = x^{-2}\frac{d^2v}{dx^2} - 4x^{-3}\frac{dv}{dx} + 6x^{-4}v.$$

Substituting in the given D.E. we get

$$\frac{d^2v}{dx^2} - 4x^{-1}\frac{dv}{dx} + 6x^{-2}v + (2x^2+4x)\left(x^{-2}\frac{dv}{dx} - 2x^{-3}v\right) + (2x^2+4x+2)x^{-2}\,v = e^{-x}\cos x,$$

that is

$$\frac{d^2v}{dx^2} + 2\frac{dv}{dx} + 2v = e^{-x}\cos x.$$

This equation has constant coefficients and may be solved by the methods of Section 23.4.6 to give

$$v = e^{-x}(\tfrac{1}{2}x\sin x + A\cos x + B\sin x).$$

Hence

$$y = x^{-2}e^{-x}(\tfrac{1}{2}x\sin x + A\cos x + B\sin x).$$

23.6.2 Change of Independent Variable Assume the relationship between x and u is given in the form

$$x = g(u). \tag{1}$$

Then

$$\frac{dy}{dx} = \frac{dy/du}{dx/du} = \frac{1}{g'(u)}\frac{dy}{du}, \tag{2}$$

$$\frac{d^2y}{dx^2} = \frac{d}{dx}\left(\frac{dy}{dx}\right) = \frac{d/du\,(dy/dx)}{dx/du} = \frac{1}{g'(u)}\frac{d}{du}\left(\frac{1}{g'(u)}\frac{dy}{du}\right). \tag{3}$$

For a second-order D.E. we can now make the required transformation by substituting from (1)–(3) in the given D.E.

Example 30 Solve the equation $\sin x\,d^2y/dx^2 - \cos x\,dy/dx = y\sin^3 x$ by using the substitution $u = \cos x$.

Here

$$x = \cos^{-1} u, \tag{1}$$

giving

$$\frac{dx}{du} = \frac{-1}{\sqrt{(1-u^2)}}.$$

Now

$$\frac{dy}{dx} = \frac{dy/du}{dx/du} = -\sqrt{(1-u^2)}\,\frac{dy}{du}. \tag{2}$$

Also

$$\frac{d^2y}{dx^2}=\frac{d}{dx}\left(\frac{dy}{dx}\right)=\frac{d/du\,(dy/dx)}{dx/du}=-\sqrt{(1-u^2)}\,\frac{d}{du}\left(-\sqrt{(1-u^2)}\,\frac{dy}{du}\right)$$

$$=(1-u^2)\frac{d^2y}{du^2}-u\frac{dy}{du}. \tag{3}$$

Substituting from (1)–(3) in the given D.E. we get

$$\sqrt{(1-u^2)}\left[(1-u^2)\frac{d^2y}{du^2}-u\frac{dy}{du}\right]+u\sqrt{(1-u^2)}\,\frac{dy}{du}=y(1-u^2)^{3/2},$$

so that

$$(1-u^2)^{3/2}\frac{d^2y}{du^2}=y(1-u^2)^{3/2},$$

that is

$$\frac{d^2y}{du^2}-y=0,$$

or

$$(D-1)(D+1)y=0, \qquad \text{where } D\equiv d/du.$$

Hence

$$y=Ae^u+Be^{-u}$$

and so

$$y=Ae^{\cos x}+Be^{-\cos x}.$$

23.6.3 Cauchy–Euler Equations This name is applied to equations of the form

$$a_0x^n\frac{d^ny}{dx^n}+a_1x^{n-1}\frac{d^{n-1}y}{dx^{n-1}}+\cdots+a_{n-1}x\frac{dy}{dx}+a_ny=b(x),$$

where $a_0, a_1, \ldots, a_n$ are constants. In particular the second-order equation has the form

$$a_0x^2\frac{d^2y}{dx^2}+a_1x\frac{dy}{dx}+a_2y=b(x),$$

as in Examples 14 and 15.

Substituting $x=e^u$, we have $dx/du=e^u$, and so

$$\frac{dy}{dx}=\frac{dy/du}{dx/du}=e^{-u}\frac{dy}{du}, \tag{1}$$

$$\frac{d^2y}{dx^2}=\frac{d}{du}\left(e^{-u}\frac{dy}{du}\right)\Big/\frac{dx}{du}=e^{-u}\left[e^{-u}\frac{d^2y}{du^2}-e^{-u}\frac{dy}{du}\right]$$

$$=e^{-2u}\left(\frac{d^2y}{du^2}-\frac{dy}{du}\right). \qquad (2)$$

From (1) we get

$$x\frac{dy}{dx}=\frac{dy}{du}$$

and from (2) we get

$$x^2\frac{d^2y}{dx^2}=\frac{d^2y}{du^2}-\frac{dy}{du}.$$

Writing $D=$ d/dx and $\mathscr{D}$ = d/du we have

$$xD=\mathscr{D},\quad x^2D^2=\mathscr{D}(\mathscr{D}-1)$$

and similar results can be verified for higher powers, e.g.

$$x^3D^3=\mathscr{D}(\mathscr{D}-1)(\mathscr{D}-2).$$

Hence, in terms of the new independent variable u, the equation has constant coefficients.

Example 31 Solve the equation

$$x^2\frac{d^2y}{dx^2}+2x\frac{dy}{dx}-2y=x^3+\ln x.$$

Using the above substitution, the equation becomes

$$[\mathscr{D}(\mathscr{D}-1)+2\mathscr{D}-2]y=e^{3u}+u,$$

that is

$$(\mathscr{D}^2+\mathscr{D}-2)y=e^{3u}+u,$$

that is

$$(\mathscr{D}+2)(\mathscr{D}-1)y=e^{3u}+u.$$

The C.F. is $Ae^{-2u}+Be^{u}$.

To find a P.I. we use the trial solution $y=ae^{3u}+bu+c$ and the equation becomes

$$10ae^{3u}-2bu+(b-2c)=e^{3u}+u.$$

Hence we require that $10a=1, -2b=1, b-2c=0$, that is

$$a=\tfrac{1}{10},\quad b=-\tfrac{1}{2},\quad c=-\tfrac{1}{4}.$$

Thus $y = \frac{1}{10}e^{3u} - \frac{1}{4}(2u + 1)$ is a P.I. and the complete solution is

$$\begin{aligned} y &= \tfrac{1}{10}e^{3u} - \tfrac{1}{4}(2u + 1) + Ae^{-2u} + Be^{u} \\ &= \tfrac{1}{10}x^3 - \tfrac{1}{4}(2\ln x + 1) + A/x^2 + Bx. \end{aligned}$$

Exercises 23(b)

For brevity, y'', y' are written for d^2y/dx^2, dy/dx respectively.

1. By the substitution $x = e^t$ transform the equation

$$x^2y'' - 3xy' + 4y = 6x^2 \ln x + 6/x$$

and hence solve it.

Find also the particular solution such that $y = 0$ and $y' = 1$, when $x = 1$.

2. Find the complete solution of the equation

$$x^2y'' - 4xy' + 6y = x^3.$$

3. If y is a function of x, $x = e^t$ and $D \equiv d/dx$, $\Delta \equiv d/dt$, prove that

$$x^2D^2y = \Delta(\Delta - 1)y.$$

Find the solution of $x^2y'' - xy' + 2y = 2x$, such that $y = 1$ and $y' = 0$ at $x = 1$.

4. (a) Solve the equation $y'' - 2y' + y = 3 \sinh x$.
   (b) Given $x = e^z$, show that $xy' = dy/dz$ and $x^2y'' = d^2y/dz^2 - dy/dz$. Hence, or otherwise, solve the equation

$$x^2y'' - 3xy' + 5y = 3x + \ln x.$$

5. (a) Solve the differential equation $y'' - 4y' + 4y = e^{2x}$.
   (b) Given that $x = e^{\theta}$, and that D and $\mathscr{D}$ denote the operators d/dx and $d/d\theta$ respectively, prove that $x^2D^2 = \mathscr{D}(\mathscr{D} - 1)$.
   Solve the differential equation $x^2y'' + y = x^2$.

Solve the differential equations given in Nos. 6–24:

6. (a) $2y'' = x^3 + y + y'$, given that $y = 0$ and $y' = 1$ when $x = 0$;
   (b) $x^2y'' + 3xy' + y = x^2$.
7. (a) $3y'' - 5y' + 2y = e^x(1 + x)$;
   (b) $x^2y'' + 5xy' - 5y = [1 + \cos(\ln x)]/x^2$.
8. (a) $y'' - 2y' + 2y = e^x(1 + \cos x)$;
   (b) $x^2y'' - xy' + y = x^2 + \ln x$.
9. (a) $y'' - 5y' + 6y = x^3e^x$;
   (b) $x^2y'' + 4xy' + 2y = (x - 1/x)^2$.
10. (a) $y'' - y = \cosh x$;
    (b) $x^2y'' - xy' - 3y = x^2 \ln x$.
11. (a) $y'' + y' + y = e^x \sin x$;
    (b) $x^2y'' - 3xy' + 4y = x^2$.

12. (a) $y'' - 2y' + 2y = e^x \cos x$;
    (b) $x^2y'' - xy' + y = 4x^3$.
13. (a) $y'' + 4y' + 5y = 4x + 6e^{-3x} + \sin x$;
    (b) $x^2y'' - xy' + y = 2x \ln x$, given that $y = 1$, $y' = 0$ when $x = 1$.
14. (a) $y'' - 6y' + 9y = e^{3x}(1 + x)$;
    (b) $x^2y'' - 2xy' - 4y = x^2 + 2 \ln x$.
15. (a) $y'' - 2y' + 5y = e^x \sin 2x$;
    (b) $x^2y'' - xy' + y = x^3$.
16. (a) $y'' + 6y' + 9y = x^4 + x + e^{-3x}$;
    (b) $x^2y'' - 2xy' + 2y = (x + 1)^2$, given that $y(1) = \frac{3}{2}$, $y(e) = \frac{1}{2}$.
17. (a) $y'' + y' + y = e^x(x + \sin x)$;
    (b) $4x^2y'' + xy' - y = x + \ln x$.
18. (a) $y'' + y = e^x \cos x + x^2$;
    (b) $x^2y'' + 4xy' + 2y = (\ln x)/x^2$.
19. (a) $y'' - y' - 2y = e^{-x}(1 + \sin x)$;
    (b) $x^2y'' + 5xy' + 3y = (1 + 1/x)^2 \ln x$.
20. (a) $y'' - 4y' + 13y = e^{2x} \sin 3x$;
    (b) $9x^2y'' + 3xy' + y = x \ln x$.
21. (a) $y'' - 4y' + 5y = e^{2x} \sin x$;
    (b) $x^2y'' - 3xy' + 4y = x^2 \ln x$.
22. (a) $y'' - 4y' + 4y = \cosh 2x + x \sinh 2x$;
    (b) $x^2y'' + 4xy' + 2y = x - 1/x$.
23. $2x^3y''' + 4x^2y'' + xy' - y = 3x$.
24. $x^3y''' - 3x^2y'' + 6xy' - 6y = 12x^4 - x^2$.
25. Given $V = x^n f(y/x)$, where f is any function, prove that

$$x\frac{\partial V}{\partial x} + y\frac{\partial V}{\partial y} = nV.$$

Show that if $\partial^2 V/\partial x\, \partial y = 0$, and u denotes y/x, then

$$(n-1)\frac{\mathrm{d}f}{\mathrm{d}u} = u\frac{\mathrm{d}^2 f}{\mathrm{d}u^2},$$

and hence verify that V is of the form $ax^n + by^n$, where a and b are constants.

26. Transform the partial differential equation $\partial^2 V/\partial x^2 + \partial^2 V/\partial y^2 = 0$ to the form

$$\frac{\partial^2 V}{\partial r^2} + \frac{1}{r}\frac{\partial V}{\partial r} + \frac{1}{r^2}\frac{\partial^2 V}{\partial \theta^2} = 0,$$

where $x = r\cos\theta$, $y = r\sin\theta$.

(a) Given that V is a function of r only, find the most general form of V satisfying the equation.
(b) Given that $V = r^n f(\theta)$, where n is a constant, find the most general form of V satisfying the equation.

27. Given $r = \sqrt{(x^2 + y^2)}$ and $u = f(r)$, show that

$$\frac{\partial^2 u}{\partial x^2} + \frac{\partial^2 u}{\partial y^2} = f''(r) + \frac{1}{r} f'(r)$$

and find u in terms of r if $\partial^2 u/\partial x^2 + \partial^2 u/\partial y^2 = 0$.

28. By changing the independent variable by means of the transformation $x = z^2$, or otherwise, solve the differential equation

$$2xy'' + y' + 2y = x^2.$$

29. Transform the differential equation

$$x^2 y'' + (3x^2 + 4x)y' + (2x^2 + 6x + 2)y = 0$$

by the substitution $x^2 y = z$.

Hence, or otherwise, solve the equation, given that $y = e^{-2}$ when $x = 1$ and $y = e^2$ when $x = -1$.

30. By means of the substitution $y = zx^{-1/2}$, transform the differential equation $4x^2 y'' + 4xy' + (4x^2 - 1)y = 0$, and obtain the solution for which $y = 0$ when $x = \frac{1}{2}\pi$ and $y = 1$ when $x = \pi$.

31. Transform the equation

$$x^2 y'' + (4x^2 + 6x)y' + (3x^2 + 12x + 6)y = 0$$

by the substitution $y = z/x^3$.

Hence, or otherwise, solve the equation, given that $y = e^{-1}$ and $y' = -4e^{-1}$ when $x = 1$.

32. By using the substitution $z = x - y$, or otherwise, solve the equation

$$y'^2 - 2y' - \frac{x^2}{1 - x^2} = 0$$

given that $y = 0$ when $x = 0$.

33. If $t = \sin x$ and if y is a function of x, prove that

$$\frac{d^2 y}{dx^2} = \frac{d^2 y}{dt^2} \cos^2 x - \frac{dy}{dt} \sin x$$

Transform the differential equation

$$y'' + y' \tan x + y \cos^2 x = 2e^{\sin x} \cos^2 x$$

into an equation connecting y and t, where $t = \sin x$. Solve the resulting equation, and hence find the solution of the given equation that satisfies the conditions $y = 1$ and $y' = 0$ when $x = 0$.

34. Show that the constant n may be chosen so that, by the substitution $y = x^n z$, the differential equation

$$x^2 y'' + 4x(x + 1)y' + (8x + 2)y = \cos x$$

reduces to the form $z'' + az' + bz = \cos x$, where a and b are constant.

Hence, or otherwise, solve the given equation.

35. Using the substitution $z = \sqrt{x}$, or otherwise, solve the equation

$$4xy'' + 2(1-\sqrt{x})y' - 6y = e^{-2\sqrt{x}}.$$

36. Prove that, in general, at every point on any curve in the xy-plane,

$$\frac{d^2x}{dy^2} + \left(\frac{dx}{dy}\right)^3 \frac{d^2y}{dx^2} = 0.$$

Transform the differential equation

$$\frac{d^2y}{dx^2} + 3\left(\frac{dy}{dx}\right)^2 = (2x-y)\left(\frac{dy}{dx}\right)^3,$$

so that y is the independent variable and x is the unknown function of y. Hence obtain the general solution of the given differential equation.

37. Given $x = \cosh z$, prove that $(x^2 - 1)y'' + xy' = d^2y/dz^2$.
    Solve the equation $(x^2 - 1)y'' + xy' - y = x$.

38. Given that y is a function of x and $x = \tan u$, prove that

$$(1+x^2)\frac{dy}{dx} = \frac{dy}{du}$$

and calculate d^2y/du^2 in terms of x, dy/dx, d^2y/dx^2.
    Find a solution involving two arbitrary constants of the equation

$$(1+x^2)^2y'' + 2x(1+x^2)y' = \tan^{-1} x.$$

What is the solution for which $y = 0$ and $y' = 1$ when $x = 0$?

39. Transform the differential equation

$$y'' \cos x + y' \sin x + 4y \cos^3 x = 8 \cos^5 x$$

into one having t as independent variable, where $t = \sin x$, and hence solve the equation.

40. By the substitution $x = \sinh t$, transform the equation

$$(1+x^2)y'' + xy' + y = 1 + x^2$$

and hence solve it.
    Find the solution for which $y = 0$ and $y' = 0$ when $x = 0$.

23.7 Simultaneous Linear Equations with Constant Coefficients

The use of operators in the solution of simultaneous linear equations with constant coefficients is demonstrated in the following examples.

Example 32 Solve the equations

$$\frac{dx}{dt} + y = \sin t + 1, \quad \frac{dy}{dt} + x = \cos t,$$

given that $x = 2$, $y = 1$ when $t = 0$.

The equations may be written in the form

$$Dx + y = \sin t + 1, \tag{1}$$

$$x + Dy = \cos t. \tag{2}$$

Operating on (1) with D, we have

$$D^2x + Dy = \cos t \tag{3}$$

and from (2) and (3)

$$(D^2 - 1)x = 0$$

which gives

$$x = Ae^t + Be^{-t}. \tag{4}$$

But from (1)

$$y = \sin t + 1 - Ae^t + Be^{-t}.$$

Since $x = 2$ when $t = 0$ we have $2 = A + B$, and since $y = 1$ when $t = 0$ we have $1 = 1 - A + B$; hence $A = B = 1$.

Hence the required solution is

$$x = e^t + e^{-t},$$

$$y = \sin t + 1 - e^t + e^{-t}.$$

Note. To avoid introducing more arbitrary constants than are required, when one of the unknowns (in this case x) has been determined, the other (in this case y) should if possible be found by means of a relation that does not involve derivatives of y.

Example 33 Solve the equations

$$2\frac{di_1}{dt} + i_1 + \frac{di_2}{dt} = \cos t, \quad \frac{di_1}{dt} + 2\frac{di_2}{dt} + i_2 = 0,$$

given that $i_2 = i_1 = 0$ when $t = 0$.

The equations are

$$(2D + 1)i_1 + Di_2 = \cos t, \tag{1}$$

$$Di_1 + (2D + 1)i_2 = 0. \tag{2}$$

We eliminate i_1 by operating on (1) with D and on (2) with $(2D + 1)$; thus

$$D(2D + 1)i_1 + D^2i_2 = -\sin t,$$

$$D(2D + 1)i_1 + (2D + 1)^2i_2 = 0,$$

giving

$$(3D^2 + 4D + 1)i_2 = \sin t = \text{Im}\,(e^{it}).$$

The C.F. is $Ae^{-t} + Be^{-t/3}$. To find a P.I. put $i_2 = ae^{it}$ and we find

$$a = \frac{1}{4i - 2} = -\frac{1}{10}(1 + 2i).$$

Hence a P.I. is given by

$$i_2 = \text{Im}\,[-\tfrac{1}{10}(1 + 2i)(\cos t + i \sin t)] = -\tfrac{1}{10}(\sin t + 2\cos t).$$

Hence the complete solution is

$$i_2 = -\tfrac{1}{10}(\sin t + 2\cos t) + Ae^{-t} + Be^{-t/3}.$$

To get i_1 in terms of i_2 and its derivatives we may now eliminate Di_1 between (1) and (2). Multiply (2) by 2 and subtract from (1) to obtain

$$i_1 - (3D + 2)i_2 = \cos t,$$

that is

$$i_1 = Be^{-t/3} - Ae^{-t} + \tfrac{1}{10}(3\cos t + 4\sin t).$$

Applying the given conditions we find that

$$0 = B - A + \tfrac{3}{10}, \quad 0 = A + B - \tfrac{1}{5},$$

that is

$$A = \tfrac{1}{4}, \quad B = -\tfrac{1}{20}.$$

Hence

$$i_1 = \tfrac{1}{20}(6\cos t + 8\sin t - e^{-t/3} - 5e^{-t}),$$

$$i_2 = \tfrac{1}{20}(5e^{-t} - e^{-t/3} - 4\cos t - 2\sin t).$$

Exercises 23(c)

For brevity, $\dot{x}$, $\dot{y}$, are written for dx/dt, dy/dt respectively.

1. Given that x and y are functions of t such that $\dot{x} = 3x - y$, $\dot{y} = x + y$, and $x = 1, y = 0$ at $t = 0$, show that $x - y = e^{2t}$.
2. Solve the simultaneous equations $\dot{y} + ay = x$, $\dot{x} + ax = y$, given that $x = 0$ and $y = 1$ when $t = 0$.
3. A point (x, y) moves in accordance with the equations

$$\dot{x} + 2y = 5e^t, \quad \dot{y} - 2x = 5e^t.$$

It is given that $x = -1$ and $y = 3$ when $t = 0$. Show that the point moves in a straight line.

4. Find x and y in terms of t, given that $\dot{y} + y = 3x$, $x + 2x = 2y$, and that $x = 0$ and $\dot{y} = \frac{1}{2}$ when $t = 0$.

Solve the pairs of simultaneous differential equations given in Nos. 5–18:

5. $\dot{x} + 2x + y = 0$, $\dot{y} + x + 2y = 0$, subject to the conditions that $x = 1$ and $y = 0$ when $t = 0$.
6. $\dot{x} = 4x - 2y + e^t$, $\dot{y} = 6x - 3y$.
7. $3\dot{x} + 3x + 2y = e^t$, $4x - 3\dot{y} + 3y = 0$.
8. $\dot{y} - 2x = \cos 2t$, $x + 2y = -\sin 2t$.
9. $\dot{x} + 3x - 2y = 1$, $\dot{y} - 2x + 3y = e^t$, given that, when $t = 0$, $x = y = 0$.
10. $\dot{x} + \dot{y} + 2x + y = e^{-3t}$, $\dot{y} + 5x + 3y = 5e^{-2t}$, given that, when $t = 0$, $x = -1$ and $y = 4$.
11. $\dot{x} + x - y = te^t$, $2y - \dot{x} + \dot{y} = e^t$. Find the particular solution for which $x = y = 0$ when $t = 0$.
12. $\dot{x} + 5x - 2y = t$, $\dot{y} + 2x + y = 0$, given that $x = 0$ and $y = 0$ when $t = 0$.
13. $2\dot{x} + \dot{y} = 2t - x - 2y$, $\dot{x} + \dot{y} = 3t + x - 3y$.

    If $x = 0$ and $y = -\frac{1}{5}$ at the initial instant $t = 0$, show that the point (x, y) describes the curve $5x + 1 = e^{x-y-1/5}$.
14. $\dot{x} + \dot{y} + x = 0$, $2\dot{x} + \dot{y} - y = 1$, subject to the conditions $x = y = 0$ at $t = 0$.
15. $2\dot{x} - \dot{y} + 3x = 2t$, $\dot{x} + 2\dot{y} - 2x - y = t^2 - t$, given that $x = 1$ and $y = 1$ when $t = 0$.
16. $\dot{x} + x - 2\dot{y} = \cos t$, $\dot{x} - \dot{y} + 6y = 0$.
17. $3x + \dot{y} - 5y = e^t$, $\dot{x} - 5x + 3y = t$, given that when $t = 0$, $x = -\frac{3}{7}$ and $y = -\frac{4}{7}$.
18. The co-ordinates of a point P moving in the xy-plane satisfy the differential equations

$$\dot{x} + y = \sin 2t, \quad \dot{y} - x = 2 \cos 2t.$$

It is given that $x = 1$, $y = 0$ when $t = 0$. Prove that the path of the point is given by

$$y^2 = (1 - x^2)(2x + 1)^2.$$

19. The co-ordinates x, y of a point P moving in the xy-plane satisfy the equations

$$\dot{x} - y = 2 \cos t, \quad \dot{x} + \dot{y} + 2x + 2y = 3 \sin t - \cos t.$$

When $t = 0$, $x = 1$ and $y = -2$. Find the co-ordinates of P when $t = \frac{1}{2}\pi$.

20. By using $\zeta = u + iv$, or otherwise, solve the simultaneous equations

$$m \frac{du}{dt} = eE - evH, \quad m \frac{dv}{dt} = euH,$$

where m, e, E, and H are constants. If $u = dx/dt$ and $v = dy/dt$, find x and y as a function of the time t from your solution for ζ, and show further that if $x = y = u = v = 0$, when $t = 0$,

$$x = (E/\omega H)(1 - \cos \omega t), \quad y = (E/\omega H)(\omega t - \sin \omega t),$$

where $\omega = eH/m$.

ANSWERS TO EXERCISES

Exercises 1

1. $x < 2$
2. $-3 < x < 1/2$
3. $-2/3 < x < 1/2$
4. $-5 < x < -2$ or $-1 < x < 2$
5. $-1 < x < 1$
6. $x < 1$ or $x > 5/2$
7. $x < 1$ or $x > 2$.

Exercises 2(a)

2. Take equation of $X'Y'$ as $x/b - y/a = k$; then XY' is $x - y/k = a \ldots$ (i) and $X'Y$ is $x/k + y = b \ldots$ (ii). Multiply (i) by x and (ii) by y and add.
4. $\lambda = 0, -\frac{585}{14}$; $3x - y = 1$, $3x + y = 14$
6. $\cos^{-1} \lambda$
7. $x^2 + y^2 - 2x - 1 = 0$,
$x^2 + y^2 + 6x + 7 = 0$
8. $x^2 + y^2 - x - y = 0$
9. $x^2 + y^2 - 3x - 7y + 14 = 0$,
$x^2 + y^2 + 2x - 4y - 3 = 0$.

Exercises 2(b)

1. $y^2 = 2a(x - a)$
2. Point is $(4a, 0)$; locus is $y^2 = 2a(x - 4a)$
4. $y^2 = a(x + a)$
7. $bx \cos \frac{1}{2}\theta + ay \sin \frac{1}{2}\theta = ab \cos \frac{1}{2}\theta$; $ay \cos \frac{1}{2}\theta - bx \sin \frac{1}{2}\theta = ab \sin \frac{1}{2}\theta$; $KK' = 2b^2/a$.
8. $x^2/a^2 + y^2/b^2 = 1$
10. $\dfrac{x^2}{36} + \dfrac{y^2}{20} = 1$
15. $(y^2 - x^2)^2 = a^2(2x^2 - y^2)$.

Exercises 2(c)

1. (a) $(-1, 1)$, $x = -1$ (b) $(4, 0)$, $y = 0$. **2.** $(-1, 1)$
3. $(-2, -1)$, $x = 2y$, $x + 2y + 4 = 0$.
4. $y = 3 + x + 2x^2$, $x + \frac{1}{4} = 0$, $(-1/4, 23/8)$, $y + 2x^2 + x = 11/4$.
5. $u = -1$, $v = 0$.

6. (a) $2X^2 + Y^2 = 1$, (b) $\frac{1}{4}X^2 + Y^2 = 1$.
7. $X^2 - Y^2 = 1, \pi/2$.
8. $\frac{(1+h)}{c^2}X^2 + \frac{(1-h)}{c^2}Y^2 = 1$. 9. $c = 1, (3/2)X^2 + \frac{1}{2}Y^2 = 1$.

Exercises 3(a)

1. $5 \times 21 \times 5 = 525$. 2. (a) $7^3 = 343$, (b) $7^2 \times 4 = 196$.
3. $90\,000 = 2^4 \times 3^2 \times 5^4$ and so factors are $2^a \times 3^b \times 5^c$ where $a = 0, 1, 2, 3$ or 4, $b = 0$, 1 or 2, $c = 0, 1, 2, 3$ or 4; hence number of factors is $5 \times 3 \times 5 = 75$ i.e. 73 excluding 1 and 90 000. Number of even factors is $4 \times 3 \times 5 = 60$, i.e. 59 if 90 000 excluded.
7. (a) $2 \times 6! = 1440$, (b) $7! - 1440 = 3600$.
8. $2 \times {}^6C_2 = 30$, (a) $2 \times {}^nC_2$, (b) ${}^nC_2 + {}^{n+1}C_2$
9. ${}^8P_8 = 40\,320$, ${}^7P_7 = 5040$.
10. (a) ${}^{52}C_3 = 22\,100$, (b) ${}^{13}C_3 = 286$, (c) $13 \times {}^4C_3 = 78$, (d) equals number of ways of selecting two of value 1 and one of value 2, i.e. $4 \times {}^4C_2 = 24$ ways.
11. $4 \times 3 \times 3 \times 3 = 108$.
12. Let S mean 'walk south for one block', E mean 'walk east for one block'. Then any sequence $SEESSS$ ...... gives a possible route if it contains $(p-1)$ of S and $(q-1)$ of E. Hence number of ways is

$$\frac{(p+q-2)!}{(p-1)!(q-1)!}.$$

Exercises 3(b)

1. -11 at (5, 3), -48 at $(-4, 0)$. 2. 12 at (0, 4).
3. 2 at (1, 1), 1/4 at (1/4, 1/2). 4. 3 at (1, 1), 6 at (4, −2).
5. 40 acres of A, 30 acres of B (30 acres unused). 6. 1 day, 3 days.

Exercises 3(c)

1. 5 at (2, 3)
2. (a) $M \cap \{(S \cap L) \cup (F \cap G) \cup (F \cap L) \cup (G \cap L)\} = C$ say,
(b) $\{(M \cap S \cap F) \cup (M \cap S \cap G) \cup (F \cap G \cap L) \cup (S \cap F \cap G) \cup (S \cap F \cap L) \cup (S \cap G \cap L)\} \cap C'$
3. $B' \cap A'$, 53. 4. $(B \cap C \cap D) \subset A$. 5. Not commutative, false.
6. Missing values are three zeros.

Exercises 3(d)

1. (a) ${}^5C_4 \times {}^4C_2 = 30$, (b) ${}^5C_4 \times {}^3C_2 + {}^6C_4 \times {}^4C_2 = 105$,
(c) ${}^{11}C_4 \times {}^7C_2 = 27\,720$. Probability $= 30/{}^9C_6 = 5/14$.
2. See Exercises 3(a), No. 10. (a) 11/850, (b) 22/425, (c) 1/425, (d) 6/5525.
3. $({}^9C_7 \times {}^3C_2)/{}^{12}C_9 = 27/55$. Ladies who can drive or are incapacitated; those who are incapacitated or are lady-drivers.
4. $3 \times {}^9C_4 \times 2 \times {}^5C_3 = 7560$, $1/(3 \times {}^9C_4) = 1/378$.

5. $n(n-1)(n-2)/n^3 > 0{\cdot}9$ if $n \geqslant 30$.
6. (a) ${}^7C_3 \times 1/2^7 = 35/128$, (b) $(1/2^7)({}^7C_5 + {}^7C_6 + {}^7C_7) = 29/128$.
7. (a) $({}^{50}C_3 \times {}^{50}C_2)/{}^{100}C_5$, (b) $({}^{20}C_3 \times {}^{80}C_2)/{}^{100}C_5$, (c) $({}^{20}C_3 \times {}^{80}C_2 + {}^{20}C_4 \times 80 + {}^{20}C_5)/{}^{100}C_5$.
8. (a) $16/20 \times 15/19 = 12/19$, (b) $4/20 \times 16/19 + 16/20 \times 4/19 = 32/95$, (c) $32/95 + 4/20 \times 3/19 = 7/19$; $4/20 \times 3/19 \times 16/18 = 8/285$.
9. (a) ${}^{17}C_2/{}^{20}C_5 = 1/114$, (b) $3/20 \times 2/19 \times 1/18 = 1/1140$.
10. (a) $7/10$, (b) $3/10 \times 7/9 = 7/30$, (c) $3/10 \times 2/9 \times 7/8 = 7/120$, (d) $3/10 \times 2/9 \times 1/8 = 1/120$.

Exercises 4(a)

1. ${}^9C_4 . 2^5 . 3^4$.
2. ${}^{20}C_{10}$.
3. (a) $32 + 80x + 80x^2 + 40x^3 + 10x^4 + x^5$, (b) $1 - 10x + 40x^2 - 80x^3 + 80x^4 - 32x^5$.
4. (a) ${}^{12}C_4 = 495$, (b) $-{}^9C_3 = -84$.
5. $x^6 + 3x^4 + \frac{15}{4}x^2 + 5/2 + \frac{15}{16}x^{-2} + \frac{3}{16}x^{-4} + \frac{1}{64}x^{-6}$, 1 030 378.
6. ${}^9C_3 . 3^3 = 2268$.
9. (a) $1{\cdot}2166$.
10. $1 + 10y + 45y^2 + 120y^3 + 210y^4 + 252y^5 + \cdots$, $a = 55$, $b = 210$, $c = 615$, 1, 10, 55.
12. (a) $64 - 192x^2 + 240x^4 - 160x^6 + 60x^8 - 12x^{10} + x^{12}$, (b) 15, (c) $1{\cdot}194$, (d) $1 + 6x + 21x^2 + 50x^3 + \cdots$, $50|x^3| < 0{\cdot}05$, which is not sufficient for accuracy to two decimal places.

Exercises 4(b)

1. $A = 6, B = 9, C = 2$.
2. $A = -3, B = 3, C = 6, D = 1$.
3. (a) $(2x + y - 4)(x - 2y + 3)$; (b) $(3x + 2y - 4)(2x - 3y + 1)$; (c) $(3x - y)(2x + y - 1)$.
4. (a) $R = \sqrt{2}, \alpha = 45°$; (b) $R = 2, \alpha = 60°$; (c) $R = 5$, $\alpha = 53° \, 8'$.
5. $a = \frac{1}{8}, b = -\frac{1}{6}, c = \frac{1}{12}$; $\Sigma n^7 = \frac{1}{24}n^2(n+1)^2(3n^4 + 6n^3 - n^2 - 4n + 2)$.

Exercises 4(c)

1. $-2(a + b + c)(b - c)(c - a)(a - b)$.
2. $-(b - c)(c - a)(a - b)(bc + ca + ab)$.
3. $(a + b + c)(a^2 + b^2 + c^2 - bc - ca - ab)$.
4. $-(b - c)(c - a)(a - b)(a^2 + b^2 + c^2 + bc + ca + ab)$.
5. $-(b - c)(c - a)(a - b)(bc + ca + ab)$.
6. $-(b - c)(c - a)(a - b)\{3(a^2 + b^2 + c^2) + 5(bc + ca + ab)\}$.
7. $(b - c)(c - a)(a - b)(a^2 + b^2 + c^2 + bc + ca + ab)$.
8. $5(b - c)(c - a)(a - b)(a^2 + b^2 + c^2 - bc - ca - ab)$.
9. $3abc(b + c)(c + a)(a + b)$.
10. $-(b - c)(c - a)(a - b)\{a^3 + b^3 + c^3 + bc(b + c) + ca(c + a) + ab(a + b) + abc\}$.

Exercises 4(d)

1. $3/(x + 4) - 2/(x + 3)$.
2. $2/(x - 4) + 1/(x + 2)$.

3. $\frac{1}{2}\{1/x + 1/(x-2) - 2/(x+1)\}$.
4. $2/(2-x) + 3/(3-x) - 2/(1-x)$.
5. $8/(4-x) - 8/(4+x) - x$.
6. $1/(x-1) + 5/(2x+1) - 7/(2x+3)$.
7. $1/(x-2) + 4/(x-2)^2 + 4/(x-2)^3$.
8. $1/(x-1) + 2/(x+1) + 3/(x+1)^2$.
9. $\frac{1}{2}\{(1/x + 1/(x+2) - 2/x^2\}$.
10. $1/(x-2) - 2/(x+1) - 4/(x+1)^2$.
11. $1/(2x-1) - 2/(x+2)^2$.
12. $2/(2x-7) + 1/(x+4)^2 - 1/(x+4)$.
13. $1/(1+x) - x/(1+x^2)$.
14. $1/x - (x+1)/(x^2+9)$.
15. $1/(x+1) + (x-1)/(x^2-x+1)$.
16. $2/x + 1/(2-x) + 1/(2-x)^2 - 1/(2+x) - 1/(2+x)^2$.
17. $1/x - x/(9+x^2) - 9x/(9+x^2)^2$.
18. $1 - 1/x^2 - \frac{1}{3}\{2/(x+1) + (1-2x)/(x^2-x+1)\}$.
19. $\frac{1}{18}\{(13x-15)/(x^2-2x+3) + 5/(x+1) - 6/(x+1)^2\}$.
20. $1/(x-1) + 2/(x-1)^2 + 2/(x-1)^3$.
21. $2x/(x^2+1) + (3-x)/(x^2+1)^2$.
22. $(1+x)/(1+x+x^2)^2 - 1/(1+x+x^2) + 1/(1-x)^2 - 1/(1-x)$.
23. $x/(x^2+2) - 4x/(x^2+2)^3$.
24. $1/(1-x)^2 - 1/(1+x)^2 + 2/(1-x)^3 - 2/(1+x)^3$.

Exercises 4(e)

1. $\frac{1}{6}n(4n^2 + 21n + 35)$.
2. $\frac{1}{2}n(6n^2 - 3n - 1)$.
3. $n(n+1)(2n^2 + 2n - 1)$.
4. $\frac{1}{6}n(n+1)(9n^2 + 19n + 8)$.
5. $\frac{1}{6}n(n+1)(3n^2 + 5n + 1)$.
6. $\{1 - (n+1)x^n + nx^{n+1}\}/(1-x)^2$.
7. $\{1 + 2x - (3n+1)x^n + (3n-2)x^{n+1}\}/(1-x)^2$.
8. $\frac{1}{4} - 1/[2(n+1)(n+2)]$.
9. $n(2n^3 + 8n^2 + 7n - 2)$.
10. $\frac{1}{2}\{n \cos A + \cos(n+2)A \sin nA \operatorname{cosec} A\}$.
15. $\Sigma_{r=n+1}^{2n} r^3 = \frac{1}{4}(2n)^2(2n+1)^2 - (1/4)n^2(n+1)^2 = (1/4)n^2(15n^2 + 14n + 3)$.
16. Must be shown that the statement holds for $n = 1$, and this is not so.
18. $\frac{1}{4}n(n^3 + 6n^2 + 5n + 4)$.
19. (a) $1 - 1/(n+1)$, (b) 348 450.
20. (a) $n/(2n+1)$, (b) 18 360, (c) $x\{1 - (1+n)x^n + nx^{n+1}\}/(1-x)^2$.

Exercises 5

1. No. 2. subtraction. 3. Yes, No.

Exercises 6(a)

1. 2, 5, 14.
4. Not closed w.r.t. addition.
7. (a) All entries in the table are members of the set, (b) the table is symmetric about the diagonal from top left to bottom right, (c) one column of entries (the first) is identical to the column listing the elements on the left; this column is below R_0 in the row listing the elements on top, so R_0 is the identity element, (d) the identity element R_0 occurs in every row of entries.

Exercises 6(b)

1. (a) Yes, (b) No, (c) No.

2. (a) Left-distributive, not right-distributive, (b) not distributive in either sense.
5. Closure, existence of identity elements and inverses need testing, the others hold automatically.
6. Not commutative. (a) Yes, (b) No. **7.** No.
8. (a) 1, 10, (b) 1, no identity element w.r.t. $\otimes$, (c) no identity elements. Yes. Inverses do not exist, so not a field in (a), (b) or (c).

Exercises 7(a)

2. (a) $3\mathbf{e}_1 + 2\mathbf{e}_2 - \mathbf{e}_3$, (b) $2\mathbf{e}_1 - 3\mathbf{e}_2 + 2\mathbf{e}_3$.
3. $(1/\sqrt{3})(\mathbf{i} + \mathbf{j} + \mathbf{k}), (1/\sqrt{5})(\mathbf{j} + 2\mathbf{k}), -(1/\sqrt{29})(2\mathbf{i} + 3\mathbf{j} + 4\mathbf{k})$; 6, −11, −18.
4. (a) $7\mathbf{i} - 3\mathbf{j} - 5\mathbf{k}$, 1, (b) −11.
5. (a) $17\mathbf{i} + 10\mathbf{k}$, 10, (b) −7/5; $\lambda = -\mathbf{u}^2/(\mathbf{u}.\mathbf{v})$.
6. −130. **7.** $g = h = 1$.
8. $\overrightarrow{AC} = 3\mathbf{i} - 2\mathbf{j} + 3\mathbf{k}$, $\overrightarrow{BD} = 2\mathbf{i} + 3\mathbf{j} - \mathbf{k}$, 80·15°.
9. (a) $\pi/2$, (b) $b = 2a$, $a = 1/3\sqrt{2}$, $b = \sqrt{2/3}$.
16. $\overrightarrow{AB} = \mathbf{b} - \mathbf{a}$, $\overrightarrow{CD} = \mathbf{d} - \mathbf{c}$. **18.** 6 lines.

Exercises 7(b)

1. (a) $\overrightarrow{DB} = \mathbf{b} - \mathbf{d}$, $\overrightarrow{DC} = \mathbf{c} - \mathbf{d}$, (b) $\overrightarrow{BF} = \frac{1}{2}(\mathbf{c} + \mathbf{d}) - \mathbf{b}$, (c) $\frac{1}{2}|\mathbf{c} \wedge \mathbf{d}|$.
2. $\angle BOC = \angle COA = \angle AOB = \pi/2$. Area $= 3\sqrt{3}$. Volume = 6.
3. $\pm(1/\sqrt{442})(3\mathbf{i} - 17\mathbf{j} - 12\mathbf{k})$.
5. $\overrightarrow{AE} = \mathbf{a} + \frac{1}{2}\mathbf{b}$, $\overrightarrow{ED} = \frac{1}{2}\mathbf{b} - \mathbf{a}$, $\overrightarrow{FE} = \frac{1}{2}(\mathbf{a} - \mathbf{b})$; $\pm(1/\sqrt{174})(11\mathbf{i} + 2\mathbf{j} - 7\mathbf{k})$.
6. $\overrightarrow{AC} = 3(\mathbf{i} + \mathbf{j} + \mathbf{k})$, $\overrightarrow{AG} = 5\mathbf{i} + 4\mathbf{j} + 3\mathbf{k}$, 11·52°, 3.
7. Identity element is $\mathbf{i} + \mathbf{j}$. No inverses for vectors with either component zero. Not true e.g. $\mathbf{i} * \mathbf{j} = \mathbf{0}$.

Exercises 8(a)

1. 7; (2, −2, −9).
2. $6x - 2y - 2z + 15 = 0$; $3x^2 + 3y^2 + 3z^2 + 18x - 20y - 26z + 102 = 0$.
3. (−2, 2, −3).
5. (3, −6, 0); 2:1 externally. **6.** 7.
7. [6/7, −2/7, 3/7]; (4, −1, 7); (−8, 3, 1).
8. 45° or 135°.
9. 3, 3, $2\sqrt{3}$; cosines of angles are $1/\sqrt{3}$, $1/\sqrt{3}$ and 1/3 respectively.
10. 2.
12. (a) (4, 12, 16), (0, 8, 23), (−6, 12, 16), (2, 2, 14), (b) 11, (c) 87, 1, 47, (d) 135·98°.

Exercises 8(b)

1. $2\sqrt{2}, \sqrt{14}, \sqrt{2}$; $[1/\sqrt{2}, 0, -1/\sqrt{2}]$, $[-2/\sqrt{14}, -1/\sqrt{14}, 3/\sqrt{14}]$, $[0, 1/\sqrt{2}, -1/\sqrt{2}]$; $x + y + z = 5$.
2. $x + 5y - 6z + 19 = 0$. **3.** $x - y - z + 4 = 0$; same side as P.
4. $2x + y - 3z = 4$. **5.** $3y + z = 5$, $6x - z = 16$, $2x + y = 7$.
6. $x/2 + y/3 + z/4 = 1$, $12/\sqrt{61}$.

7. $2x + 3y + 6z = 38$; sine of angle is 3/7.

8. $4y - z = 0$.

9. $x - 3y - 2z + 11 = 0$.

11. $\left[\dfrac{-a, b, 0}{\sqrt{(a^2 + b^2)}}\right]$.

Exercises 8(c)

1. (a) 3, (b) $\left[\dfrac{-2, -2, -1}{3}\right]$, (c) $(x + 1)/2 = (y - 1)/2 = (z - 1)/1$; (3, 1, 3) not on AB.

2. (a) $\sqrt{11}$, (b) $\left[\dfrac{1, 3, -1}{\sqrt{11}}\right]$, (c) $(x - 3)/1 = (y - 2)/3 = (z - 1)/(-1)$; (4, 5, 0).

3. (−7, 11, −2).

4. (a) $x = 2t + 1$, $y = 2t$, $z = t - 1$,
(b) $\left[\dfrac{2, 2, 1}{3}\right]$,
(c) (1/3, −2/3, −4/3), (d) 1.

5. $\left[\dfrac{1, 2, -3}{\sqrt{14}}\right]$.

6. $x - 2 = y - 3 = z - 4$.

7. $x + 3y - 2z + 17 = 0$; [1 : 1 : 2].

8. $2x - 2y - z + 3 = 0$; 2.

9. $4x - 5y - 5z = 0$.

10. $4x - 10y + z + 13 = 0$.

11. (5, 0, 6), $(3514)^{1/2}/7$.

14. (14/3, 4/3, −8)

15. $\left[\dfrac{1, 4, 1}{3\sqrt{2}}\right]$; 4.

16. 14.

17. $x + y + z = 11$; $7\sqrt{3}$.

18. $(1/ap, 1/bp, 1/cp)$ where $p = 1/a^2 + 1/b^2 + 1/c^2$.

19. $x + 2y + 2z = 11$; (1, 2, 3).

20. $\left[\dfrac{5, -2, -3}{\sqrt{38}}\right]$.

21. 18 : 5 externally.

22. $2\sqrt{17}$; (−2, 3, −5).

23. (3, 1, 1); $5x - 17y + 19z = 17$.

24. Q is (−1, 1, 5), R is (3, 5, 3).

26. $x/5 = y/(-10) = z/9$; (10/7, −20/7, 18/7).

27. $(la' + mb' + nc')(ax + by + cz + d)$
$= (la + mb + nc)(a'x + b'y + c'z + d')$.

28. (a) $(x - 5)/2 = (y - 4)/3 = (z - 4)/4$; $2\sqrt{29}$,
(b) $(x - 8)/5 = (y - 1)/(-3) = (z - 3)/5$; $\sqrt{59}$,
(c) $(x - 1)/3 = (y - 1)/4 = (z - 2)/12$; 13,
(d) $(x - 2)/2 = (y - 1)/(-1) = (z + 1)/2$; 3.

Exercises 8(d)

1. (−1, 3/2, 0), 5/2, (a) 4π, (b) $\sqrt{(37/3)}$.

2. (−1, 2, −2), 3, (a) $\left[\dfrac{1, -2, 2}{3}\right]$, (b) $x - 2y + 2z = 0$,
(c) $x - 2y + 2z + 18 = 0$.

3. (29/9, 2/9, −35/9), 12.

4. (3, 7, 4); $3y + 4z = 27$, $3y + 4z + 3 = 0$.

5. (1, 0, 0), (0, 1, 0), (0, 0, 1); centre (1/3, 1/3, 1/3), radius $\sqrt{6}/3$.
8. $(au + bv + cw - d)^2 = (a^2 + b^2 + c^2)(u^2 + v^2 + w^2 - k)$.

Exercises 9(a)

1. (a) $-(29 + 2i)/5$, (b) $(-29 + 28i)/25$.
2. $-i$.
3. 0·7, 0·9.
4. $x = \frac{2}{5}, y = -\frac{1}{5}$.
5. $x = 1, y = 2$.
6. $z = i, w = 1 + i$.
7. $z = 1 + i, w = 2 - 3i$.

Exercises 9(b)

1. $2, \sqrt{2}, -\pi/3$.
2. $\sqrt{2}, \pi/4$.
3. (a) $(46 + 9i)/13$, (b) $\sqrt{13}/13$, (c) $1 \angle -\pi/4$, $1 \angle -\pi/2$, $1 \angle -3\pi/4$.
4. u inside, v and w outside.
5. (a) modulus = 1, argument = $-\theta$; (b) modulus = $\sec\theta$, argument = θ.
6. (a) $\sqrt{2}, \frac{1}{4}\pi; \sqrt{2}, \frac{3}{4}\pi; \sqrt{2}, -\frac{1}{4}\pi$.
7. $1, \frac{1}{2}\pi$; $1, \frac{1}{4}\pi$.
8. $x - y = 1$.
9. $4x + 8y = 3$.
12. Centre $(3R, -4R)$, radius $5R$.
13. $3x^2 + 3y^2 - 34x + 96 = 0$.
15. (a) $x^2 + y^2 - x - 2 = 0$.
17. $P_1 \equiv r(-\cos\theta + i\sin\theta)$; the circle is $c(u^2 + v^2) + 4u = 0$.
18. (b) $|z_1| = |z_2|$ and $\arg z_1 - \arg z_2 = \frac{1}{3}\pi$. Hence $\arg z_1^3 - \arg z_2^3 = \pi$ $\therefore z_1^3 = -z_2^3$, etc.
20. $\rho^2 + r^2 - 2\rho r\cos(\theta - \alpha)$.
21. (a) Divide AB at C so that $AC:CB = \mu:\lambda$. Then $\lambda\overrightarrow{OA} + \mu\overrightarrow{OB} = (\lambda + \mu)\overrightarrow{OC}$ where O is the origin. Thus $\lambda z_1 + \mu z_2 = z = k\overrightarrow{OC}$ so that P is the point on OC such that $OP = kOC$. Hence P lies on a line parallel to AB.
22. $B \equiv (1 + \frac{5}{3}\sqrt{2}) - i(1 + \frac{1}{3}\sqrt{2})$; $D \equiv (1 - \frac{5}{3}\sqrt{2}) - i(1 - \frac{1}{3}\sqrt{2})$.

Exercises 9(c)

1. (a) $-i$, (b) $0{\cdot}9659 + 0{\cdot}2588i$.
3. $1, -1/2 \pm (\sqrt{3}/2)i$.
4. $e^{i\{\pi/10 + k(2\pi/5)\}}$ $(k = 0, 1, 2, 3, 4)$.
6. $2^{1/6}\, e^{i(8k+1)\pi/12}$ $(k = 0, 1, 2)$.
7. $\sqrt{2}\, e^{i(8k+3)\pi/12}$ $(k = 0, 1, 2)$.
8. $\pm(0{\cdot}9808 + 0{\cdot}1951i), \pm(0{\cdot}1951 - 0{\cdot}9808i)$.
9. $\sqrt{2}[\cos(\pi/4) + i\sin(\pi/4)]$.

Exercises 10

1. (a) $2(1 + i)$, $1 + 3i$, (b) $\frac{1}{2}[(3\sqrt{2} - 4) + i(6 + \sqrt{2})]$, $\frac{1}{2}[-(3\sqrt{2} + 4) + i(6 - \sqrt{2})]$ (c) $2^{1/6}\, e^{i(8k\pm1)\pi/12}$ $(k = 0, 1, 2)$.
2. $x = 2, 5, -3, -6$.
3. (a) $2, \frac{1}{2}, 2 \pm \sqrt{3}$, (b) $\frac{1}{2}, 2, (5 \pm i\sqrt{11})/6$.
4. $w = \pm 2, \pm 2i$, $z = 3, 1/3, (3 \pm 4i)/5$.
5. (a) $\pm i, \pm(1 \pm i)/\sqrt{2}$, (b) $-1/2, \frac{1}{2}(-1 \pm i)$.
6. $2 - i, -15/4$.
7. $-1/2 - i\sqrt{3}/2$, $z^2 + z + 1$, $z^2 - z + 5$.
8. $1 - 2i, -3/2$; $(2x + 3)(x^2 - 2x + 5)$.
9. $e^{i2k\pi/5}$ $(k = 0, 1, 2, 3, 4)$;

$(z-1)[z-\cos(2\pi/5)-i\sin(2\pi/5)][z-\cos(2\pi/5)+i\sin(2\pi/5)]$
$\times[z-\cos(4\pi/5)-i\sin(4\pi/5)][z-\cos(4\pi/5)+i\sin(4\pi/5)]$;
$(z-1)(z^2-2z\cos(2\pi/5)+1)(z^2-2z\cos(4\pi/5)+1)$.

10. $\frac{1}{2}[1-i\cot(4k+1)\pi/12]$, $k=1, 2, 3$.
11. $z=\frac{1}{2}[1\pm 3i(\sqrt{2}+1)]$, $\frac{1}{2}[1\pm 3i(\sqrt{2}-1)]$.
12. 1, $e^{\pm i2k\pi/5}$ $(k=1, 2)$.
13. (a) $1+3i$, $-1+i$;
(b) $\pm(1+i)$, $\pm\frac{1}{2}[(\sqrt{3}+1)-i(\sqrt{3}-1)]$, $\pm\frac{1}{2}[(\sqrt{3}-1)+i(\sqrt{3}+1)]$.
14. 1, $e^{\pm i2k\pi/5}$ $(k=1, 2)$.
16. p^2, $-2pr$.
17. $p(3q-p^2)-3r$.
18. $q^3=p^3r$; $x=2/3, 2, 6$.
19. $(2pr-q^2)/r$.

Exercises 11(a)

1. $(4, 7)=5(2, 5)-2(3, 9)$.
2. $\mathbf{r}_1-2\mathbf{r}_2+\mathbf{r}_3=\mathbf{0}$.
3. $\mathbf{i}=-\frac{16}{5}\mathbf{r}_1+\frac{12}{5}\mathbf{r}_2-\frac{1}{5}\mathbf{r}_3$, $\mathbf{j}=\frac{12}{5}\mathbf{r}_1-\frac{9}{5}\mathbf{r}_2+\frac{2}{5}\mathbf{r}_3$,
$\mathbf{k}=-\frac{1}{5}\mathbf{r}_1+\frac{2}{5}\mathbf{r}_2-\frac{1}{5}\mathbf{r}_3$.
4. $5\mathbf{r}_1+\mathbf{r}_2-4\mathbf{r}_3=\mathbf{0}$, $\dim\{\mathbf{r}_1, \mathbf{r}_2, \mathbf{r}_3\}=2$.
5. Altered values are 0, 2.
6. $(1, 2, 0)=-\frac{1}{2}(1, -1, 3)+\frac{3}{4}(2, 0, 1)+\frac{3}{4}(0, 2, 1)$.
7. Required scalars are 3, 2; 3, −2; 7, −3.

Exercises 11(b)

1. (a) 1800, (b) 24.
6. $8abc$ (Add first column to second and third columns).
7. a, b, $-(a+b)$.
8. 2, 3, 6.
9. $3, (1\pm\sqrt{561})/10$.
10. (a) 2, −5/2, 1/2, (b) 1, 2, −3.
11. ±1, −2.
12. (a) $-a$(triple root), (b) $-\frac{1}{2}a$(triple root).
13. $(\lambda\mu\nu+1)D$.
14. (a) 24, (b) 0 (add first row to second, third row to fourth), (c) 505.
17. Add $a^2\times$ row 1 + ab row 2 + ac row 3 + ad row 4.

Exercises 11(c)

3. $\mathbf{i}=\mathbf{r}_4-\mathbf{r}_3$, $\mathbf{j}=\mathbf{r}_3-\mathbf{r}_2$, $\mathbf{k}=\mathbf{r}_2-\mathbf{r}_1$, $\mathbf{l}=\mathbf{r}_1$.
4. $\mathbf{i}=(1/3)(\mathbf{r}_1+\mathbf{r}_2+\mathbf{r}_3-2\mathbf{r}_4)$, $\mathbf{j}=(1/3)(\mathbf{r}_1+\mathbf{r}_2-2\mathbf{r}_3+\mathbf{r}_4)$,
$\mathbf{k}=(1/3)(\mathbf{r}_1-2\mathbf{r}_2+\mathbf{r}_3+\mathbf{r}_4)$, $\mathbf{l}=(1/3)(-2\mathbf{r}_1+\mathbf{r}_2+\mathbf{r}_3+\mathbf{r}_4)$.
5. $\mathbf{r}_1-\mathbf{r}_2+\mathbf{r}_3-\mathbf{r}_4=\mathbf{0}$.
6. 4.
7. 2.

Exercises 11(d)

1. (a) $x=3, y=5, z=6$, (b) $x=-0{\cdot}6, y=1{\cdot}1, z=2{\cdot}7$,
(c) $x=1, y=2, z=2$, (d) $x=2, y=3, z=4$,
(c) $x=-5, y=7, z=2$.
2. $x=12, y=-60, z=60$.
3. (a) Inconsistent, (b) consistent; $x=4, y=1, z=0$.

4. $\lambda = 3, x = \frac{1}{6}, y = \frac{3}{2}; \lambda = 14, x = -\frac{1}{5}, y = \frac{2}{5}$.
5. $\lambda = 3, x = 4, y = -2$.
6. $\lambda = 1, x = -5, y = 1; \lambda = -1, x = -1/11, y = -15/11$;
$\lambda = 12, x = 1/2, y = 1$.
7. $a + b + c = 0$.
9. $\lambda = 1; 1:-1:0; \lambda = -2; 1:1:\frac{1}{2}\sqrt{2}; \lambda = 3; 1:1:-2\sqrt{2}$.
10. $a + b + c = 0; 5:3:-8$.
11. $t = 0, x = y = z; t = 3$, the three equations are identical.
12. $abc + 2fgh - af^2 - bg^2 - ch^2 = 0$.

Exercises 12(a)

1. (a) No synonyms in language B.
(b) Same as (a), and also a word in language A corresponding to each word in B.
(c) Same as (b) and also no synonyms in A.
2. Domain: all lines not parallel to the y-axis. Range: R^2.
3. Domain: all lines not passing through the origin. Range: all points in R^2 except points on the axes.
4. Restrict a, b to positive (or negative) numbers. Domain: all ellipses, centre O. Range: all points in the positive (or negative) quadrant. (Also $a > 0, b < 0$ or $a < 0, b > 0$ would do).
5. Set of points on the surface of a unit sphere, centre O. (1, 0, 0), (0, 1, 0), (0, 0, 1).
6. Set of all vectors in V_3 with magnitude a, where a is the radius of the sphere.
7. Set of all points outside the sphere.

Exercises 12(b)

1. f^{-1} exists. g^{-1} does not exist since many rectangles have the same area.
2. No. Parallel lines have the same image.
3. No. All points on OP have the same image as P.
4. All points excluding the origin and the positive x-axis. Image rule for f^{-1} similar to that for f but with *double* replaced by *half*.
5. Same as image rule for f.

Exercises 12(c)

1. $f \circ g = g \circ f$ when lines at right-angles, otherwise $f \circ g \neq g \circ f$.
3. $\theta = \alpha + \beta$. Yes.
5. Yes.

Exercises 12(d)

1. $\mathbf{t} = -a\mathbf{i} - b\mathbf{j}$.
2. $\theta = -\alpha$.

Exercises 12(e)

1. Translation of axes, $x' = x - a, y' = y - b$.
3. $a = \cos\alpha + i\sin\alpha, b = \cos\alpha - i\sin\alpha$.

4. Reflection in the x-axis, y-axis, and the line $y = x$ respectively.
7. $\{z: \arg z \neq 0\}$; f^{-1} exists but g is not a (1, 1) mapping.

Exercises 12(f)

1. $x \mapsto V$, $V = x(18 - 2x)^2$, $0 < x < 9$ (or $0 \leqslant x \leqslant 9$ if limiting cases included).
2. $x \mapsto A$, $A = \frac{1}{8}(x^2 - 48x + 1152)$, $0 < x < 48$ (or $0 \leqslant x \leqslant 48$ if limiting cases included).
3. $x \mapsto A$, $A = x\sqrt{(36 - x^2)}$, $0 < x < 6$ (or $0 \leqslant x \leqslant 6$ if limiting cases included).
4. 3/5, undefined.
5. 20, 31, 91, 43, 13/7, 19/4.
6. (a) $\{x: |x| \leqslant 2\}$, $\{y: 0 \leqslant y \leqslant 6\}$,
   (b) $\{x: |x| \leqslant 2 \text{ or } |x| \geqslant 3\}$, $\{y: 0 \leqslant y < \infty\}$.
7. (a) $\{x: x \neq -2, 3\}$, (b) $\{x: x < -2 \text{ or } x > 3\}$, (c) R,
   (d) $\{x: x > 0\}$, (c) $\{x: x \neq 0\}$.
8. $\{x: x \neq (2n + 1)\pi/2\}$, $\{x: x \neq n\pi\}$, $\{x: x \neq n\pi\}$ where $n = 0, \pm 1, \pm 2, \ldots$
9. (a) $\{x: x \neq n\pi/3\}$, (b) $\{x: x \neq (2n + 1)\pi\}$ where $n = 0, \pm 1, \pm 2, \ldots$

Exercises 12(g)

2. $\{x: |x| \leqslant 2\}$.
5. $\{x: x \neq 1, x \geqslant -\sqrt{2}\}$.
9. (a) polynomial, (b) rational function, (c) rational function.
10. (a) polynomial, (b) rational function, (c) rational function.

Exercises 12(h)

2. $(1 - 2x)/(1 + x)$.
4. $\log_{10} x$, $(\log_{10} x)/\log_{10} 2$.
5. $x \mapsto \sqrt{x} - 1$; $x \mapsto -\sqrt{x} - 1$
8. $f^{-1}: x \mapsto y$, $x^2 + xy + y^2 = 4$, $y \geqslant -2x$.

Exercises 12(i)

1. $\pi/6$, $\pi/4$, $\pi/3$, 0, $\pi/2$; $\pi/6$, $\pi/3$, $\pi/4$, 0.
2. 56/33.
4. $x = \pm 2$.
5. $x = -\sqrt{3} \pm \sqrt{5}$.
9. Definitions reasonable since (a) if $\cot \alpha = x$, then $\tan \alpha = 1/x$, (b) if $\cot \alpha = x$, then $\tan(\pi/2 - \alpha) = x$. (a) Range is $\{y: -\pi/2 < y < \pi/2, y \neq 0\}$, (b) Range is $\{y: 0 < y < \pi\}$.

Exercises 12(j)

1. $\dfrac{1 - 5x^2}{2x + x^2}$, $\dfrac{x^2 - 5}{1 - 2x}$, $\dfrac{x^2 - 2x - 4}{2x - 1}$, $\dfrac{\sin^2 x - 5}{2 \sin x + 1}$.
6. (a) $f(x) = x + b$, (b) $f(x) = ax$, (c) $f(x) = a(x + 1) - 1$.
10. (a) not true, (b) true.
11. (a) false, (b) true, (c) false, (d) true, (e) false, (f) true.

Exercises 13(a)

1. $A + B = \begin{bmatrix} 3 & 1 & 0 \\ 5 & 1 & 5 \end{bmatrix}$, $A + C$ impossible, $C + D = \begin{bmatrix} 4 & 3 \\ 1 & 4 \end{bmatrix}$,

$CA = \begin{bmatrix} 6 & 6 & 2 \\ 10 & 2 & 14 \end{bmatrix}$, AB and AC impossible, $A'C' = \begin{bmatrix} 6 & 10 \\ 6 & 2 \\ 2 & 14 \end{bmatrix}$.

2. $\begin{bmatrix} 3 & 2 \\ 3 & 6 \end{bmatrix}, \begin{bmatrix} 4 & 2 \\ 3 & 7 \end{bmatrix}, \begin{bmatrix} 18 & 20 \\ 30 & 48 \end{bmatrix}$.

3. $\begin{bmatrix} 9 & 8 \\ 5 & 8 \end{bmatrix}, \begin{bmatrix} 13 & 11 \\ 6 & 10 \end{bmatrix}, \begin{bmatrix} 165 & 179 \\ 113 & 135 \end{bmatrix}$.

4. $\lambda = 1$.

5. $a = \pm 1, b = 0$, or $a = 1/3, b = -2/3$, or $a = -1/3, b = 2/3$.

10. Multiplication non-commutative, system not closed w.r.t. either operation since $A + B, AB$ are not defined for every pair A, B.

Exercises 13(b)

1. $\begin{bmatrix} 1 & 1 & 1 \\ -1 & 1 & -2 \\ 0 & 1 & 1 \end{bmatrix}, \frac{1}{3}\begin{bmatrix} 3 & 0 & -3 \\ 1 & 1 & 1 \\ -1 & -1 & 2 \end{bmatrix}$.

2. $A^{-1} = -\frac{1}{2}\begin{bmatrix} 0 & 1 & 1 \\ 1 & 0 & 1 \\ 1 & 1 & 0 \end{bmatrix}$.

3. $A^{-1} = \begin{bmatrix} -7 & 5 & 3 \\ 3 & -2 & -2 \\ 3 & -2 & -1 \end{bmatrix}$, $A^{-1}A' = \begin{bmatrix} 3 & 4 & -2 \\ -4 & -5 & 0 \\ 0 & 0 & 1 \end{bmatrix}$.

4. $AB = \begin{bmatrix} 1 & 3(a+1) & 3b \\ 0 & 6+5a & 5b \\ 0 & 6(a+1) & 1+6b \end{bmatrix}$, $a = -1, b = 0$.

5. $A^{-1} = -\frac{1}{4}\begin{bmatrix} 1 & 3 \\ 2 & 2 \end{bmatrix}$, $B^{-1} = -\frac{1}{4}\begin{bmatrix} -3 & 1 & 1 \\ 1 & -3 & 1 \\ 1 & 1 & -3 \end{bmatrix}$.

7. $c = -2$, $\frac{1}{9}\begin{bmatrix} 1 & -2 & -2 \\ -2 & 1 & -2 \\ -2 & -2 & 1 \end{bmatrix}$, $-\frac{1}{2}\begin{bmatrix} 0 & 1 & 1 \\ 1 & 0 & 1 \\ 1 & 1 & 0 \end{bmatrix}$.

9. (a) $\frac{1}{12}\begin{bmatrix} 8 & 0 & 4 \\ -1 & 3 & -5 \\ -5 & 3 & -1 \end{bmatrix}$, (b) $\mu = 1, 5$.

Exercises 13(c)

1. $A^{-1}=\begin{bmatrix}-5 & 1 & 9\\ -1 & 0 & 2\\ 6 & -1 & -10\end{bmatrix}$, $X=\begin{bmatrix}6\\ 1\\ -6\end{bmatrix}$, $Y=\begin{bmatrix}6 & 4\\ 1 & 1\\ 6 & -4\end{bmatrix}$.

2. $A^{-1}=\begin{bmatrix}0 & 1 & -1\\ -1 & -1 & 2\\ 2 & 0 & -1\end{bmatrix}$, $X=\begin{bmatrix}0 & -1\\ 0 & 1\\ 1 & 1\end{bmatrix}$.

3. $A^{-1}=\begin{bmatrix}7 & -3 & -3\\ -1 & 1 & 0\\ -1 & 0 & 1\end{bmatrix}$, $X=\begin{bmatrix}2 & 18\\ 1 & -2\\ -1 & -3\end{bmatrix}$, $\begin{bmatrix}x\\ y\\ z\end{bmatrix}=\begin{bmatrix}1\\ -1\\ 2\end{bmatrix}$.

4. $A^{-1}=\frac{1}{2}\begin{bmatrix}1 & -1 & 1\\ 1 & 1 & -1\\ -1 & 1 & 1\end{bmatrix}$, $x=\frac{1}{2}(a-b+c), y=\frac{1}{2}(a+b-c)$, $z=\frac{1}{2}(b+c-a)$.

5. $x_1=a-b, x_2=\frac{1}{3}(3b-c-d), x_3=\frac{1}{3}(-c+2d), x_4=\frac{1}{3}(2c-d)$.

6. (a) $x=-1/3, y=-1, z=-1$. (b) If $A\neq 0, B\neq 0$ both are singular.

7. $A^{-1}=\begin{bmatrix}1 & -1 & 0\\ 0 & 1 & -1\\ 0 & 0 & 1\end{bmatrix}$, $(A^n)^{-1}=\begin{bmatrix}1 & -n & \frac{1}{2}n(n-1)\\ 0 & 1 & -n\\ 0 & 0 & 1\end{bmatrix}$.

Exercises 14(a)

1. $\frac{2}{3}$. **2.** $-\frac{1}{2}$. **3.** $\frac{4}{5}$. **4.** 2. **5.** $\frac{1}{2}$. **6.** $1/2a$. **7.** $-5/7$.
8. -9. **9.** 2. **10.** -1. **11.** 1. **12.** 0. **13.** k. **14.** k.
15. 0. **16.** 1. **17.** 80. **18.** (a) $-1, 1$. (b) $-1, 1$.
19. $\lim_{x\to 0-} f(x)=-1$, $\lim_{x\to 0+} f(x)=1$, $\lim_{x\to 0} f(x)$ does not exist.

Exercises 14(b)

1. $x=2$. **2.** $x=0$.
3. $x=2, x=-1$.
4. $x=(2k+1)\pi/4, k=0, \pm 1, \pm 2, \ldots$.
5. $x=k\pi+\pi/4, k=0, \pm 1, \pm 2, \ldots$.
6. $x=k\pi/3, k=0, \pm 1, \ldots$.
7. $x=2k\pi, k=0, \pm 1, \ldots$. **8.** $x=5$.
9. $x=k\pi, k=1, 2, \ldots$.
10. $x=-(2k+1)\pi/2, x=k\pi, k=0, 1, 2, \ldots$.
11. $x=k/4, k=1, 2, \ldots$.

Exercises 14(c)

1. $2x - 4$.
2. $-3/(2\sqrt{x^3})$.
3. $-6x/(3x^2 + 5)^2$.
4. $1/\sqrt{(2x + 7)}$.
5. $1/(x + 1)^2$.
6. $x \cos x + \sin x$.
7. $\cos x - x \sin x$.
8. $2x \cos (x^2)$.

10. $-\frac{1}{2}$; $f'(1) = -\frac{1}{2}$; $f(0)$ and $f'(0)$ undefined.

Exercises 14(d)

1. $x = 2$ and $x = 2/9$.

3. (a) $-71° 34'$. (b) $-26° 34'$.
4. $2x^2(5x^2 + 16x + 12)$.
5. $(15x^2 - 6x + 1)/(2\sqrt{x})$.
6. $3/\{2(1 - 3x)^{3/2}\}$.
7. $12/(1 - 2x)^3$.
8. $-11/(2x - 1)^2$.
9. $\frac{1}{2}(1 - 2x)/\{x(1 - x)\}^{1/2}$.
10. $\frac{1}{2}(x^5 + 4)/x^3(x^5 - 1)^{1/2}$.
11. $n[x + \sqrt{(1 + x^2)}]^n/\sqrt{(1 + x^2)}$.
12. $-\frac{1}{2}(x^3 - 3x^2 + 2)/(1 - x^3)^{3/2}$.
13. $\frac{1}{2}(9x^2 + 2x - 25)/(3x - 1)^{3/2}$.
14. $5(3x + 8)(3x + 1)^4/(2 - x)^{11}$.
15. $3 \tan^2 x \sec^2 x$.
16. $15 \sin^4 3x \cos 3x$.
17. $\cos x/[2\sqrt{(\sin x)}]$.
18. $(\sin x - 2)/(1 + \sin x)^2$.
19. $(1 - \sin x - \cos x)/(1 - \cos x)^2$.
20. $(x + \sin x)/(1 + \cos x)$.
21. $\frac{1}{2} \cos x\,(2 + 2 \sin x - \sin^2 x)/[(2 + \sin^2 x)(1 - \sin x)^3]^{1/2}$.
22. $x^{-2} \operatorname{cosec}^2 (1/x)$.
23. $[2x/(x^2 + 1)^2] \operatorname{cosec}^2 [1/(x^2 + 1)]$.
24. $\frac{1}{2}/\sqrt{(3x - 2 - x^2)}$.
25. $-1/\sqrt{(1 - x^2)}$.
26. -1.
27. $2a/\sqrt{(1 - a^2x^2)}$.
28. $2ax/[(a^2 + x^2)\sqrt{(a^2 + 2x^2)}]$.
29. $-2/\sqrt{(1 - x^2)}$.
30. $2x/(2 - 2x^2 + x^4)$.
31. $ab/(a^2 + b^2x^2)$.
32. $m \sec^2 x/(1 + m^2 \tan^2 x)$.
33. $1/(1 + x)\sqrt{x}$.
34. $-1/\{2(1 + x)\sqrt{x}\}$.
35. $2x/[(x^2 + 2)\sqrt{(1 + x^2)}]$.
36. $2/(1 + 4x)\sqrt{x}$.
37. $3/[2(1 + x)\sqrt{x}]$.
38. -1.
39. $2x \sin^{-1} (\frac{1}{2}x)$.
40. $\sqrt{(a^2 - b^2)}/(a + b \cos x)$; $\frac{1}{2}\sqrt{(a^2 - b^2)}/(a + b \cos x)$.

Exercises 14(e)

1. $-x^2/y^2$.
2. $-(y/x)^{1/2}$.
3. $-(2xy + y^2)/(2xy + x^2)$.
4. $-(my)/(nx)$.
5. $na^m x^{n-1}/(mb^n y^{m-1})$.
6. $(ax - x - y)/(x + y - by)$.
7. $\{ay - (x + y)^2\}/\{(x + y)^2 - ax\}$.
8. $-(ax + hy + g)/(hx + by + f)$.
9. $2(x + y) \cos (x + y)^2/\{1 - 2(x + y) \cos (x + y)^2\}$.

12. $(78x^4 - 2x^2y^2 - y^4)/\{xy(x^2 + 2y^2)\}$; 3, -3.
13. $\{1 + y \cos (xy)\}/\{1 - x \cos (xy)\}$; tangent is $y = x$, and meets curve again at the points $(\sqrt{(n\pi)}, \sqrt{(n\pi)}), (-\sqrt{(n\pi)}, -\sqrt{(n\pi)})$ where $n = 1, 2, 3, \ldots$.

15. $20x + 17y + 12a = 0$.

Exercises 14(f)

1. $(t^2 - 1)/2t^3$.
2. $-\operatorname{cosec} t$.
3. $-1/t$.
4. $\operatorname{cosec} t$.

5. $-\cot t$.
6. $\cot t$.
7. $-\frac{1}{2}(1 + 2\cos 2t)\operatorname{cosec} 2t$.
12. $(\frac{1}{4}at^2, -\frac{1}{8}at^3)$.

Exercises 14(g)

10. (a) $-6/(x + 2y)^3$,
    (b) $-\cot^3 t$.
16. -2536.
18. $\frac{1}{3}\sec^3 2\theta \operatorname{cosec} \theta$.
19. $\tan \frac{5}{7} \theta$.
20. $b = -5/4$, $c = 3/8$, $d = -1/64$.

Exercises 14(h)

1. (a) $(-2, -1)$ Min. T.P., (b) $(2, 9)$ Max. T.P.,
   (c) $(-5, 0)$ Stat. pt, also a point of inflexion,
   (d) $(3/2, 3/2^{2/3})$ Min. T.P., (e) $(20, -16)$ Min. T.P.
2. $a = b = -2$.
4. Min. T.P.
5. Stat. point is also a point of inflexion.
6. $x = \frac{1}{3}$ (max.), $x = 1$ (min.), $x = 0$ (inflexion).
8. $dy/dx = -y/(4y^3 + x)$, $d^2y/dx^2 = (2xy - 4y^4)/(4y^3 + x)^3$.
9. $x = (27 + 3\sqrt{78})^{1/3}$ gives max.; $x = (27 - 3\sqrt{78})^{1/3}$ gives min.
10. $y = 6x + 8$, $y = 6x - 8$.

Exercises 14(i)

1. (a) $4/27, 0$, (b) $2, 0$.
2. (a) $-14/3, -6$, (b) $0, -5$, (c) $321/32, 6$.
3. (a) Max. turning value $= 3$, Min. turning value $= 0$, (b) $3, -1$.
4. Greatest value $\frac{6}{7}$, at $x = -\frac{1}{3}$. Least value $\frac{2}{5}$, at $x = 1$. On $-2 \leqslant x \leqslant 2$ there is no greatest and no least value.
5. (a) $\sqrt{2}$, (b) 3.

Exercises 15(a)

1. 8/3 m.
2. $2 < t < 3$ and $t > 4$; 28 units.
3. 0·4760 cm²/min.
4. 1·5 m/s.
5. 3·054 cm/min.
6. $14\sqrt{2}$ knots.
9. 576π cm²/min.
10. 25 cm/s.
11. $28\,900\pi$ cm³/min.
12. (a) 12 km/h, (b) 16/5 radians/h.
13. 360 m/s, 180 m/s.
14. 3π m/s, 54 degrees/s.
15. $(36\pi + 10)$ cm³/s.
17. $9\sqrt{2}/2$ units/s; rate of change of slope is $-\frac{3}{2}$.

Exercises 15(b)

3. 5000 m.
4. $a\sqrt{2}/2$.
7. 4 km.
10. P at mid-point of BC.
11. $\sqrt{2}a, \sqrt{2}b$.
15. $a, \sqrt{3}a$.
16. $I = (k \cos\theta \sin^2\theta)/d^2$, k a constant; $\sqrt{2}d/2$.

17. Clearly the shortest ladder will touch the top of the wall and so the length of the ladder is h cosec $x + d$ sec x, where x radians is the inclination of the ladder to the horizontal. Minimising w.r.t. x we get $(h^{2/3} + d^{2/3})^{3/2}$.

18. $4\sqrt{3}$ m.

27. 80 at $x = 12$, 180 at $x = 6$.

29. $3a^2\sqrt{3}/8$.

31. $t(3 + t^2)x + 2y = 3t$.

33. Time $= (a^2 + x^2)^{1/2}/u + [(c - x)^2 + b^2]^{1/2}/v$.

37. h.

39. $x = a^{2/3}b^{2/3}/(a^{2/3} + b^{2/3})^{1/2}$.

40. The circles are orthogonal and cut at (0, 0), $[-2\lambda/(1 + \lambda^4), -2\lambda^3/(1 + \lambda^4)]$; area of quadrilateral = 1; length of common chord $= 2\lambda/(1 + \lambda^4)^{1/2}$.

41. $4a^3/3\sqrt{3}$ = max. value, 0 = min. value. For max. value $r_1 = a(1 + 1/\sqrt{3})$; but r_1 must lie between the values a and $a(1 + e)$, where e is the eccentricity of the ellipse. Hence if $e > 1/\sqrt{3}$, $a(1 + 1/\sqrt{3})$ is a possible value of r_1 and it gives $r_1r_2(r_1 - r_2)$ its max. value. If $e < 1/\sqrt{3}$, $a(1 + 1/\sqrt{3})$ is not a possible value of r_1, and the max. value of $r_1r_2(r_1 - r_2)$ occurs when r_1 takes its highest possible value $a(1 + e)$.

42. $a + b$.

Exercises 15(c)

1. 4·6570.

2. 5·0008, 1·0349.

3. −7%.

5. 1/6%.

6. 26·46 cm; 0·28 cm.

7. $R = r$.

10. 0·4.

11. 1·86.

12. 4·493 (radians).

13. 4·275.

14. 1·547.

15. 1·47.

Exercises 15(d)

1. $-\sqrt{2}$.

3. (c, c) and $(-c, -c)$.

4. $a/6$.

Exercises 15(e)

1. 2.

2. ∞.

3. 0.

4. 0.

5. $\frac{3}{4}$.

6. $\frac{3}{4}$.

7. $y = x + 3$.

8. $4y = 6x - 9$.

9. $y = x$.

Exercises 15(f)

5. (1/4, 23/16) Max; (2/3, 31/27) Min.; 11/24.

6. $(-1/2, -13/8), (0, -1), (1, -5); (1 \pm \sqrt{7})/6$.

9. $(-1, -\frac{1}{2})$ Min.; $(1, \frac{1}{2})$ Max.; $(0, 0)$, $(-\sqrt{3}, -\sqrt{3}/4)$, $(\sqrt{3}, \sqrt{3}/4)$.

14. $a = 1$, $b = 0$; -2.

Exercises 15(g)

2. (a) $\phi = \frac{1}{2}(\pi - \theta)$, (b) $\phi = \frac{1}{2}\pi - 2\theta$, (c) $\phi = 2\theta$, (d) $\phi = \theta/n$, (e) $\phi = n\theta$, (f) $\phi = \frac{1}{2}\pi + 3\theta$, (g) $\phi = \frac{1}{2}(\theta + \pi)$.

7. $\frac{1}{2}\sqrt{2}$.
8. $a[(\theta - \sin\theta)^2 + (1 - \cos\theta)^2]^{1/2}$; $4\pi^2 a/(\pi^2 + 4)^{1/2}$.

Exercises 16(b)

1. (a) $x^2 + 1/x + c$, (b) $\frac{1}{5}x^5 + 2x^3 + 9x + c$,
   (c) $\frac{2}{5}x^{5/2} - \frac{4}{3}x^{3/2} - 2x^{1/2} + c$.
2. $y = 1/x$.
3. 164/9 units.
6. $dV/dt = -kt^{-2}$ and $k = r$.
7. $dV/dt = -kt^{-1/2}$ and $k = r$. Empty when $t = 49$.

Exercises 16(c)

1. (a) $4x^4 + c$, (b) $-8/x^2 + c$, (c) $\frac{3}{29}x^{29/3} + c$,
   (d) $-\frac{5}{2}.10^{1/5}.x^{-2/5} + c$, (e) $-4(3x)^{-1/4} + c$,
   (f) $2(2x + 1)^4 + c$, (g) $-4/(2x + 1)^2 + c$, (h) $\frac{3}{145}(5x - 7)^{29/3} + c$,
   (i) $-\frac{5}{2}.10^{1/5}.(\sin x)^{-2/5} + c$, (j) $-4(3\tan x)^{-1/4} + c$.
2. $\frac{1}{12}(2x + 3)^6 + c$.
3. $\frac{1}{24}(2x^2 + 3)^6 + c$.
4. $-1/3(3x + 5) + c$.
5. $\frac{1}{3}\sqrt{(5 + 6x)} + c$.
6. $-\frac{1}{5}\cot 5x + c$.
7. $\frac{1}{12}(x^2 + 1)^6 + c$.
8. $-\frac{1}{12}(9 - 4x^2)^{3/2} + c$.
9. $\sec x + c$.
10. $-\frac{1}{6}\operatorname{cosec}^6 x + c$.
11. $2(\sqrt{3} - 1)$.
12. 8/3.
13. 1/6.
14. 1/3.
15. $\frac{1}{6}\tan^{-1}\frac{1}{3}(5 + 2x) + c$.
16. $\frac{1}{4}\sin^{-1}\frac{1}{3}(3 + 4x) + c$.
17. $\frac{1}{3}(\sin^{-1} x)^3 + c$.
20. $(24y - 1)(5 - 3x)^2 = 4$.

Exercises 16(d)

1. $\frac{1}{14}(2x + 1)(x - 3)^6 + c$.
2. $\frac{2}{5}(x - 2)(3 + x)^{3/2} + c$.
3. $\frac{2}{15}(3x - 2)(x + 1)^{3/2} + c$.
4. $-\frac{1}{8}(4x + 1)(2x + 1)^{-2} + c$.
5. $\frac{1}{3}\tan^{-1}(x^3) + c$.
6. $\frac{1}{6}\tan^{-1}(\frac{1}{3}x^2) + c$.
7. $\frac{1}{4}$.
8. $\pi/2$.

Exercises 16(f)

1. 1/3.
2. Does not exist.
3. Does not exist.
4. $\frac{1}{2}$.
5. $\pi/2$.
6. 1.
7. $\pi/8$.
8. $\pi/2$.
9. $\pi/2ab$.
10. π.
11. $\pi/20$.
12. 2.

Exercises 17(a)

1. $2\sec x$.
2. $2/[x\sqrt{(1 + x^2)}]$.
3. $\sec x$.
4. $2x/(x^4 - 1)$.
5. $(2/a)\tan(x/a)$.
6. $1/(x \ln x)$.
7. $\{x/(a^2 + x^2)\}\sec\{\frac{1}{2}\ln(a^2 + x^2)\}\tan\{\frac{1}{2}\ln(a^2 + x^2)\}$.
9. 0. The given expression is constant; its value is ln 2.

15. $x = \pi/3$ gives minimum value.
16. $-\frac{1}{2}x^{-3/2}(\sqrt{x} - 1)^2$.
17. (a) $\frac{1}{2}(9x^2 + 2x - 25)/(3x - 1)^{3/2}$, (b) $5(3x + 8)(3x + 1)^4/(2 - x)^{11}$,
(c) $-\frac{1}{2}(x^3 - 3x^2 + 2)/(1 - x^3)^{3/2}$,
(d) $\frac{1}{2} \cos x\,(2 + 2 \sin x - \sin^2 x)/\sqrt{\{(2 + \sin^2 x)(1 - \sin x)^3\}}$.
18. $x(n + 1 + x)^n/(n + x)^{n+1}$.
19. $x = 2$, Max. T.P.; $x = 3$, Min. T.P.; when $x = 2$, $y = 0{\cdot}70$.
20. Max. T.P.

Exercises 17(b)

1. $\frac{1}{2} \ln (1 + x^2) + c$.
2. $\frac{1}{2} \ln (x^2 + 4x - 5) + c$.
3. $\ln (\sin x) + c$.
4. $-\frac{1}{3} \ln (2 + 3 \cos x) + c$.
5. $\frac{1}{2} \ln (1 + 2 \tan x) + c$.
11. $-(2 \ln x + 1)/4x^2 + c$.
12. $2\sqrt{x}\,(\ln x - 2) + c$.
13. $(\ln x)^{n+1}/(n + 1) + c$.
14. $\frac{1}{2}n(\ln x)^2 + c$.

Exercises 17(c)

9. $(1, -6 \ln 7)$ Max. T.P.; $(3, 6 \ln (3/7) - 8)$ Min. T.P.; $x = 7$.

Exercises 17(d)

1. $2x\, e^{x^2}$.
2. $e^x(1 + x)$.
3. $e^{2x}(2 \sin 3x + 3 \cos 3x)$.
4. $e^{-x}\,(3x - x^2 - 1)$.
5. $-e^x/(e^x + 1)^2$.
6. $-8xe^{-2x}/\sqrt{(1 + 4x)}$.
7. $2e^x/(e^x + 1)^2$.
8. $x\, 10^x(2 + x \ln 10)$.
9. $x^{(1/x)-2}(1 - \ln x)$.
10. $-e^{-x}(21x^2 + 32x + 33)/(7x - 1)^2$.
11. $-2e^{\cos x} \sin x/(e^{\cos x} + 1)^2$.
12. $4e^{2x}/(1 - e^{2x})^2$.
13. $(\ln x)^x\{\ln (\ln x) + 1/\ln x\}$.
14. $\frac{1}{2} - 1/2\sqrt{x}$.
15. $x^2/(x^2 - 1)$.
16. $\frac{1}{2}e$.
17. 10.
20. $x = \pi/4$, Max. T.P.; $x = 5\pi/4$, Min. T.P.
23. -10, $8 \ln 2 - 16$.

Exercises 17(e)

2. (a) $x = \ln 3$ twice, (b) $x = \ln 2$.
3. $\cosh 2x = 1$, i.e. $x = 0$ (double root).
5. $\frac{1}{8}(\cosh 4x - 4 \cosh 2x + 3)$.
6. (a) $-(1/x^2) \cosh (1/x)$, (b) $(1/x) \operatorname{sech}^2 (\ln x)$, (c) $2/\sinh 2x$,
(d) $\operatorname{sech}^2 x \cosh (\tanh x)$.
8. $\sinh [\ln(4/e^2)]$.

Exercises 17(f)

4. $2/(1 - x^2)\sqrt{x}$.
7. $\dfrac{dy}{dx} = \dfrac{\sinh^{-1} y - y/\sqrt{(1 + x^2)}}{\sinh^{-1} x - x/\sqrt{(1 + y^2)}}$.
9. $\theta = \pi/4$ or $3\pi/4$.

Exercises 18(a)

1. $\frac{1}{3}e^{3x+2} + c.$
2. $-\frac{1}{2}e^{-2x} + c.$
3. $\frac{1}{2}\cosh(2x + 3) + c.$
4. $-\frac{1}{5}\sinh(3 - 5x) + c.$
5. $\frac{1}{3}\sinh^{-1}[(3x - 1)/\sqrt{2}] + c.$
6. $-\cosh^{-1}[(2 - x)/\sqrt{3}] + c.$
7. $\frac{1}{2}e^{x^2} + c.$
8. $\frac{1}{3}e^{x^3} + c.$
9. $e^{\sin x} + c.$
10. $e^{(x^2+5x+4)} + c.$
11. $\frac{1}{2}\cosh(x^2 + 4) + c.$
12. $\sinh^{-1}(\frac{1}{2}\sin x) + c.$
13. $\sinh^{-1}(e^x) + c.$
14. $\cosh^{-1}(e^x) + c.$
15. $\frac{1}{2}\sinh^{-1}(e^{2x}) + c.$
16. $-\cos(\ln x) + c.$
17. $2e^{\sqrt{x}} + c.$
18. $\frac{1}{4}(\ln x)^4 + c.$

Exercises 18(b)

1. $\frac{1}{3}\ln[(x - 2)/(x + 1)] + c.$
2. $\ln[(x - 3)/(x - 2)] + c.$
3. $\ln[(x + 2)/(x + 3)] + c.$
4. $\frac{1}{4}\ln[(x - 3)/(x + 1)] + c.$
5. $(1/\sqrt{2})\tan^{-1}[(x + 1)/\sqrt{2}] + c.$
6. $\frac{1}{5}\ln[(x - 3)/(x + 2)] + c.$
7. $\ln[x/(x + 1)] + c.$
8. $-1/(x + 4) + c.$
9. $\frac{1}{4}\tan^{-1}[(x + 3)/4] + c.$
10. $(1/2\sqrt{13})\ln[(x + 3 - \sqrt{13})/(x + 3 + \sqrt{13})] + c.$
11. $\frac{1}{5}\tan^{-1}[(x + 1)/5] + c.$
12. $\frac{1}{6}\tan^{-1}[(2x + 4)/3] + c.$
13. $(1/2\sqrt{15})\ln[(x + 1 - \sqrt{(5/3)})/(x + 1 + \sqrt{(5/3)})] + c.$
14. $(1/\sqrt{22})\ln[(\sqrt{11} + 2x - 3)/(\sqrt{11} - 2x + 3)] + c.$

Exercises 18(c)

1. $\sin^{-1}[(x + 1)/2] + c.$
2. $\sin^{-1}[(x - 1)/2] + c.$
3. $\sin^{-1}[(x + 1)/3] + c.$
4. $\sin^{-1}[(x + 3)/4] + c.$
5. $\sinh^{-1}[(x + 1)/5] + c.$
6. $\cosh^{-1}[(x - 2)/5] + c.$
7. $\sin^{-1}[(x - 2)/2] + c.$

Exercises 18(d)

1. $\frac{1}{10}\ln[(2x - 3)^3(x + 1)^2] + c.$
2. $\frac{1}{12}\ln[(3x + 5)^7/(x - 1)^3] + c.$
3. $-\frac{1}{2}\ln[(3x - 1)(x - 1)] + c.$
4. $\frac{1}{8}\ln[x^3(x - 4)^7/(x - 2)^{10}] + c.$
5. $\frac{1}{2}\ln[x^6(x^2 + 4)] - \frac{1}{2}\tan^{-1}(\frac{1}{2}x) + c.$
6. $-\frac{3}{2}[\ln(9 - x^2) + 4/(3 + x)] + c.$
7. $x - 4\ln(x + 2) - 5/(x + 2) + c.$
8. $\ln[(x - 1)/x] + 2/x - 2/(x - 1) + c.$
9. $-2\ln(12 - 6x - x^2) - (7/\sqrt{21})\ln[(\sqrt{21} + x + 3)/(\sqrt{21} - x - 3)] + c.$
10. $\ln[\sqrt{(x^2 + 1)}/(1 - x)] + c.$
11. $\frac{1}{8}\ln[(4x^2 + 1)/(2x - 1)^4] + \frac{5}{4}\tan^{-1}(2x) + c.$
12. $\ln(x^2 + 2x + 26) - \frac{1}{5}\tan^{-1}[(x + 1)/5] + c.$
13. $\frac{1}{2}\ln(4x^2 + 16x + 25) - (7/6)\tan^{-1}[(2x + 4)/3] + c.$
14. $(5/6)\ln(3x^2 + 6x - 2) - (2/\sqrt{15})\ln[(3x + 3 - \sqrt{15})/(3x + 3 + \sqrt{15})] + c.$
15. $-\frac{3}{4}\ln(1 + 6x - 2x^2) + (4/\sqrt{11})\ln[(\sqrt{11} + 2x - 3)/(\sqrt{11} - 2x + 3)] + c.$
23. $\ln[e^x/(1 + e^x)] + c.$

Exercises 18(e)

1. $2/3$.
2. $3\pi/16$.
3. $(1/10)\sin^5 2x + c$.
4. $\ln(\cos x + \sin x) + c$.
5. $1/12$.
6. $\sin^5 x(63 - 90\sin^2 x + 35\sin^4 x)/315 + c$.
7. Put $\sin x = z$; $\frac{1}{2}\sqrt{2}\ln(1 + \sqrt{2})$.
8. $(3\sqrt{2} - 4)/10$.
9. $\frac{1}{2}\ln(\operatorname{cosec} 2x - \cot 2x) + c$.
10. $\frac{1}{3}\ln(\sec 3x + \tan 3x) + c$.
11. $2\tan\frac{1}{2}x - x + c$.
12. $\frac{1}{3}\sec^3 x - \sec x + c$.
13. $\tan x + \frac{1}{3}\tan^3 x + c$.
14. $-\frac{1}{3}\cot^3 x + c$.
15. $\frac{1}{3}\tan^3 x + 2\tan x - \cot x + c$.
16. $\tan\frac{1}{2}x + c$.
17. $\tan(\frac{1}{2}x - \pi/4) + c$.
18. $(5\sin 2x - \sin 10x)/20 + c$.
19. $(4\sin 2x + \sin 8x)/16 + c$.
20. $\frac{1}{5}\ln[(5 + \tan\frac{1}{2}x)/(5 - \tan\frac{1}{2}x)] + c$.
21. $(1/60)\ln[(11\tan\frac{1}{2}x + 1)/(\tan\frac{1}{2}x + 11)] + c$.
22. $\frac{1}{12}\tan^{-1}(\frac{3}{4}\tan x) + c$.
23. $4x + \ln(\sin x + 2\cos x) + c$.
24. $x + \ln(2 + 2\cos x - \sin x) + \ln(2 - \tan\frac{1}{2}x) + c$.
25. $2\pi - 4$.
26. 1.
27. $[(9x^2 - 2)\sin 3x + 6x\cos 3x]/27 + c$.
28. $e^{2x}(2\sin 3x - 3\cos 3x)/13 + c$.
29. $\pi^2/4$.
30. $\frac{1}{2}$.
31. $\frac{1}{2}$.
32. $(2e + \pi)/(4 + \pi^2)$.
33. $x\tan x + \ln(\cos x) + c$.
34. $\frac{1}{4}(\sinh 2x - 2x) + c$.
35. $\ln(\cosh x) - \frac{1}{2}\tanh^2 x + c$.
36. $[\tan^{-1}(\sqrt{6}) - \tan^{-1}(\frac{1}{2}\sqrt{6})]/\sqrt{6}$.
37. $x\sinh x - \cosh x + c$.
38. $\tan^{-1}(\sinh x) + c$, or $\tan^{-1}(e^x) + c$.
39. Express denominator as $5\cos[\theta - \tan^{-1}(4/3)]$; $\frac{1}{5}\ln 3$.
40. $\pi(8\sqrt{3} - 9)/36$.

Exercises 18(f)

1. $\frac{2}{5}x^{5/2} + \frac{4}{3}x^{3/2} + 2x^{1/2} + c$.
2. $\frac{2}{5}(x + 1)^{5/2} - \frac{2}{3}(x + 1)^{3/2} + c$.
3. $\frac{4}{3}(2 - \sqrt{2})$.
4. $1 - 1/\sqrt{3}$.
5. $\frac{3}{16}(2x^2 + 5)^{4/3} + c$.
6. $\frac{2}{9}(x^3 + 1)^{3/2} + c$.
7. $\frac{4}{9}$.
8. $-\frac{1}{3}(9 - 2x)^{3/2} + c$.
9. $\frac{1}{2}\sin^{-1}(2x/3) + c$.
10. $\sqrt{(x^2 + 2x + 10)} + 2\sinh^{-1}[(x + 1)/3] + c$.
11. $\sqrt{(x^2 - 7x + 12)} + \frac{13}{2}\cosh^{-1}(2x - 7) + c$.
12. $(x - 1)/\sqrt{(x^2 + 1)} + c$.
13. $\frac{1}{2}(\sqrt{26} - \sqrt{10}) + \ln[(3 + \sqrt{10})/(5 + \sqrt{26})]$.
14. $10 + 9\pi/2$.
15. $22{\cdot}67$.
16. $3{\cdot}77$.
17. $\sin^{-1}\sqrt{x} - \sqrt{[x(1 - x)]} + c$ (e.g. put $x = \sin^2 z$).
18. $\frac{1}{2}\pi - 1$.
19. 3π.
20. $\sqrt{(x^2 - 1)} + \ln[x + \sqrt{(x^2 - 1)}] + c$.

Exercises 18(g)

1. $\frac{16}{[illegible]}$.
2. $5\pi/32$.
3. $\frac{2}{15}$.
4. $\pi/32$.

5. $35\pi/8$.
6. $3\pi/16$.
7. $\frac{1}{4}\tan^4 x - \frac{1}{2}\tan^2 x - \ln(\cos x) + c$.
8. $\frac{10}{3}$.
9. $5\pi/256$.
10. $\frac{1}{4}(2\ln 2 - 1)$.
11. $(16e - 38)/e^2$.
12. $(n - 1)!$
13. $5(\pi/2)^4 - 60(\pi/2)^2 + 120$.
14. $\frac{3}{5}\cosh(\pi/2)$.
17. Write $I_n = \int_0^{\pi/2} (x\cos^{n-1} x)(\cos x)\,dx$; $I_4 = (3\pi^2 - 16)/64$; $I_5 = (60\pi - 149)/225$.

Exercises 18(h)

1. $\frac{7}{3}$; in this case we are approximating, with a parabolic arc, to a parabola.
2. 0·5184.
4. 0·4055.
5. 0·6045; 0·6046.
7. 7·514; 7·525.
8. 0·248; 0·247.
9. $\pi^2(4 + \sqrt{3})/18$; π.

Exercises 19(a)

1. (a) π, (b) π, (c) 0, (d) 0.
2. $\pi - 2$.
3. $4\frac{1}{2}$.
4. $\frac{128}{5}$, $x = 16^{2/5}$.
5. $\frac{29}{7}$ m.
6. 0·071.
7. 256/15.
8. $\frac{4}{3}$.
9. $\frac{1}{6}$.
10. 9.
11. $8\sqrt{2}$.
12. $\frac{1}{6}ab^3$, $y = \frac{1}{8}x(6 - x)$.
13. (a) $\pi/4$, (b) $\pi/3$, (c) $21\sqrt{3}/2\pi$, (d) $\pi/2 - 4/\pi$.
14. 16/3.
15. 32/3.
16. 8/35.
19. $a^2(\ln 2 - \frac{2}{3})$.
20. $y - 4a^2x + 8a^3 = 0$.
23. $(22\pi + 15\sqrt{3})/6$.
24. $3\pi/16$.

Exercises 19(b)

1. (a) $512\pi/9$, (b) $56\pi/3$.
2. (a) $384\pi/7$, (b) $\pi^2/4$.
3. $\frac{1}{3}\pi r^2 h$; $\frac{1}{3}\pi r^2(3h' - 3h'^2/h + h'^3/h^2)$.
4. (a) $\pi^2/2$, (b) $\pi^2/2$.
5. 414 cm^3.
6. 148 200 cm^3.
7. $\pi/10$.
9. $\frac{32}{3}$, $128\pi/15$.
10. $512\pi/15$.
11. $256/\sqrt{3}$.
12. $\frac{1}{12}\pi b^2 h$.
14. $\frac{4}{3}a^3$.
15. 10 cm^3.
16. $16a^3/3$.
17. $\frac{1}{3}\pi abh$.
18. $\frac{4}{3}l^2$, $\frac{16}{9}ab^2$.
19. $\pi[(h + 1)^5 - 1]/5$, $dV/dt = \pi(h + 1)^4\, dh/dt$, $\pi(h + 1)^4 k$ where $k = dh/dt$.
20. $8a^2/15$; $\pi a^3/12$.
21. $8a^2/15$; $\pi a^3/4$.
26. $3\pi a^2/4$; $\pi^2 a^3/4$.
27. $\pi a^3(\ln 4 - 1)$.
28. $\pi(\frac{8}{3} - \sqrt{2})$.

Exercises 19(c)

1. $\frac{1}{2}(e - 1/e)$.
2. 17/12.

3. $\sqrt{3}$.
4. $33/16$.
7. $59/24$.
9. $(13^{3/2} - 8)/27$.
10. $(13^{3/2} - 8)a/27$.
11. $3/4$.
12. $16a/3$.
13. 4.
15. $\frac{1}{3}\sqrt{2}$.
17. length $= \ln[(1 + x_1)/(1 - x_1)] - x_1$;
area $= (1 - x_1)\ln(1 - x_1) - (1 + x_1)\ln(1 + x_1) + 2x_1$.
18. $3\pi a/2$.
19. $(r_2 - r_1)\sec\alpha$.

Exercises 19(d)

1. $56\pi a^2/3$.
2. (a) $4\pi a^2$, (b) $2\pi ah$.
3. $\frac{1}{4}\pi(e^2 - e^{-2} + 4)$.
4. $141\pi/48$.
5. $\frac{6}{5}\pi a^2$.
7. $\frac{1}{3}\pi a^2\{67\sqrt{5} - 8 - \frac{3}{2}\ln(2 + \sqrt{5})\}$; $96\pi a^3/5$.
8. $2\pi[\ln(\sqrt{2} + 1) + \sqrt{2}]$.
9. $(0, \frac{2}{3}a)$, $(0, -\frac{2}{3}a)$; two tangents at $(-2a, 0)$; $3\pi a^2$.

Exercises 20(a)

1. $2x/y + y^2/x^2$, $-(x^2/y^2 + 2y/x)$.
2. $2y/(x + y)^2$, $-2x/(x + y)^2$.
3. $(1 + 2x)e^{2x+3y}$, $3xe^{2x+3y}$.
4. $xy/\sqrt{(x^2 - y^2)}$, $(x^2 - 2y^2)/\sqrt{(x^2 - y^2)}$.
5. $y/(x^2 + y^2)$, $-x/(x^2 + y^2)$.
6. $e^{x-y}[\ln(y - x) - 1/(y - x)]$, $e^{x-y}[1/(y - x) - \ln(y - x)]$.
7. $2\sin y - y^2\cos x$, $2\cos x - x^2\sin y$.
8. $y(2\ln x - 3)/x^3$, 0.
9. $y(2x^2 - y^2)/x^2(x^2 - y^2)^{3/2}$, $y/(x^2 - y^2)^{3/2}$.
24. $w = xy(x + y)e^{x-y}$.

Exercises 20(b)

1. $\pm 1{\cdot}28\pi$.
2. $\pm 3{\cdot}8\%$.
4. 324π, $\pm 0{\cdot}47\%$.
5. $\pm 2\frac{1}{3}\%$.
6. $\pm 4\frac{4}{15}\%$.
7. $\Delta A \simeq \cos C\,\Delta b + \cos B\,\Delta c$; the triangle should be right-angled at C.
8. $\Delta b/b + \Delta c/c + \cot A\,\Delta A$; $0{\cdot}015$.
10. area $= \frac{1}{2}a^2\sin B\sin C\operatorname{cosec}(B + C)$.
11. $D = a^2/\sqrt{(a^2 - b^2)}$; $\Delta D \simeq [a(a^2 - 2b^2)\Delta a + a^2 b\,\Delta b]/(a^2 - b^2)^{3/2}$.
14. $(4 - \sqrt{3})\pi a/1080\sqrt{2}$.

Exercises 20(c)

14. $c = a^2b/7$, $d = 8b/7$.
15. $\sin\theta\cos\theta/r^2$.

Exercises 20(d)

1. x and y both odd integers or both even integers.
2. Min. T.P. when $x = -\frac{5}{3}$, $y = -\frac{4}{3}$.
3. Saddle-points when $x = \frac{2}{3}$, $y = \frac{1}{3}$ and when $x = -\frac{2}{3}$, $y = -\frac{1}{3}$.

4. Max. T.P. when $x = y = 0$, saddle-points when $x = \pm 1, y = \pm 1$.
5. Saddle-points when $x = 1/\sqrt{6}, y = \sqrt{6}/2$ and when $x = -1/\sqrt{6}, y = -\sqrt{6}/2$.
6. Min. T.P. when $x = \frac{1}{8}, y = \frac{1}{2}$, saddle-points when $x = -1, y = -1$ and when $x = -1, y = 2$.
7. Max. T.P. when $x = 1, y = 0$, saddle-points when $x = 2, y = \pm 2$.
8. Max. T.P. when $x = -3, y = -2$, saddle-point when $x = 1, y = 2$.
9. Max. T.P. when $x = y = \frac{2}{3}$.

Exercises 20(f)

1. 4.
2. 256.
3. $\frac{1}{6}a^3$.
4. $\frac{1}{10}$.
5. $\frac{31}{13}$.
6. $\frac{1}{2}\ln(1 + \sqrt{2})$.
7. $\frac{1}{6}$.
8. $\frac{1}{3}(3\sqrt{2}/2 - 2)$.
9. $\pi/8$.
10. $\frac{1}{9}\ln\frac{5}{4}$.
11. π.
12. $\frac{1}{4}(e - 1)$.
13. $\frac{1}{24}$.
14. $\frac{1}{2}\ln 17 + \frac{1}{16}\tan^{-1} 4 - \frac{1}{4}$.
15. $\frac{1}{8}$.
16. $\pi/8$.

Exercises 21(a)

1. $r(d^2\phi/dr^2) + 2(d\phi/dr) = 0$.
2. $(d^2y/dx^2) - \tan x\,(dy/dx) - y\sec^2 x = 0$.
3. $(dT/dt) = h(T_f - T)$, T = temp. of object, T_f = temp. of flame.
4. $(dA/dt) = kA$, A = area of substance alight.
5. $v(dv/ds) = -h(s + v^3)$.
6. $dV/dt = -k(V_0 - V)$, V = volume of liquid in the vessel at time t.
7. $dy/dx = (x^2 + y^2)^{3/2}$.
8. $dy/dx = -1/y^2$.

Exercises 21(b)

1. $xy - e^x = C$.
2. $y = x(x^2 + C)$.
3. $ax^2 + 2hxy + by^2 + 2gx + 2fy + C = 0$.
4. $y(1 - \cos 2x) = C$.
5. $x^3 + 2y^3 + 2xy + x + 2y = C$.

Exercises 21(c)

1. $y = (C + e^x)\sin^2 x$.
2. $y = C\sec^4 x - \cos^2 x$.
3. $xy = \frac{1}{2}(\ln x)^2 + C$.
4. $y\sin^2 x = x\sin x + \cos x + C$.
5. $y(1 + x) = x(C + x^3)$.
6. $y = (C - e^{\cos x})\sqrt{(1 - x^2)}$.
7. $y = x^2 - 2 + 2e^{-x^2/2}$.
8. $y(x - 3)^2 = (x - 2)\{(x - 1)^2 + C(x - 2)\}$.
9. $2xy = (1 + x)^2 + C(1 - x^2)^2$; $2y = (1 + x)^2(2 - x)$.
10. $y = \frac{1}{4}(x - 1)^2\{(x - 1)\sin 2x + \frac{1}{2}\cos 2x + x^2 - 2x + C\}$.
11. $2x^3y = e^{-2x}\{x^2 - 2e^{-x}(x + 1) + C\}$.
12. $2xy = (1 - x)\cos x + x\sin x + Ce^{-x}$.
13. $2y = x(1 + x^2) + (C - \sinh^{-1} x)\sqrt{(1 + x^2)}$.
14. $y(1 + 3x)^2 = Ce^{3x} - (2 + 3x)$.
15. $y(x + 1) = x^4 + x$.
16. $y/x = 1 + x\sin x + \cos x$.

Exercises 21(d)

1. $3y^2 = 2x^3 + C$.
2. $\ln \sin y = C + e^x$.
3. $Cy^4 = (x - 2)/(x + 2)$.
4. $2 \tan^{-1} x - \tan^{-1}(\frac{1}{2}y) = C$.
5. $1 + y^2 = Cx^2e^{x^2}$.
6. $2e^x(\sin x - \cos x) + e^{-2y}(2y + 1) = C$.
7. $4x^2 = y(1 + 6x - 3x^2)$.
8. $y = \sin(x \sin x + \cos x - 1)$.
9. $y = 1/\{2 - e^{-x}(1 + x)\}$.
10. $dx/dt = x(\cot t - 1)$, $x = e^{\pi/2-t} \sin t$; π.
11. 30·4°C.
12. $ds/dp = -ks/(A + p)$, $s = C(p + A)^{-k}$.
13. $dA/dt = kA$ where A is the area infected at time t, or $dr/dt = kr$ where r is the ratio of the infected area to the whole area, $\ln r = (\frac{1}{12} \ln 2{\cdot}8)t + \ln \frac{1}{4}$.
14. $dp/dt = kp$, $p = 10^{(3/2)t+2}$.
15. $dq/dt = kq$ where q is the amount of active ferment; $q = Ae^{kt}$, $t = 2$.
16. $dP/dt = kP$, $P = Ae^{kt}$.
17. $dx/dt = -kx^3$ where x is the concentration; $x = \sqrt{[a^3/(a + 2bt)]}$.
18. 12 min.
19. $dy/dx = y/k$, $y = 5e^{x/k}$.
20. $\sqrt{(d - x)}(\sec\theta + \tan\theta) = C$, $C = \sqrt{d}$.

Exercises 21(e)

1. $Cx^4 = y^2 - 6xy + 8x^2$.
2. $y = C(x^2 - y^2)$.
3. $y^3 = x^3 \ln Cx^3$.
4. $x^4y^4 = C(x + y)^3$.
5. $4x/(y - 2x) = \ln\{C(y - 2x)\}$.
6. $x^2 - xy - y^2 + x + 3y = C$.
7. $y - x = \ln C(x + y)^2$.
8. $(x + y)^3 = C(x^2 - xy + y^2)$.
9. $2x + y = C(x + y - 3)^2$.
10. $3x - 4y - 6 = C(x - y - 1)^2$.
11. $y = 2x \tan^{-1}(Ce^x)$.
12. $x(x + y)^2 = 9(x - y)^2$.
13. $y = 2x^2 \ln x + x^2$.
14. $y^2 = x^4 - x^2$.
15. $y(3 - \sqrt{x}) = x(3 + \sqrt{x})$.

Exercises 21(f)

1. $y \sin x = 1/(C + \cos x)$.
2. $2 \sec^2 x = y^2(C - \tan^4 x)$.
3. $y^2 = Ce^{2x} - (x + \frac{1}{2})$.
4. $y^2(C - \sin^2 x) = 2 \sin x$; $y + \{(2 \sin x)/(3 - \sin^2 x)\}^{1/2} = 0$.
5. $2/y^2 = 2x + 1 + Ce^{2x}$.
6. $y(\sin^3 x + C \cos^3 x) + 3 \cos x = 0$.
7. $y^2\{x^3 + 3x^2 + 9x + 9 \ln(x - 1) + C\} = 3(x^2 + x + 1)$.
8. $y^3(x^2 - 1)^{3/2} = 3e^x(x - 1)^2 + C$.
9. $(1/y^2) - (1/x^2) = 1$.

Exercises 21(g)

1. (a) $(x - 2y)^5(3x + 2y)^3 = C$, (b) $y = 8x^2 - x - \frac{1}{2}$.
2. (a) $(x + y)^2(x - y)^3 = C$, (b) $y = \sin x(C - \cos x)$.
3. (a) $x = C\sqrt{(1 - t^2)} + t^2 - 1$, (b) $(9x - 2y)^7(x + 2y)^3 = C$.
4. (a) $y \cos x = \frac{1}{2} \cos 2x + c$, (b) $(x^2 + y^2)^2 = C(x^2 - y^2)$.
5. (a) $y = \cos x(C - 2 \cos x)$, (b) $(x - y)^2e^{y/x} = Cx$,
   (c) $y \cos x - x \sin y = C$.

6. (a) $y(x^2+1)^2 = x^5 - 5x + C$, (b) $(x^2 - y^2)^2 = Cxy$.
7. (a) $y = 1 + x + C \sin x$, (b) $(2x - y)^2 = Cx^2(x - y)$.
8. (a) $y = \sin 2x + \cos x \ln \{C(\sec x - \tan x)\}$,
   (b) $8 \tan^{-1}(y/x) = 3 \ln \{C(x^2 + y^2)\}$.
9. (a) $(y - x)^2/(y + x)^2 = Cx$, (b) $y = (1 + x^2) \ln \{C(1 + x^2)\}$,
   (c) $2(x + 4y) = \tan (2x + C)$.
10. (a) $\ln (xy^{-1/2} - 1) + \frac{1}{2}x^{n+1}/(n+1) = C$, (b) $(y - x + 4)^2(x + y + 2) = C$.
11. (a) $y - x = Ce^{-(x^2+x)}$, (b) $x + 2y + 1 = C(2x + y - 1)^2$.
12. (a) $\ln (10x + 5y - 1)^7 = 5x - 10y + C$,
    (b) $1/(y \sin x) = \ln (\operatorname{cosec} x + \cot x) - \cos x + C$.
13. (a) $y^2(C - 2x^5) = 5x^2(1 - x^2)$, (b) $y + (x^2 + y^2)^{1/2} = Cx^2$.
14. (a) $(x - 2y + 5)^4 = C(x + y - 1)$, (b) $y^2 \ln (C \operatorname{cosec} x) = 2 \sin x$.
15. (a) $2\sqrt{3} \tan^{-1}\{(2y + x)/x\sqrt{3}\} = \ln \{C(x - y)^3(x^3 - y^3)\}$,
    (b) $\sqrt{2} \tan^{-1} \{(y - 1)/(x - 1)\sqrt{2}\} = \ln C + \ln \{2(x - 1)^2 + (y - 1)^2\}$,
    (c) $2y\sqrt{(1 + x)} = Ce^x - (x^2 + 2x + 2)$.
16. (a) $3 \tan^{-1} \{(y + 1)/(x - 2)\} + \ln \{(x - 2)^2 + (y + 1)^2\} = C$,
    (b) $(\cos^2 x)/y^2 + 2 \tan x + \frac{2}{3} \tan^3 x = C$.
17. (a) $x(1 - x)y + 1 = C(1 + x)$, (b) $2x/(x + y) + \ln \{(x + y)^3(x - y)\} = C$.
18. (a) $y = Ce^{-x} + 2(x - 1)$, (b) $y^2 + xy - x^2 - 5y + 5 = C$.
19. (a) $y = x^2 \sin x + C \cos x$, (b) $(x - 1)^2 - 2(x - 1)(y - 1) + 8(y - 1)^2 = C$.
20. (a) $(x + 3y)^3(2x - y)^4 = C$, (b) $1/y^2 = Ce^{2x} + x + \frac{1}{2}$.
21. (a) $(\sqrt{2} - 1) \ln (x + y\sqrt{2}) - (\sqrt{2} + 1) \ln (x - y\sqrt{2}) - 2 \ln x = C$,
    (b) $y = \frac{1}{4}(C - \cos x)^2 \operatorname{cosec} x$.
    If y is finite when $x \to 0$, $C = 1$ and $y = \frac{1}{2} \sin^2 \frac{1}{2}x \tan \frac{1}{2}x$.
22. (a) $(x - 1)y = x\{x \ln x - x - \frac{1}{2}(\ln x)^2 + C\}$, (b) $y + (x^2 + y^2)^{1/2} = Cx^2$,
    (c) $x^2y(1 + Cx) = 1$.
23. (a) $Cx^3 + 6x^2y + 3xy^2 + 2y^3 + 6x^3 \ln \{(y - x)^2/x^3\} = 0$,
    (b) $y \sin^2 x = C - \frac{1}{4} \cos 2x$.
24. (a) $(x + y)e^{-y/x} = x \ln (C/x)$, (b) $x^2y = x(1 - x^2) + 1 + C(1 + x)$.
25. (a) $2x^3 + 3x^2y + y^3 = C$, (b) $y = \frac{1}{2}e^{-x} + Ce^x$.
26. (a) $y = Cx^2e^{1/x} - (1 + 2x + 2x^2)$,
    (b) $x^4 + 2x^3y + 3x^2y + x^2y^2 + 6xy^2 + 3y^3 = C$.
27. $y^3 = 3x^6 - 2x^3$.
28. (a) I.F. $= x^2$; $x^3(x^2 + 3xy - y^2) = C$, (b) $4(x - 1)y = (2x - 3)e^x + 3e^{-x}$.
29. (a) $e^{2y/x} = Gx^3(x + 2y)$, (b) $2xy + 1 = Cx^3(xy - 1)$.
30. (a) $xy = 2 + (x + 1)\{2 \ln (x + 1) + C\}$,
    (b) $\tan^{-1} \{(y^2 + x)/x\} = \ln \sqrt{\{C(2x^2 + 2xy^2 + y^4)\}}$.
31. (a) $y^2 = 2x^2 + 2x + 1$, (b) $x + y = (x - y)^3$.

Exercises 21(h)

1. G.S. $y = Cx + a/C$; S.S. $y^2 = 4ax$.
2. G.S. $y = Cx + 2\sqrt{(aC)}$; S.S. $xy + a = 0$.
3. G.S. $y = Cx + C^3$; S.S. $27y^2 + 4x^3 = 0$.
4. G.S. $y = Cx + a\sqrt{(1 + C^2)}$; S.S. $x^2 + y^2 = a^2$.
5. G.S. $y = Cx + C - C^2$; S.S. $4y = (x + 1)^2$.

6. G.S. $y = C(x - 1) + \ln(1 + C)$; S.S. $y + x + \ln(1 - x) = 0$.
7. G.S. $2y = 2Cx - \ln\sec^2 C$; S.S. $2y = 2x\tan^{-1} x - \ln(1 + x^2)$.
8. $Y = PX + c^2P/(P - 1)$ where $P = dY/dX$; G.S. $y^2 = ax^2 + c^2a/(a - 1)$ where a is an arbitrary constant;
S.S. $(y + x + c)(y + x - c)(y - x + c)(y - x - c) = 0$.
9. $v = Px + 1/P$ where $P = dv/dx$; G.S. $y = Cx^2 + x/C$; S.S. $y^2 = 4x^3$.
10. The given equation may be written in the form $(y - px)^2 = a^2(1 + p^2) - c^2$.

Exercises 21(i)

1. $y^3 = C(y^2 - x^2)$. **2.** $x - y = C(x + y)^3$.
3. $y' = (y^2 - x^2)/2xy$.
4. $y = \frac{1}{2}x(y' + 1/y')$; $x^2 = C(2y - C)$; $y = -\frac{1}{2}x(y' + 1/y')$.
5. $y'^2 = a^2/x^2 - 1$; $\pm(y + C) = \sqrt{(a^2 - x^2)} - a\ln\{a + \sqrt{(a^2 - x^2)}\} + a\ln x$.
6. $y' = \pm 1/\sqrt{(a^2y^2 - 1)}$; $\pm 2(x + C) = y\sqrt{(a^2y^2 - 1)} - (1/a)\cosh^{-1} ay$;
$2x = y\sqrt{(a^2y^2 - 1)} - (1/a)\cosh^{-1} ay$; $ay = \cosh ax$.

Exercises 22(a)

1. (a) Divergent, (b) convergent, (c) convergent, (d) convergent, (e) divergent, (f) convergent, (g) divergent, (h) convergent, (i) convergent, (j) divergent, (k) divergent, (l) convergent.
2. (a) A.C. when $|x| < 1$; C.C. when $x = -1$, (b) A.C. for all values of x, (c) A.C. when $|x| < 1$; C.C. when $x = -1$, (d) A.C. for all values of x, (e) A.C. when $|x| < 1$; C.C. when $|x| = 1$.
4. (d) $u_n \to 0$ since $u_n < \frac{1}{3}(\frac{1}{2})^{n-1}$, (e) $u_n < \frac{1}{3}(\frac{2}{3})^{n-1}$.
7. Convergent if $|x| \leqslant 1$; divergent if $|x| > 1$.

Exercises 22(b)

1. (a) $2\left(\frac{1}{1!} + \frac{1}{3!} + \frac{1}{5!} + \frac{1}{7!} + \cdots\right)$, (b) $2\left(1 + \frac{4}{2!} + \frac{16}{4!} + \frac{64}{6!} + \cdots\right)$,

(c) $1 + \frac{3}{1!} + \frac{9}{2!} + \frac{27}{3!} + \cdots$.

2. (a) $1 + \frac{2x}{1!} + \frac{4x^2}{2!} + \frac{8x^3}{3!} + \cdots$; $2^r x^r/r!$,

(b) $1 + 2x + \frac{3x^2}{2!} + \frac{4x^3}{3!} + \cdots$; $(r + 1)x^r/r!$,

(c) $1 - x - \frac{x^2}{2!} + \frac{5x^3}{3!} + \cdots$; $(-1)^{r-1}(r^2 - r - 1)x^r/r!$.

3. $1 + x + \frac{3x^2}{2} + \frac{7x^3}{6} + \frac{25}{24}x^4 + \cdots$.

5. Last part: put $m/n = e^{2\theta}$.
8. (a) $-\frac{1}{3}$, (b) Put $x = 1 + y$; 1, (c) $-1/48$, (d) $1/18$, (e) 1, (f) $\frac{1}{4}$, (g) 1, (h) $\frac{1}{4}$, (i) 0, (j) $\frac{1}{2}$.

9. $x^2-(4/3)x^4+(32/45)x^6+\cdots; 22/45$.
10. $1/2560$.
14. $2^{-1/3}[1-x^2/36+x^4/6480+\cdots]$.
21. $x+(1/6)x^3+\cdots; t-t^2+(7/6)t^3+\cdots$.
22. $1+x^3/3!+14x^4/6!+\ldots$.
24. (a) $\ln 2-\frac{1}{2}x^2$, (b) $-\frac{1}{2}$.

Exercises 22(c)

1. (a) $\ln 2-i(\pi/3)$, (b) $(1/\sqrt{2})\cosh t-(i/\sqrt{2})\sinh t$, (c) $-e^2$,
(d) $\frac{1}{2}\sinh t-i(\sqrt{3}/2)\cosh t$, (e) $-5/8+i(3\sqrt{3}/8)$, (f) $-\cosh t$.
2. $2n\pi \pm i\ln 2\ (n=0, \pm 1, \pm 2, \ldots)$.
4. e^z is real if $z=x+ik\pi$, $\cos z$ is real if $z=k\pi+iy$, $k=0, \pm 1, \pm 2, \ldots$.
5. 2π apart.
6. (a) $(4k+1)\pi/2 \pm i\ln(3+2\sqrt{2})$;
(b) $k\pi-\frac{1}{2}\tan^{-1}2+\frac{1}{4}i\ln 5\ (k=0, \pm 1, \ldots)$.
7. $x=\tanh 2u$, $y=\operatorname{sech} 2u$. **8.** $x=\sin u\cosh v$, $y=\cos u\sinh v$.
12. If $|z|=r$, $\arg z=\theta$, then $|e^z|=e^{r\cos\theta}$, $\arg(e^z)=r\sin\theta$.
15. $(\sqrt{3}/4\sqrt{2})(\cosh\theta+\sinh\theta), (1/4\sqrt{2})(3\sinh\theta-\cosh\theta)$.

Exercises 23(a)

1. $30x=11\sin 3t-3\cos 3t+3e^{-t}$.
2. $y=A\cos 2x+B\sin 2x+\frac{1}{4}x+\frac{1}{5}\sin 3x$.
3. $y=A\sin x+B\cos x+\frac{1}{5}e^{-x}(\sin x+2\cos x)$.
4. $x=Ae^t+Be^{3t}+\frac{1}{9}(3t+4)-e^{2t}$.
5. $y=A+Be^{-x}+\frac{1}{3}x^3-x^2+3x$. **6.** $y=e^{-2x}(A+Bx+3x^3)$.
7. $y=e^x(A+Bx-\cos x)$.
8. $y=Ae^{-x}+Be^{-3x}+(\sin x-2\cos x)/10$.
9. $y=Ae^{4x}+Be^x-e^x(3x^2+2x)/18$.
10. $y=Ae^{-x}+Be^{-2x}+xe^{-x}(x-2)$.
11. $y=e^{-2x}(A\sin 3x+B\cos 3x+\frac{1}{8}\cos x)$.
12. $y=Ae^x+Be^{2x}-e^x(\frac{1}{2}x^2+2x)$.
13. $y=Ae^{5x}+Be^{-x}-(2\sin x+3\cos x)/26$.
14. $y=(e^{(1/2)t}+1)(\cos t-4\sin t)$.
15. $y=Ae^{4x}-e^x(B+3x^2+2x)/18$.
16. $y=e^x(A\cos x+B\sin x)+(2\cos 2x-\sin 2x)/10$.
17. $y=e^{bx/a}\{A\cos(bx\sqrt{3}/a)+B\sin(bx\sqrt{3}/a)\}+x(bx+a)$.
18. $y=e^{-2x}(A+Bx)+(e^{2x}+8x^2e^{-2x})/32$.
19. $y=e^x(A+Bx)+e^{2x}(x^2-2x+3)$.
20. $y=e^{-x}\{A\cos(x\sqrt{2})+B\sin(x\sqrt{2})\}+\frac{1}{2}e^{-x}+(4\sin 2x-\cos 2x)/17$.
21. $y=e^{-x}(A\cos x+B\sin x)+\frac{1}{2}(x-1)+\frac{1}{5}(\sin x-2\cos x)$.
22. $y=Ae^{3x}+Be^{-3x}+(27x\sinh 3x-18x^2-4)/162$.
23. $s=e^{-3t}(A\cos 4t+B\sin 4t)+(24\cos 4t+64\sin 4t)/73$.
24. $y=\frac{1}{4}e^{2x}(A\cos x+B\sin x-x^2\cos x+x\sin x)$.
25. $y=\frac{1}{2}e^x\{(2\pi+1)\sin x-x\cos x\}$.
26. $y=e^{-2x}(A\cos x+B\sin x)+\frac{1}{8}(\cos x+\sin x+4xe^{-2x}\sin x)$.

27. $y = Ae^{(1/2)x} + Be^{-2x} - (5 \sin 2x + 3 \cos 2x)/68 - \frac{1}{4}(2x^2 + 6x + 13)$.
28. $y = A + B \cos 2x + C \sin 2x - \frac{1}{3} \cos x - \frac{1}{8}x^2$.
29. $y = Ae^{-2x} + B \cos 2x + C \sin 2x + \frac{1}{3} \sin x + \frac{1}{4} + \frac{3}{4}x$.
30. $y = A + Be^{2x} + Ce^{-2x} + E \cos x + F \sin x -$ $(36x + \cos 2x + 6x^4 - 54x^2)/48$.
31. (a) $y = e^x(Ax + B) + C \cos 2x + E \sin 2x + 2e^{2x} - \sin x$;
(b) $y = Ae^{-x} + B \cos x + C \sin x + \frac{1}{2}xe^{-x} + x^4 - 4x^3 + 24$.
32. $y = Ae^x + Be^{-x} + C \cos 2x + E \sin 2x + \frac{17}{2}xe^x + \frac{5}{8}e^x(5 \cos 2x + 3 \sin 2x)$.
33. $x = \frac{1}{4}\{\cos t - e^{-t} + t(\sin t - \cos t)\}$.
34. $y = e^x \sin^2 x - x^2$.
35. (a) $y = Ae^x + Be^{-x} + C \cos x + E \sin x + \frac{1}{4} x \cosh x + \frac{1}{15} \sin 2x$;
(b) $y = Ax + B + Ce^{-x} + (x^3 - 3x^2 + 3 \cos x - 3 \sin x)/6$.
36. (a) $y = e^t(A \cos 2t + B \sin 2t) + (\cos 2t - 4 \sin 2t)/17$;
(b) $y = Ae^t + B \cos 2t + C \sin 2t + te^t$.
37. (a) $y = Ae^x + Be^{-x} + C \cos 2x + E \sin 2x - \frac{3}{8} - \frac{1}{4}x^2 - \sin x + e^{2x}$;
(b) $x = a(3e^{-2t} - 2e^{-4t})$; $x_{\max} = 9a/8$.
38. (a) $y = A + Be^x + Ce^{-x} + E \cos x + F \sin x - x + (6e^{2x} - \cos 2x)/30$;
(b) $x = (\cos 3nt - 9 \cos nt)a/12$; max. speed $= an$.
39. $x = (kt/2p)e^{-ht} \sin pt$.
40. $x = e^{-t}(A \sin 2t + B \cos 2t) + te^{-t} \sin 2t$; the particular solution is $x = te^{-t} \sin 2t$.
41. $\sin (\theta + k)$ where k is a constant.

Exercises 23(b)

1. $y = x^2(A + B \ln x) + x^2(\ln x)^3 + 2/3x$;
$y = \frac{1}{3}x^2\{3(\ln x)^3 + 9(\ln x) - 2 + 2/x^3\}$.
2. $y = Ax^2 + Bx^3 + x^3 \ln x$. **3.** $y = x\{2 - \sin (\ln x) - \cos (\ln x)\}$.
4. (a) $y = e^x(Ax + B) + \frac{3}{8}(2x^2e^x - e^{-x})$;
(b) $y = x^2 \{A \sin (\ln x) + B \cos (\ln x)\} + (75x + 10 \ln x + 8)/50$.
5. (a) $y = \frac{1}{2}e^{2x}(A + Bx + x^2)$;
(b) $y = \frac{1}{3}x^2 + \sqrt{x}\{A \cos (\frac{1}{2}\sqrt{3} \ln x) + B \sin (\frac{1}{2}\sqrt{3} \ln x)\}$.
6. (a) $y = \frac{2}{3}(4e^x - 49e^{-(1/2)x}) - x^3 + 3x^2 - 18x + 30$;
(b) $y = (A \ln x + B)/x + \frac{1}{9}x^2$.
7. (a) $y = Ae^x + Be^{2x/3} + \frac{1}{2}e^x(x^2 - 4x)$;
(b) $y = Ax + B/x^5 - \{10 + 9 \cos (\ln x)\}/90x^2$.
8. (a) $y = \frac{1}{2}e^x(A \cos x + B \sin x + x \sin x + 2)$;
(b) $y = x(A + x) + (1 + Bx) \ln x + 2$.
9. (a) $y = Ae^{2x} + Be^{3x} + \frac{1}{8}e^x(4x^3 + 18x^2 + 42x + 45)$;
(b) $y = A/x + (B - \ln x)/x^2 + (x^2 - 12)/12$.
10. (a) $y = Ae^x + Be^{-x} + \frac{1}{2}x \sinh x$;
(b) $y = Ax^3 + B/x - x^2(3 \ln x + 2)/9$.
11. (a) $y = e^{-(1/2)x}\{A \cos (\frac{1}{2}x\sqrt{3}) + B \sin(\frac{1}{2}x\sqrt{3})\} - e^x(3 \cos x - 2 \sin x)/13$;
(b) $y = x^2(A \ln x + B) + \frac{1}{2}x^2 (\ln x)^2$.
12. (a) $y = e^x(A \cos x + B \sin x + \frac{1}{2}x \sin x)$; (b) $y = x^3 + x(A \ln x + B)$.
13. (a) $y = e^{-2x}(A \cos x + B \sin x) + 4(5x - 4)/25 + 3e^{-3x} + (\sin x - \cos x)/8$;
(b) $y = x(1 - \ln x) + \frac{1}{3}x(\ln x)^3$.

14. (a) $y = \frac{1}{6}e^{3x}(A + Bx + 3x^2 + x^3)$;
(b) $y = Ax^4 + B/x + (9 - 12 \ln x - 4x^2)/24$.

15. (a) $y = e^x(A \cos 2x + B \sin 2x - \frac{1}{4}x \cos 2x)$;
(b) $y = x(A + B \ln x) + \frac{1}{4}x^3$.

16. (a) $y = e^{-3x}(\frac{1}{2}x^2 + Ax + B) + (27x^4 - 72x^3 + 108x^2 - 69x + 22)/243$;
(b) $y = \frac{1}{2} - x(x - 2)(1 - \ln x)$.

17. (a) $y = e^{-(1/2)x}\{A \cos(\frac{1}{2}x\sqrt{3}) + B \sin(\frac{1}{2}x\sqrt{3})\}$
$+ e^x\{13(x - 1) + 3(2 \sin x - 3 \cos x)\}/39$;
(b) $y = Ax + Bx^{-1/4} + 3 + \frac{1}{5}(x - 5) \ln x$.

18. (a) $y = A \sin x + B \cos x + \frac{1}{5}e^x(2 \sin x + \cos x) + x^2 - 2$;
(b) $2x^2y = Ax + B - (\ln x)(2 + \ln x)$.

19. (a) $y = Ae^{2x} + e^{-x}(B + 9 \cos x - 10x - 3 \sin x)/30$;
(b) $y = A/x + B/x^3 - \frac{4}{9} + \frac{1}{6}(2 - 3/x - 6/x^2) \ln x + \frac{1}{2}(\ln x)^2/x$.

20. (a) $y = e^{2x}(A \cos 3x + B \sin 3x) - \frac{1}{6}xe^{2x} \cos 3x$;
(b) $y = x^{1/3}(A + B \ln x) + \frac{1}{4}x(\ln x - 3)$.

21. (a) $y = e^{2x}(A \cos x + B \sin x) - \frac{1}{2}xe^{2x} \cos x$;
(b) $y = x^2(A + B \ln x) + \frac{1}{6}x^2(\ln x)^3$.

22. (a) $y = e^{2x}(Ax + B + 3x^2 + x^3)/12 + e^{-2x}(1 - 2x)/64$;
(b) $y = A/x + B/x^7 + \frac{1}{8}x - (\ln x)/x$.

23. $y = Ax + B \cos(\frac{1}{2}\sqrt{2} \ln x) + C \sin(\frac{1}{2}\sqrt{2} \ln x) + x \ln x$.

24. $y = Ax + Bx^2 + Cx^3 + 2x^4 + x^2 \ln x$.

26. (a) $V = \ln ar^n$ where a and n are constants;
(b) $V = r^n(A \cos n\theta + B \sin n\theta)$, where A and B are constants.

27. $u = \ln ar^n$ where a and n are constants.

28. $y = A \sin 2\sqrt{x} + B \cos 2\sqrt{x} + \frac{1}{4}(2x^2 - 6x + 3)$.

29. $z'' + 3z' + 2z = 0$; $x^2y = e^{-2x}$. **30.** $z'' + z = 0$; $y \sec x = -\sqrt{(\pi/x)}$.

31. $z'' + 4z' + 3z = 0$; $y = e^{-x}/x^3$. **32.** $y = x - \sin^{-1} x$.

33. $(\mathrm{d}^2y/\mathrm{d}t^2) + y = 2e^t$; $y = e^{\sin x} - \sin(\sin x)$.

34. $n = -2$; $x^2y = A + Be^{-4x} + (4 \sin x - \cos x)/17$.

35. $y = Ae^{3\sqrt{x}} + Be^{-2\sqrt{x}} - \frac{1}{5}\sqrt{x}e^{-2\sqrt{x}}$.

36. the given equation is found from the relation $\dfrac{\mathrm{d}^2x}{\mathrm{d}y^2} = \dfrac{\mathrm{d}}{\mathrm{d}x}\left(1\Big/\dfrac{\mathrm{d}y}{\mathrm{d}x}\right)\dfrac{\mathrm{d}x}{\mathrm{d}y}$;

$(\mathrm{d}^2x/\mathrm{d}y^2) - 3(\mathrm{d}x/\mathrm{d}y) + 2x = y$; $x = Ae^y + Be^{2y} + \frac{1}{4}(2y + 3)$.

37. $y = Ax + B\sqrt{(x^2 - 1)} + \frac{1}{2}(x^2 - 1)^{1/2} \ln\{x + \sqrt{(x^2 - 1)}\}$.

38. $(\mathrm{d}^2y/\mathrm{d}u^2) = (1 + x^2)^2(\mathrm{d}^2y/\mathrm{d}x^2) + 2x(1 + x^2)(\mathrm{d}y/\mathrm{d}x)$;
$y = \frac{1}{6}(\tan^{-1} x)^3 + A \tan^{-1} x + B$; $y = \frac{1}{6}(\tan^{-1} x)^3 + \tan^{-1} x$.

39. $(\mathrm{d}^2y/\mathrm{d}t^2) + 4y = 8(1 - t^2)$; $y = A \sin(2 \sin x) + B \cos(2 \sin x)$
$+ 3 - 2 \sin^2 x$.

40. $(\mathrm{d}^2y/\mathrm{d}t^2) + y = \cosh^2 t$; $y = A \sin(\sinh^{-1} x) + B \cos(\sinh^{-1} x) + \frac{1}{3}(3 + x^2)$;
$y = \frac{1}{3}[3 + x^2 - 3 \cos(\sinh^{-1} x)]$.

Exercises 23(c)

1. $x = e^{2t}(t + 1)$; $y = te^{2t}$.

2. $x = e^{-at} \sinh t$; $y = e^{-at} \cosh t$.

3. The line is $3x + y = 0$.

4. $x = (e^{-4t} - e^t)/5;\ y = -(3e^t + 2e^{-4t})/10.$
5. $x = \frac{1}{2}(e^{-3t} + e^{-t});\ y = \frac{1}{2}(e^{-3t} - e^{-t}).$
6. $x = e^t(A + 4t) + B;\ y = \frac{3}{2}e^t(4t + A - 1) + 2B.$
7. $x = Ae^{(1/3)t} + Be^{-(1/3)t};\ y = \frac{1}{2}e^t - 2Ae^{(1/3)t} - Be^{-(1/3)t}.$
8. $x = A \cos 2t - B \sin 2t - t \sin 2t;\ y = A \sin 2t + B \cos 2t + t \cos 2t.$
9. $x = (36 + 10e^t - 45e^{-t} - e^{-5t})/60;\ y = (24 + 20e^t - 45e^{-t} + e^{-5t})/60.$
10. $x = 3 \sin t - 2 \cos t + e^{-2t};\ y = \frac{1}{2}(9 \cos t - 7 \sin t - e^{-3t}).$
11. $x = e^{-t}(A \cos t + B \sin t) + e^t(15t - 2)/25;$
    $y = e^{-t}(B \cos t - A \sin t) + e^t(5t + 11)/25.$
    For the particular solution $A = 2/25$, $B = -11/25$.
12. $x = [(3t + 1) - e^{-3t}(6t + 1)]/27;\ y = 2[(2 - 3t) - e^{-3t}(3t + 2)]/27.$
13. $x = Ae^{-5t} + Be^{-t} - \frac{1}{5};\ y = t - 3Ae^{-5t} + Be^{-t} - \frac{2}{5}.$
14. $x = \sin t;\ y = \cos t - \sin t - 1.$
15. $x = \frac{1}{8}(7e^{3t/5} + 9e^{-t} - 8);\ y = \frac{1}{8}(49e^{3t/5} - 9e^{-t} - 8t^2 - 24t - 32).$
16. $x = 4Ae^{-2t} + 3Be^{-3t} + (5 \cos t + 7 \sin t)/10;$
    $y = Ae^{-2t} + Be^{-3t} + (\sin t - \cos t)/10.$
17. $x = (16e^{2t} + e^{8t} - 40t - 17)/128 - \frac{3}{7}e^t;$
    $y = (16e^{2t} - e^{8t} - 24t - 15)/128 - \frac{4}{7}e^t.$
19. $x = 1 + e^{-\pi/2};\ y = -e^{-\pi/2}.$
20. $u = A \cos \omega t + B \sin \omega t;\ v = (E/H) + A \sin \omega t - B \cos \omega t;$
    $x = (A/\omega) \sin \omega t - (B/\omega) \cos \omega t + C;$
    $y = (E/H)t - (A/\omega) \cos \omega t - (B/\omega) \sin \omega t + C'.$

INDEX